Introduction to the Chemical Analysis of Foods

Introduction to the Chemical Analysis of Foods

Editor
S. Suzanne Nielsen

Purdue University
West Lafayette, Indiana

Jones and Bartlett Publishers
Boston London

Editorial, Sales, and Customer Service Offices
Jones and Bartlett Publishers
One Exeter Plaza
Boston, MA 02116
1-800-832-0034
1-617-859-3900

Jones and Bartlett Publishers International
PO Box 1498
London W6 7RS
England

Library of Congress Cataloging-in-Publication Data

Introduction to the chemical analysis of foods / S.S. Nielsen, editor.
p. cm.
Includes bibliographical references and index.
ISBN 0-86720-826-0
1. Food—Analysis. 2. Food adulteration and inspection. I. Nielsen, S. S. (S. Suzanne)
TX545.I58 1994
664′.07—dc20 93-44871
CIP

Acquisitions Editor: Joseph E. Burns
Manufacturing Buyer: Dana L. Cerrito
Editorial Production Service: Colophon
Typesetting: Modern Graphics, Inc.
Cover Design: Hannus Design Associates
Printing and Binding: Braun-Brumfield, Inc.

Printed in the United States of America

98 97 96 95 94 10 9 8 7 6 5 4 3 2 1

Contents

Contributing Authors

Jörg Augustin
Department of Food Science and Toxicology
University of Idaho
Moscow, Idaho 83843

Maurice R. Bennink
Department of Food Science and Human Nutrition
Michigan State University
East Lansing, Michigan 48824-1224

Robert L. Bradley, Jr.
Department of Food Science
University of Wisconsin
Madison, Wisconsin 53706

Sam K. C. Chang
Department of Food and Nutrition and Department of Cereal Science and Food Technology
North Dakota State University
Fargo, North Dakota 58105

Genevieve L. Christen
Department of Food Science and Technology
University of Tennessee
Knoxville, Tennessee 37901-1071

Eugenia A. Davis
Department of Food Science and Nutrition
University of Minnesota
St. Paul, Minnesota 55108-6099

Deborah E. Dixon
Becton Dickinson Advanced Diagnostics
Sparks, Maryland 21152

Thomas M. Eads
Department of Food Sciences
Purdue University
West Lafayette, Indiana 47907-1160

Jesse F. Gregory, III
Department of Food Science and Human Nutrition
University of Florida
Gainsville, Florida 32611-0370

Yong D. Hang
Department of Food Science and Technology
Cornell University
Geneva, New York 14456

Leniel H. Harbers
Department of Animal Sciences and Industry
Kansas State University
Manhattan, Kansas 66506-1600

Deloy G. Hendricks
Department of Nutrition and Food Sciences
Utah State University
Logan, Utah 84322-8700

Y. H. Hui
American Food and Nutrition Center
Cutten, California 95534

Dick H. Kleyn (deceased)
Department of Food Science
Rutgers University
New Brunswick, New Jersey 08903

Nicholas H. Low
Department of Applied Microbiology and Food Science
University of Saskatchewan
Saskatoon, Saskatchewan, Canada S7I 0W0

William D. Marshall
Department of Food Science and Agricultural Chemistry
MacDonald Campus of McGill University
St.-Anne-de-Bellevue, Quebec, Canada H9X 3V9

Dennis D. Miller
Department of Food Science
Cornell University
Ithaca, NY 14853-7201

David B. Min
Department of Food Science and Technology
The Ohio State University
Columbus, Ohio 43210

S. Suzanne Nielsen
Department of Food Science
Purdue University
West Lafayette, Indiana 47907-1160

John R. Pedersen
Department of Grain Science and Industry
Kansas State University
Manhattan, Kansas 66506-2201

Michael H. Penner
Department of Food Science and Technology
Oregon State University
Corvallis, Oregon 97331-6602

Oscar A. Pike
Department of Food Science and Nutrition
Brigham Young University
Provo, Utah 84602

Joseph R. Powers
Department of Food Science and Human Nutrition
Washington State University
Pullman, Washington 99164-5184

Barbara A. Rasco
Institute of Food Science and Technology
School of Fisheries
University of Washington
Seattle, Washington 98195

Gary A. Reineccius
Department of Food Science and Nutrition
University of Minnesota
St. Paul, Minnesota 55108-6099

Mary Ann Rounds
Department of Food Science
Purdue University
West Lafayette, Indiana 47907-1160

Gerald F. Russell
Department of Food Science and Technology
University of California
Davis, California 95616-8598

George D. Sadler
National Center for Food Safety and Technology
Illinois Institute of Technology
Summit-Argo, Illinois 60501

Steven J. Schwartz
Department of Food Science
North Carolina State University
Raleigh, North Carolina 27695-7624

Denise M. Smith
Department of Food Science and Human Nutrition
Michigan State University
East Lansing, Michigan 48824-1224

J. Scott Smith
Department of Animal Sciences and Industry
Kansas State University
Manhattan, Kansas 66506-1600

Randy L. Wehling
Department of Food Science and Technology
University of Nebraska
Lincoln, Nebraska 68583-0919

James M. Zdunek
Kraft General Foods, Inc.
Glenview, Illinois 60025

List of Abbreviations

AACC	American Association of Cereal Chemists
AAS	atomic absorption spectroscopy
ADC	analog-to-digital converter
ADP	adenosine-5′-diphosphate
AES	atomic emission spectroscopy
AI	artificial intelligence
AOAC	Association of Official Analytical Chemists
AOCS	American Oil Chemists' Society
AOM	active oxygen method
APHA	American Public Health Association
ASCII	American Standard for Information Interchange
ASTM	American Society for Testing Materials
ATP	adenosine-5′-triphosphate
ATR	attenuated total reflectance
BCA	bicinchoninic acid
BCD	binary coded decimal
Be	degrees Baumé
BGG	bovine gamma globulin
BHA	butylated hydroxyanisole
BHT	butylated hydroxytoluene
BOD	biochemical oxygen demand
BSA	bovine serum albumin
BV	biological value
CAST	calf antibiotic and sulfa test
CFR	Code of Federal Regulations
CGC	capillary gas chromatography
CI	chemical ionization
CI	confidence interval
CID	Commercial Item Description
COD	chemical oxygen demand
CPC	cetylpyridinium chloride
C-PER	calculated protein efficiency ratio
CPU	central processing unit
CQC	2,6-dichloroquinonechloroimide
CV	cofficient of variation
CVM	Center for Veterinary Medicine
CW	continuous wave
DAL	defect action level
DC	direct current
DC-PER	discriminatt calculated protein efficiency ratio
DHHS	Department of Health and Human Services
DMF	dimethylformamide
DMD	D-malate dehydrogenase
DMTA	dynamic mechanical thermal analysis
DNFB	1-fluoro-2,4-dinitrobenzene
DNP	dinitrophenyl
DSC	differential scanning colorimetry
DTA	differential thermal analysis
DTNB	5,5′-dithiobis-2-nitrobenzoic acid
dwb	dry weight basis
ECD	electron capture detector
EDTA	ethylenediaminetetraacetic acid
EI	electron impact
EIA	enzyme immunoassay
ELISA	enzyme linked immunosorbent assay
EMF	electromotive force
EPA	Environmental Protection Agency
ERH	equilibrium relative humidity
ESR	electron spin resonance
FAB	fast atom bombardment
FAME	fatty acid methyl esters
FAO/WHO	Food and Agricultural Organization/ World Health Organization
FDA	Food and Drug Administration
FD&C	Food, Drug and Cosmetic
FDNB	1-fluoro-2,4-dinitrobenzene
FFA	free fatty acid
FGIS	Federal Grain Inspection Service
FIA	fluoroimmunoassay
FID	flame ionization detector
FID	free induction decay
FIFRA	Federal Insecticide, Fungicide, and Rodenticide Act
FNB/NAS	Food and Nutrition Board of the National Academy of Sciences
F-6-P	fructose-6-phosphate
FPD	flame photometric detector
FT	Fourier transform
FTC	Federal Trade Commission
FT-ESR	Fourier transform - electron spin resonance
FTIR	Fourier transform infrared
FT-NMR	Fourier transform - nuclear magnetic resonance
Gal-DH	β-galactosidase dehydrogenase or galactose dehydrogenase
GC	gas chromatography
GC-AED	gas chromatography - atomic emission detector
GC-FTIR	gas chromatography - Fourier transform infrared

GC-MS	gas chromatography - mass spectrometry
GFC	gel-filtration chromatography
GHz	gigahertz
GMP	Good Manufacturing Practices
G-6-P	glucose-6-phosphate
GPC	gel-permeation chromatography
G6P-DH	glucose-6-phosphate dehydrogenase
HACCP	Hazard Analysis Critical Control Point
HETP	height equivalent to a theoretical plate
HK	hexokinase
HMDS	hexamethyldisilazane
HMF	5-hydroxymethyl-2-furfural
HPCE	high-performance capillary electrophoresis
HPLC	high-performance liquid chromatography
HPLC-MS	high-performance liquid chromatography - mass spectroscopy
HPTLC	high-performance thin-layer chromatography
HRGC	high-resolution gas chromatography
HRP-B	horseradish peroxidase-benzidine
IC	integrated circuit
ICP	inductively coupled plasma
ICP-AES	inductively coupled plasma - atomic emission spectroscopy
ICTA	International Confederation for Thermal Analysis
IDEA	immobilized digestive enzyme assay
IMS	Interstate Milk Shippers
INT	indonitrotetrazolium
IR	infrared
ISA	ionic strength adjustor
ISE	ion-selective electrode
IU	International Units
IUPAC	International Union of Pure and Applied Chemistry
KFR	Karl Fischer reagent
KFReq	Karl Fischer reagent water equivalence
KHL	keyhole limpet hemocyanin
KHP	potassium acid phthalate
LALLS	low-angle laser light scattering
LAN	local area network
LC	liquid chromatography
LC-MS	liquid chromatography - mass spectroscopy
LDH	lactate dehydrogenase
LIMS	laboratory information management system
MAS	magic angle spinning
MCL	maximum contaminant level
MECC	micellar electrokinetic capillary chromatography
MFL	million fibers per liter
MHz	megahertz
MPIP	Meat Poultry Inspection Program
MRI	magnetic resonance imaging
MRMs	multiresidue methods
MS	mass spectrometry (or spectrometer)
MW	molecular weight
NAD	nicotinamide-adenine dinucleotide
NADP	nicotinamide-adenine dinucleotide phosphate
NADPH	reduced NADP
NCWM	National Conferences on Weights and Measures
NIR	near-infrared
NIST	National Institute of Standards Technology
NMFS	National Marine Fisheries Service
NMR	nuclear magnetic resonance
nOe	nuclear Overhauser enhancement
NPR	net protein ratio
NPU	net protein utilization
NRC	National Research Council
NSSP	National Shellfish Sanitation Program
OCls	organochlorines
ODS	octadecylsilyl
OPs	organophosphates
PAGE	polyacrylamide gel electrophoresis
PAM I	Pesticide Analytical Manual, Volume I
PAM II	Pesticide Analytical Manual, Volume II
PCBs	polychlorinated biphenyls
PDCAAS	protein digestibility - corrected amino acid score
PEEK	polyether ether ketone
PER	protein efficiency ratio
PGI	phosphoglucose isomerase
pI	isoelectric point
PID	photoionization detector
PMO	Pasteurized Milk Ordinance
ppb	parts per billion
PPD	Purchase Product Description
ppm	parts per million
PUFA	polyunsaturated fatty acids
PVPP	polyvinylpolypyrrolidone
RAC	raw agricultural commodity
RF	radio frequency
RIA	radioimmunoassay
RPAR	Rebuttable Presumption Against Registration
RPER	relative protein efficiency ratio
RCS	rapid scan correlation
SD	standard deviation
SDH	sorbitol dehydrogenase
SDS	sodium dodecyl sulfate
SDS-PAGE	sodium dodecyl sulfate - polyacrylamide gel electrophoresis
SEC	size-exclusion chromatography
SFC	solid fat content
SFC	supercritical-fluid chromatography

SFI	solid fat index
SIM	selected ion monitoring
S/L	solid/liquid
SNF	solids-not-fat
SO	sulfite oxidase
SRMs	single residue methods
STOP	swab test on premises
TBA	thiobarbituric acid
TCD	thermal conductivity detector
TEMED	tetramethylethylenediamine
TEPA	tetraethylenepentamine
TGA	thermogravimetric analysis
TIC	total ion current
TLC	thin-layer chromatography
TMCS	trimethylchlorosilane
TMS	tri-methylsilyl
TNBS	trinitrobenzenesulphonic acid
TOC	total organic carbon
TS-MS	thermospray - mass spectrometry
TSS	total soluble solids
TSUSA	Tariff Schedules of the United States of America
USCS	United States Customs Service
USDA	United States Department of Agriculture
UV	ultraviolet
UV-Vis	ultraviolet-visible
Vis	visible
VPP	vegetable protein product
wwb	wet weight basis

Preface and Acknowledgments

This book was designed for use as a text primarily for undergraduate students majoring in food science who are currently studying the chemical analysis of foods. It should also be useful to workers in the food industry who do food analysis. Chapter authors are primarily university faculty members who are teaching or have taught a food analysis course, and who are very familiar with the specific topic of the chapters by nature of their research programs. Each chapter has been reviewed by persons working in the food industry who are familiar with and utilize that technique.

The book is not a laboratory manual but instead is designed to provide the lecture materials in an easy-to-follow outline format, with a brief discussion for each section. It provides much of the information on techniques that students must possess before they are able to conduct those laboratory experiments that normally accompany a food analysis course.

This book covers only the analysis of chemical properties of foods, and not physical properties. It is not intended as a detailed reference, but as a general introduction to the techniques used in food analysis. The course instructor can expand on the information in lecture, as desired, providing more details for any particular technique.

General information on sampling and data handling provides the background for discussing specific methods to determine the chemical composition and characteristics of foods. Large sections on spectroscopy and chromatography are included to explain principles of the techniques themselves and how they relate to methods of food analysis. Other methods and instrumentation such as ion selective electrodes, enzymes, immunoassays, and thermal analysis are also covered from the perspective of their use in the chemical analysis of foods. A chapter is included that relates food analysis to government regulations and recommendations.

All topics include information on the basic principles, procedures, advantages, limitations, and applications of food analysis. All chapters have summaries and study questions, and key words or phrases are identified. Many also have practice problems and a comparison of possible methods.

Most of the material covered in this book requires an understanding of general, organic, analytical and food chemistry, as well as biochemistry. With this basic knowledge, along with a food analysis course and other relevant food science courses, it is hoped that students can function within the food industry as necessary relevant to food analysis. This book will provide a good basis for food scientists and technologists as they begin work in the food industry, whether or not they are the persons directly involved in analysis of the food products. I would greatly appreciate comments from students, instructors and food industry professionals as to how well this book meets their needs, as well as any suggestions for later editions.

I wish to thank the persons who prepared each of the chapters in this book. Those of us who teach or have taught food analysis are indebted to our former students, who gave us the needed perspective as we prepared our chapters, intended to meet the needs of future students. I also wish to thank the authors of articles and books, as well as the publishers and industrial companies, for their permission to reproduce materials used here. Special thanks is extended to Dr. Y. H. Hui who advised me throughout this project. Becky Hitt-Otkinson is acknowledged for providing exceptional secretarial assistance. I thank the many persons from the food industry, government, and academia who kindly reviewed one or more chapters in this book. They offered their assistance on this project with the hope that this book might meet the needs of present employees in their companies, and that it might also meet the needs of students who will be their future employees. They are:

Sami M. Al-Hasani, Ph.D.
ConAgra Frozen Foods Analytical Laboratory
Columbia, Missouri

Yousef H. Atallah, Ph.D., DABT
Sandoz Agro Inc.
Des Plaines, Illinois

Douglas Bark, B.S.
Woodson-Tenent Laboratories, Inc.
Memphis, Tennessee

Allen E. Blaurock, Ph.D.
Kraft General Foods, Inc.
KGF Technology Center
Glenview, Illinois

David G. Cunningham, Ph.D.
Ocean Spray Cranberries, Inc.
Middleboro, Massachusetts

Virginia L. Carlson, M.S.
Nabisco Foods Group
E. Hanover, New Jersey

Francis J. Farrell, Ph.D.
Thomas J. Lipton Co.
Englewood Cliffs, New Jersey

Norman E. Fraley, Jr.
Express Analytic
A ConAgra Company
Downers Grove, Illinois

Beverly A. Friend, Ph.D.
Friend Consulting Services
Collinsville, Illinois

Bonita L. Funk, B.S.
Gerber Products Co.
Nutritional Regulatory Affairs
Fremont, Michigan

J. Richard Gorham, Ph.D.
Food & Drug Administration
Washington, District of Columbia

Bruce R. Hamaker, Ph.D.
Purdue University
West Lafayette, Indiana

Erich Heftmann, Ph.D.
Elsevier Science Publishers
Amsterdam, The Netherlands

Kevin B. Hicks, Ph.D.
Eastern Regional Research Ctr.
ARS/USDA
Philadelphia, Pennsylvania

Clyde B. Hoskins, DVM
Office of the Surgeon General
U.S. Army
Falls Church, Virginia

W. Jeffrey Hurst, Ph.D.
Editor, Laboratory Robotics & Automation
Hershey, Pennsylvania

Edward J. Kikta, Jr., Ph.D.
FMC Corporation
AgriChem Group
Princeton, New Jersey

Sandra Pfeiffer
Gerber Products Co.
Fremont, Michigan

Frank J. Sasevich
Kraft General Foods, Inc.
Glenview, Illinois

Shyamala Subramanian, M.S.
Shaklee Corporation
Hayward, California

Robert L. Wade, Ph.D.
The Procter & Gamble Co.
Cincinnati, Ohio

Loren J. Wagner
Bohdan Automation, Inc.
Mundelein, Illinois

Durward Ray Walker
E & J Gallo Winery
Modesto, California

Edward H. Waysek
Hoffmann-LaRoche Inc.
Vitamin Research & Development
Nutley, New Jersey

James E. Woodbury
Santa Ana, California

PART I

General Information

CHAPTER

1

Introduction to Food Analysis

S. Suzanne Nielsen

1.1 INTRODUCTION

Investigations in food science and technology, whether in universities, governmental agencies, or the food industry, often require determination of food composition. Various types of samples may require analysis as part of a research program, as new food products are developed, or as part of a **quality assurance** program for existing products. The chemical composition of foods is often determined to establish the acceptability or nutritive value of the food product. The nature of the sample and the specific reason for the analysis commonly dictate the choice of analysis method. Speed, **precision,** and **accuracy** are often key factors that determine the choice of method. In addition to actually performing the assay, the success of any analysis method relies on the proper selection and preparation of the food sample and on the appropriate calculations and interpretation of the data. **Official methods** of analysis developed by several nonprofit scientific organizations allow for comparison of results between different laboratories, and for comparisons to less standard procedures. Such official methods are critical in the analysis of foods, to ensure that they meet the legal requirements established by governmental agencies. **Government regulations** most relevant to the chemical analysis of foods will be covered in Chapter 2.

1.2 TYPES OF SAMPLES ANALYZED

The chemical analysis of foods is an important part of the quality assurance program in food processing, from ingredients and raw materials, through processing, to the finished products (1–3). It is also important in formulating and developing new products and evaluating new processes for making food products, and in identifying the source of the problem for unacceptable products (Table 1-1). For each type of product to be analyzed, it may be necessary to determine either just one or many components. The nature of the sample and the way in which the information obtained will be used may dictate the specific method of analysis. For example, process control samples are usually analyzed by rapid methods, whereas nutritive value information for **nutritional labeling** generally requires the use of official methods of analysis. Critical questions, including those listed in Table 1-1, can be answered by analyzing various types of samples in a food processing system.

TABLE 1-1. Types of Samples Analyzed in a Quality Assurance Program for Food Products

Sample Type	*Critical Questions*
Raw Materials	Do they meet your specifications? Do they meet required legal specifications? Will a processing parameter have to be modified because of any change in the composition of raw materials? Are the quality and composition the same as for previous deliveries? How does the material from a potential new supplier compare to that from the current supplier?
Process Control Samples	Did a specific processing step result in a product of acceptable composition or characteristics? Does a further processing step need to be modified to obtain a final product of acceptable quality?
Finished Product	Does it meet the legal requirements? What is the nutritive value, so that label information can be developed? or Is the nutritive value as specified on an existing label? Will it be acceptable to the consumer? Will it have the appropriate shelf life?
Competitor's Sample	What are its composition and characteristics? How can we use this information to develop new products?
Complaint Sample	How do the composition and characteristics of a complaint sample submitted by a customer differ from a sample with no problems?

Adapted from (4,5)

1.3 STEPS IN ANALYSIS

1.3.1 Select and Prepare Sample

In analyzing food samples of the types described above, all results depend on obtaining a representative sample and converting the sample to a form that can be analyzed. Neither of these is as easy as it sounds! **Sampling** and **sample preparation** are covered in detail in Chapter 3.

1.3.2 Perform the Assay

Performing the assay is the step in food analysis that is unique for each component or characteristic to be analyzed and may be unique to a specific type of food product. Single chapters in this book address the other two steps in analysis described here (sections 1.3.1 and 1.3.3), while the remainder of the book addresses this step of actually performing the assay. The descriptions of the various specific procedures are meant to be overviews of the methods. To actually perform the assays, details regarding chemicals, reagents, apparatus, and step-by-step instructions should be found in the referenced books and articles.

1.3.3 Calculate and Interpret the Data

To make decisions and take action based on the results obtained from performing the assay to determine the chemical composition or characteristics of a food product, one must make the appropriate calculations to correctly interpret the data. *Data handling* is covered in Chapter 4.

1.4 CHOICE OF METHODS

Numerous methods are often available to assay food samples for a specific characteristic or component. To select or modify methods used to determine the chemical composition and characteristics of foods, one must know something about the chemical composition and physical properties of the food product. One must also be familiar with the properties of the methods, including the principles of the procedures and the critical steps in the procedures. Selection of a method depends largely on the objective of the measurement. Methods used for quality assurance measurements may be less accurate than those official methods used for nutritional labeling purposes. Recent advances in food analysis have resulted in the increased use of instrumental techniques. Some analyses may be too difficult and costly to be done on site. It may be better economically to send certain samples to an analytical laboratory that has the necessary equipment and trained personnel. Some analyses can be done using commercially available kits as opposed to the standard method of analysis. Such kits may save time, effort, and capital investment in terms of bulk reagents and special equipment. The criteria described in Table 1-2 are useful to evaluate the appropriateness of a method in current use or a new method being considered.

1.5 OFFICIAL METHODS OF ANALYSIS

The choice of method for a specific characteristic or component of a food sample is often made easier by the availability of **official methods** of analysis. Several nonprofit scientific organizations have compiled and published official methods of analysis for food products. These methods have been carefully developed and are standardized. They allow for the comparison of results between different laboratories that follow the same procedure, and for comparison of results to those obtained using new or more rapid procedures.

1.5.1 Association of Official Analytical Chemists International

The **Association of Official Analytical Chemists International** (AOAC) is an organization begun in 1884 to serve the needs of government regulatory and research agencies for analytical methods. Its goal is to provide methods that will perform with the necessary accuracy and precision under usual laboratory conditions.

The organization functions as follows:

1. Selects method of analysis from published literature or develops new methods.
2. Collaboratively tests methods through interlaboratory studies.
3. Approves methods.
4. Publishes approved methods for a wide variety of materials related to foods, drugs, cosmetics, agriculture, forensic science, and products affecting the public health and welfare.

Methods refereed by the AOAC International and the data supporting the method are published in the *Journal of the Association of Official Analytical Chemists*. Such methods must be successful in a formal interlaboratory collaborative study before being accepted by **first action** as an official method by the AOAC International. The first action methods are subject to scrutiny and general testing by other scientists and analysts for one year before **final action** adoption. Approved first action and final action methods are compiled in books published and updated every four to five years in the *Official Methods of Analysis* (6) of the AOAC International. Supplements to the book are published yearly and contain new and revised methods. The *Official Methods of Analysis* of the AOAC includes methods appropriate for a wide variety of products and other materials (Table 1-3). These methods are often those specified by the **Food and Drug Administration** (FDA) with regard to legal requirements for food products. They are generally the methods followed by the FDA and USDA **(United States Department of Agriculture)** to check the nutritional labeling information on foods and to check foods for the presence of residues.

1.5.2 American Association of Cereal Chemists

The **American Association of Cereal Chemists** (AACC) publishes a set of approved laboratory methods, most applicable to cereal products (Table 1-4). The *AACC Approved Methods of Analysis* (7) are consistent with official methods set by the AOAC and the **American Oil Chemists' Society** (AOCS). The *Methods* are continuously reviewed, critiqued, and updated. They are printed in a looseleaf format and are contained in ring binders. Supplements containing new and revised procedures are developed annually.

The AACC has a **check sample** service in which a subscribing laboratory receives specifically prepared

TABLE 1-2. Criteria for Choice of Food Analysis Methods

Characteristic	*Critical Questions*
Inherent Properties	
• Specificity	Is the property being measured the same as that claimed to be measured? Is it what you really need to be measuring? What steps are being taken to ensure a high degree of specificity?
• Precision	What is the precision of the method? Is it within-batch, batch-to-batch, or day-to-day variation? What step in the procedure contributes the greatest variability?
• Accuracy	How does the new method compare in accuracy to the old or a standard method? What is the percent recovery?
Applicability of Method to Laboratory	
• Sample size	How much sample is needed? Is it too large or too small to fit your needs? Does it fit your equipment and/or glassware?
• Reagents	Can you properly prepare them? What equipment is needed? Are they stable? For how long and under what conditions? Is the method very sensitive to slight or moderate changes in the reagents?
• Equipment	Do you have the appropriate equipment? Are personnel competent to operate equipment?
• Cost	What is the cost in terms of equipment, reagents, and personnel?
Usefulness	
• Time required	How fast is it? How fast does it need to be?
• Reliability	How reliable is it from the standpoints of precision and stability?
• Need	Does it meet a need or better meet a need? Is any change in method worth the trouble of the change?
Personnel	
• Safety	Are special precautions necessary?
• Procedures	Who will prepare the written description of the procedures and reagents? Who will do any required calculations?

TABLE 1-3. Table of Contents of 1990 *Official Methods of Analysis* of the Association of Official Analytical Chemists (AOAC) International (6)

Chapter	*Title*
1	Agriculture Liming Materials
2	Fertilizers
3	Plants
4	Animal Feed
5	Drugs in Feeds
6	Disinfectants
7	Pesticide Formulations
8	Hazardous Substances
9	Metals and Other Elements at Trace Levels in Foods
10	Pesticide and Industrial Chemical Residues
11	Waters; and Salt
12	Microchemical Methods
13	Radioactivity
14	Veterinary Analytical Toxicology
15	Cosmetics
16	Extraneous Materials: Isolation
17	Microbiological Methods
18	Drugs: Part I
19	Drugs: Part II
20	Drugs: Part III
21	Drugs: Part IV
22	Drugs: Part V
23	Drugs and Feed Additives in Tissues
24	Forensic Sciences
25	Baking Powders and Baking Chemicals
26	Distilled Liquors
27	Malt Beverages and Brewing Materials
28	Wines
29	Nonalcoholic Beverages and Concentrates
30	Coffee and Tea
31	Cacao Bean and Its Products
32	Cereal Foods
33	Dairy Products
34	Eggs and Egg Products
35	Fish and Other Marine Products
36	Flavors
37	Fruits and Fruit Products
38	Gelatin, Dessert Preparations, and Mixes
39	Meat and Meat Products
40	Nuts and Nut Products
41	Oils and Fats
42	Vegetable Products, Processed
43	Spices and Other Condiments
44	Sugars and Sugar Products
45	Vitamins and Other Nutrients
46	Color Additives
47	Food Additives: Direct
48	Food Additives: Indirect
49	Natural Poisons

TABLE 1-4. Table of Contents of 1983 *Approved Methods of the American Association of Cereal Chemists* (7)

Chapter	Title	Chapter	Title
2	Acidity	48	Oxidizing, Bleaching, and Maturing Agents
4	Acids	50	Particle Size
6	Admixture of Flours	52	Pentosans
7	Amino Acids	54	Physical Dough Tests
8	Total Ash	55	Physical Tests
10	Baking Quality	56	Physicochemical Tests
12	Carbon Dioxide	58	Special Properties of Fats, Oils, and Shortenings
14	Color and Pigments	60	Residues
16	Cooking Characteristics	61	Rice
18	Drugs	62	Preparation of Sample
20	Egg Solids	64	Sampling
22	Enzymes	66	Semolina Quality
26	Experimental Milling	68	Solids
28	Extraneous Matter	70	Solutions
30	Crude Fat	71	Soybean Protein
32	Fiber	72	Specific Volume
33	Flavor	74	Staleness/Texture
36	Glossary	76	Starch
38	Gluten	78	Statistical Principles
39	Infrared Analysis	80	Sugars
40	Inorganic Constituents	82	Tables
42	Microorganisms	86	Vitamins
44	Moisture	88	Water Hydration Capacity
45	Mycotoxins	89	Yeast
46	Nitrogen		

test samples from AACC. The laboratory performs the specified analyses on the samples and returns the results to AACC. The AACC then provides a statistical evaluation of the analytical results and compares the subscribing laboratory's data with those of other laboratories to inform the subscribing laboratory of its degree of accuracy. Check samples are offered in areas such as flours and cereals, vitamins and minerals, sugars, sodium, total dietary fiber, soluble and insoluble dietary fiber, near infrared analysis, sanitation, and microbiology. The AACC also provides reference food fiber sources in bulk quantities for experimental research studies.

1.5.3 American Oil Chemists' Society (AOCS)

The AOCS publishes a set of official methods and recommended practices, most applicable to fat and oil analysis (Table 1-5)(8). This is a widely used methodology source on the subjects of edible fats and oils, oilseeds and oilseed proteins, soaps and synthetic detergents, industrial fats and oils, fatty acids, oleochemicals, glycerin, and lecithin. The AOCS also has a check sample program for fats, oils, oilseed meals, and related substances. Laboratories from many countries participate in the program for the purpose of checking the accuracy of their work, their reagents, and their laboratory apparatus against the statistical norm derived from the entire group.

TABLE 1-5. Table of Contents of 1993 *Official Methods and Recommended Practices of the American Oil Chemists' Society* (8)

Section	Title
A	Vegetable Oil Source Materials
B	Oilseed By-Products
C	Commercial Fats and Oils
D	Soap and Synthetic Detergents
E	Glycerin
F	Sulfonated and Sulfated Oils
G	Soap Stocks
H	Specifications for Reagents, Solvents, and Apparatus
J	Lecithin
M	Evaluation and Design of Test Methods
R	Official Listings
S	Recommended Practices for Testing Industrial Oils and Derivatives
T	Test Methods for Industrial Oils and Derivatives

1.5.4 Other Official Methods

Standard Methods for the Examination of Dairy Products (9), published by the American Public Health Association, includes methods for the chemical analysis of products (Table 1-6). *Standard Methods for the Examination of Water and Wastewater* (10) is published jointly by the American Public Health Association, American Water Works Association, and the Water Pollution Control Federation. *Food Chemical Codex* (11), published by

TABLE 1-6. Contents of Chapter 15 on Chemical and Physical Methods in *Standard Methods for the Examination of Dairy Products* (9)

15.1	Introduction
15.2	Acid Degree Value (Hydrolytic Rancidity)
15.3	Acidity Acidity: titratable Acidity: potentiometric, pH Acidity: titratable, potentiometric endpoint Acidity: pH, gold electrode/quinhydrone method
15.4	Ash and Alkalinity of Ash Ash gravimetric Alkalinity of ash
15.5	Chloride (Salt) Mohr method Volhard method Coulometric titration
15.6	Chloride, Available
15.7	Extraneous Material Cheese Milk, quantitative, laboratory analysis Milk, qualitative, Sani-guide
15.8	Fat Babcock method Pennsylvania modified Babcock method Roccal Babcock method Gerber method Ether extraction (Roese-Gottlieb) method Mojonnier method Automated turbidimetry Vegetable oil in milkfat (β-sitosterol)
15.9	Lactose in Milk Polarimetric method HPLC method
15.10	Moisture and Solids Vacuum oven Vacuum oven, sand pan Forced-draft oven Moisture, microwave oven Moisture balance: dry milk products Refractometer: whey and whey products Solids in milk: lactometric method
15.11	Multicomponent Methods Infrared milk analysis: fat, protein, lactose, total solids Near infrared analysis: fat, protein, total solids in milk Modified Kohman method: fat, moisture and salt in butter and margarine
15.12	Protein Kjeldahl standard Kjeldahl (block digester) Dye binding: acid orange 12 Undenatured whey protein nitrogen in nonfat dry milk
15.13	Water: Added to Milk Thermistor cryoscope
15.14	Iodine: Selective Ion Procedure
15.15	Vitamins A and D in Milk Products Vitamin A; HPLC method Vitamins D_2 and D_3, HPLC method
15.16	Pesticide Residues in Milk
15.17	Radionuclides
15.18	References

the Food and Nutrition Board of the National ResearchCouncil/National Academy of Science, contains methods for the analysis of certain food additives. The American Spice Trade Association publishes standard methods for the analysis of spices (12).

1.6 SUMMARY

Food scientists and technologists often determine the chemical composition of foods as part of research, food product development, or quality assurance activities. For example, the types of samples analyzed in the quality assurance program for a food company may include raw materials, process control samples, finished products, competitors' samples, and complaint samples. Chemical analysis of foods is often done to determine the acceptability or nutritive value. To successfully base decisions on results of any analysis, one must correctly conduct all three major steps in the analysis: (1) select and prepare samples, (2) perform the assay, and (3) calculate and interpret the data. The choice of the analysis methods is usually based on the nature of sample, the specific reason for the analysis, and characteristics of the method itself, such as specificity, accuracy, precision, speed, cost of equipment, and training of personnel. Methods used for quality assurance may be less accurate but much faster than official methods used for nutritional labeling. Official methods for the chemical analyses of foods have been compiled and published by the AOAC International, American Association of Cereal Chemists, American Oil Chemists' Society, and certain other nonprofit scientific organizations. These methods allow for the comparison of results between different laboratories and for comparison to new or more rapid procedures.

1.7 STUDY QUESTIONS

1. Identify six reasons you might need to determine certain chemical characteristics of a food product as part of a quality assurance program.
2. You are considering the use of a new method to measure Compound X in your food product. List six factors you will consider before adopting this new method in your quality assurance laboratory.

3. In your work at a food company, you mentioned to a coworker something about the *Official Methods of Analysis* published by the AOAC International. The coworker asks you what the term "AOAC" refers to, what it does, and what the *Official Methods of Analysis* is. Answer your coworker's questions.
4. For each type of product listed below, identify a publication of standard methods of analysis that would be appropriate to consult:
   a. ice cream
   b. enriched flour
   c. wastewater (from food processing plant)
   d. margarine
   e. ground cinnamon

1.8 REFERENCES

1. Herschdoefer, S.M. (Ed.) 1984. *Quality Control in the Food Industry*, Vol. 1, 2nd ed. Academic Press, New York, NY.
2. Stauffer, J.E. 1988. *Quality Assurance of Food Ingredients, Processing and Distribution*. Food & Nutrition Press. Inc., Westport, CT.
3. Gould, W.A., and Gould, R.W. 1988. *Total Quality Assurance for the Food Industries*. CTI Publications, Inc., Baltimore, MD.
4. Pearson, D. 1973. Introduction—Some basic principles of quality control. Ch. 1, in *Laboratory Techniques in Food Analysis*, 1–26. John Wiley & Sons, New York, NY.
5. Pomeranz, Y., and Meloan, C.E. 1987. *Food Analysis: Theory and Practice*, 2nd ed. Van Nostrand Reinhold Co., New York, NY.
6. AOAC. 1990. *Official Methods of Analysis*, 15th ed. Association of Official Analytical Chemists, Washington, DC.
7. AACC. 1983. *Approved Methods of Analysis*, 8th ed. American Association of Cereal Chemists, St. Paul, MN.
8. AOCS. 1990. *Official Methods of Recommended Practices*, 4th ed. 2nd printing (Additions and revisions through 1993.) American Oil Chemists' Society, Champaign, IL.
9. Marshall, R.T. (Ed.) 1992. *Standard Methods for the Examination of Dairy Products*, 16th ed. American Public Health Association, Washington, DC.
10. Greenberg, A. E., Clesceri, L. S., and Eaton, A. D. (Eds.) 1992. *Standard Methods for the Examination of Water and Wastewater*, 18th ed. American Public Health Association, Washington, DC.
11. National Academy of Sciences. 1981. *Food Chemical Codex*, 3rd ed. Food and Nutrition Board, National Research Council. National Academy Press, Washington DC.
12. American Spice Trade Association. 1986. *ASTA Analytical Methods*, 3rd ed. American Spice Trade Association, Inc., Englewood Cliffs, NJ.

CHAPTER

2

Government Regulations and Recommendations Related to Food Analysis

S. Suzanne Nielsen and Y. H. Hui

2.1 INTRODUCTION

Knowledge of government regulations relevant to the chemical analysis of foods is extremely important to persons working in the food industry. Government laws and regulations reinforce the efforts of the food industry to provide wholesome foods, to inform consumers about the nutritional composition of foods, and to eliminate economic frauds. In some cases, they dictate what ingredients a food must contain, what must be tested, and the procedures used to analyze foods for safety factors and quality attributes. This chapter describes the federal and state regulations that affect the composition of foods and the latest federal regulations on nutritional labeling. The reader is referred to references 1–4 for comprehensive coverage of U.S. food laws and regulations. Many of the regulations referred to in this chapter are published in the various Titles of the ***Code of Federal Regulations*** (CFR) (5).

2.2 FEDERAL AND STATE REGULATIONS AFFECTING FOOD COMPOSITION

2.2.1 U.S. Food and Drug Administration

The **Food and Drug Administration** (FDA) is a U.S. government agency within the **Department of Health and Human Services** (DHHS). The FDA is responsible for regulating, among other things, the safety of foods, cosmetics, drugs, medical devices, biologics, and radiological products. It acts under laws passed by the U.S. Congress to monitor the affected industries and assure the consumer of safety of such products.

2.2.1.1 Legislative History

2.2.1.1.1 Food and Drug Act of 1906 The Food and Drug Act of 1906, reenacted in 1907 to extend the provisions for an indefinite period, was the first federal law governing the food supply in the United States. It made illegal the interstate commerce of misbranded or adulterated manufactured or natural foods, beverages, drugs, medicines, or stock feeds. It stated that only substances not likely to render a food injurious to health could be added to foods.

The **U.S. Department of Agriculture** (USDA) was responsible for administering the 1906 Act until 1931, when the FDA was created to administer it. Although the FDA was originally a part of the USDA, it now operates within the DHHS. The USDA and then the FDA were both ineffective at enforcing the 1906 Act because the Act lacked fines or other penalties for violators. This led to eventual passage of the **Federal Food, Drug, and Cosmetic (FD&C) Act** of 1938.

2.2.1.1.2 Federal Food, Drug, and Cosmetic Act of 1938 The FD&C Act of 1938 broadened the scope of the 1906 Act, intending to assure consumers that foods are safe and wholesome, produced under sanitary conditions, and packaged and labeled truthfully. The law further defined and set regulations on adulterated and misbranded foods. The FDA was given power to seize illegal products and to imprison and fine violators. An important part of the 1938 Act relevant to food analysis is the section that authorizes food definitions and standards of identity, as further described below.

2.2.1.1.3 Amendments and Additions to the 1938 FD&C Act The 1938 FD&C Act has been amended several times to increase its power. It was amended in 1954, with the **Miller Pesticide Amendment,** to specify the acceptable amount of pesticide residues on fresh fruits, vegetables, and other raw agricultural products when they enter the marketplace. This Amendment, then under the authority of the FDA, is now administered by the **Environmental Protection Agency** (EPA).

The **Food Additives Amendment** to the 1938 Act was enacted in 1958. This amendment was designed to protect the health of consumers by requiring a food additive to be proven safe before addition to a food and to permit the food industry to use food additives that are safe at the intended level of use. The highly controversial **Delaney Clause,** attached as a rider to this amendment, prohibits the FDA from setting any tolerance level as a food additive for substances known to be carcinogenic.

The **Color Additives Amendment,** passed in 1960, defines color additives, sets rules for both certified and uncertified colors, provides for the approval of color additives that must be certified or are exempt from certification, and empowers the FDA to list color additives for specific uses and set quantity limitations. Similar to the Food Additives Amendment, the Color Additives Amendment contains a Delaney Clause.

2.2.1.1.4 Other FDA Regulations The FDA has developed many administrative rules, guidelines, and action levels, in addition to the regulations described above, to implement the FD&C Act of 1938. Most of them are published in Title 21 of the CFR. They include the **Good Manufacturing Practice** (GMP) **Regulations** (21 CFR 110), regulations regarding **food labeling** (21 CFR 101), **recall guidelines** (21 CFR 7.40), **nutritional quality guidelines** (21 CFR 104), and **nutritional labeling guidelines** (discussed in section 2.6).

The FDA administers several other federal statutes related to foods. The **Fair Packaging and Labeling Act** of 1966 requires that the net weight of a food product, among other information, be accurately stated on the label in a specific manner. Among the provisions enforced by the FDA in the **Public Health Service Act** of 1944 are the safety of pasteurized milk and shellfish, as discussed in sections 2.3 and 2.4, respectively. The importation of milk and cream into the United States is regulated by the **Import Milk Act** of 1927 for economic and public health considerations. The **Tea Importation Act** of 1887, as amended, assures consumers that imported teas meet the quality standards set by the U.S. Board of Tea Experts. Certain aspects of the amended **Federal Meat Inspection Act** of 1967 and the amended **Poultry Products Inspection Act** of 1957 (discussed in sections 2.2.2.1.1 and 2.2.2.2.3) are administered by the FDA.

2.2.1.2 Food Definitions and Standards

The food definitions and standards established by the FDA are published in 21 CFR 100-169 and include standards of identity, quality, and fill. The **standards of identity,** which have been set for a wide variety of food products, are most relevant to the chemical analysis of foods since they specifically establish which ingredients a food must contain. They limit the amount of water permitted in certain products. The minimum levels for expensive ingredients are often set, and maximum levels for inexpensive ingredients are sometimes set. The kind and amount of certain vitamins and minerals that must be present in foods labeled "enriched" are specified. The standards of identity for some foods include a list of optional ingredients. The standard of identity for frozen concentrated orange juice (21 CFR 146.146) is given in Fig. 2-1. Table 2-1 summarizes the standards of identity relevant to food analysis for a number of other foods. Note that the standard of identity often includes the recommended analytical method for determining chemical composition.

Although standards of quality and fill are less related to the chemical analysis of foods than are standards of identity, they are important for economic and quality control considerations. **Standards of quality,** established by the FDA for some canned fruits and vegetables, set minimum standards and specifications for factors such as color, tenderness, weight of units in the container, and freedom from defects. The **standards of fill** established for some canned fruits and vegetables, tomato products, and seafoods state how full a container must be to avoid consumer deception.

2.2.2 U.S. Department of Agriculture

The USDA, created in 1862, is now one of the cabinet-level federal agencies within the executive branch of the U.S. government. The USDA administers several federal statutes relevant to food composition and analysis; some programs are mandatory and others voluntary.

2.2.2.1 Mandatory Inspection Programs for Fresh and Processed Food Commodities: Standards and Composition

2.2.2.1.1 Meat and Poultry The **Meat Poultry Inspection Program** (MPIP) administered by the USDA provides for inspecting the slaughter of certain domestic livestock and poultry and the processing of meat and poultry products. Such inspection applies to all meat and poultry products in interstate or foreign commerce, to prevent the sale and distribution of adulterated or misbranded products. The MPIP reviews foreign inspection systems and packing plants that export meat and poultry to the United States. Imported products are reinspected at ports of entry.

The MPIP derives its authority from the Federal Meat Inspection Act of 1906, updated in 1967; the Poultry Products Inspection Act of 1957; the **Agricultural Marketing Act** of 1946; the **Humane Slaughter Act** of 1958; and the **Imported Meat Act** (a part of the 1930 Tariff Act).

The regulations relating to the inspection and certification of meat and poultry products are published in Title 9 of the CFR. Comprehensive inspection manuals, such as the *Meat and Poultry Inspection Manual* (6), have been developed to assist MPIP and industry personnel to interpret and utilize the regulations. Standards of identity have been established for many meat products (9 CFR 319), commonly specifying percentages of meat, fat, and water. Analyses are to be conducted using AOAC methods, if available. The standard of identity for breakfast sausage is given in Fig. 2-2 as an example.

2.2.2.1.2 Grains The **Federal Grain Inspection Service** (FGIS) within the USDA administers the mandatory requirements of the **U.S. Grain Standards Act** of 1976 as amended. The regulations to enforce this act and provide for a national inspection system for grain are published in 7 CFR 800. Mandatory official grade standards exist for a number of grains, including barley, oats, wheat, corn, rye, flaxseed, sorghum, soybeans, and triticale. The FGIS has issued many handbooks and instructions for its inspectors, such as the *Grain Inspection Handbook* (7). Grades are determined by factors such as test weight per bushel and percentages of heat-damaged kernels, broken kernels, and foreign material. A grade limit is commonly set for moisture, which is as specified by contract or load order grade.

2.2.2.2 Voluntary Grading and Inspection Programs for Fresh and Processed Food Commodities: Standards and Composition

2.2.2.2.1 General Information on Grade Standards Although **grade standards** developed for foods

§ 146.146 Frozen concentrated orange juice.

(a) Frozen concentrated orange juice is the food prepared by removing water from the juice of mature oranges as provided in § 146.135, to which juice may be added unfermented juice obtained from mature oranges of the species *Citrus reticulata*, or hybrids thereof, or of *Citrus aurantium*, or both. However, in the unconcentrated blend the volume of juice from *Citrus reticulata* shall not exceed 10 percent and from *Citrus aurantium* shall not exceed 5 percent. The concentrate so obtained is frozen. In its preparation, seeds (except embryonic seeds and small fragments of seeds that cannot be separated by good manufacturing practice) and excess pulp are removed, and a properly prepared water extract of the excess pulp so removed may be added. Orange oil, orange pulp, orange essence (obtained from orange juice), orange juice and other orange juice concentrate as provided in this section or concentrated orange juice for manufacturing provided in § 146.153 (when made from mature oranges), water, and one or more of the optional sweetening ingredients specified in paragraph (b) of this section may be added to adjust the final composition. The juice of *Citrus reticulata* and *Citrus aurantium*, as permitted by this paragraph, may be added in single strength or concentrated form prior to concentration of the *Citrus sinensis* juice, or in concentrated form during adjustment of the composition of the finished food. The addition of concentrated juice from *Citrus reticulata* or *Citrus aurantium*, or both, shall not exceed, on a single-strength basis, the 10 percent maximum for *Citrus reticulata* and the 5 percent maximum for *Citrus aurantium* prescribed by this paragraph. Any of the ingredients of the finished concentrate may have been so treated by heat as to reduce substantially the enzymatic activity and the number of viable microorganisms. The finished food is of such concentration that when diluted according to label directions the diluted article will contain not less than 11.8 percent by weight of orange juice soluble solids, exclusive of the solids of any added optional sweetening ingredients. The dilution ratio shall be not less than 3 plus 1. For the purposes of this section and § 146.150, the term "dilution ratio" means the whole number of volumes of water per volume of frozen concentrate required to produce orange juice from concentrate having orange juice soluble solids of not less than 11.8 percent by weight exclusive of the solids of any added optional sweetening ingredients.

(b) The optional sweetening ingredients referred to in paragraph (a) of this section are sugar, sugar sirup, invert sugar, invert sugar sirup, dextrose, corn sirup, dried corn sirup, glucose sirup, and dried glucose sirup.

(c) If one or more of the sweetening ingredients specified in paragraph (b) of this section are added to the frozen concentrated orange juice, the label shall bear the statement "——— added", the blank being filled in with the name or an appropriate combination of names of the sweetening ingredients used. However, for the purpose of this section, the name "sweetener" may be used in lieu of the specific name or names of the sweetening ingredients.

(d) The name of the food concentrated to a dilution ratio of 3 plus 1 is "frozen concentrated orange juice" or "frozen orange juice concentrate". The name of the food concentrated to a dilution ratio greater than 3 plus 1 is "frozen concentrated orange juice, ——— plus 1" or "frozen orange juice concentrate, ——— plus 1", the blank being filled in with the whole number showing the dilution ratio; for example, "frozen orange juice concentrate, 4 plus 1". However, where the label bears directions for making 1 quart of orange juice from concentrate (or multiples of a quart), the blank in the name may be filled in with a mixed number; for example, "frozen orange juice concentrate, 4⅓ plus 1". For containers larger than 1 pint, the dilution ratio in the name may be replaced by the concentration of orange juice soluble solids in degrees Brix; for example, a 62° Brix concentrate in 3½-gallon cans may be named on the label "frozen concentrated orange juice, 62° Brix".

(e) Wherever the name of the food appears on the label so conspicuously as to be easily seen under customary conditions of purchase, the statements specified in this section for naming the optional ingredients used shall immediately and conspicuously precede or follow the name of the food, without intervening written, printed, or graphic matter.

(f) Nothing in this section is intended to interfere with the adoption and enforcement by any State, in regulating the production of frozen concentrated orange juice in such State, of State standards, consistent with this section, but which impose higher or more restrictive requirements than those set forth in this section.

FIGURE 2-1.
Standard of identity for frozen concentrated orange juice. From 21 CFR 146.146 (1993).

TABLE 2-1. Selected Chemical Composition Requirements of Some Foods with Standards of Identity

			AOAC METHOD[2]		
Section in 21 CFR[1]	*Food Product*	*Requirement*	*Number in 13th Ed.*	*Number in 15th Ed.*	*Name/ Description*
131.110	Milk	Milk Solids Non Fat ≥ 8¼%	16.032	925.23A	Total solids, by hot air oven
		Milkfat ≥ 3¼%	16.059	905.02	Roese-Gottlieb
		Vit. A (if added) ≥ 2000 IU[3]/qt[4]			
		Vit. D (if added) - 400 IU[3]/qt[4]	43.195–43.208	936.14	Bioassay line test with rats
131.125	Nonfat Dry Milk	Moisture ≤ 5% by wt.	16.192	927.05	Vacuum oven
		Milkfat ≤ 1½% by wt.	16.199–16.200	932.06A, 932.06B	Roese-Gottlieb
131.160	Sour Cream	Milkfat ≥ 18%[5]	16.172	945.48G	Roese-Gottlieb
		Titratable Acidity ≥ 0.5%, calculated as lactic acid	16.023	947.05	Titration with NaOH
133.113	Cheddar Cheese	Milkfat ≥ 50% by wt. of solids	16.255	933.05	Digest with HCl; Roese-Gottlieb
		Moisture ≤ 39% by wt.	16.233	926.08	Vacuum oven
		Phosphatase Level ≤ 3 µg phenol equivalent/0.25 g[6]	16.275–16.277	946.03–946.03C	Residual Phosphatase
135.110	Ice Cream and Frozen Custard	Total Solids ≥ 1.6 lb/gal	16.287, 16.054	952.06, 953.08D	Roese-Gottlieb
		Milkfat ≥ 10%			
		Nonfat Milk Solids ≥ 10%[7]			
137.165	Enriched Flour	Moisture ≤ 15%	14.002, 14.003	925.09, 925.09B	Vacuum oven
		Ascorbic Acid ≤ 200 ppm (if added as dough conditioner)			
		Ash[8] ≤ (0.35 + 1/20 of the percent of protein, calculated on dwb[9])	14.006	923.03	Dry ashing
		(Protein)	2.057	955.04C	Kjeldahl, for nitrate-free samples
		Thiamine, 2.9 mg/lb			
		Riboflavin, 1.8 mg/lb			
		Niacin, 24 mg/lb			
		Iron, 20 mg/lb			
		Calcium (if added), 960 mg/lb			
137.230	Corn Grits	Crude Fiber ≤ 1.2% dwb[9]	14.062, 14.065	945.38A, 945.38D (modified)	Crude fiber
		Fat ≤ 2.25%	14.062, 14.067	945.38A, 945.38F	Ether extraction
		(Moisture)	14.062, 14.063	945.38A, 945.38B	Vacuum oven
145.110	Canned Applesauce	Soluble Solids ≥ 9%[10]	22.024	932.12	Refractometer
146.185	Pineapple Juice	Soluble Solids ≥ 10.5°Brix[11]	31.009	932.14A	Hydrometer (Brix spindle)
		Total Acidity ≤ 1.35g/100ml (as anhydrous citric acid)	[12]		Titration with NaOH

by the USDA are not legal requirements, they are widely used voluntarily by food processors and distributors as an aid in wholesale trading, since the quality of a product affects its price. The USDA has issued grade standards for more than 300 food products under authority of the Agricultural Marketing Act of 1946 and related statutes. Grade standards exist for many types of meats, poultry, dairy products, fruits, vegetables, and grains, along with eggs, domestic rabbits, certain preserves, dry beans, rice, and peas. Additional information about each of these is given in the next several sections, except for dairy products, which is given in section 2.3. The USDA has published for consumers a summary of these standards, in *USDA Grade Standards for Food and Farm Products* (8). Complete information regarding the standards is published in the CFR.

Grade standards, issued by the USDA for agricultural products and by the Department of Commerce for fishery products, must not be confused with standards of quality or standards of identity set by the FDA, as discussed above. A **standard of identity** estab-

TABLE 2-1. *Continued*

			AOAC METHOD[2]		
Section in 21 CFR[1]	*Food Product*	*Requirement*	*Number in 13th Ed.*	*Number in 15th Ed.*	*Name/ Description*
		Brix/Acid Ratio ≥ 12	[13]		Calculated
		Insoluble Solids ≥ 5 and ≤ 30%	[14]		Calculated from volume of sediment
158.170	Frozen Peas	Alcohol Insoluble Solids ≤ 19%	[15]		Extract with alcohol solution; dry insoluble material
163.113	Cocoa	Cocoa Fat < 22% and ≥ 10%	13.031	925.07 (modified)	Extraction with petroleum ether
164.150	Peanut Butter	Fat ≤ 55%	27.006(a)	948.22	Ether extraction with Soxhlet unit
168.120	Glucose Syrup	Total Solids ≥ 70% mass/mass (m/m)	31.208–31.209	941.14A, 941.14B	Vacuum oven, with diatomaceous earth
		Reducing Sugar ≥ 20% m/m (dextrose equivalent, dwb)	31.220(a)	945.66(a)	Lane-Eynon
		Sulfated Ash ≤ 1% m/m, dwb	31.216	945.63B	Dry ashing
		Sulfur Dioxide ≤ 20 mg/kg	20.106–20.111	962.16A–963.20C	Modified Monier-Williams
169.175	Vanilla Extract	Ethyl Alcohol ≥ 35% by volume			
		Vanilla Constituent ≥ 1 unit[16]/gal			

[1]CFR, Code of Federal Regulations
[2]*Official Methods of Analysis* of the Association of Official Analytical Chemists (AOAC) International
[3]IU = International Units
[4]Within limits of good manufacturing practice
[5]Exceptions allowed are explained
[6]If pasteurized dairy ingredients are used
[7]Exceptions clarified
[8]Excluding ash resulting from any added iron or salts of iron or calcium or wheat germ
[9]dwb = moisture-free or dry weight basis
[10]Exclusive of the solids of any added optional nutritive carbohydrate sweeteners; expressed as % sucrose, °Brix, with correction for temperature to the equivalent at 20°C
[11]Exclusive of added sugars, without added water. As determined by refractometer at 20°C uncorrected for acidity and read as °Brix on International Sucrose Scales. Exception stated for juice from concentrate.
[12]Detailed titration method given in 21 CFR 145.180 (b)(2)(ix)
[13]Calculated from °Brix and total acidity values, as described in 21 CFR 146.185 (b)(2)(iii)
[14]Detailed method given in 21 CFR 146.185 (b)(2)(iv)
[15]Detailed method given in 21 CFR 158.170 (b)(3)
[16]Defined in 21 CFR 169.3(c); requires measurement of moisture content, according to modification of AOAC Method 7.004 and 7.005

lishes or defines what a given food product is; it establishes for some foods which ingredients they must contain. **Standards of quality** are the minimum standards for some canned fruits and vegetables. **Standards for grades** may classify products from average to excellent in quality. Standards for grades are not required to be stated on the label, but if they are stated, the product must comply with the specifications of the declared grade. Official USDA grading services are provided, for a fee, to packers, processors, distributors, and others who seek official certification of the grades of their products. Such grade standards are often used as quality control tools.

2.2.2.2.2 Fruits and Vegetables The USDA is responsible for assuring the quality of fruits, vegetables, and related products sold in the United States. The Fresh Products Branch and the Processed Products Branch of the **Fruits and Vegetables Quality Division of the USDA** standardize, grade, and inspect fruits and vegetables under various voluntary programs. The regulations promulgated for fresh fruits and vegetables are given in 7 CFR 51 and those for processed fruits and vegetables are given in 7 CFR 52. Standards for grades of processed fruits and vegetables often include factors such as color, texture or consistency, defects, size and shape, tenderness, maturity, flavor, and a variety of

§ 319.143 Breakfast sausage.

"Breakfast Sausage" is sausage prepared with fresh and/or frozen meat; or fresh and/or frozen meat and meat byproducts, and may contain Mechanically Separated (Species) in accordance with § 319.6, and may be seasoned with condimental substances as permitted in Part 318 of this subchapter. The finished product shall not contain more than 50 percent fat. To facilitate chopping or mixing, water or ice may be used in an amount not to exceed 3 percent of the total ingredients used. Extenders or binders as listed in Part 318 of this subchapter may be used to the extent of 3½ percent of the finished sausage as permitted in § 319.140.

[35 FR 15597, Oct. 3, 1970, as amended at 43 FR 26424, June 20, 1978; 47 FR 28257, 28258, June 29, 1982]

FIGURE 2-2.
Standard of identity for breakfast sausage. From 9 CFR 319.143 (1993).

chemical characteristics. Sampling procedures and methods of analysis are commonly given. As examples, partial information about the standards for grades of frozen concentrated orange juice and sugarcane molasses are given in Tables 2-2 and 2-3, respectively.

2.2.2.2.3 Meat and Poultry The **Meat Quality Division of the USDA** provides voluntary grading and certification services, as described in 7 CFR 53 (Livestock) and 54 (Meat, Prepared Meats, and Meat Products). The **Poultry Quality Division of the USDA** provides voluntary inspection and grading services for egg products (7 CFR 55), voluntary grading of shell eggs (7 CFR 56), voluntary grading of poultry products and rabbit products (7 CFR 70), voluntary inspection of poultry (9 CFR 362), and voluntary inspection of rabbits and their edible products (9 CFR 354).

The voluntary inspection and grading program for egg products covers services such as laboratory analyses required but not covered by the mandatory inspection program that exist for eggs and egg products (7 CFR 59) under the **Egg Products Inspection Act.** Similarly, the voluntary poultry inspection program exists for poultry products not covered by the mandatory regulations of the Poultry Products Inspection Act (9 CFR 381).

2.2.2.4.4 Other Agricultural Commodities The FGIS implements voluntary regulations and standards for inspection and certification of certain agricultural commodities and their products. Such regulations and standards for beans, rough rice, brown rice, milled rice, whole dry peas, split peas, and lentils are given in 7 CFR 68. Grade standards for these products are commonly determined by factors such as defects, presence of foreign material, and insect infestation. The standard for beans (7 CFR 68.101-68.142) also specifies a moisture content of 18 percent. Beans with more than 18 percent moisture are graded as "high moisture." The regulations state that moisture is to be determined by the use of equipment and procedures prescribed by the FGIS, or by any method that gives equivalent results. Laboratory test services are made available (9) (Table 2-4), at a fee, for these agricultural commodities, as they are for other food products through inspection and grading programs.

2.2.3 U.S. National Marine Fisheries Service

The **National Marine Fisheries Service** (NMFS) is one of the major services administered under the National Oceanic and Atmospheric Administration, which is part of the **U.S. Department of Commerce.** The Service assures the safety and quality of seafoods consumed in the United States through **voluntary grading, standardization, and inspection programs,** as described in 50 CFR 260-267. The NMFS has also published a very comprehensive manual on these subjects entitled *Fishery Products Inspection Manual* (10). The **U.S. Standards for Grades of Fishery Products** are intended to help the fishing industry maintain and improve product quality and to thereby increase consumer confidence in seafoods. Standards are based on attributes such as color, size, texture, flavor, odor, workmanship defects, and consistency.

A mandatory seafood system inspection is expected within the next few years. The U.S. Congress in 1986 requested NMFS to design a new **mandatory seafood inspection program** based on the **Hazard Analysis Critical Control Point** (HACCP) concept. Legislative proposals have been introduced recently into Congress to require mandatory federal inspection of all seafood and seafood products (11).

The FDA and the EPA work with the NMFS for the assurance of seafood safety. The FDA, under the FD&C Act, is responsible for ensuring that seafood shipped or received in interstate commerce is safe, wholesome, and not misbranded or deceptively packaged. The FDA has primary authority in setting and enforcing allowable levels of contaminants and pathogenic microorganisms in seafood. The EPA assists the FDA in identifying the range of chemical contaminants that pose a human health risk and are most likely to accumulate in seafood. A tolerance of 2.0 parts per million (ppm) for total polychlorinated biphenyls (PCBs) is the only formal tolerance specified by the FDA to date to mitigate human health impacts in seafood (12).

TABLE 2-2. USDA Standards for Grades of Frozen Concentrated Orange Juice

Factors	*Grade A*	*Grade B*
QUALITY:[1]		
Appearance	Fresh orange juice	Fresh orange juice
Reconstitution	Reconstitutes properly	Reconstitutes properly
Color	Very good (Equal to or better than USDA OJ 5)	Good (Not as good as USDA OJ 5, but not off color)
Score points	36–40	32–35
Defects	Practically free	Reasonably free
Score points	18–20	16–17
Flavor	Very good	Good
Score points	36–40	32–35
Total score points	Minimum, 90	Minimum, 80

ANALYTICAL:	*Grade A Unsweetened*		*Grade A Sweetened*		*Grade B Unsweetened*		*Grade B Sweetened*	
Concentrate:								
Brix value: Minimum	41.8[2]		42.0[2]		41.8[2]		42.0[2]	
Brix value/acid ratio:	*Min*	*Max*	*Min*	*Max*	*Min*	*Max*	*Min*	*Max*
California/Arizona	11.5:1	19.5:1	12.0:1	19.5:1	10.0:1		10.0:1	
Outside California/Arizona	12.5:1	19.5:1	13.0:1	19.5:1	10.0:1		10.0:1	
Reconstituted juice:								
Brix: Minimum	11.8[2]				11.8[2]			
Soluble orange solids, exclusive of sweetener (percent by weight of finished product): Minimum			11.8				11.8	
Recoverable oil (percent by volume): Maximum		0.035				0.040		

From 7 CFR 52.1557 (1993)

[1]Reconstituted prior to grading

[2]Definitions of terms (7 CFR 52.1553) (1993). In these U.S. standards, unless otherwise required by the context, the following terms shall be construed, respectively, to mean:

(a) *Acid* means the percent, by weight, of total acidity (calculated as anhydrous citric acid).

(b) *Appearance* means the physical properties of orange juice which are evaluated by the human eye.

(c) *Brix* means the total soluble solids as determined when tested with a Brix hydrometer and applying the applicable temperature correction. The Brix may be determined by any other method that gives equivalent results.

(d) *Brix/acid ratio* means the ratio of the degrees Brix of the juice to the grams of anhydrous citric acid per 100 grams of the juice.

(e) *Brix value* means the refractometric sucrose value determined in accordance with the "International Scale of Refractive Indices of Sucrose Solutions" and to which the applicable correction for acid is added. The Brix value is determined in accordance with the refractometric method outlined in the "Official Methods of Analysis of the Association of Official Analytical Chemists."

(f) *Brix value/acid ratio* means the ratio of the Brix value of the concentrate, in degrees Brix, to the grams of anhydrous citric acid per 100 grams of concentrate.

2.2.4 U.S. Bureau of Alcohol, Tobacco, and Firearms

2.2.4.1 Regulatory Responsibility for Alcoholic Beverages

Beer, wines, liquors, and other alcoholic beverages are termed "food" according to the FD&C Act of 1938. However, regulatory control over their quality, standards, manufacture, and other related aspects is specified by the **Federal Alcohol Administration Act,** which is enforced by the **Bureau of Alcohol, Tobacco, and Firearms** of the **U.S. Department of the Treasury.** Issues regarding the composition and labeling of most alcoholic beverages are handled by the Bureau. However, the FDA has jurisdiction over certain other alcoholic beverages and cooking wines. The FDA also deals with questions of sanitation, filth, and the presence of deleterious substances in alcoholic beverages.

TABLE 2-3. USDA Standards for Grades of Sugarcane Molasses. Required Minimum Brix Solids and Total Sugar and Maximum Ash and Total Sulfite[1]

	MINIMUM				MAXIMUM		
	Brix solids (percent)		*Total sugar (percent)*		*Ash (percent)*		*Total sulfites (ppm)*
Grade Designation	*Average from all containers*	*Limit for individual containers*	*Average from all containers*	*Limit for individual containers*	*Average from all containers*	*Limit for individual containers*	*Average from all containers*
Grade A	79.0	78.5	63.5	63.0	5.00	5.25	200
Grade B	79.0	78.5	61.5	61.0	7.00	7.50	250
Grade C	79.0	78.5	58.0	57.0	9.00	10.00	250
SStd	Under 79.0	—	Under 58.0	—	Over 9.00	—	Over 250

From 7 CFR 52.3651 (1993).
[1]Definitions and Methods of Analysis (7 CFR 52.3663–3668) (1993).
52.3663 *Brix solids.* Brix solids means the applicable solids content of sugarcane molasses or the Brix value as determined by the double dilution method by means of a Brix hydrometer corrected to 20°C (68°F).
52.3664 *Ash.* Percent ash means the ash content of sugarcane molasses determined as sulfated ash.
52.3665 *Sulfur dioxide, p.p.m.* Sulfur dioxide, p.p.m., means the total sulfites determined by the Monier-Williams method calculated as parts per million of sulfur dioxide (SO_2).
52.3666 *Reducing sugars.* The percent of reducing sugars is determined by the Lane-Eynon volumetric method for reducing sugars.
52.3667 *Sucrose.* The percentage of sucrose is determined by the Clerget or double polarization method, using invertase as the inverting agent.
52.3668 *Total sugar.* The percent of total sugar is the sum of the percent of reducing sugars and the percent of sucrose.

TABLE 2-4. Laboratory Test Services Available Through Federal Grain Inspection Service

1. Alpha monoglycerides
2. Aflatoxin test (other than TLC or minicolumn method)
3. Aflatoxin (TLC)
4. Aflatoxin (minicolumn method)
5. Appearance and color
6. Ash
7. Bacteria count
8. Baking test (cookies)
9. Bostwick (cooked)
10. Bostwick (uncooked/cook test/dispersibility)
11. Brix
12. Calcium
13. Carotenoid color
14. Cold test (oil)
15. Color test (syrups)
16. Cooking test (other than corn soy blend)
17. Crude fat
18. Crude fiber
19. Dough handling (baking)
20. *E. coli*
21. Falling number
22. Fat (acid hydrolysis)
23. Fat-stability (A.O.M.)
24. Flash point (open and closed cup)
25. Free fatty acid
26. Hydrogen ion activity-pH
27. Iron enrichment
28. Iodine number/value
29. Linolenic acid (fatty acid profile)
30. Lipid phosphorous
31. Lovibond color
32. Margarine (nonfat solids)
33. Moisture
34. Moisture average (crackers)
35. Moisture and volatile matter
36. Performance test (prepared bakery mix)
37. Peroxide value
38. Pesticide residue (carbon tetrachloride, methyl bromide, and ethylene dibromide)
39. Phosphorus
40. Popcorn kernels (total defects)
41. Popping ratio/value popcorn
42. Potassium bormate
43. Protein
44. Rope spore count
45. *Salmonella*
46. Salt or sodium content
47. Sanitation (filth light)
48. Sieve test
49. Smoke point
50. Solid fat index
51. Specific volume (bread)
52. *Staphylococcus aureus*
53. Texture
54. Tilletia controversa kuhn (TCK)
55. Unsaponifiable matter
56. Urease activity
57. Visual exam (hops pellet)
58. Visual exam (insoluble impurities, oils and shortening)
59. Visual exam (pasta)
60. Visual exam (processed grain products)
61. Visual exam (total foreign material other than cereal grains)
62. Vitamin enrichment
63. Water activity
64. Wiley melting point
65. Other laboratory tests

Adapted from 7 CFR 68.90 (1993) and (9).

2.2.4.2 *Standards and Composition of Beer, Wine, and Distilled Beverage Spirits*

Information related to **definitions, standards of identity,** and certain **labeling requirements** for beer, wine, and distilled beverage spirits is given in 27 CFR 1-299. Standards of identity for these types of beverages stipulate the need for analyses such as percent alcohol by volume, total solids content, volatile acidity, and °Brix. For example, the fruit juice used for the production of wine is often specified by its °Brix and total solids content. The maximum volatile acidity (calculated as acetic acid and exclusive of sulfur dioxide) for grape wine must not be more than 0.14g/100 ml (20°C) for natural red wine and 0.12g/100 ml for other grape wine (27 CFR 4.21). The percent alcohol by volume is often used as a criterion for class or type designation of alcoholic beverages. For example, dessert wine is grape wine with an alcoholic content in excess of 14 percent but not in excess of 24 percent by volume, while table wines have an alcoholic content not in excess of 14 percent alcohol by volume (27 CFR 4.21). No product with less than 0.5 percent alcohol by volume is permitted to be labeled "beer," "lager beer," "lager," "ale," "porter," "stout," or any other class or type designation normally used for malt beverages with higher alcoholic content (27 CFR 7.24).

2.2.5 U.S. Environmental Protection Agency

The EPA was established as an independent agency in 1970 through a reorganization plan to consolidate certain federal government environmental activities. The EPA regulatory activities most relevant to this book are control of pesticide residues in foods, drinking water safety, and the composition of effluent from food processing plants.

2.2.5.1 *Pesticide Residues*

Pesticides are chemicals intended to protect our food supply by controlling harmful insects, diseases, rodents, weeds, bacteria, and other pests. However, most pesticide chemicals can have harmful effects on people, animals, and the environment if they are improperly used. The two federal laws relevant to protection of food from pesticide residues are the **Federal Insecticide, Fungicide, and Rodenticide Act** (FIFRA), as amended, and certain provisions of the Federal FD&C Act. FIFRA, supplemented by the FD&C Act, authorizes a comprehensive program to regulate the manufacturing, distribution, and use of pesticides, along with a research effort to determine the effects of pesticides.

Section 408 of the FD&C Act is a special pesticide amendment that authorizes the EPA to establish an **allowable limit** or **tolerance** for any detectable pesticide residues that might remain in or on a harvested food or feed crop. The tolerance level is often many times the level expected to produce undesirable health effects in man or animal. While the EPA establishes the tolerance levels, the FDA enforces the regulations by collecting and analyzing food samples, mostly agricultural commodities. Livestock and poultry samples are collected and analyzed by the USDA. Pesticide residue levels that exceed the established tolerances are considered in violation of the FD&C Act.

Regulations regarding pesticide tolerances in raw agricultural chemicals are given in 40 CFR 180. The **pesticide tolerance levels** in foods is included in 40 CFR 185, with such pesticides described as "food additives permitted in food for human consumption." The 40 CFR 180 specifies general categories of products and specific commodities with tolerances or exemptions, and in some cases which part of the agricultural product is to be examined. Agricultural products covered include a wide variety of both plants (e.g., fruits, vegetables, grains, legumes, nuts) and animals (e.g., poultry, cattle, hogs, goats, sheep, horses, eggs, milk). Unless otherwise noted, the specific tolerances established for the pesticide chemical apply to residues resulting from their application prior to harvest or slaughter. Tolerances are expressed in terms of parts by weight of the pesticide chemical per 1 million parts by weight of the raw agricultural commodity (i.e., ppm). For example, the residue tolerance for the pesticide picloram is 0.05 ppm for eggs, milk, and poultry; 0.2 ppm for meat from cattle, hogs, goats, sheep, and horses; 0.5 ppm for the grain of barley, oats, and wheat; and 1 ppm for the green forage and straw of these plants (40 CFR 180.292) (Table 2-5). Tolerance levels for selected pesticides and insecticides permitted in foods as food additives are given in Table 2-6.

The analytical methods to be used for determining whether pesticide residues are in compliance with the tolerance established are identified among the methods contained or referenced in the *Pesticide Analytical Manual* (13) maintained by and available from the FDA. The method must be sensitive and reliable at and above the tolerance level. Pesticides are generally detected and quantitated by gas chromatographic methods (see Chapters 19 and 30).

2.2.5.2 *Drinking Water Standards and Contaminants*

The EPA administers the **Safe Drinking Water Act** of 1974, which is to provide for the safety of drinking water supplies in the United States and to enforce national drinking water standards. The EPA has identi-

TABLE 2-5. Tolerances for Residences of the Pesticide Picloram

Commodity	*Parts per Million*
Barley, grain	0.5
Barley, green forage	1
Barley, straw	1
Cattle, fat	0.2
Cattle, kidney	5
Cattle, liver	0.5
Cattle, mbyp (exc kidney and liver)	0.2
Cattle, meat	0.2
Eggs	0.05
Flax, seed	0.5
Flax, straw	0.5
Goats, fat	0.2
Goats, kidney	5
Goats, liver	0.5
Goats, mbyp (exc kidney and liver)	0.2
Goats, meat	0.2
Grasses, forage	80
Hogs, fat	0.2
Hogs, kidney	5
Hogs, liver	0.5
Hogs, mbyp (exc kidney and liver)	0.2
Hogs, meat	0.2
Horses, fat	0.2
Horses, kidney	5
Horses, liver	0.5
Horses, mbyp (exc kidney and liver)	0.2
Horses, meat	0.2
Milk	0.05
Oats, grain	0.5
Oats, green forage	1
Oats, straw	1
Poultry, fat	0.05
Poultry, mbyp	0.05
Poultry, meat	0.05
Sheep, fat	0.2
Sheep, kidney	5
Sheep, liver	0.5
Sheep, mbyp (exc kidney and liver)	0.2
Sheep, meat	0.2
Wheat, grain	0.5
Wheat, green forage	1
Wheat, straw	1

From 40 CFR 180.292 (1993).

fied potential contaminants of concern and established their maximum acceptable levels in drinking water. The EPA has primary responsibility to establish the standards, while the states enforce them and otherwise supervise public water supply systems and sources of drinking water. **Primary and secondary drinking water regulations** have been established; enforcement of the former is mandatory, whereas enforcement of the latter is optional. The national primary and secondary drinking water regulations are given in 40 CFR 141 and 143, respectively. Recently, concerns have been expressed regarding the special standardization of water used in the manufacturing of foods and beverages.

Maximum contaminant levels (MCL) for primary drinking water are set for **inorganic and organic chemicals** (Table 2-7), **turbidity, certain types of radioactivity,** and **microorganisms.** Sampling procedures and analytical methods for the analysis of chemical contaminants are specified, with common reference to *Standard Methods for the Examination of Water and Wastewater* (14), published by the American Public Health Association; *Methods of Chemical Analysis of Water and Wastes* (15), published by the EPA; and *Annual Book of ASTM Standards* (16), published by the American Society for Testing Materials. Methods commonly specified for the analysis of inorganic contaminants in water include atomic absorption (direct aspiration or furnace technique), inductively coupled plasma (see Chapter 25), ion chromatography (see Chapter 29), and ion selective electrode (see Chapter 31) (Table 2-8).

2.2.5.3 Effluent Composition from Food Processing Plants

In administering the **Federal Water Pollution and Control Act,** the EPA has developed effluent guidelines and standards that cover various types of food processing plants. Regulations promulgated under 40 CFR 402-699 prescribe effluent limitation guidelines for existing sources, standards of performance for new sources, and pretreatment standards for new and existing sources. Point sources of discharge of pollution are required to comply with these regulations, where applicable. Regulations are prescribed for specific foods under the appropriate point source category: dairy products processing (40 CFR 405), grain mills (40 CFR 406), canned and preserved fruits and vegetables processing (40 CFR 407), canned and preserved seafood processing (40 CFR 408), sugar processing (40 CFR 409), and meat products (40 CFR 432). **Effluent characteristics** commonly prescribed for food processing plants are **biochemical oxygen demand** (BOD) (see Chapter 21), **total soluble solids** (TSS) (see Chapter 7), and **pH** (see Chapter 31), as shown in Table 2-9 for effluent from a plant that makes natural and processed cheese. The test procedures for measurement of effluent characteristics are prescribed in 40 CFR 136.

2.2.6 U.S. Customs Service

Over 100 countries export food, beverages, and related edible products to the United States. The **U.S. Customs Service** (USCS) assumes the central role in ensuring that imported products are taxed properly, safe for human consumption, and not economically deceptive. The USCS receives assistance from the FDA and USDA as it assumes these responsibilities. The major regulations promulgated by the USCS are given in Title 19 of the CFR.

TABLE 2-6. Tolerances for Selected Pesticides and Insecticides Classified as Food Additives Permitted in Foods for Human Consumption

Section	Food Additive	Food	Tolerance
185.1000	Chlorpyrifos	Citrus oil	25.0
		Corn oil	3.0
		Mint oil	10.0
		Peanut oil	1.5
185.1050	Chlorpyrifos-methyl	Barley, milled fractions[1]	90
		Oats, milled fractions[1]	130
		Sorghum, milled fractions	90
		Rice, milled fractions[1]	30
		Wheat, milled fractions[1]	30
185.1580	Deltamethrin	Tomato products (concentrated)	1.0
185.3300	Flucythrinate	Cottonseed oil	0.2
185.4850	Picloram	Barley, milled fractions	3
		Oats, milled fractions	3
		Wheat, milled fractions	3
185.5000	Propargite	Figs, dried	9
		Hops, dried	30
		Raisins	25
		Tea, dried	10

Adapted from 40 CFR 185 (1993).
[1]Except flour

2.2.6.1 Harmonized Tariff Schedule of the United States (TSUSA)

All goods imported into the United States are subject to duty or duty-free entry according to their classification under applicable items in the ***Harmonized Tariff Schedule of the United States*** (TSUSA). The TSUSA can be purchased in an annotated looseleaf edition from the U.S. Government Printing Office (17). The U.S. tariff system has official tariff schedules for over 400 edible items exported into the United States. The TSUSA specifies the food product in detail and gives the general rate of duty applicable to that product coming from most countries and any special higher or lower rates of duty for certain other countries.

2.2.6.2 Food Composition and the TSUSA

The **rate of duty** for certain food products is determined by their chemical composition. The rate of duty on some dairy products is determined in part by the fat content, as shown for milk and cream in Table 2-10. The tariff for some syrups is determined by the fructose content, for some chocolate products by the sugar or butterfat content, for butter substitutes by the butterfat content, and for some wines by their alcohol content (percent by volume).

2.2.7 U.S. Federal Trade Commission

The **Federal Trade Commission** (FTC) is the most influential of the federal agencies that have authority over various aspects of advertising and sales promotion practices for foods in the United States. The major role of the FTC is to keep business and trade competition free and fair.

2.2.7.1 Enforcement Authority

The **Federal Trade Commission Act** of 1914 authorizes the FTC to protect both the consumer and the businessperson from anticompetitive behavior and unfair or deceptive business and trade practices. The FTC periodically issues industry guides and trade regulations rules that tell businesses what they can and cannot do. These issuances are supplemented with advisory opinions given to corporations and individuals upon request. The proposal of any new rules, guides, or regulations is preceded by widespread notice or announcement in the Federal Register, and comments are invited. The FTC not only has guidance and preventive functions but is also authorized to issue complaints or shut-down orders and sue for civil penalties for violation of trade regulation rules. The **Bureau of Consumer Protection** is one of the FTC bureaus that enforce and develop trade regulation rules.

2.2.7.2 Food Labels, Food Composition, and Deceptive Advertising

While the **Fair Packaging and Labeling Act** of 1966 is administered by the FTC, that agency does not have specific authority over the packaging and labeling of

TABLE 2-7. Maximum Contaminant Levels for Inorganic and Organic Chemicals in Primary Drinking Water

Contaminant	*Level Milligrams per Liter*
INORGANIC	
Arsenic	0.05
Barium	1
Cadmium	0.010
Chromium	0.05
Lead	0.05
Mercury	0.002
Nitrate (as N)	10
Selenium	0.01
Silver	0.05
ORGANIC	
(a) Chlorinated hydrocarbons:	
Endrin (1,2,3,4,10, 10-hexachloro-6, 7-epoxy-1,4, 4a,5,6,7,8,81-octahydro-1,4-endo, endo-5,8-dimethano naphthalene)	0.0002
Lindane (1,2,3,4,5,6-hexachlorocyclo-hexane, gamma isomer)	0.004
Methoxychlor (1,1,1-Trichloro-2, 2-bis [p-methoxyphenyl] ethane)	0.1
Toxaphene ($C_{10}H_{10}Cl_8$-Technical chlorinated camphene, 67–69 percent chlorine)	0.005
(b) Chlorophenoxys:	
2,4-D, (2,4-Dichlorophenoxyacetic acid)	0.1
2,4,5-TP Silver (2,4,5-Trichlorophenoxy-propionic acid)	0.01
(c) Total trihalomethanes (the sum of the concentrations of bromodichloromethane, dibromochloromethane, tribromomethane (bromoform) and trichloromethane (chloroform))	0.10

Adapted from 40 CFR 141.11 and 141.12 (1993).

foods. The FTC and FDA have agreed upon responsibilities: FTC has primary authority over advertising of foods and FDA has primary authority over labeling of foods.

Grading, standards of identity, and labeling of foods regulated by several federal agencies as described above have eliminated many of the potential problems in the advertising of foods. Such federal regulations and voluntary programs have reduced the scope of advertising and other forms of product differentiation. Misleading, deceptive advertising is less likely to be an issue and is more easily controlled. For example, foods such as ice cream, mayonnaise, and peanut butter have standards of identity that set minimum ingredient standards. If these standards are not met, the food must be given a different generic designation (e.g., salad dressing instead of mayonnaise) or be labeled "imitation." Grading, standards, and labeling of food aid consumers in making price-quality comparisons. Once again, analyses of chemical composition play an important role in developing and setting these grades, standards, and labels. In many cases in which the FTC intervenes, data from a chemical analysis become central evidence for all parties involved.

2.3 REGULATIONS AND RECOMMENDATIONS FOR MILK

The safety and quality of milk and dairy products in the United States are the responsibility of both federal (FDA and USDA) and state agencies. The FDA has regulatory authority over the dairy industry in interstate commerce, while the USDA involvement with the dairy industry is voluntary and service oriented. Each state has its own regulatory office for the dairy industry within that state. The various regulations for milk involve several types of chemical analyses.

2.3.1 FDA Responsibilities

The FDA has responsibility under the FD&C Act, the Public Health Service Act, and the Import Milk Act to assure consumers that the U.S. milk supply and imported dairy products are safe, wholesome, and not economically deceptive. Processors of both Grade A and Grade B milk are required under FDA regulations to take remedial action when conditions exist that could jeopardize the safety and wholesomeness of milk and dairy products being handled. As described above in section 2.2.1.2, the FDA also promulgates standards of identity and labeling, quality, and fill-of-container requirements for milk and dairy products moving in interstate commerce.

For **Grade A milk and dairy products,** each state shares with the FDA the responsibility of assuring safety, wholesomeness, and economic integrity. This is done through a **Memorandum of Understanding with the National Conference on Interstate Milk Shipments,** which is comprised of all 50 states. In cooperation with the states and the dairy industry, the FDA has also developed for state adoption model regulations regarding sanitation and quality aspects of producing and handling Grade A milk. These regulations are contained in the ***Grade A Pasteurized Milk Ordinance*** (PMO) (18), which all states have adopted as minimum requirements.

The standards for Grade A pasteurized milk and milk products and bulk-shipped heat-treated milk products under the PMO are given in Table 2-11. The PMO specifies that "all sampling procedures and required laboratory examinations shall be in substantial compliance with the . . . Edition of *Standard Methods for the Examination of Dairy Products* of the American

TABLE 2-8. Detection Limits for Inorganic Contaminants in Drinking Water

Contaminant	*MCL*[1] *(mg/l)*	*Analytical Method*	*Detection Limit (mg/l)*
Asbestos	7 MFL[2]	Transmission electron microscopy	0.01 MFL
Barium	2	Atomic absorption; furnace technique	0.002
		Atomic absorption; direct aspiration	0.1
		Inductively coupled plasma	0.002(0.001)[3]
Cadmium	0.005	Atomic absorption; furnace technique	0.0001
		Inductively coupled plasma	0.001[3]
Chromium	0.1	Atomic absorption; furnace technique	0.001
		Inductively coupled plasma	0.007(0.001)[3]
Mercury	0.002	Manual cold vapor technique	0.0002
		Automated cold vapor technique	0.0002
Nitrate	10 (as N)	Manual cadmium reduction	0.01
		Automated hydrazine reduction	0.01
		Automated cadmium reduction	0.05
		Ion selective electrode	1
		Ion chromatography	0.01
Nitrite	1 (as N)	Spectrophotometric technique	0.01
		Automated cadmium reduction	0.05
		Manual cadmium reduction	0.01
		Ion chromatography	0.004
Selenium	0.05	Atomic absorption; furnace	0.002
		Atomic absorption; gaseous hydride	0.002

From 40 CFR 141.23 (1993).
[1]MCL = maximum contaminant level; maximum permissible level of a contaminant in water as specified in 40 CFR 141.2 (1993)
[2]MFL = million fibers per liter > 10 $_{\mu}$m
[3]Using concentration technique in Appendix A to EPA Method 200.7

TABLE 2-9. Effluent Limitations for Plants Processing Natural and Processed Cheese

	Effluent Characteristics					
	*Metric Units**			*English Units***		
Effluent Limitations	*BOD5*	*TSS*	*pH*	*BOD5*	*TSS*	*pH*
Processing *more than* 100,000 lb/day of milk equivalent						
Maximum for any 1 day	0.716	1.088	(1)	0.073	0.109	(1)
Avg of daily values for 30 consecutive days shall not exceed	0.290	.435	(1)	0.029	0.044	(1)
Processing *less than* 100,000 lb/day of milk equivalent						
Maximum for any 1 day	0.976	1.462	(1)	0.098	0.146	(1)
Avg of daily values for 30 consecutive days shall not exceed	0.488	.731	(1)	0.049	.073	(1)

Adapted from 40 CFR 405.62 (1993).
[1]Kilograms per 1,000 kg of BOD5 input
[2]Pounds per 100 lbs of BOD5 input
[3]Within the range 6.0–9.0

Public Health Association, and the . . . Edition of *Official Methods of Analysis* of the Association of Official Analytical Chemists. (Insert edition number current at time of adaption.)" (18–20).

The FDA monitors state programs for compliance with the PMO and trains state inspectors. To facilitate movement of Grade A milk in interstate commerce, a federal-state certification program exists: the **Interstate Milk Shippers** (IMS) **Program.** This program is maintained by the **National Conference on Interstate Milk Shipments,** which is a voluntary organization that includes representatives from each state, the FDA, the USDA, and the dairy industry. In this program, the producers of Grade A pasteurized milk are required to pass inspections and be rated by cooperating state agencies, based on PMO sanitary standards, requirements, and procedures. The ratings appear in the **IMS List,** which is published by the FDA, and made avail-

TABLE 2-10. U.S. Harmonized Tariff Schedule for Milk and Cream

Article Description	Units of Quantity	Rates of Duty 1 General[1]	Rates of Duty 1 Special[2]	Rates of Duty 2[3]
Milk and cream, not concentrated nor containing added sugar or other sweetening matter:				
Of a fat content, by weight, not exceeding 1%	liters	0.4¢/liter	Free (E, IL) 0.2¢/liter (CA)	0.5¢/liter
Of a fat content, by weight, exceeding 1% but not exceeding 6%:				
For not over 11,356,236 liters entered in any calendar year	liters	0.5¢/liter	Free (E, IL) 0.3¢/liter (CA)	1.7¢/liter
Other	liters	1.7¢/liter	Free (E, IL) 1¢/liter (CA)	1.7¢/liter
Of a fat content, by weight, exceeding 6%:				
Of a fat content, by weight, not exceeding 45%:				
Described in additional U.S. note 1(a) to this chapter *1/*	liters	3.2¢/liter	Free (E, IL) 1.9¢/liter (CA)	15¢/liter
Other	liters	15¢/liter	Free (E, IL) 9¢/liter (CA)	15¢/liter

Adapted from (17) (Section 0401).
[1]General tariff rate that applies to most countries
[2]Special lower tariff treatment, with symbols referring to specific programs
[3]Special tariff rate applying to certain countries

TABLE 2-11. Pasteurized Milk Ordinance Standards for Grade A Pasteurized Milk and Milk Products and Bulk-Shipped Heat-Treated Milk Products

Criteria	Requirement
Temperature	Cooled to 7°C (45°F) or less and maintained thereat.
Bacterial limits[1]	20,000 per ml.
Coliform	Not to exceed 10 per ml: Provided that, in the case of bulk milk transport tank shipments, shall not exceed 100 per ml.
Phosphatase[2]	Less than 1 microgram per ml by the Scharer Rapid Method or equivalent.
Antibiotics	No zone greater than or equal to 16 mm with the *Bacillus sterothermophilus* disc assay method specified.

Adapted from (18).
[1]Not applicable to cultured products
[2]Not applicable to bulk-shipped heat-treated milk products

able to state authorities and milk buyers to ensure the safety of milk shipped from other states.

2.3.2 USDA Responsibilities

Under authority of the Agricultural Marketing Act of 1946, the **Dairy Quality Division of the USDA** offers **voluntary grading services** for manufactured or processed dairy products (7 CFR 58). If USDA inspection of a dairy manufacturing plant shows that good sanitation practices are being followed to meet the requirements in the *General Specifications for Dairy Plants Approved for USDA Inspection and Grading Service* (21), the plant qualifies for the USDA services of grading, sampling, testing, and certification of its products. The laboratory analyses available, for a fee, through the inspection and grading service for various types of dairy products are listed in Table 2-12. A product such as nonfat dry milk is graded based on flavor, physical appearance, and various laboratory analyses, the last of which is given in Table 2-13.

As with the USDA voluntary grading programs for other foods described in section 2.2.2.2, the USDA has no regulatory authority regarding dairy plant inspections and cannot require changes in plant operations. The USDA can only decline to provide the grading services, which are available to the dairy plants for a fee. The USDA, under an arrangement with FDA, assists states in establishing safety and quality regulations for manufacturing-grade milk. Much as described above for the FDA with Grade A milk, the USDA has developed model regulations for state adoption regarding the quality and sanitation aspects of producing and handling manufacturing-grade milk. These regulations are given in the *Milk for Manufacturing Purposes*

TABLE 2-12. Laboratory Analysis Services Available Through Inspection and Grading and Services for Manufactured or Processed Dairy Products

Dry Milk and Related Products
Total fat (ether extraction)
Moisture
Titratable acidity
Solubility index
Scorched particles
Bacterial plate count
Bacterial direct microscopic count
Whey protein nitrogen
Vitamin A
Alkalinity of ash
Dispersibility
Coliform (solid media)
Salmonella
Phosphatase
Oxygen
Density
Antibiotic
Condensed Milk and Related Products
Fat (fat extraction)
Total solids
Sugar (sucrose)
Net weight (per can)
Cheese and Related Products
Moisture
Moisture in duplicate
Total fat (ether extraction)
Moisture and fat (dry basis) complete
Meltability (Process cheese)
Butter and Related Products
Moisture
Fat
Salt
Complete Kohman analysis
Fat and moisture (same sample)
Peroxide value
Free fatty acid
Yeast and mold
Proteolytic count
Meat and Related Products
Fat (hamburger)

Adapted from 7 CFR 58.44 (1993).

and Its Production and Processing, Recommended Requirements (22). The states that have **Grade B milk** have essentially adopted these model regulations.

2.3.3 State Responsibilities

As described above, individual states have enacted safety and quality regulations for Grade A and manufacturing-grade milk that are essentially identical to those in the PMO and the USDA Recommended Requirements, respectively. The department of health or agriculture in each state normally is responsible for enforcing these regulations. The states also establish their own standards of identity and labeling requirements for milk and dairy products, which are generally similar to the federal requirements.

TABLE 2-13. U.S. Standards for Grades of Nonfat Dry Milk (Spray Process): Classification According to Laboratory Analysis

Laboratory Tests	*U.S. Extra Grade*	*U.S. Standard Grade*
Bacterial estimate, standard plate count per gram	50,000	100,000
Milkfat content, percent	1.25	1.50
Moisture content, percent	4.0	5.0
Scorched particle content, mg	15.0	22.5
Solubility index, ml	1.2	2.0
U.S. High Heat	2.0	2.5
Titratable acidity, percent	.015	0.17

From 7 CFR 58.2528 (1993).

2.4 REGULATIONS AND RECOMMENDATIONS FOR SHELLFISH

Shellfish include fresh or frozen oysters, clams, or mussels. They may transmit intestinal diseases such as typhoid fever or be carriers of natural or chemical toxins. This makes it very important that they be obtained from unpolluted waters and be handled and processed in a sanitary manner.

2.4.1 State and Federal Shellfish Sanitation Programs

Shellfish must comply not only with the general requirements of the FD&C Act but also with the requirements of state health agencies cooperating in the **National Shellfish Sanitation Program** (NSSP) administered by the FDA (23). The FDA has no regulatory power over shellfish sanitation unless the product is shipped interstate. However, the Public Health Service Act authorizes the FDA to make recommendations and to cooperate with state and local authorities to ensure the safety and wholesomeness of shellfish. Under special agreement, Canada, Japan, Korea, Iceland, and Mexico are in the NSSP and are subject to the same sanitary controls as required by states in the United States. Through the NSSP, state health personnel continually inspect and survey bacteriological conditions in shellfish-growing areas. Any contaminated location is supervised or patrolled so that shellfish cannot be harvested from the area. State inspectors check harvesting boats and shucking plants before issuing approval **certificates,** which serve as operating licenses. The certification number of the approved plant is placed on each shellfish package shipped.

2.4.2 Natural and Environmental Toxic Substances in Shellfish

A major concern is the ability of shellfish to concentrate radioactive material, insecticides, and other chemicals from their environment. Thus, one aspect of the NSSP is to ensure that shellfish-growing areas are free from sewage pollution and/or toxic industrial waste. **Pesticide residues** in shellfish are usually quantitated by **gas chromatographic techniques,** and heavy metals such as mercury are commonly quantitated by **atomic absorption spectroscopy** (e.g., AOAC Method 977.15). Another safety problem of shellfish is the control of **natural toxicity,** which is a separate issue from sanitation. Paralytic shellfish poisoning is caused by a poison, which is tested by mouse bioassays and produced by planktonic organisms known as dinoflagellates. Control of this toxicity is achieved by a careful survey followed by prohibition of harvesting from locations inhabited by toxic shellfish.

2.5 VOLUNTARY FEDERAL AND STATE RECOMMENDATIONS AFFECTING FOOD COMPOSITION

2.5.1 Food Specifications, Food Purchase, and Government Agencies

Large amounts of food products are purchased by federal agencies for use in domestic (e.g., school lunch) and foreign programs, prisons, veterans hospitals, the armed forces, and others. Specifications or descriptions developed for many food products are used by federal agencies in procurement of foods, to be assured of the safety and quality of the product specified. Such specifications or descriptions often include information that requires assurance of chemical composition.

2.5.1.1 Federal Specifications

A **Federal Specification** serves as a document for all federal user agencies to procure essential goods and services on a competitive basis. All such specifications for foods should include at least the following information:

1. Name of product
2. Grade or quality designation
3. Size of container or package
4. Number of purchase units
5. Any other pertinent information

2.5.1.2 Commercial Item Descriptions

Commercial Item Descriptions (CIDs) are a series of federal specifications that usually contain the same basic components, with certain optional components. CIDs are used in lieu of federal specifications to purchase commercial off-the-shelf products of good commercial quality. These products must adequately serve government requirements and have an established commercial acceptability. The Agricultural Marketing Service of the USDA has management authority for all food federal standardization documents, including CIDs. The basic format of a CID follows:

1. Title—name of product
2. Code—number
3. Salient characteristics—classification (style, form, type, variety, container size), acceptable quality criteria
4. Regulatory requirements—federal and state mandatory requirements and regulations, where applicable
5. Quality assurance—USDA grade standards and requirements for inspection and certification, if available; product specifications if no standards available
6. Preservation, packaging, packing, labeling, and marking
7. Notes—any special notes
8. Sources of government documents—titles and sources of relevant state and federal documents

2.5.1.3 Other Specifications

In addition to Federal Specifications and CIDs, federal agencies use other terms for the specifications they use in the purchase of foods. These include **Purchase Product Description** (PPD), **USDA Specifications, Commodity Specifications,** and **Military Specifications.**

2.5.1.4 Examples of Specifications for Food Purchase

The CID for canned fruit nectars (apricot, pear, and peach) specifies the acceptable ranges for pH, soluble solids (°Brix, by refractometer), and total acidity (g/100 ml juice, calculated as anhydrous citric acid), all according to AOAC methods. The CID for canned dried beans (in tomato sauce or brine, with or without meat or meat product) specifies the moisture content of the beans at time of fill and the percent lean of the pork, if used (3).

Various CIDs, PPDs, Federal Specifications, or USDA Specifications are used by the USDA (Commodity Procurement Branch, Livestock and Seed Division, Agricultural Marketing Service) to purchase meat products for programs such as school lunch. For example, the CID for canned tuna (24) specifies salt/sodium levels, with analysis to be done by the AOAC flame photometric method for Na and K in seafood. The PPD for canned pink salmon (25) also specifies salt content, but the analysis is to be done by the AOAC indicating strip method for salt (Cl as NaCl) in seafood. The Federal Specification for canned (ready-to-eat) luncheon meat (26) specifies fat and salt contents, with analysis to be done by AOAC methods (unspecified). The USDA Specifications for canned pork (27) or beef (28) with natural juices, frozen ground pork (29), and low-fat beef patties (30), and the Federal Specification for frozen ground beef products (31) all state maximum allowable fat contents, with the fat content to be determined by an AOAC method. In many cases, a specific method is given for withdrawal of samples for fat content analysis. In some cases, the purchaser may specify discount ranges for fat content analysis below the maximum allowable fat content, such that a premium price is paid for a product with a lower fat content. Such specification information for frozen ground turkey (32) is given in Table 2-14.

The Federal Specification for frozen ground beef products (31) includes as an option ground beef with added **vegetable protein product** (VPP). The protein, moisture, fat, and ash contents of the VPP used are specified, along with the contents of certain vitamins and minerals and the biological quality of the protein in the VPP (compared to that of casein).

TABLE 2-14. Fat Content Requirements, USDA Specifications for Purchase of Frozen Ground Turkey for Distribution to Eligible Outlets

Fat Content (Average for Lot) USDA Laboratory Analysis	*Discount Applicable*
No more than 11.0%[1]	None
Greater than 11.0% but not more than 11.5%	2.5% of contract price
Greater than 11.5% but not more than 12.0%	5.0% of contract price
Greater than 12.0% but not more than 12.5%	7.5% of contract price

Adapted from (32).

[1]Lot of packaged ground turkey must contain no more than 11.0% fat (average for the lot), but will be accepted with more than 11.0% but not more than 12.5% fat with the deviations subject to the price discount indicated.

The USDA Specification for low-fat beef patties (30) specifies that the product shall contain either added carrageenan or added oat bran and oat fiber. The oat bran used is to meet the oat bran definition developed by the American Association of Cereal Chemists (AACC) and the American Oat Association; the protein, fat, ash, moisture, crude fiber, total dietary fiber, and soluble dietary fiber contents of the oat bran are specified. The pH, calories, moisture, and total dietary fiber contents of oat fiber are specified, determined according to official test methods of the AACC. Enzyme activity (peroxidase) in both products is to be negative.

Commodity Specifications for a variety of poultry products have been issued by the USDA (Commodity Procurement Branch, Poultry Division, Agricultural Marketing Service). Samples for analyses may be submitted to USDA laboratories. Specifications generally state how the USDA laboratory will sample the product and report the results and in some cases, what method will be used to do the assay. For example, the fat content of frozen turkey burgers (33) will be determined by the American Oil Chemists' Society (AOCS) method for fat (crude) or ether extract in meat. The moisture content of dried egg mix (34) will be analyzed in accordance with *Laboratory Methods for Egg Products* (35). Such dried egg mix is to consist of liquid whole eggs, nonfat dry milk, vegetable oil, and salt. Cottonseed, corn, or soybean oil can be used as the vegetable oil, with specifications given for the following, as determined by AOCS test methods: free fatty acid value, peroxide value, cold test, linolenic acid, moisture and volatile matter, fat stability (active oxygen methods), and Lovibond color values (see Chapter 13 for some of these tests).

Commodity Specifications have been developed for bulk dairy products purchased by the Commodity Credit Corporation of the USDA under the Dairy Price Support Program (Dairy Division or Processed Commodities Division, Agricultural Stabilization and Conservation Service) (36). For example, the moisture content and vitamin A content of nonfat dry milk are specified, as are the moisture content and pH of cheese. Pasteurized process American cheese (37) and mozzarella cheese (38) "for use in domestic donation programs" have specifications on moisture content and milk fat content (on a solids basis).

The Defense Personnel Support Center of the Defense Logistics Agency, Department of Defense, utilizes a variety of specifications, standards, and notes in the purchase of food for the military: USDA Notices or Purchase Specifications (Schedules), CIDs, Federal Specifications, Military Specifications, Military Standards, and Non Governmental Standards (e.g., ASTM Document). For example, they use the Federal Specification for frozen ground beef products (32) and syrup

(39), CID for instant tea (40) and frozen frankfurters (41), USDA Specification for catsup (7 CFR 52) and slab or sliced bacon (42), and Military Specification for pork steak (flaked, formed, breaded, frozen) (43) and beef stew (dehydrated, cooked) (44). The analytical requirements for instant tea according to its CID (40) are given in Table 2-15.

2.5.2 National Conference on Weights and Measures: State Food Packaging Regulations

Consumers assume that the weighing scale for a food product is accurate and that a package of flour, sugar, meat, or ice cream contains the amount claimed on the label. While this assumption is usually correct, city or county offices responsible for weights and measures need to police any unfair practices. Leadership in this area is provided by the **National Conference on Weights and Measures** (NCWM) (45).

2.5.2.1 National Conference on Weights and Measures

The NCWM was established in 1905 by the **National Institute of Standards and Technology** (NIST) (formerly the National Bureau of Standards), which is part of the U.S. Department of Commerce. This came from a need to bring about uniformity in state laws referring to weights and measures and to create close cooperation between the state measurement services and NIST. The NCWM has no regulatory power, but it develops many technical, legal, and general recommendations in the field of weights and measures administration and technology. The NCWM is a membership organization comprised of state and local weights and measures regulatory officers, other officials of federal, state, and local governments, and representatives of manufacturers, industry, business, and consumer organizations. It assembles for an annual meeting of decision-making officials and generates uniformity in the regulations issued by these officials concerning weights and measures.

2.5.2.2 NIST Handbook 133

The NIST Handbook 133, *Checking the Net Contents of Packaged Goods* (46), gives model state **packaging and labeling regulations** that have been adopted by a majority of states. The Handbook provides detailed procedures for (1) testing packages labeled by liquid or dry volume, length, area, count, and combinations of labeled quantities, (2) testing certain hard-to-measure prepackaged goods, and (3) sampling to determine compliance with regulations. The Handbook specifies that the average quantity of contents of packages must at least equal the labeling quantity, with the variation between the individual package net contents and the labeled quantity not be too "unreasonably large." Variations are permitted within the bounds of good manufacturing practices and are due to gain or loss of moisture (within the bounds of good distribution practice). For certain products (e.g., flour, pasta, rice) this requires careful monitoring of moisture content and control of storage conditions by the manufacturer.

2.6 NUTRITIONAL LABELING

The FDA was authorized under the 1906 Federal Food and Drug Act and the 1938 FD&C Act to require certain types of food labeling. This includes the amount of food in a package, its common or usual name, and its ingredients. Major modifications to the FD&C Act have been made twice to regulate nutritional labeling. A reference manual that explains nutritional labeling regulation (with continual updating) can be purchased from the National Food Processors Association (a nonprofit organization) (47) and several commercial publishers.

TABLE 2-15. Analytical Requirements for Instant Tea According to Commercial Item Description[1]

Type	*Moisture (%) (Max.)*	*Caffeine (%) (Min.)*	*Sugar (%) (Max.)*	*Titratable Acidity (Dry Basis) (Min.)*	*Ascorbic Acid (Class B Product) (Min.)*
I	4.0	2.5	63[2]	—	15 mg/1.3 g product[5]
II	4.0	5.0	26[3]	—	15 mg/0.65 g product[5]
III	0.40	0.110	96[4]	2.4	—

Adapted from (40).
[1]AOAC method specified for each constituent to be tested
[2]After hydrolysis, for reducing sugars on a dry basis
[3]For reducing sugars on a dry basis
[4]The total sugar (sucrose plus reducing sugar)
[5]Lot average

2.6.1 1973 Regulations on Nutritional Labeling

The FDA promulgated regulations in 1973 that permitted, and in some cases required, foods to be labeled for their nutritional value. Nutrition labeling was required only if a food contained an added nutrient or if a nutrition claim was made for the food on the label or in advertising. In 1984, the FDA adopted regulations to include sodium content on the nutritional label (effective 1985). A nutritional label under the 1973 regulations, with the addition of sodium content, included the information given in Table 2-16.

2.6.2 Nutrition Labeling and Education Act of 1990

Since the nutrition label was established in 1973, dietary recommendations for better health have focused more on the role of calories and macronutrients (fat, carbohydrates, etc.) in chronic diseases and less on the role of micronutrients (minerals and vitamins) in deficiency diseases. Therefore, the FDA revised the content of the nutritional label to make it more consistent with dietary concerns. The **Nutritional Labeling and Education Act of 1990** (48) amended the FD&C Act with regard to three primary changes:

1. Mandatory nutrition labeling for almost all food products
2. Federal regulation of nutrient content claims and health claims
3. National uniformity for most food labeling requirements

TABLE 2-16. Nutritional Labeling Format, 1973–1993.

Serving size
Number of servings per container
Amount of the following nutrients per serving:
  Calories
  Protein (g)
  Carbohydrate (g)
  Fat (g)
  Sodium[1] (mg)
Percentage of U.S. Recommended Daily Allowance per serving
  Protein
  Vitamin A
  Vitamin C
  Thiamine
  Riboflavin
  Niacin
  Calcium
  Iron

[1]Required effective 1985.

2.6.2.1 Mandatory Nutrition Labeling

The FDA regulations implementing the Nutrition Labeling and Education Act of 1990 require nutrition labeling for most foods offered for sale and regulated by the FDA (49–53). Certain information is required on the label, and other information is voluntary (Table 2-17). The standard format for nutrition information on food labels is given in Fig. 2-3 and consists of the following: (1) quantitative amount per serving of each nutrient except vitamins and minerals, (2) amount of each nutrient, except sugars and protein, as a percent of the **Daily Value** (i.e., the new label reference values) for a 2,000 calorie diet, and (3) footnote with Daily Values for selected nutrients based on 2,000 calorie and 2,500 calorie diets. A Daily Value for sugars has not been established. Reporting the amount of protein as a percent of its Daily Value is optional (49). Caloric conversion information on the label for fat, carbohydrate, and protein is optional (53). The FDA regulations specify five methods by which caloric content may be calculated (49).

A simplified format for nutrition information may be used if seven or more of the 13 required nutrients

TABLE 2-17. Mandatory (Bold) and Voluntary Components for Food Label Under Nutrition Labeling and Education Act of 1990[1]

Total calories
Calories from fat
Calories from saturated fat
Total fat
Saturated fat
Polyunsaturated fat
Monounsaturated fat
Cholesterol
Sodium
Potassium
Total carbohydrate
Dietary fiber
Soluble fiber
Insoluble fiber
Sugars
Sugar alcohol (e.g., sugar substitutes xylitol, mannitol, and sorbital)
Other carbohydrate (the difference between total carbohydrate and the sum of dietary fiber, sugars, and sugar alcohols, if declared)
Protein
Vitamin A
Vitamin C
Calcium
Iron
Other essential vitamins and minerals

From (49).
[1]Nutrition panel will have the heading "Nutrition Facts." Only components listed are allowed on the nutrition panel, and they must be in the order listed. Components are to be expressed as amount and/or as percent of an established "Daily Value."

Serving sizes are now more consistent across product lines, stated in both household and metric measures, and reflect the amounts people actually eat.

The **list of nutrients** covers those most important to the health of today's consumers, most of whom need to worry about getting too much of certain items (fat, for example), rather than too few vitamins or minerals, as in the past.

Nutrition Facts

Serving Size 1/2 cup (114g)
Servings Per Container 4

Amount Per Serving

Calories 90 Calories from Fat 30

	% Daily Value*
Total Fat 3g	**5%**
Saturated Fat 0g	**0%**
Cholesterol 0mg	**0%**
Sodium 300mg	**13%**
Total Carbohydrate 13g	**4%**
Dietary Fiber 3g	**12%**
Sugars 3g	
Protein 3g	

Vitamin A	80%	•	Vitamin C	60%
Calcium	4%	•	Iron	4%

* Percent Daily Values are based on a 2,000 calorie diet. Your daily values may be higher or lower depending on your calorie needs:

	Calories	2,000	2,500
Total Fat	Less than	65g	80g
Sat Fat	Less than	20g	25g
Cholesterol	Less than	300mg	300mg
Sodium	Less than	2,400mg	2,400mg
Total Carbohydrate		300g	375g
Fiber		25g	30g

Calories per gram:
Fat 9 • Carbohydrate 4 • Protein 4

New title signals that the label contains the newly required information.

Calories from fat are now shown on the label to help consumers meet dietary guidelines that recommend people get no more than 30 percent of their calories from fat.

% Daily Value shows how a food fits into the overall daily diet.

Daily Values are also something new. Some are maximums, as with fat (65 grams or less); others are minimums, as with carbohydrate (300 grams or more). The daily values for a 2,000- and 2,500-calorie diet must be listed on the label of larger packages. Individuals should adjust the values to fit their own calorie intake.

FIGURE 2-3.
An example of the new nutrition label, Nutrition Labeling and Education Act of 1990. Exact specifications are in the final regulations (49-53). Note that final regulations were modified to make optional giving the caloric conversion information for fat, carbohydrate, and protein (53). Figure courtesy of the Food and Drug Administration.

are present in only insignificant amounts (e.g., soft drinks). Certain foods are exempt from mandatory nutrition labeling requirements (Table 2-18) unless a nutrient content claim or health claim is made. Special labeling provisions apply to certain other foods (e.g., foods in small packages; foods for children; raw fruits, vegetables, and fish; foods sold from bulk containers; unit containers in multiunit packages; foods in gift packs).

The numerical expression of quantity per serving is specified for all mandatory nutrients. For example, calories are to be reported to the nearest 5 calories up

TABLE 2-18. Foods Exempt from Mandatory Nutrition Labeling Requirements of the Nutrition Labeling and Education Act of 1990

Food offered for sale by small business
Food sold in restaurants or other establishments in which food is served for immediate human consumption
Foods similar to restaurant foods that are ready to eat but are not for immediate consumption, are primarily prepared on site, and are not offered for sale outside that location
Foods that contain insignificant amounts of all nutrients subject to this rule, e.g., coffee and tea
Dietary supplements, except those in conventional food form
Infant formula
Medical foods
Custom-processed fish organ meats
Foods shipped in bulk form
Donated foods

From (49).

to and including 50 calories, and to the nearest 10 calories above 50 calories. Calories can be reported as zero if there are less than 5 calories per serving.

2.6.2.2 Nutrient Content Claims

The FDA has defined **nutrient descriptors,** which are claims that characterize the level of a nutrient but do not include nutrient labeling information or disease prevention claims. These include the terms "free," "low," "lean," "light," "reduced," "less," "more," and "high" (Table 2-19). The new requirements on nutrient content claims do not apply to infant formulas and medical foods. Also, restaurant foods are not affected by the limitations on cholesterol, fat, and fiber content claims.

Only nutrient descriptors defined by the FDA may be used. The terms "less" (or "fewer"), "more," "reduced," "added" (or "fortified" and "enriched"), and "light/lite" are relative terms and require label information about the food product that is the basis of the comparison. The percentage difference between the original food and the food product being labeled must be listed on the label for comparison. These and the other nutrient descriptors are based on definitions (48–50,52,54) that require certain types of nutrient analysis.

2.6.2.3 Health Claims

The FDA has defined and will allow as part of the 1990 Act claims for seven relationships between a nutrient or a food and the risk of a disease or health-related condition:

1. Calcium and osteoporosis
2. Sodium and hypertension
3. Dietary saturated fat and cholesterol and risk of coronary heart disease
4. Dietary fat and cancer
5. Fiber-containing grain products, fruits, and vegetables and cancer
6. Fruits, vegetables, and grain products that contain fiber, particularly soluble fiber, and risk of coronary heart disease
7. Fruits and vegetables and cancer

Such claims can be made through third-party references (e.g., National Cancer Institute), statements, symbols (e.g., heart), and vignettes or descriptions. The claim must meet the requirements for authorized health claims, and it must state that other factors play a role in that disease.

2.6.2.4 National Uniformity and Preemption

To provide for national uniformity, the 1990 Act authorizes federal preemption of certain state and local labeling requirements that are not identical to federal requirements. This pertains to requirements for food standards, nutrition labeling, claims of nutrient content, health claims, and ingredient declaration. States may petition the FDA for exemption of state requirements from federal preemption.

2.6.2.5 Miscellaneous Provisions

The 1990 Act amends the FD&C Act to allow a state to bring, in its own name in state court, an action to enforce the food labeling provisions of the FD&C Act that are the subject of national uniformity. The rulemaking procedure for standards of identity was modified by the 1990 Act. The FD&C Act is also amended to impose several new requirements concerning **ingredient labeling,** intended to make this aspect of labeling more useful to consumers.

2.6.2.6 Adequacy of Methods for Nutritional Labeling

The **AOAC Task Force on Nutrient Labeling Methods** has considered the adequacy of AOAC methods to meet nutritional labeling requirements in the 1990 Nutritional Labeling and Education Act. Adequacy was judged on the basis of a survey of users of nutrient methods and on the basis of collaboratively validated and officially approved status of methods published in the AOAC *Official Methods of Analysis* (20). The task

TABLE 2-19. Nutrient Content Descriptors that May Be Used on Food Labels

Descriptor[1]	*Definition*[2]
Free	A serving contains no or a physiologically inconsequential amount: <5 calories; <5 mg of sodium; <0.5 g of fat; <0.5 g of saturated fat; <2 mg of cholesterol; or <0.5 g of sugar
Low	A serving (and 50 g of food if the serving size is small) contains no more than 40 calories; 140 mg of sodium; 3 g of fat; 1 g of saturated fat and 15% of calories from saturated fat; or 20 mg of cholesterol; not defined for sugar; for "very low sodium," no more than 35 mg of sodium
Lean	A serving (and 100 g) of meat, poultry, seafood, and game meats contains <10 g of fat, <4 g of saturated fat, and <95 mg of cholesterol
Extra lean	A serving (and 100 g) of meat, poultry, seafood, and game meats contains <5 g of fat, <2 g of saturated fat, and <95 mg of cholesterol
High	A serving contains 20% or more of the daily value (DV) for a particular nutrient
Good source	A serving contains 10–19% of the DV for the nutrient
Reduced	A nutritionally altered product contains 25% less of a nutrient or 25% fewer calories than a reference food; cannot be used if the reference food already meets the requirement for a "low" claim
Less	A food contains 25% less of a nutrient or 25% fewer calories than a reference food
Light	1. An altered product contains one-third fewer calories or 50% of the fat in a reference food (if 50% or more of the calories come from fat, the reduction must be 50% of the fat); or 2. The sodium content of a low-calorie, low-fat food has been reduced by 50% (the claim "light in sodium" may be used); or 3. The term describes such properties as texture and color, as long as the label explains the intent (e.g., "light brown sugar," "light and fluffy")
More	A serving contains at least 10% of the DV of a nutrient more than a reference food. Also applies to fortified, enriched, and added claims for altered foods
% Fat Free	A product must be low-fat or fat-free, and the percentage must accurately reflect the amount of fat in 100 g of food. Thus, 2.5 g of fat in 50 g of food results in a "95% fat-free" claim
Healthy[3]	A food is low in fat and saturated fat, and a serving contains no more than 480 mg of sodium and no more than 60 mg of cholesterol
Fresh	1. A food is raw, has never been frozen or heated, and contains no preservatives (irradiation at low levels is allowed); or 2. The term accurately describes the product (e.g., "fresh milk" or "freshly baked bread")
Fresh frozen	The food has been quickly frozen while still fresh; blanching is allowed before freezing to prevent nutrient breakdown

From (54), used with permission.
[1]See the regulations (49,50) for acceptable synonyms.
[2]These definitions have been simplified for this table; see the regulations (49,50,52,53) for specific restrictions and additional requirements.
[3]Proposed by USDA Food Safety and Inspection Service for meat and poultry products.

force compiled a list of methods judged to be adequate relative to the 1990 Act (55). Methods were identified by specific combinations of analyte (e.g., ash, calcium, fat, protein) and matrix (see Table 2-20). The task force has also recommended collaborative studies to provide methodology in cases in which (1) no adequate method exists for specific analyte/matrix combinations, (2) methods for analyte/matrix combinations are deemed adequate based on common use, but collaborative studies have not been conducted, or (3) updated methods have been recommended for analyte/matrix combinations (56). Persons involved in analysis of food products to meet nutritional labeling requirements specified in the 1990 Act will need to follow carefully the development of new and improved methods for specific analyte/matrix combinations.

TABLE 2-20. Matrixes Considered by AOAC International as Representative of All Food Types

Baby foods (fruits)	Infant formula/medical diets
Baby foods (meats)	Meat (beef/pork/fowl)
Baby foods (vegetables)	Mixed dinners (TV dinners)
Beverages and juices	Nuts
Candy	Oils/fats (dressings)
Cereals and products	Potatoes and products
Cheese	Shellfish
Dairy products	Sweet mixes (cake/pie/etc.)
Fish	Spices
Fruits	Vegetables

Adapted from (55).

2.7 SUMMARY

Various kinds of standards set for certain food products by federal agencies make it possible to get essentially the same food product whenever and wherever purchased in the United States. The standards of identity set by the FDA and USDA define what certain food products must consist of. The USDA and National Marine Fisheries Service of the Department of Commerce have specified grade standards to define attributes for certain foods. Grading programs are voluntary, while inspection programs may be voluntary or mandatory, depending on the specific food product.

While the FDA has broadest regulatory authority over most foods, responsibility is shared with other regulatory agencies for certain foods. The USDA has significant responsibilities for meat and poultry, the National Marine Fisheries Service for seafood, and the Bureau of Alcohol, Tobacco, and Firearms for alcoholic beverages. The FDA, the USDA, state agencies, and the dairy industry work together to ensure the safety, quality, and economic integrity of milk and milk products. The FDA, the EPA, and state agencies work together in the National Shellfish Sanitation Program to ensure the safety and wholesomeness of shellfish. The EPA shares responsibility with the FDA for control of pesticide residues in foods and has responsibility for drinking water safety and the composition of effluent from food processing plants. The Customs Service receives assistance from the FDA and USDA in its role to ensure the safety and economic integrity of imported foods. The Federal Trade Commission works with the FDA to prevent deceptive advertising of food products, as affected by food composition and labels.

The chemical composition of foods is often an important factor in determining the quality, grade, and price of a food. Government agencies that purchase foods for special programs often rely on detailed specifications that include information on food composition. Nutritional labeling requirements for foods are based on their chemical composition. The various types of chemical analyses required to obtain such food composition information are described in the remainder of this book.

2.8 STUDY QUESTIONS

1. Define the abbreviations FDA, USDA, and EPA, and give two examples for each of what they do and/or regulate relevant to food analysis.
2. Differentiate "standards of identity," "standards of quality," and "grade standards" with regard to what they are and which federal agency establishes and regulates them.
3. Government regulations regarding the composition of foods often state the official or standard method by which the food is to be analyzed. Give the full names of three organizations that publish commonly referenced sources of such methods.
4. For each type of product listed below, identify the governmental agency (or agencies) that has regulatory and/or other responsibility for quality assurance. Specify the nature of that responsibility.
   a. frozen fish sticks
   b. contaminants in drinking water
   c. dessert wine
   d. Grade A milk
   e. frozen oysters
   f. imported chocolate products
   g. residual pesticide on wheat grain
   h. corned beef
5. Food products purchased by federal agencies often have specifications that include requirements for chemical composition. Give the names of four such specifications.
6. For the nutritional labeling of foods, what new food components and characteristics are required according to the 1990 regulations (finalized in 1993) that were not required according to the 1973 regulations? What food components and characteristics are not required on the new label that were required on the old label?

2.9 REFERENCES

1. Hui, Y.H. 1979. *United States Food Laws, Regulations, and Standards*. John Wiley & Sons, New York, NY.
2. Hui, Y.H. 1986. *United States Food Laws, Regulations, & Standards*, Vol. I and II, 2nd ed. John Wiley & Sons, New York, NY.
3. Hui, Y.H. 1988. *United States Regulations for Processed Fruits and Vegetables*. John Wiley & Sons, New York, NY.
4. Aurand, L.W., Woods, A.E., and Wells, M.R. 1987. Food laws and regulations. Ch. 1, in *Food Composition and Analysis*. Von Nostrand Reinhold Co., New York, NY.
5. Anonymous. 1992. Code of Federal Regulations. Titles 7, 9, 21, 27, 40, 50. U.S. Government Printing Office, Washington, DC.
6. USDA. 1987. *Meat and Poultry Inspection Manual*. Meat and Poultry Inspection Program, Animal and Plant Health Inspection Services, U.S. Dept. of Agriculture. U.S. Government Printing Office, Washington, DC.
7. USDA. 1979. *Grain Inspection Handbook*. Federal Grain Inspection Service, U.S. Dept. of Agriculture, Washington, DC.
8. USDA. 1981. *USDA Grade Standards for Food and Farm Products*. Agriculture Handbook No. 533. Food Safety and Quality Service, U.S. Dept. of Agriculture. U.S. Government Printing Office, Washington, DC.
9. USDA. 1993. Fees for FGIS Commodity Inspection Services (Pulses, Hops, and Miscellaneous Processed Commodities). Administrative Notice. October 29, 1993. Federal Grain Inspection Service, U.S. Dept. of Agriculture, Washington, DC.
10. National Marine Fisheries Service (NMFS). *Fishery Products Inspection Manual*. (Updated continuously.) National Seafood Inspection Laboratory, Pascagoula, MS.
11. Garrett, E.S., III and Hudak-Roos, M. 1991. Developing an HACCP-based inspection system for the seafood industry. *Food Tech.* 45(12):53–57.
12. Ahmed, F.E. (Ed.) 1991. *Seafood Safety*. Food and Nutrition Board, Institute of Medicine. National Academy Press, Washington, DC.

13. FDA. 1985. *Pesticide Analytical Manual,* Vol. 1 *(Methods Which Detect Multiple Residues)* and Vol. 2 *(Methods for Individual Pesticide Residues).* Public Records and Documents Center, Food and Drug Administration, HFI-35, Rockville, MD.
14. Greenberg, A.E., Clesceri, L.S., and Eaton, A.D. (Eds) 1992. *Standard Methods for the Examination of Water and Wastewater,* 18th ed. American Public Health Association, Washington, DC.
15. EPA. 1979. *Methods of Chemical Analysis of Water and Wastes.* EPA-600/4-79-020, March 1979. EPA Environmental Monitoring and Support Laboratory, Cincinnati, OH.
16. American Society for Testing Materials (ASTM). 1976. *Annual Book of ASTM Standards,* Part 31, Water. ASTM, Philadelphia, PA.
17. U.S. International Trade Commission. 1992. *Harmonized Tariff Schedule of the United States.* USITC Publication 2449. U.S. Government Printing Office, Washington, DC.
18. U.S. Department of Health and Human Services, Public Health Service, Food and Drug Administration. 1989. *Grade "A" Pasteurized Milk Ordinance.* Publication No. 229. U.S. Government Printing Office, Washington, DC.
19. Marshall, R.T. (Ed.) 1992. *Standard Methods for the Examination of Dairy Products,* 16th ed. American Public Health Association, Washington, DC.
20. AOAC. 1990. *Official Methods of Analysis,* 15th ed. Association of Official Analytical Chemists, Washington, DC.
21. USDA. 1975. *General Specifications for Dairy Plants Approved for USDA Inspection and Grading Service.* 40 Federal Register (F.R.) 198, October 10, 1975. U.S. Government Printing Service, Washington, DC. (Amendments: 50 F.R. 166, August 27, 1985; 55 F.R. 39912, October 1, 1990; 56 F.R. 33854, July 24, 1991).
22. USDA. 1972. *Milk for Manufacturing Purposes and Its Production and Processing, Recommended Requirements.* April 7, 1972 (updated July 1, 1986). Dairy Division, Agricultural Marketing Service, U.S. Dept. of Agriculture, Washington, DC.
23. FDA. 1991. *National Shellfish Sanitation Program Manuals.* Food and Drug Administration, Washington, DC.
24. Anonymous. 1989. Commercial Item Description. Tuna, Canned. A-A-20155. April 28, 1989. General Services Administration, Specifications Unit, Washington, DC.
25. Anonymous. 1991. Product Purchase Description. Salmon, Pink, Canned. PPD-05-S-003E. August 22, 1991. General Services Administration, Specifications Unit, Washington, DC.
26. Anonymous. 1992. Federal Specification. Luncheon Meat, Canned (Ready to Eat). PP-L-800E. Amendment-1. March 13, 1992. Amendment to PP-L-800E. September 9, 1986. General Services Administration, Specifications Unit, Washington, DC.
27. USDA. 1991. USDA Specification of Pork and Natural Juices, Canned. Schedule PJ—October 1991. Livestock and Seed Division, Agricultural Marketing Service, U.S. Dept. of Agriculture, Washington, DC.
28. USDA. 1990. USDA Specification of Beef with Natural Juices, Canned. Amendment No. 3 to Schedule BJ—December 1984. August 1990. Livestock and Seed Division, Agricultural Marketing Service, U.S. Dept. of Agriculture, Washington, DC.
29. USDA. 1991. USDA Specification for Frozen Ground Pork. Amendment No. 2 to Schedule GP—August 1990. Livestock and Seed Division, Agricultural Marketing Service, U.S. Dept. of Agriculture, Washington, DC.
30. USDA. 1991. USDA Specification for Low-Fat Beef Patties. Interim Schedule LP—December 1991. Livestock and Seed Division, Agricultural Marketing Service, U.S. Dept. of Agriculture, Washington, DC.
31. Anonymous. 1991. Federal Specification. Ground Beef Products, Frozen. PP-B-2120B. Amendment-1. November 14, 1991. Amendment to PP-B-2120B, May 31, 1991. General Services Administration, Specifications Unit, Washington, DC.
32. USDA. 1991. Purchase of Frozen Ground Turkey for Distribution to Eligible Outlets. Announcement PY-146. July 1991. Poultry Division, Agricultural Marketing Service, U.S. Dept. of Agriculture, Washington, DC.
33. USDA. 1991. Purchase of Frozen Turkey Burgers for Distribution to Eligible Outlets. Announcement PY-149. Poultry Division, Agricultural Marketing Service, U.S. Dept. of Agriculture, Washington, DC.
34. USDA. 1992. Purchase of Dried Egg Mix for Distribution to Eligible Outlets. Announcement/Invitation PY-153, April 1992. Poultry Division, Agricultural Marketing Service, U.S. Dept. of Agriculture, Washington, DC.
35. USDA. 1984. *Laboratory Methods for Egg Products.* Grading Brand Poultry Division, Agricultural Marketing Service, U.S. Dept. of Agriculture, Washington, DC.
36. USDA. 1988. Purchase of Bulk Dairy Products. Announcement Dairy-4. December 28, 1988. Dairy Division, Agricultural Stabilization and Conservation Service, U.S. Dept. of Agriculture, Washington, DC.
37. USDA. 1989. Purchase of Pasteurized Process American Cheese by Competitive Offers for Use in Donation Programs. Announcement KC-C-11. July 7, 1989. Processed Commodities Division, Agricultural Stabilization and Conversation Service, U.S. Dept. of Agriculture, Kansas City, MO.
38. USDA. 1985. Purchase of Low Moisture Part-Skim Mozzarella Cheese for Use in Domestic Donation Programs. Announcement KC-MC-3. January 25, 1985. Processed Commodities Division, Agricultural Stabilization and Conservation Service, U.S. Dept. of Agriculture, Kansas City, MO.
39. Anonymous. 1981. Federal Specification. Syrup. JJJ-S-351H. October 20, 1981. General Services Administration, Specifications Unit, Washington, DC.
40. Anonymous. 1990. Commercial Item Description. Tea, Instant. A-A-20183. November 1, 1990. General Services Administration, Specifications Unit, Washington, DC.
41. Anonymous. 1986. Commercial Item Description. Frankfurters, Frozen. A-A-20132A. 11 February 1986. General Services Administration, Specifications Unit, Washington, DC.
42. USDA. 1979. USDA Specification for Slab or Sliced Bacon. Schedule SB. August 1979. Livestock and Seed Division, Agriculture Marketing Service, U.S. Dept. of Agriculture, Washington, DC.

43. Department of Defense. 1987. Military Specification. Pork Steak, Flaked, Formed, Breaded, Frozen. MIL-P-44131A. 30 March 1987. Defense Quality and Standardization Office, Falls Church, VA.
44. Department of Defense. 1991. Military Specification. Beef Stew, Dehydrated, Cooked. MIL-B-43404E. Notice 2. 20 September 1991. Reference to MIL-B-43404E, 29 June 1984. Defense Quality and Standardization Office, Falls Church, VA.
45. National Conference on Weights and Measures. *The National Conference on Weights and Measures, Its Organization Procedures and Membership Plan.* NCWM Publication 6. NCWM, P.O. Box 3137, Gaithersburg, MD.
46. U.S. Dept. of Commerce, National Bureau of Standards. 1988. *Checking the Net Contents of Packaged Goods.* NBS Handbook 133, 3rd ed. U.S. Government Printing Office, Washington, DC.
47. National Food Processors Association. (Updated continuously.) *Food Labeling: A User's Manual.* National Food Processors Association, Washington, DC.
48. U.S. Public Law 101-535. Nutrition Labeling and Education Act of 1990, November 8, 1990. U.S. Congress, Washington, DC.
49. Federal Register. 1993. 21 CFR Part 1, et al. Food Labeling; General Provisions; Nutrition Labeling; Label Format; Nutrient Content Claims; Health Claims; Ingredient Labeling; State and Local Requirements; and Exemptions; Final Rules. January 6, 1993. 58(3):2066–2941. Superintendent of Documents. U.S. Government Printing Office, Washington, DC.
50. FDA. 1993. The new food label summaries. January 6, 1993. Food and Drug Administration, Dept. of Health and Human Services, Washington, DC.
51. Federal Register. 1993. 21 CFR Parts 1 and 101. Food Labeling and Nutrient Content Revision, Format for Nutrition Label; Correction. April 2, 1993. 58(62):17328–17340. Superintendent of Documents. U.S. Government Printing Office, Washington, DC.
52. Federal Register. 1993. 21 CFR Parts 5 and 101. Food Labeling: Nutrient Content Claims, General Principles, Petitions, Definition of Terms; Definitions of Nutrient Content Claims for the Fat, Fatty Acid, and Cholesterol Content of Food; Correction. April 2, 1993. 58(62):17341–17346. Superintendent of Documents. U.S. Government Printing Office, Washington, DC.
53. Federal Register. 1993. 21 CFR Part 101. Food Labeling: Mandatory Status of Nutrition Labeling and Nutrient Content Revision. Format for Nutrition Label; Technical Amendments. August 18, 1993. 58(158):44063–44077. Superintendent of Documents. U.S. Government Printing Office, Washington, DC.
54. Mermelstein, N.H. 1993. The new era in food labeling. *Food Tech.* 47(2):81–86, 88–92, 94, 96.
55. AOAC International. July 1992. Nutrient labeling task force report. *The Referee* 16(7):1–12.
56. AOAC International. October 1992. Nutrient labeling task force update. *The Referee* 16(10):5–7.

CHAPTER

3

Sampling and Sample Preparation

Genevieve L. Christen

3.1 INTRODUCTION

In the food industry, characteristics of products are of concern at many points. To monitor these characteristics we have two alternatives: We can inspect 100 percent of the product to ensure that each piece is acceptable, or we can select a portion of the whole and examine that portion for the characteristic of interest. Only rarely is 100 percent inspection feasible. Weighing each individual unit is possible with high-speed checkweighing devices. Usually 100 percent inspection is not feasible because it is time consuming and expensive; it is limited to characteristics that can be determined nondestructively, or we would have no product to sell; it is highly dependent on the inspector's skill; and if done manually, it is only 85–95 percent effective because of fatigue (1).

Because 100 percent inspection is frequently not possible, a subsection of the whole is evaluated. **Sampling** consists of obtaining a portion of a larger group (referred to as the **population**). From this sample, one hopes to obtain an estimate of the true value of the parameter of interest with sufficient accuracy for the intended purposes. Sampling permits a reduction in cost and personnel while allowing information to be obtained quickly and comprehensively. It is, however, only an estimate of the true value for the population. If sampling plans are created carefully, the value is a very good estimate, and any inaccuracies can be estimated. Various sampling procedures and aspects of their selection are discussed in sections 3.2 and 3.3.

A **laboratory sample** is anything that is sent to a laboratory for analysis. It can be of any size or quantity (2). Considerations in determining the size and characteristics of the laboratory sample are described in this chapter along with potential problems (sections 3.3 and 3.4). The laboratory sample must be reduced to test sample size while maintaining the essential properties of the original sample. Preparation of test samples from laboratory samples is discussed in section 3.5.

3.2 SELECTION OF SAMPLING PROCEDURES

3.2.1 General Information

The first requirement in sampling is to clearly define the population to be sampled. Is it a lot, one day's production, all like material in the warehouse, or all the material of this type in the world? Extrapolating information from a sample taken of a particular lot to other samples in that same lot can be done with accuracy. Expanding information on a sample from a particular lot to all like units in the warehouse or the world is inaccurate and should not be attempted.

Populations may be finite or infinite (1). For **finite populations,** sampling provides an estimate of lot quality. For **infinite populations,** sampling determines characteristics about the process. **Acceptance sampling** provides information about a particular lot or process. This information is compared to specific, predetermined values to determine if the population sampled is within specification.

The final results obtained from any analytical technique follows a series of steps. The uncertainty associated with the final result is cumulative and depends on the uncertainty at each step (3,4). **Variance** is an estimate of the uncertainty, with total variance for the overall testing procedure equal to the sum of the variances associated with the sampling process, the subsampling process, and the analytical method (3). The variance is equal to the square of the standard deviation and represents a measure of the **precision** of the process. The most efficient way to improve the precision of the final result is to improve the reliability of the step with the greatest variance. Frequently, this is the initial sampling step. A characteristic of sampling statistics is that the reliability of a random sample does not depend so much on the size of the lot as on the size of the sample (1). The larger the sample size, the better, within the constraints of time, cost, and facilities available for collecting the samples and analyzing the data.

3.2.2 Factors That Affect Choice of Sampling Plan

Factors that affect the choice of sampling plans are given in Table 3-1. Each must be considered in the selection of the sampling plan. Once information is obtained about the purpose for inspection and about the nature of the product, test method, and lot, a sampling plan can be created that will provide the best information possible under the given constraints.

3.2.3 Sampling for Attributes or Variables

Sampling plans are usually described as sampling for attributes or sampling for variables (1). In **attribute sampling,** the inspector makes a decision, based on the sample, about whether the population is acceptable or unacceptable. The characteristic is present or it is not. The statistical distribution of such a sampling plan is hypergeometric, binomial, or Poisson.

In **sampling for variables,** the characteristic is measured on a continuous scale, and the actual value obtained is compared to the expected value and the deviation determined. Such information is usually normally distributed. Fill of container and percent total solids are examples of characteristics that are measured

TABLE 3-1. Factors that Affect the Choice of Sampling Plans

Factors to Be Considered	*Questions*
Purpose of the inspection	Is it to accept or reject the lot? Is it to measure the average quality of the lot? Is it to determine the variability of the product?
Nature of the product	Is it homogeneous or heterogeneous? (See section 3.3.2.) What is the unit size? How consistently have past populations met specifications? What is the cost of the material being sampled?
Nature of the test method	Is the test critical or minor? Will someone become sick or die if the population fails to pass the test? Is the test destructive or nondestructive? How much does the test cost to complete?
Nature of the population being investigated	Is the lot large but uniform? Does the lot consist of smaller, easily identifiable sublots? What is the distribution of the units within the population?

Adapted from (1).

on a continuous scale. Variables sampling plans require smaller sample sizes for an equal level of protection compared to attributes sampling plans (1). Under a variables sampling plan, each characteristic of the product must be treated separately.

There are three basic types of sampling plans: **single, double,** or **multiple** (5). Each may be used for evaluation of attributes or variables, or a combination of both. Selection of the appropriate sampling plan depends on the expected overall quality of the lot and the cost of sampling. **Single sampling plans** permit accept/reject decisions to be made on inspection of one sample of specified size drawn from the lot of interest. **Double sampling plans** require selection of two sample sets in many cases. However, if the lot is extremely good or extremely bad, acceptance or rejection may be determined after evaluation of the first set of samples. If the lot does not fall within the very good or very bad category, another set of samples is taken and a decision made on the total sample. Double sampling plans have a positive psychological impact since samples are selected twice from many lots. **Multiple sampling plans** also reduce sampling costs by rejecting low-quality lots or accepting high-quality lots quickly. The amount of sampling depends on the overall quality of the lot. A multiple sampling chart must be developed relating the cumulative number of defects to the size of the sample. When the cumulative number of defects falls above the reject line or below the accept line, the appropriate decision is made regarding the lot. If the number of defects falls between the two lines, sampling and inspection continue.

For a contaminate such as aflatoxin, a normal distribution cannot be assumed (2). Aflatoxin is distributed broadly and erratically within a population. Distributions such as this require combining many randomly selected portions homogenously to obtain a reasonable average estimate of the variable of interest (e.g., ppb aflatoxin). Horowitz (2) points out that there is no need to have methods of analysis that are extremely precise to determine aflatoxin concentrations when sampling error is many times larger than analytical error. He further advises that "the best use of laboratory time would be to ensure adequate comminution and mixing before size reduction," rather than in the analysis per se. For additional information on sampling for aflatoxins, consult Chapter 19.

3.2.4 Risks in Selection of Sampling Plan

Aside from the considerations just described, there are certain risks involved in selection of sampling plans. The first is described as the **consumer's risk** (1). This is the chance of accepting a population of poor quality. With this risk, we are concerned with the possibility of substandard material or product passing inspection. Usually we want this to happen rarely (less than 5 percent of the time). The actual probability associated with consumer's risk depends on the consequences of failure: Will someone get sick and/or die if we fail to detect a defect that should have been rejected, or will the product appear slightly darker orange than the consumer is accustomed to? The former demands a low probability of occurring, while the latter could occur more frequently. The **vendor's risk** is the chance of rejecting acceptable product (1). Again, the actual probability associated with this risk depends on the consequences of a wrong decision. Vendor's risk is usually set at 5–10 percent.

A good sampling plan should protect both the vendor and the consumer, be simple and easy to use, take into consideration the end use of rejected lots, and be flexible (1). In section 3.3.3, several sampling plans will be described.

3.3 SAMPLING PROCEDURES

3.3.1 Introduction

The results of any analytical determination are only as good as the sample upon which the results are based. As pointed out in Table 3-1, the anticipated use of the results from the sample will determine the sampling procedure. Specific details for some products are given in the *Official Methods of Analysis* of the Association of Official Analytical Chemists (AOAC) International (6) and in the Code of Federal Regulations (CFR) (7). For example, AOAC Method 925.08 (6) describes in detail the method for sampling flour in sacks. The number of sacks to be sampled is determined by the square root of the total number in the lot. Sacks to be sampled are selected according to their exposure, with those that are most exposed sampled more frequently and those least exposed sampled less frequently. Each sack is to be sampled by drawing a core from one corner of the top, diagonally, to the center of the sack. The sampling device is a cylindrical, pointed, polished metal trier, 13 mm in diameter, with a slit at least one third of the circumference of the trier. A second sample is taken from the opposite corner in a similar manner. These cores are to be delivered to a clean, dry, airtight container that has stood open near the lot of flour to be sampled for a few minutes. A separate container is to be used for each sack, and the sample container is to be sealed immediately after the sample is placed inside. The type of container to be used is specified. Procedures to ensure a homogenous subsample are also described.

Title 21 CFR specifies sampling procedures required for certain products to ensure that they conform with the standard of identity. As an example, 21 CFR 145.3, in describing canned fruits, defines a sample unit as "a container, a portion of the contents of a container, or a composite mixture of product from small containers that is sufficient for the examination or testing as a single unit" (7). Further, a sampling plan is specified for containers of different net weights. Under each container size category, lots of different sizes are listed. For the various size lots, a specified number of containers is to be examined, and the lot is to be rejected if the specified number of defective units are identified. For a lot containing 48,001 to 84,000 units each weighing 1 kg or less, 48 sample units should be selected. If six of these units fail to conform with the criteria under evaluation (net weight, soluble solids, extraneous material, etc.), the lot is to be rejected. Under this sampling plan, a defective lot will be successfully rejected 95 percent of the time (consumer's risk of 5 percent, discussed in section 3.2.4 above) (7).

The following sections describe some general considerations in taking the laboratory sample.

3.3.2 Homogeneous versus Heterogeneous Populations

The ideal population would be exactly the same at every location. Such a population would be **homogeneous.** Sampling from a homogeneous population is simple. One can select a portion from any location and obtain results that are representative of the whole. Unfortunately, in the real world such populations are rare. Even a seemingly homogeneous product such as sugar syrup may not be homogeneous if it has suspended particles or sediment present only in a few locations. Most populations are **heterogeneous.** Results obtained with samples taken from heterogeneous populations will depend on the location of sampling. Fortunately, there are several methods to make a heterogeneous population more homogeneous. Some depend on the sampling plan (section 3.2.3) and some depend on sample preparation prior to analysis (section 3.5).

3.3.3 Manual versus Continuous Sampling

Manual sampling is done by humans. The person taking the sample physically performs some process to select the sample. Manual sampling may be as simple as selecting finished packages from the processing line and as complicated as using an auger to sample a rail car. Regardless of the process, it is imperative that the unit being sampled be as homogeneous as possible prior to sampling. For liquids in small containers, this is achieved by shaking prior to sampling. For liquids in silos, aeration maintains a homogeneous unit. For grain in rail cars, mixing prior to sampling is impossible. Samples are probed from several points at random and a composite sample prepared to represent the whole.

Liquids may be sampled by pipetting, pumping, or dipping (Fig. 3-1) a portion from the whole. Granular or powdered samples may be taken with the aid of triers or probes that are inserted into the material. The sample is removed and transferred into a sample container. Some solid products may be sampled by cutting representative portions from specific areas.

Manual sampling of powdered or granular materials is subject to many errors (8). Round particles flow into sampler compartments more easily than do angu-

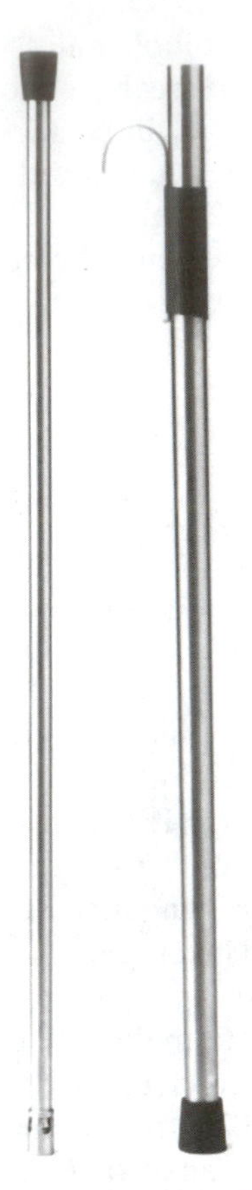

FIGURE 3-1.
Manual core sampler for fluids. Sampler takes 4 mL of product for each inch (2.54 cm) of depth. Sampler is shown with sanitizing/carrying case (right). Figure courtesy of Liquid Sampling Systems, Inc., Cedar Rapids, IA.

FIGURE 3-2.
Automatic sampling device for liquids that uses high-pressure air to collect a large number of 1.5 mL portions. The control box (left) regulates the frequency of sampling. Figure courtesy of Liquid Sampling Systems, Inc., Cedar Rapids, IA.

lar particles of similar size. Uncoated hygroscopic materials flow into the sampler compartment more readily than do similar nonhygroscopic materials. Horizontal cores were found to contain a higher proportion of smaller-sized particles than vertical cores (8).

Continuous sampling is performed by mechanical sampling devices. Fig. 3-2 shows an automatic sampling device for liquids flowing through lines. Figures 3-3, 3-4, and 3-5 are of automatic sampling devices for dry materials flowing through various types of lines.

3.3.4 Statistical Considerations

3.3.4.1 Nonprobability Sampling

Nonprobability sampling may be of several types, and it should be used with the understanding that the probability of inclusion of any portion of the whole in the sample is not equal. In these sampling plans, the investigator determines which samples will be selected. Accurate estimates of the entire population are not possible because sampling error cannot be determined.

Judgment sampling depends on the person choosing the sample. Frequently this method is the only practical and feasible way to obtain a sample. If the investigator is experienced in sample selection and the limitations in extrapolation of the results are understood, this method may better represent the true state of the population than would random sampling (1). **Convenience sampling** is often referred to as "chunk sampling" or "grab sampling." The first pallet or easiest-to-get-to box is selected. Such a sample is not representative of the whole. **Restricted sampling** may be required when the entire population is not accessible. A heavily loaded boxcar may of necessity be sampled in this manner. Results are not representative of the entire population. **Haphazard sampling,** selection of any portion, should be avoided. **Quota sampling** is the division of the lot into groups representing various categories. A specific number of samples are selected from each group by judgment. The sampling plan is less costly than random sampling but is also less reliable.

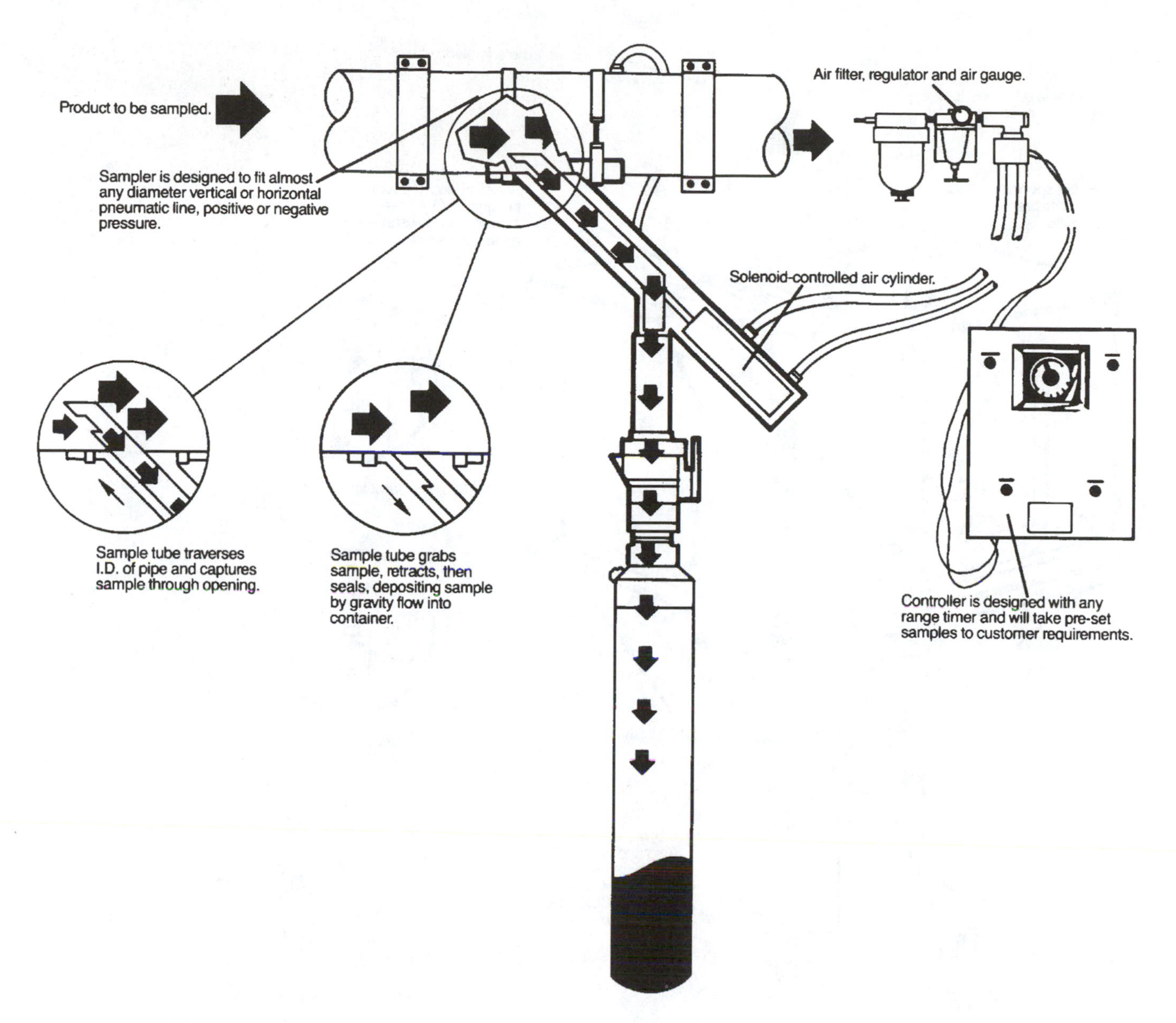

FIGURE 3-3.
Automatic sampling device for powders, granules, and pellets flowing in positive or negative pressure, in horizontal or vertical pneumatic conveying systems. Figure courtesy of Gustafson, Inc., Dallas, TX.

3.3.4.2 Probability Sampling

Probability sampling plans are of several types and provide a scientific method for selection of samples according to a statistical plan. The plan provides for automatic selection of the elements with no choice by the inspector (1). The chance of including each item is known, and sampling error may be calculated.

Simple random sampling requires that the number of units in the population to be sampled is known. Each unit is assigned (in order) a number. A specific quantity of random numbers are selected between one and the total number of units in the lot. Sample size depends on lot size and the seriousness of committing a consumer or vendor error. Sampling plans are available in reference 10. Random number tables may be used; computer programs are available to perform this task. Units are chosen corresponding to the random numbers and evaluated for the characteristics of interest.

Systematic sampling is applied when a complete list of sample units is not available. The first sample unit is selected at random, and every *n*th unit after that is selected. This plan is used when materials are continuously distributed over time or space. It is most frequently used in production-line sampling. The variance is difficult to measure.

In **stratified random sampling,** the population to

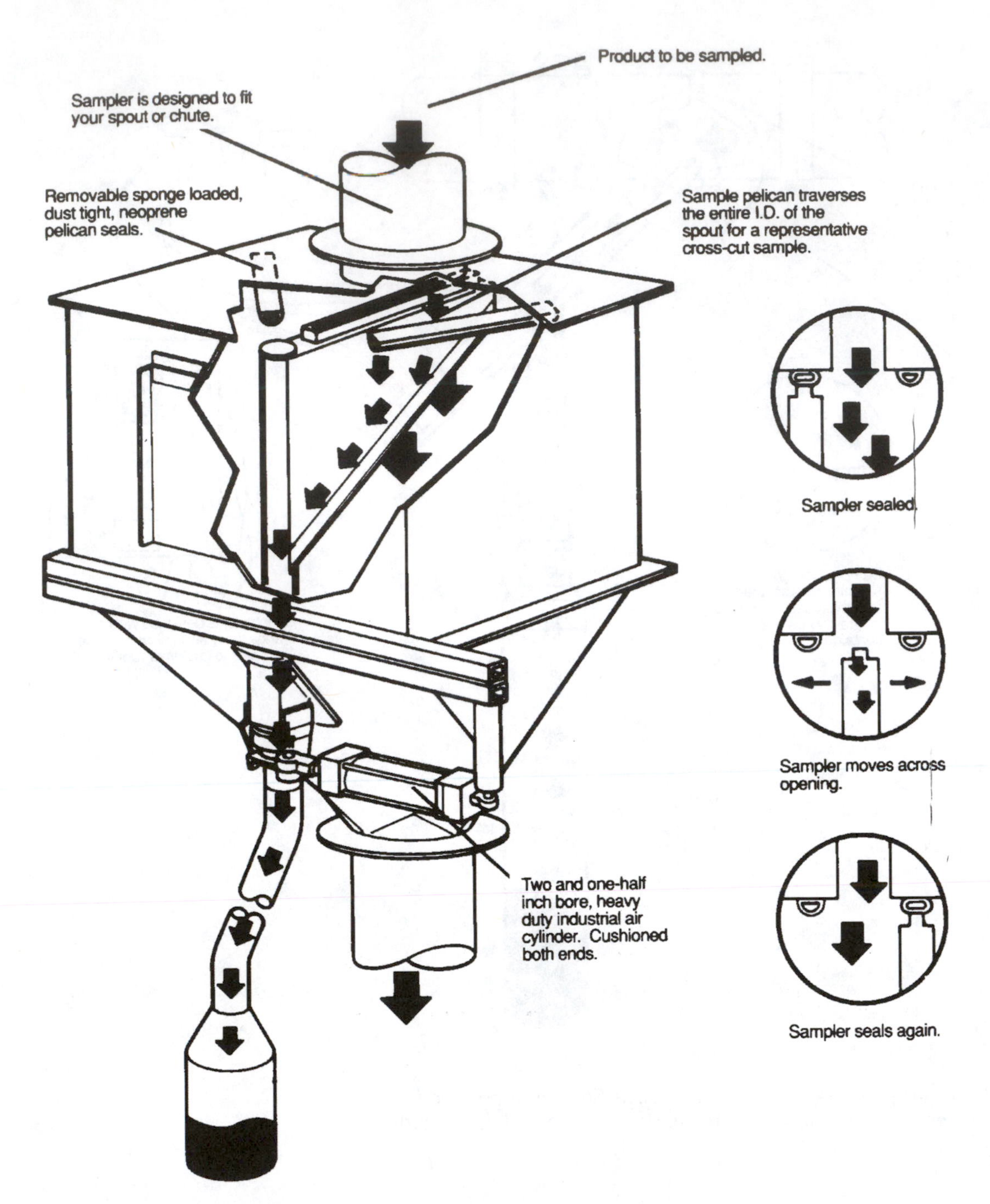

FIGURE 3-4.
Cross-cut sampler for automatic sampling from vertical spouts or chutes. Sampler is designed for powders, granules, flakes, pellets, and slurries. Figure courtesy of Gustafson, Inc., Dallas, TX.

be sampled is divided into subgroups such that units within each group are as homogeneous as possible. Group means are as widely different as possible. Samples are taken randomly from each subgroup. This procedure provides the most representative cross section of the entire population because no part is excluded. It also provides an estimate on each group within the population and of the whole population. It is less expensive than simple random sampling.

Another sampling variation is **cluster sampling.** The population is divided into subgroups termed "clusters" such that each subgroup is as similar to all others as possible. Heterogeneity is within the cluster. Clusters are chosen randomly and the selected cluster

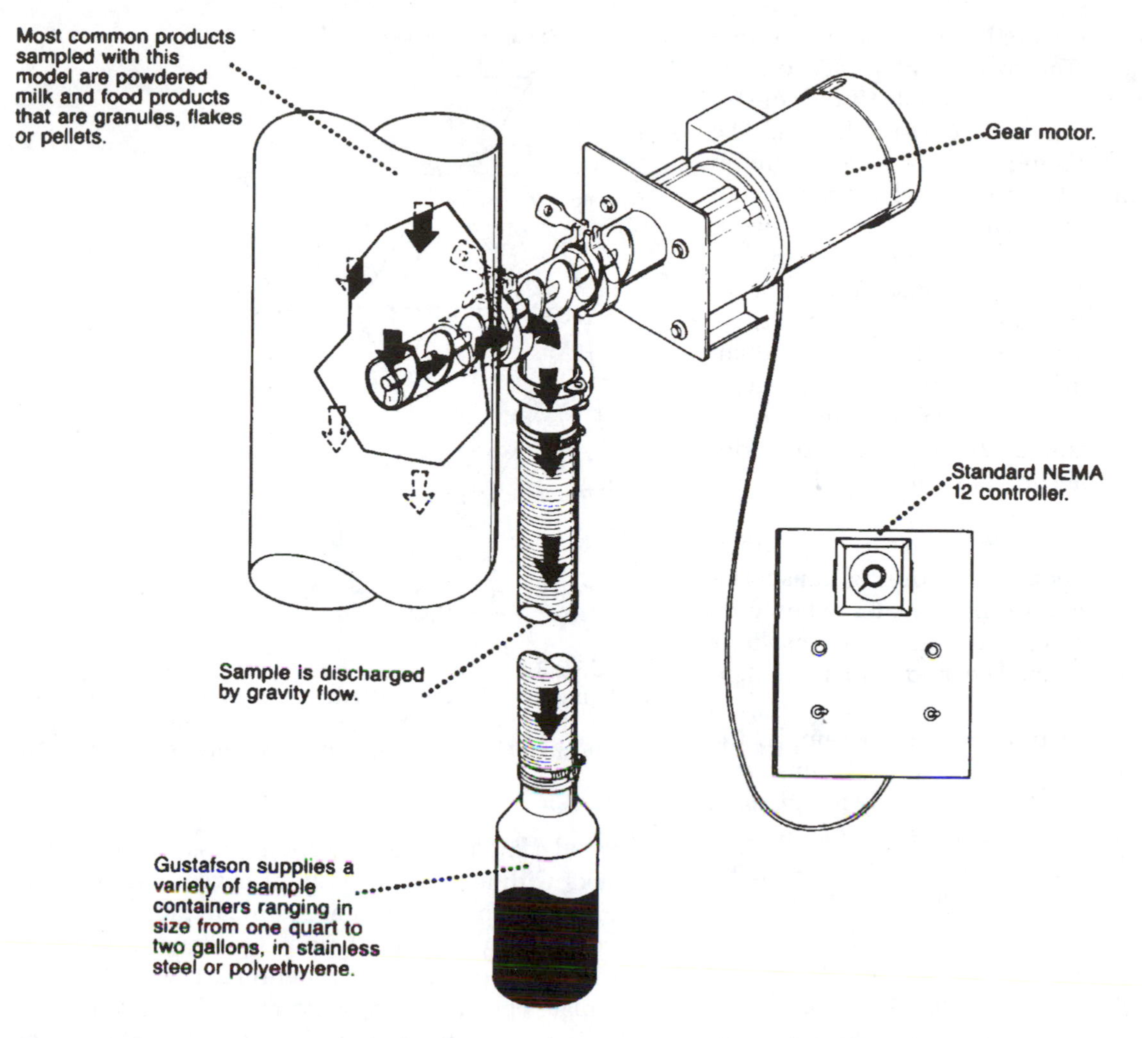

FIGURE 3-5.
Automatic sampler designed for vertical gravity or low-pressure lines carrying powders, granules, flakes, or pellets. Figure courtesy of Gustafson, Inc., Dallas, TX.

is either 100 percent inspected or subsampled randomly. This process is more efficient and less costly than simple random sampling for populations that easily can be divided into homogeneous groups. Clusters should be small and the number of sample units from each cluster about the same.

Composite sampling is commonly used for flour, seed, and other items in bags. It is also useful for solid samples in bulk. Two or more random samples are combined to give one sample for analysis. This procedure averages differences within the population.

3.3.4.3 Mixed Sampling

Mixed sampling combines random and purposeful sampling. The lot is subdivided based on purposeful sampling methods and items from within the group are selected randomly.

3.4 PROBLEMS IN SAMPLING

The analytical results obtained for any sample are only as good as the sample. Among the problems associated with sampling is sample bias. Bias may be introduced by nonprobability sampling plans such as purposeful or convenience sampling. Bias may also result from substitution of a more conveniently acquired sample by the sampler (or perhaps he or she forgot to pull the sample when required and took one a little later in the process). One major source of error results from a lack of understanding of the population distribution. Incorrect choice of a sampling plan not applicable to the true distribution may result.

Some often overlooked problems in sampling are not in sampling but in identification of the sample and its storage once obtained. The sample should be placed in an appropriate container that protects it from moisture loss or absorption during transport and storage.

The sample should be stored under conditions that prevent degradation. The method of protection depends upon the final use of the sample. If the constituent of interest is light sensitive, the sample should be protected from light by being wrapped with aluminum foil or placed in an amber or otherwise opaque container. If oxygen can cause changes in the sample, it should be stored under nitrogen or other inert gas. If temperature is a problem, it should be controlled. Freezing protects many samples, but it should be avoided in products consisting of unstable or potentially unstable emulsions. Preservatives (e.g., mercuric chloride, potassium dichromate, chloroform) may be used under some conditions (2). Reference 2 provides information on the protection of samples during transport and storage.

Sample containers should be clearly identified with markings that are not affected during transportation and storage. Plastic bags that are later to be stored in ice water should be marked with water-insoluble inks. Otherwise, samples may be unidentifiable at their destination.

If the sample is an official or legal sample, the sample container must be sealed to ensure against tampering. The sealing mark should be easily identifiable. Official samples also must include the date of sampling and the name and signature of the sampling agent. Chain of custody must be identified at all times.

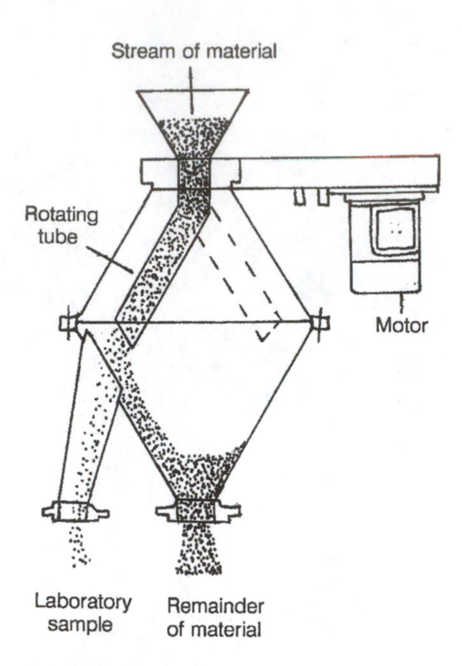

FIGURE 3-6.
Rotating tube divider for reducing a large sample (ca. 880 kg) of dry, free-flowing material to a laboratory size sample (ca. 0.2 kg). Figure courtesy of Glen Mills, Inc., Clifton, NJ.

3.5 PREPARATION OF SAMPLES

3.5.1 General Size Reduction Considerations

Upon arrival at the laboratory, the sample is usually too large for analysis (2). It must be reduced in bulk size and/or particle size. A simple way to prepare a smaller sample from a bulk sample of small, solid-particle size is to pile the sample evenly on a clean surface, flatten the pile, and divide it into quarters (2). Combine two opposite quarters and discard the other two. Repeat this operation until the sample is reduced to a manageable size. A similar process can be done with homogeneous liquids and four containers. The process may be automated (Fig. 3-6). The objective is to prepare a sample homogeneous enough to ensure that there is negligible difference between repeated test portions (2). Grinders, blenders, and food processors are useful equipment to reduce particle size.

AOAC (6) provides details on sample preparation. Preparation depends on the nature of the food and the type of analysis. For example, in Method 983.18 for meat and meat products, AOAC specifies that small samples should not be used, as this leads to moisture loss during preparation and subsequent handling. Ground samples are to be kept in glass or similar containers with air- and watertight covers. Fresh, dried, cured, or smoked meats and similar products are to be completely separated from the bone and passed rapidly through a food chopper three times. The chopper plate opening must be no more than 3 mm wide. The sample is to be mixed thoroughly after being ground, and determinations are to begin immediately. Samples not analyzed immediately are to be chilled and, for long-term storage, are to be dried. Another example is that described for solid sugar products in AOAC Method 920.175 (6). AOAC specifies that the sugar should be ground, if necessary, and mixed to uniformity. Raw sugars are to be thoroughly and quickly mixed with a spatula. Lumps are to be broken either by placing the sample on a glass plate and crushing it with a glass or iron rolling pin, or by placing the sample in a large, clean, dry mortar and crushing the lumps with a pestle. In the following sections, some considerations of sample preparation will be discussed.

3.5.2 Grinding

Various types of mills are available to reduce sample particle size and provide a homogeneous mass from a heterogeneous mixture (9). Bowl cutters, meat mincers, tissue grinders, mortars and pestles, or blenders are most useful for moist samples. Mortars and pestles or mills are best for dry samples. Mills are classified according to their mode of action. The Buhler mill is a burr mill, the Udy mill is a hammer mill, the Culatti mill is an impeller mill, the Cyclotec is a cyclone mill, the Wiley mill is a cutter mill, and the Retsch mill is a centrifugal mill (9). Particle size is controlled in some mills by adjusting the distance between burrs or blades and/or by screen mesh size. Final particle size of dry foods should be 20-mesh (openings per linear inch) if the sample is to be used for moisture, total protein, or mineral determination. A final particle size of 40-mesh is used for assays involving extraction, such as for lipid, carbohydrate, and various forms of protein.

Some foods are better ground after drying in a desiccator or vacuum oven. Grinding wet samples may cause significant losses of moisture and/or chemical changes. Grinding foods when they are frozen reduces undesirable chemical changes.

Grinding should not heat the product. This can be controlled by not overloading the grinder. Bare-metal mills should be avoided and must not be used if the sample is to be analyzed for trace metals.

3.5.3 Enzymatic Inactivation

Food materials are rich in enzymes. Under many conditions of analysis, the enzymes are still active and may interfere with final results. If the action of the enzyme alters the compound being analyzed, it must be controlled or inactivated. The control method varies widely depending on the size, consistency, and composition of the material; the enzyme(s) present; and the intended analyses (9). Some enzymes are particularly heat labile and can be inactivated with minimal heating. Storage at low temperatures (−20 to −30°C) protects many, but not all, foods from the action of enzymes. Some enzymes are inactivated by inorganic compounds, by a shift in pH, or by salting out (9). Oxidative enzymes may be controlled by reducing agents.

3.5.4 Lipid Protection

Lipids create a particular problem in sample preparation. Foods high in fat are difficult to grind at room temperature. These foods may need to be ground in a frozen state. Unsaturated lipid components may be altered by oxygen exposure. Samples should be protected by storage under nitrogen. Antioxidants may be helpful if they do not interfere with final analysis. Light can initiate oxidation of unsaturated fatty acids. Lipid materials appear to be more stable during frozen storage in intact tissue rather than in extracted tissue (9). Low-temperature storage under nitrogen is usually recommended to protect most foods.

3.5.5 Microbial Growth and Contamination

Microorganisms are present in nearly all foods and if not controlled can alter the composition of the sample. Likewise, they are present on all but sterilized surfaces and can easily contaminate samples. The former is a problem in all samples; the latter is more of a problem in samples destined for microbiological examination. Freezing, drying, and chemical preservatives can control microbial growth. Frequently, a combination is employed. The method of preservation depends on the nature of the food, the expected contamination, the storage period and conditions, and the analyses that are to be performed (9).

3.6 SUMMARY

Quality characteristics of foods are monitored at many points. For this monitoring, 100 percent inspection is rarely possible. To ensure that the portion monitored is representative of the whole, sampling and sample reduction procedures must be developed. The sampling procedure selected depends on the purpose of the inspection, the nature of the product, the nature of the test method, and the nature of the population being investigated. In general, increasing the sample size will increase the reliability of the final results; sampling is usually the most variable step in the overall analysis process.

Sampling may be for attributes or variables. Attributes are those characteristics that are present or not present. Variables are those characteristics that are measured on a continuous scale. Sampling plans are developed for either characteristic and may be single, double, or multiple. Multiple sampling plans reduce sampling costs by rejecting low-quality lots or accepting high-quality lots quickly. Intermediate quality lots require more sampling. Every sampling plan has inherent risks associated with it. Consumer's risk is the chance of accepting a poor-quality lot, while vendor's risk is the chance of rejecting acceptable product. The probability acceptable for either depends on the seriousness of the occurrence.

Sampling plans depend on whether the population is homogeneous or heterogeneous. Sampling of homogeneous populations is simple. Unfortunately, few populations in real life are homogeneous. Heterogeneous populations must be sampled to achieve the most homogeneous sample representative of the whole. Sampling may be manual or continuous. Selection of the sample to be evaluated may be determined statistically. Nonprobability sampling is less desirable but sometimes is unavoidable. Nonprobability sampling does not give each portion an equal chance of being included. The person doing the sampling determines the portions selected. Probability sampling is statistically sound, and each portion has an equal probability of being selected. The chance of including any item and the sampling error may be calculated.

Sample identity must be clearly marked, and the sample must be protected from changes from the time of selection to the completion of the analysis. Official samples must be sealed and chain of custody identified at all times.

Prior to laboratory analysis, samples usually must be reduced in size. This must be done in such a manner as to ensure that the final sample is representative of the whole. Considerations must be given to the potential activity of enzymes in the sample, protection of lipid material from oxidation, and microbial growth and/or contamination of the sample.

3.7 STUDY QUESTIONS

1. As part of your job as supervisor in a quality assurance laboratory, you need to give a new employee instructions regarding choice of a sampling plan. What general factors should your employee consider in choosing a sample plan? Differentiate sampling for attributes versus sampling for variables. Differentiate the three basic types of sampling plans. What risks are associated with selection of a sampling plan?
2. Differentiate nonprobability sampling and probability samples. Which is preferable? Why?
3. Identify (a) a piece of equipment that would be useful in *collecting* a representative sample for analysis and (b) a piece of equipment that would be useful in *preparing* a sample for analysis. For each piece of equipment, give an example of a type of product suitable for its use. Give any appropriate precautions in the use of the equipment, to be certain a representative sample is obtained and sample composition is not changed during sample preparation.
4. List two problems that can be associated with the *collection* of samples for analysis and two problems that can be associated with the *preparation* of samples for analysis. For each problem, state how it can be at least partially overcome.
5. The instructions for an assay procedure you are to follow were left by the person previously in your position. The instructions specify to grind a cereal sample to 10-mesh before you extract the protein using a series of solutions.
   a. What does "10-mesh" mean?
   b. Would you question the use of a 10-mesh for your specific application? Why or why not?

3.8 REFERENCES

1. Puri, S.C., Ennis, D., and Mullen, K. 1979. *Statistical Quality Control for Food and Agricultural Scientists.* G. K. Hall and Co., Boston, MA.
2. Horwitz, W. 1988. Sampling and preparation of sample for chemical examination. *J. Assoc. Off. Anal. Chem.* 71:241–245.
3. Harris, D.C. 1991. *Quantitative Chemical Analysis,* 3rd ed. W. H. Freeman and Co., New York, NY.
4. Miller, J.C., and Miller, J.N. 1988. Basic statistical methods for analytical chemistry, Part I, Statistics of repeated measurements. A review. *Analyst* 113:1351–1355.
5. Springer, J.A., and McClure, F.D. 1988. Statistical sampling approaches. *J. Assoc. Off. Anal. Chem.* 71:246–250.
6. AOAC. 1990. *Official Methods of Analysis,* 15th ed. Association of Official Analytical Chemists, Washington, DC.
7. Anonymous. 1992. Code of Federal Regulations. Title 21. U.S. Government Printing Office, Washington, DC.
8. Baker, W.L., Gehrke, C.W., and Krause, G.F. 1967. Mechanisms of sampler bias. *J. Assoc. Off. Anal. Chem.* 50: 407–413.
9. Pomeranz, Y., and Meloan, C.E. 1987. *Food Analysis: Theory and Practice,* 2nd ed. Van Nostrand Reinhold, New York, NY.
10. U.S. Dept. of Defense. 1963. MIL-STD-105D. U.S. Government Printing Office, Washington, DC.

CHAPTER

4

Evaluation of Analytical Data

J. Scott Smith

4.1 INTRODUCTION

The field of food analysis, or any type of analysis, involves a considerable amount of time spent learning principles, methods, and instrument operations and perfecting various techniques. Although these areas are extremely important, much of our effort would be for naught if there were not some way for us to evaluate the data obtained from the various analytical assays. Several mathematical treatments are available that provide an idea of how well a particular assay was performed or how well we can reproduce an experiment. Fortunately, the statistics are not too involved and apply to most analytical determinations.

The focus in this chapter will be primarily on how to evaluate replicate analysis of the same sample for accuracy and precision. In addition, considerable focus will be on the determination of best line fits for standard curves type data.

4.2 MEASURES OF CENTRAL TENDENCY

To increase accuracy and precision, as well as to evaluate these parameters, the analysis of a sample is usually performed (repeated) several times. At least three assays are typically performed, though often the number can be much higher. Because we are not sure which value is closest to the true value, we determine the mean (or average) using all the values obtained and report the results of the **mean.** The mean is designated by the symbol $\bar{x}$ and calculated according to the equation below.

$$\bar{x} = \frac{x_1 + x_2 + x_3 + \ldots + x_n}{n} = \frac{\sum x_i}{n} \quad (1)$$

where: $\bar{x}$ = mean
x_1, x_2, etc. = individually measured values
n = number of measurements

For example, suppose we measured a sample of uncooked hamburger for percent moisture content four times and obtained the following results: 64.53 percent, 64.45 percent, 65.10 percent, and 64.78 percent.

$$\bar{x} = \frac{64.53 + 64.45 + 65.10 + 64.78}{4} = 64.72\% \quad (2)$$

Thus, the results would be reported as 64.72 percent moisture. When we report the mean value, we are indicating that this is the best experimental estimate of the value. We are not saying anything about how accurate or true this value is. Some of the individual values may be closer to the true value, but there is no way to make that determination, so we report only the mean.

Another determination that can be used is the **median,** which is the midpoint or middle number within a group of numbers arrayed in numerical order. Basically, half of the experimental values will be less than the median and half will be greater. The median is not used often, because the mean is such a superior experimental estimator.

4.3 RELIABILITY OF ANALYSIS

Returning to our previous example, recall that we obtained a mean value for moisture. However, we did not have any indication of how repeatable the tests were or of how close our results were to the true value. The next several sections will deal with these questions and some of the relatively simple ways to calculate the answers.

4.3.1 Accuracy and Precision

One of the most confusing aspects of data analysis for students is grasping the concepts of accuracy and precision. These terms are commonly used interchangeably in society, which only adds to this confusion. If we consider the purpose of the analysis, then these terms become much clearer. If we look at our experiments, we know that the first data obtained are the individual results and a mean value ($\bar{x}$). The next questions should be How close were our individual measurements? and How close were they to the true value? Both questions involve accuracy and precision. Now, let us turn our attention to these terms.

Accuracy refers to how close a particular measure is to the true or correct value. In the moisture analysis for hamburger, recall that we obtained a mean of 64.72 percent. Let us say the true value was actually 65.05 percent moisture. By comparing these two numbers, you could probably make a guess that your results were pretty accurate because they were very close to the correct value. (The calculations of accuracy will be discussed later.)

The problem in determining accuracy is that most of the time we are not sure what the true value is. For certain types of materials we can purchase known samples from, for example, the National Institute of Standards and Technology, and then check our assays against these samples. Only then can we have an indication of the accuracy of the testing procedures. Another approach is to compare our results with those of other labs to see how well they agree, assuming the other labs are accurate.

A term that is much easier to deal with and determine is **precision.** This parameter is a measure of how reproducible or how close replicate measurements become. If repetitive testing yields very similar results, then we would say the precision of that test was good.

The difference between precision and accuracy can best be illustrated with the figure below. Imagine shooting a rifle at a target that represents experimental values. The bull's-eye would be the true value, and where the bullets hit would represent the individual experimental values. As you can see in Fig. 4-1a, the values can be tightly spaced (good precision) and close to the bull's-eye (good accuracy). Or in some cases, there can be good precision but poor accuracy (Fig. 4-1b). The worst situation, as illustrated with Fig. 4-1d, is one in which both accuracy and precision are poor. Because of errors in the determination, in this case, interpretation of the results becomes very difficult. Later, we will discuss the practical aspects of the various types of error.

When evaluating data, several tests are commonly used to give some appreciation of how much the experimental values would vary if we were to repeat the test (indicators of precision). An easy way to look at dispersion or scattering is to report the range of the experimental values. The **range** is simply the difference between the largest and smallest observations. This measurement is not too useful and, thus, is seldom used in evaluating data.

Probably the best and most commonly used statistical evaluation of the precision of analytical data is the standard deviation. The **standard deviation** measures the spread of the experimental values and gives a good indication of how close the values are to each other. When evaluating the standard deviation, one has to remember that we are never able to analyze the entire food product. That would be difficult, if not impossible, and very time consuming. Thus, the calculations we use are only estimates of the unknown true value.

If we have many samples, then the standard deviation is designated by the Greek letter sigma (σ). It is calculated according to the equation below, assuming all the food product was evaluated (which would be an infinite amount of assays).

$$\sigma = \frac{\sqrt{\Sigma (x_i - \mu)^2}}{n} \tag{3}$$

where: σ = standard deviation
x_i = individual sample values
μ = true mean
n = total population of samples

Because we do not know the value for the true mean, the equation becomes somewhat simplified so that we can use it with real data. In this case, we now call the σ term the standard deviation of the sample and designate is by SD or s. It is determined according to the calculation below, where $\bar{x}$ replaces the true mean term μ, and n represents the number of samples.

$$SD = \frac{\sqrt{\Sigma (x_i - \bar{x})^2}}{n} \tag{4}$$

If the number of replicate determinations is small (about 30 or less), which is common with most assays, the n is replaced by the n-1 term, and the equation below is used. Unless you know otherwise, the equation below is always used in calculating the standard deviation of a group of analyses.

$$SD = \frac{\sqrt{\Sigma (x_i - \bar{x})^2}}{n - 1} \tag{5}$$

Depending on which of the equations above is used, the standard deviation may be reported as SD_n or σ_n and SD_{n-1} or σ_{n-1}. (Different brands of scientific calculators sometimes use different labels for the keys, so you must be careful.) Table 4-1 shows an example of the determination of standard deviation. The sample results would be reported to average 64.72 percent moisture with a standard deviation of 0.293.

Once we have a mean and standard deviation, we must next determine how to interpret these numbers. One easy way to get a feel for the standard deviation is to calculate what is called the **coefficient of variation** (CV), also known as the relative standard deviation. This calculation is shown below for our example of the moisture determination of uncooked hamburger.

$$\text{Coefficient of Variation (CV)} = \frac{SD}{\bar{x}} \times 100\% \tag{6}$$

$$CV = \frac{0.293}{64.72} \times 100\% = 0.453\% \tag{7}$$

The coefficient of variation tells us that our standard deviation is only 0.453 percent as large as the

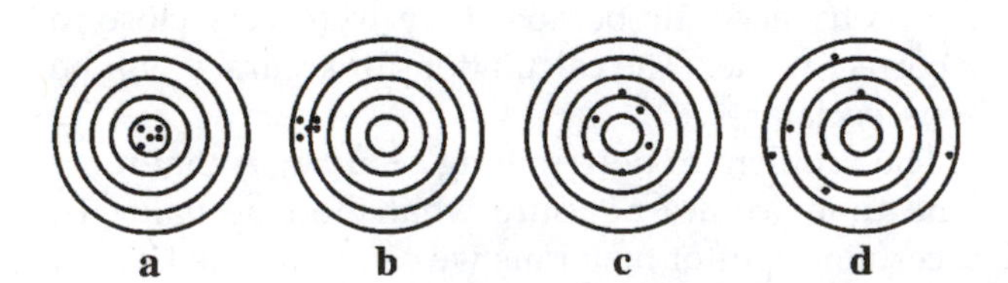

FIGURE 4-1.
Comparison of accuracy and precision: (a) good accuracy and good precision, (b) good precision and poor accuracy, (c) good accuracy and poor precision, and (d) poor accuracy and poor precision.

TABLE 4-1. Determination of the Standard Deviation of Percent Moisture in Uncooked Hamburger

Measurement	*Observed % Moisture*	*Deviation from the Mean* $(x_i-\bar{x})$	$(x_i-\bar{x})^2$
1	64.53	−0.19	0.0361
2	64.45	−0.27	0.0729
3	65.10	+0.38	0.1444
4	64.78	+0.06	0.0036
	$\Sigma x_i = 258.86$		$\Sigma (x_i-\bar{x})^2 = 0.257$

$$\bar{x} = \frac{\Sigma x_i}{n} = \frac{258.86}{4} = 64.72 \qquad SD = \sqrt{\frac{\Sigma (x_i-\bar{x})^2}{n-1}} = \sqrt{\frac{0.257}{3}} = 0.2927$$

mean. For our example, that number is small, which indicates a high level of precision or reproducibility of the replicates. As a rule, a CV below 5 percent is considered acceptable, although it depends on the type of analysis.

Another way to evaluate the meaning of the standard deviation is to examine its origin in statistical theory. Many populations (in our case, sample values or means) that exist in nature are said to have a normal distribution. If we were to measure an infinite number of samples, we would get a distribution similar to that represented by Fig. 4-2. In a population with a **normal distribution,** 68 percent of those values would be within ± 1 standard deviation from the mean; 95 percent would be within ± 2 standard deviations, and 99.7 percent would be within ± 3 standard deviations. In other words, there is a probability of only 1 percent that a sample in a population would fall outside ± 3 standard deviations.

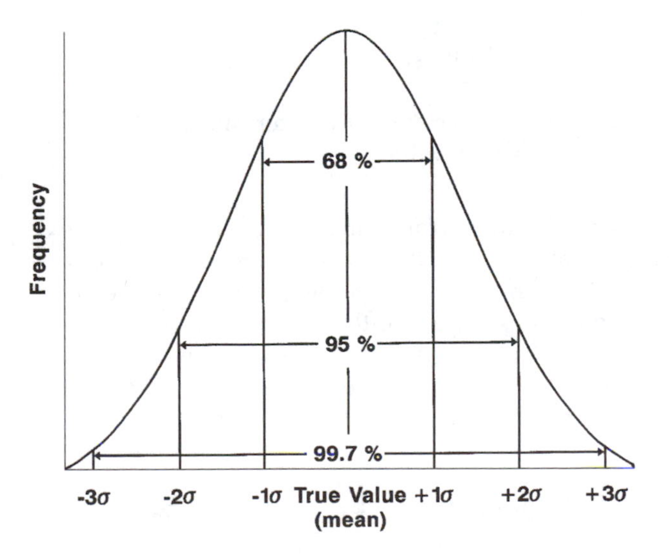

FIGURE 4-2.
A normal distribution curve for a population or a group of analyses.

Another way of understanding the normal distribution curve is to realize that the probability of finding the true mean is within certain confidence intervals as defined by the standard deviation. For large numbers of samples, we can determine the **confidence limit** or **interval** using the statistical parameter called the **Z value.** We do this calculation by first looking up the Z value from statistical tables once we have decided the desired degree of certainty. Some of the Z values are listed in Table 4-2.

The confidence limit (or interval) for our moisture data, assuming a 95 percent probability, is calculated according to the equation below. Since this calculation is not valid for small numbers, assume we had run 25 samples instead of four.

$$\text{Confidence Interval (CI)} = \bar{x} \pm Z \text{ value} \times \frac{\text{standard deviation}}{\sqrt{n}} \tag{8}$$

$$\begin{array}{c}\text{Confidence Interval} \\ \text{(at 95\%)}\end{array} = 64.72 \pm 1.96 \times \frac{0.2927}{\sqrt{25}} = 64.72 \pm 0.115\% \tag{9}$$

Because our example had only four values for the moisture levels, the confidence interval should be calculated using statistical t tables. In this case, we have to look up the t score from Table 4-3 based on the degrees of freedom, which is the sample size minus one (n−1), and the desired level of confidence.

TABLE 4-2. Values for Z for Checking Both Upper and Lower Levels

Degree of Certainty (Confidence)	*Z Value*
80%	1.29
90%	1.64
95%	1.96
99%	2.58
99.9%	3.29

TABLE 4-3. Values of t for Various Levels of Probability[1]

Degrees of Freedom (n-1)	LEVELS OF CERTAINTY		
	95%	*99%*	*99.9%*
1	12.7	63.7	636
2	4.30	9.93	31.60
3	3.18	5.84	12.9
4	2.78	4.60	8.61
5	2.57	4.03	6.86
6	2.45	3.71	5.96
7	2.36	3.50	5.40
8	2.31	3.56	5.04
9	2.26	3.25	4.78
10	2.23	3.17	4.59

[1]More extensive t-tables can be found in statistics books.

The calculation for our moisture example with four samples (n) and 3 degrees of freedom (n−1) is given below.

$$\text{Confidence Interval} = \bar{x} \pm \text{t value} \times \frac{\text{standard deviation (SD)}}{\sqrt{n}} \quad (10)$$

$$\begin{array}{l}\text{Confidence}\\ \text{Interval}\\ \text{(at 95\%)}\end{array} = 64.72 \pm 3.18 \times \frac{0.2927}{\sqrt{4}} \quad (11)$$

$$= 64.72 \pm 0.465\%$$

To interpret this number, we can say, with 95 percent confidence, that the true mean for our moisture will be between 64.72 ± 0.465 percent or fall between 65.185 and 64.255 percent.

The expression $SD/\sqrt{n}$ is often reported as the **standard error of the mean.** It is then left to the reader to calculate the confidence interval based on the desired level of certainty.

Other quick tests of precision sometimes used are the **relative deviation from the mean** and the **relative average deviation from the mean.** The **relative deviation from the mean** is useful when only two replicates have been performed. It is calculated according to the equation below, with values below 2 percent considered acceptable.

$$\begin{array}{c}\text{Relative}\\ \text{deviation}\\ \text{from the mean}\end{array} = \frac{x_i - \bar{x}}{\bar{x}} \times 100 \quad (12)$$

The x_i represents the individual sample value, and $\bar{x}$ is the mean.

If there are several experimental values, then the **relative average deviation from the mean** becomes a useful indicator of precision. It is calculated similarly to the relative deviation from the mean, except the average deviation is used instead of the individual deviation. It is calculated according to the equation below:

$$\begin{array}{c}\text{Relative}\\ \text{average}\\ \text{deviation}\\ \text{from the mean}\end{array} = \frac{\dfrac{\Sigma|x_i - \bar{x}|}{n}}{\bar{x}} \times 1000 \quad (13)$$

$$= \text{parts per thousand}$$

Using the moisture values discussed in Table 4-1, the $x_i - \bar{x}$ terms for each determination are −0.19, −0.27, +0.38, +0.06. Thus, the calculation becomes:

$$\begin{array}{l}\text{Rel.}\\ \text{Avg.}\\ \text{Dev.}\end{array} = \frac{\dfrac{0.19 + 0.27 + 0.38 + 0.06}{4}}{64.72} \times 1000$$

$$= \frac{0.225}{64.72} \times 1000 \quad (14)$$

$$= 3.47 \text{ parts per thousand}$$

Up to now, our discussions and calculations have involved ways to evaluate precision. If the true value is not known, we can only calculate precision. A low degree of precision would make it difficult to predict a value for the sample.

However, we occasionally may have a sample in which we know the true value and can compare our results with the known value. In this case, we can actually calculate the error for our test, compare it to the known value, and determine the accuracy. One term that can be calculated is the **absolute error,** which is simply the difference between the experimental value and the true value.

$$\text{Absolute error} = E_{abs} = x - T \quad (15)$$

where: x = experimentally determined value
T = true value

The absolute error term can have either a positive or a negative value. If the experimentally determined value is from several replicates, then the mean ($\bar{x}$) would be substituted for the x term. This is not a very good test for error, because the value is not related to the magnitude of the true value. A more useful measurement of error is **relative error.**

$$\text{Relative error} = E_{rel} = \frac{E_{abs}}{T} = \frac{x - T}{T} \quad (16)$$

The results are reported as a negative or positive value, which represents a fraction of the true value.

If desired, the relative error can be expressed as percent relative error by multiplying by 100 percent. Then the relationship becomes the following, where x can be either an individual determination or the mean ($\bar{x}$) of several determinations:

$$\% \, E_{rel} = \frac{E_{abs}}{T} \times 100\% \qquad (17)$$
$$= \frac{x - T}{T} \times 100\%$$

Using the data for the percent moisture of uncooked hamburger, suppose the true value of the sample is 65.05 percent. The percent relative error then is calculated using our mean value of 64.72 percent and equation 17.

$$\% \, E_{rel} = \frac{\bar{x} - T}{T} \times 100\%$$
$$= \frac{64.72 - 65.05}{65.05} \times 100\% \qquad (18)$$
$$= -0.507\%$$

Note that we keep the negative value, which indicates the direction of our error: Our results were 0.507 percent lower than the true value.

4.3.2 Sources of Errors

As you may recall from our discussions on accuracy and precision, error can be quite important in analytical determinations. Although we strive to obtain correct results, it is unreasonable to expect an analytical technique to be entirely free of error. The best we can hope for is that the error is small and, if possible, at least consistent. As long as we know about the error, the analytical method will often be satisfactory. There are several sources of error, which can be classified as random error (indeterminate), systematic error (determinate), and gross error or blunders.

Blunders are easy to eliminate, since they are so obvious. The experimental data are usually scattered, and the results are not close to an expected value. This type of error is a result of using the wrong reagent or instrument or of grossly sloppy technique. Some people have called it the "Monday morning syndrome" error. Fortunately, blunders are easily identified and corrected.

Random errors produce data that vary in a nonreproducible way from one measurement to another, although the mean of the individual measurements may be close to the expected value. This type of error would have a poor level of precision, as represented by Fig. 4-1c, and may or may not be accurate. Generally, this type of error can be caused by several factors at the same time and can be hard to avoid. For example, misreading an analytical balance, misjudgment of the end-point change in a titration, and improper use of a pipette can all contribute to random error. Background instrument noise, which is always present to some extent, is often a factor in random error. Both positive and negative errors are equally possible. Although this type of error is difficult to avoid, it fortunately is usually small.

Systematic error produces results that consistently deviate from the expected value in one direction or the other. As illustrated in Fig. 4-1b, the results will be closely spaced together, but they are always off the mark. Identifying the source of this serious type of error can be difficult and time consuming, because it often involves inaccurate instruments or measuring devices. For example, a pipette that consistently delivers the wrong volume of reagent will produce a high degree of precision yet inaccurate results. Sometimes impure chemicals or the analytical method itself are the cause. Generally, we can overcome systematic errors by proper calibration of instruments, running blank determinations, or using a different analytical method.

4.3.3 Specificity

Specificity of a particular analytical method means that it only detects the component of interest. Analytical methods can be very specific for a certain food component or, in many cases, can analyze for a broad spectrum of components. Quite often, it is desirable for the method to be somewhat broad in its detection. For example, the determination of food lipid (fat) is actually the crude analysis of any compound that is soluble in an organic solvent. Some of these compounds are glycerides, phospholipids, carotenes, and free fatty acids. Since we are not concerned about each individual compound when considering the crude fat content of food, it is desirable that the method be broad in scope. On the other hand, determining the lactose content of ice cream would require a specific method. Since ice cream contains other types of simple sugars, without a specific method, we would tend to overestimate the amount present.

There are no hard rules for what specificity is required. Each situation is different and depends on the desired results and type of assay used. However, it is something to keep in mind as the various analytical techniques are discussed.

4.3.4 Sensitivity and Detection Limit

Although often used interchangeably, sensitivity and detection limit should not be confused. They are different terms, yet they are closely related. **Sensitivity** relates to the magnitude of change of a measuring device (instrument) with changes in compound concentration. It is an indicator of how little change we can make in the unknown material before we will be able to note a difference on a needle gauge or a digital readout. We are all familiar with the process of tuning in a radio station on our stereo and know how, at some point, once the station is tuned in, we can move the dial without disturbing the reception. This is sensitivity. In many situations, we can adjust the sensitivity of an assay to fit our needs, that is, whether we desire more or less sensitivity. We may even desire a lower sensitivity so that samples with widely varying concentration can be analyzed at the same time.

Detection limit, in contrast to sensitivity, is the lowest possible increment that we can detect with some degree of confidence (or statistical significance). With every assay, there is a lower limit at which point we are not sure if something is present or not. Obviously, the best choice would be to concentrate the sample so we are not working close to the detection limit. However, this may not be possible, and we may need to know the detection limit so we can work away from that limit.

There are several ways to measure detection limit, depending on the apparatus that is used. If we are using something like a spectrophotometer, gas chromatograph, or high-performance liquid chromatograph, the limit of detection often is reached when the signal-to-noise ratio is 2 or greater. In other words, when the sample gives a value that is twice the magnitude of the noise detection, it is at the lowest limit possible. Noise is the random signal fluctuation that occurs with any instrument.

A more general way to define the limit of detection is to approach the problem from a statistical viewpoint, in which the variation between samples is considered. A common mathematical definition of detection limit is given below.

$$X_{LD} = X_{Blk} + 3 \times SD_{Blk} \quad (19)$$

where: X_{LD} = the minimum detectable concentration

X_{Blk} = the signal of a blank

SD_{Blk} = the standard deviation of the blank readings

In this equation, the variation of the blank values (or noise, if we are talking about instruments) determines the detection limit. High variability in the blank values decreases the limit of detection.

4.4 STANDARD CURVES: REGRESSION ANALYSIS

Quite often in analytical determinations, we use standard curves to determine concentrations of an unknown sample. This procedure involves making a group of standards in increasing concentration and then recording the particular analytical parameter (for example, absorbance). What results is a curve (actually a straight line) relating concentration versus observed instrument value (absorbance). Once we know how the absorbance changes with concentration, we can estimate the concentration of an unknown by extrapolation from the standard curve.

4.4.1 Linear Regression

The figure below illustrates differences in standard curves; Fig. 4-3a shows a good correlation of the data, and Fig. 4-3b shows a less than desirable correlation. In both cases, we can draw a straight line through the data points, but it would be more difficult for the latter. Both curves yield the same straight line, but the precision is poorer for the latter. In working with these standard curves, two questions arise: How do we draw the line through the data points? How well do the data fit to the straight line?

Consider the following example. Data for a Lowry protein assay standard curve are shown in Table 4-4. Basically, the protein solutions (in this case, standards) are reacted with several reagents to yield a blue-colored complex. The intensity of the blue color increases with protein concentration and can be measured as absorbance at 650 nm. (Note that we actually measure the absorbed radiation, not the reflected blue color.) If you plot absorbance (y-axis) versus concentration (x-axis), a straight line is generated that goes through (or

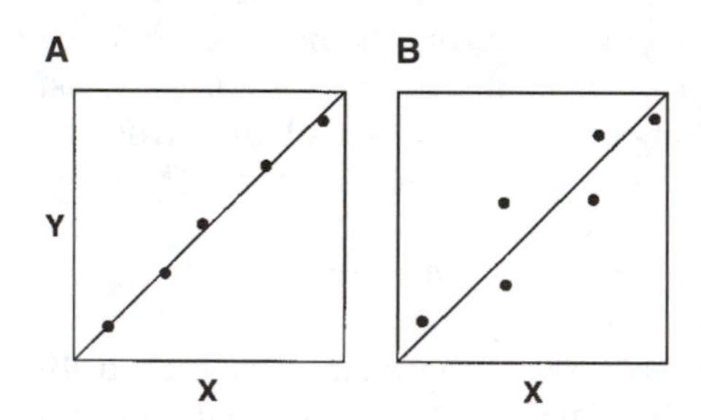

FIGURE 4-3.
Examples of standard curves showing the relationship between the X and Y variables when there is (A) a high amount of correlation and (B) a lower amount of correlation. Both lines have the same equation.

TABLE 4-4. Lowry Protein Standard Curve Experimental Data

Protein Concentration (μg/ml)	*Absorbance at 650 nm*
18	0.100
38	0.250
60	0.380
85	0.510

almost through) the origin. However, the line is not perfectly straight and never is.

We can mathematically determine the best fit of the line by use of linear regression. Keep in mind the equation for a straight line, which is: $y = ax + b$, where a = slope, and b = y-intercept. To determine the regression slope and y-intercept, a table is set up (Table 4-5). (Note n = sample size = 4). We determine a and b and, thus, for any value of Y (absorbance) we can determine the concentration (X) in μg/ml.

EXAMPLE. An unknown gives an absorbance at 650 nm of 0.310. What is the concentration of protein in the sample?

$$y = ax + b \tag{20}$$

or

$$x = \frac{y - b}{a} \tag{21}$$

$$x \text{ (conc)} = \frac{0.310 - 0.0025}{0.0061} = 50.4\ \mu\text{g/ml protein} \tag{22}$$

The above equations and calculations assume that the y-intercept is valid (i.e., it has some value). However, a standard curve should theoretically pass through the origin (i.e., $x = 0$ and $y = 0$). This makes sense, because a sample that has zero protein should have no absorbance. In this case, the straight line equation becomes

$$y = ax \tag{23}$$

(There is no "b" term, because $b = 0$.)

The regression equation for this line is:

$$\text{slope} = a = \frac{\Sigma XY}{\Sigma X^2} \tag{24}$$

For our example:

$$a = \frac{77.5}{12{,}593} = 0.006154 \tag{25}$$

As you can see, there are differences in the slopes of the two lines, but they are rather small, less than 1 percent. Most standard curves do not go through the origin for a multitude of reasons. Sometimes the relationship between Y and X is curved near zero, resulting in a line that is slightly off the origin (i.e., it has a small y-intercept value). Unless we know otherwise, the standard line equation is used.

4.4.2 Correlation Coefficient

The **correlation coefficient** defines how well the data fit to a straight line. For a standard curve, the ideal situation would be that all data points lie perfectly on a straight line. However, this is never the case, because errors are introduced in making standards and reading absorbance values. The correlation coefficient and coefficient of determination are defined below.

$$\text{correlation coefficient} = r = \frac{\Sigma XY}{\sqrt{(\Sigma X^2)(\Sigma Y^2)}} \tag{26}$$

TABLE 4-5. Linear Regression Analysis of the Lowry Protein Standard Curve Data

	X(conc)	*Y(abs)*	*X^2*	*Y^2*	*XY*
	18	0.10	324	0.010	1.8
	38	0.25	1444	0.063	9.5
	60	0.38	3600	0.144	22.8
	85	0.51	7225	0.260	43.4
Σ	201	1.24	12,593	0.477	77.5

slope

$$a = \frac{n\Sigma XY - \Sigma X \Sigma Y}{n\Sigma X^2 - (\Sigma X)^2}$$

$$a = \frac{4(77.5) - (201)(1.24)}{4(12{,}593) - (201)^2}$$

$$a = \frac{60.8}{9971}$$

$$a = 0.006098$$

y-intercept

$$b = \frac{\Sigma Y - a\Sigma X}{n}$$

$$b = \frac{1.24 - (0.0061)(201)}{4}$$

$$b = \frac{1.24 - 1.23}{4} = \frac{0.01}{4}$$

$$b = 0.0025$$

for our example:

$$r = \frac{77.5}{\sqrt{(12{,}593)(0.477)}}$$
$$= \frac{77.5}{77.503942} \qquad (27)$$
$$r = 0.9999491$$

(Values are usually reported to four significant figures.)

Note: This number differs from values obtained when using a calculator or computer software because of rounding off (calculator value, $r = 0.9965$). The **coefficient of determination** $= r^2 = 0.9930$.

For standard curves, we want the value of r as close to 1.0000 as possible, because this value is a perfect correlation (perfect straight line). Generally, in analytical work, the r should be 0.9970 or better. (This does not apply to biological studies.)

The **coefficient of determination** (r^2) is also used quite often because it gives a better feel for what happens with the straight line. The r^2 for the example presented is 0.9930, which represents the proportion of the variance of absorbance (Y) that can be attributed to its linear regression on concentration (X). This means about 0.70 percent of the straight line variation ($1.000 - 0.9930 = 0.0070 \times 100\% = 0.70\%$) does not vary with changes in X and Y and, thus, is due to indeterminate variation. A small amount of variation is normally expected.

The calculations presented above can be very time consuming if you do them by hand. Most low-priced scientific calculators will readily perform these calculations in minutes. In addition, most computer graphics/plotting software packages will perform the regression analysis and provide an actual graph of the data.

4.5 REPORTING RESULTS

In dealing with experimental results, we are always confronted with reporting the data in a way that indicates the sensitivity and precision of the assay. Ideally, we do not want to overstate or understate the sensitivity of the assay and, thus, we strive to report a meaningful value, be it a mean, standard deviation, or some other number. The next three sections will look at how we can evaluate experimental values so as to be precise when reporting results.

4.5.1 Significant Figures

The term **significant figure** is used rather loosely to describe some judgment of the number of reportable digits in a result. Often, the judgment is not soundly based, and meaningful digits are lost or meaningless digits are retained.

Proper use of significant figures is meant to give an indication of the sensitivity and reliability of the analytical method. Thus, reported values should contain only significant figures. A value is made up of significant figures when it contains all digits known to be true and one last digit that is in doubt. For example, a value reported as 64.72 contains four significant figures, of which three digits are certain (64.7) and the last digit that is uncertain. Thus, the 2 is somewhat uncertain and could be either 1 or 3. As a rule, numbers that are presented in a value represent the significant figures, regardless of the position of any decimal points. This is also true for values containing zeros, provided they are bounded on either side by a number. For example, 64.72, 6.472, 0.6472, and 6.407 all contain four significant figures. Note that the zero to the left of the decimal point is used only to indicate that there are no numbers above 1. We could have reported the value as .6472, but using the zero is better, since we know that a number was not inadvertently left off our value.

Special considerations are necessary for zeros that may or may not be significant:

1. Zeros after a decimal point are always significant figures. For example, 64.720 and 64.700 both contain five significant figures.
2. Zeros before a decimal point with no other preceding digits are not significant. As indicated before, 0.6472 contains four significant figures.
3. Zeros after a decimal point are not significant if there are no digits before the decimal point. For example, 0.0072 has no digits before the decimal point; thus, this value contains two significant figures. In contrast, the value 1.0072 contains five significant figures.
4. Final zeros in a number are not significant unless indicated otherwise. Thus, the value 7,000 only contains one significant figure. However, adding a decimal point and another zero gives the number 7,000.0, which has five significant figures.

A good way to measure the significance of zeros, if the above rules become confusing, is to convert the number to the exponential form. If the zeros can be omitted, then they are not significant. For example, 7,000 expressed in the exponential form is 7×10^3 and contains one significant figure. With 7,000.0, the zeros are retained and the number becomes 7.0000×10^3. If we were to convert 0.007 to exponent form, the value is 7×10^{-3}, and only one significant figure is indicated. As a rule, determining significant figures in arithmetic operations is dictated by the value with the least number of significant figures. The easiest way to avoid any confusion is to perform all the calculations and then round off the final answer to the appropriate digits. For example, 36.54 × 238 × 1.1 = 9566.172, and because 1.1 contains only two significant figures, the answer would be reported as 9,600 (remember, the two zeros are not significant). This method works fine for most calculations, except when adding or subtracting numbers containing decimals. In those cases, the number of significant figures in the final value is determined by the numbers that follow the decimal point. Thus, adding 7.45 + 8.635 = 16.175; because 7.45 has only two numbers after the decimal point, the sum is rounded to 16.18. Likewise, 433.8 − 32.66 gives 402.14, which rounds off to 402.1

A word of caution is warranted when using the simple rule stated above, for there is a tendency to underestimate the significant figures in the final answer. For example, take the situation in which we determined the Lowry protein in an unknown solution to be 21.1 μg/ml. We had to dilute the sample fiftyfold using a volumetric flask in order to fit the unknown within the range of our method. To calculate the protein in the original sample, we multiply our result by 50, or 21.1 μg/ml × 50 = 1055 μg/ml in the unknown. Based on our rule above, we would then round the number to one significant figure (because 50 contains one significant figure) and report the value as 1,000. However, doing this actually underestimates the sensitivity of our procedure, because we ignore the accuracy of the volumetric flask used for the dilution. A Class-A volumetric flask has a tolerance of ± 0.05 ml; thus, a more reasonable way to express the dilution factor would be 50.0 instead of 50. We have now increased the significant figures in the answer by two, and the value becomes 1,060 μg/ml.

As you can see, an awareness of significant figures and how they are adopted requires close inspection. The guidelines can be helpful but they do not always work unless each individual value or number is closely inspected.

4.5.2 Rounding Off Numbers

Rounding off numbers is an important and necessary operation in all analytical areas. However, premature or incorrect rounding off can produce serious errors in the final results. It is usually desirable to carry extra numbers during calculations and perform the rounding off on the final answers. A good example of a rounding-off error is the correlation coefficient (r value) that was obtained from the example covered under the section on linear regression analysis (section 3.3). There we saw a considerable difference between the values obtained from the table and by using a calculator (which carries eight digits through all calculator steps). So the advice is to be careful and carry extra numbers throughout calculations, if in doubt.

Rounding-off procedures are fairly straightforward and commonly used by most everyone. Even the Internal Revenue Service allows taxpayers to round off fractions of a dollar to the whole dollar when filling out income tax forms. However, analytical data require a little more accuracy than the IRS, and, thus, the rules are slightly different.

The basic rules of **rounding off** are listed below:

1. If the figure following those numbers to be retained is less than 5, the figure is dropped and the retained numbers are kept unchanged. For example, 64.722 is rounded off to 64.72.
2. If the figure following those numbers to be retained is greater than 5, the figure is dropped, and the last retained number is increased by 1. For example, 64.727 is rounded off to 64.73.
3. If the number following those to be retained is a 5, and there are no figures other than zeros beyond the 5, the figure is dropped and the last retained figure is increased by 1 if it is an odd number, or it is kept unchanged if it is an even number. For example, 64.725 is rounded off to 64.72 and 64.705 is rounded off to 64.70, whereas 64.715 is rounded off to 64.72.

A simplified version of rule 3 above is to increase by 1 if the 5 is followed by numbers other than zeros and ignore the even-odd method. If the 5 is followed by zeros, then it is simply dropped. For example, 64.715 would round off to 64.71, whereas 64.715001 would round off to 64.72.

Remember, it is best to round off after performing any mathematical operations.

4.5.3 Rejecting Data

Invariably, in the course of working with experimental data we will come across a value that does not match the others. Can you reject that value and, thus, not use it in calculating the final reported results?

The answer is "sometimes," but only after careful consideration. If you are routinely rejecting data to help make your assay look better, then you are misrepresenting the results and the precision of the assay. If the bad value resulted from an identifiable mistake in that particular test, then it is probably safe to drop the value. Again, caution is advised, because you may be rejecting a value that is closer to the true value than some of the other values.

Consistently poor accuracy and/or precision indicates that an improper technique or incorrect reagent was used or that the test was not very good. It is best to make changes in the procedure or change methods rather than try to figure out ways to eliminate undesirable values.

There are several tests for rejecting an aberrant value. One of these, the **Q-test,** is commonly used. In this test, a **Q-value** is calculated as shown below and compared to values in a table. If the calculated value is larger than the table value, then the questionable measurement can be rejected at the 90 percent confidence level.

$$\text{Q-value} = \frac{X_2 - X_1}{W} \quad (28)$$

where: X_1 = the questionable value
X_2 = the next closest value to X_1
W = the total spread of all values, obtained by subtracting the lowest value from the highest value

Table 4-6 provides the rejection Q-values for a 90 percent confidence level.

The example below shows how the test is used for the moisture level of uncooked hamburger for which four replicates were performed giving values of 64.53, 64.45, 64.78, and 55.31. The 55.31 value looks as if it is too low compared to the other results. Can that value be rejected? For our example, X_1 = the questionable value = 55.31, and X_2 is the closest neighbor to X_1 (which is 64.45). The spread (W) is the high value minus the low measurement, which is 64.78 − 55.31.

$$\text{Q-value} = \frac{64.45 - 55.31}{64.78 - 55.31} = \frac{9.14}{9.47} = 0.97 \quad (29)$$

From Table 4-6, we see that the calculated Q-value must be greater than 0.76 to reject the data. Thus, we make the decision to reject the 55.31 percent moisture value and do not use it in calculating the mean.

TABLE 4-6 Q-Values for the Rejection of Results

Number of Observations	*Q of Rejection (90% level)*
3	0.94
4	0.76
5	0.64
6	0.56
7	0.51
8	0.47
9	0.44
10	0.41

Reprinted with permission from Dean, R.B., and Dixon, W.J. 1951. Simplified statistics for small numbers of observations. *Anal. Chem.* 23:636–638. Copyright 1951, American Chemical Society.

4.6 SUMMARY

This chapter has focused on the basic mathematical treatment that will most likely be used to evaluate a group of data. For example, it should be almost second nature to determine a mean, standard deviation, and coefficient of variation when evaluating replicate analysis of an individual sample. In evaluating linear standard curves, best line fits should always be determined along with the indicators of the degree of linearity (correlation coefficient or coefficient of determination). Fortunately, most computer spreadsheet and graphics software will readily perform the calculations for you. Guidelines are available to enable one to report analytical results in a way that tells something about the sensitivity and confidence of a particular test. These include the proper use of significant figures, rules for rounding off numbers, and use of the Q-test to reject grossly aberrant individual values.

4.7 STUDY QUESTIONS

1. Method A to quantitate a particular food component was reported to be more specific and accurate than method B, but method A had lower precision. Explain what this means.
2. You are considering adopting a new analytical method in your lab to measure moisture content of cereal products. How would you determine the precision of the new method and compare it to the old method? Include any equations to be used for any needed calculations.
3. Differentiate "standard deviation" from "coefficient of variation," "standard error of the mean," and "confidence interval."
4. Differentiate the terms "absolute error" versus "relative error." Which is more useful? Why?
5. For each of the errors described below in performing an analytical procedure, classify the error as random error, systematic error, or blunder, and describe a way to overcome the error.
   a. automatic pipettor consistently delivered 0.96 ml rather than 1.00 ml
   b. substrate was not added to one tube in an enzyme assay
6. Differentiate the terms "sensitivity" and "detection limit."
7. The correlation coefficient for standard curve A is reported as 0.9970. The coefficient of determination for standard curve B is reported to 0.9950. In which case do the data better fit a straight line?

4.8 PRACTICE PROBLEMS

1. How many significant figures are in the following numbers: 0.0025, 4.50, 5.607?
2. What is the correct answer for the following calculation expressed in the proper amount of significant figures?

$$\frac{2.43 \times 0.01672}{1.83215} =$$

3. Given the following data on dry matter (88.62, 88.74, 89.20, 82.20), determine the mean, standard deviation, and coefficient of variation. Is the precision for this set of data acceptable? Can you reject the value 82.20 since it seems to be different than the others? What is the 95 percent confidence level you would expect your values to fall within if the test were repeated? If the true value for dry matter is 89.40, what is the percent relative error?
4. Compare the two groups of standard curve data below for sodium determination by atomic emission spectroscopy. Draw the standard curves using graph paper or a computer software program. Which group of data provides a better standard curve? Note that the absorbance of the emitted radiation at 589 nm increases proportionally to sodium concentration. Calculate the amount of sodium in a sample with a value of 0.555 for emission at 589 nm. Use both standard curve groups and compare the results.

GROUP A—SODIUM STANDARD CURVE

Sodium Concentration (μg/ml)	*Emission at 589 nm*
1.00	0.050
3.00	0.140
5.00	0.242
10.0	0.521
20.0	0.998

GROUP B—SODIUM STANDARD CURVE

Sodium Concentration (μg/ml)	*Emission at 589 nm*
1.00	0.060
3.00	0.113
5.00	0.221
10.0	0.592
20.0	0.917

Answers:

1. 2, 3, 4. 2. 0.0222. 3. Mean = 87.19, SD_{n-1} = 3.34, CV = 3.83 percent; thus the precision is acceptable. Q-value = 0.92; therefore the value 82.20 can be rejected. CI = 87.19 ± 4.64. Percent E_{rel} = −2.47 percent. 4. Group A is the better standard curve (group A, R^2 = 0.9990; group B, R^2 = 0.9708). Sodium in the sample using group A standard curve = 11.07 µg/ml; with group B curve = 11.48 µg/ml.

4.9 RESOURCE MATERIALS

Bender, F.E., Douglass, L.W., and Kramer, A. 1989. *Statistical Methods for Food and Agriculture*. Haworth Press, Binghamton, NY. This book is broad in scope and has very readable sections on quality control and regression analysis, with many practical examples.

McCormick, D., and Roach, A. 1987. *Measurement, Statistics and Computation*. John Wiley & Sons, New York, NY. This is a very good book covering a wide variety of topics of interest to an analytical chemist. The best feature of this book is that it covers the topics at an introductory level yet with sufficient detail and examples.

Miller, J.C., and Miller, J.N. 1988. (Reprinted with corrections 1989.) *Statistics for Analytical Chemistry*, 2nd. ed. Ellis Horwood Ltd., distributed by John Wiley & Sons, New York, NY. (Also available from Aldrich Chemical Co., Milwaukee, WI.) This is another excellent introductory text for beginning analytical chemists. It contains a fair amount of detail, sufficient for most analytical statistics, yet works through the material starting at a basic introductory level. The authors also discuss the Q-test used for rejecting data.

Phillips, J.L., Jr. 1982. *Statistical Thinking*. W.H. Freeman, San Francisco, CA. This is a small, easy-to-read book on general statistics that covers the material from a conceptual point of view.

Skoog, D.A., and West, D.M. 1982. *Fundamentals of Analytical Chemistry*, 4th ed. Holt-Saunders, New York, NY. Chapter 3 does an excellent job of covering, in an easy-to-read style, most of the statistics needed by an analytical chemist.

Willard, H.H., Merritt, L.L., Jr., Dean, J.A., and Settle, F.A., Jr. 1988. *Instrumental Methods of Analysis*, 7th. ed. Wadsworth Publishing, Belmont, CA. This gives a rigorous treatment of instrumentation and has a very useful chapter (Chapter 2, Measurements, Signals and Data) on types of error generated by instruments.

Contribution No. 92-12-B from the Kansas Agricultural Experiment Station, Kansas State University, Manhattan, KS.

CHAPTER

5

Computerization and Robotics

Gerald F. Russell and James M. Zdunek

5.1 COMPUTERS FOR DATA ACQUISITION

The promise of computers, robotics, and automation is to relieve the scientist from tedious and undesirable tasks. These might include the three Ds—dirty, dull, and dangerous—to say nothing of demeaning and debilitating. Apart from the need for computers to do repetitive tasks quickly and efficiently, many tasks can be accomplished with great speed, accuracy, and precision. The determination of the type of computer and robotics systems needed is the responsibility of the food scientist.

5.1.1 Historical Overview

The advent of modern computers in the food scientist's laboratory has been made possible by rapidly decreasing costs of computer hardware packaged in smaller physical size, but with ever-increasing computational power. Vendors of analytical equipment have incorporated microprocessors in most modern instrumentation and robotics systems. The microprocessor is usually dedicated to the control and operation of an instrument such as a spectrophotometer, gas chromatograph (GC), or robotic autosampler, among many others. The vendor supplies a system as a ready-to-use or **turnkey** product. Often vendors make available optional interfaces to attach to computers for customized needs in the laboratory. A typical example of such a system might be a GC system in which a microprocessor monitors and controls operations of the GC and then sends a filtered and conditioned signal to a recording device. Typically, this is in the form of a recording integrator, in which the chromatographic peaks are integrated by another dedicated microprocessor and a tracing of the chromatogram is provided along with a printed copy of analytical results. The user often has very little optional, direct control of such a system unless further connection and interfacing with a separate computer are possible and desirable.

Differing from the dedicated microprocessor environment above, the next step in automation is to make connection directly to the experiment with a laboratory computer. This is often desirable for long-term storage of data and further treatment of experimental data. An example would be to interface a Macintosh computer to a Hewlett-Packard GC and integrator. The computer controls all operations and captures the GC signal along with the results of the integrator's calculations of retention times and peak areas. An example is shown in Fig. 5-1 of experimental results from such an experimental configuration that was used to capture all relevant data for GC analyses of tomato headspace volatiles. Advantages to such a system include permanent long-term and easily retrievable data for both the signal and the integrator report. Additionally, the signal can be regraphed with differing amplitudes in a customized form by a graphics program such as Igor™, from which Fig. 5-1 was taken.

5.1.2 Hardware Requirements

5.1.2.1 Computers

Choices of computers for the laboratory can be overwhelming. Computers for laboratories first became available in the 1970s largely through minicomputer products from Digital Equipment Corporation (DEC). In the mid 1980s, the availability of IBM PC microcomputers became widespread, especially with the introduction of low-cost clones. More recently, the availability of high-quality Apple Macintosh computers has added a large product base with the promise of increased ease of use. Choices of computers should be based on the food scientist's experimental needs, such as the type of direct connections or interfacing necessary to monitor and control instruments or experiments in the laboratory.

5.1.2.2 Transducers

If an analytical apparatus is not directly computer compatible, it is necessary to use the correct **transducer** to change into an electrical signal each physical parameter to be measured. For example, to measure temperature, thermocouples are often employed. The resulting thermocouple electrical signals need to amplified, filtered, and conditioned to provide a reproducible and stable electrical signal. Fortunately, there are many vendors to whom the food scientist needs merely to describe the experimental needs to find an appropriate transducer.

If the transducer produces an analog electrical signal, this signal must be converted to a digital form to be appropriate input for modern digital computers. Modern **analog-to-digital converters** (ADCs) are available in convenient form as **boards** or **cards** that merely need be inserted into the backplane or bus slots on most computers. Often, a single multipurpose interface card will serve all the experimental needs of an analytical laboratory; however, appropriate software must be available to effectively use it.

5.1.2.3 Interfaces with the Computer

If an instrument or experiment's transducers already provide a digitized signal, interfacing can be accomplished directly through a serial port provided on most computers. The exact **protocol** for sending and receiving this digital information needs to be defined, and there exist many protocols for the lab computer. The protocols defined as EIA RS-232c and RS-423 are com-

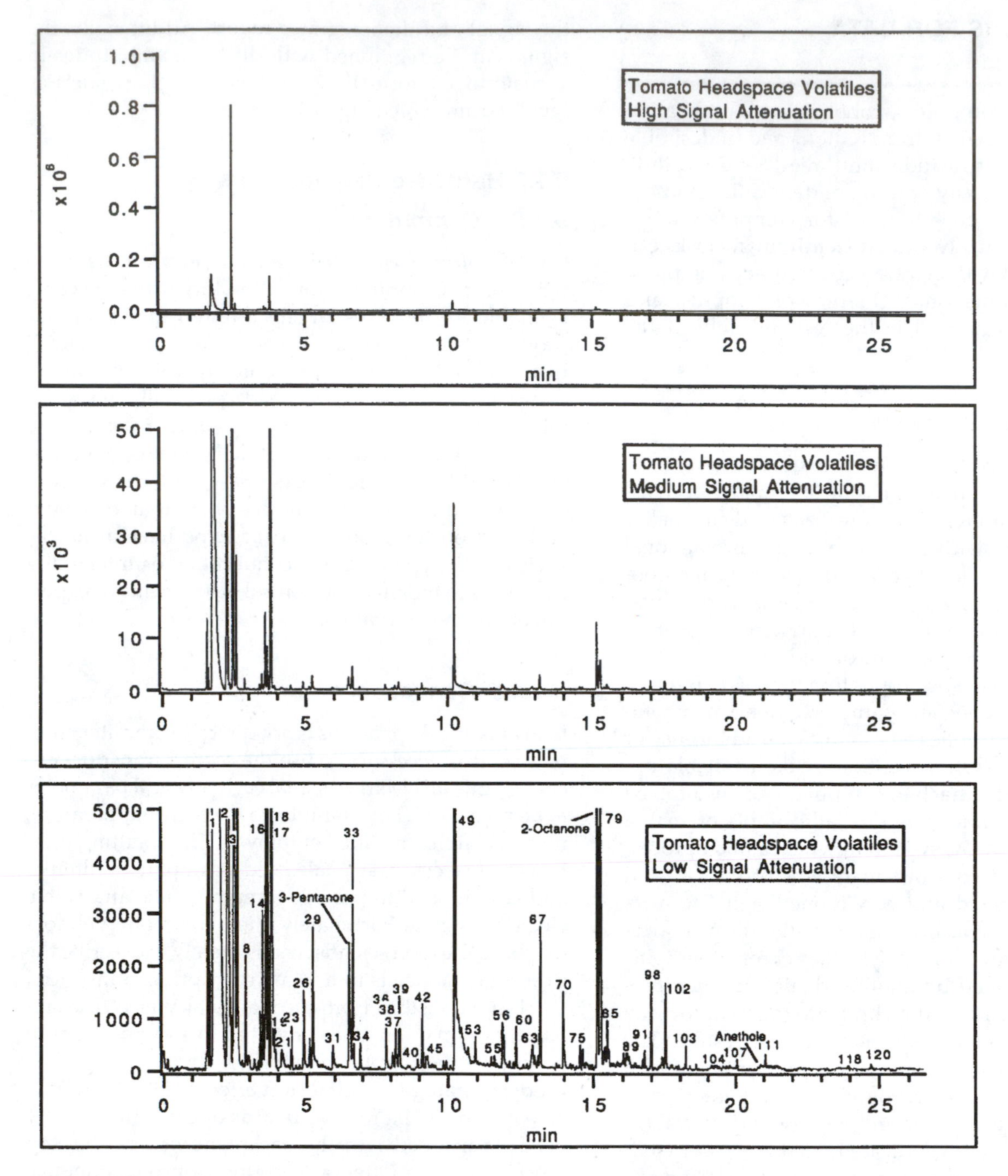

FIGURE 5-1.
Layout of three chromatograms produced by the graphics program Igor™ that displays three different signal amplification levels or degrees of attenuation.

mon for IBM and Macintosh computers. Another data exchange protocol termed either IEEE-488 or GPIB requires a separate interface card. This later protocol has been adapted for thousands of instruments, and many computer programs have been written to accommodate this standard protocol, independent of specific computers or instruments.

Often, the microcomputers themselves are interfaced together and can thus acquire and transmit data simultaneously. The use of **local area networks** (LANs) is becoming common. Standards for LANs are not yet firmly established, but Ethernet is rapidly becoming a de facto standard for high-speed data transmission.

5.1.3 Software Requirements

5.1.3.1 Real-Time and Post-Run

If computers are used to analyze data after the physical measurements are completed, the analysis is said to be post-run. This is a common application of computers; the data are either typed by hand, input through a digitizing tablet, or perhaps entered in graphic form from a commercial scanner/digitizer. No matter what the form of data entry, the scientist must be vigilant to monitor errors during the process. (Estimates of typed error rates are commonly given as from 3 to 15 percent.)

A compelling reason for directly interfacing a computer to an instrument or experimental apparatus is to reduce such data-recording errors. Ideally, a computer is able to input the data immediately and at the same time as it is acquired experimentally; this is termed **real-time data acquisition.** Since the data ultimately must be transmitted in digital form as bytes or words of information, the rate at which computers can do real-time acquisition may be limiting to the experiment; however, it is not uncommon to acquire data points at rates of 50 kHz or more in instruments such as those used for nuclear magnetic resonance experiments or imaging.

5.1.3.2 Interface Hardware and Drivers

The data can arrive at the computer interface in different formats and needs to be converted into an appropriate form for further use and storage. Incoming data may be binary, binary coded decimal (BCD), American Standard for Information Interchange (ASCII), or Gray code, among others. **Interface cards** can do required conversions directly or otherwise transmit the data for further handling and storage by software in the computer.

Drivers are specialized software programs that provide the link between the hardware interface and more advanced software or programs that handle, manipulate, display, and store the data. Driver programs are designed to provide the most efficient data exchange possible and at the highest possible speed. They are the first link between the hardware and computer software. Drivers are written in the most concise computer code possible (assembly or machine language) and are not usually changed or reprogrammed by users of the driver who wish only to link their more sophisticated programs to the driver.

Many software vendors can provide the laboratory scientist with the software and drivers necessary for their experiments. For example, LabTech NoteBook™ is popular with the IBM DOS and Windows™ platforms. For Macintosh computers, National Instruments provides a state-of-the-art program, LabVIEW™, that has recently also been released for IBM and Sun workstation platforms as well. Both of these products have used a development environment in which the user designs a customized interface to the experiment using **graphical icons** that simply describe the needs of the interface. The difficult computer programming is transparent. An example of a LabVIEW program application is shown in Fig. 5-2; it shows the computer screen display for the interface and controller of an oxygen electrode apparatus used to measure oxygen uptake during accelerated lipid oxidation experiments. The user simply clicks a mouse pointer on the appropriate controls to change settings, and instantly the entire experiment is under computer control. Graphical results and calculations are displayed in real-time and data are simultaneously stored for archive purposes.

5.1.3.3 Integration into Software Programs

Directly incoming (real-time) data or stored data needs to be treated with software programs. The use of once-formidable programming languages such as Basic, Fortran, and Pascal is no longer the rule for analyzing data with software. Many specialized programs for handling data exist that do not require extensive programming skills. The concept and development of user-friendly programs has taken the laboratory computer to new levels of ease of use.

5.1.4 Ergonomics and Economics

The modern computer system can be purchased as part of packaged commercial instruments and apparatus. When commercial products do not meet the exact needs of the food scientist, easily adaptable computer systems and programs are available that need not be intimidating to the nonprogrammer of computers. The products discussed above, along with a myriad of others, offer easy-to-use products with graphical and visually aesthetic programs. Ultimately, the promise of error-free data acquisition and the speed of computers make the use of computers in the laboratory very cost effective and can free the food analyst for other tasks.

5.2 COMPUTERS FOR DATA ANALYSIS AND DISPLAY

The power of computers has become available in surprisingly small physical packages that engender tremendous calculation power. Important uses of laboratory computers include needs for computations, graphics, and database management.

For calculations and computational needs, the use of **spreadsheet programs** can serve most needs of the food scientist. Two popular spreadsheet programs, Lotus 1-2-3 and Microsoft Excel, are available for both IBM and Macintosh computers and contain virtually all the scientific functions needed for most physical

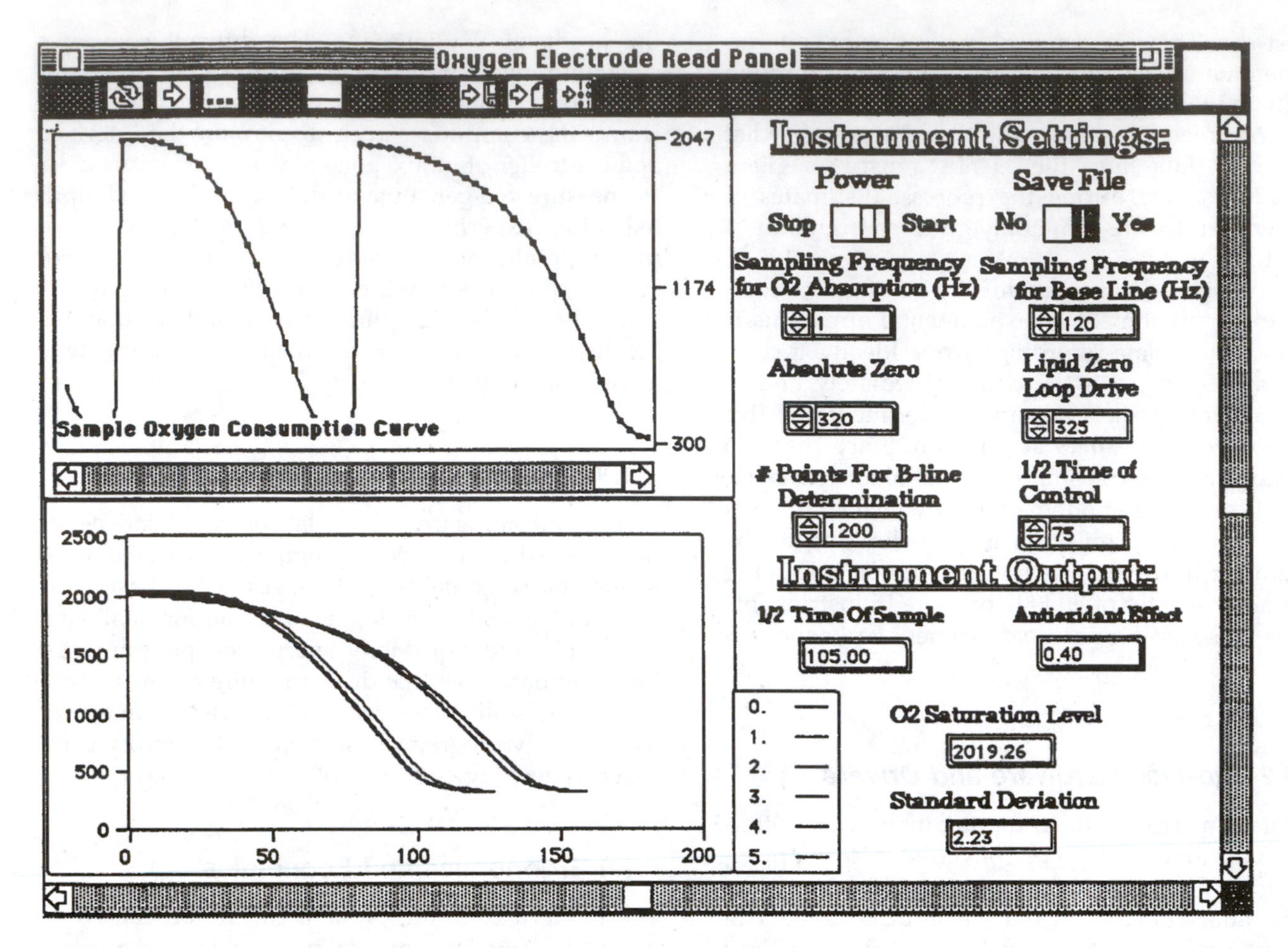

FIGURE 5-2.
A graphical Macintosh screen of a LabVIEW™ interface to an oxygen electrode experimental apparatus. The experimental parameters are set directly from this screen display. The results of up to six experiments are displayed in separate colors on the computer display.

and chemical analyses. The food scientist needs very little computer programming expertise to create custom applications from spreadsheet programs that handle all needed calculations and formatting for reports.

If a picture is worth a thousand words, the use of graphics is one of the most valued reasons for the food scientist to use computers in the laboratory. From the graphing of experimental results to programs that draw chemical structures for reports, specialized programs abound. Most spreadsheets, as discussed above, also have limited graphing capabilities, but they may not be totally adequate. For example, results of an enzyme assay could easily be calculated and displayed from a spreadsheet program. However, as an alternate method, a specific product written for enzyme experiments, such as EnzymeKinetics™ for the Macintosh, can instantaneously do required calculations and print graphs such as shown in Fig. 5-3.

Calculations and graphics for very large data sets are especially easily handled on Macintosh computers with an Igor™ program. With Igor, data sets are limited

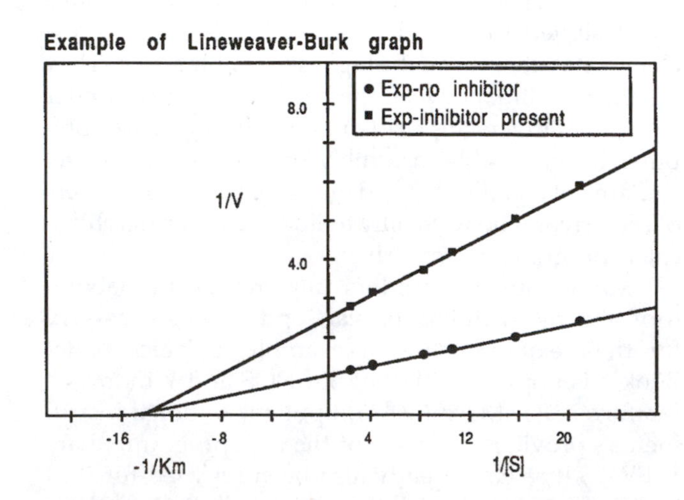

FIGURE 5-3.
Graphical presentation of calculations from the EnzymeKinetics™ program for the Macintosh computer.

in size only by computer memory and are easily manipulated and graphed in forms designed for scientists, as opposed to the business-oriented applications of many graphics programs. Fourier transformations, curve fitting, data smoothing, and plotting of multiple data sets are all quickly accomplished by clicking on appropriate pull-down menus. The GC graph or chromatogram shown in Fig. 5-1 is an example of a plot of over 32,000 data points.

Database management can be accomplished efficiently with many programs for the lab computer; the use of dBASE™ for the IBM environment is very popular. Database management programs are also bundled into **laboratory information management systems** (LIMS), discussed later. Statistical calculations are easily accomplished in spreadsheets for all but the most sophisticated applications. For the more demanding statistical needs, commercial programs bring the traditional mainframe computer power of programs such as SAS, MINITAB, and MATLAB to the laboratory microcomputer.

5.3 COMPUTERS FOR AUTOMATION AND ROBOTICS

An ever-increasing analytical sample load and the need for improved information quality will require solutions that are highly productive and cost effective. The food laboratory of the future will be a highly integrated, information-intensive domain that will combine sophisticated analytical techniques with automated robotic procedures.

Advancements of computer technology in both hardware and software have made a considerable impact in the way laboratory procedures are performed today. The usefulness of the computer in the laboratory has been extended not only to include the acquisition and processing of analytical data, but in many cases, it has helped to replace the physical work of sample preparation as well.

5.3.1 Laboratory Automation

The analytical food scientist of today is faced with ever-increasing quantities of samples, tests, and test replicates required to complete a desired experiment. Specialized tools are needed to overcome the mundane and repetitive tasks that often confront the food scientist.

To help meet these needs, analytical instrument makers were challenged by the task of converting simple analytical tools into sophisticated computer-driven analytical data stations.

5.3.1.1 Analytical Methods

Automation opportunities were created by industries in which the demand for high sample throughput provided the economic justification to purchase or build special instrumentation that met the demand. The areas of industry that had the most to gain by laboratory automation included industrial quality control.

Steel mills, for instance, required fast and accurate analysis of the molten steel before it was poured into ingots. Without automated multielement analyzers, a proper assay of the metal would take several hours instead of seconds to perform.

Medical and pharmaceutical testing laboratories quickly caught on to laboratory automation. Obtaining fast and accurate results from a hospital's medical laboratory are important, and much of today's medical laboratory instrumentation is geared for multiple testing of biologicals with emphasis on data system integration.

The analytical methods that were commonly automated included spectrometric [ultraviolet-visible, atomic absorption spectroscopy (AAS), inductively coupled plasma-atomic emission spectroscopy (ICP-AES) (see Chapters 23 and 25)] and chromatographic (see Chapters 28–30), in which much of the automation simply meant introducing a large number of prepared samples to the instrument.

5.3.1.2 Instrument Interfaces

The introduction of the computer to analytical instrumentation has changed forever the way the analytical chemist approaches the bench. In many cases, much of the time-consuming sample preparation and measurement is performed by a machine under the control of a computer.

The most basic instrument interfaces are sample delivery systems or **autosamplers** that eliminate the repetitive work required to introduce a large number of samples. Usually there is some preparation work to be done before analysis. After groups of samples are arranged in a logical order, the analysis of each sample can be allowed to continue without the analyst's intervention. Complete analytical computer systems have been linked to these delivery systems so that once the samples are added to a sample tray or carousel, the system automatically performs the analysis. Options for data reduction and report generation can also be included.

An example of an automated analytical system is shown in Fig. 5-4. The analytical instrument is a gas chromatograph equipped with a multiple sample carousel. The instrument can select each sample from the carousel, inject it into the column, and run the chromatogram for a specified time. The computer in-

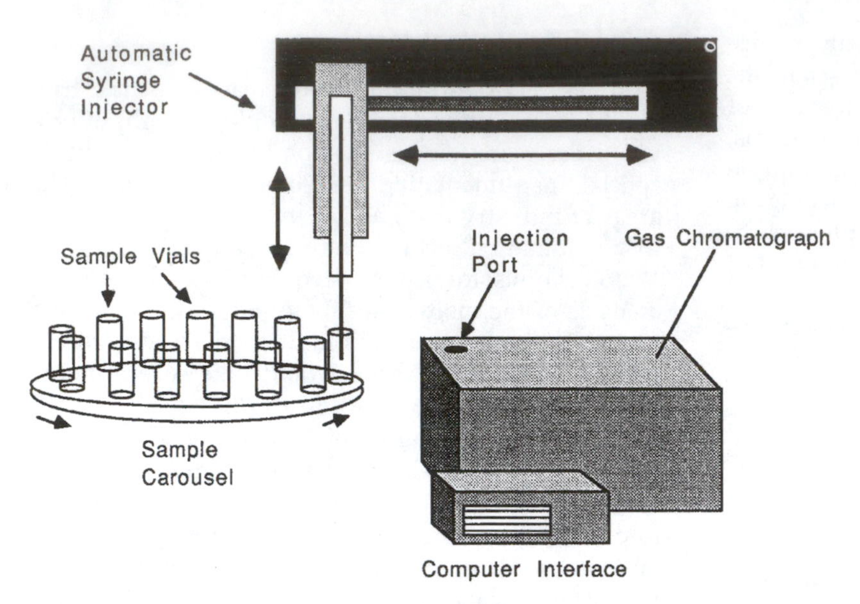

FIGURE 5-4.
Multiple sample gas chromatographic analyzer.

terface is installed on the instrument to interpret chromatographic peak information (data reduction) and enter this information in a database for reporting purposes.

Automation systems like the one in the example are available for many analytical techniques including high-performance liquid chromatography (HPLC), AAS, and ICP-AES. Somewhat more advanced instrument interfaces may include an additional step prior to analysis. One such device, known as an **autoanalyzer,** utilizes spectrophotometric techniques of liquid samples by pumping the liquid to be measured through a special flow cell. Before the liquid reaches the cell, various preparation steps can be made to the flowing liquid sample, including the introduction of specific reagents within the stream.

5.3.2 Laboratory Robotics

Although the automation tools available to the analytical chemist have reduced some of the manual labor, many additional methods, especially in the food industry, require far more manual preparation steps that are not performed by any of today's analytical instrumentation. As advances in technologies continue into the twenty-first century, there is a clear need to automate entire procedures and give the scientist more time to continue the thinking part of the analytical procedure.

There are three steps in an analytical method that can be automated (see Fig. 5-5). The analytical chemist follows these steps in this order: (1) Prepare the sample, (2) analyze the sample, and (3) reduce and report the data. In the early days of laboratory automation, the third step was automated first. In time, the second step involving automated analytical equipment became more sophisticated, and direct linkage between analysis and data reduction in one instrument became a reality. Finally, when analysis and data reduction are completely automated, the next logical step will be to automate sample preparation. As more sample preparation and analysis techniques become fully automated, more emphasis can be focused on the **decision-making process.**

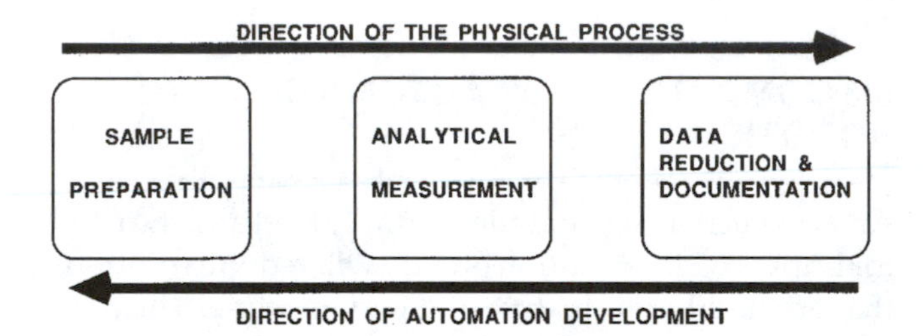

FIGURE 5-5.
Steps in an analytical method that can be automated.

5.3.2.1 Early Development

A **laboratory robot** is defined as a programmable manipulator that is an extension of the capabilities present in laboratory autosamplers and autoanalyzers. It adds versatility because it can be programmed to perform a number of preparation steps over a range of samples and methods. A more generic class of autosamplers was introduced to perform a wider range of methods, but until that time, none of these systems could have been called a robot.

One of the first commercially successful laboratory robotic systems was introduced in the mid-1970s. The robot was specifically designed for the laboratory, utilized a **modular system** for building laboratory methods, and used a robot program language designed with the chemist in mind. Other robotic systems followed, and many laboratories designed their own specific automation requirements around one or more of the commercially available robot arms.

5.3.2.2 Mechanical Considerations

An increased use of laboratory robotics over the past 20 years is unquestionably the result of the advancement of the **integrated circuit** (IC) or micro chip. The major IC advances have flourished not only in the function of the central processing units, or computers on a chip, but also for other IC devices. Many kinds of sophisticated integrated circuits have been introduced to aid in getting the computer to do physical labor.

Unfortunately, the advancement of mechanical robotic technology has not kept pace with that of the computer. Many of the robotic arm mechanisms on the market still utilize the electromechanical technology that was common over 50 years ago. Smaller and more powerful drive mechanisms need to be developed for robot arms. Also, robots need to be more aware, as it were, of their surroundings; robot sensors have to be more advanced in order to move about in the real world.

5.3.2.2.1 Robotic Drive Mechanisms The drive method behind the laboratory robot arm is an important factor in choosing a specific robot application. The four main drive methods currently available are hydraulic, pneumatic, DC electric motor, and electric stepping motor drives.

Hydraulic and **pneumatic drives** are seldom used in analytical laboratories and are primarily used in large industrial applications where heavy lifting or explosive atmospheres are factors. The two remaining methods are the drives that make up virtually all of the laboratory robots in use today and are based on electric motor drive technology.

DC electric motors are particularly well-known motor drive mechanisms. They are feedback loop controlled motors that have exceptionally high torque-to-volume ratios that make them ideal for use in robot wrist and finger mechanisms. Many small drive systems of this type are used to drive the pen in the familiar laboratory strip chart recorders; larger versions are now being used to move robot arms for automated chemical analysis.

Electric stepper motors have several advantages over DC motors; they are inexpensive and do not require costly feedback loop control hardware. Their low torque-to-volume ratio makes them less desirable for small- and medium-size robot applications, but these motors are ideal for driving positioning tables and gantries as part of the automation system.

5.3.3.2.2 Sensors A robot that does not know where it is going in a laboratory will have trouble performing an analysis. Sensors of various types must tell the robot system the obvious: Is the hand open or closed? Is the arm near the balance? Is the test tube rack full or empty?

Position sensors are the most basic of the robotic sensors. All robotic systems have some means of determining where one part of the robot is relative to the other robot parts. This position information, although helpful, still does not give an adequate picture of the outside world; additional sensor information is necessary to help guide the robot through its motions.

True robotic vision is not yet perfected, so the robot must "see" using other means. Force detection and proximity switch activation, for example, can give a sense of touch to the robot. Sensors determine the position of the robot and the position of the items around it, but it is also necessary to incorporate sensors that will recognize failed attempts during certain critical activities. Questions such as Did the robot add enough solvent? or Is the tube placed properly in the centrifuge? are common questions answered by sensors.

5.3.2.3 Laboratory Devices

The laboratory robot is not a substitute for a laboratory scientist. It is merely a tool to help the scientist be more productive. Like any sophisticated tool, a robot must have a built-in relationship with its surroundings. One cannot simply install a robot arm onto a laboratory bench and expect it to perform its job using laboratory equipment and methods that were designed for humans to use. This laboratory equipment or laboratory device must be manufactured specifically for automation purposes.

An example of a **laboratory device** is an analytical balance with automated doors that open when the robot is ready to weigh a sample. Laboratory devices also include laboratory glassware and disposables that the robots must use. It is important that the glassware and disposables are absolutely free of defects and are made to exacting tolerances. The robot may not be able to detect a cracked tube or a poorly molded pipette tip, so it is important that these items are properly made and easy for the robot to use.

5.3.2.4 Computer Hardware

The computer acts as the central control for the laboratory robot system. More sophisticated robot systems may have more than one computer, but there is one central unit that has control over the rest.

The major job for the computer system is data acquisition and control. The computer has control over the robot arm and must have the necessary hardware to fulfill this function. This hardware may function as an output to a motor drive in a robot arm or provide serial data information to control a smaller subunit of the robot system. An example of a typical control setup is shown in Fig. 5-6.

Along with the **central processing unit** (CPU), the robot control computer must have additional hardware to allow it to talk to the robot. This hardware exists as digital input/output printed circuit boards that are plugged into a socket mounted near the CPU.

5.3.2.5 Robot Software

The key to the versatility of a laboratory robot system is its software. The computer is under the control of a program, and this program can be modified or changed completely to make the robot do some other kind of job. The program can be saved outside the system and provide documentation for the method. It can also be duplicated and loaded onto other similar robot configurations, aiding in quicker set-up of additional robot systems.

Robot programming languages are usually designed for easy control of the robot and laboratory devices. The following examples of robot programs (Table 5-1) show how two different robot languages accomplish the same task.

It is important to realize that although the programs in the above example are fairly easy to comprehend, the robot must be programmed to point out the exact location of each object. The command to MOVE to the FLASK will not work until the robot has learned where the flask was originally placed. Once all of the objects have been defined within the robot work space, the programmer must teach the robot where each object can be found.

In an effort to keep the amount of robotic programming to a minimum, a number of manufacturers have offered some excellent solutions to help make this task easier. The Zymark Corporation's PyTechnology™ helps modularize individual robot tasks around a centralized robot base. In the Py configuration, each laboratory device is attached to the robot on a laboratory bench like a slice of pie. Each pie slice contains a laboratory module: balance, vortex, rack of tubes, and the like. The software for each slice is prepackaged at the factory so that the programming required for the completed system is minimal.

Hewlett-Packard's ORCA (Optimized Robot for Chemical Analysis) exhibits a high degree of integration with laboratory instrumentation as well as interaction with spreadsheets and other software applications. The methods development software, with its easy-to-

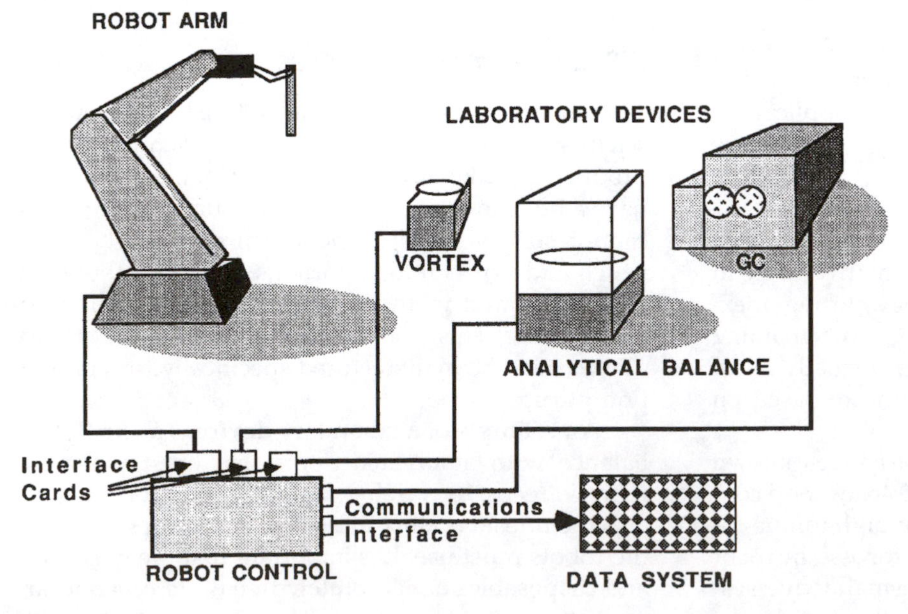

FIGURE 5-6.
Laboratory robot system.

TABLE 5-1. Examples of Robot Programs

RAPL[1] *Program GET FLASK*	*EASYLAB*[2] *Procedure Get.Flask*	*Comments*
100 SPEED 100	SPEED = 1	;Select high speed
120 APPRO FLASK,5	APPROACH.FLASK	;Move near flask
140 SPEED 20	SPEED = .2	;Select slow speed
160 MOVE FLASK	OVER.FLASK	;Move up to flask
180 GOSUB FLASK_UP	PICK.UP.FLASK	;Pick flask up
200 MOVE SAFE	APPROACH.SAFE	;Move flask away

[1]RAPL (Robot Automation Programming Language) is a trademark of CRS Plus Inc., Burlington, Ontario, Canada.
[2]EASYLAB is a trademark of Zymark Corp., Hopkinton, MA.

use **graphical user interface,** offers a sophisticated approach to laboratory automation.

5.3.2.6 Applications

Robots have successfully found their way into the food analysis laboratory. The applications are few, but much attention has been placed on a number of methods that have been duplicated in several food laboratories. These methods include vitamin, sugars, fiber, and fat analyses. (Figure 5-7 shows a bench layout for a typical laboratory robot system.)

In addition to the above applications, many preanalysis sample preparation robots routinely prepare samples for analysis by GC, HPLC, and ICP-AES.

5.3.3 Systems Integration

Systems integration means taking all the pieces of the automated laboratory and linking them together so that they work like one large smoothly running machine. A fully integrated laboratory is able to access incoming samples, determine the appropriate action, and follow through with timely and accurate analytical data.

5.3.3.1 Laboratory Information Management

Food laboratories must keep track of incoming samples and be able to store the analytical data in a way that is easy to access at a later date. This analytical information is stored on a **laboratory information management system,** or LIMS, which is a database program that is available on any size computer depending on the number of samples and number of users of the system.

The LIMS can be accessed by the user community or those involved with the analytical laboratory in some way, so that the exchange of information is rapid and efficient. Analysts can check the LIMS to see what kind of work is required on the samples. Clients can check the LIMS to review the analytical results. Administrators can check the LIMS to see how many samples pass through the laboratory.

From a systems integration standpoint, the LIMS computer and the automated system computers can be linked together to form an integrated analytical system. The result of this integration refines data handling and eliminates data-entry errors from the analytical instrumentation.

5.3.3.2 Artificial Intelligence and Expert Systems

The vast power available from computer technology has helped bring forth a new age in programming capability. One area that has received a great deal of interest recently is AI, or artificial intelligence, and its more application-oriented sibling, expert systems.

Expert systems involve gaining experience in a specific area of knowledge. From this base of knowledge can be formed a method for properly coordinating and selecting bits and pieces of this knowledge information. When a particular problem needs to be solved, the expert system can rely on this knowledge to help provide a solution.

The use of expert systems within laboratory automation, especially robotics, takes the power of the computer one step further. Instead of merely controlling a robot or automated analyzer under a rigid program regime, the expert system has the ability to modify the activity of the system. The knowledge that the expert system would use may simply be a list of rules for checking the types of analytical samples with available methods to help schedule the best way to run several analyses. Another possible knowledge base could help

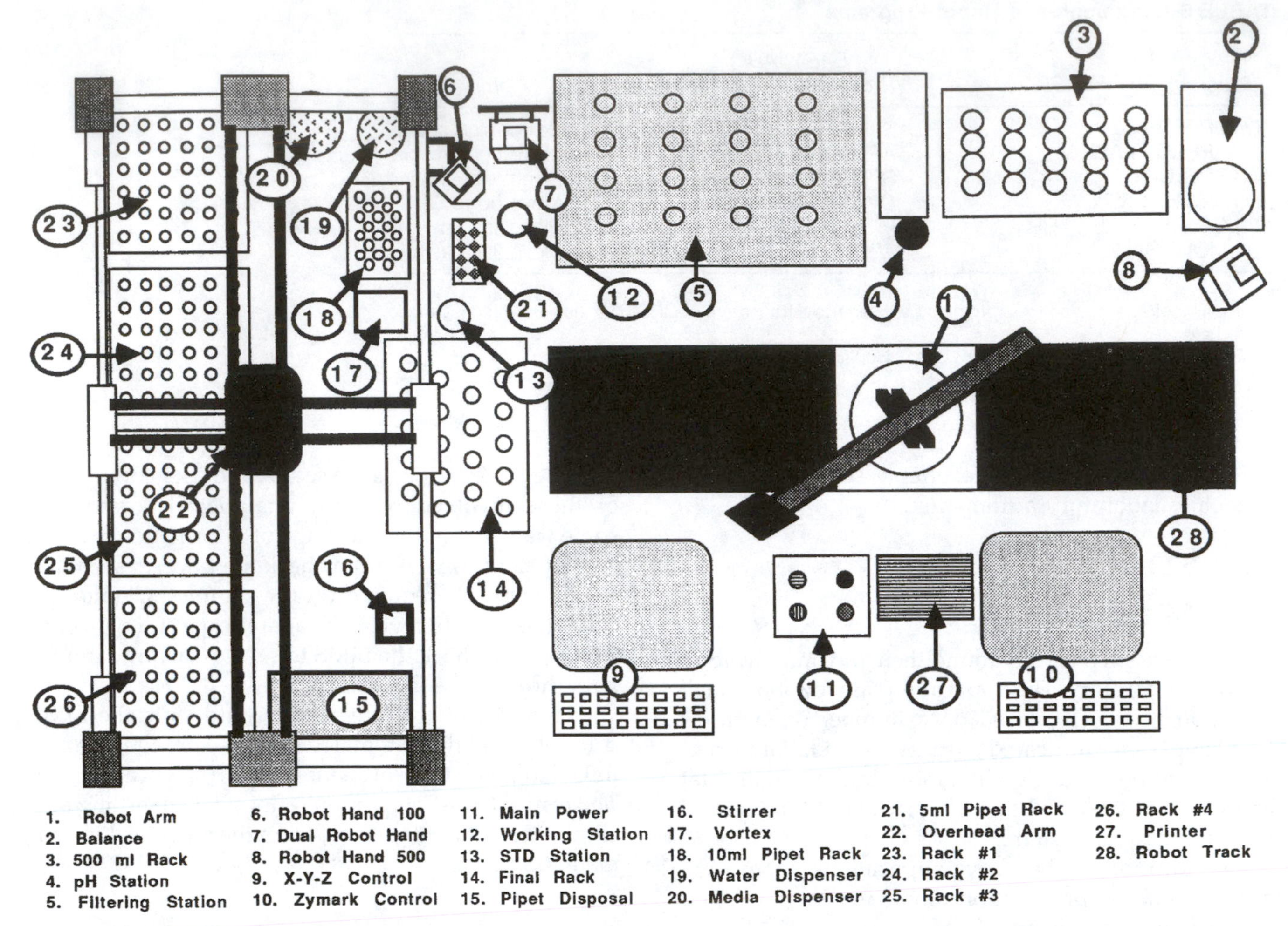

FIGURE 5-7.
Vitamin assay robot.

evaluate robotic sense inputs to develop a more accurate picture of the world around it. As the range of applications for LIMS and laboratory automation increase, the role of the expert system will become more evident.

5.4 SUMMARY

Significant progress has been made through the use of computers and automation over the last 20 years. This chapter details some of these recent advances in the food sciences with new applications for data acquisition and control as well as new laboratory robotics. New approaches to laboratory software are discussed, with emphasis on ease of use in practical applications. Examples are given for several analytical methods used in food laboratories. The background and use of laboratory robotics are covered, with emphasis on systems integration and future trends in enhanced data management.

5.5 STUDY QUESTIONS

1. Some advantages of automated data acquisition by computer might include less chance of errors in recording results, greater reproducibility of measurements, possibility for operations unattended by technician, ease in obtaining permanent records, ease in treating data with graphics and report-generating programs. List others that would be of importance to you.
2. Some disadvantages of automated data acquisition by computer might include loss of understanding of the process when it is treated as a black box and the fact that reliance on a computer-controlled protocol may not facilitate insight into limitations to the information obtained. List others that would be of importance to you.
3. A vendor has just sold you a new automated balance. It

has a built in RS-232c port and is claimed to be computer compatible. What additional items might be required to have a computer automate a series of weighings?

4. You read a report generated by a new computer-controlled infrared analyzer. The result you are checking shows "Fat content = 43.7146%." What are some potential limitations to this result? Is it wise to believe, without question, results printed by a computer?
5. Differentiate between the concepts of real-time and post-run data acquisition and analysis.
6. Suggest alternative steps for automating a robotic method for titratable acidity. List the laboratory devices required to perform the method and then suggest possible modifications to each device (e.g., specialized glassware) that would increase the efficiency of the automated method.
7. Review some analytical methods that would be good candidates for robotic automation. Give specific reasons for each; include human and environmental factors.
8. A large analytical laboratory will generally have a LIMS, or laboratory information management system, to store and recover analytical data. What other functions would be desirable in a food laboratory LIMS?

5.6 RESOURCE MATERIALS

Hawk, G.L., and Strimaitis, J.R. (Eds.). *Advances in Laboratory Automation—1984.* Zymark Corporation, Hopkinton, MA.

Hewlett-Packard Co. 1992. *The HP ORCA System Specifications Guide.* Hewlett Packard Co., Naperville, IL.

Hurst, W.J., and Mortimer, J.W. 1987. *Laboratory Robotics.* VCH Publishers, Inc., New York, NY.

Schoeny, D.E., and Rollheiser, J.J. 1991. The automated analytical laboratory: Introduction of a new approach to laboratory robotics. *American Laboratory* 23:14.

Zymark Corporation. 1988. *Laboratory Robotics Handbook.* Zymark Corporation, Hopkinton, MA.

PART

II

Chemical Composition and Characteristics of Foods

CHAPTER

6

Titratable Acidity

George O. Sadler

6.1 INTRODUCTION

The most common naturally occurring food acids are citric, malic, lactic, and tartaric acids. In the carbonated beverage industry, CO_2 and phosphoric acid also contribute to acidity. Food acids can affect food quality in many ways. They may act as nutrients, flavor components, flavor enhancers, gelling agents and chelates for certain destabilizing ions, suppressors of enzymatic browning, and reducing entities for stabilizing such redox reactions as ascorbic acid oxidation.

Organic acids can dictate the dominant microflora in foods. Many pathogenic and food spoilage microorganisms are unable to grow in high-acid foods. The ratio of sugars to acids gives an accurate prediction of tartness for many high-acid foods. This ratio value is one of the few objective tests that correlates well with organoleptic perception.

Food acids are typically quantified by titration with a standard base. This process provides **total acidity,** commonly referred to as **titratable acidity.** Because titratable acidity is such a determining factor in food quality, it is one of the most frequently run analytical tests in the food industry. Although tests for many food components have changed along with advances in analytical technology, procedures for titratable acidity have not changed substantially in over 100 years. This attests to the simplicity and time-honored importance of the procedures presented in this chapter.

This chapter describes the theory and procedures for measuring titratable acidity in foods. Most students have already encountered titratable acidity in general chemistry. Although this chapter has sought to provide sufficient detail to instruct the beginner, it was also designed to provide resource and review information for the veteran food technologist.

6.2 DEFINITIONS

Terminology is often the greatest obstacle to mastering a concept. To assist in learning, the key definitions presented below are stated in terms appropriate to food science. Although broader definitions are sometimes possible, they encompass situations that are of minor importance to food analysis.

Acid A proton donor. The hydrogen ion is the only donated proton of importance in foods.

Base A proton acceptor. The hydrogen ion is the only accepted proton of importance in foods.

Titratable acidity The percent acid in a sample determined by titration with a standard base and stated in terms of the predominant acid in the sample.

Standard acid A solution of precise normality made from a pure, dry, accurately weighed organic acid (usually potassium acid phthalate).

Standard base An alkali solution (usually sodium hydroxide) whose normality has been precisely determined by titration against a standard acid.

Conjugated acid The free hydrogen ion produced during the disassociation of an acid.

Conjugated base The free anion (negatively charged ion) resulting from the disassociation of an acid.

Equivalent weight The molecular weight of an acid or base divided by the number of ionizable hydrogens (acids) or hydroxyl groups (bases) on the molecule.

Normality The number of equivalents per liter or milli-equivalents per milliliter.

Molarity The number of moles of a compound per liter of a solution.

Equivalence point The point in a titration where the equivalents of acid exactly equal the equivalents of base.

pH The negative log of hydrogen ion concentration.

pK_a The negative log of the ionization constant for a weak acid.

Endpoint In an acid-base titration, the apparent equivalence point as marked by some secondary indicator (usually phenolphthalein).

Potentiometric method Instrumental method of pH determination using a pH meter and electrode.

Brix/acid ratio The soluble solids (w/w) determined by refractometer or hydrometer divided by the grams acid determined by titration.

6.3 CONTENT IN FOOD

6.3.1 Predominant Acid

Foods, for the most part, are as chemically complex as life itself. As such, they contain the full complement of Krebs cycle acids (and their derivatives), fatty acids, and amino acids. Theoretically, all of these contribute to titratable acidity. Since routine titration cannot differentiate between individual acids, a compromise must be struck. Usually, titratable acidity is stated in terms of the predominant acid. For most foods, this is unambiguous. In some cases, two acids are present in large concentrations and the predominant acid may change with maturity. In grapes, malic acid often predominates prior to maturity, while tartaric acid typically predominates in the ripe fruit. A similar phenomenon is observed with malic and citric acids in pears. Fortunately, the equivalent weights of common food acids are fairly similar. Therefore, percent titratable acidity is not substantially affected by mixed predominance or incorrect selection of the predominant acid.

6.3.2 Acid Composition

The range of acids in foods is very broad. Acids can exist at levels below detection limits or they can be the principal substance in certain fruits. In most cases, acid content only tells part of the story. The tartness of acids is reduced by sugars. Consequently, the **Brix/acid ratio** (often simply called **ratio**) is usually a better predictor of flavor quality than Brix or acid alone. Acids tend to decrease with the maturity of fruit, while sugar content increases. Therefore, the Brix/acid ratio is often used as an index of fruit maturity. The ratio can also be affected by climate, variety, and horticultural practices. Table 6-1 gives typical acid composition and sugar levels for many commercially important fruits. Citric and malic acid are the most common acids in fruits. Citric and malic acids also predominate in most vegetables; however, leafy vegetables may also contain significant quantities of oxalic acid. In the dairy industry, titratable acidity is commonly used to monitor the progress of lactic acid fermentations in cheese and yogurt production.

Organic acids contribute to the refractometer reading of soluble solids. When foods are sold on the basis of pounds solids, Brix readings are sometimes corrected for acid content. For citric acid, 0.20 °Brix are added for each percent titratable acidity.

6.4 GENERAL CONSIDERATIONS

6.4.1 Normality

By definition, an **acid** is a proton donor. The only proton of concern in food systems is the hydrogen ion (H^+), which is hydrated in aqueous environments to produce the hydronium ion (H_3O^+). Common food acids have one, two, or three acid groups per molecule. These are mono-, di-, and triprotic acids, respectively. **Bases** by definition are proton acceptors. Complete neutralization of a triprotic acid requires three times as much titrant as an equimolar concentration of a monoprotic acid. Titration problems could be made much easier if one unit of base always reacted with one unit of acid. This can easily be achieved by dividing the molecular weight of the acid by the number of acid groups on the molecule. The resulting unit describes the fraction of molecular weight equivalent to one acid group. The value is consequently called the **equivalent weight.** Mono-, di-, and triprotic acids have respectively one, two, and three equivalents per mole. Similar to **molarity,** which is stated in terms of moles/liter, **normality** is expressed in terms of equivalents/liter. Normality has the advantage that one equivalent of acid always reacts with one equivalent of base. Table 6-2 provides a list of molecular and equivalent weights for acids important to food analysis.

6.4.2 Acidity

The pH is used to determine the endpoint of acid/base titration. This can be achieved directly with a pH meter (refer to Chapter 31 for a complete discussion of pH and the pH meter), but more commonly, an indicator dye is used. In some cases, the way pH changes during a titration can lead to subtle problems. Some background in acid theory is necessary to fully understand titration and to appreciate the occasional problems that might be encountered.

Acidity is a function of the free hydrogen ion con-

TABLE 6-1. Acid Composition and °Brix of Some Commercially Important Fruits

Fruit	*Principal Acid*	*Typical Percent Acid*	*Typical Brix*
Apples	Malic	0.27–1.02	9.12–13.5
Bananas	Malic/Citric (3:1)	0.25	16.5–19.5
Cherries	Malic	0.47–1.86	13.4–18.0
Cranberries	Citric	0.9–1.36	
	Malic	0.70–0.98	12.9–14.2
Grapefruit	Citric	0.64–2.10	7–10
Grapes	Tartaric/Malic (3:2)	0.84–1.16	13.3–14.4
Lemons	Citric	4.2–8.33	7.1–11.9
Limes	Citric	4.9–8.3	8.3–14.1
Oranges	Citric	0.68–1.20	9–14
Peaches	Citric	1–2	11.8–12.3
Pears	Malic/Citric	0.34–0.45	11–12.3
Pineapples	Citric	0.78–0.84	12.3–16.8
Raspberries	Citric	1.57–2.23	9–11.1
Strawberries	Citric	0.95–1.18	8–10.1
Tomatoes	Citric	0.2–0.6	4

TABLE 6-2. Molecular and Equivalent Weights of Common Food Acids

Acid	*Chemical Formula*	*Molecular Weight*	*Equivalents per Mole*	*Equivalent Weight*
Citric (anhydrous)	$H_3C_6H_5O_7$	192.12	3	64.04
Citric (hydrous)	$H_3C_6H_5O_7 \cdot H_2O$	210.14	3	70.05
Acetic	$HC_2H_3O_2$	60.06	1	60.05
Lactic	$HC_3H_5O_3$	90.08	1	90.08
Malic	$H_2C_4H_4O_5$	134.09	2	67.05
Oxalic	$H_2C_2O_4$	90.04	2	45.02
Tartaric	$H_2C_4H_4O_6$	150.09	2	75.05
Ascorbic	$H_2C_6H_6O_6$	176.12	2	88.06
Hydrochloric	HCl	36.47	1	36.47
Sulfuric	H_2SO_4	98.08	2	49.04
Phosphoric	H_3PO_4	98.00	3	32.67
Potassium acid phthalate	$KHC_8H_4O_4$	204.22	1	204.22

centration. Theoretically, in aqueous solution, the hydrogen ion concentration can range from approximately 10^1 to 10^{14} moles/liter. The term pH (the negative log of hydrogen ion concentration) was developed to compress this broad continuum of concentrations into a simple value. Although pH can hypothetically range from −1 to 14, pH readings below 1 are difficult to obtain. This is due to incomplete disassociation of hydrogen ions at high-acid concentrations. At 0.1 N, strong acids are assumed to be fully disassociated. Since full disassociation exists when a strong base is used to titrate a strong acid, the pH at any point in the titration is equal to the hydrogen ion concentration of the remaining acid (Fig. 6-1).

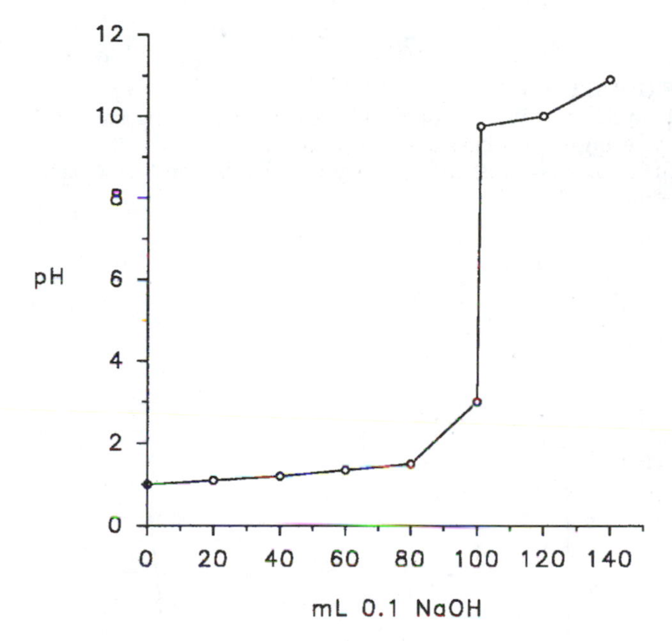

FIGURE 6-1.
Titration of a strong acid with strong base. The pH at any point in the titration is dictated by the hydrogen ion concentration of the acid remaining after partial neutralization with base.

All food acids are weak acids. Less than 1 percent of their ionizable hydrogens are disassociated from the parent molecule. When free hydrogen ions are removed through titration, new hydrogen ions can arise from other previously undisassociated parent molecules. This tends to cushion the solution from abrupt changes in pH. This property of a solution to resist change in pH is termed **buffering.** Buffering occurs in foods whenever a weak acid and its salt (i.e., its conjugated base) are present in the same medium. Due to buffering, a graph of pH versus titrant is more complex for weak acids than strong acids. However, this relationship can be predicted by the Henderson-Hasselbach equation.

$$pH = pK_a + \log \frac{[A^-]}{[HA]} \qquad (1)$$

HA represents some acid. A^- represents its conjugated base. The **conjugated base** is equal in concentration to the **conjugated acid** $[H^+]$. The **pK_a** is the pH at which equal quantities of acid and conjugated base are present. The equation indicates maximum buffering capacity will exist when the pH equals the pK_a. A graph showing the titration of 0.1 N acetic acid with 0.1 N NaOH illustrates this point (Fig. 6-2).

Di- and triprotic acids will have two and three buffering regions, respectively. A pH versus titrant graph of citric acid is given in Fig. 6-3. If the pK_a steps in polyprotic acids differ by three or more pK_a units, then the Henderson-Hasselbach equation can predict the plateau corresponding to each step. However, the transition region between steps is complicated by the presence of protons and conjugate bases arising from other disassociation state(s). Consequently, the Henderson-Hasselbach equation breaks down near the

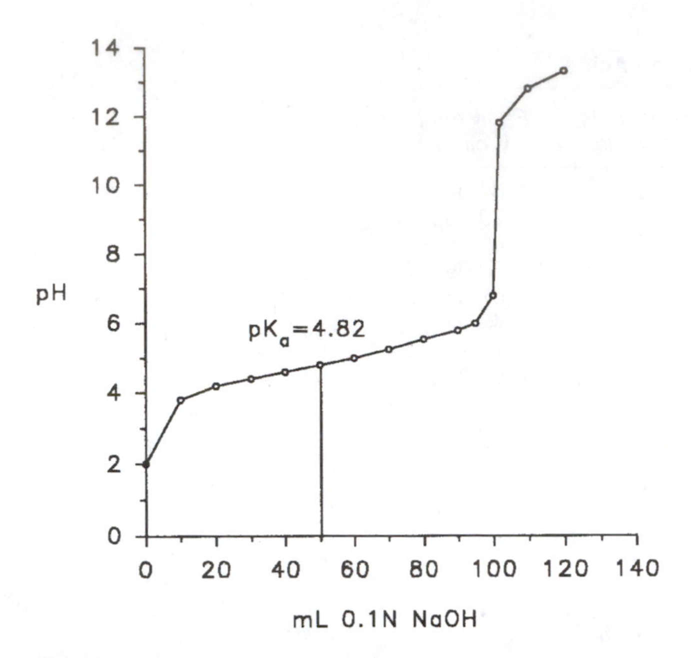

FIGURE 6-2.
Titration of a weak monoprotic acid with a strong base. A buffering region is established around the pK_a (4.82). The pH at any point is described by the Henderson-Hasselbach equation.

FIGURE 6-3.
Titration of a weak polyprotic acid with a strong base. Buffering regions are established around each pK_a. The Henderson-Hasselbach equation can predict the pH for each pK_a value if pK_a steps are separated by more than three units. However, complex transition mixtures between pK_a steps make simple calculations of transition pHs impossible.

equivalence point between two pK_a steps. However, the pH at the equivalence point is easily calculated. The pH is simply $(pK_a1 + pK_a2)/2$. Table 6-3 gives pK_a's of acids important to food analysis.

6.4.3 Equivalence Point

At the **equivalence point** in a titration, the number of acid equivalents exactly equals the number of base equivalents, and total acid neutralization is achieved. As the equivalence point is approached, the denominator [HA] in the Henderson-Hasselbach equation becomes insignificantly small and the quotient $[A^-]/[HA]$ increases exponentially. As a result, the solution pH rapidly increases and ultimately approaches the pH of the titrant. The exact equivalent point is the halfway mark on this slope of abrupt pH increase. The advantage of determining the equivalence point by pH change is that the precise equivalent point is identified. Since change in pH (and not some final pH value) signals the end of titration, accurate calibration of the pH meter is not even essential. However, in order to identify the equivalent point, a record of pH versus titrant must be kept. This and the physical constraints of pH probes and slow response with some electrodes make the potentiometric approach somewhat cumbersome.

6.4.4 Indicators

For simplicity in routine work, an indicator solution is often used to approximate the equivalent point. This approach tends to overshoot the equivalent point by a small amount. When indicators are used, the term **endpoint** is substituted for equivalent point. This emphasizes that the resulting values are approximate and dependent on the specific indicator. Phenolphthalein is the most common indicator for food use. It changes from clear to red in the pH region 8.0–9.6. Significant

TABLE 6-3. pK_a Values for Some Acids Important in Food Analysis

Acid	*pK_a1*	*pK_a2*	*pK_a3*
Oxalic	1.19	4.21	—
Phosphoric	2.12	7.21	12.30
Tartaric	3.02	4.54	—
Malic	3.40	5.05	—
Citric	3.06	4.74	5.40
Lactic	3.86	—	—
Ascorbic	4.10	11.79	—
Acetic	4.76	—	—
Potassium acid phthalate	5.40	—	—
Carbonic	6.10	10.25	—

color change is usually present by pH 8.2. This pH is termed the **phenolphthalein endpoint.**

A review of pK_a values in Table 6-3 indicates that naturally occurring food acids do not buffer in the region of the phenolphthalein endpoint. However, phosphoric acid (used as an acidulant in some soft drinks) and carbonic acid (carbon dioxide in aqueous solution) do buffer at this pH. Consequently, taking the solution from the true equivalent point to the endpoint may require a large amount of titrant for these acids. Indistinct endpoints and erroneously large titration values may result. When these acids are titrated, potentiometric analysis is usually preferred. Interference by CO_2 can be removed by boiling the sample and titrating the remaining acidity to a phenolphthalein endpoint.

Deeply colored samples also present a problem for endpoint indicators. When colored solutions obscure the endpoint, a **potentiometric method** is normally used. For routine work, pH versus titrant data are not collected. Samples are simply titrated to an 8.2 pH (the **phenolphthalein endpoint**). Even though this is a potentiometric method, the resulting value is an endpoint and not the true equivalent point since it simply echoes the pH value of phenolphthalein. A pH of 7 may seem to be a better target for a potentiometric endpoint than 8.2. This would appear to identify the point of full acid neutralization and should, consequently, represent the actual equivalent point. However, once all acid has been neutralized, the conjugate base remains. As a result, the pH at equivalent point is usually greater than 7. Confusion might also arise if pH 7 was standard for colored samples and pH 8.2 was standard for noncolored samples.

Dilute acid solutions (e.g., vegetable extracts) require dilute solutions of standard base for optimal accuracy in titration. However, a significant volume of dilute alkali may be required to take a titration from the equivalence point to pH 8.2. Bromthymol blue is sometimes used as an alternative indicator in low-acid situations. It changes from yellow to blue in the pH range 6.0–7.6. The endpoint is usually a distinct green. However, endpoint detection is somewhat more subjective than the phenolphthalein endpoint.

All indicators are either weak acids or weak bases. In excessive amounts, they can influence the titration. Indicator solutions should be held to the minimum necessary to impart effective color. Typically, two to three drops of indicator are added to the solution to be titrated. Indicator solutions rarely contain over a few tenths percent dye (w/v). The lower the indicator concentration, the sharper will be the endpoint.

In acetic acid fermentations, it is sometimes desirable to know how much acidity comes from the acetic acid and how much is contributed naturally by other acids in the product. This can be achieved by first performing an initial titration to measure total acidity. The acetic acid is then boiled off, the solution is allowed to cool, and a second titration is performed to determine the **fixed acidity.** The difference between fixed and total acidity is **volatile acidity.**

6.5 PREPARATION OF REAGENTS

6.5.1 Standard Alkali

Sodium hydroxide (NaOH) is the most commonly used base in titratable acidity determinations. In some ways, it appears to be a poor candidate for a **standard base.** Reagent grade NaOH is very hygroscopic and often contains significant quantities of insoluble sodium carbonate (Na_2CO_3). However, NaOH's economy, availability, and long tradition of use outweigh its shortcomings. Standard base solutions are normally made from a stock solution containing 50 percent sodium hydroxide in water (w/v). Sodium carbonate is essentially insoluble in concentrated alkali and gradually precipitates out of solution over approximately the first 10 days of storage.

NaOH can react with dissolved and atmospheric CO_2 to produce new Na_2CO_3. This reduces alkalinity and sets up a carbonate buffer that can obscure the true endpoint of a titration. Therefore, CO_2 should be removed from water prior to making the stock solution. This can be achieved by purging water with CO_2-free gas for 24 hours or by boiling distilled water for 20 minutes and allowing it to cool before use. During cooling and long-term storage, air (with accompanying CO_2) will be drawn back into the container. Carbon dioxide can be stripped from reentering air with a soda-lime (20% NaOH, 65% CaO, 15% H_2O) or ascarite (NaOH-impregnated asbestos) trap.

Stock alkali solution is approximately 18 N. A working solution is made by diluting stock solution with CO_2-free water. It is best to store alkali solution in a glass container with a rubber or thick plastic cap. Storing alkali in plastic containers is not advised since carbon dioxide permeates freely through most common plastics. Over time, alkali can react with glass. Working solutions should be restandardized weekly to correct alkalinity for losses arising from interactions with glass and CO_2.

6.5.2 Standard Acid

Sodium hydroxide's impurities and hygroscopic nature make it unsuitable as a primary standard. Therefore, NaOH titrating solutions must be standardized against a **standard acid. Potassium acid phthalate** (KHP) is commonly used for this purpose.

KHP's single ionizable hydrogen (pK_a = 5.4) provides very little buffering at pH 8.2. It can be manufactured in very pure form, it is relatively nonhygroscopic, and it can be dried at 120°C without decomposition or volatilization. Its high molecular weight favors accurate weighing.

KHP should be dried for two hours at 120°C and allowed to cool to room temperature in a dessicator immediately prior to use. An accurately measured quantity of KHP solution is titrated with a base of unknown normality. The base is always the titrant. CO_2 is relatively insoluble in acidic solutions. Consequently, stirring an acid sample to assist in mixing does not significantly alter the titration.

6.6 SAMPLE PREPARATION AND ANALYSIS

A number of standard or official methods exist for determining titratable acidity in various foods. (See *Official Methods of Analysis*, published by the Association of Official Analytical Chemists International, and *Standard Methods for the Examination of Dairy Products*, published by the American Public Health Association.) However, determining titratable acidity on most samples is relatively routine, and various procedures share many common steps. An aliquot of sample (often 10 mL) is titrated with a standard alkali solution (often 0.1 N NaOH) to a phenolphthalein endpoint. Potentiometric endpoint determination is used when sample pigment makes a color indicator impractical. Problems may arise, however, when concentrates, gels, or particulate-containing samples are titrated. These matrices prevent rapid diffusion of acid from densely packed portions of sample material. This slow diffusion process results in a fading endpoint. Concentrates can simply be diluted with CO_2-free water. Titration is then performed and the original acid content calculated from dilution data. Starch and similar weak gels can often be mixed with CO_2-free water, stirred vigorously, and titrated in a manner similar to concentrates. However, some pectin and food gum gels require mixing in a blender to adequately disrupt the gel matrix. Thick foams are occasionally formed in mixing. Antifoam or vacuum can be used to break the foams.

The pH of particulate samples often varies from one particulate piece to another. Acid equilibration throughout the entire mass may require several months. As a result, particulate-containing foods should be finely comminuted in a blender before titrating. The comminuting process may incorporate large quantities of air. Air entrapment makes volumetric measurements questionable. Aliquots are often weighed when air incorporation is a problem.

NaOH has a low surface tension. This predisposes it to leakage around the stopcock. Leakage during titration will produce erroneously high acid values. If leakage occurs when the burette is not in use, concentrated alkali can form on the outside of the burette. This concentration may be reflected in a higher titrant normality within the stopcock, with resulting titration errors for the initial titrations. The possibility of caustic burns also exists.

6.7 CALCULATIONS

6.7.1 Calculating Titratable Acidity

The student will encounter several steps in determining titratable acidity. The standard acid must be made, concentrated alkali must be diluted to working strength, and the normality of the working solution must be precisely determined. The exact normality of the stock alkali solution can also be determined for future reference. Samples can then be analyzed for titratable acidity. When a large number of samples will be run on a routine basis, it is possible to adjust the base normality so percent acid can be read directly from the burette. The Practice Problems at the end of the chapter are designed to present these steps in their logical order of execution.

6.7.2 Other Methods

High-performance liquid chromatography (HPLC) and electrochemistry have both been used to measure acids in food samples. Both methods allow identification of specific acids. HPLC uses refractive index or UV detection. Ascorbic acid has strong absorbance at 265 nm. Significant absorbance of other prominent acids does not occur until 200 nm or below.

Many acids can be measured with such electrochemical techniques as voltammetry and polarography. In ideal cases, the sensitivity and selectivity of electrochemical methods are exceptional. However, interfering compounds frequently hinder identification of other compounds of interest.

Chromatographic and electrochemical techniques do not differentiate between an acid and its conjugate base. Both species inevitably exist side by side as part of the inherent food-buffer system. As a result, acids determined by instrumental methods may be 50 percent higher than values determined by titration. It follows that Brix/acid ratios can only be based on titration values.

6.8 SUMMARY

When acids compose more than 0.2 percent of a food, they have a strong impact on flavor perception. Although titratable acidity usually cannot quantify individual acids, it does provide a general prediction of an acid's impact on flavor.

Although titratable acidity is one of the easiest tests to perform, constant restandardization of the NaOH is essential for accuracy. Accuracy is also influenced by inconsistent or over addition of indicator. Most foods are titrated to a pH 8.2 endpoint. Whenever practical, phenolphthalein is used as the indicator; however, a pH meter is often required for deeply colored samples.

6.9 STUDY QUESTIONS

1. For each of the food products listed below, what acid should be used to express the titratable acidity?
   a. orange juice
   b. yogurt
   c. apple juice
   d. grape juice
2. Why is the Brix/acid ratio often used as an indicator of flavor quality for certain foods, rather than simply Brix or acid alone?
3. How would you recommend determining the endpoint in the titration of tomato juice to determine the titratable acid? Why?
4. The titratable acidity was determined by titration to a phenolphthalein endpoint for boiled and unboiled carbonated beverage (clear-colored) samples. Which sample would you expect to have a higher calculated titratable acidity? Why? Would you expect one of the samples to have a fading endpoint? Why?
5. Why and how is an ascarite trap used in the process of determining titratable acidity?
6. Why is volatile acidity useful as a measure of quality for acetic acid fermentation products, and how is it determined?
7. What factors make potassium acid phthalate a good choice as a standard acid for use in standardizing NaOH solutions to determine titratable acidity?
8. Could a sample that is determined to contain 1.5 percent acetic acid also be described as containing 1.5 percent citric acid? Why or why not?

6.10 PRACTICE PROBLEMS

1. How would you make 100 mL of a 0.1 N solution of potassium acid phthalate (KHP)?
2. How would you make 100 mL of a citrate buffer that is 0.1 N in both citric acid (anhydrous) and potassium citrate $KH_2C_6H_5O_7$ (MW 230.22)?
3. What would be the pH of the 0.1 N citrate buffer described in Problem 2?
4. How would you make 1 L of 0.1 N NaOH solution from an 18 N stock solution?
5. A stock base solution assumed to be 18 N was diluted to 0.1 N. KHP standardization indicated the normality of the working solution was 0.088 N. What was the actual normality of the solution?
6. A 20 mL sample of juice requires 25 mL of 0.1 N NaOH titrant. What would be the percent acid if the juice is (1) apple juice, (2) orange juice, (3) grape juice?
7. A lab analyzes a large number of orange juice samples. All juice samples will be 10 mL. It is decided that 5 mL of titrant should equal 1 percent citric acid. What base normality should be used?
8. A lab wishes to analyze apple juice. They would like each milliliter of titrant to equal 0.1 percent malic acid. Sample aliquots will all be 10 mL. What base normality should be used?

Answers

1. From Table 6-2, the equivalent weight of KHP is 204.22 g/Eq. The weight of KHP required can be calculated from the equation.

$$\text{Acid Wt.} = \frac{\text{Desired volume (mL)}}{1000\ \text{mL/L}} \times \text{Eq. Wt. (g/Eq)} \times \text{desired N (Eq/L)} \tag{2}$$

   Therefore,

$$\text{KHP Wt.} = \frac{100\ \text{mL}}{1000\ \text{mL/L}} \times 204\ \text{g/Eq} \times 0.1\ \text{Eq/L} = 2.0422\ \text{g} \tag{3}$$

   The solution can be made by weighing exactly 2.0422 g of cool, dry KHP into a 100 mL volumetric flask and diluting to volume.
2. This problem is the same as Problem 1 above, except two components are being added to 100 mL of solution. From Table 6-2, the equivalent weight of citric acid (anhydrous) is 64.04 g/Eq. Therefore, the weight of citric acid (CA) would be

$$\text{CA Wt.} = \frac{100\ \text{mL}}{1000\ \text{mL/L}} \times 64.04\ \text{g/Eq} \times 0.1\ \text{Dq/L} = 0.6404\ \text{g} \tag{4}$$

   Potassium citrate (PC) is citric acid with one of its three hydrogen ions removed. Consequently, it has one less equivalent per mole than CA. The equivalent weight of PC would be its molecular weight (230.22) divided by its two remaining hydrogen ions, or 115.11 g per equivalent. Therefore, the weight contribution of PC would be

$$\text{PC Wt.} = \frac{100\ \text{mL}}{1000\ \text{mL}} \times 115.11\ \text{g/Eq} \times 0.1\ \text{Eq/L} = 1.511\ \text{g} \tag{5}$$

   100 mL of a 0.1 N citrate buffer would be made by disso!ving 0.6404 g of anhydrous CA and 1.1511 g of PC into a 100 mL volumetric flask and filling to volume.
3. The relationship between pH and conjugate acid/base pair concentrations is given by the Henderson-Hasselbach equation.

$$\text{pH} = \text{pK}_a + \log \frac{[\text{A}^-]}{[\text{HA}]} \tag{6}$$

When acid and conjugate base concentrations are equal $[A^-]/[HA] = 1$. Since the log of 1 is 0, the pH will equal the pK_a of the acid. Because CA and PC are both 0.1 N, the pH will equal the pKa1 of citric acid given in Table 6-3 (pH = 3.2).

4. In general,

$$\text{beginning normality} \times \text{ml of concentrated solution} = \text{final normality} \times \text{final mL of dilute solution} \quad (7)$$

Solving for volume of concentrate, we get

$$\text{mL concentrated solution} = \frac{\text{final N} \times \text{final mL}}{\text{beginning N}} = \frac{0.1\ \text{N} \times 1000\ \text{mL}}{18\ \text{N}} = 5.55\ \text{mL} \quad (8)$$

Consequently, 5.55 mL would be dispensed into a 1 L volumetric flask. The flask would then be filled to volume with distilled CO_2-free water.

The normality of this solution will only be approximate since NaOH is not a primary standard. Standardization against a KHP solution or some other primary standard is essential. It is sometimes useful to back-calculate the true normality of the stock solution. Even under the best circumstances, the normality will decrease with time, but back-calculating will permit a closer approximation of the target normality the next time a working standard is prepared.

5. This answer is a simple ratio.

$$\frac{0.088}{0.100} \times 18 = 15.84\ \text{N} \quad (9)$$

In general chemistry, acid strength is frequently reported in normality. Normality is useful in preparing standard solutions for measurement of titratable acidity in foods. However, normality is not a very useful term for reporting results of food analysis. Customarily, food acids are reported as percent of total sample weight. Trace acids may also be reported in milligram percent (mg/100 g of sample).

Titratable acidity measurements require relatively small acid and titrant volumes. Normalities for small volumes are typically reported in milli-equivalents per mL (mEq/mL). Normalities reported in Eq/L are numerically identical to normalities reported in mEq/mL, only the unit portions differ. Therefore, a 1 normal solution of NaOH contains 1 equivalent of NaOH per L or 1 milli-equivalent per mL. The percent acid can be calculated by the equation

$$\%\ \text{acid} = \frac{\text{base normality (mEq/mL)} \times \text{mL base} \times \text{Eq. Wt. of acid (mg/Eq)}}{\text{sample weight (mg)}} \times 100 \quad (10)$$

It is often awkward in routine work to cite sample weights in milligrams. A modification of eq. 10 allows sample weights to be reported directly in grams.

$$\%\ \text{acid} = \frac{\text{base normality (mEq/mL)} \times \text{mL base} \times \text{Eq. Wt. of acid (mg/Eq)}}{\text{sample weight (g)} \times 10} \quad (11)$$

For routine titration of single-strength juice, milliliters can often be substituted for sample weight in grams. Depending on the soluble solids content of the juice, the resulting acid values will be high by 1 to 6 percent.

6. Table 6-1 indicates the principal acids in apple, orange, and grape juice are malic, citric, and tartaric acids, respectively. Table 6-2 indicates the equivalent weight of these acid are malic (67.05), citric (64.04), and tartaric (75.05). The percent acid for each of these juices would be

$$\text{malic acid} = \frac{0.1\ \text{mEq/mL NaOH} \times 25\ \text{mL} \times 67.05\ \text{mg/mEq}}{20\ \text{mL (10) mg/mL}} = 0.84 \quad (12)$$

$$\text{citric acid} = \frac{0.1\ \text{mEq/mL NaOH} \times 25\ \text{mL} \times 64.04\ \text{mg/mEq}}{20\ \text{mL (10) mg/mL}} = 0.80 \quad (13)$$

$$\text{tartaric acid} = \frac{0.1\ \text{mEq/mL NaOH} \times 25\ \text{mL} \times 75.05\ \text{mg/mEq}}{20\ \text{mL (10) mg/mL}} = 0.94 \quad (14)$$

The three values obtained for the different acids are closer than one might expect from a casual examination of molecular weights. When two acids predominate in a juice, they are usually malic and citric or malic and tartaric acids. Choosing the wrong predominant acid does not seriously affect the calculated percent acid since equivalent weights of naturally occurring food acid are similar.

Notice that the equivalent weight of anhydrous citric acid was used. The anhydrous form will always be used in calculating and reporting the results of titration. Pure citric acid has a tendency to absorb water. Some manufacturers of pure citric acid intentionally hydrate the molecule to stabilize it against further hydration. The equivalent weight of citric acid monohydrate will only be used when making solutions from hydrated starting materials. Hydrated chemicals should never be dried in a drying oven. Total dehydration is rarely possible. The resulting compound would have some intermediate (and unknown) hydration number, and solutions made from the compound would be inaccurate.

Quality control laboratories often analyze a large number of samples having a specific type of acid. Speed and accuracy are increased if acid concentration can be read directly from the burette. It is possible to adjust the normality of the base to achieve this purpose. The proper base normality can be calculated from the equation

$$N = \frac{10 \times A}{B \times C} \quad (15)$$

where: A = weight (or volume) of the sample to be titrated
B = volume (mL) of titrant you want to equal 1 percent acid
C = equivalent weight of the acid

7.

$$N = \frac{10 \times 10}{5 \times 64.04} = 0.3123 \text{ N} \quad (16)$$

In actuality, the standard alkali solution used universally by the Florida citrus industry is 0.3123 N.

8. Since each milliliter will equal 0.1 percent malic acid, 1 percent malic acid will equal 10 mL. Therefore,

$$N = \frac{10 \times 10}{10 \times 67.05} = 0.1491 \text{ N} \quad (17)$$

6.11 RESOURCE MATERIALS

AOAC. 1990. *Official Methods of Analysis*, 15th ed. Association of Official Analytical Chemists, Washington, DC.

Christian, G.D. 1977. *Analytical Chemistry*, 2nd ed. John Wiley & Sons, New York, NY.

Gardner, W.H. 1966. *Food Acidulants.* Allied Chemical Co., New York, NY.

Harris, D.C. 1991. *Quantitative Chemical Analysis*, 3rd ed. W.H. Freeman, Inc., New York, NY.

Nelson, P.E., and Tressler, D.K. 1980. *Fruit and Vegetable Juice Processing Technology*, 3rd ed. AVI Publishing Co., Westport, CT.

Marshall, R.T. (Ed.) 1992. *Standard Method for Examination of Dairy Products*, 16th ed. American Public Health Assn., Washington, DC.

Segel, I. 1976. *Biochemical Calculation.* John Wiley & Sons, Inc., New York, NY.

CHAPTER

7

Moisture and Total Solids Analysis

Robert L. Bradley, Jr.

7.1 INTRODUCTION

Moisture determination can be one of the most important analyses performed on a food product and yet one of the most difficult from which to obtain accurate and precise data. This chapter describes various methods for moisture analysis—their principles, procedures, applications, cautions, advantages, and disadvantages. Water activity measurement is also described, since it parallels the measurement of total moisture as an important quality factor. With an understanding of techniques described, one can apply appropriate moisture analyses to food products.

7.1.1 Importance of Moisture Assay

One of the most fundamental and important analytical procedures that can be performed on a food product is an assay for the amount of moisture (1–3). The dry matter that remains after moisture removal is commonly referred to as **total solids.** This analytical value is of great economic importance to a food manufacturer because water is an inexpensive filler. The following listing gives some examples in which moisture content is important to the food processor.

1. Moisture is a quality factor in the preservation of some products and affects stability in
   a. dehydrated vegetables and fruits
   b. dried milks
   c. powdered eggs
   d. dehydrated potatoes
   e. Spices and herbs
2. Moisture is used as a quality factor for
   a. jams and jellies, to prevent sugar crystallization
   b. sugar syrups
   c. prepared cereals—conventional, 4–8 percent; puffed, 7–8 percent
3. Reduced moisture is used for convenience in packaging and/or shipping of
   a. concentrated milks
   b. liquid cane sugar (67 percent solids) and liquid corn sweetener (80 percent solids)
   c. dehydrated products (these are difficult to package if too high in moisture)
   d. concentrated fruit juices
4. Moisture (or solids) content is often specified in compositional standards (i.e., Standards of Identity)
   a. Cheddar cheese must be ≤ 39 percent moisture
   b. enriched flour must be ≤ 15 percent moisture
   c. pineapple juice must have soluble solids of ≥ 10.5 °Brix (conditions specified)
   d. glucose syrup must have ≥ 70 percent total solids (mass/mass)
   e. processed meats' percentage of added water is commonly specified
5. Computations of the nutritional value of foods require that you know the moisture content.
6. Moisture data are used to express results of other analytical determinations on a uniform basis (i.e., dry weight basis).

7.1.2 Moisture Content of Foods

The moisture content of foods varies greatly, as shown in Table 7-1. Water is a major constituent of most food products. The approximate, expected moisture of content of a food can affect the choice of the method of mea-

TABLE 7-1. Moisture Content of Selected Foods

Food	*Moisture Content (%)*[1]
Fruits	
watermelons	92.6
oranges	86.0
apples	84.4
grapes	81.6
dried, e.g., raisins	18.0
Vegetables	
cucumbers	95.1
potatoes, white	79.8
snap beans, green	90.1
Meat, Poultry, and Fish	
ground beef (10% fat)	68.3
chicken, fryer — light meat	59.5
chicken, fryer — dark meat	57.5
flounder, fillet	58.1
Dairy Products	
milk, whole (3.5% fat)	87.4
yogurt	89.0
cottage cheese (4.2% milk fat)	78.3
Cheddar cheese	37.0
ice cream	63.2
Eggs	
chicken eggs (whole, fresh)	73.7
Nuts	
walnuts, black (in shell)	3.1
peanuts, roasted in shell	1.8
peanut butter	1.6
Dry Legumes	
blackeye peas (dry)	10.5
Bread, Cereal, and Pasta	
flour, wheat	12.0
white bread, enriched (firm)	35.0
corn flakes cereal	3.8
crackers, saltines	4.3
macaroni	10.4
Sweeteners	
sugar, granulated	0.5
sugar, brown	2.1
honey	17.2
Fats and Oils	
margarine (regular)	15.5
butter	15.5
oils, salad or cooking	0

[1]Values from (4).

surement. It can also guide the analyst in determining the practical level of accuracy required when measuring moisture content, relative to other food constituents.

7.1.3 Forms of Water in Foods

The ease of water removal from foods depends on how it exists in the food product. The three states of water in food products are:

1. **Free water**—This water retains its physical properties and thus acts as the dispersing agent for colloids and the solvent for salts.
2. **Adsorbed water**—This water is held tightly or is occluded in cell walls or protoplasm and is held tightly to proteins.
3. **Water of hydration**—This water is bound chemically, for example, lactose monohydrate; also some salts like $Na_2SO_4 \cdot 10H_2O$.

Depending on the form of the water present in a food, the method used for determining moisture may measure more or less of the water present. This is one of the reasons it is often necessary to use official methods with stated procedures (5–7). However, several official methods may exist for a particular product. For example, the Association of Official Analytical Chemists (AOAC) International methods for cheese include: method 926.08, vacuum oven; 948.12, forced draft oven; 977.11, microwave oven; 969.19, distillation. Usually, the first method listed by AOAC is preferred over others in any section.

7.1.4 Sample Collection and Handling

General procedures for sampling, sample handling and storage, and sample preparation are given in Chapter 3. These procedures are perhaps the greatest potential source of error in any analysis. Precautions must be taken to minimize inadvertent **moisture losses or gains** that occur during these steps. Obviously, any exposure of a sample to the open atmosphere should be as short as possible. Any heating of a sample during grinding should be minimized. Headspace in the sample storage container should be minimal because moisture is lost from the sample to equilibrate the container environment against the sample.

To illustrate the need for optimum efficiency and speed in weighing samples for analysis, Vanderwarn (8) showed using shredded Cheddar cheese (2–3 g in a 5.5 cm aluminum foil pan) that moisture loss within an analytical balance was a straight line function, and the rate of loss was related to the relative humidity. At 50 percent relative humidity, it required only 5 sec to lose 0.01 percent moisture. This time doubled at 70 percent humidity, or 0.01 percent moisture loss in 10 sec. While one might expect a curvilinear loss, the moisture loss was actually linear over a 5 min study interval. These data demonstrate the necessity of absolute control during collection of samples through weighing before drying.

7.2 OVEN DRYING METHODS

In **oven drying methods,** the sample is heated under specified conditions, and the loss of weight is used to calculate the moisture content of the sample. The moisture content value obtained is highly dependent on the type of oven used, conditions in the oven, and the time and temperature of drying. Various drying oven methods are AOAC approved for many food products. The methods are simple, and many ovens allow for simultaneous analysis of large numbers of samples. The time required may be from a few minutes to over 24 hr.

7.2.1 General Information

7.2.1.1 Removal of Moisture

Any oven method used to evaporate moisture has as its foundation the fact that the boiling point of water is 100°C; however, this considers only pure water at sea level. Free water is the easiest of the three forms of water to remove. However, consider if 1 gram molecular weight (1 mol) of a solute is dissolved in 1.0 liter of water, the boiling point would be raised by 0.512°C. This boiling point elevation continues throughout the moisture removal process as more and more concentration occurs.

Moisture removal is sometimes best achieved in a two-stage process. Liquid products (e.g., juices, milk) are commonly predried over a **steam bath** before drying in an oven. Products such as bread and field-dried grain are often air dried, then ground and oven dried, with the moisture content calculated from moisture loss at both air and oven drying steps. Particle size, particle size distribution, sample sizes, and surface area during drying influence the rate and efficiency of moisture removal.

7.2.1.2 Decomposition of Other Food Constituents

Moisture loss from a sample during analysis is a function of time and temperature. **Decomposition** enters the picture when time is extended too much or temperature is too high. Thus, most methods for food moisture involve a compromise between time at a particular temperature at which limited decomposition might be a factor. One major problem exists in that the physical process must separate all the water without decomposing any of the constituents that could release water. For example, carbohydrates decompose at 100°C according to the following reaction:

$$C_6H_{12}O_6 \rightarrow 6C + 6H_2O \quad (1)$$

The water generated in carbohydrate decomposition is not the moisture that we want to measure. Certain other chemical reactions (e.g., sucrose hydrolysis) can result in utilization of water, which would reduce the moisture for measurement. A less serious problem, but one that would be a consistent error, is the loss of **volatile constituents,** such as acetic, propionic, and butyric acids; and alcohols, esters, and aldehydes among flavor compounds. While weight changes in oven drying methods are assumed to be due to moisture loss, weight gains can also occur due to oxidation of unsaturated fatty acids and certain other compounds.

Nelson and Hulett (9) determined that moisture was retained in biological products to at least 365°C, which is coincidentally the critical temperature for water. Their data indicate that among the decomposition products at elevated temperatures were CO, CO_2, CH_4, and H_2O. These were not given off at any one particular temperature but at all temperatures and at different rates at the respective temperature in question.

By plotting moisture liberated against temperature, curves were obtained that show the amount of moisture liberated at each temperature (Fig. 7-1). Distinct breaks were shown that indicated the temperature at which decomposition became measurable. None of these curves showed any break before 184°C. Generally, proteins decompose at temperatures somewhat lower than those required for starches and celluloses. Extrapolation of the flat portion of each curve to 250°C gave a true moisture content based on the assumption that there was no adsorbed water present at the temperature in question.

7.2.1.3 Temperature Control

Drying methods utilize specified drying temperatures and times, which must be carefully controlled. Moreover, there may be considerable variability of temperature, depending on the type of oven used for moisture analysis. One should consider how much temperature variation exists within an oven before relying on data collected from its use.

Consider the temperature variation in three types of ovens: **convection, (atmospheric) forced draft,** and **vacuum.** The greatest temperature variation exists in a **convection oven.** This is because hot air slowly circulates without the aid of a fan. Air movement is obstructed further by pans placed in the oven. When the oven door is closed, the rate of temperature recovery is generally slow. This is dependent also upon the load placed in the oven and upon the ambient temperature. A 10°C temperature differential across a convection oven is not unusual. This must be considered in view of anticipated analytical accuracy and precision. A convection oven should not be used when precise and accurate measurements are needed.

Forced draft ovens have the least temperature differential across the interior of all ovens, usually not greater than 1°C. Air is circulated by a fan that forces air movement throughout the oven cavity. Forced draft ovens with air distribution manifolds appear to have added benefit where air movement is horizontal across shelving. Thus, no matter whether the oven is filled with moisture pans or only half filled, the result would be the same for a particular sample. This has been demonstrated using a Lab-Line oven (Melrose Park, IL) in which three stacking configurations for the pans were used (8). In one configuration, the oven was filled with as many pans holding 2–3 grams of Cheddar cheese as the forced draft oven could hold. In the two others, one half of the full load of pans with cheese was used with the pans (1) in orderly vertical rows with the width of one pan between rows, or (2) staggered such that pans on every other shelf were in vertical alignment. The results after drying showed no difference in the mean value or the standard deviation.

Two features of some **vacuum ovens** contribute to a wider temperature spread across the oven. One feature is a glass panel in the door. While from an educational point of view it may be fascinating to observe some samples in the drying state, the glass is a heat sink. The second feature is the way by which air is bled into the oven. If the air inlet and discharge are on opposite sides, conduct of air is virtually straight across the oven. Some newer models (Lab-Line model 3623) have air inlet and discharge manifolds mounted top and bottom. Air movement in this style of vacuum oven is upward from the front, then backward to the discharge in a broad sweep. The effect is to minimize cold spots as well as to exhaust moisture in the interior air.

7.2.1.4 Types of Pans for Oven Drying Methods

Pans used for moisture determinations are varied in shape and may or may not have a cover. The AOAC (5) moisture pan is about 5.5 cm in diameter with an insert cover. Other pans have covers that slip over the outside edge of the pan. These pans, while reusable, are expensive, particularly when reviewing labor costs to clean appropriately to allow reuse.

Pan covers are necessary to control loss of sample by spattering during the heating process. If the cover is metal, it must be slipped to one side during drying to allow for moisture evaporation. However, this slip-

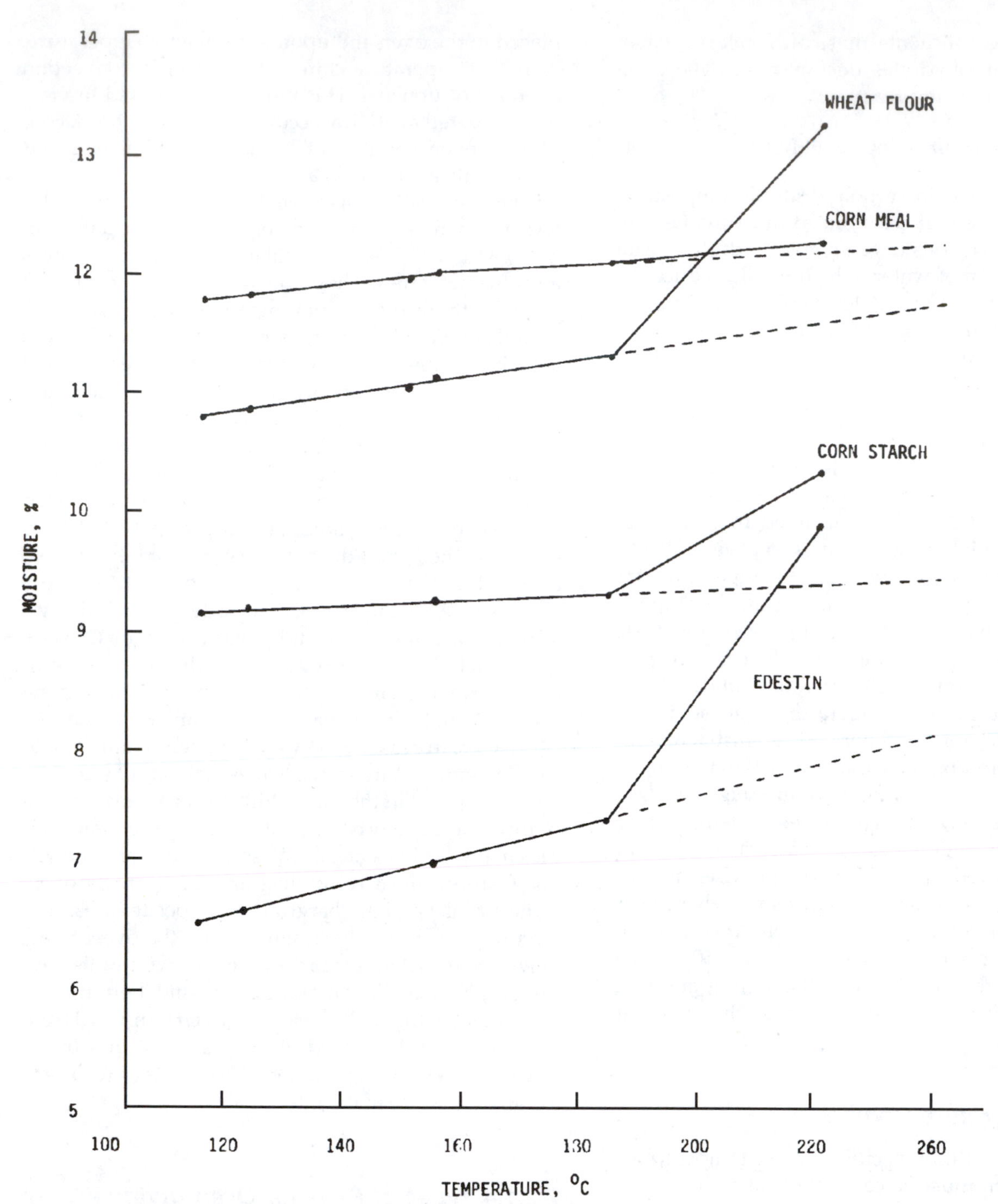

FIGURE 7-1.
Moisture content of several foods held at various temperatures in an oven. The hyphenated line extrapolates data to 250°F, the true moisture content. Reprinted with permission from (9) Nelson, O.A., and Hulett, G.A. 1920. The moisture content of cereals. *J. Ind. Eng. Chem.* 12:40–45. Copyright 1920, American Chemical Society.

ping of the cover also creates an area where spattering will result in product loss. Examine the interior of most moisture ovens and you will detect odor and deposits of burned-on residue. Such happenings, while undetected at the time of occurrence, produce erroneous results and large standard deviations (8).

Consider the use of **disposable pans** whenever possible; then purchase **glass fiber discs** for covers. At 5.5 cm in diameter, these covers fit perfectly inside a disposable aluminum foil pan and prevent spattering while allowing the surface to breathe. Paper filter discs foul with fat and thus do not breathe effectively. Consider the evidence presented in Fig. 7-2, derived from at least 10 replicate analyses of the same cheese with various pans and covers. These data prove two points: (1) fat does spatter from pans with slipped covers and (2) fiberglass is a very satisfactory cover.

7.2.1.5 Handling and Preparation of Pans

The handling and preparation of pans before use requires consideration. Use only **tongs** to handle any pan. Even fingerprints have weight. *All* pans must be oven treated to prepare them for use. This is a factor of major importance unless disproved by the technologist doing moisture determinations with a particular type of pan. To illustrate the weight change that occurs with disposable aluminum pans, consider the examples in Fig. 7-3. Disposable aluminum pans must be vacuum oven dried for 3 hr before use. At 3 hr and 15 hr in either a vacuum or forced draft oven at 100°C, pans varied in their weight within the error of the balance, or 0.0001 g (8). Store dry moisture pans in a functioning **desiccator.** The glass fiber covers do not need drying before use.

7.2.1.6 Surface Crust Formation (Sand Pan Technique)

Some food materials tend to form a semipermeable crust or to lump together during drying. Such an occurrence will contribute to erratic and erroneous results. To control this problem, analysts use the **sand pan technique.** Clean, dry sand and a short glass stirring rod are preweighed into a moisture pan. Subsequently, after weighing in a sample, the sand and sample are admixed with the stirring rod left in the pan. The remainder of the procedure is by a standardized procedure if available; otherwise the sample is dried to constant weight. The purpose of the sand is twofold: to prevent **surface crust** from forming and to disperse the sample so evaporation of moisture is less impeded.

7.2.1.7 Calculations

Moisture and total solids contents of foods can be calculated as follows using oven drying procedures:

$$\text{\% Moisture (wt/wt)} = \frac{\text{wt } H_2O \text{ in sample}}{\text{wt of wet sample}} \times 100 \qquad (2)$$

$$\text{\% Moisture (wt/wt)} = \frac{\text{wt of wet sample} - \text{wt of dry sample}}{\text{wt of wet sample}} \times 100 \qquad (3)$$

$$\text{\% Total Solids (wt/wt)} = \frac{\text{wt of dry sample}}{\text{wt of wet sample}} \times 100 \qquad (4)$$

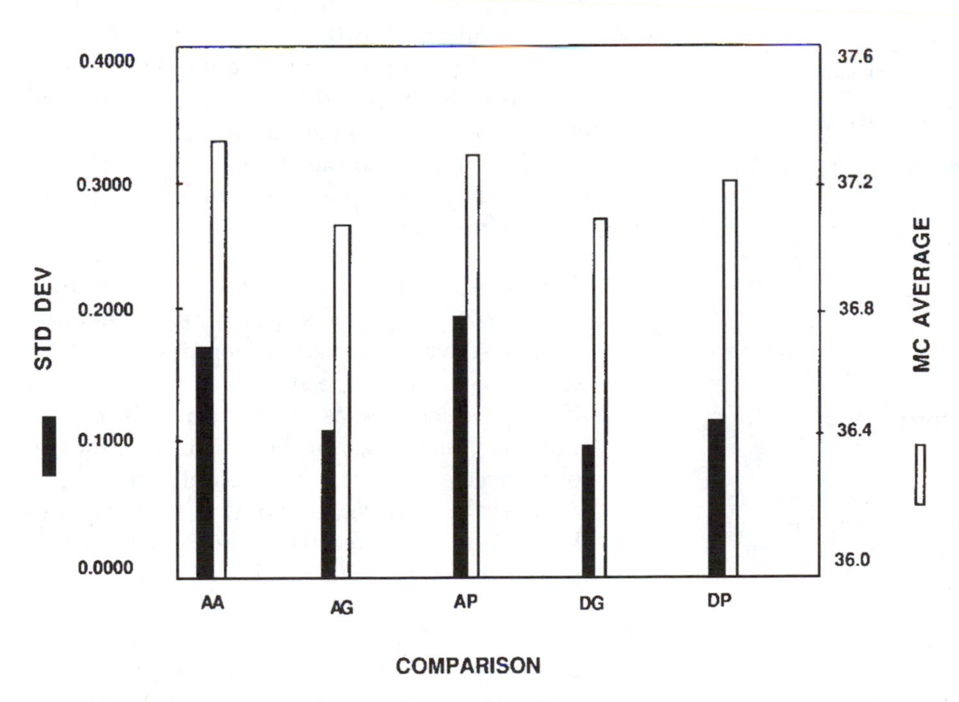

FIGURE 7-2.
Effect of various pan and cover combinations on the moisture content (MC) of Cheddar cheese. Standard deviations show precision of the analysis. Pans: A = AOAC, D = disposable; Covers: A = AOAC, G = glass fiber disc, P = filter paper disc. From (8), used with permission.

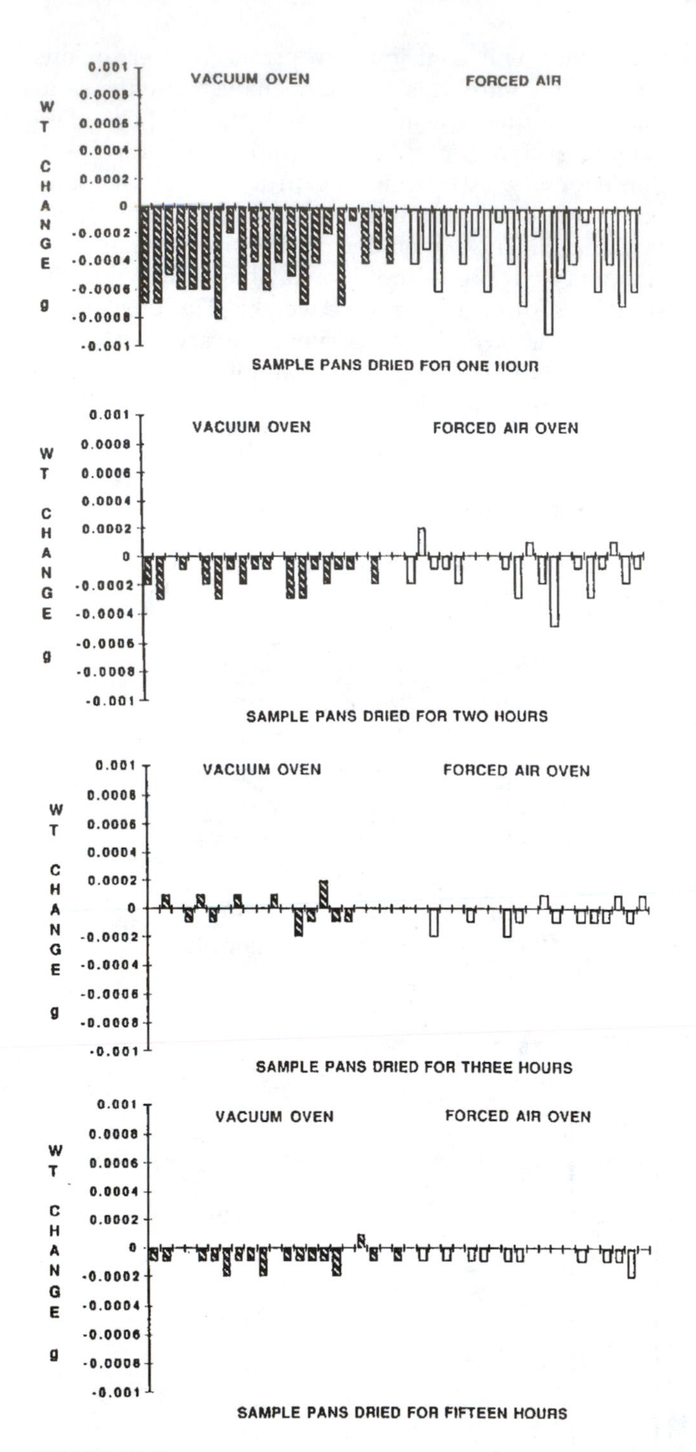

FIGURE 7-3.
Effect of drying new disposable aluminum moisture pans in a vacuum and forced draft oven at 100°C. The sensitivity of the balance was 0.0001 g. From (8), used with permission.

7.2.2 Forced Draft Oven

When using a forced draft oven, the sample is rapidly weighed into a moisture pan and placed in the oven for an arbitrarily selected time if no standardized method exists. Drying time periods for this method are .75–24 hr (Table 7-2), depending on the food sample and its pretreatment; some liquid samples are dried initially on a steam bath at 100°C to minimize spattering. In these cases, drying times are shortened to .75–3 hr. A forced draft oven is used with or without a steam table predrying treatment to determine the solids content of fluid milks (AOAC Method 990.19, 990.20).

An alternative to selecting a time period for drying is to weigh and reweigh the dried sample and pan until two successive weighings taken 30 min apart agree within a specified limit, for example, .1–.2 mg for a 5 g sample. The user of this second method must be aware of sample transformation, such as browning that suggests moisture loss of the wrong form. Samples high in carbohydrates should not be dried in a forced draft oven but rather in a vacuum oven at a temperature no higher than 70°C. Lipid oxidation and a resulting sample weight gain can occur at high temperatures in a forced draft oven.

7.2.3 Vacuum Oven

By drying under reduced pressure (25–100 mm Hg), one is able to obtain a more complete removal of water and volatiles without decomposition within a 3–6 hr drying time. Vacuum ovens need a dry air purge in addition to temperature and vacuum controls to operate within method definition. In older methods, a vacuum flask is used, partially filled with concentrated sulfuric acid as the desiccant. One or two air bubbles per second are passed through the acid. Recent changes now stipulate an air trap that is filled with calcium sulfate containing an indicator to show moisture saturation. Between the trap and the vacuum oven is an appropriately sized rotameter to measure air flow (100–120 ml per min) into the oven.

The following are important points in the use of a vacuum drying oven:

1. **Temperature** used depends on the product, such as 70°C for fruits and other high-sugar products. Even with reduced temperature, there can be some decomposition.
2. If the product to be assayed has a high concentration of **volatiles,** you should consider the use of a correction factor to compensate for the loss.
3. Analysts should remember that in a **vacuum** heat is not conducted well. Thus pans must be placed directly on the metal shelves to conduct heat.
4. **Evaporation** is an endothermic process; thus, a pronounced cooling is observed. Because of the cooling effect of evaporation, when several samples are placed in an oven of this type, you will note that the temperature will drop. Do not attempt to compensate for the cooling effect by increasing the temperature, otherwise samples

TABLE 7-2. Forced Draft Oven Temperature and Times for Selected Foods

Product	*Dry on Steam Bath*	*Oven Temperature (°C ± 2)*	*Time in Oven (Hrs)*
Buttermilk, liquid	X[1]	100	3
Cheese, natural type only		100	16.5 ± .5
Chocolate and cocoa		100	3
Cottage cheese		100	3
Cream, liquid and frozen	X	100	3
Egg albumin, liquid	X	130	0.75
Egg albumin, dried	X	100	0.75
Ice cream and frozen desserts	X	100	3.5
Milk: whole, low fat, and skim	X	100	3
condensed skim		100	3
evaporated		100	3
Nuts: almonds, peanuts, walnuts		130	3

From (6) p. 492, with permission, *Standard Methods for the Examination of Dairy Products*, Robert T. Marshall, ed. Copyright 1993 by the American Public Health Association.
[1]X = samples must be partially dried on steam bath before being placed in oven.

during the last stages of drying will be overheated.

5. The **drying time** is a function of the total moisture present, nature of the food, surface area per unit weight of sample, whether sand is used as a dispersant, and the relative concentration of sugars and other substances capable of retaining moisture and/or decomposing. The drying interval is determined experimentally to give reproducible results.

7.2.4 Microwave Oven

Microwave oven drying in its infancy was looked upon as a great boon to moisture determination. It was the first precise and rapid technique that allowed some segments of the food industry to make in-process adjustment of moisture in the food before final packaging. For example, process cheese could be analyzed and the composition adjusted before the blend was dumped from the cooker. Such control could effectively pay for the microwave oven within a few months. Methods texts indicated that users must check results against the AOAC vacuum oven procedure to determine how much microwave energy was needed. Adjustment of megatron output on the original Apollo oven was done by turning a control knob.

The original manufacturer sold his invention, and then came a new model from CEM Corporation (Matthews, NC). The new model had significant improvements over the original Apollo oven. A particular microwave oven, or equivalent, is specified in the AOAC procedures for moisture analysis of cheese (AOAC Method 977.11), total solids analysis of processed tomato products (AOAC Method 985.26), and moisture analysis of meat and poultry products (AOAC Method 985.14).

With the CEM microwave oven, the user controls the microwave output by setting the microprocessor controller to a percentage of full power. Next the internal balance is tared with two fiberglass pads on the balance. As rapidly as possible, a sample is placed between the two pads and weighed against the tare weight. Time for the drying operation is set by the operator and "start" is activated. The microprocessor controls the drying procedure, with percentage moisture indicated in the controller window.

The procedure described above suffers somewhat from a few inherent difficulties. The focus of the microwave energy is such that unless the sample is centrally located and evenly distributed, some portions may burn while other areas are underprocessed. The amount of time needed for an inexperienced operator to place an appropriate sample weight between the pads results in too much moisture loss before weighing. Newer models will likely eliminate these problems.

7.3.5 Infrared Drying

Infrared drying involves penetration of heat into the sample being dried, as compared to heat conductivity and convection with conventional ovens. Such heat penetration to evaporate water from the sample can significantly shorten the required drying time, to

10–25 min. The infrared lamp used to supply heat to the sample results in a filament temperature of 2000–2500°K. Factors that must be controlled include distance of the infrared source from the dried material and thickness of the sample. The analyst must be careful that the sample does not burn or case harden while drying. Infrared drying ovens may be equipped with forced ventilation to remove moisture air and an analytical balance to read moisture content directly. No infrared moisture analysis techniques are approved by AOAC currently. However, because of the speed of analysis, this technique is suited for qualitative in-process use.

7.3 DISTILLATION PROCEDURES

7.3.1 Overview

Distillation techniques involve co-distilling the water in a food sample with a high boiling point solvent that is immiscible in water, collecting the mixture that distills off, and then measuring the volume of water. Two distillation procedures are in use today: **direct** and **reflux distillations,** with a variety of solvents. For example, in direct distillation with immiscible solvents of higher boiling point than water, the sample is heated in mineral oil or liquid with a flash point well above the boiling point for water. Other immiscible liquids with boiling point only slightly above water can be used (e.g., toluene, xylene, and benzene). However, reflux distillation with an immiscible solvent is the most widely used method.

Distillation techniques were originally developed as rapid methods for quality control work, but they are not adaptable to routine testing. The distillation method is an AOAC-approved technique for moisture analysis of spices (AOAC Method 986.21), cheese (AOAC Method 969.19), and animal feeds (AOAC Method 925.04). It can also give good accuracy and precision for nuts, oils, soaps, and waxes.

Distillation methods cause less thermal decomposition of some foods than oven drying at high temperatures. Adverse chemical reactions are not eliminated but can be reduced by using a solvent with a lower boiling point. This, however, will increase the distillation times. Water is measured directly in the distillation procedure (rather than by weight loss), but reading the menicus of a receiving tube to determine the volume of water may be less accurate than using a weight measurement.

7.3.2 Reflux Distillation with Immiscible Solvent

Reflux distillation uses either a solvent less dense than water (e.g., toluene, with a boiling point of 110.6°C; or xylene, with a boiling range of 137–140°C) or a solvent more dense than water (e.g., tetrachlorethylene, with a boiling point of 121°C). The advantage of using this last solvent is that material to be dried floats; therefore it will not char or burn. In addition, there is no fire hazard with this solvent.

A **Bidwell-Sterling moisture trap** (Fig. 7-4) is commonly used as part of the apparatus for reflux distillation with a solvent less dense than water. The distillation procedure using such a trap is described in Fig. 7-5, with emphasis placed on dislodging adhering water drops, thereby minimizing error. When the toluene in the distillation just starts to boil, the analyst will observe a hazy cloud rising in the distillation flask. This is a vaporous emulsion of water in toluene. As the vapors rise and heat the vessel, the Bidwell-Sterling trap, and the bottom of the condenser, condensation occurs. It is also hazy at the cold surface of the condenser, where water droplets are visible. The emulsion inverts and becomes toluene dispersed in water. This turbidity clears very slowly on cooling.

Three potential sources of error with distillation should be eliminated if observed:

1. Formation of **emulsions** that will not break. Usually this can be controlled by allowing the apparatus to cool after distillation is completed and before reading the amount of moisture in the trap.
2. Clinging of **water droplets** to dirty apparatus. Clean glassware is essential, but water seems to cling even with the best cleaning effort. A

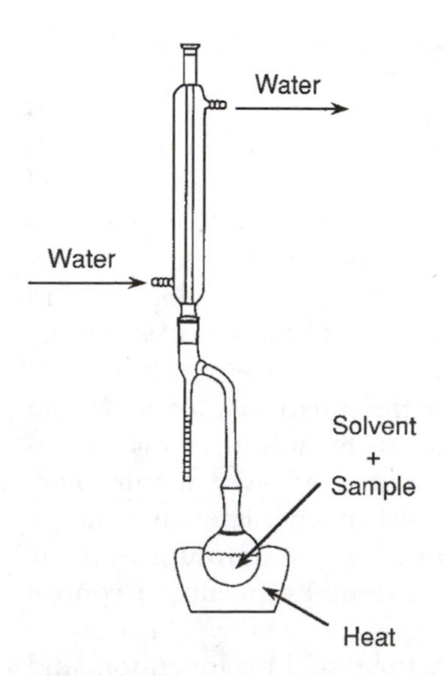

FIGURE 7-4.
Apparatus for reflux distillation of moisture from a food. Key to this setup is the Bidwell-Sterling moisture trap. This style can only be used where the solvent is less dense than water.

Place sample in distillation flask and cover completely with solvent.
⇓
Fill the receiving tube (e.g., Bidwell-Sterling trap) with solvent, by pouring it through the top of the condenser.
⇓
Bring to a boil and distill slowly at first then at increased rate.
⇓
After the distillation has proceeded for approximately 1 hr, use an adapted buret brush to dislodge moisture droplets from the condenser and top part of the Bidwell-Sterling trap.
⇓
Slide the brush up the condenser to a point above the vapor condensing area.
⇓
Rinse the brush and wire with a small amount of toluene to dislodge adhering water drops.
⇓
If water has adhered to the walls of the calibrated tube, invert the brush and use the straight wire to dislodge this water so it collects in the bottom of the tube.
⇓
Return the wire to a point above the condensation point, and rinse with another small amount of toluene.
⇓
After no more water has distilled from the sample, repeat the brush and wire routine to dislodge adhering water droplets.
⇓
Rinse the brush and wire with toluene before removing from the condenser.
⇓
Allow the apparatus to cool to ambient temperatures before measuring the volume of water in the trap.
⇓
Volume of water × 2 (for a 50 g sample) = % moisture

FIGURE 7-5.
Procedure for reflux distillation with toluene using a Bidwell-Sterling trap. Steps to dislodge adhering moisture drops are given.

burette brush, with the handle end flattened so it will pass down the condenser, is needed to dislodge moisture droplets.

3. **Decomposition** of the sample with production of water. This is principally due to carbohydrate decomposition to generate water ($C_6H_{12}O_6 \rightarrow 6H_20 + 6C$). If this is a measurable problem, discontinue method use and find an alternative procedure.

7.4 CHEMICAL METHODS

7.4.1 Karl Fischer Titration

The **Karl Fischer titration** is particularly adaptable to food products that show erratic results when heated or submitted to a vacuum. This is the method of choice for determination of water in many low-moisture foods like dried fruits and vegetables (AOAC Method 967.19 E-G), candies, chocolate (AOAC Method 977.10), roasted coffee, oils and fats (AOAC Method 984.20), or any low-moisture food high in sugar or protein. The method is quite rapid and sensitive and uses no heat.

This method is based on the fundamental reaction described by Bunsen in 1853 (10) involving the reduction of iodine by SO_2 in the presence of water:

$$2H_2O + SO_2 + I_2 \rightarrow H_2SO_4 + 2HI \tag{5}$$

This was modified to include methanol and pyridine in a four-component system to dissolve the iodine and SO_2:

$$C_5H_5N \cdot I_2 + C_5H_5N \cdot SO_2 + C_5H_5N + H_2O \rightarrow 2C_5H_5N \cdot HI + C_5H_5N \cdot SO_3 \tag{6}$$

$$C_5H_5N \cdot SO_3 + CH_3OH \rightarrow C_5H_5N(H)SO_4 \cdot CH_3 \tag{7}$$

These reactions show that for each mol of water, 1 mol of iodine, 1 mol of SO_2, 3 mols of pyridine, and 1 mol of methanol are used. For general work, a methanolic solution is used that contains these components in the ratio of 1 iodine: 3 SO_2:10 pyridine, and at a concentration so that 3.5 mg water = 1 ml reagent. A procedure for standardizing this reagent is given below.

In a **volumetric titration** procedure (Fig. 7-6), iodine and SO_2 in the appropriate form are added to the sample in a closed chamber protected from atmospheric moisture. The excess of I_2 that cannot react with the water can be determined **visually.** The end point color is dark red-brown. Some instrumental systems are improved by the inclusion of a **potentiometer** to electronically determine the endpoint, which increases the sensitivity. Instruments are also available to automatically perform the Karl Fischer moisture analysis by the **conductometric method.**

The volumetric titration procedure described above is appropriate for samples with a moisture content greater ~0.03 percent. A second type of titration, referred to as **coulometric titration,** is ideal for products with very low levels of moisture, from 0.03 percent down to parts per million (ppm) levels. In this method, iodine is electrolytically generated to titrate the water. The amount of iodine required to titrate the water is determined by the current needed to generate the iodine.

In a Karl Fischer volumetric titration, the **Karl Fischer reagent** (KFR) is added directly as the titrant if the water in the sample is accessible. However, if water in a solid sample is inaccessible to the reagent, the moisture is extracted from the food with an appropriate solvent (e.g., methanol). Then the methanol extract is titrated with KFR.

The obnoxious odor of pyridine makes it an undesirable reagent. Therefore, researchers have experimented with other amines capable of dissolving iodine

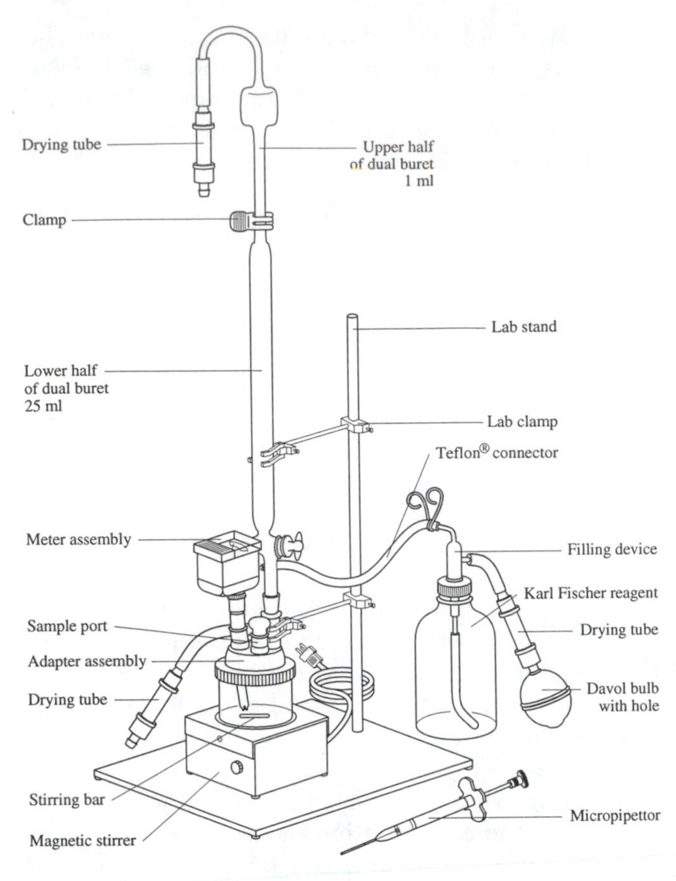

FIGURE 7-6.
Karl Fischer titration unit. Figure courtesy of Labindustries, Inc., Berkeley, CA.

and sulfur dioxide. Some aliphatic amines and several other heterocyclic compounds were found suitable. On the basis of these new amines, **one-component reagents** (solvent and titrant components together) and **two-component reagents** (solvent and titrant components separate) have been prepared. The one-component reagent may be more convenient to use, but the two-component reagent has greater storage stability.

Before the amount of water found in a food sample can be determined, a **KFR water equivalence** (KFReq) must be determined. The KFReq value represents the equivalent amount of water that reacts with 1 ml of KFR. Standardization must be checked before each use because the KFReq will change with time.

The KFReq can be established with **pure water,** a **water-in-methanol standard,** or **sodium tartrate dihydrate.** Pure water is a difficult standard to use because of inaccuracy in measuring the small amounts required. The water-in-methanol standard is premixed by the manufacturer and generally contains 1 milligram water per milliliter of solution. This standard can change over prolonged storage periods by absorbing atmospheric moisture. Sodium tartrate dihydrate ($Na_2C_4H_4O_6 \cdot 2H_2O$) is a primary standard for determining KFReq. This compound is very stable, contains 15.66 percent water under all conditions expected in the laboratory, and is the material of choice to use.

The KFReq is calculated as follows using sodium tartrate dihydrate:

$$\text{KFReq (mg H}_2\text{O/ml)} = \frac{36 \text{ g H}_2\text{O/mol Na}_2\text{C}_4\text{H}_4\text{O}_6 \cdot 2\text{H}_2\text{O} \times S \times 1000}{230.08 \text{ g/mol} \times A} \quad (8)$$

where: KFReq = Karl Fischer Reagent water equivalence
S = weight of sodium tartrate dihydrate (g)
A = ml of KFR required for titration of sodium tartrate dihydrate

Once the KFReq is known, the moisture content of the sample is determined as follows:

$$\%\text{H}_2\text{O} = \frac{\text{KFReq} \times \text{Ks}}{S} \times 100 \quad (9)$$

where: KFReq = Karl Fischer Reagent water equivalence
Ks = ml of KFR used to titrate sample
S = weight of sample (mg)

The major difficulties and sources of error in the Karl Fischer titration methods are:

1. **Incomplete water extraction**—For this reason, fineness of grind is very important in preparation of cereal grains and some foods.
2. **Atmospheric moisture**—External air must not be allowed to infiltrate the reaction chamber.
3. **Moisture adhering** to walls of unit—All glassware and utensils must be carefully dried.
4. **Interferences** from certain food constituents—**Ascorbic acid** is oxidized by KFR to dehydroascorbic acid to overestimate moisture content; **carbonyl compounds** react with methanol to form acetals and release water, to overestimate moisture content (this reaction may also result in fading endpoints); **unsaturated fatty acids** will react with iodine, so moisture content will be overestimated.

7.4.2 Gas Production Procedure

The **calcium carbide method** is the only gas production procedure reported here. It is a simple but rapid method that requires no elaborate apparatus. In fact, many people might recall the nonanalytical form of this chemical reaction being used for Fourth of July celebrations where gas-generating combustion cannons were involved.

$$CaC_2 + 2H_2O \rightarrow Ca(OH)_2 + C_2H_2$$

In this carbide and water reaction, one only needs to accurately collect the generated gas, acetylene. The amount of acetylene may be determined in one of three ways:

1. Measure the volume liberated using an inverted graduated cylinder filled with water and a tubing connection.
2. Measure the loss in weight of a mixture after treatment.
3. Determine the pressure developed in a closed system after the reaction is completed (a calibration table is needed).

7.5 PHYSICAL METHODS

7.5.1 Electrical Methods

7.5.1.1 Dielectric Method

Water content of certain foods can be determined by measuring the change in **capacitance** or **resistance to an electric current** passed through a sample. These instruments require calibration against samples of known moisture content as determined by standard methods. Sample density or weight/volume relationships and sample temperature are important factors to control in making reliable and repeatable measurements by dielectric methods. These techniques can be very useful for process control measurement applications, where continuous measurement is required. These methods are limited to food systems that contain no more than 30–35 percent moisture.

The moisture determination in dielectric-type meters is based on the fact that the dielectric constant of water (80.37 at 20°C) is higher than that of most solvents. The **dielectric constant** is measured as an index of **capacitance.** As an example, the dielectric method is used widely for cereal grains. Its use is based on the fact that water has a dielectric constant of 80.37, whereas starches and proteins found in cereals have dielectric constants of 10. By determining this properly on samples in standard metal condensers, dial readings may be obtained and the percentage of water determined from a previously constructed standard curve for a particular cereal grain.

7.5.1.2 Conductivity Method

The **conductivity method** functions because the conductivity of an electric current increases with the percentage of water in the sample. A modestly accurate and rapid method is created when one measures resistance. Ohm's law states that the strength of an electricity current is equal to the electromotive force divided by the resistance. The electrical resistance of wheat with 13 percent water is seven times as great as that with 14 percent water and 50 times that with 15 percent water. Temperature must be kept constant, and 1 min is necessary for a single determination.

7.5.2 Hydrometry

Hydrometry is the science of measuring **specific gravity** or **density,** which can be done using several different principles and instruments. While hydrometry is considered archaic in some analytical circles, it is still widely used and, with proper technique, is highly accurate. Specific gravity measurements with a **pycnometer,** various types of **hydrometers,** or a **Westphal balance** are commonly used for routine testing of moisture (or solids) content of numerous food products. These include beverages, salt brines, and sugar solutions.

7.5.2.1 Pycnometer

One approach to measuring specific gravity is a comparison of the weights of equal volumes of a liquid and water in standardized glassware, a **pycnometer.** This will yield density of the liquid compared to water. In some texts and reference books, *20/20* is given after the specific gravity number. This indicates that the temperature of both fluids was 20°C when the weights were measured. Using a clean, dry pycnometer at 20°C, the analyst weighs it empty, fills it to the full point with distilled water at 20°C, inserts the thermometer to seal the fill opening, and then touches off the last drops of water and puts on the cap for the overflow tube. The pycnometer is wiped dry in case of any spillage from filling and is reweighed. The density of the sample is calculated as follows:

$$\frac{\text{weight of sample-filled pycnometer} - \text{weight of empty pycnometer}}{\text{weight of water-filled pycnometer} - \text{weight of empty pycnometer}} = \text{density of sample} \quad (10)$$

This method is used for determining alcohol content in alcoholic beverages (e.g., distilled liquor, AOAC

Method 930.17), solids in sugar syrups (AOAC Method 932.14B), and solids in milk (AOAC Method 925.22).

7.5.2.2 Hydrometer

A second approach to measuring specific gravity is based on **Archimedes' principle,** which states that a solid suspended in a liquid will be buoyed by a force equal to the weight of the liquid displaced. The weight per unit volume of a liquid is determined by measuring the volume displaced by an object of standard weight. A **hydrometer** is a standard weight on the end of a spindle, and it displaces a weight of liquid equal to its own weight. For example, in a liquid of low density, the hydrometer will sink to a greater depth, whereas in a liquid of high density, the hydrometer will not sink as far. Hydrometers are available in narrow and wide ranges of specific gravity. The spindle of the hydrometer is calibrated to read specific gravity directly at 15.5°C or 20°C. A hydrometer is not as accurate as a pycnometer, but the speed with which you can do an analysis is a decisive factor.

The rudimentary but surprisingly accurate hydrometer comes equipped with various modifications depending on the fluid to be measured:

1. The Quevenne and New York Board of Health **lactometer** is used to determine the density of milk. The Quevenne lactometer reads from 15 to 40° and corresponds to 1.015 to 1.040. For every degree above 60°F, 0.1 is added to the reading, and 0.1 is subtracted for every degree below 60°F.
2. The **Baumé hydrometer** was used originally to determine the density of salt solutions (originally 10% salt), but it has come into much wider use. From the value obtained in the Baumé scale, you can convert to specific gravity of liquids heavier than water. For example, it is used to determine the specific gravity of milk being condensed in a vacuum pan.
3. The **Brix hydrometer** is used for sugar solutions, and one usually reads directly the percentage of sucrose at 20°C.
4. **Alcoholometers** are used to estimate the alcohol content of beverages. Such hydrometers are calibrated in 0.1 or 0.2° proof to determine percentage of alcohol in distilled liquors (AOAC Method 957.03).
5. The **Twaddell hydrometer** is only for liquids heavier than water.

7.5.2.3 Westphal Balance

The **Westphal balance** functions on Archimedes' principle such that the plummet on the balance will be buoyed by the weight of liquid equal to the volume displaced. This is more accurate than a hydrometer but less accurate than a pycnometer. It provides measurements to four decimal places. The balance has a plummet that displaces exactly 5 g of water at 15.5°C. If the specific gravity is 1, as would be the case with water at 15.5°C, a gravity weight hung at the 10 mark would bring this device into balance.

The specific gravity measurement of solid objects is made as described below, with the determination of frozen pea maturity given as the example:

1. Weigh peas in air.
2. Immerse peas in solvent.
3. Obtain weight in this solvent.

$$\text{Specific gravity} = \frac{\begin{array}{c}\text{weight in air}\\ \times \text{ specific gravity}\\ \text{of liquid}\end{array}}{\begin{array}{c}\text{weight in liquid}\\ - \text{ weight in air}\end{array}} \qquad (11)$$

The difference between the weight in air and the weight in liquid equals the weight of a volume of the liquid, which equals the volume of peas. Industry grade standards may be based on specific gravity values (Scott Rambo, personal communication, Dean Foods, Rockford, IL).

Suggested standards for frozen peas:

Fancy, 1.072 and lower
Standard, 1.073–1.084
Substandard, 1.085 and higher

Whole kernel corn can be assayed similarly with the following specific gravity standards:

Fancy, 1.080–1.118
Reject immature, 1.079 and lower
Reject overmature, 1.119 and higher

7.5.2.4 Golding Beads

Occasionally an innovative technology appears that makes some determinations somewhat easier. Golding (11) developed such a technique using a sample jar with plastic beads of different colors and known densities. Milk was added to the beads in the jar, warmed to 40°C for 5 min, cooled to 20°C to temper milkfat from a liquid state, and then allowed to stand several minutes. The number of beads on the bottom and floating should equal 10, although one bead occasionally will equal the exact specific gravity of the milk and remain suspended. The beads are standardized so that they will just sink in milk at their respective densities. Milk solids-not-fat (SNF) can be calculated as follows:

$$\text{SNF} = 9.133 - (0.279 \times \#\text{ beads down}) + (0.307 \times \%\text{ fat}) \quad (12)$$

7.5.3 Refractometry

Moisture in liquid sugar products and condensed milks can be determined using a Baumé hydrometer (solids), a Brix hydrometer (sugar content), gravimetric means, or a **refractometer.** If it is performed correctly and no crystalline solids are evident, the refractometer procedure is rapid and surprisingly accurate (AOAC Method 9.32.14C, for solids in syrups). The refractometer has been valuable in determining the soluble solids in fruits and fruit products (AOAC Method 932.12; 976.20; 983.17). The refractive index of an oil, syrup, or other liquid is a dimensionless constant that can be used to describe the nature of the food. While some refractometers are designed only to provide results as refractive indices, others, particularly hand-held, quick-to-use units, are equipped with scales calibrated to read percent solids, percent sugars, and the like, depending on the products for which they are intended. Tables are provided with the instruments to convert values and adjust for temperature differences.

When a beam of light is passed from one medium to another and the density of the two differs, then the beam of light is bent. Bending of the light beam is a function of the media and the sines of the angles of incidence and refraction at any given temperature and pressure, and is thus a constant. The **refractive index** (η) is a ratio of the sines of the angles:

$$\eta = \frac{\text{sine incident ray angle}}{\text{sine refracted ray angle}} \quad (13)$$

Instruments are designed to give a reading by passing a light beam through a glass prism into a liquid, the sample. Monochromatic light must be used since refractive index varies with wavelength. Bench-top or hand-held units use **Amici prisms** to obtain the **D line of the sodium spectrum** or 589 nm from white light. Whenever refractive indices of standard fluids are given, these are prefaced with η_D^{20} = a value from 1.3000 to 1.7000. The Greek letter η is the symbol for refractive index, the 20 refers to temperature in °C, and D is the wavelength of the light beam, the D line of the sodium spectrum.

Bench-top instruments are more accurate compared to hand-held units mainly because of temperature control. These former units have provisions for water circulation through the head where the prism and sample meet. **Abbe refractometers** are the most popular for laboratory use. Care must be taken when cleaning the prism surface following use. Wipe the contact surface clean with lens paper and rinse with distilled water and then ethanol. Close the prism chamber and cover the instrument with a bag when not in use to protect the delicate prism surface from dust or other debris that might lead to scratches and inaccuracy.

7.5.4 Infrared Milk Analysis

Infrared spectroscopy and the use of mid-infrared milk analyzers to determine total solids in milk (AOAC Method 972.16) are covered in Chapter 24 of this text. The midrange spectroscopic method does not yield moisture or solids results except by computer calculation because these instruments do not monitor at wavelengths where water absorbs. The instrument must be calibrated using a minimum of eight milk samples that were previously analyzed for fat (F), protein (P), lactose (L), and total solids (TS) by standard methods. Then, a mean difference value, *a*, is calculated for all samples used in calibration:

$$a = \Sigma(\text{TS} - \text{F} - \text{P} - \text{L})/n \quad (14)$$

where: a = solids not measurable by the F, P, and L methods
n = number of samples
F = fat percentage
P = protein percentage
L = lactose percentage
TS = total solids percentage

Total solids can then be determined from any infrared milk analyzer results by using the formula

$$\text{TS} = a + \text{F} + \text{P} + \text{L} \quad (15)$$

The *a* value is thus a standard value mathematically derived. Newer instruments have the algorithm in their computer software to ascertain this value automatically.

7.5.5 Freezing Point

When water is added to a food product, many of the physical constants are altered. Some properties of solutions depend on the number of solute particles as ions or molecules present. These properties are vapor pressure, freezing point, boiling point, and osmotic pressure. Measurement of any of these properties can be used to determine the concentration of solutes in a solution. However, the most commonly practiced assay of this variety for milk is the change of the **freezing point** value. It has economic importance both from the raw and pasteurized milk viewpoints. The freezing point of milk is its most constant physical property. The secretory process of the mammary gland is such that the osmotic pressure is kept in equilibrium with blood and milk. Thus, with any decrease in the synthesis of lactose, there is a compensating increase in the concentration of Na^+ and Cl^-. While termed a physical constant, the freezing point varies within narrow limits, and the vast majority of samples from individual cows fall between −0.525 and −0.565°H (temperature in °H or Hortvet) (−0.503°C and −0.541°C). The average is very close to −0.540°H (−0.517°C). Herd or bulk milk will exhibit a narrower range unless the supply was watered intentionally or accidentally. **Hortvet** is the surname of the inventor of the first freezing point apparatus used for many years before automated equipment forced its obsolescence. All values today are given in °C by agreement. The following is used to convert °H to °C, or °C to °H (5, 6):

$$°C = 0.9623\ °H - 0.0024 \tag{16}$$

$$°H = 1.03916\ °C + 0.0025 \tag{17}$$

The principle utility of freezing point is to measure for **added water.** However, the freezing point of milk can be alternated by mastitis infection in cows and souring of milk. In special cases, nutrition and environment of the cow, stage of lactation, and processing operations for the milk can affect the freezing point. If the solute remains constant in weight and composition, the change of the freezing point varies inversely with the amount of solvent present. Therefore, we can calculate the percent H_2O added:

$$\%H_2O \text{ added} = \frac{0.540 - T}{0.540} \times 100 \tag{18}$$

where: 0.540 = freezing point in °H of all milk entering a plant
T = freezing point in °H of a sample

The AOAC cryoscopic method for water added to milk (AOAC Method of 961.07) assumes a freezing point for normal milk of −0.550°H (−0.527°C). The Food and Drug Administration will reject all milk with freezing points above −0.525°H (−0.503°C). Since the difference between the freezing points of milk and water is slight and since the freezing point is used to calculate the amount of water added, it is essential that the method be precise as possible. The thermister used can sense temperature change to 0.001°H (0.001°C). The general technique is to supercool the solution and then induce crystallization by a vibrating reed. The temperature will rise rapidly to the freezing point or eutectic temperature as the water freezes. In the case of pure water, the temperature remains constant until all the water is frozen. With milk, the temperature is read when there is no further temperature rise.

Instrumentation available is the old Hortvet cryoscope and the new Advanced Instruments (Fig. 7-7) and Fiske cryoscopes. Time required for the automated instruments is 1–2 min per sample using a prechilled sample.

7.6 WATER ACTIVITY

Water content alone is not a reliable indicator of food stability, since it has been observed that foods with the same water content differ in their perishability (12). This is at least partly due to differences in the way that water associates with other constituents in a food. Water tightly associated with other food constituents is less available for microbial growth and chemical reactions to cause decomposition. **Water activity** (a_w) is a better indication of food perishability than is water content. However, it is also an important quality factor for organoleptic properties such as hard/soft, crunchy/chewy, and the like.

Water activity is defined as follows:

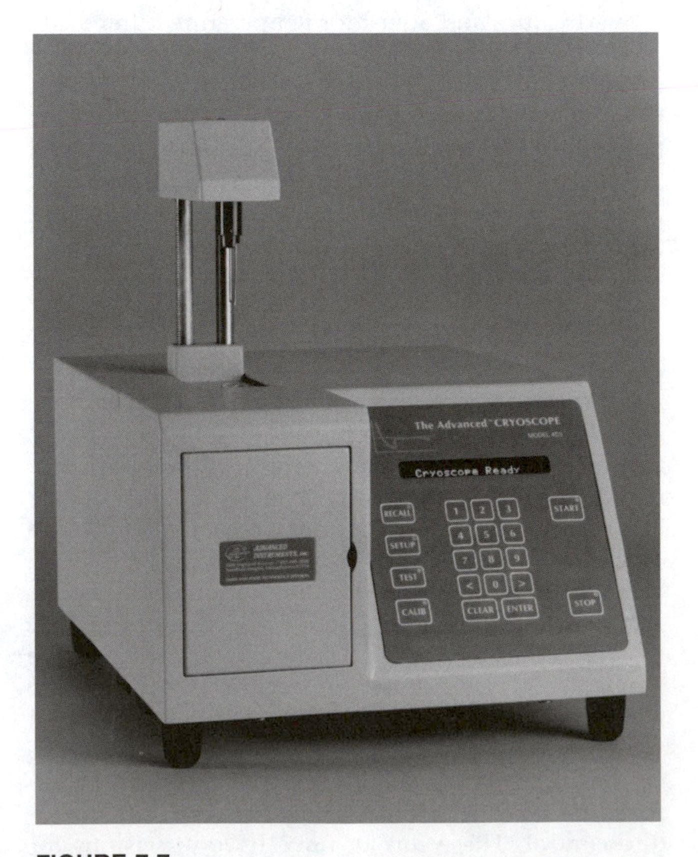

FIGURE 7-7.
A model 4D3 Advanced Instruments cryoscope for freezing point determination in milk. Figure courtesy of Advanced Instruments, Inc., Norwood, MA.

$$a_w = \frac{P}{P_o} \quad (19)$$

$$a_w = \frac{ERH}{100} \quad (20)$$

where: a_w = water activity
P = partial pressure of water above the sample
P_o = vapor pressure of pure water at the same temperature (specified)
ERH = equilibrium relative humidity (%) surrounding the product

There are various techniques to measure a_w. A commonly used approach relies on measuring the amount of moisture in the equilibrated headspace above a sample of the food product, which correlates directly with sample a_w. A sample for such analysis is placed in a small closed chamber at constant temperature, and a relative humidity sensor is used to measure the ERH of the sample atmosphere after equilibration. A simple and accurate variation of this approach is the chilled mirror technique, in which the water vapor in the headspace condenses on the surface of a mirror that is cooled in a controlled manner. The dew point is determined by the temperature at which condensation takes place, and this determines the relative humidity in the headspace. Two other general approaches to measuring a_w are (1) using the sample freezing point depression and moisture content to calculate a_w and (2) equilibrating a sample in a chamber held at constant relative humidity (by means of a saturated salt solution) and then using the water content of the sample to calculate a_w (12).

7.7 COMPARISON OF METHODS

7.7.1 Principles

Oven drying methods involve the removal of moisture from the sample and then a weight determination of the solids remaining to calculate the moisture content. Non-water volatiles can be lost during drying, but their loss is generally a negligible percentage of the amount of water lost. Distillation procedures also involve a separation of the water from the solids, but the water is quantitated directly by volume. Karl Fischer titration and the gas production method are both based on chemical reactions of the water present, reflected as either the amount of titrant used (Karl Fischer) or the amount of acetylene generated (gas production).

Dielectric and conductivity methods are based on electrical properties of water. Hydrometric methods are based on the relationship between specific gravity and moisture content. The refractive index method is based on how water in a sample affects the refraction of light. Near infrared analysis of water in foods is based on measuring the absorption at wavelengths characteristic of the molecular vibration in water. Freezing point is a physical property of milk that is changed by a change in solute concentration.

7.7.2 Nature of Sample

While most foods will tolerate oven drying at high temperatures, some foods contain volatiles that are lost at such temperatures. Some foods have constituents that undergo chemical reactions at high temperatures to generate or utilize water or other compounds, to affect the calculated moisture content. Vacuum oven drying at reduced temperatures may overcome such problems for some foods. However, a distillation technique is necessary for some food to minimize volatilization and decomposition. For foods very low in moisture and/or high in fats and sugars, Karl Fischer titration is often the method of choice. The use of a pycnometer, hydrometer, and refractometer requires liquid samples, ideally with limited constituents.

7.7.3 Intended Purposes

Moisture analysis data may be needed quickly for quality control purposes, and high accuracy may not be necessary. Of the oven drying methods, microwave and infrared drying are fastest. Some forced draft oven procedures require less than 1 hr drying, but most forced draft oven and vacuum oven procedures require much longer. The electrical, hydrometric, refractive index, and near infrared analyses methods are very rapid but often require correlation to less empirical methods. Oven drying procedures are official methods for a variety of food products. Reflux distillation is an AOAC method for chocolate, dried vegetables, and oils and fats. Such official methods are used for regulatory and nutritional labeling purposes.

7.8 SUMMARY

The moisture content of foods is important to food processors and consumers for a variety of reasons. While moisture determination may seem simplistic, it is often one of the most difficult assays in obtaining accurate and precise results. The free water present in food is generally more easily quantitated as compared to the adsorbed water and the water of hydration. Some moisture analysis methods involve a separation of water in the sample from the solids and then quantitation by weight or volume. Other methods do not involve such a separation but instead are based on some physical and/or chemical property of the water in the sample. A major difficulty with many methods is attempting to remove or otherwise quantitate all water present. This is often complicated by decomposition or interference by other food constituents. For each moisture analysis method, there are factors that must be controlled or precautions that must be taken to ensure accurate and precise results. Careful sample collection and handling procedures are extremely important and cannot be overemphasized. The choice of moisture analysis method is often determined by the expected moisture content, nature of other food constituents (e.g., highly volatile, heat sensitive), equipment available, speed necessary, accuracy and precision required, and intended purpose (e.g., regulatory or in-plant quality control).

7.9 STUDY QUESTIONS

1. Identify five factors that one would need to consider when choosing a moisture analysis method for a specific food product.
2. Why is standardized methodology needed for moisture determinations?
3. What are the potential advantages of using a vacuum oven rather than a forced draft oven for moisture content determination?
4. In each case specified below, would you likely overestimate or underestimate the moisture content of a food product being tested? Explain your answer.
   a. hot air oven
      - particle size too large
      - high concentration of volatile flavor compounds present
      - lipid oxidation
      - sample very hygroscopic
      - alteration of carbohydrates (e.g., Maillard browning)
      - sucrose hydrolysis
      - surface crust formation
      - splattering
      - desiccator with dried sample not sealed properly
   b. toluene distillation
      - emulsion between water in sample and solvent not broken
      - water clinging to condenser
   c. Karl Fischer
      - very humid day when weighing original samples
      - glassware not dry
      - sample ground coarsely
      - food high in Vitamin C
      - food high in unsaturated fatty acids
5. The procedure for an analysis for moisture in a liquid food product requires the addition of 1–2 ml of deionized water to the weighed sample in the moisture pan. Why should you add moisture to an analysis in which moisture is being determined?
6. A new instrument based on infrared principles has been received in your laboratory to be used in moisture analysis. Briefly describe the way you would ascertain if the new instrument would meet your satisfaction and company standards.
7. A technician you supervise is to determine the moisture content of a food product by the Karl Fischer method. Your technician wants to know what is this "Karl Fischer Reagent Water Equivalence" that is used in the equation to calculate percentage of water in the sample, why is it necessary, and how is it determined. Give the technician your answer.
8. You are fortunate to have available in your laboratory the equipment for doing moisture analysis by essentially all methods—both official and rapid quality control methods. For each of the food products listed below (with the purpose specified as rapid quality control or official), indicate (a) the name of the method you would use, (b) the principle (not procedure) for the method, (c) a justification for use of that method (as compared to using a hot air drying oven), and (d) two cautions in use of the method to ensure accurate results.
   a. ice cream mix (liquid)—quality control
   b. milk chocolate—official
   c. spices—official
   d. syrup for canned peaches—quality control
   e. oat flour—quality control

7.10 PRACTICE PROBLEMS

1. As an analyst, you are given a sample of condensed soup to analyze to determine if it is reduced to the correct concentration. By gravimetric means, you find that the concentration is 26.54 percent solids. The company standard reads 28.63 percent. If the starting volume were 1000 gallons at 8.67 percent solids, and the weight is 8.5 pounds per gallon, how much more water must be removed?
2. Your laboratory just received several sample containers of peas to analyze for moisture content. There is a visible condensate on the inside of the container. What is your procedure to obtain a result?
3. You have the following gravimetric results: weight of dried pan and glass disc = 1.0376 grams, weight of pan and liquid sample = 4.6274 grams, and weight of the pan and dried sample = 1.7321 grams. What was the moisture content of the sample and what is the percent solids?

Answers

1. The weight of the soup initially is superfluous information. By condensing the soup to 26.54 percent solids from 8.67 percent solids, the volume is reduced to 326.7 gallons [(8.67/26.54) $\times$ 1000]. You need to reduce the volume further to obtain 28.63 percent solids [(8.67/28.63) $\times$ 1000], or 302.8 gallons. The difference in the gallons obtained is 23.9, or the volume of water that must be removed from the partially condensed soup to comply with company standards (326.7–302.8).

2. This problem focuses on a real issue in the food processing industry—when do you analyze a sample and when don't you? It would appear that the peas have lost water that should be within the vegetable for correct results. You will need to grind the peas in a food mill or blender. If the peas are in a Mason jar or one that fits an Oster blender head, no transfer is needed. Blend the peas to a creamy texture. If a container transfer was made, then put the blended peas back into original container. Mix with the residual moisture to a uniform blend. Collect a sample for moisture analysis. You should note on the report form containing the results of the analysis that the pea samples had free moisture on container walls when they arrived.
3. Note eqs. 2–4 in section 7.2.1.7. To use any of the equations, you must subtract the weight of the dried pan and glass disc. Then you obtain 3.5898 g of original sample and 0.6945 g when dried. By subtracting these results, you have water removed or 2.8953 g. Then (0.6945/3.5898) × 100 = 19.35 percent solids and (2.8953/3.5898) × 100 = 80.65 percent water.

7.11 REFERENCES

1. Pomeranz, Y., and Meloan, C. 1987. *Food Analysis: Theory and Practice,* 2nd ed. Van Nostrand Reinhold, New York, NY.
2. Aurand, L.W., Woods, A.E., and Wells, M.R. 1987. *Food Composition and Analysis*. Van Nostrand Reinhold, New York, NY.
3. Josyln, M.A. 1970. *Methods in Food Analysis,* 2nd ed. Academic Press, New York, NY.
4. USDA. 1975. *Nutritive Value of American Foods.* Agricultural Handbook No. 456. Agricultural Research Service, United States Dept. of Agriculture, Washington, DC.
5. AOAC 1990. *Official Methods of Analysis,* 15th ed. Association of Official Analytical Chemists, Washington, DC.
6. Marshall, R.T. (Ed.) 1992. *Standard Methods for the Examination of Dairy Products,* 16th ed. Amer. Public Health Assoc., Washington, DC.
7. AACC. 1983. *Approved Methods of Analysis,* 8th ed. American Association of Cereal Chemists, St. Paul, MN.
8. Vanderwarn, M.A. 1989. Analysis of cheese and cheese products for moisture. M.S. Thesis, Univ. of Wisconsin–Madison.
9. Nelson, O.A., and Hulett, G.A. 1920. The moisture content of cereals. *J. Ind. Eng. Chem.* 12:40–45.
10. Mitchell, J., Jr., and Smith, D.M. 1948. *Aquametry.* John Wiley & Sons, New York, NY.
11. Golding, N.S. 1964. A new method for estimating total solids and solids-not-fat in milk and liquid milk products by hydrometric means. *Washington Agr. Expt. Sta. Bull.,* 650:1–24.
12. Fennema, O.R. 1985. Water and Ice. Ch. 2, in *Food Chemistry.* O.R. Fennema (Ed.), Marcel Dekker, Inc., New York, NY.

CHAPTER

8

Ash Analysis

Leniel H. Harbers

8.1 INTRODUCTION

Ash refers to the inorganic residue remaining after either ignition or complete oxidation of organic matter in a foodstuff. A basic knowledge of the characteristics of various ashing procedures and types of equipment is essential to ensure reliable results. Three major types of ashing are available: dry ashing for the majority of samples, wet ashing (oxidation) for samples with high fat content (meats and meat products) as a preparation for elemental analysis, and low-temperature plasma dry ashing (also called simply plasma ashing or low-temperature ashing) for preparation of samples when volatile elemental analyses are conducted. Most dry samples (i.e., whole grain, cereals, dried vegetables) need no preparation, while fresh vegetables need to be dried prior to ashing. High-fat products such as meats may need to be dried and fat extracted before ashing. Fruits and vegetables may be subjected to additional ashing procedures such as soluble ash in water and alkalinity of ash. The ash content of foods can be expressed on either a wet weight (as is) or on a dry weight basis.

8.1.1 Definitions

Dry ashing refers to the use of a muffle furnace capable of maintaining temperatures of 500 to 600°C. Water and volatiles are vaporized and organic substances are burned in the presence of oxygen in air to CO_2 and oxides of N_2. Most minerals are converted to oxides, sulfates, phosphates, chlorides, and silicates. Elements such as Fe, Se, Pb, and Hg may partially volatilize with this procedure, so other methods must be used if ashing is a preliminary step for specific elemental analysis.

Wet ashing is a procedure for oxidizing organic substances by using acids and oxidizing agents or their combinations. Minerals are solubilized without volatilization. Wet ashing is often preferable to dry ashing as a preparation for specific elemental analysis. Nitric and perchloric acids are preferable, but a special perchloric acid hood is essential. This procedure *must be* conducted in a perchloric acid hood and caution must be taken when fatty foods are used.

Low-temperature plasma ashing refers to a specific type of dry ashing method whereby foods are oxidized in a partial vacuum by nascent oxygen formed by an electromagnetic field. Ashing occurs at a much lower temperature than with a muffle furnace, preventing volatilization of most elements. The crystalline structures usually remain intact.

Acid insoluble ash generally refers to insoluble mineral contaminants in foods. Soil minerals (largely silicates and opaline silica) soluble only in HBr or HF comprise the major portion of this ash.

Alkalinity of ash is a useful measurement to determine the acid-base balance of foods and to detect adulteration of foods with minerals.

8.1.2 Importance of Ash in Food Analysis

Ash content represents the total mineral content in foods. Determining the ash content may be important for several reasons. It is a part of the proximate analysis for nutritional evaluation. Ashing is the first step in the preparation of a food sample for specific elemental analysis. Because certain foods are high in particular minerals, ash content becomes important. We can usually expect a constant elemental content from the ash of animal products, but that from plant sources is variable.

8.1.3 Ash Contents in Foods

The average ash content of the various food groups is given in Table 8-1. The ash content of most fresh foods rarely is greater than 5 percent. Pure oils and fats generally contain little or no ash, products such as cured

TABLE 8-1. Ash Content of Selected Foods

Food	*Percent Ash (wet weight basis)*
Milk and Dairy Products	
Butter	2.5
Cream	2.9
Evaporated milk	1.6
Margarine	2.5
Milk	0.7
Yogurt	0.8
Meat, Poultry, and Fish	
Eggs	1.0
Fish fillet	1.3
Ham, fresh	0.8
Hamburger, cooked	1.1
Poultry	1.0
Roast beef	3.0
Fruits and Vegetables	
Apples	0.3
Bananas	0.8
Cherries	0.5
Dried fruits	2.3
Potatoes	1.0
Tomatoes	0.6
Cereals	
Brown rice	1.0
Corn meal	1.3
Hominy	0.4
White rice	0.7
Whole wheat flour	1.7

From Wooster, H.A. 1956. *Nutritional Data*, 3rd ed. H.J. Heinz Co., Pittsburg, PA. Used with permission.

bacon may contain 6 percent ash, and dried beef may be as high as 11.6 percent (wet weight basis).

Fats, oils, and shortenings vary from 0.0 to 4.09 percent ash, while dairy products vary from 0.5 to 5.1 percent. Fruits, fruit juice, and melons contain 0.2 to 0.6 percent ash, while dried fruits are higher (2.4 to 3.5 percent). Flours and meals vary from 0.3 to 1.4 percent ash. Pure starch contains 0.3 percent and wheat germ 4.3 percent ash. It would be expected that grain and grain products with bran would tend to be higher in ash content than such products without bran. Nuts and nut products contain 0.8 to 3.4 percent ash, while meat, poultry, and seafoods contain 0.7 to 1.3 percent ash.

8.2 METHODS

Principles, materials, instrumentation, general procedures, and applications are described below for various ash determination methods. Refer to methods cited for detailed instructions of the procedures.

8.2.1 Sample Preparation

It cannot be overemphasized that the small sample used for ash, or other determinations, needs to be very carefully chosen so that it represents the original materials. A 2–10 g sample is generally used for ash determination. For that purpose, milling, grinding, and the like probably will not alter the ash content much; however, if this ash is a preparatory step for specific mineral analyses, contamination by microelements is of potential concern. Remember, most grinders and mincers are of steel construction. Repeated use of glassware can be a source of contaminants as well. The water source used in dilutions may also contain contaminants of some microelements. Distilled-deionized water should be used.

8.2.1.1 Plant Materials

Plant materials are generally dried by routine methods prior to grinding. The temperature of drying is of little consequence for ashing. However, the sample may be used for multiple determinations—protein, fiber, and so on—which require consideration of temperature for drying. Fresh stem and leaf tissue probably should be dried in two stages (i.e., first at a lower temperature of 55°C, then a higher temperature) especially to prevent artifact lignin. Plant material with 15 percent or less moisture may be ashed without prior drying.

8.2.1.2 Fat and Sugar Products

Animal products, syrups, and spices require treatments prior to ashing because of high fat and moisture (spattering, swelling) or high sugar content (foaming) that may result in loss of sample.

Meats, sugars, and syrups need to be evaporated to dryness on a steam bath or with an infrared (IR) lamp. One or two drops of olive oil (which contains no ash) are added to allow steam to escape as a crust is formed on the product.

Smoking and burning may occur upon ashing for some products (e.g., cheese, seafood, spices). Allow this to finish slowly by keeping the muffle door open prior to the normal procedure. Ashing of the same sample may follow drying and fat extraction. In most cases, mineral loss is minimal during drying and fat extraction. Under no circumstances should fat-extracted samples be heated until all the ether has been evaporated.

8.2.2 Dry Ashing

8.2.2.1 Principles and Instrumentation

Dry ashing is incineration at high temperature (525°C or higher). Incineration is accomplished with a muffle furnace. Several models of muffle furnaces are available, ranging from large-capacity units requiring either 208 or 240 voltage supplies to small bench-top units utilizing 110 volt outlets. A microwave muffle furnace is available that uses quartz fiber crucibles. Ashing time is drastically reduced with microwaving; however, microwave ovens are generally of small capacity.

Crucible selection becomes critical in ashing because type depends upon the specific use. **Quartz crucibles** are resistant to acids and halogens, but not alkali, at high temperatures. **Vycor® brand crucibles** are stable to 900°C, but **Pyrex® Gooch crucibles** are limited to 500°C. Ashing at a lower temperature of 500–525°C may result in slightly higher ash values because of less decomposition of carbonates and loss of volatile salts. **Porcelain crucibles** resemble quartz crucibles in their properties but will crack with rapid temperatures changes. Porcelain crucibles are relatively inexpensive and usually the crucible of choice. **Steel crucibles** are resistant to both acids and alkalies and are inexpensive, but they are composed of chromium and nickel, which are possible sources of contamination. **Platinum crucibles** are very inert and are probably the best crucibles but they are currently far too expensive for routine use for large numbers of samples.

All crucibles should be **marked for identification.** Marks on crucibles with a felt-tip marking pen will disappear during ashing in a muffle furnace. Laboratory inks scribed with a steel pin are available commercially. Crucibles may also be etched with a diamond point and marked with a 0.5M solution of $FeCl_2$ in 20 percent HCl. An iron nail dissolved in concentrated HCl forms a brown goo that is a satisfactory marker. The crucibles should be fired and cleaned prior to use.

The *advantages* of conventional dry ashing are that it is a safe method, it requires no added reagents or blank subtraction, and little attention is needed once ignition begins. Usually a large number of crucibles can be handled at once, and the resultant ash can be used for such other analyses as most individual elements, acid insoluble ash, and water soluble and insoluble ash. The *disadvantages* are the length of time required (12–18 hrs, or overnight) and expensive equipment. There will be a loss of the volatile elements and interactions between mineral components and crucibles. Volatile elements at risk of being lost include As, B, Cd, Cr, Cu, Fe, Pb, Hg, Ni, P, V, and Zn.

8.2.2.2 Procedures

The Association of Official Analytical Chemists (AOAC) International has several dry ashing procedures (e.g., AOAC Methods 900.02 A or B, 920.117, 923.03) for certain individual foodstuffs.

The general procedure includes the following steps:

1. Weigh a 5–10 g sample into a tared crucible. Pre-dry if the sample is very moist.
2. Place crucibles in cool muffle furnace. Use tongs, gloves, and protective eyeware if the muffle furnace is warm.
3. Ignite 12–18 hrs (or overnight) at about 550°C.
4. Turn off muffle furnace and wait to open it until the temperature has dropped to at least 250°C, preferably lower. Open door carefully to avoid losing ash that may be fluffy.
5. Using safety tongs, quickly transfer crucibles to a desiccator with a porcelain plate and desiccant. Cover crucibles, close desiccator, and allow crucibles to cool prior to weighing.

Note: Warm crucibles will heat air within the desiccator. With hot samples, a cover may bump to allow air to escape. A vacuum may form on cooling. At the end of the cooling period, the desiccator cover should be removed gradually by sliding to one side to prevent a sudden inrush of air. Covers with a ground glass sleeve or fitted for a rubber stopper allow for slow release of a vacuum.

The ash content is calculated as follows:

$$\begin{array}{c}\%\ \text{ash} \\ \text{(dry basis)}\end{array} = \frac{\begin{array}{c}\text{wt after ashing} \\ -\ \text{tare wt of crucible}\end{array}}{\begin{array}{c}\text{original sample wt} \\ \times\ \text{dry matter coefficient}\end{array}} \qquad (1)$$

where: dry matter coefficient = % solids/100

For example, if corn meal is 87 percent dry matter, the dry matter coefficient would be 0.87. If ash is calculated on an as-received or wet-weight basis (includes moisture), delete the dry matter coefficient from the denominator. If moisture was determined in the same crucible prior to ashing, the denominator becomes (dry sample wt − tared crucible wt).

8.2.2.3 Special Applications

Some of the AOAC procedures recommend steps in addition to those listed above. If carbon is still present following the initial incineration, several drops of H_2O or HNO_3 should be added; then the sample should be re-ashed. If the carbon persists, such as with high-sugar samples, follow this procedure:

1. Suspend the ash in water.
2. Filter through ashless filter paper because this residue tends to form a glaze.
3. Dry the filtrate.
4. Place paper and dried filtrate in muffle furnace and re-ash.

Other suggestions that may be helpful and accelerate incineration:

1. High-fat samples should be extracted either by using the crude fat determination procedure or by burning off prior to closing the muffle furnace. Pork fat, for example, can form a combustible mixture inside the furnace and burn with the admission of oxygen if the door is opened.
2. Glycerin, alcohol, and hydrogen will accelerate ashing.
3. Samples such as jellies will spatter and can be mixed with cotton wool.
4. Salt-rich foods may require a separate ashing of water-insoluble components and salt-rich water extract. Use a crucible cover to prevent spattering.
5. An alcoholic solution of magnesium acetate can be added to accelerate ashing of cereals. An appropriate blank determination is necessary.

8.2.3 Wet Ashing

8.2.3.1 Principle, Materials, and Applications

Wet ashing is sometimes called **wet oxidation** or **wet digestion.** Its primary use is preparation for specific mineral analysis and metallic poisons.

There are several *advantages* to using the wet ashing procedure. Minerals will usually stay in solution, and there is little or no loss from volatilization because of the lower temperature. The oxidation time is short and requires a hood, hot plate, and long tongs, plus safety equipment.

The *disadvantages* of wet ashing are that it takes virtually constant operator attention, corrosive reagents are necessary, and only small numbers of samples can be handled at any one time. All work needs to be carried out in a special fume hood that may be washed. Those hoods are generally called **perchloric acid hoods.**

Unfortunately, a single acid used in wet ashing does not give complete and rapid oxidation of organic material. Nitric acid with either sulfuric or perchloric acids and potassium chlorate or sulfate are used in varying combinations. Different combinations are recommended for different samples. Sulfur and nitric oxides are expelled for complete oxidation. The nitric-perchloric combination is generally faster than the sulfuric-nitric procedure. Perchloric acid has a tendency to explode, so a special perchloric acid hood that has wash-down capabilities is recommended. The hood does not contain plastic or glycerol base caulking compounds.

8.2.3.2 Procedures

The following wet ash procedure for nitric-perchloric oxidation is similar to that of the AOAC Method 975.03 for metals in plants:

1. A dried, ground 1 g sample is accurately weighed into a 150 ml Griffin beaker.
2. Add 10 ml HNO_3 and allow to soak. If the material has a high fat content, allow it to soak overnight.
3. Add 3 ml of 60 percent $HClO_4$ (*Precaution:* Place a beaker under pipette tip during transport) and slowly heat on a hot plate up to 350°C until frothing stops and HNO_3 is almost evaporated.
4. Continue boiling until perchloric reaction occurs (copious fumes), and then place watch glass on beaker. Sample should become colorless or light straw in color. *Do not let liquid in beaker reduce to dryness.*
5. Remove beaker from hot plate and let cool.
6. Wash watch glass with a minimum of distilled, deionized water and add 10 ml 50 percent HCl.
7. Transfer to appropriate volumetric flask (usually 50 ml) and dilute with distilled, deionized water.
8. Start wash-down procedure for hood after last sample.

An alternate procedure (i.e., as a preparation for iron analysis in meats) that could be used would be to use a 2 g sample boiled in 30 ml HNO_3 on a 350°C hotplate until 10 ml remain. Then add 10 ml of 60 percent perchloric acid and proceed as in step 4 above. Dilute to 100 ml in a volumetric flask following oxidation.

The wet ashing technique described above is very hazardous. Precautions for its use are found in the AOAC methods under "Safe Handling of Special Chemical Hazards." Perchloric acid interferes in the assay for iron by reacting with iron in the sample to form ferrous perchlorate, which forms an insoluble complex with the o-phenanthrolene in the procedure. It should not interfere with atomic absorption spectrophotometry.

The following procedure for a modified dry-wet ash oxidation may be used. It is listed under "Minerals in Ready-to-Feed Milk-Based Infant Formula" (AOAC Method 985.35).

1. Evaporate moist samples (25–50 mL) at 100°C overnight or in a microwave drying oven.
2. Heat on a hot plate until smoking ceases.
3. Ash at 525°C for 3–5 hrs.
4. Cool and wet with deionized distilled water plus 0.5–3.0 mL HNO_3.
5. Dry on a hot plate or steam bath and incinerate at 525°C for 1–2 hrs.
6. Weigh sample after cooling in a desiccator.
7. Repeat steps 4 and 5 if carbon persists. (*Caution:* some K may be lost with repeated ashing.)

8.2.4 Low-Temperature Plasma Ashing

8.2.4.1 Principles and Instrumentation

The equipment used for low-temperature plasma ashing consists of a glass system with a variable number of chambers for samples that may be evacuated by a vacuum pump. A small amount of oxygen is introduced that is broken into nascent oxygen by a radiofrequency electromagnetic field generator. A variable power frequency adjusts the rate of incineration. Air may be introduced as a gentler incineration procedure to preserve microscopic and structural components such as calcium oxalate crystals in various leaf tissues.

8.2.4.2 Procedures

The specific procedures for each type of low-temperature plasma ashing instrument may vary. The operator's manual should be consulted for proper operation.

The low-temperature plasma ashers usually contain two or more separate glass chambers with glass boats for holding samples. Intact or ground material is placed in individual boats, which are inserted into individual chambers. The chambers are sealed and a vacuum is applied. Once a vacuum is satisfactory (1 Torr or less), a small flow of oxygen or air is introduced into the system while maintaining a specific minimum vacuum. The frequency generator is then activated at a frequency slightly less than 14 mHz and adjusted by the amount of wattage applied (50–200 watts) to control the rate of incineration. Some models contain shaking devices to stir the sample. The progress of ashing may be viewed through the chambers.

The instrument is not without operational problems. These are usually due to leaks in the vacuum system. Either the seals around the chambers develop a leak and need to be replaced or breaks occur in the T-joints (usually plastic in the vacuum system). Those joints need to be replaced by glass material.

8.2.4.3 Applications

Low-temperature plasma ashing is a variation of dry ashing. The major *advantage* of this method is that there is less chance of losing trace elements by volatilization than with classical dry ashing techniques. The low temperature used with plasma ashers (150°C or less) generally allows the microscopic and crystalline structures to remain unaltered. The major *disadvantages* are small sample capacity and the expense of the equipment. However, it may be the equipment of choice under certain circumstances, especially for volatile salts.

8.2.5 Other Ash Measurements

8.2.5.1 Soluble and Insoluble Ash in Water

These measurements are an index of the fruit content of preserves and jellies. A lower ash in the water-soluble fraction is an indication that extra fruit is added to fruit and sugar products. Use the following procedure to measure soluble and insoluble ash:

1. Weigh the total ash.
2. Add 10 mL distilled H_2O.
3. Cover the crucible and heat nearly to boiling.
4. Filter on ashless filter paper and rinse with hot distilled water several times.
5. Dry and re-ash filter paper at least 30 min.
6. Weigh and calculate as percent H_2O-insoluble ash.
7. Calculate soluble ash by subtracting insoluble ash from total ash or dry the filtrate, re-ash, and weigh.

8.2.5.2 Ash Insoluble in Acid

This ash determination is a useful measure of the surface contamination of fruits and vegetables and wheat and rice coatings. Those contaminants are generally silicates and remain insoluble in acid, except HBr.

Use the following procedure:

1. Add 25 mL 10 percent HCl to total ash or H_2O-insoluble ash.
2. Cover and boil 5 min.
3. Filter on ashless filter paper and wash several times with hot distilled water.
4. Re-ash dried filter paper and residue at least 30 min.
5. Weigh and calculate as a percentage.

8.2.5.3 Alkalinity of Ash

The ash of fruits and vegetables is alkaline (Ca, Mg, K, Na), while that of meats and some cereals is acid (P, S, Cl). The alkalinity of ash has been used as a quality index of fruit and fruit juices. The salts of citric, malic, and tartaric acids yield carbonates upon combustion. Phosphates may interfere with this procedure. The procedure has been used for calculating acid-base balance, but its value in dietary calculations is questionable.

The following procedure is used to determine alkalinity of ash:

1. Place ash (total or water-insoluble ash) in platinum dish and accurately add 10 mL 0.1N HCl.
2. Add boiling H_2O if necessary and warm on a steam bath.
3. Cool and transfer to an Erlenmeyer flask.
4. Titrate the excess HCl with 0.1N NaOH using methyl orange as an indicator.
5. Express in terms of mL 1N acid/100 g sample.

Alkalinity of insoluble ash can be determined by titrating directly with 0.1N HCl using methyl orange. Express as described above.

8.3 COMPARISON OF METHODS

Ashing by any one of three methodologies (dry ashing, wet ashing, low-temperature plasma ashing) requires expensive equipment, especially if a large number of samples is analyzed. The muffle furnace may have to be placed in a heat room along with drying ovens and requires a 220-volt outlet. It is important to make sure large furnaces of that type are equipped with a double-pole, single-throw switch. Heating coils are generally exposed, and care must be taken when taking samples in and out with metal tongs. Desk-top furnaces (110 volts) are available for fewer samples. Wet ashing by the nitric acid method or nitric-sulfuric acid combination requires a hood and corrosive reagents. It also requires constant operator attention. There are several digesters currently available for wet ashing, including microwave ovens and bomb colorimetry. Those may be viable alternatives to a perchloric acid hood (which is expensive) even though the nitric-perchloric acid method is rapid. The low-temperature plasma asher requires a large vacuum pump in addition to the investment in the asher. Obviously, the type of further elemental analyses will dictate the equipment. Some micro- and most volatile elements will require special equipment and procedures. While wet oxidation and plasma ashing cause little volatilization, dry ashing will result in the loss of volatile elements. Refer to Chapter 9 for specific preparation procedures for elemental analyses.

8.4 SUMMARY

Three major types of ashing have been described: dry ashing, wet oxidation (ashing), and low-temperature plasma ashing. The procedure of choice depends upon the use of ash following its determination. Dry ashing, the method most commonly used, is based upon incineration at high temperatures in a muffle furnace. Except for certain elements, the residue may be used for further specific mineral analyses. Wet ashing (oxidation) is used in meat and meat products as a preparation for specific elemental analysis by simultaneously dissolving minerals and oxidizing all organic material. Low-temperature plasma ashers incinerate organic matter in a partial vacuum by forming nacent oxygen with a radio-frequency electromagnetic field generator. Highly volatile elements are preserved by this method. Wet ashing and low-temperature plasma ashing conserve volatile elements but are expensive, require operator time, and are limited to a small number of samples. Three post-ashing procedures (soluble and insoluble ash in water, ash insoluble in acid, and alkalinity of ash) are special measurements for certain foods.

8.5 STUDY QUESTIONS

1. Identify four potential sources of error in the preparation of samples for ash analysis, and describe a way to overcome each source of error.
2. You are determining the total ash content of a product using the dry ashing method. Your boss asks you to switch to a wet ashing method because he has heard it takes less time than dry ashing.
   a. Do you agree or disagree with your boss concerning the time issue, and why?
   b. Not considering the time issues, why might you want to continue using dry ashing, *and* why might you change to wet ashing?
3. Your lab technician was to determine the ash content of buttermilk by dry ashing. The technician weighed 5 g of buttermilk into one weighed platinum crucible, immediately put the crucible into the muffle furnace using a pair of all stainless steel tongs, and ashed the sample for 48 hours at 800°C. The crucible was removed from the muffle furnace and set on a rack in the open until it was cool enough to reweigh. Itemize the instructions you should have given your technician before beginning, so there would not have been the mistakes made as described above.
4. Differentiate low-temperature dry ashing from conventional dry ashing with regard to principle and applications.
5. How would you recommend to your technician to overcome the following problems that could arise in dry ashing various foods?
   a. You seem to be getting volatilization of phosphorus, when you want to later determine the phosphorus content.
   b. You are getting incomplete combustion of a product high in sugar after a typical dry ashing procedure (i.e., the ash is dark colored, not white or pale gray).
   c. The typical procedure takes too long for your purpose. You need to speed up the procedure, but you do not want to use the standard wet ashing procedure.
   d. You have reason to believe the compound you want to measure after dry ashing may be reacting with the porcelain crucibles being used.
   e. You want to determine the iron content of some foods but cannot seem to get the iron solubilized after the dry ashing procedure.
6. Explain two special ash measurements that can be useful for estimating the quality of fruits and fruit products.

8.6 PRACTICE PROBLEMS

1. A grain was found to contain 11.5 percent moisture. A 5.2146 g sample was placed into a crucible (28.5053 g tare). The ashed crucible weighed 28.5939 g. Calculate the percentage ash on (a) an as-received basis and (b) a dry-matter basis.
2. A vegetable (23.5000 g) was found to have 0.0940 g acid insoluble ash. What is the percentage acid insoluble ash?
3. You wish to have at least 100 mg ash from a cereal grain. Assuming 2.5 percent ash on average, how many grams should be weighed for ashing?
4. You wish to have 95 percent relative precision with your ash analyses. The following ash data are obtained: 2.15 percent, 2.12 percent, 2.07 percent. Are these data acceptable, and what is the relative precision?

5. The following data were obtained on a sample of hamburger: sample wt, 2.034 g; wt after drying, 1.0781 g; wt after ether extraction, 0.4679 g; and wt of ash, 0.0223 g. What is the percentage ash on (a) a wet weight basis and (b) a fat-free basis?

Answers

1. (a) 1.70%, (b) 1.95%. 2. 0.4 percent. 3. 4 g. 4. Yes, 96.3 percent. 5. (a) 1.1 percent, (b) 1.57 percent.

8.7 RESOURCE MATERIALS

Analytical Methods Committee. 1960. Methods for the destruction of organic matter. *Analyst* 85:643–656. This report gives a number of methods for wet and dry combustion and their applications, advantages, disadvantages, and hazards.

AOAC. 1990. *Official Methods of Analysis,* 15th ed. Association of Official Analytical Chemists, Washington, DC. This two-volume series contains the official methods for each specific food ingredient. It may be difficult for the beginning student to follow.

Aurand, L.W., Woods, A.E., and Wells, M.R. 1987. *Food Composition and Analysis.* Van Nostrand Reinhold, New York, NY. The chapters that deal with ash are divided by foodstuffs. General dry procedures are discussed under each major heading.

Ockerman, H.W. 1991. *Food Science Source Book,* Part 2, 2nd ed. Van Nostrand Reinhold, New York, NY.

Pomeranz, Y., and Meloan, C. 1987. *Food Analysis: Theory and Practice,* 2nd ed. Van Nostrand Reinhold, New York, NY. Chapter 34 on ash and minerals gives an excellent narrative on ashing methods and is easy reading for a student in food chemistry. A good reference list of specific mineral losses is given at the end of the chapter. No stepwise procedures are given.

Marshall, R.T. (Ed.) 1992. *Standard Methods for the Examination of Dairy Products,* 16th ed. American Public Health Assn., Washington, DC. This text gives detailed analytical procedures for ashing dairy products.

Smith, G.F. 1953. The wet ashing of organic matter employing hot concentrated perchloric acid. The liquid fire reaction. *Analytica Chimica Acta* 8:397–421. The treatise gives an in-depth review of wet ashing with perchloric acid. Tables on reaction times with foodstuffs and color reactions are informative. It is easy for the food scientist to understand.

Wooster, H.A. 1956. *Nutritional Data,* 3rd ed. H.J. Heinz Co., Pittsburgh, PA.

Contribution # 92-57-B from the Kansas Agricultural Experiment Station, Kansas State University, Manhattan, KS.

CHAPTER

9

Mineral Analysis by Traditional Methods

Deloy G. Hendricks

9.1 INTRODUCTION

Modern instrumentation has made it possible to quantitate an entire spectrum of minerals in one process. Some instruments are capable of detecting mineral concentrations in the parts-per-billion range. Instrumentation capable of such analysis is beyond the financial resources of many quality assurance laboratories. Large numbers of samples to be analyzed may justify the automation of some routine analyses and perhaps the expense of some of the modern pieces of equipment. The requirements for only occasional samples to be analyzed for a specific mineral, however, will not justify the overhead of much instrumentation. This leaves the options of (1) sending samples out to certified laboratories for analysis or (2) utilizing one of the more traditional methods for analysis. Traditional methods generally require chemicals and equipment that are routinely available in an analytical laboratory.

In this chapter, the nutritional need for minerals, their roles in food, and methods for analysis of minerals involving gravimetric, titrimetric, and colorimetric procedures are described. Procedures for analysis of minerals of major nutritional or food processing concern are used for illustrative purposes. For additional examples of traditional methods currently in use, refer to references 1–3. Slight modifications of these methods are often needed for specific foodstuffs in order to minimize interferences or to be in the range of analytical accuracy. Methods for plant or animal foodstuffs are reported here. For analytical requirements for specific foods, see the *Official Methods of Analysis* of the Association of Official Analytical Chemists (AOAC) International, or other official methods.

9.1.1 Importance of Minerals in the Diet

Approximately 98 percent of the calcium and 80 percent of the phosphorus in the body are found in the skeleton. Sodium, potassium, calcium, and magnesium are minerals involved in neural conduction and muscle contraction. Hydrochloric acid in the stomach greatly influences solubility and consequently absorbability of many minerals from foods in the diet. Calcium, phosphorus, sodium, potassium, magnesium, chlorine, and sulfur make up the dietary macro minerals, those minerals required at more than 100 mg per day by the adult (4). Each of these minerals has a specific function in the body. The body malfunctions if these minerals are not provided in the diet on a regular basis.

An additional 10 minerals are required in milligram quantities per day and are referred to as **trace minerals** (4). These include iron, iodine, zinc, copper, chromium, manganese, molybdenum, fluoride, selenium, and silica. These minerals have specific biochemical roles in maintaining body functions. Iron, for example, is part of the hemoglobin and myoglobin molecules involved in oxygen transport to and within the cells.

There is also a group of minerals called **ultra trace minerals** that are currently being investigated for possible biological function, but that do not currently have clearly defined biochemical roles. These include vanadium, tin, nickel, arsenic, and boron.

Some mineral elements have been documented to be toxic to the body and should, therefore, be avoided in the diet. Some metals of concern because of toxicity include lead, mercury, cadmium, and aluminum. Essential minerals such as fluoride and selenium are also known to be harmful if consumed in excessive quantities, even though they do have beneficial biochemical functions at proper dietary levels.

9.1.2 Minerals in Food Processing

Some minerals are inherent in natural foodstuffs. For example, milk is a good source of calcium, containing about 300 mg of calcium per 8-ounce cup. In some cases, salt is added in processing to decrease water activity and act as a preservative, thus increasing significantly the sodium content of products such as bacon, pickles, and Cheddar cheese. The enrichment law for flour requires that iron be replaced in white flour to the level it occurred naturally in the wheat kernel before removal of the bran. Fortification of foods has allowed addition of minerals into some foods above levels ever expected naturally. Prepared breakfast cereals are often fortified with minerals such as calcium, iron, and zinc, formerly thought to be limited in the diet. Fortification of salt with iodine has almost eliminated goiter in the United States.

Some processing of foods results in decreased mineral content. A large portion of the phosphorus, zinc, manganese, chromium, and copper found in a grain kernel is in the bran layer. When the bran layer is removed in processing, these minerals are removed. Direct acid cottage cheese is very low in calcium because of the action of the acid causing the calcium bound to the casein to be freed and consequently lost in the whey fraction.

The mineral content of foodstuffs is, therefore, important because of nutritional value, toxicological potential, and proper processing function and safety of some foods.

9.2 BASIC CONSIDERATIONS

Some sample preparation is required for traditional methods of mineral analysis. Methods used in sample preparation can remove interference for some analyses, add contaminants, or cause a loss of volatile elements. Proper handling of samples prior to the final analysis is very important for obtaining reliable analytical results for mineral content of foodstuffs.

9.2.1 Sample Preparation

Methods such as near infrared and neutron activation allow for mineral estimation without destruction of the carbon matrix of carbohydrates, fats, protein, and vitamins that make up foods. However, traditional methods generally require that the minerals be freed from this organic matrix in some manner. Chapter 8 describes the various methods used to ash foods in preparation for determination of specific mineral components of the food.

A major concern in mineral analysis is **contamination.** Solvents such as water can contain significant quantities of minerals. Therefore, all procedures involving mineral analysis require the use of the purest reagents available. In some cases, the cost of ultrapure reagents may be prohibitive. When this is the case, the alternative is to always work with a reagent blank. A **reagent blank** is a sample of reagents used in the sample analysis, quantitatively the same as used in the sample but without the material being analyzed. This reagent blank, representing the sum of the mineral contamination in the reagents, is then subtracted from the samples as they are quantitated.

9.2.2 Interferences

Factors such as **pH, sample matrix, temperature,** and **other analytical conditions** and **reagents** influence the ability of an analytical method to accurately be used to quantify a mineral. This is very clearly illustrated by the Parks, Hood, Hurwitz, and Ellis scheme of 12 inorganic elements as reviewed by Winton and Winton (5). In this scheme, molybdenum, manganese, iron, and phosphorus are determined on a dilute hydrochloride acid solution of a nitric acid-perchloric acid wet digest sample of food sample. An alkaline dithizone extraction is then used to separate sulfur, calcium, magnesium, potassium, and sodium for their further individual analysis. An acid dithizone extraction is then used for separation of zinc, cobalt, and copper, which can be individually quantitated.

If interferences are suspected, it is a common practice to use a sample matrix for standard curve preparation. A **sample matrix** is made up of elements known to be in the sample at the same level they exist in the sample. For example, if a food sample were to be analyzed for calcium content, a solution of the known levels of sodium, potassium, magnesium, and phosphorus would be used to make up the calcium standards for developing the standard curve. If the major minerals known to exist are used to make up the background solution for the standards, then the standard solutions more closely resemble the samples in solution. If there are interferences among the major minerals, the impact in the standards and the samples should be similar if a sample matrix is used. For some minerals, there are specific interfering substances that must be suppressed for accurate analysis.

9.3 METHODS

9.3.1 Gravimetric Analysis

9.3.1.1 Principles

Insoluble forms of minerals are precipitated, rinsed, dried, and weighed to estimate mineral content using gravimetric procedures. **Gravimetric analysis** is based on the fact that the constituent elements in any pure compound are always in the same proportions by weight. For example, NaCl is always 39.3 percent sodium. In gravimetric analysis, the desired constituent is separated from contaminating substances by selective precipitation and then rinsing to minimize any adhering or trapped elements. The precipitated compound is then dried and weighed. The weight of the mineral element is the same proportion of the weight of the compound as it is of the compound formed in the precipitated complex. Chloride, for example, is often precipitated as silver chloride. The silver chloride is rinsed, dried, and weighed. The weight of the chloride can then be calculated from the weight of the silver chloride, because chloride is 24.74 percent of the molecular weight of silver chloride.

9.3.1.2 Procedure—Modified Gravimetric Determination of Calcium (AOAC Method 910.01)

Calcium can be determined by precipitation as the oxalate (Fig. 9-1). The precipitate may then be converted to CaO by ignition and reported as calcium weight per sample weight.

9.3.1.3 Applications

Gravimetric procedures are best suited to large sample sizes and are generally limited to foods that contain large amounts of the element to be determined. Proce-

Weigh 10 g of sample to be analyzed into a crucible.

⇓

Ash in a muffle furnace at 500–550°C for 24 hr. (If there is any black or grey remaining indicating carbon particles, add 2 ml of 6 N HCl to the cooled samples. Dry at low heat in an oven or on a hot plate and ash for another 6 hr in the muffle furnace.)

⇓

If the material being analyzed is plant material, which may be high in silica, dissolve the ash in 10 ml of 2 N HCl and transfer to a 100 ml beaker.

⇓

Add 5 ml conc. HCl and evaporate to dryness in a steam bath to dehydrate SiO_2.

⇓

Moisten the residue with 5 ml of 2 N HCl and add about 50 ml of d H_2O.

⇓

Heat for a few minutes in a steam bath.

⇓

Transfer to a 100 ml vol. flask.

⇓

Cool rapidly to room temp.

⇓

Dilute to volume.

⇓

Filter.

⇓

Pipet a 10 ml aliquot[1] into a conical tip centrifuge tube.

⇓

Add to each sample in the centrifuge tubes 2 ml of saturated ammonium oxalate solution.

⇓

Add 2 drops of methyl red indicator (0.5% methyl red in ethanol) to each tube and adjust the pH to 4.5 by dropwise adding 3 N ammonium hydroxide while mixing. (Methyl red is a faint pink at pH 4.5.)

⇓

Allow sample to stand at least 4 hr then centrifuge at 1500 × g rpm for 15 min.

⇓

Carefully remove the supernatant by decantation or by suction, taking care not to disturb the precipitate.

⇓

Wash the precipitate by adding 2 ml of .3 N ammonium hydroxide.

⇓

Break up the precipitate for thorough washing by creating a vortex in the tube or a sharp shaking of the tube.

⇓

Repeat the centrifuging and washing 3 times.

⇓

After removing the last supernatant, resuspend the calcium oxalate precipitate in 2 ml of d H_2O and quantitatively transfer to a carefully dried and weighed crucible.

⇓

Rinse 3 times, adding the rinse to the crucible.

⇓

Dry carefully on a hot plate avoiding splattering.

⇓

Completely ash at 600°C in a muffle furnace for 12 hr.

⇓

Cool in a desiccator and weigh.

Express results as:
weight of the calcium (in a 10 ml aliquot[1] of the sample) = mg calcium oxide × 0.7147[2]

[1]The 10 ml aliquot of sample is from 100 ml total volume generated from a 10 g sample.
[2]M.W. Ca/M.W. CaO = .7147

FIGURE 9-1.
Procedure for modified gravimetric determination of calcium. AOAC Method 910.01. Adapted from (1).

dures using silver nitrate have been used to quantitate chloride. Most trace elements are in such low quantities in foods that gravimetric procedures are too insensitive to be of analytical value.

A disadvantage of the gravimetric procedure is the extra time involved in the second ignition, where the CaC_2O_4 is converted to CaO. However, simple concentration calculations show that an increased percent calcium of the total weight is worth the extra step. Repeated washing tends to solubilize some of the CaC_2O_4. However, co-precipitation of other minerals necessitates the rinsing steps.

9.3.2 EDTA Complexometric Titration

9.3.2.1 Principles

There are a number of carboxylic acids containing tertiary amines that form stable complexes with a variety of metal ions. **Ethylenediaminetetraacetic acid** (EDTA) is the most important of this class of reagents referred to as **versenes.** The disodium salt, usually written as Na_2H_2Y, is available in high purity as the dihydrate. Since EDTA has both donor nitrogen and donor oxygen atoms, it can form as many as six five-membered chelate rings and forms complexes with practically all metals except the alkali metals of group I.

In general, 1:1 complexes are formed between EDTA and metallic ions. Typical reaction can be summarized as:

$$m^{2+} + H_2Y^{2-} \rightarrow mY^{2-} + 2H^+ \qquad (1)$$

$$m^{3+} + H_2Y^{2-} \rightarrow mY^{-} + 2H^+ \qquad (2)$$

$$m^{4+} + H_2Y^{2} \rightarrow mY + 2H^+ \qquad (3)$$

Obviously, pH will greatly influence the complex formation. The EDTA complexes are highly stable and can therefore be used for volumetric analysis.

9.3.2.2 Procedure—Calcium Determination Using EDTA Titration (AOAC Method 968.31)

Calcium content of foods can be determined by complexing calcium with EDTA in a titration procedure (Fig. 9-2). The standard curve should be established before running a sample to be certain of the endpoint color.

9.3.2.3 Applications

EDTA complexometric titration is suitable for fruits and vegetables and other foods that have calcium without appreciable magnesium or phosphorous. Phosphorus may be removed by passing the ashed material through an Omberlite IR-4B resin bed at pH 3.5 prior to adjusting the base for titration. Using calmagite as an indicator, magnesium content of the sample can be calculated by difference (AOAC Method 967.30).

Pipette an aliquot of dissolved ashed sample expected to contain 2–10 mg of calcium into a 250 ml beaker.

⇓

Dilute to 50 ml and adjust pH to 12.5–13.0 by dropwise adding KOH-KCN solutions (dissolve 28 g KOH and 6.6 g KCN in 100 ml of d H_2O) while stirring with a magnetic stirrer.

⇓

Add 100 mg of ascorbic acid and about 250 mg of hydroxynaplithol blue indicator.

⇓

Titrate immediately with 0.01 M EDTA solution (Dissolve 3.72 g of Na_2H_2 EDTA · $2H_2O$, 99 + % purity, in d H_2O in a 1 L vol. flask. Dilute to volume and mix). The endpoint in this titration is a deep blue end point.

Standard Curve

Prepare and titrate standard solutions containing 2.5, 5.0, 7.5, and 10.0 mg of calcium to develop a standard curve:

Dry primary standard grade calcium carbonate for 2 hr at 285°F.

⇓

Weigh out 2.5 g of the dried calcium carbonate and transfer quantitatively into a 1 L vol. flask.

⇓

Dissolve the calcium carbonate in 50 ml of 3 N HCl.

⇓

Dilute to volume with d H_2O and thoroughly mix. Samples of 2.5, 5.0, 7.5 and 10 ml of this solution represent 2.5, 5.0, 7.5 and 10 mg of calcium, respectively.

FIGURE 9-2.
Procedure for calcium determination by EDTA titration. AOAC Method 968.31. Adapted from (1).

9.3.3 Redox Reactions

9.3.3.1 Principles

The basis of many analytical methods is oxidation-reduction reactions. The reaction of a substance with oxygen is defined as **oxidation.** Therefore, a **reduction** is the removal of oxygen. As we now know, oxidation is actually the **removal of electrons** from an atom, while reduction is the **gain of electrons.** Other reactions that do not involve oxygen involve the loss or gain of electrons. Any reaction that results in an increase in positive charge is termed oxidation, while any reaction that decreases positive charge is termed reduction whether or not oxygen is involved.

Since electrons cannot be created or destroyed in ordinary chemical reactions, any oxidation must be accompanied by a corresponding reduction. All oxidation-reduction reactions can be considered to be the reaction of an **oxidizing agent** with a **reducing agent.** Such reactions cause the oxidizing agent to be reduced and the reducing agent to be oxidized.

In some oxidation-reduction titrations, a colored reactant or product can act as the indicator. Permanganate ion is a deep purple, while manganous ion is a very pale pink. Thus, permanganate titrations have a built-in indicator.

9.3.3.2 Procedures

9.3.3.2.1 Calcium Determination Using Redox Titration (AOAC Method 921.01) Phosphates and magnesium tend to interfere with the analysis for calcium. Therefore, calcium is often precipitated as an oxalate to minimize the presence of interfering minerals in the final titration solution. This method, like weighing the precipitated calcium as the oxide, has the disadvantage of requiring the precipitation and washing of the oxalate. A diagram that describes the determination of calcium by redox titration is given in Fig. 9-3.

9.3.3.2.2 Iron Determination Using Redox Reaction and Colorimetry (AOAC Method 944.02) There are a number of organic compounds that effectively function as redox indicators. These compounds form stable colors that can be quantitated colorimetrically by measuring light absorbance at characteristic wavelengths.

Iron is quantitated by its ability to complex with organic compounds, resulting in formation of colored products proportional to the iron content (Fig. 9-4). All glassware must be acid washed and triple rinsed in distilled water to avoid iron contamination. Because many reagents contain small amounts of iron, it is important to always use a reagent blank for iron determinations.

Generate calcium oxalate precipitate as in the gravimetric procedure (Fig. 9-1).

⇓

Wash the precipitate.

⇓

Solubilize the precipitate with 2 ml of 3 N H_2SO_4.

⇓

Heat to 80–90°C in a hot H_2O bath.

⇓

Titrate with 0.02 N potassium permanganate to a slight pink endpoint.

⇓

Titrate 2 ml of 3 N H_2SO_4 heated to 80–90°C to determine the blank value for the potassium permanganate titrant.

Express results as:
wt. of calcium (per 10 ml aliquot[1] of sample) = ml of 0.02 N potassium permanganate × 0.0004[2]

[1]The 10 ml aliquot of sample is from 100 ml total volume generated from a 10 g sample as described in Fig. 9-1.

[2]$0.0004 = \frac{40.08}{1000 \times 100}$

40.08 = M.W. of calcium
1000 = conversion of L to ml
100 = difference in equivalence between 1 N calcium and .02 N potassium permanganate

FIGURE 9-3.
Procedure for calcium determination using redox titration. AOAC Method 921.01. Adapted from (1).

9.3.3.3 Applications

Generally, redox reactions have been of limited use for quantitating metals in foods. Calcium, iron, copper, and iodine concentrations have been determined using this approach. Determination of iron in foods using this method appears to have some advantages over atomic absorption spectroscopy (see Chapter 25). Higher recovery from spiked samples and closer match to a wider variety of National Institute of Standards and Technology (formerly known as the National Bureau of Standards) samples have been observed for iron analyzed using this redox colorimetric method compared to atomic absorption spectroscopy.

9.3.4 Precipitation Titration

9.3.4.1 Principles

When at least one product of a titration reaction is an insoluble precipitate, it is referred to as **precipitation titrimetry.** Few of the many gravimetric methods, however, can be adapted to yield accurate volumetric methods. Some of the major factors blocking the adaptation are (1) time to complete a reaction, resulting in a complete precipitation of the compound being formed; (2) failure of the reaction to yield a single product of definite composition; and (3) lack of an endpoint indicator for the reaction.

The potential of precipitation titration has resulted in at least two methods that are widely used in the food industry today. The **Mohr method** for chloride determination is based on the formation of an orange-colored solid, silver chromate, after silver from silver nitrate has complexed with all the available chloride.

$$Ag^+ + Cl^- \rightarrow AgCl \quad \text{(until all } Cl^- \text{ is complexed)} \tag{4}$$

$$2Ag^+ + CrO_4^{2-} \rightarrow Ag_2CrO_4 \quad \text{(orange only after } Cl^- \text{ is all complexed)} \tag{5}$$

Weigh into a clean, dried crucible a food sample expected to contain 50–500 μg of iron.

⇓

Add 10 ml of a glycerol-ethanol (1:1) mixture and dry over a low heat to avoid splattering.

⇓

Ash for 24 hr at 600°C.

⇓

After cooling add 1 ml conc. nitric acid then evaporate to dryness.

⇓

Return to the muffle furnace at 600°C for 1 hr to completely eliminate carbon particles.

⇓

Cool and add 5 ml of 6 N HCl to the ash.

⇓

Heat in a steam bath for 15 min.

⇓

Filter through a hardened filter paper into a 100 ml vol. flask with at least 3 rinsings of hot d H_2O.

⇓

Dilute to volume after allowing to reach room temp.

⇓

Pipette a 10 ml aliquot of the dissolved ash solution into a clean 25 ml vol. flask.

⇓

Add 1 ml of a 10% solution of hydroxylamine hydrochloride.

⇓

Allow to stand after mixing for a few minutes.

⇓

Add 5 ml of acetate buffer. (Made by dissolving 8.3 g of sodium acetate in 20 ml of water in a 100 ml vol. flask, adding 12 ml of acetic acid, and diluting to volume.)

⇓

Add as a color developing agent 1 ml of 0.1% orthophenanthroline or 2 ml of 0.1% alpha alpha-dipyridyl solution.

⇓

Dilute to volume with mixing.

⇓

Allow to stand for 30 minutes.

⇓

Read absorbance at 510 nm.

Standard Curve

Make an iron standard stock solution of 100 ppm by dissolving 0.1 g of analytical grade iron wire in 20 ml conc. HCl and diluting to 1 L.

⇓

Prepare working standards by pipetting 0, 2.0, 5.0, 10.0, 15.0, 20.0, 25.0, 30.0, 35.0, 40.0, and 45.0 ml of the standard stock solution plus 2 ml conc. HCl into 100 ml vol. flasks and diluting to volume.

⇓

10 ml of each of these working standards should be treated as the 10 ml of samples in the analytical procedure above.

⇓

Plot the standard curve and use it to calculate iron concentration in your sample.

FIGURE 9-4.
Procedure for iron determination using redox reaction and colorimetry. AOAC Method 944.02. Adapted from (1).

The **Volhard method** is an indirect or **back titration** method in which an excess of a standard solution of silver nitrate is added to a chloride-containing solution. The excess silver nitrate is then back-titrated using a standardized solution of potassium or ammonium thiocyanate with ferric ion as an indicator. In a back titration such as the Volhard method, the excess silver is then back-titrated to calculate the amount of chloride that precipitated with the silver in the first steps of the reaction.

$$Ag^+ + Cl^- \rightarrow AgCl \quad \text{(until all } Cl^- \text{ is complexed)} \qquad (6)$$

$$Ag^+ + SCN^- \rightarrow AgSCN \quad \text{(to quantitate silver not complexed with chloride)} \qquad (7)$$

$$SCN^- + Fe^{+3} \rightarrow FeSCN \quad \text{(red when there is any } SCN^- \text{ not complexed to } Ag^+\text{)} \qquad (8)$$

9.3.4.2 Procedures

9.3.4.2.1 Mohr Titration of Salt in Butter (AOAC Method 960.29) Salt in foods may be estimated by titrating the chloride ion with silver (Fig. 9-5). The orange endpoint in this reaction occurs only when all chloride ion is complexed, resulting in an excess of silver to form the colored silver chromate. The endpoint of this reaction is therefore at the first hint of an orange color. When preparing reagents for this assay use boiled water to avoid interferences from carbonates in the water.

9.3.4.2.2 Volhard Titration of Chloride in Plant Material (AOAC Method 915.01) In the Volhard method (Fig. 9-6), water must be boiled to minimize errors due to interfering carbonates, since the solubility product of silver carbonate is greater than the solubility product of silver chloride. Once chloride is determined by titration, the chloride weight is multiplied by 1.648 to obtain salt weight, if salt content is desired.

9.3.4.3 Applications

Gravimetric titration methods are well suited for any foods high in chlorides. Because of added salt in processed cheeses and meats, these products should certainly be considered for using this method to detect chloride; then salt content is estimated by calculation. The Quantab chloride titration used in AOAC Method 971.19 is an adaptation of the principles involved in the Mohr titration methods. This adaptation allows for very rapid quantitation of salt in food products. The Quantab adaptation is accurate to ± 10 percent over a range of 0.3 to 10 percent NaCl in food products.

Weigh about 5 g of butter into 250 ml Erlenmeyer flask and add 100 ml boiling H_2O.

⇓

Let stand 5 to 10 min with occasional swirling.

⇓

Add 2 ml of a 5% solution of K_2CrO_4 in d H_2O.

⇓

Titrate with 0.1 N $AgNO_3$ standardized as below until an orange-brown color persists for 30 sec.

Standardization of 0.1 N $AgNO_3$

Accurately weigh 300 mg of recrystallized dried KCl and transfer to a 250 ml Erlenmeyer flask with 40 ml of water.

⇓

Add 1 ml of K_2CrO_4 solution and titrate with $AgNO_3$ solution until first preciptable pale red-brown appears.

⇓

From the titration volume subtract the ml of the $AgNO_3$ solution required to produce the endpoint color in 75 ml of water containing 1 ml K_2CaO_4.

⇓

From the net volume of $AgNO_3$ calculate normality of the $AgNO_3$ as:

$$\text{Normality } AgNO_3 = \frac{\text{mg KCl}}{\text{ml } AgNO_3 \times 74.555 \text{ g KCl/mole}}$$

Calculating Salt in Butter

$$\text{Percent salt} = \frac{\text{ml .1 N } AgNO_3 \times 0.585}{\text{g of sample}}$$

[0.585 = (58.5 g NaCl/mole)/100]

FIGURE 9-5.
Procedure for Mohr titration of salt in butter. AOAC Method 960.29. Adapted from (1).

Moisten 5 g sample in crucible with 20 ml of 5% Na_2CO_3 in water.

⇓

Evaporate to dryness.

⇓

Char on a hot plate under a hood until smoking stops.

⇓

Combust at 500°C for 24 hr.

⇓

Dissolve residue in 10 ml of 5 N HNO_3.

⇓

Dilute to 25 ml with d H_2O.

⇓

Titrate with standardized $AgNO_3$ solution (from the Mohr method) until white AgCl stops precipitating and then add a slight excess.

⇓

Stir well, filter through a retentive filter paper, and wash AgCl thoroughly.

⇓

Add 5 ml of a saturated solution of $FeNH_4(SO_4)_2 \cdot 12\ H_2O$ to the combined titrate and washings.

⇓

Add 3 ml of 12 N HNO_3 and titrate excess silver with 0.1 N potassium thiocyanate.

Standardization of Potassium Thiocyanate Standard Solution

Determine working titer of the 0.1 N potassium thiocyanate standard solution by accurately measuring 40–50 ml of the standard $AgNO_3$ and adding it to 2 ml of $FeNH_4(SO_4)_2 \cdot 12\ H_2O$ indicator solution and 5 ml of 9 N HNO_3.

⇓

Titrate with thiocyanate solution until solution appears pale rose after vigorous shaking.

Calculating Cl Concentration

Net volume of the $AgNO_3$ = Total volume $AgNO_3$ added − Volume titrated with thiocyanate

1 ml of 0.1 M $AgNO_3$ = 3.506 mg chloride

FIGURE 9-6.
Procedure for Volhard titration of chloride in plant material. AOAC Method 915.01. Adapted from (1).

9.3.5 Colorimetric Methods

9.3.5.1 Principles

In the visible region of the electromagnetic spectrum, certain wavelengths are absorbed and others are reflected from an object. The reflected wavelength range is the color we see. In the colorimetric methods, a chemical reaction must result in a stable color that develops rapidly and is the result of a single colored product. The color-forming reaction should be selective for the mineral being analyzed.

As a color intensity increases, less light is able to pass through a solution. As the light passes through a longer pathway of the solution, there is also less light transmitted. **Beer's law,** which defines these relationships, is stated as:

$$A = -\log T = abc \tag{9}$$

where: A = absorbance
T = transmittance
a = an absorptivity constant for the reaction
b = the optical path length in centimeters
c = concentration in grams per liter

With the ability to quantitate light transmitted through a solution or, conversely, light absorbed by a solution, it is possible to determine concentrations of reacting substances. This principle has been used to develop methods for determining concentration of many minerals.

9.3.5.2 Procedure—Determination of Phosphorus by Colorimetry (AOAC Method 986.24)

The color intensity of phosphomolybdovanadate can be quantitated spectrophotometrically as described in Fig. 9-7. This is only one of many methods described

Ash a 2 g sample for 4 hr at 600°C.

⇓

Cool; add 5 ml of 6 N HCl and several drops of nitric acid.

⇓

Heat to dissolve the ash completely.

⇓

Cool and transfer to a 100 ml vol. flask and dilute to volume with d H_2O.

⇓

Pipette an aliquot expected to contain 0.5–1.5 mg of phosphorus into a 100 ml vol. flask.

⇓

Add 20 ml of molybdovanadate reagent. (This reagent is prepared by dissolving 20 g of ammonium molybdate in 200 ml of hot H_2O, then dissolving 1 g of ammonium meta vanadate in 125 ml of hot H_2O to which when cooled is added 140 ml conc. nitric acid. The cooled molybdate and vanadate solutions are then combined and diluted to 1 L.)

⇓

Dilute the sample and the molybdovanatate reagent to 100 ml.

⇓

Allow the color to develop for 10 min.

⇓

Read the absorbance at 400 nm against a phosphorus standard curve.

Preparation of Standard Curve

Make a stock standard solution of 2 mg P/ml by weighing 8.7874 g of KH_2PO_4 that has been dried at 105°C for 2 hr.

⇓

Quantitatively transfer to a 1 L vol. flask and add about 750 ml of d H_2O to dissolve.

⇓

Dilute to volume with H_2O.

⇓

Store refrigerated until use.

⇓

Make a working standard solution containing 0.1 mg phosphorus/ml by diluting 50 ml of the stock solution to 1 L with d H_2O.

⇓

Transfer aliquots of the working standard solution of 0, 5, 8, 10, and 15 ml to freshly rinsed 100 ml vol. (These represent 0, 0.5, 0.8, 1.0 and 1.5 mg of phosphorus, respectively.)

⇓

Add 20 ml of the molybdovanadate reagent to each flask containing the standards

⇓

Dilute to volume with H_2O and mix well.

⇓

Let flasks stand for 10 min. to complete color development.

⇓

Read absorbance at 400 nm. Use the 0.0 standard (blank) to zero the spectrophotometer.

FIGURE 9-7.
Procedure for determination of phosphorus by colorimetry. AOAC Method 986.24. Adapted from (1).

using the phosphomolybdate reaction. This procedure has the advantage of producing a more stable color than most and is therefore preferred.

9.3.5.3 Applications

Colorimetry is used for a wide variety of minerals. The example of iron determination given as an example of an oxidation-reduction reaction is quantitated using colorimetry. An oxidation-reduction reaction is, however, involved in the color development.

Some detergents contain phosphorus. It is necessary to thoroughly rinse all glassware carefully at least three times with distilled water to avoid contamination in determination of phosphorus by colorimetry.

9.4 COMPARISON OF METHODS

All minerals of concern nutritionally, for food processing, and toxicologically cannot be assessed by any single method with an equal degree of analytical accuracy. For labeling, processing, and even practical nutrition, we are only concerned with a few minerals, which generally can be analyzed by traditional methods. The traditional methods available for mineral analysis are varied; a very limited number of examples have been given.

Generally, for a small laboratory with skilled analytical personnel, the traditional methods can be carried out rapidly, with accuracy and at minimal costs. If a large number of samples of a specific element are to be run, there is certainly a time factor in favor of using atomic absorption spectroscopy or emission spectroscopy (see Chapter 25), depending on the mineral being analyzed. The graphite furnace on the atomic absorption spectrophotometer is capable of sensitivity in the parts-per-billion range. This is beyond the limits of the traditional methods. However, for most minerals of practical concern in the food industry, this degree of sensitivity is not required.

Individual choice of methods for mineral analysis must be made based on cost per analysis completed. Equipment cost, analytical time, analytical volume, and requirements for sensitivity should all be considered in making the final decision on which methods to use.

9.5 SPECIAL CONSIDERATIONS

The **Nutrition Labeling and Education Act of 1990** has made health claims on food labels legal under some conditions. Two minerals that are specifically identified as relating to health claims are calcium and sodium. Sodium analysis is also important in making claims for low-sodium-content food items that are being promoted for people with hypertension. With full implementation of the Labeling Law in 1994, more attention will be paid to the rapid and accurate analysis of these elements. Traditional methods described in this chapter will be used for quality assurance work and labeling compliance by companies with few specialized products. Methods such as atomic absorption spectrometry and emission spectrometry described in Chapter 25 will be utilized by laboratories specializing in providing large quantities of mineral data for labeling purposes and for compliance checks.

9.6 SUMMARY

The mineral content of foodstuffs is important because of nutritional value, toxicological potential, and proper processing function and safety of some foods. Traditional methods for mineral analysis include gravimetric, titrimetric, and colorimetric procedures. Foods are typically ashed prior to these analyses, since the methods generally require that the minerals be freed from the organic matrix of the foods. Sample preparation must include steps necessary to prevent contamination or loss of volatile elements and must deal with any potential interferences. The basic principles of gravimetric, titrimetric, and colorimetric methods for mineral analyses are described in this chapter, with procedures given for some minerals of concern in the food industry.

Gravimetric, titrimetric, and colorimetric procedures for mineral analyses generally require chemicals and equipment routinely available in an analytical laboratory and do not require expensive instrumentation. These methods may be suited to a small laboratory with skilled analytical personnel and a limited number of samples to be analyzed. Adequate sample must be available, and a high degree of sensitivity must not be required.

Traditional methods for mineral analysis are being kit-adapted for rapid analysis. Tests for water hardness and the Quantab for salt determination are examples currently being used. The basic principles involved in these methods will continue to be utilized to develop inexpensive rapid methods for screening mineral content of foods and beverages.

9.7 STUDY QUESTIONS

1. What are the major concerns in sample preparation for specific mineral analysis? How can each concern be addressed?
2. Calcium can be quantitated by gravimetric analysis, EDTA complexometric titration, and redox titration. Differentiate these techniques with regard to the principles involved.
3. The Mohr and Volhard titration methods are often used to determine the NaCl content of foods. Compare and contrast these two methods, as you explain the principles involved.
4. In a back-titration procedure, would overshooting the endpoint in the titration cause an over- or underestimation of the compound being quantitated? Explain your answer.
5. What is the function of a matrix standard? How is it prepared?
6. Describe analytical conditions that may call for the use of a reagent blank.
7. What factors should be considered in selecting a specific method for mineral analysis for a food product?

9.8 PRACTICE PROBLEMS

1. If a given sample of food yields 0.750 g of silver chloride in a gravimetric analysis, what weight of chloride is present?
2. A 10 g food sample was dried, then ashed, and analyzed for salt (NaCl) content by the Mohr titration method ($AgNO_3 + Cl^- \rightarrow AgCl$). The weight of the dried sample was 2 g, and the ashed sample weight was 0.5 g. The entire ashed sample was titrated using a standardized $AgNO_3$ solution. It took 6.5 ml of the $AgNO_3$ solution to reach the endpoint, as indicated by the red color of Ag_2CO_4 when K_2CrO_4 was used an an indicator. The $AgNO_3$ solution was standardized using 300 mg of dried KCl as described in Fig. 9-5. The corrected volume of $AgNO_3$ solution used in the titration was 40.9 ml. Calculate the salt (NaCl) content of the original food sample in terms of percent (w/w) NaCl.
3. A 25 g food sample was dried, then ashed, and finally analyzed for salt (NaCl) content by the Volhard titration method. The weight of the dried sample was 5 g, and the ashed sample weighed 1 g. Then 30 ml of 0.1 N $AgNO_3$ was added to the ashed sample, the resultant precipitate was filtered out, and a small amount of ferric ammonium sulfate was added to the filtrate. The filtrate was then titrated with 3 ml of 0.1 M KSCN to a red endpoint.
   a. What was the moisture content of the sample, expressed as percent H_2O (w/w)?
   b. What was the ash content of the sample, expressed as percent ash (w/w) on a dry-weight basis?
   c. What was the salt content of the original sample in terms of percent (w/w) NaCl? (molecular weight Na = 23, molecular weight Cl = 35.5)
4. Compound X in a food sample was quantitated by a colorimetric assay. Use the following information and Beer's law to calculate the content of Compound X in the food sample, in terms of mg Compound X/100 g sample:
   a. A 4 g sample was ashed.
   b. Ashed sample was dissolved with 1 ml of acid and the volume brought to 250 ml.
   c. A 0.75 ml aliquot was used in a reaction in which the total volume of the sample to be read in the spectrophotometer was 50 ml.
   d. Absorbance at 595 nm for the sample was 0.543.
   e. The absorptivity constant for the reaction (i.e., molar extinction coefficient) was known to be 1754 $L\ g^{-2}\ cm^{-1}$.
   f. Inside diameter of cuvette for spectrophotometer was 1 cm.
5. Colorimetric Analysis
   a. You are using a colorimetric method to determine the concentration of Compound A in your liquid food sample. This method allows a sample volume of 5 ml. This volume must be held constant but can be comprised of diluted standard solution and water. For this standard curve, you need standards that contain 0, 0.25, 0.50, 0.75, and 1.0 mg of Compound A. Your stock standard solution contains 5 g/l of Compound A.

Devise a dilution scheme(s) for preparing the samples for this standard curve that could be followed by a lab technician. Be specific. In preparing the dilution scheme, use no volumes less than 0.2 ml.

b. You obtain the following absorbance values for your standard curve.

SAMPLE	ABS (500nm)
.00 mg	0.00
.25 mg	0.20
.50 mg	0.40
.75 mg	0.60
1.00 mg	0.80

On a sheet of graph paper, construct a standard curve and determine the equation of the line.

c. A 5 ml sample is diluted to 500 ml, and 3 ml of this solution is analyzed as per the standard samples; the absorbance is 0.50 units at 500 nm. Use the equation of the line calculated in part (b) and information about the dilutions to calculate what the concentration is of Compound A in your original sample in terms of g/L.

Answers

1.
$$\frac{x \text{ g Cl}}{0.750 \text{ g AgCl}} = \frac{35.45 \text{ g/mol}}{143.3 \text{ g/mol}}$$
$$x = 0.1855 \text{ g Cl}$$

2.
$$\text{N AgNO}_3 = \frac{0.3 \text{ g KCl}}{\text{ml AgNO}_3 \times 74.555 \text{ g KCl/mole}}$$
$$0.0984 \text{ N} = \frac{0.3 \text{ g}}{40.9 \text{ ml} \times 74.555}$$
$$(0.0984 \text{ M AgNO}_3)(0.0065 \text{ L}) = 0.0006396 \text{ mole Ag}^+$$
$$\Rightarrow 0.0006396 \text{ mole Cl}^-$$
$$\Rightarrow 0.006396 \text{ mole NaCl}$$
$$(.0006396 \text{ mole NaCl}) \times \frac{58.5 \text{ g NaCl}}{\text{mole}} = .0374 \text{ g NaCl}$$

3. a.
$$\frac{25 \text{ g wet sample} - 5 \text{ g dry sample}}{25 \text{ g wet sample}} \times 100 = 80\%$$

b.
$$\frac{1 \text{ g ash}}{5 \text{ g dry sample}} \times 100 = 20\%$$

c. moles Ag added = mole Cl^- in sample + moles SCN^- added

moles Ag = (0.1 mole/L) × (0.03L) = 0.003 mole

moles SCN^- = (0.1 mole/L) × (0.003 L) = 0.0003 mole

0.003 mole = moles Cl^- + 0.003 mole

0.0027 mole = mole Cl^-

$$(0.0027 \text{ mole Cl}^-) \times \frac{58.5 \text{ g NaCl}}{\text{mole}} = 0.1580 \text{ g NaCl}$$
$$\frac{0.1580 \text{ g NaCl}}{25 \text{ g wet sample}} = \frac{0.00632 \text{ g NaCl}}{\text{g wet sample}} \times 100$$
$$= 0.632\% \text{ NaCl (w/w)}$$

4. A = abc
$$0.543 = (1.574 \text{ L g}^{-1} \text{ cm}^{-1} (1 \text{ cm})\ c$$
$$c = 3.4498 \times 10^{-4} \text{ g/L}$$
$$c = 3.44498 \times 10^{-4} \text{ mg/ml}$$
$$\frac{3.4498 \times 10^{-4} \text{ mg}}{\text{ml}} \times 50 \text{ ml} = 1.725 \times 10^{-2} \text{ mg}$$
$$\frac{1.725 \times 10^{-2} \text{ mg}}{0.75 \text{ ml}} \times \frac{250 \text{ ml}}{4 \text{ g}} = 1.437 \text{ mg/g}$$
$$= 143.7 \text{ mg/100g}$$

5. a. Say that you want to know what dilution to do on the 5 mg/ml stock solution to pipette 0.2 ml for the lowest point on the std. curve (0.25 mg/5 ml). What dilution must you do?
$$\frac{0.25 \text{ mg}}{5 \text{ ml}} \times \frac{5 \text{ ml}}{0.2 \text{ ml}} \times \frac{X \text{ ml}}{1 \text{ ml}} = \frac{5 \text{ mg}}{\text{ml}}$$
$$X = 4 \text{ ml}$$

∴ Dilution of 1 ml to 4 ml would give 0.25 mg in a 0.2 ml sample.

Say that you have diluted the 5 mg/ml solution by 1 to 10. How much of this diluted solution do you pipette to get a concentration of 0.25 mg/5 ml (in the total sample volume)?
$$\frac{0.25 \text{ mg}}{5 \text{ ml}} \times \frac{5 \text{ ml}}{X \text{ ml}} \times \frac{10 \text{ ml}}{1 \text{ ml}} = \frac{5 \text{ mg}}{\text{ml}}$$
$$X = 0.5 \text{ ml}$$

∴ Dilution of 1 ml to 10 ml would give 0.25 mg in a 0.5 ml sample.

Using diluted stock solution of 0.25 mg/0.5 ml:
$$\frac{0.5 \text{ mg}}{\text{ml}} \times X \text{ ml} = 1.0 \text{ mg}$$
$$X = 2 \text{ ml}$$

etc. for 0.75, 0.50, 0.25 mg

mg A/5 ml	ml DILUTED STOCK SOLUTION	ml H_2O
0	0	5.0
.25	.5	4.5
.50	1.0	4.0
.75	1.5	3.5
1.0	2.0	3.0

b. *Draw graph:*

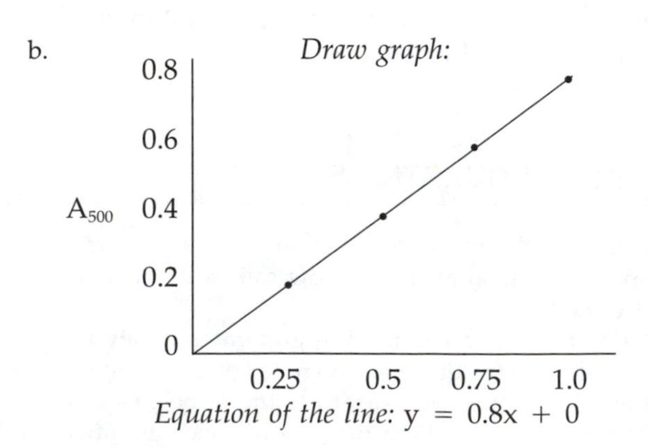

Equation of the line: y = 0.8x + 0

c.
$$A_{500} = 0.50 = y$$
$$0.50 = 0.8x + 0$$
$$x = 0.625$$
$$\frac{0.625 \text{ mg}}{5 \text{ ml}} \times \frac{5 \text{ ml}}{3 \text{ ml}} \times \frac{500 \text{ ml}}{5 \text{ ml}} = 20.83 \text{ mg/ml}$$
$$= 20.83 \text{ g/l}$$

9.6 REFERENCES

1. AOAC. 1990. *Official Methods of Analysis*, 15th ed. Association of Official Analytical Chemists, Washington, DC.
2. Kenner, C.T. 1971. *Analytical Separations and Determinations.* MacMillan Company, New York, NY.
3. Kirk, R.S., and Sawyer, R. 1991. *Pearson's Composition and Analysis of Foods*, 9th ed. Longman Scientific and Technical, Essex, England.
4. National Research Council. 1989. *Recommended Dietary Allowances*, 10th ed. National Academy Press, Washington, DC.
5. Winton, A.L., and Winton, K.B. 1945. *The Analysis of Foods.* John Wiley & Sons, Inc., London, England.

CHAPTER

10

Carbohydrate Analysis

Nicholas H. Low

10.1 INTRODUCTION

10.1.1 Importance of Carbohydrates

Carbohydrates as a group comprise some of the most important compounds known to humans. These molecules occur wherever carbon dioxide and air combine in the leaves of plants in the presence of water. Carbohydrate biosynthesis by plants (photosynthesis) releases the by-product oxygen, which supports the existence of all other organisms.

The multitude of roles played by carbohydrates in nature is unique. Glucose (a simple monosaccharide) is converted into many other natural substances, such as fats, proteins, and vitamins. The fossil fuels used today, such as petroleum and coal, were formed by carbohydrate conversion of primeval forests.

Carbohydrates play an important role in human nutrition as energy reserves. Simple household sugars such as beet and cane sugar can be used directly as foods. In nature, these compounds play only a minor role as energy stores, since their excellent solubility in water results in the inability of many plants to enrich their concentration. Starch, cellulose, and glycogen, as polymers of glucose, can act as storage forms much more readily. Cellulose, which is distinguished from starch only by the linkage of the glucose monomers, is an ideal support material and is the main constituent of wood. Another polysaccharide, chitin (N-acetylglucosamine monomers), forms the highly resistant shell and exoskeletal material of all insects and many marine animals.

An additional aspect of the function of carbohydrates in nature is in the transport of substances. Numerous active and auxiliary substances occur in combination with carbohydrates (glycoconjugates). It is by these mechanisms that natural barriers such as cell membranes are overcome and improved solubilities are realized. Recognition mechanisms between cells also rely on cell surface carbohydrates. Sugar-containing structures embedded in cell membranes (glycoproteins and glycolipids) are responsible for extremely specific interactions with other cell surfaces. Examples include the AIDS virus (HIV), which can only attack certain cells (T-lymphocytes, macrophages), and blood-group proteins that contain oligosaccharide side chains that determine blood-group specificity. Many organisms employ the same principle for immune defense by means of antibodies.

Due to the importance of these molecules, carbohydrate detection methods have received and continue to receive a great deal of attention. This chapter will cover the wide assortment of qualitative and quantitative methods of carbohydrate analysis.

Initially it is important to introduce to some readers, and review with others, some of the basic features and nomenclature of carbohydrates.

10.1.2 Carbohydrate Classification

Carbohydrates are classified into three major groups: **monosaccharides,** also called simple sugars, **oligosaccharides,** and **polysaccharides.** In this chapter, the rules of carbohydrate nomenclature as published by the American Chemical Society (1) will be followed. However, as the trivial names for the important food carbohydrates are familiar to carbohydrate and food chemists, they will also be used.

10.1.2.1 Monosaccharides

Monosaccharides are water-soluble crystalline compounds. These compounds are generally aliphatic aldehydes, or ketones, which contain one carbonyl group and one or more hydroxyl groups. A broader description is that monosaccharides are polyhydroxy aldehydes, ketones, acids, alcohols, amines, and their simple derivatives. These compounds can be further classified as an **aldose** (aldehyde-containing carbohydrate) or a **ketose** (ketone carbohydrate). Depending on the number of carbon atoms, a monosaccharide is known as a **triose, tetrose, pentose, hexose,** and so on. An aldohexose, for example, is a six-carbon monosaccharide containing an aldehyde group. Most naturally occurring monosaccharides are pentoses or hexoses.

The reactive centers of monosaccharides are the carbonyl and hydroxyl groups. Carbohydrates that reduce **Fehling's** (or **Benedict's**) or **Tollens's reagent** are known as **reducing sugars.** Reducing sugars differ from most other organic compounds in one characteristic property. When a pure organic compound that is not a reducing sugar is dissolved in a solvent, it will usually contain only one compound. However, when a reducing sugar is dissolved in water, a solution is obtained that may contain up to six compounds: the two pyranoses, the two furanoses, and the acyclic (open chain) carbonyl form and its hydrate (2) (Fig. 10-1). These forms, often referred to as **tautomeric forms,** are distinct compounds that differ from the other forms in their chemical, physical, and biological properties. A number of modern carbohydrate analytical methods can measure each of these forms. Therefore, when choosing a method for carbohydrate analysis, it is important to define the question(s) you are trying to answer.

The most important monosaccharides in food analysis are the hexoses, D-glucose (also called dextrose), D-fructose (also called levulose), and D-galactose. Of minor importance are D-mannose, and the pentoses, D-ribose, D-arabinose, and D-xylose.

10.1.2.2 Oligosaccharides

Oligosaccharides [Greek *oligos,* meaning a few (3)] are relatively low molecular weight (340–1600 daltons)

FIGURE 10-1.
Tautomeric forms of D-glucose.

polymers that yield monosaccharides upon hydrolysis that are covalently bonded through glycosidic linkages with resultant loss of water. Arrows such as 1→4 indicate that the #1 carbon atom of one monomeric unit is attached to the #4 carbon atom of the other monomeric unit via a glycosidic linkage.

The standard convention for the number of monomer units making up an oligosaccharide is 10 (4). Oligosaccharides comprised of two monomer units are called **disaccharides,** three monomer units are **trisaccharides,** and so on.

Oligosaccharides occur naturally in plants, animals, and microorganisms. These compounds can be synthesized enzymatically (5) (i.e., by transferase or reverse hydrolase reactions) or by acid hydrolysis (6) (reversion reaction).

Oligosaccharides comprised of D-glucose, D-fructose, and D-galactose units are the most important in food analysis. The most important food oligosaccharide is sucrose, which is a nonreducing disaccharide. Other important disaccharides include lactose, maltose, and isomaltose. Other important food oligosaccharides include raffinose, maltotriose, stachyose, verbascose, and the kestoses (nystose, 1-, 3-, and 6-kestose).

10.1.2.3 Polysaccharides

The majority of the carbohydrates found in nature occur as polysaccharides. Polysaccharides are high molecular weight (up to 420 million daltons!) polymers that yield monosaccharides upon acid or specific enzyme hydrolysis.

Polysaccharides containing identical carbohydrate monomers are called **homopolysaccharides** (or homoglycans), whereas those that are comprised of more than one type of monomer are called **heteropolysaccharides** (or heteroglycans).

The most important food homopolysaccharides contain D-glucose and include starch, glycogen, cellulose, and dextrins. Important food heteropolysaccharides include pectin (comprised of D-galacturonic acid, its methylester, neutral sugars, etc.), hemicellulose (comprised of at least six different monomers), and numerous plant and microbial gums.

10.1.3 Carbohydrate Content of Selected Foods

Table 10-1 lists a number of common foods and their carbohydrate content. It is interesting to note that in most of the foods shown, the major chemical constituent of each of these foods (after water) is carbohydrate. This fact further emphasizes the importance of analytical techniques to accurately determine carbohydrate content (both qualitatively and quantitatively).

10.1.4 Importance of Carbohydrate Analysis

Heightened consumer interest and expectations in conjunction with the rapid increases in food choices and products have placed a heavy demand on analytical methods to provide complete and accurate food compositional information.

Carbohydrate analysis of raw materials and processed foods can be used to provide a wealth of important information. For example, so-called fingerprint oligosaccharide patterns can be used to detect the addition of one food material to another (food adulteration) (7). In addition, carbohydrate breakdown products can be used to determine if a food has been irradiated (8).

TABLE 10-1. Carbohydrate Content of Selected Foods

Food	*Percent Carbohydrate (wet weight basis)*
Yogurt	5.60
Milk (2%)	4.78
Starch (potato)	83.10
Potato	15.40
Carrot	3.59
Broccoli	2.30
Tomato	3.63
Apple	12.39
Grape	16.11
Orange	9.19
Honey	75.10
Beer (light)	2.90

10.2 METHODS OF ANALYSIS

In this section, a variety of analytical methods will be presented that attempt to cover the myriad of techniques currently available for carbohydrate analysis. Some methods referenced are official methods of the Association of Official Analytical Chemists (AOAC) International. Refer to these methods and other references cited for detailed information about procedures.

10.2.1 Sample Preparation

Due to the complex nature of foods, analysis of carbohydrates is normally preceded by sample purification. This step is routinely employed to remove possible interferences that may affect analysis.

10.2.1.1 Extraction of Monosaccharides and Oligosaccharides

The extraction of simple sugars (monosaccharides) and oligosaccharides from food materials without polysaccharide removal is accomplished employing **80 percent alcohol** (AOAC Method 922.02, 925.05). The method consists of addition of the finely ground or chopped solid material to hot redistilled alcohol to which enough precipitated $CaCO_3$ has been added to neutralize acidity, using enough alcohol so that the final concentration is 80 percent. The mixture is then heated nearly to boiling on a hot plate or steam bath for 30 min so that carbohydrate extraction can take place. The alcohol solution is then passed through filter paper or an extraction thimble with retention of the filtrate. The insoluble residue is then returned to an appropriate container, covered with hot 80 percent alcohol, heated on a hot plate or steam bath for 1 hr, cooled, and filtered through the same filter. If the second filtrate is highly colored, the extraction is repeated. The solid residue is allowed to dry and is ground so that all particles will pass through a 1 mm sieve. This material is transferred to an extraction thimble and extracted for 12 hrs in a Soxhlet apparatus with 80 percent alcohol. The residue is dried and saved for starch determination. The alcohol filtrates are combined and diluted to a known volume with 80 percent alcohol for carbohydrate determination.

If starch is not to be determined, the sample preparation is the same as that followed in AOAC Method 922.02, but the sample is boiled on a steam bath or hot plate for 1 hr. Decant the solution into a volumetric flask and comminute solids in a high-speed blender with 80 percent alcohol. Boil the blended material for 0.5 hr, cool, transfer to a volumetric flask, dilute to volume with 80 percent alcohol at room temperature, filter, and take an aliquot for analysis.

It is important to point out that the efficacy of the aforementioned method relies on the complete solubility of monosaccharides and oligosaccharides in aqueous alcohol. Various aqueous alcoholic treatments have been employed for the isolation of food carbohydrates; these include 80 percent ethanol (9), 85 percent methanol (10), and 20 percent ethanol/40 percent isopropanol (11). In each case, the alcoholic extracts contain, in addition to carbohydrates, some lipids, pigments, and the free amino and organic acids present in the sample. Of more concern to the food analyst is the fact that many important food carbohydrates (specifically oligosaccharides) are only partially soluble in aqueous alcohol (12).

10.2.1.2 Removal of Interfering Compounds

Depending on the analytical method used to determine carbohydrates (specifically colorimetric methods), the interfering chemical compounds (enzymes, amino acids, organic acids, metals, pigments, and other small molecules) require removal. These compounds can be removed by treatment with either **lead acetate** or **ion-exchange resins.**

10.2.1.2.1 Clarification with Lead Acetate Clarification with lead acetate (AOAC Method 925.46B): Place an aliquot of alcohol extract (AOAC Method 922.02, 925.05) on a steam bath and evaporate the alcohol. Avoid evaporation to dryness by the addition of water if necessary. When the odor of alcohol disappears, add ca. 100 mL of water and heat to 80°C to soften any gummy precipitates and to break up insoluble masses. Cool to room temperature and proceed as in (1) or (2) below:

1. Transfer the solution to a volumetric flask, rinse the beaker thoroughly with water, and add rinsing to the flask. Add enough saturated neutral $Pb(OAc)_2$ solution to produce a flocculent precipitate (ppt), shake thoroughly, and let stand for 15 min. Test the supernate with a few drops of the lead solution. If more ppt forms, shake and let stand again; if no further ppt forms, dilute to volume with water, mix thoroughly, and filter through dry paper. Add enough solid sodium oxalate (Na oxalate) to the filtrate to ppt all the Pb, and refilter through dry paper. Test the filtrate for the presence of Pb with a small amount of solid Na oxalate.
2. Add twice the minimum amount of saturated neutral $Pb(OAc)_2$ solution required to cause complete precipitation, as found by testing a portion of the supernate with a few drops of

dilute Na oxalate solution. Let the mixture stand only a few minutes; then filter into a beaker containing an estimated excess of Na oxalate crystals. Let Pb ppt drain on the filter and wash with cold water until the filtrate no longer gives ppt in oxalate solution. Ensure excess of oxalate by testing with one drop of $Pb(OAc)_2$. Filter and wash pptd Pb oxalate, catching the filtrate and washings in a volumetric flask. Dilute to volume with water and mix.

10.2.1.2.2 Clarification with Ion Exchange Resins

Place an aliquot of the alcohol extract (AOAC Method 922.02, 925.05) in a beaker and heat on a steam bath to evaporate the alcohol. Avoid evaporation to dryness by adding water. When the odor of alcohol disappears, add ca. 15–25 mL of water and heat to 80°C to soften gummy ppts and break up insoluble masses. Cool to room temperature. Prepare a thin mat of celite on filter paper in a buchner or a fritted glass filter and wash until the water comes through clear. Filter the sample through the celite mat, wash the mat with water, dilute the filtrate and washings to appropriate volume in a volumetric flask, and mix well.

Place a 50 mL aliquot in a 250 mL Erlenmeyer; add 2 g of Amberlite IR-120(H) analytical grade cation [replace by REXYN 101(H) resin, Fisher Scientific Co.] and 3g Duolite A-4(OH) anion ion exchange resins (the company that produced this material no longer exists, and this resin can be replaced by the use of AG 3-X4, Bio-Rad Laboratories) (AOAC Method 931.02C). Let stand 2 hrs with occasional swirling. Take a 5 mL aliquot of the deionized solution and determine reducing sugars as glucose as in 31.054.

10.2.1.3 Other Sample Preparation and Clarification Procedures

We have found some of the sample preparation and clarification schemes described above to be quite laborious and in many cases unsuitable for analysis when the exact structures of the food carbohydrates are desired. Considerable problems are encountered when employing the ion exchange procedure. Treatment of a carbohydrate sample with a strong cation exchange resin (stirring for 2 hr) will result in hydrolysis of oligosaccharides and may result in the formation of a variety of compounds from hexoses, such as 5-hydroxymethyl-2-furfural (HMF), anhydro-sugars, and epimerization. Carbohydrates (monosaccharides, oligosaccharides) can also interact with anion exchange resins (especially strong anion resins), resulting in binding of the carbohydrate to the resin.

Simpler sample preparation schemes avoid partial solubilization, hydrolysis, and binding problems (13). This sample preparation involves simple extraction/dilution with water (following grinding or macerating if applicable) and rapid passage of the resulting solution through a strong cation exchange resin column (approximately 5 mL of resin-AG 50W-X8, Bio-Rad Laboratories) at room temperature, or slow passage through the column at reduced temperatures (4°C), followed by passage of the eluent through a weak anion exchange resin column (5 mL of resin-AG1-X4 formate form, Bio-Rad Laboratories). The sample is then passed through a hydrophobic column (C_{18} sep pak cartridge, Waters Associates). These treatments can be performed rapidly (<10 minutes) and result in a sample free of proteins/amino acids, organic acids, trace minerals, and hydrophobic compounds.

Sample preparation techniques and analytical methods for the analysis of polysaccharides will be covered later in this chapter.

10.2.2 Calculation by Difference

Basic food analysis consists of the determination of moisture, protein, fat/lipid, and ash (i.e., minerals). Each of these methods of analysis is covered in depth in this book. When the results from these analyses are tabulated, that remaining is considered to be total carbohydrate. Therefore, one can quantitate the amount of total carbohydrate in a food by difference: 100% − (%moisture + %protein + %fat/lipid + %ash).

The obvious problems with this method involve incomplete digestion/extraction of each of these major food constituents and the experimental error involved in their determination. This error, even though it may be small in each case, will accumulate and can lead to inaccurate results for carbohydrate content.

In addition, this method does not differentiate between available and nonavailable carbohydrates. The complete carbohydrate profile of a food (reducing sugars, oligosaccharides, polysaccharides, fiber, etc.) relates to and influences the quality of the product. Therefore, as foods become more complex and as new carbohydrates and their derivatives are used in food formulations, more specific analyses (than the by-difference method) are necessary.

10.2.3 Chemical Methods for Monosaccharides and Oligosaccharides

The majority of chemical methods for the analysis of carbohydrates are based on the reaction of reducing sugars with chemical reagents to yield precipitates (such as cuprous oxide) or a colored complex (as in the anthrone reaction). The final stages of these methods involve gravimetric analysis or solubilization of the precipitate followed by titration or spectrophotometric determination at a specific wavelength (colored complex). Of great importance is the fact that a number of these methods (where strong acid treatment is not a part of the reaction) are only suitable for reducing sugars and that nonreducing sugars must be hydrolyzed prior to analysis. An alternate method is to carry out the reaction prior to hydrolysis and then re-analyze following hydrolysis to determine the amount of reducing and nonreducing sugar in the sample. In addition, the reaction between the chemical reagent and the carbohydrate is dependent upon the structure of the carbohydrate. Therefore, standard curves or specific tables (9) must be used. In samples in which a variety of carbohydrates are present (as in honey), these methods can yield inaccurate results.

10.2.3.1 Lane-Eynon Method

The Lane-Eynon method (AOAC Method 923.09, 920.183b) is based on the reaction of **reducing sugars** with a solution of **copper sulfate** followed by reaction with **alkaline tartrate** (or by treatment with the Soxhlet solution: equal volume mixture of copper sulfate solution and alkaline tartrate solution).

The approximate carbohydrate concentration of the sample must be known so that this sample solution can be added (by burette) to reduce almost all of the copper in the Soxhlet solution. This mixture is then boiled for a very specific time period (2 min, crucial step) followed by the addition of methylene blue. This colored solution is then titrated with the original sample solution, again within a specific time period (3 min) until decoloration of the indicator.

The problem associated with any reducing sugar method is that the method cannot be used to distinguish between different reducing sugars. As previously mentioned, this method lacks the ability to determine the concentration of nonreducing sugars. It has been shown that many biological compounds can interfere with this determination (14). In addition, this method is a time-specific assay that is dependent upon the heating time, temperature, and reactant concentration; in the hands of the untrained (and even trained hands!) it can lead to inaccurate results.

This method is used routinely for the determination of reducing sugars in honey and other high-reducing sugar syrups.

10.2.3.2 Munson and Walker Method

The Munson and Walker method can be applied to all **reducing sugars,** and it is has been used extensively for determining carbohydrate concentration in foods [AOAC Method 906.03 (refers to Methods 31.037-31.044, 14th ed.)]. The results from this test are reproducible, and if the proper conversion table is used (to convert cuprous oxide to weight of a particular carbohydrate), it is quite accurate. The method is not very sensitive (>5.0 mg of D-glucose/5 mL of clarified sample solution), but this is normally not a significant problem in foods. Conversion tables are available for most of the common reducing sugars (9).

The Munson and Walker method involves the oxidation of the carbohydrates in the presence of heat and an excess of **cupric sulfate** and **alkaline tartrate,** under carefully controlled conditions. The basic conditions are required to keep the copper in solution as copper hydroxide (Cu^+). Upon heating, water is driven off and copper oxide is converted to cuprous oxide. **Cuprous oxide** precipitates as the carbohydrates are oxidized and can be determined

1. Gravimetrically (AOAC Method 31.039, 14th ed.)
2. By electrolytic deposition from a nitric acid solution (AOAC Method 31.044, 14th ed.), where the copper oxide is dissolved in nitric acid and then deposited on platinum electrodes; the weight gain of the electrode is related to the reducing sugar content
3. By titration with either
   a. sodium thiosulfate (AOAC Method 31.040, 14th ed.), in which cuprous oxide is dissolved in nitric acid, it undergoes oxidation to cupric nitrate, potassium iodide is added, and the iodide is oxidized to iodine, which is titrated with thiosulphate using a starch indicator
   b. Potassium permanganate (AOAC Method 31.042, 14th ed.), in which cuprous oxide is reacted with ferric sulfate and the ferric ion is reduced to the ferrous ion (+3 to +2); the ferrous ion is then titrated with permanganate, resulting in a color change due to permanganate (+3, pink → +2, colorless)

The mechanism of this reaction is quite complex, as carbohydrates in basic solutions undergo a variety of reactions including keto-enol tautomerization (this explains why keto sugars can be determined), unsaturation, and base elimination. A simplified representation of the reaction (and subsequent titration with permanganate) is shown below:

$$\text{reducing sugar} + Cu^{+2} + \text{base} \rightarrow \text{oxidized sugar} + Cu_2O \quad (1)$$

Titration of the cuprous oxide with permanganate:

$$Cu_2O + Fe_2(SO_4)_3 \rightarrow 2FeSO_4 + CuSO_4 + CuO \quad (2)$$

$$10\,FeSO_4 + 2KMnO_4 + 8H_2SO_4 \rightarrow 5Fe_2(SO_4)_3 + K_2SO_4 + 2MnSO_4 + 8H_2O \quad (3)$$

The Munson and Walker method depends on the ability of reducing sugars to react with the copper solution. This reaction (formation of cuprous oxide) depends on the rate and time of heating, carbohydrate concentration, type of carbohydrate, and alkalinity of the reaction mixture. Another major problem is that other biological molecules may interfere with the reaction. In addition, this method lacks the ability to distinguish carbohydrates and requires a thorough knowledge of the type and concentration of the carbohydrate in the sample prior to analysis (because carbohydrates differ in their ability to reduce the copper solution).

A modification of this method (15) involves the use of an excess of alkaline copper citrate in place of tartrate with sodium carbonate (as a base). Following the reduction, the excess copper citrate is reacted with excess potassium iodide and the liberated iodine is titrated with sodium thiosulfate. The advantage of this modified method is that glucose, fructose, and invert sugar can be determined using the same reference tables.

10.2.3.3 Nelson-Somogyi Method

The Nelson-Somogyi method (16,17) to measure **reducing sugars** is a modification of the Munson and Walker and Lane-Eynon methods, which are applicable for samples that contain low concentrations of carbohydrate (0.3–3 mg/tested aliquot).

The method is based on the reaction of **cuprous oxide** with **arsenomolybdate reagent,** which is prepared by reacting ammonium molybdate ($(NH_4)_6Mo_7O_{24}$) and sodium arsenate (Na_2HAsO_7) in sulfuric acid. Oxidation of cupric to cuprous by reducing sugars, with concomitant reduction of the arsenomolybdate complex, produces an intense blue-colored solution (due to the reduced arsenomolybdate) that is very stable. The absorbance of this solution is determined at either 500 or 520 nm.

This method requires preparation of a standard curve (using D-glucose at levels of 50, 100, and 150 μg/mL). The same problems associated with the other methods involving cuprous oxide are encountered when employing this method of analysis.

10.2.3.4 Alkaline Ferricyanide Method

The alkaline ferricyanide method, first introduced in 1962, is based on the principle that carbohydrates in a basic solution (pH > 10.5) can reduce **ferricyanide** to **ferrocyanide.** Ferrocyanide can then react with **ferric ions** to produce Prussian blue, which can be read spectrophotometrically at 700 nm. The resulting blue color is quite stable, and the reaction obeys **Beer's law.** However, standards must be employed to obtain accurate results.

10.2.3.5 Phenol-Sulfuric Acid Method

The phenol-sulfuric acid method is a simple, rapid, and universal test for the measurement of total carbohydrate in a food (18). Because sulfuric acid is used in this method, both reducing and nonreducing carbohydrates can be quantitated. The test is based on the reaction of carbohydrate with phenol in the presence of a **strong acid** (which produces heat), followed by heating (25–30°C for 20 min). Under the strong acidic conditions, dehydration of the carbohydrates occurs to form **furfural** and **hydroxy methyl furfural.** These products then condense with **phenol.** The overall reaction results in color formation (yellow-orange) that can be read spectrophotometrically at 490 nm (hexose) and 480 nm (pentose). Standard reference curves must be prepared for individual sugars, and blanks must be carried through with each batch of samples analyzed.

10.2.3.6 Anthrone Method

Although most carbohydrates react with anthrone, the strong reactivity of hexoses to this reagent provides a reasonably specific colorimetric method for these carbohydrates (19). Carbohydrates react with 9,10-dihydro-9-oxoanthracene **(anthrone)** under **acidic conditions** (concentrated sulfuric acid) to yield a blue-green color. The reaction mixture is heated in boiling water for 15 min and then allowed to cool in the dark (for color formation, 20–30 min) before measuring the absorbance at 620 nm. Standards and reagent blanks must be run simultaneously with unknowns.

Because the method involves the addition of concentrated sulfuric acid, both reducing and nonreducing carbohydrates (hexoses) are determined. The method can only tolerate small amounts of alcohol without significant interference (up to 5 percent w/v); therefore, alcohol must be removed prior to analysis. The reaction is dependent upon the type of carbohydrate present in the sample, the concentration of the reagent, and the time/temperature of the actual reaction.

10.2.3.7 Other Colorimetric Methods

Reducing sugars react with 2,3,5-triphenyl tetrazolium bromide or chloride above pH 12.5 to produce a colored complex, the triphenyl formazon (pink-violet-blue, measured at 485 nm). The final color of the reaction mixture depends on the carbohydrate (20). Reducing sugars also react with 3,5-dinitrosalicylate under alkaline conditions to produce red-brown colors that can be measured spectrophotometrically (21).

Carbohydrates react with resorcinol under strongly acidic conditions (22) to give colored complexes. All hexoses react under these conditions; however, the color yield with ketoses is much greater than that for aldoses. Therefore, this method can be used to provide analytical information on keto sugars.

Carbohydrates react with orcinol in strongly acidic conditions (23) to give colored complexes. This method has been used to determine pentoses in the presence of hexoses because the color response for a hexose is approximately 5 percent of that for a pentose. The color formed is quite stable, and the absorbance can be measured at 670 nm.

10.2.4 Enzymatic Methods for Monosaccharides and Oligosaccharides

10.2.4.1 Introduction

Enzymes are proteins with catalytic activity due to their power of specific activation. These chemical compounds have a wide range of molecular masses (4,000 to >1,000,000 daltons). Because these compounds are proteins, they are susceptible to denaturation (loss of catalytic activity) by pH, temperature, and other environmental factors. In most cases, enzymes that are used for analytical purposes have a maximum temperature range of 20–60°C and pH values from 4 to 10.

Due to their high specificity and sensitivity, enzyme assays are ideal for the analysis of food carbohydrates. Assays may be performed on an aliquot of the food itself without partial or total separation of the carbohydrates prior to analysis. These enzymatic reactions are usually conducted at temperatures close to room temperature, at neutral pH, and in minutes (typically <20 min), thus minimizing changes in the compounds during analysis.

Enzymatic assays for carbohydrates in biological materials have been available since the end of the nineteenth century. However, the routine use of enzymes for carbohydrate analysis has only received widespread acceptance in the last 15 years. This surge in the use of enzymes in carbohydrate analysis is due in part to the numerous **enzyme assay kits** available, which in turn relates to the availability of sufficiently pure enzymes at reasonable cost. In addition, instrumentation such as microtitre plates and readers (UV or fluorescent) affords multiple (96 sample/well microtitre plate) and rapid (<3 seconds per well) sample analysis. Enzyme kits for carbohydrate analysis are available from a number of companies, which provide detailed outlines on their specific use (e.g., Boehringer-Mannheim Corp., Sigma Chemical Co., Worthington Biochemical Corp., Calbiochem Corp.).

There are two methods used in the analysis of foods with enzymes: the **total change method** and the **rate assay (or kinetic) method.** In the total change method, sufficient enzyme is added to convert all of the substrate in the sample to product within a short period of time. The amount of substrate in the sample is then determined from the total change in the sample either as substrate disappearance or as product formation. In the rate assay method, the initial rate of the enzyme substrate reaction is determined, and from this relation the concentration of enzyme, substrate, activator, or inhibitor may be determined.

Regardless of the enzymatic method used in food analysis, the **enzyme concentration, substrate concentration, activator concentration, inhibitor concentration, pH,** and **temperature** all affect the rate of enzyme catalyzed reaction. Therefore, it is necessary to control all of these parameters (especially in the rate assay method) if accurate results are to be obtained.

Further information on the use of enzymes in food analysis (not limited to carbohydrate analysis) can be found in Chapter 32 of this text and a number of excellent books and articles (24–27).

10.2.4.2 D-Glucose/D-Fructose/D-Sorbitol Method

D-glucose and D-fructose can be phosphorylated to glucose-6-phosphate (G-6-P) and fructose-6-phosphate (F-6-P), respectively, by the enzyme **hexokinase** (HK) and adenosine-5′-triphosphate (ATP) with the simultaneous formation of adenosine-5′-diphosphate (ADP) (28).

Glucose-6-phosphate is oxidized by nicotinamide-adenine dinucleotide phosphate (NADP) in the presence of the enzyme **glucose-6-phosphate dehydrogenase** (G6P-DH) to gluconate-6-phosphate and reduced NADP (NADPH):

$$\text{G-6-P} + \text{NADP}^+ \xrightarrow{\text{G6P-DH}} \text{gluconate-6-phosphate} + \text{NADPH} + \text{H}^+ \qquad (4)$$

The amount of NADPH formed is stoichiometric with the amount of D-glucose, and the increase in NADPH can be measured spectrophotometrically at 334, 340, or 365 nm.

Upon completion of this reaction, **phosphoglucose isomerase** (PGI) is used to convert F-6-P to G-6-P. The G-6-P formed then reacts with NADP to form gluconate-6-phosphate and NADPH. The amount of

NADPH formed in this reaction is stoichiometric with D-fructose.

In the presence of the enzyme **sorbitol dehydrogenase** (SDH), D-sorbitol is oxidized to D-fructose by nicotinamide adenine dinucleotide (NAD):

$$\text{D-sorbitol} + \text{NAD}^+ \xrightarrow{\text{SDH}} \text{D-fructose} + \text{NADH} + \text{H}^+ \quad (5)$$

In this reaction, the equilibrium lies far to the right, completely in favor of D-fructose. The compound NADH is removed from the reaction mixture by **lactate dehydrogenase** (LDH) and pyruvate. The D-fructose can then be determined using the aforementioned methods. The amount of NADPH formed is now stoichiometric with the D-fructose formed from D-sorbitol.

In this method, it is important that proteins and free amino acids are removed because they can interfere in the spectroscopic measurements. In addition, SDH can also catalyze the oxidation of other polyols (although at slower rates) with the formation of NADH, which can lead to erroneous results for the concentration of D-sorbitol.

10.2.4.3 Lactose/D-Galactose Method

Lactose can be enzymatically hydrolyzed to its component monosaccharides (D-glucose and D-galactose) by the enzyme **β-galactosidase:**

$$\text{lactose} + H_2O \xrightarrow{\beta\text{-galactosidase}} \text{D-glucose} + \text{D-galactose} \quad (6)$$

D-galactose is then oxidized to D-galactonic acid by NAD in the presence of **β-galactosidase dehydrogenase** (Gal-DH):

$$\text{D-galactose} + \text{NAD}^+ \xrightarrow{\text{Gal-DH}} \text{D-galactonic acid} + \text{NADH} + \text{H}^+ \quad (7)$$

The amount of NADH (which can be measured at 334, 340, or 365 nm) formed is stoichiometric with the amount of lactose.

The D-galactose concentration of a sample can be measured by treatment of the sample with NAD and Gal-DH.

A problem with this method is the fact that Gal-DH is not specific for D-galactose because it also oxidizes L-arabinose. As the concentration of L-arabinose in foods is small, this lack of specificity is considered to be insignificant.

10.2.4.4 Maltose/Sucrose/D-Glucose Method

In the presence of the enzyme **α-glucosidase,** maltose and sucrose can be hydrolyzed to two molecules of D-glucose or to D-glucose and D-fructose, respectively:

$$\text{maltose} + H_2O \xrightarrow{\alpha\text{-glucosidase}} \text{2D-glucose} \quad (8)$$

$$\text{sucrose} + H_2O \xrightarrow{\alpha\text{-glucosidase}} \text{D-glucose} + \text{D-fructose} \quad (9)$$

Sucrose can also be enzymatically hydrolyzed employing the enzyme **β-fructosidase.**

The D-glucose and D-fructose produced from these enzymatic reactions can then be determined using the method in section 10.2.4.2.

Because the enzymes α-glucosidase and β-fructosidase are not specific for a single oligosaccharide, the usefulness of these assays can be limited. For example, maltotriose and other dextrose polymers can be hydrolyzed by α-glucosidase. In addition, ubiquitous carbohydrates such as raffinose can be hydrolyzed by β-fructosidase. Therefore, for optimum accuracy, the carbohydrate profile of a food sample should be known prior to using this method.

10.2.4.5 Raffinose Method

Raffinose (O-α-D-galactopyranosyl-(1->6)-α-D-glucopyranosyl β-D-fructofuranoside) can be hydrolyzed to D-galactose and sucrose by the enzyme **α-galactosidase:**

$$\text{raffinose} + H_2O \xrightarrow{\alpha\text{-galactosidase}} \text{D-galactose} + \text{sucrose} \quad (10)$$

The resulting D-galactose is oxidized by NAD to galactono-lactone in the presence of **galactose dehydrogenase** (Gal-DH):

$$\text{D-galactose} + \text{NAD}^+ \xrightarrow{\text{Gal-DH}} \text{galactono-lactone} + \text{NADH} + \text{H}^+ \quad (11)$$

The amount of NADH produced in this reaction is stoichiometric with the concentration of raffinose in the sample.

The enzyme α-galactosidase also has the ability to hydrolyze other commonly occurring food α-galactosides, such as melibiose (O-α-D-galactopyranosyl-(1->6)-D-glucopyranose) and stachyose (O-α-D-galactopyranosyl-(1->6)-O-α-D-galactopyranosyl-(1->6)-α-D-glucopyranosyl β-D-fructofuranoside). The presence of these carbohydrates in a sample will adversely affect the determination of raffinose.

Raffinose can also be hydrolyzed by the enzyme β-fructosidase to yield D-fructose:

$$\text{raffinose} + H_2O \xrightarrow{\beta\text{-fructosidase}} \text{D-fructose} + \text{melibiose} \quad (12)$$

The D-fructose formed can then be analyzed using the method in section 10.2.4.2, and the amount of NADH formed is stoichiometric with the concentration of raffinose in the sample.

10.2.4.6 Oxidase Method

The oxidase method is a combined enzymic/instrumental method based on the enzymatic oxidation of the carbohydrate and the subsequent measurement of the hydrogen peroxide formed by electrochemical oxidation at a platinum electrode. The general reaction is as follows:

$$\text{carbohydrate} + O_2 \longrightarrow \text{oxidized carbohydrate} + H_2O_2 \quad (13)$$

The enzyme is an immobilized **oxidase** contained in a membrane support. Each carbohydrate requires a specific oxidase enzyme/membrane support, and the hydrogen peroxide produced is proportional to the carbohydrate concentration. The oxygen required for the reaction is provided by the buffer solution. The amount of oxygen uptake during the enzymatic oxidation reaction can be determined in place of measuring hydrogen peroxide (29).

Carbohydrates that can be analyzed by this method include D-glucose (glucose oxidase), D-fructose (galactose oxidase), lactose (galactose oxidase), and sucrose. Sucrose must be hydrolyzed prior to oxidation, and the glucose produced is oxidized with glucose oxidase:

$$\text{sucrose} + \text{invertase} \longrightarrow \text{D-glucose} + \text{D-fructose} \quad (14)$$

This assay is fast (< 1 min) and has been automated (Yellow Springs Instruments, Yellow Springs, OH) for use in the food and beverage industries. Reagent strips (Destrostix; Ames Division of Miles Laboratories Ltd.) containing immobilized glucose oxidase/peroxidase/chromogen indicator are used in the biomedical industry for the rapid determination of blood glucose.

Most of the problems associated with the oxidase method are due to interfering carbohydrates. For example, raffinose and melibiose interfere at levels of 2 and 8 percent, respectively, in sucrose determination. The analysis of lactose is affected by fructose, erythrose, glycerol, and mannose. Conversely, fructose analysis is affected by the presence of lactose. An additional significant problem related to D-glucose determination is the fact that the enzyme glucose oxidase only reacts with the β anomer. Therefore, for reproducible, accurate results, the sample must be completely equilibrated prior to analysis.

10.2.5 Physical Methods

10.2.5.1 Introduction

There are many methods available to determine the carbohydrate concentration of foods. The most facile and rapid of these methods are those based on the physical properties of carbohydrates. These methods include **polarimetry, refractometry,** and **specific gravity.** Each of these aforementioned methods suffers from the limitation that it is only suitable for pure carbohydrate syrups (honey, dextrose syrups, etc.) and requires clarified solutions. In addition, these cases require the use of conversion tables to obtain accurate results.

10.2.5.2 Polarimetry

Most compounds containing an **asymmetric carbon atom** have the ability to rotate the plane of polarization of polarized light. A **polarimeter** measures the rotatory power exerted by a compound in solution on plane polarized light. The polarimeter consists of a source of monochromatic radiation; a polarizer for converting the light waves in the beam of monochromatic light into plane polarized light (light that vibrates in only one plane); a tube of known length, with parallel faces, that contains the sample solution; and an analyzer assembly for measuring the extent of rotation of the plane polarized light.

Polarimetry makes use of the **optical activity** of carbohydrates. This means that carbohydrates can rotate plane polarized light through an angle that is dependent on the **wavelength of the light, concentration of the sample, temperature,** and **amount of sample.** The wavelength most often used for this determination is the **sodium D line** (589.3 nm); the cell length is usually one decimeter (10 cm); and the operating temperature is 20°C. The formula that relates these parameters is:

$$(\alpha)_D^{20} = 100\ \alpha/lc \qquad (15)$$

where:
α = observed angular rotation
l = length of the tube in decimeters
c = concentration of solution in g/100 mL
$(\alpha)_D^{20}$ = specific rotation for the sugar at the sodium D line at 20°C

The above relationship holds well in most cases, but with some carbohydrates (especially monosaccharides), the relationship between rotation and concentration is not linear: $(\alpha)_D^{20}$ is not a constant. This variation is slight and can be accounted for by introducing mathematical factors that have been determined for a number of carbohydrates (see Table 10-2).

Polarimetry can be used to determine the concentration of a single carbohydrate in a food sample. If a known weight of the food is dissolved in water (Y g/100 mL), the percentage of the carbohydrate (N) in the sample may be calculated by

$$N\ (\%\ wt/vol) = (\alpha)/(\alpha)_D^{20} \times 100/l \times 100/Y \qquad (16)$$

where:
α = observed angular rotation
l = length of the tube in decimeters
$(\alpha)_D^{20}$ = specific rotation for the sugar at the sodium D line at 20°C

Prior to analysis by polarimetry, the solution to be analyzed must be clarified. Another major concern with polarimetry is that the carbohydrate is in equilibrium. All reducing carbohydrates display mutarotation between α and β isomers. If the carbohydrate solution is freshly prepared or if the sample solution has not equilibrated, there may be errors due to this phenomenon. To alleviate this problem, the solution should first be allowed to stand for several hours or a few drops of ammonia should be added to establish the equilibrium rapidly. The major drawback with this procedure is the fact that in most cases, mixtures of carbohydrates cannot be analyzed by polarimetry. An exception to this is the determination of sucrose in the presence of other carbohydrates. The principle in this determination is to take a reading at the beginning and after sucrose hydrolysis by acid or enzymes. The difference between these readings can then be related to the percentage sudrose:

$$\%\ sucrose = ((\alpha_b)_D - (\alpha_a)_D)/Q \qquad (17)$$

where:
$(\alpha_b)_D$ = specific rotation before inversion
$(\alpha_a)_D$ = specific rotation after inversion
Q = constant (change in rotation due to inversion from +66.5 to −22.15°), which is 88.65°

Other carbohydrates that are hydrolyzed (by acid or enzyme) during this procedure will adversely affect the accuracy of this measurement.

Specific rotation has also been shown (25) to be indirectly related to the dextrose equivalent in glucose syrups.

TABLE 10-2. Formula to Calculate Specific Rotation at the Sodium D Line at 20°C

Carbohydrate	*Formula*
Glucose	$52.50 + 0.0188p + 0.000517p^2$
Fructose	$-113.96 + 0.258q$
Sucrose	$66.462 + 0.0087c - 0.000235c^2$
Maltose	$138.475 - 0.01837p$
Invert sugar	$-(19.415 + 0.07065c - 0.00054c^2)$

Where: c = concentration in grams per 100 mL
p = percentage by weight
q = percentage of water

Temperature corrections can also be made employing mathematical formulas, for example, the temperature correction for invert sugar:

$$(\alpha)_D^t = (\alpha)_D^{20} - (0.283 + 0.0014c)\ (t - 20°C)$$

Where: c = concentration in grams per 100 mL
t = temperature (°C)

From (104) (p. 270), used with permission.

10.2.5.3 Refractive Index Measurement

When electromagnetic radiation passes from one medium to another, it can change direction; it is either bent or refracted. The ratio of the sine of the angle of incidence to the sine of the angle of refraction is termed the **index of refraction** (n). Because all chemical compounds have an index of refraction, this measurement has been used for the qualitative identification of an unknown compound by comparing its refractive index with literature values.

Refractive index varies with **concentration, temperature,** and the **wavelength of light.** Standard refractive index values in literature are given for **monochromatic sodium light** (589 nm) at a temperature of **20°C.** These values are indicated as n. The refractive index for white light is only slightly different than that for the sodium D line. However, the use of white light introduces difficulties due to light dispersion that can only be solved by the incorporation of specific prisms **(Amici prism).**

Refractometers used in carbohydrate analysis are constructed so that the critical angle (the angle of incidence for which the angle of refraction is ≥90) is such that total reflection is obtained. An example is the **Abbe refractometer,** which covers the refractive index range from 1.3 to 1.7 with a precision of 0.0003 units. In this type of refractometer, the whole surface of the prism is illuminated with light, the telescope collects all the rays travelling parallel to each other into one line in the focal plane, and the critical boundary shows up in the telescope as a sharp line of demarcation between dark and brightly illuminated fields.

As mentioned previously, the refractive index of a solution increases with concentration (see Table 10-3). This phenomenon has been exploited in the analysis of food carbohydrates (actually the total soluble solids) in a variety of products. Examples include sugar syrups (honey, maple syrup, molasses), fruit products (juices, jams, jellies, etc.), and tomato products. These measurements are performed on a refractometer that has been calibrated with sucrose, and the readings are normally expressed as percent sugar wt/wt (as sucrose with reference to the appropriate table and with corrections for the presence of other soluble compounds in the sample). Alternately, most refractometers are also calibrated in **°Brix** (grams of sucrose/100 grams of sample), which is numerically equivalent to the percent sucrose on a wt/wt basis. These refractive index measurements are accurate only for pure sucrose solutions but are widely used to approximate values in these foods.

10.2.5.4 Specific Gravity Measurement

10.2.5.4.1 Introduction **Specific gravity** (S; also referred to as **relative density**) is defined as the ratio of the density of a substance (D_a) to the density of a reference substance (D_b, usually water):

$$S = D_a/D_b = \frac{\text{wt of X mL of substance a}}{\text{wt. of X mL of water}} \quad (18)$$

Because the volume of a given weight of substance varies with temperature, specific gravity is not defined unless the temperatures at which the densities are measured are given. The density of water at a variety of temperatures is shown in Table 10-4 (30). Because these values are all quite close to unity, the specific gravity can be taken to be numerically equal to the density.

Specific gravity measurements are best applied to the analysis of solutions consisting of only one component (in a medium of water). Dr. F. Plato used this method to accurately determine the concentration of pure cane sugar solutions at 20°C (31). These published values are routinely used as a standard worldwide.

The specific gravity for a number of important food carbohydrates (at the same concentration) have similar values (Table 10-5). Therefore, it is possible to determine the concentration of a pure carbohydrate solution or a mixture of pure carbohydrates by referring to appropriate specific gravity tables.

Specific gravity measurements can be made employing a variety of methods. The most important of these are the **pycnometer, Westphal balance,** and **hydrometer.**

10.2.5.4.2 Pycnometer With pycnometers, the relative weights of equal volumes of the sample liquid

TABLE 10-3. Refractive Indices of Carbohydrate Solutions at 20°C

Concentration (Percent)	D-Glucose	D-Fructose	Sucrose
10	1.34775	1.34762	1.3478
20	1.36356	1.36335	1.3638
30	1.38051	1.38030	1.3811
40	1.39872	1.39857	1.3998
50	1.41826	1.41819	1.4200

From (30). Reprinted with permission from *CRC Handbook of Chemistry and Physics,* 59th ed., R.C. Weast (Ed.). Copyright CRC Press, Boca Raton, FL.

TABLE 10-4. Density of Water at Various Temperatures

Density (g/mL)	Temperature (°C)
0.99987	0
1.00000	3.98
0.99999	5
0.99973	10
0.99823	20
0.99567	30

TABLE 10-5. Specific Gravities of Glucose, Fructose, and Sucrose at 20°C

Concentration (Percent by Wt)	SPECIFIC GRAVITY		
	Glucose	Fructose	Sucrose
5	1.01769	1.01803	1.01824
10	1.03769	1.03853	1.03901
20	1.07981	1.08162	1.08270
30	1.12475	1.12760	1.1300

From (104) (p. 207), used with permission.

and of water are determined. These measurements are made in either a pycnometer flask or an Ostwald-Sprengel tube. The **pycnometer flask** is a vessel that contains a fixed volume of liquid at a specific temperature. The flask is then weighed prior to and following both sample and water addition. The specific gravity can be determined as follows:

$$\text{specific gravity}^t = \frac{\text{wt. pycnometer} + \text{sample}^t - \text{wt. pycnometer}}{\text{wt. pycnometer} + \text{water}^t - \text{wt. pycnometer}} \quad (19)$$

where: t = temperature

The **Ostwald-Sprengel tube** is simply a pycnometer flask with side arms made of small-bore capillaries.

Pycnometers suffer from a number of problems that affect accuracy. One of the most obvious is the fact that the accuracy of the method is dependent upon the gravimetric equipment used, and this gravimetric measurement in turn is affected by humidity (or lack thereof!). In addition, maintaining a constant fixed temperature without stirring can be difficult. Because this piece of equipment is made from glass, volume changes may be encountered if it is exposed to excessive temperature or pressure changes. Errors can also be created by the introduction of bubbles. In general, pycnometers are accurate to the fourth decimal place and only when duplicate or triplicate measurements are made.

10.2.5.4.3 Westphal Balance Use of a Westphal balance is based on **Archimedes's principle:** A solid submerged in a liquid will lose in weight equal to the weight of the volume of liquid displaced. The **Westphal balance** consists of a glass plummet suspended from a knife-edge balanced beam that is supported on a stand equipped with a leveling screw. The plummet is commercially produced so that at 15.5°C it will displace exactly 5.0 g of water. The beam consists of 10 equal divisions. Placement of a 5.0 g weight (rider) on the tenth division (the division from which the plummet is suspended) results in the balance being in equilibrium in water (at 15.5°C), and the resultant specific gravity reading is 1.0. Movement of this 5.0 g rider to any of the other divisions indicates tenths of specific gravity. In a similar manner, riders of 0.5, 0.05, and 0.005 determine 10^{-1}, 10^{-2}, and 10^{-3} of the specific gravity reading, respectively.

The Westphal balance is calibrated to read specific gravities at 15.5°C only. The instrument cannot be used to read specific gravities at other temperatures without calibration or a table of temperature corrections.

10.2.5.4.4 Hydrometer The hydrometer is also based on **Archimedes's principle** that the same body displaces equal weights of all liquids in which it floats. The weight of the displaced liquid is equal to the product of its density and volume:

$$V_1D_1 = V_2D_2 \quad (20)$$

where: V_1 and V_2 denote the volumes of the liquids replaced by the same floating body, and D_1 and D_2 are their densities.

When the floating body is an upright cylinder (of uniform diameter), the displaced liquid volumes are proportional to the depth to which the body sinks:

$$D_1/D_2 = H_2/H_1 \quad (21)$$

The **hydrometer** consists of a weighted spindle whose stem is calibrated to read specific gravity when used at a specified temperature.

Hydrometers covering a wide range of specific gravities are normally used to determine the approximate value of the specific gravity of a liquid. More accurate specific gravity measurements can then be conducted by choosing a hydrometer more accurately calibrated in the desired region.

Special hydrometers (saccharometers) have been constructed to give readings other than specific gravity. **Baumé hydrometers** read in **degrees Baumé** (Be). This value can be converted to specific gravity:

$$\text{Baumé} = 145 - 145/\text{specific gravity}_{60/60} \quad (22)$$

Saccharometers are graduated to indicate the percentage of sucrose by weight. Balling, who first introduced the use of a hydrometer to directly read the percent sucrose in a water solution, calibrated his scale at 60°F. Therefore, **Balling saccharometers** are graduated to indicate percentage of sugar by weight at 60°F. **Brix saccharometers** are graduated to indicate the percent sucrose by weight at 17.5°C. The terms **Brix** and **Balling** are interpreted as the weight percent of pure sucrose. When applied to fruit juices and syrups, the hydrometer indicates the percentage composition of a sucrose solution that has the same specific gravity as the solution being tested. To obtain a true measurement of the solid content of these "real solutions" requires the use of Brix correction factors.

10.2.6 Modern Analytical Methods

In the last 25 years carbohydrate analysis has undergone dramatic change. With the advent/application and optimization of techniques such as high performance liquid chromatography, gas chromatography, nuclear magnetic resonance spectroscopy, and mass spectroscopy, separation, identification and quantitation of food carbohydrates (to parts-per-billion levels) has become routine. Powerful new techniques such as immunochemistry, supercritical fluid chromatography, capillary electrophoresis, and capillary high performance liquid chromatography will ensure continued evolution of rapid, precise, and accurate carbohydrate analysis.

10.2.6.1 High-Performance Liquid Chromatography (HPLC)

10.2.6.1.1 Introduction Although the author can be considered biased in this assessment, HPLC is currently the most facile, reproducible, accurate, and sensitive method for carbohydrate analysis in foods. In general, HPLC is the method of choice for the analysis of food carbohydrates.

The basic principles of HPLC are covered in this text (see Chapter 29) and will not be duplicated here. However, direct application of column chemistries and detectors to the analysis of food carbohydrates will be examined.

As mentioned previously, HPLC is a widely used technique for the analysis of food carbohydrates. There are four major reasons responsible for this universal acceptance:

1. HPLC provides a significant reduction in analysis time in comparison with conventional liquid chromatographic techniques (i.e., paper chromatography, thin-layer chromatography). This reduction in analysis time is accomplished over a wide range of sample concentrations, including trace levels, with a high degree of accuracy and precision.
2. HPLC is capable of generating high column efficiencies. Plate heights are very small in comparison with conventional liquid chromatographic techniques; therefore a large number of theoretical plates are possible for a given column length. This equates to shorter columns and/or the separation of complex sample mixtures.
3. HPLC, unlike gas chromatography, does not require the sample to have an appreciable vapor pressure. HPLC can be applied to the separation of monosaccharides, oliogosaccharides, and polysaccharides.
4. The growth and design in HPLC instrumentation has paralleled the demand for its application. Separations based on adsorption, partition, ion exchange, and size exclusion are possible.

The three important parameters in the separation and quantitation of carbohydrates by HPLC are the **stationary phase,** the **mobile phase,** and the **detector.**

10.2.6.1.2 Stationary Phases for HPLC A number of stationary phases have been successfully employed in carbohydrate analysis. Each of these phases will be discussed individually.

1. Normal Phase. Normal-phase chromatography, in which the stationary phase is polar and elution is accomplished by employing a mobile phase of increasing polarity, is a widely used HPLC method for carbohydrate analysis. Silica gel that has been derivatized with aminopropyltriethoxysilane, ethylenediaminopropyltrimethoxy-silane, or diethylenetriamino-propyltriemethoxysilane comprise amine bonded aminopropyl silica gels. These stationary phases when employed with acetonitrile/water (50 to 85% acetonitrile) as the eluent have been extremely effective in carbohydrate separations. In addition, aminopropyl/cyanopropyl silica gels have also been employed as effective stationary phases for carbohydrates. The elution order of carbohydrates employing these techniques is monosaccharides and polyhydric alcohols, then disaccharides, followed by oligosaccharides (Fig. 10-2). These columns have been successfully used to analyze the carbohydrate content of honey, beverages, breakfast cereals, ice cream, cakes, snacks, infant foods, fruits, vegetables, meats, and so on (41).

A severe disadvantage of amino bonded silica gel is the tendency for glycosylamine formation between reducing carbohydrates and the amino groups on the stationary phase, which results in a deterioration of column performance over time. This situation can be partially alleviated through the use of amine modified silica gel columns.

Amine modified or in-situ modified silica gel columns are different from amine bonded columns because the modification chemical is added to the mobile phase. This modifier must have at least two amino functions, as one is needed to adsorb to the silica gel and the other must be free for the specific carbohydrate. Because the modifier is in the eluent, the column is continuously regenerated. Modifiers such as piperazine, ethylenediamine, 1,4-diaminobutane, and tetraethylenepentamine (TEPA) have all found use in the separation of carbohydrates when used at low levels (0.01–0.1%).

2. Reversed Phase. The use of reversed-phase chromatography with water as the mobile phase is fairly recent (33) in carbohydrate analysis (Fig. 10-3). This system is applicable to the group separation of mono-, di-, and trisaccharides. Reversed-phase chromatography has been used to analyze the sucrose, raffinose, and stachyose content of soybeans and soybean products (34). In addition, juices, syrups, and brewery wort have

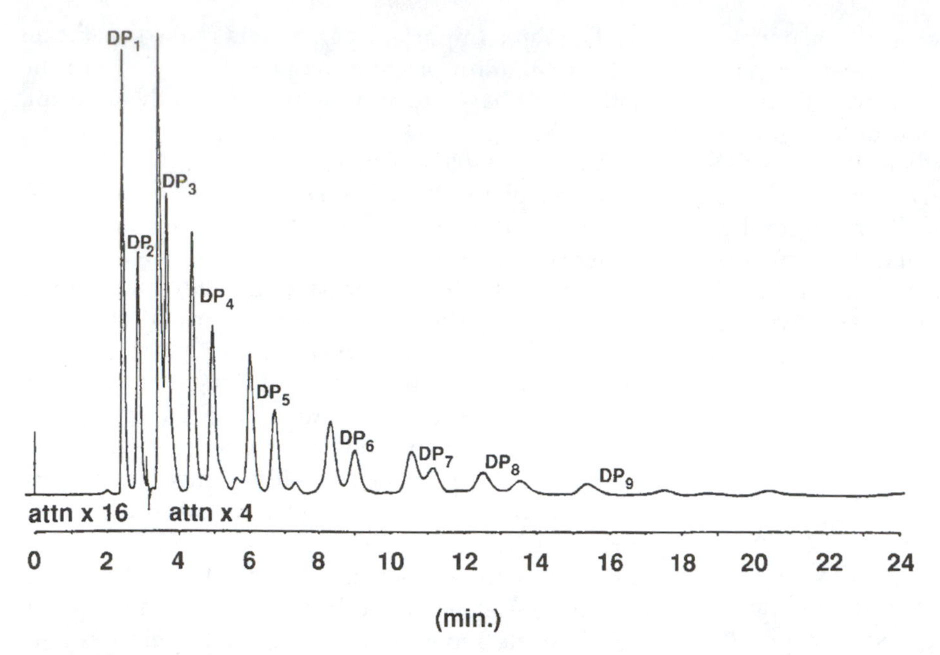

FIGURE 10-2.
High performance liquid chromatogram of dextrose polymers (DP1–DP9) employing a C_{18} stationary phase (reversed-phase). From (105), used with permission.

been analyzed for invert sugar, sucrose, maltose, and maltotriose (35, 36). Further information on the use of reverse-phase columns for unsubstituted carbohydrate separation can be found (37).

A major disadvantage of this stationary phase is the short retention times of monosaccharides, which result in elution as a single unresolved peak. The addition of salts (such as sodium chloride) can increase retention on the stationary phase and may increase the utility of this method for monosaccharide analysis (38). Retention times can also be increased by using a shorter alkyl chain or by using an octadecyl stationary phase with low carbon loading (12%) (39). Reversed-phase chromatography is complicated by peak doubling and/or peak broadening due to the presence of anomers. This problem can be alleviated by the addition of an amine (e.g., triethylamine) to the mobile phase, which will accelerate the mutarotation process. This addition has the unfortunate side effect of decreasing the retention time of the carbohydrate, which may affect separation.

Both normal- and reversed-phase columns have long life, high stability over a wide range of solvent composition and pH (>2 to <10), are suitable for the separation of a wide range of carbohydrates, and are of relatively low cost. All silica-based stationary phases share the disadvantage that the silica dissolves to a small extent in water-rich eluents.

3. Cation Exchange. Microparticulate spheres (5–12 μm) of sulphonated polystyrene-divinylbenzene, with 6 to 12 percent crosslinking, comprise the resin-based cation-exchange stationary phases. The resin is loaded with a variety of metal counter ions, depending on the type of separation desired. Usually Ca, Pb, or Ag are used as the counter ion; however, Zn has also been employed. The mobile phase used with these columns is aqueous with varying amounts (typically < 40%) of organic solvents such as acetonitrile and/or methanol. As with reversed-phase chromatography, an amine can be added to the mobile phase to avoid anomeric peak broadening. These columns are normally operated at elevated temperatures (>80°C) that increase column efficiency by increasing the mass transfer rate between the stationary and mobile phase, resulting in less band broadening and improved resolution (36). In addition, elevated column temperatures are effective against anomeric peak broadening.

Carbohydrate elution from cation-exchange resins takes place in order of decreasing molecular weight, the size-exclusion mechanism being predominant. The higher oligosaccharides elute first (>DP 3), followed by trisaccharides, disaccharides, monosaccharides, and polyhydric alcohols. There is some resolution of disaccharides, but the real strength of this stationary phase is in the separation of individual monosaccharides.

The obvious drawback with this cation-exchange stationary phase is the lack of resolution of oligosaccharides. Although there has been considerable activity recently (40) in the use of lower cross-linked resins (2–4%) for oligosaccharide separation, these resins are very soft and must be used at low flow rates (<0.5 mL/min) to avoid permanent damage of the stationary

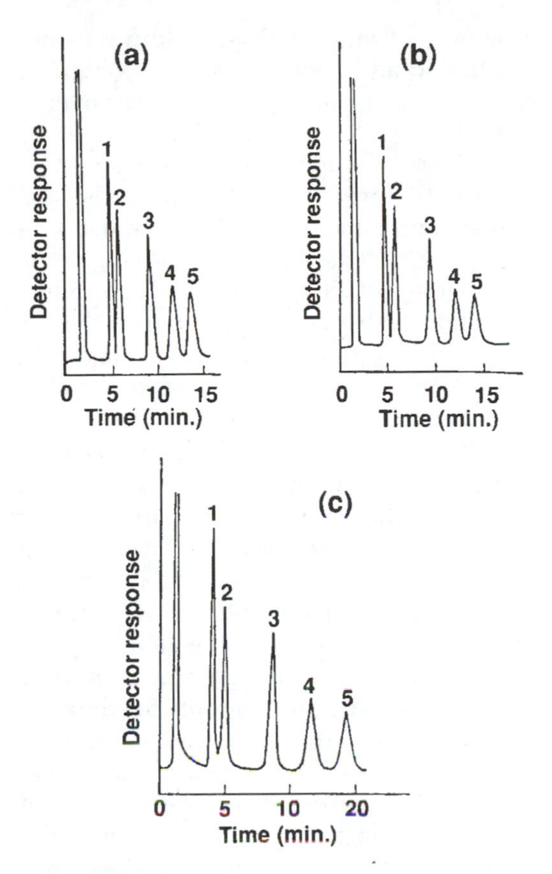

FIGURE 10-3.
High performance liquid chromatograms of (1) fructose, (2) glucose, (3) sucrose, (4) maltose, (5) lactose, employing different amino stationary phases. (a) Waters μ-Bondapak/Carbohydrate column. (b) Merck LiChrosorb NH_2 column. (c) Dupont Zorbax NH_2 column. Reprinted from (106), p. 184, by courtesy of Marcel Dekker, Inc.

phase. Cation-exchange resins can also be easily poisoned by the presence of Na and K ions in the sample, which can only be removed from the column by strong acid followed by reloading of the exchanger with the desired counter ion.

4. Anion Exchange. Carbohydrates are weak acids and have pK values in the range of pH 12 to 14. In a high pH solution such as 100 mM sodium hydroxide, the hydroxyl groups of carbohydrates are partially ionized and can be separated by anion-exchange resins. Special column packings have been developed employing 5 to 10 μm pellicular latex beads coated with a strong cation exchanger. These beads are electrostatically bound to small (approximately 0.1 μm) submicron ion exchange beads coated with a strong anion exchanger (42). The general elution sequence employing this stationary phase with sodium hydroxide as the mobile phase is polyhydric alcohols, monosaccharides, disaccharides, and oligosaccharides. The hydroxyl ions in the mobile phase displace the carbohydrate anions from the resin sites, and the pH at which this dissociation occurs is dependent on the pK of the carbohydrate. Sodium acetate can be added to the eluent to increase the ionic strength (without affecting the overall pH) of the mobile phase, resulting in sharper peaks and elution of more strongly retained oligosaccharides.

The use of this packing material in conjunction with a pulsed amperometric detector (see Chapter 29 and section 10.2.6.1.3) has proven to be one of the great breakthroughs in carbohydrate separation and detection. This technique has been used to examine the complex oligosaccharide patterns of honey (43), brewing syrups (44), beet sugar hydrolyzates, and orange juice (45).

A potential problem with this technique relates to the strongly alkaline conditions that are employed for elution. Reducing sugars are known to undergo enolization, and interconversion via a 1,2-enediol intermediate commonly known as the Lobry de Bruyn–Alberda van Ekenstein reaction (46). In addition, under extreme conditions, base and temperature base catalyzed hydrolysis of glycosides can occur (47). However, research has shown that under the conditions normally employed for HPLC analysis, no transformation products are formed (48). This method has the advantage of being applicable to baseline separation within each class of carbohydrates with rapid separation of dextrose polymers up to DP 85. Because the mobile phase is aqueous sodium hydroxide, the system is inexpensive to run.

5. Others. Other stationary phases that have been used in carbohydrate separation are diol/polyol bonded silica. These packing materials exhibit similar elution patterns to amino bonded phases (with acetonitrile/water as the mobile phase), but without the problem of glycosylamine formation. Cyclodextrin bonded silicas (both α and β) have also been used for carbohydrate separation and may be extremely useful for the separation of enantiomers (41).

10.2.6.1.3 Detection Systems for HPLC **1. Refractive Index.** The most common detector employed for carbohydrate analysis is the refractive index (RI) detector. RI measurements are linear over a wide range of carbohydrate concentrations and can be universally applied to all carbohydrates.

As previously mentioned in this chapter (see section 10.2.5.3), RI is a bulk physical property that is sensitive to changes in flow, pressure, and temperature. With modern HPLC equipment and a temperature-controlled detector, problems arising from these changes are minimal. The most significant limiting factor with RI detection is that the application of gradient elution is not readily feasible. However, gradient programming using RI has been performed by running the gradient program prior to sample addition. With computer-controlled instruments, the resulting electronic signal can then be subtracted from the gradient elution of the actual sample. Although this method is cumbersome and lengthy, it may have applications in the future. Another serious drawback is the fact that RI detection is nonspecific. Therefore noncarbohydrate compounds may co-elute with carbohydrates, causing inaccurate results. In addition, refractivity is not sensitive enough to detect samples of less than 10 nmol.

More recent advancements in the use of a laser-based refractive index detector with microbore (0.25 mm internal diameter) HPLC columns have been employed to detect ng quantities of mono-, di-, and trisaccharides (49). The authors point out that the detector is simple to use and that the only critical step is in the alignment of the cuvette (perpendicular to the laser beam). Although this laser-based detector is not yet commercially available, the authors estimate the total cost to be less than $1,200.

2. Ultraviolet. Generally, carbohydrates cannot be detected by absorption in the ultraviolet (UV), visible regions or by fluorescence because they lack chromophores/flurophores. However, carbohydrates do exhibit absorption maxima in the near-ultraviolet region of 180–220 nm (50) due to the carbonyl functional group. Unfortunately, a number of mobile phase (solvent) contaminants also absorb electromagnetic radiation at this wavelength. Therefore, expensive high-purity solvents must be used in the direct application of UV to carbohydrate detection, and this is the main reason for the paucity of published information on this technique.

The formation of strong UV-absorbing chromophores of carbohydrates can be readily achieved via precolumn or postcolumn derivatization. A number of derivatives are commonly employed for precolumn derivatization. These include benzylation, benzoylation, arylosazone, O-methyloximes, p-toluenesulfonylamides, and phenyldimethylsilyl. These derivatives may be detected by absorption in the region of 230–280 nm. Fluorescent chromophores include dansylhydrazones, dansylamides, and 1-(N-2-pyridyl-amino)deoxyalditols, which can be measured at (excitation/emission) 350 nm/500 nm, 340 nm/470 nm, and 310 nm/380 nm, respectively. Postcolumn derivatization techniques include color reactions (orcinol, resorcinol, and anthrone); flurogen production, which is the reaction of reducing carbohydrates with aliphatic amines (such as ethylenediamine, ethanolamine, arginine, and 2-cyanoacetamide); and lutidine formation, which is the condensation reaction between the carbohydrate and periodate to form formaldehyde, which is then reacted with ammonia and 2,4-pentanedione (50).

The advantage of pre- and postcolumn derivatization of carbohydrates is the greatly enhanced detection limits obtained, typically to levels of 1 ng. Although many of these derivatization techniques can be automated, they still add additional steps to carbohydrate analysis, with the result being an increase in the overall analysis time.

3. Electrochemical. Electrochemical detectors based on the oxidation/reduction of compounds (specifically organic compounds) have been used extensively with HPLC. Problems with the poisoning of the electrode due to accumulation of the oxidized product on the surface of the electrode severely limited the use of these types of detectors for carbohydrate analysis. In 1983, Rocklin and Pohl (51) introduced a triple-pulsed electrochemical detector that overcame this problem of electrode poisoning. They employed a gold electrode with a sequence of three potentials. The first potential oxidized the carbohydrate, the second cleaned the electrode surface, and the third reduced the gold oxide formed back to gold. The use of this detector in conjunction with anion exchange chromatography, employing strong base as the mobile phase, has revolutionized the use of HPLC for carbohydrate analysis (see Fig. 10-4). The optimum pH for the electrochemical detection of carbohydrates is 13. Because electrochemical detectors of this type are sensitive to slight pH changes (which result in baseline fluctuations), postcolumn addition of 300 mM sodium hydroxide is often employed to alleviate baseline drift. The drawback to this postcolumn addition of sodium hydroxide is dilution of the sample prior to detection.

This system is compatible with gradient elution, and the solvents employed are inexpensive (e.g., sodium hydroxide, sodium acetate). On column detection, limits are approximately 1.5 ng for monosaccharides and 5 ng for di-, tri-, and tetrasaccharides. The detector is suitable for both reducing and nonreducing carbohydrates; however, the detection limits are slightly lower for reducing sugars.

4. Mass. The mass detector is reasonably sensitive and suitable for nonvolatile solutes. The principle of operation of this detector is based on the evaporation of solvent following nebulization by a heated column. The resulting finely divided solute particles pass through a light beam. Light scattered from the solute particles is detected by a photomultiplier that is placed at an angle (120°) to the light beam. The signal is amplified and can be recorded on a conventional chart recorder. In most commercial instruments, internal reflection is minimized by a light trap (52).

This detector has been employed for the analysis of glucose and maltose in a glucose syrup and of stachyose, raffinose, and sucrose levels in wheat germ and soybean extracts (53). The mass detector is quite sensitive, 2 to 3 mg for mono-, di-, and trisaccharides, and is linear over a fairly wide concentration range (10–200 mg).

5. Others. Other HPLC detectors have been and are

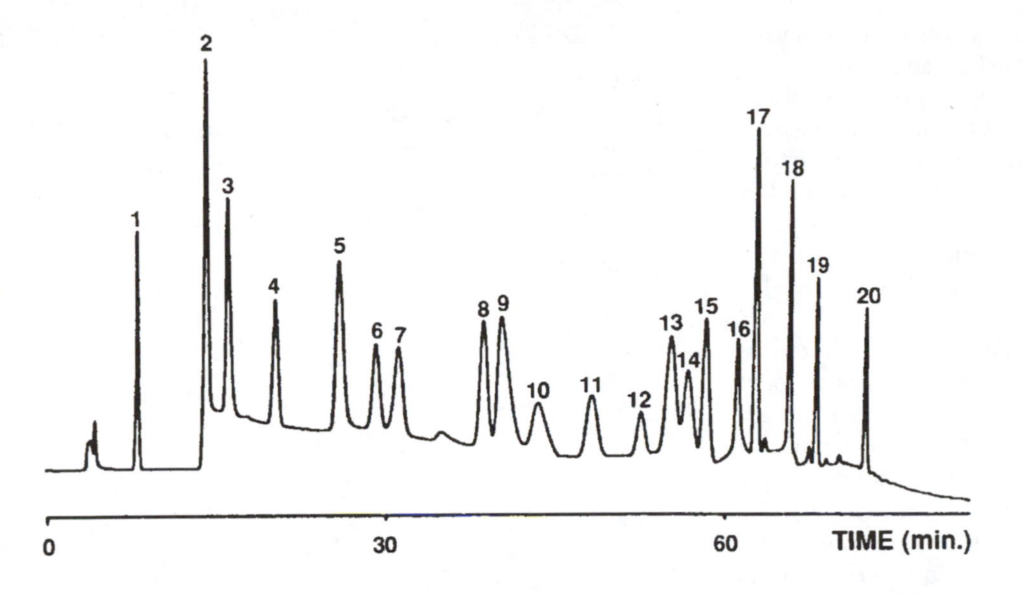

peak	carbohydrate	T_r, min
1	neotrehalose	7.8 ± 0.3
2	glucose	13.7 ± 0.1
3	fructose	15.5 ± 0.1
4	melibiose	19.9 ± 0.1
5	isomaltose	24.9 ± 0.3
	maltulose	
6	sucrose	28.2 ± 0.4
7	kojibiose	30.0 ± 0.4
8	turanose	37.9 ± 0.4
	gentiobiose	
9	palatinose	39.3 ± 0.4
10	melezitose	43.2 ± 0.5
11	isomaltotriose	47.6 ± 0.5
12	nigerose	51.2 ± 0.4
13	maltose	53.2 ± 0.5
	1-kestose	
14	theanderose	55.4 ± 0.6
15	laminaribiose	57.2 ± 0.7
16	isopanose	60.3 ± 0.5
17	erlose	61.4 ± 0.4
18	panose	63.9 ± 0.5
19	maltotriose	66.4 ± 0.6
20	laminaritriose	71.1 ± 0.5

FIGURE 10-4.
High-performance liquid chromatogram of honey monosaccharides and oligosaccharides employing anion exchange chromatography with pulsed amperometric detection. Reprinted with permission from (43). Copyright 1990, American Chemical Society.

being used for carbohydrate detection. These include nuclear magnetic resonance (54), conductivity (55), electrochemical/platinum electrode in conjunction with immobilized fructose 5-dehydrogenase (56), optical activity, and a pressure ionized mass spectrometer (57), just to name a few. The application of these instruments to the analysis of carbohydrates is indicative of the enormous interest in this field.

10.2.6.2 Gas Chromatography (GC)

10.2.6.2.1 Introduction As was the case with HPLC, the general principles of gas chromatography are covered elsewhere in this text (Chapter 30) and will not be introduced here. However, the direct application of column chemistries, reducing techniques, and derivatization of carbohydrates will be presented.

Due to progress in HPLC technology, the application of gas chromatography to carbohydrate analysis is rapidly becoming obsolete. This is partially due to the fact that carbohydrates are nonvolatile; therefore, they must be derivatized prior to GC analysis. In addition, the analysis times (that is, the instrument time necessary to actually perform the analysis) are typically longer for GC than for HPLC. These drawbacks became easy to overlook a few years ago because of the increased sensitivity of the GC flame ionization detector (FID; detection limits of 5 ng) when compared to the HPLC RI detector (detection limits of approximately 10 μg). However, with the introduction of more sensitive HPLC detectors (e.g., pulsed amperometric detector), detection limits are no longer a factor. This trend is unfortunate because GC has some significant separatory advantages over HPLC, specifically in the area of structurally similar di- and trisaccharides (58).

10.2.6.2.2 Carbohydrate Derivatives for GC Derivatization of the carbonyl group of carbohydrates is of particular importance. If this group is not derivatized, multiple peaks will be observed chromatographically for each reducing carbohydrate. Reducing sugars differ from most other organic compounds in one characteristic property. When a reducing sugar is dissolved in a solvent, the solution can contain up to six compounds (2). The six compounds in solution include two pyranoses (α and β), two furanoses (α and β), and the acyclic carbonyl form and its hydrate (Fig. 10-1). Therefore, reduction of a reducing carbohydrate with either sodium borohydride (to yield the alditol) or hydroxylamine-HCl (to yield the oxime) is necessary.

Problems arise during **oxime formation** because two products are formed, corresponding to the syn and anti forms of the reduced carbohydrates. This problem is not severe with respect to aldoses because subsequent derivatization to the acetate or trifluroacetate converts the oxime to the nitrile. The nitrile imparts no stereochemistry, and each aldose yields a unique aldononitrile (Fig. 10-5). However, this situation does not hold for ketoses or oxime derivatives that are subsequently silylated. In these cases, two compounds are produced upon reduction, and each must be quantitated individually. Obvious problems can arise during analysis because the quantity of each compound produced is related to carbohydrate structure, reaction time, solvent, and temperature of the reaction.

D-Glucose → (NH_2OH HCl) → Oxime → ($-H_2O$) → Nitrile

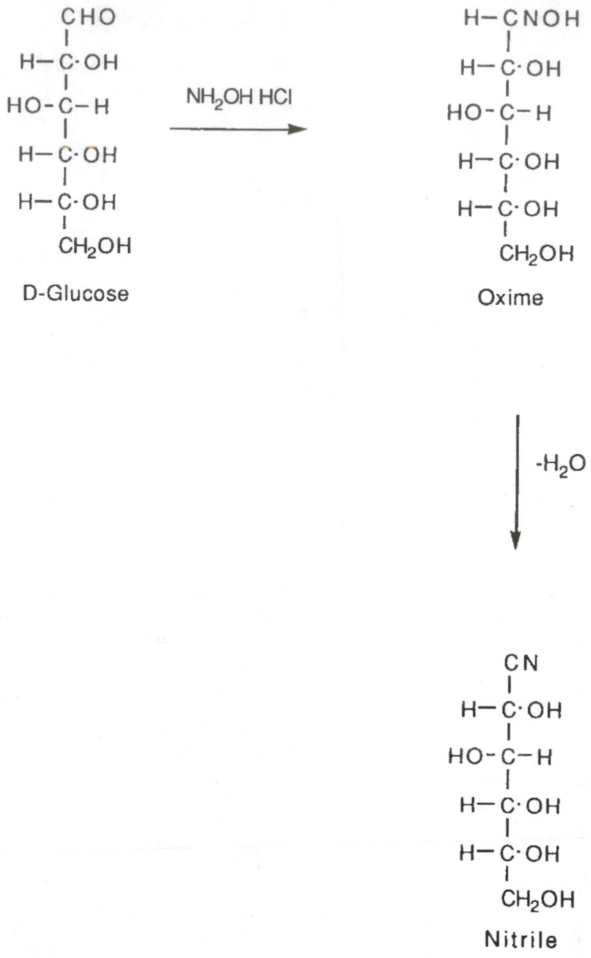

FIGURE 10-5.
Aldononitrile formation.

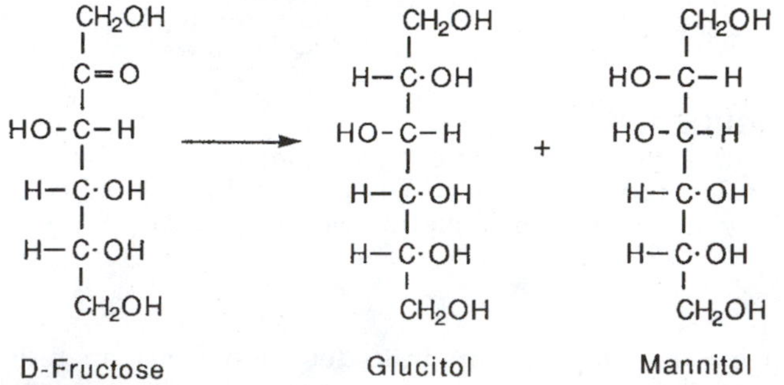

D-Fructose → Glucitol + Mannitol

FIGURE 10-6.
Reduction of D-fructose.

Alditol formation does not result in any stereochemistry being introduced at the anomeric carbon. Reduction of aldoses results in the formation of a unique alditol. However, keto sugars yield two compounds upon reduction. For example, when fructose is reacted with sodium borohydride, glucitol and mannitol are formed (Fig. 10-6). This production of two compounds from a ketose sugar can lead to significant problems when the sample being analyzed contains both glucose and fructose. In most cases this is not a problem in food, as the amount of fructose in the sample can be determined by quantifying the mannitol peak produced (with comparison to a known quantity of reduced fructose). The amount of glucose is then determined by subtracting the contribution of glucitol produced from fructose. Solutions of fructose upon reduction produce ratios of glucitol/mannitol that are reproducible. Real difficulties are only observed when the sample being analyzed contains a significant quantity of mannose.

The polar groups of carbohydrates, specifically the hydroxyl groups, make these compounds nonvolatile. Derivatizing these groups greatly increases the volatility, affording the analysis of mono- and oligosaccharides (up to hexasaccharides) by GC techniques (59).

Hydroxyl groups can be silylated, acetylated,

trifluor-acetylated, methylated, or ethylated. The two most common of these derivatives are silylation and acetylation. **Silylation** of carbohydrates is usually carried out by employing a mixture of trimethylsilylchloride and hexamethyldisilazane in pyridine. The reaction time is normally quite short, <5 min at 60 to 80°C, and the derivatization can even be carried out at room temperature (60). The sample must be relatively free from moisture because water hydrolyzes both the final product and the reactants. The resulting product from carbohydrate silylation is the trimethylsilylated derivative. The addition of the trimethylsilyl group to each hydroxly moiety greatly increases the molecular weight of the final product and limits the usefulness of this derivative for higher oligosaccharides.

The **acetate derivatives** are prepared by reacting the carbohydrate with either a mixture of acetic anhydride/anhydrous acetic acid/sulfuric acid or pyridine/acetic anhydride. Derivatization is carried out at elevated temperatures (>60°C) for 10 to 30 min (depending on the carbohydrate). More recent derivatization techniques employ catalysts such as 4-dimethylaminopyridine or N-methylimidazole (59). As is the case with silylation, the sample must be relatively water free to avoid inactivation of the reactants. The resulting molecular weight of the derivatized carbohydrate is much less than that obtained in silylation and facilitates GC analysis of higher oligosaccharides.

10.2.6.2.3 Capillary GC (CGC) A major improvement in the analysis of carbohydrates by GC was the development (61) of capillary columns. Capillary columns with an internal diameter of 0.100, 0.25, 0.32, and 0.53 mm and lengths of 15, 30, and 60 m are commercially available. These columns are made from fused silica (metal-free silica) and are coated with a polymeric material for flexibility. The liquid (stationary) phase is coated (film thickness of 0.2–1.0 μm) and, in most cases for carbohydrate work, covalently bonded to the capillary walls. The choice of film thickness and column diameter is dependent on the amount of sample to be injected on the column. Sample capacity in units of μg per compound injected is approximately 50 for 0.25 mm i.d., 500 for 0.32 mm i.d., 1,500 for 0.53 mm i.d., and 5 ng for 0.100 i.d. These columns have replaced traditional packed columns of glass or stainless steel that had internal diameters of 2–4 mm and were 0.9–2 m in length. There are three main *advantages* to the use of capillary columns in place of traditional packed columns for carbohydrate analysis:

1. Much shorter analysis times. Because the columns are open tubular, there is rapid mobile phase movement through the column.
2. Increased resolution of carbohydrates, specifically structurally similar carbohydrates. Capillary columns have the ability to separate α and β anomers, and with chiral stationary phases D and L sugars can be resolved. A 30 m (0.25 mm i.d.) capillary column has > 100,000 theoretical plates.
3. Improved quantitation of trace carbohydrates due to minimal stationary phase bleeding.

A variety of stationary phases have been used for the analysis of carbohydrates, which include polar (up to 50% cyanopropylsilicone) and nonpolar (dimethylsilicone) (62, 63). Employing a stationary phase of intermediate polarity (5% phenylmethylsilicone), the quantitative analysis of simple monosaccharides and structurally complex oligosaccharides has been accomplished (58) (Fig. 10-7).

10.2.6.2.4 GC Detection of Carbohydrates For most samples, the **flame ionization detector** (FID) is the detector of choice for carbohydrates. The detector is quite sensitive, with detection limits of approximately 5 ng for monosaccharides, and has a broad linear range. The **electron capture detector** is more sensitive (100–1000x) than the FID detector but is only suitable for halogenated derivatives such as the pertrifluoroacetates. **Specific detectors,** such as the nitrogen detector, can be used for certain classes of carbohydrates (amino sugars) but offer only a minor increase (10x) in sensitivity over the FID detector.

Combined GC and **mass spectrometry** (MS) (where the mass spectrometer is the actual detection system) affords both qualitative and quantitative anal-

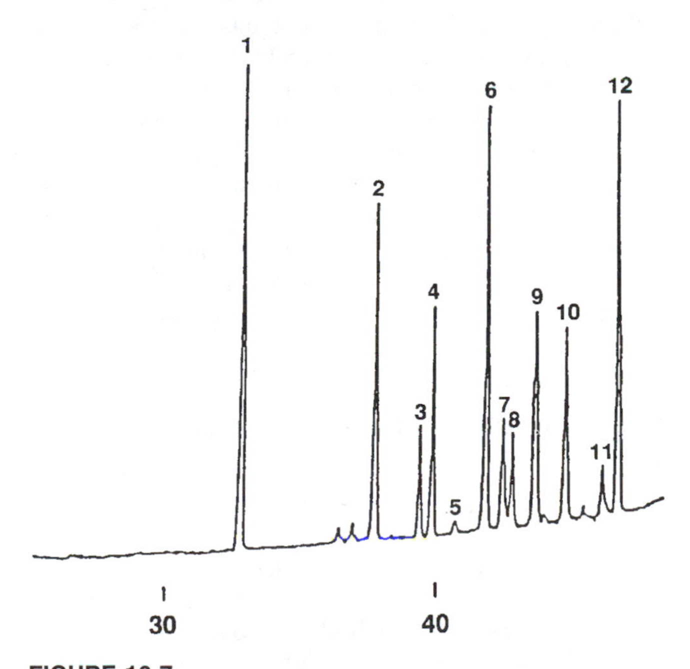

FIGURE 10-7.
Capillary gas chromatogram of silylated structurally similar disaccharides. From (58) with permission.

ysis of carbohydrates. The principles and instrumentation of GC-MS are covered in Chapter 26.

10.2.6.3 Mass Spectroscopy

Molecular weight determination is of considerable importance in determining the structure of an organic compound. A **mass spectrometer** is an instrument that affords the determination of the molecular weight of an organic compound. A mass spectrometer bombards the substance of interest with an electron beam and records the result as a spectrum of positive ion fragments (see Chapter 26 for a more complete explanation of mass spectroscopy). Separation of the positive ion fragments is on the basis of mass—strictly, mass/charge. However, the majority of ions are singly charged.

Underivatized carbohydrates (specifically mono- and oligosaccharides) are not sufficiently volatile to be analyzed directly by MS and require conversion to more volatile compounds such as their acetal, ester, or ether derivatives. A second technique for the molecular weight determination of carbohydrates is reaction of the sample with **secondary ions** (i.e., methane ions CH_5^+ and $C_2H_5^+$, or ammonia, NH_4^+), which are produced by electron impact on methane or ammonia. These secondary ions react with the sample ion (RH) to produce RH_2^+ and/or $(RH + NH_4)^+$. This technique is called **chemical ionization.**

Another ionization technique called **fast atom bombardment** (FAB) (64) has been successfully employed to determine the molecular weight of biomolecules (specifically carbohydrates) directly. In this technique, an accelerated beam of xenon atoms is fired from an atom gun toward a metal target that has been coated with a viscous liquid, called the matrix, which contains the carbohydrate sample. The matrix most commonly used for carbohydrates is glycerol. When the beam impacts the matrix, kinetic energy is transferred to the sample molecules and gas phase ions are generated (this process is referred to as **sputtering**). The sputtered ions, which include molecular ions, are then accelerated from the ion source and analyzed (65).

10.2.6.4 Nuclear Magnetic Resonance Spectrometry

Nuclear magnetic resonance (NMR) spectroscopy (see Chapter 27 for a complete description of this method) is simply another form of absorption spectroscopy. Under the appropriate conditions, a sample can absorb electromagnetic radiation in the radio frequency range. The frequencies absorbed are governed by the three-dimensional structure of the sample. A plot of the frequencies of the absorption peaks versus peak intensities constitutes an NMR spectrum.

NMR is an extremely powerful tool for both qualitative and quantitative analysis of food carbohydrates. The technique is nondestructive and is quite rapid if a sufficient quantity of the carbohydrate is present. In addition, NMR allows elucidation of carbohydrates in both their static and dynamic structures.

A number of modern NMR techniques are available to determine the absolute structure of carbohydrates. If the carbon assignments are known (by ^{13}C spectroscopy), two-dimensional heteronuclear correlation experiments can be conducted to determine proton assignments. Proton coupling can be determined using two-dimensional J-correlated (COSY) experiments. The basic carbon structure can be elucidated with two-dimensional ^{13}C-correlation. A complete explanation of these techniques is not possible here but may be found in the literature (66, 67).

Routine nuclear magnetic resonance spectroscopic experiments can provide a wealth of information on carbohydrate samples. This information, based on ^{1}H and ^{13}C data, includes:

1. In most instances, the chemical shifts of individual ring protons can be determined in the ^{1}H-NMR spectrum, and structural assignments may be made from these values. In general, anomeric protons resonate at lower field than other methine or methylene protons. Anomeric ring protons in the equatorial position generally resonate at lower field than their axial counterparts.
2. For ^{1}H-NMR spectroscopy, the area under the peaks is proportional to the number of protons present in the sample. Therefore, with the use of a suitable internal standard, the quantity of the compound can be determined.
3. The decoupled ^{13}C-NMR spectrum provides the total number of carbon atoms in the carbohydrate molecule. Therefore, in general, hexoses will have six resonance peaks, disaccharides (comprised of two hexoses) 12 and so on.
4. The chemical shifts observed in the decoupled ^{13}C-NMR spectrum of an oligosaccharide can be used to identify both the position of attachment of the monomeric units and the stereochemistry of this attachment (i.e., if the monomers are attached in an α or β conformation).
5. The area under the ^{13}C-NMR resonances can be used to quantitate the amount of carbohydrate present. This fact holds when a standard curve for the carbohydrate being analyzed is determined. In addition, a paramagnetic relaxation agent (such as chromium acetylacetonate) must be added to the sample to normalize the nOe

(nuclear Overhauser enhancement) and to allow spin-lattice relaxation to be more efficient, which results in the ability to do more frequent pulsing.

6. The decoupled ^{13}C-NMR spectrum of the anomeric carbon region can be used to identify the structure and concentration of a mixture of carbohydrates in solution.

Some examples of the use of high resolution ^{1}H and ^{13}C-NMR for carbohydrate analysis include measurement of the sucrose content in raw and processed sugar beets (68); the glucose, fructose, sucrose, and maltose content in wheat (69); qualitative and quantitative analysis of carbohydrates in *Pinus radiata* (70) and honey (71) (Fig. 10-8); analysis of carbohydrates in tropical root crops (72); and the analysis of Amadori- and Heyns-rearrangement products formed during processing and storage of foods (73).

Low resolution pulsed ^{1}H-NMR has also been used to study carbohydrates in food systems. This technique has been especially useful in the study of carbohydrate hydration (74) and solvation (75).

10.2.6.5 Immunoassays

Immunochemical techniques are based on the specific interactions of antibodies with antigens. There are a number of methods by which these interactions can be detected, and the principles of these techniques and their application to food analysis are covered in this text (Chapter 33) and elsewhere (76, 77).

Low molecular weight carbohydrates must be covalently linked to proteins in order to produce an immune response. This small molecule-protein complex is referred to as a **hapten** and when introduced into an animal can result in antibody formation.

In food systems, analysis employing immunoassays is certainly the method of the future. These techniques are compound specific and are extremely sensitive (ng levels are routinely attained), and make immunoassays the method of choice for the future.

With respect to carbohydrates, a number of specific antibodies have been produced commercially. These include raffinose (Westinghouse Bio-Analytical); blood group antigens, which are oligosaccharides (Chembiomed); and specific oligosaccharide linkages such as Gal a-(1→4) Gal (Monocarb), as examples. In addition, antibodies can also be produced to carbohydrate reaction products, such as the chemical products produced from the reaction of sulfathiazole and reducing carbohydrates (76).

Immunoassays are not without their drawbacks, one of which is cross reactivity. The commercial antibody for raffinose also reacts with melibiose (at levels

disaccharide	chem shift, ppm	disaccharide	chem shift, ppm
O-methyl β-D-ribofuranoside	108.53	maltose	100.77
laminaribiose	104.30	turanose	100.32
sucrose	103.87		99.57
	91.64	nigerose	99.59
gentiobiose	103.61	kojibiose	99.20
cellobiose	103.59	palatinose	98.70
neotrehalose	103.56		98.61
	100.68	isomaltose	98.62
maltulose	100.79	trehalose	92.93
	99.56		

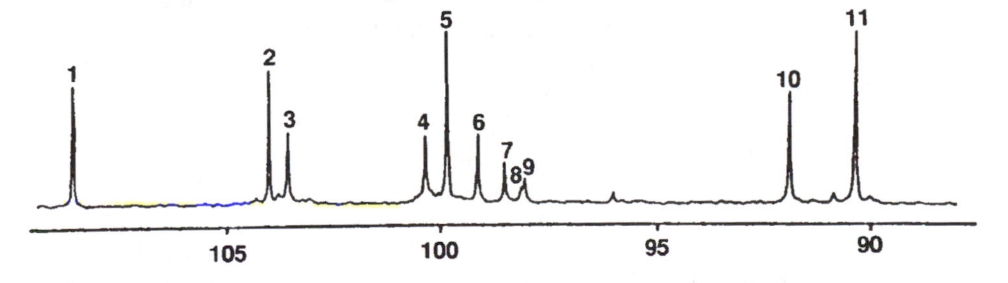

FIGURE 10-8.
^{13}C-Nuclear magnetic resonance spectroscopy (including chemical shifts) of reduced disaccharides. Reprinted with permission from (71). Copyright 1988, American Chemical Society.

of 0.8%) and sucrose. However, the major problem with immunoassays is in the preparation stage because development and purification of antibodies is both a science and an art.

10.2.6.6 Other Modern Carbohydrate Analysis Methods

As mentioned previously, the analysis of carbohydrates is rapidly changing. This section will briefly describe three different chromatographic techniques that can be used for carbohydrate analysis. These techniques were chosen for the following reasons: (1) They are not specifically covered elsewhere in this text, (2) they are fairly recent and have had very little literature exposure in the field of carbohydrate analysis, and (3) the equipment is commercially available.

10.2.6.6.1 Microscale HPLC The internal diameter of the packed columns generally used in HPLC is 4–6 mm. Micro-HPLC, as introduced by Horvath et al. (77) and Ishii et al. (78), has column internal diameters ranging from 0.2–0.01 mm. The *advantages* of employing small diameter columns are:

1. Reduced usage of both the mobile and stationary phase. This facilitates the use of expensive and novel mobile and stationary phases.
2. High resolution. With open tubular columns (similar to capillary GC), a large number of theoretical plates are possible, which results in increased resolution.
3. Increased mass sensitivity. Assuming the same column efficiency, the concentration of solutes eluting from the column is inversely proportional to the square of the inner diameter. This means that micro-HPLC will gain mass sensitivity if a concentration sensitive detector is employed. In addition, the small diameter columns will allow direct coupling to mass spectrometers.
4. The application of temperature programming.
5. Greater choice in operating conditions and reduced equilibrium times.

This technique has been successfully to separate food components, including carbohydrates (Fig. 10-9). An excellent treatment of micro-HPLC has recently been published (79).

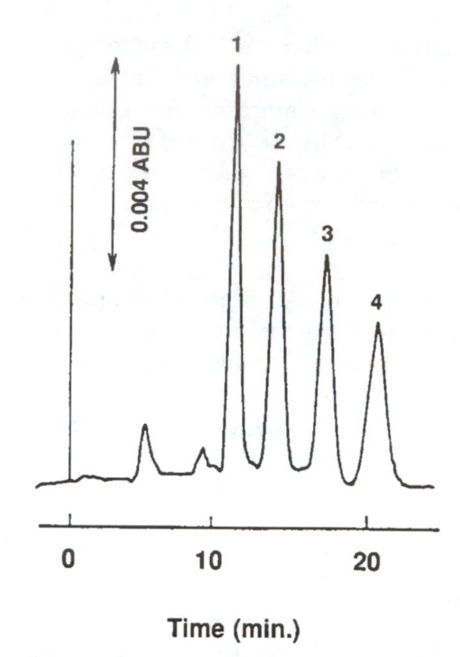

FIGURE 10-9.
Microscale high performance liquid chromatogram of (1) xylose, (2) glucose, (3) fructose, (4) lactose, on a 15 cm × 0.5 mm i.d. polytetrafluoroethylene (PTFE) tube packed with Nucleosil NH_2. Each peak corresponds to 9.5 μg of carbohydrate. From (79), reprinted with permission by VCH Publishers © 1988.

10.2.6.6.2 Supercritical-Fluid Chromatography (SFC) Although supercritical fluid extraction was first introduced in the late 1800s (80), the first chromatographic use of this principle was not demonstrated until 1962 (81). The application of SFC to compounds that are nonvolatile, polar, thermally labile, or of high molecular weight is very recent (82).

A supercritical fluid is one that is above its critical pressure and critical temperature. In general, these supercritical liquids exhibit characteristics between those of liquids and gases. A number of supercritical fluids have been used in food applications, including carbon dioxide, nitrous oxide, trifluoromethane, sulphur hexafluoride, pentane, and ammonia. The solvent of choice for most food applications is carbon dioxide; however, this solvent is not useful for carbohydrate analysis (as carbohydrates are insoluble in this solvent). The addition of a small amount of an entrainer, that is, a compound of intermediate volatility (such as methanol), dramatically affects compound solubility and selectivity (i.e., CO_2 with an appropriate entrainer is suitable for carbohydrate extraction and analysis).

When employed as mobile phases, supercritical fluids confer chromatographic properties intermediate to liquid (LC) and gas chromatography. The low densities and high diffusivities of gases give GC superior resolution; conversely, high liquid densities are responsible for the excellent solvating power of LC. SFC overcomes the disadvantages of GC, that is, the limited volatility and thermal stability of many organic compounds, and the disadvantages of LC, longer analysis times and reduced resolution.

SFC offers flexibility in chromatographic techniques that can be used for the separation and detection of organic compounds. This technique makes possible the separation of nonvolatile, thermally labile organic compounds outside the volatility and stability range of GC. Although this technique has not traditionally been employed for carbohydrate separation and analysis (the technique has mainly been used for nonpolar compounds), it is certainly applicable to these compounds (83).

10.2.6.6.3 High-Performance Capillary Electrophoresis (HPCE) High-performance capillary electrophoresis is recognized by many researchers as a breakthrough in the science of compound separation. The basic equipment setup of HPCE is shown in Fig. 10-10. In principle, a small plug of sample solution is introduced into one end of a fused silica capillary tube that contains an appropriate electrolyte (carrier). When high voltage is applied between the ends of the capillary tube, the surface of the inner wall of the tube is negatively charged and the boundary monolayer of the carrier that is in contact with the inner wall becomes positively charged. This layer then moves towards the cathode in conjunction with the medium of the carrier to generate electroosmotic flow. When the sample is introduced at the anode, all components of the sample, both charged and neutral, will migrate to the cathode at a uniform velocity. Anionic components are held back by electrostatic forces (84).

Under optimum conditions, HPCE can analyze less than a nanoliter of sample with a theoretical plate number approaching or exceeding 1 million and with a detection sensitivity at the attamole level.

Unfortunately, carbohydrates in general are neutral molecules and travel as a single band at the electroosmotic flow rate and are therefore unresolved. This problem has been recently resolved by:

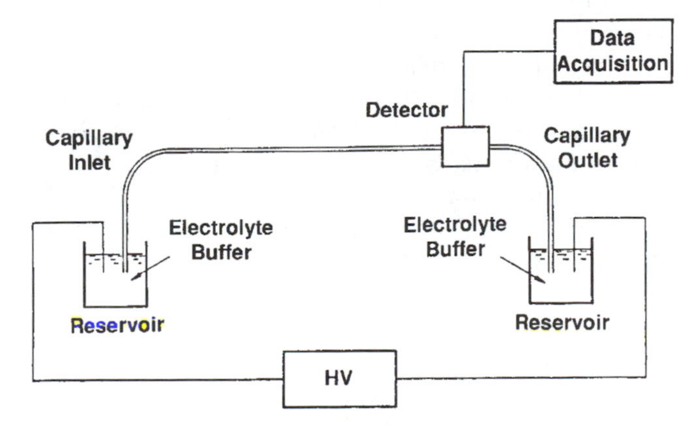

FIGURE 10-10.
Diagram of a high performance capillary electrophoresis unit.

1. Interaction of carbohydrate molecules with micelles that have been incorporated into the electrolyte solution. This mode of HPCE separation is known as micellar electrokinetic capillary chromatography (MECC) (85).
2. Conversion of the carbohydrate in situ to anionic borate complexes (84).

The commercially available detectors for HPCE are based on absorption. Therefore, the carbohydrates must be derivatized prior to analysis (84, 86, 87). Recently, HPCE employing indirect UV has been used as a method for carbohydrate detection (88).

10.2.7 Starch Analysis

All natural food starch contains two types of homopolysaccharide materials: **amyloses,** which are essentially linear (although some starch amylose is minimally branched) polymers of D-glucose, α1->4 linked; and **amylopectins,** which are highly branched polysaccharides made up of D-glucose, α1->4 and α1->6 linked. In general, the molecular weight of amylopectin is much greater than that of amylose and results in amylopectins being much less dispersable in water.

Lower molecular weight carbohydrates can be removed from starch by extraction with 80 percent ethanol. Further treatment with hot water will remove most of the amylose and the dextrins, leaving mainly amylopectin as the residue. Starch can be extracted from foods (more specifically, cereals) by direct treatment of the food with a hot calcium chloride solution, treatment of a gelatinized sample with perchloric acid, or extraction with dimethyl sulfoxide.

Following extraction, starch can be quantitatively determined by **polarimetry** [research has shown that a specific optical rotation of +203 could be taken for all cereal starches (89)]; by acid hydrolysis (AOAC Method 920.44), employing sulfuric or hydrochloric acid; by enzymatic hydrolysis (AOAC Method 979.10), employing glucoamylase (amyloglucosidase); followed by reducing sugar analysis (see sections 10.2.3.1 and 10.2.3.4). These methods were recently employed and evaluated in the analysis of cassava starch (90).

More recently, the glucose produced from starch hydrolysis has been determined using **cerium (IV) oxidation.** The reaction involves the colorimetric conversion of Ce (IV) to Ce (III), which can be monitored at 445 nm. This analysis is simple and rapid and was used to determine the starch content of a number of commercial starch hydrolyzates (91). The glucose produced from starch hydrolysis (either by acid or enzymes) can also be determined chromatographically by HPLC or GC methods.

Recently a great deal of interest has been shown

in the use of NMR for starch analysis. Proton and ^{13}C NMR has been successfully used to analyze modified starches for both their structure and quantity (92).

The accuracy of starch analysis can be compromised by interfering substances such as phenolics, lipids, and proteins. Lipids and phenolics can be easily removed by extraction with alcohol or a mixture of methanol, chloroform, and water prior to hydrolysis (93). Samples with high protein levels can be treated with **Carrez reagent,** which results in protein precipitation. It has been shown that there is a gradual increase in apparent starch content with increasing Carrez reagent addition, and the recommendation is for the addition of 5 mL each of Carrez I and Carrez II solutions (94).

The purity of commercial enzyme preparations is also important when examining the accuracy of starch analysis. These preparations may contain trace levels of hemicellulases, pectinases, and cellulase that can produce overestimations in starch content.

Treatment of starch with perchloric acid can solubilize pectic substances. This interference can be avoided by treatment of the starch with iodine, which precipitates the starch fragments (95). However, this procedure is unreliable because the glucose produced by perchloric acid treatment of starch is not precipitated by iodine.

Each method employed for starch analysis has some inherent error associated with it, and care must be taken to ensure that this error is minimized. Recent reviews on starch analysis methods are available for interested readers (96, 97).

10.2.8 Structural Polysaccharide Analysis

Structural polysaccharide includes **pectic substances, hemicelluloses,** and **cellulose.**

Initially, the sample is dried and finely ground prior to extraction to remove simple carbohydrates, lipids/pigments, and proteins. It is essential to monitor each extraction step to ensure that no polysaccharides have been removed during these preliminary treatments (note: 90%, v/v, ethanol will precipitate virtually all polysaccharides). The resulting residue can then be further manipulated for specific polysaccharide extraction.

10.2.8.1 Pectic Substances

Pectic substances are widely distributed in plants. Pectin is mainly a polymer consisting of D-galacturonic monomers joined by $\alpha 1 \rightarrow 4$ linkages. The main chain of pectin also contains approximately 10 percent rhamnose. Pectin also contains small amounts of galactans, arabinans, fucose, and D-xylose as short side chains. Many of the D-galacturonic acid residues along the main chain are esterified to yield the methyl ester, while the hydroxyl groups in the 2 and 3 positions may be acylated to a small extent.

Pectic substances can be extracted employing hot water (90–100°C) or ammonium oxalate (0.5%, w/v). Following filtration, the supernatant is slightly acidified with hydrochloric acid and the pectic substances are precipitated with 70 percent ethanol. The precipitate is then transferred to a preweighed crucible and washed with ethanol and acetone. The crucible and contents (pectic substances) are dried at 100°C, cooled, and then weighed. The precipitates usually formed are fine gelatinous masses that are quite difficult to recover quantitatively. In addition, these precipitates are not completely pure (98).

The most accurate method to quantitatively determine the filtrate from the initial extraction is by treatment with strong acid (sulfuric)/heat to hydrolyze the polymer to its constituent monomeric units. The resulting monosaccharides can then be qualitatively and quantitatively determined employing HPLC or GC.

10.2.8.2 Hemicelluloses

Hemicelluloses are structurally diverse polysaccharides. These compounds typically have molecular masses ranging from 60,000–600,000 daltons. They can be comprised of a number of monomeric units, including D-galacturonic, L-arabinose, 4-O-methyl-D-glucuronic, D-aldohexoses, D-pentoses, and a small quantity of L-rhamnose.

Hemicelluloses can be extracted after removal of pectic substances by treatment of the residue with dilute alkali (5% potassium hydroxide, w/v) under an atmosphere of nitrogen. Neutralization of this fraction results in the precipitation of a class of hemicelluloses (hemicellulose A) that is mainly comprised of xylans and uronic acid residues. This material can be collected, dried, and weighed. Treatment of the supernatant with ethanol results in the precipitation of hemicelluose B, which can be collected, dried, and weighed. The supernatant can then be treated with strong alkali (24% potassium hydroxide), which results in the precipitation of hemicelluose C upon neutralization. The precipitates formed are normally gelatinous masses that are difficult to recover completely.

The extracted hemicelluoses can also be treated with strong acid (sulfuric)/heat to completely hydrolyze the polymers. The resulting monosaccharides can be determined employing HPLC or GC by comparison to appropriate standards.

10.2.8.3 Cellulose

Cellulose is a high molecular weight polymer of D-glucopyranosyl units linked by $\beta 1\text{->}4$ glycosidic bonds. This polymer is relatively free from bound water (i.e., amorphous), and its resistance to chemical hydrolysis is due to its ability to establish crystalline microfibrils that are stabilized by hydrogen bonding.

The residue from hemicellulose extraction can be washed with water and acetone and weighed. This residue is referred to as α-cellulose. An alternate method involves treatment of the residue with strong sulfuric acid (72%, w/w) followed by reducing sugar determination, or D-glucose measurement (HPLC, GC), which can yield quantitative results for cellulose content.

These methods for structural polysaccharide extraction and quantitation require a number of sample manipulations. Each manipulation of a sample will result in some loss of material, and the extraction techniques described are not polysaccharide specific. Therefore, there will be some degree of uncertainty in each of these measurements. An excellent treatment on the extraction techniques for polysaccharides is contained in reference (98).

10.2.9 Non-Structural Polysaccharide Analysis

This class of compounds includes polysaccharides and their derivatives that form viscous solutions or dispersions in cold or hot water (99). Included in this class are **seaweed extracts** (agar, alginates, carrageenan), **plant exudates** (gum arabic, ghatti, karaya, tragacanth), **seed and seed root gums** (guar, locust bean, tamarind), and **gums obtained from microbial fermentations** (xanthan, dextrans).

In general, the concentrations of these compounds in foods are quite low (0.1–0.2% when used as a stabilizer; up to 2.2% when employed as gelling or thickening agents). There are few standard procedures for the isolation and analysis of these polysaccharides.

The AOAC has outlined a method (AOAC Method 935.61) for the extraction and analysis of gums in food dressings. This method involves analysis on a sample that is free of starch. The sample is first treated with hot water (60–70°C) followed by treatment with 50 percent trichloroacetic acid. The mixture is then centrifuged and the aqueous layer is removed and filtered. Ethanol is added to the filtrate and allowed to stand overnight to precipitate the polysaccharide gums. Most of the alcohol is then removed and the residue (gums) is dissolved in hot water. Acetic acid is added to reprecipitate the gums. This precipitate is then hydrolyzed in hot hydrochloric acid and the resulting monosaccharides analyzed employing reducing sugar tests or chromatography.

More recent techniques involve precipitation of hydrocolloids and gellan gum with cetylpyridinium chloride (CPC). These materials can be trapped on a celite column and the gums eluted with sodium hexametaphosphate. Reaction of an aliquot of the eluent with thiourea (0.1%)/sulfuric acid/cysteine hydrochloride results in color development (due to the rhamnose moiety of the gum) that can be monitored at 455 nm. Comparison to a standard curve allows for quantitation (100).

Galactomannans can be analyzed following CPC precipitation by specific oxidation with galactose oxidase. This assay has been applied to the identification of guar and locust bean gums in polysaccharide blends (101).

A specific staining technique for anionic gums, such as algin, carrageenan, gum arabic, and pectin has proven useful for the analysis of food gums. In this technique, solutions of food grade gums are applied to polyamide strips followed by staining with horseradish peroxidase-benzidine (HRP-B). The strips are then scanned with a laser densitometer and compared to standard curves. Detection limits of 20 ng are possible, and the technique is linear over a wide concentration range (0.008–0.4%) (102). This method could prove very useful to the food industry, as samples can be analyzed directly.

A lectin specific for D-galactose [*Bandeiraea simplicifolia* lectin (BS-lectin)] has been successfully employed for the detection and quantitation of the galactomannans of guar and locust bean gum. Although this enzyme-linked lectin assay is not able to distinguish between these two polysaccharides, it was able to specifically detect guar and locust bean gums in the presence of other food polysaccharides (such as xanthan, carrageenan, pectin, and alginate). The detection limit for this assay was approximately 10 ng/mL (103).

10.3 SUMMARY

The purpose of this chapter is to introduce to students the myriad of analytical techniques that are currently available for carbohydrate analysis. Regardless of the age of these analytical techniques, each has an application for food. However, it is important that the user of an analytical technique be cognizant of both the advantages and limitations of the method.

Many analytical techniques for carbohydrates require sample preparation that includes extraction and clarification procedures. Most of the chemical techniques for the analysis of carbohydrates are based on the reaction of reducing sugars with chemical reagents to yield precipitates or colored complexes, which are quantitated by solubilization, then titration, or by spectrophotometric determination. The enzymatic analysis of specific carbohydrates has greatly increased with the availability of enzyme assay kits. The physical methods of polarimetry, refractometry, and specific gravity are useful as rapid quality control techniques for products such as syrups and juices. While there are few standard procedures for the isolation and analysis of nonstructural polysaccharides, starch and the structural polysaccharides (pectic substances, hemicelluloses, and cellulose) can be quantitated by extraction techniques followed by further treatment and often GC or HPLC analysis.

Modern analytical methods have revolutionized carbohydrate analysis. Techniques such as HPLC, CGC, and HPCE afford the rapid and accurate analysis of structurally similar carbohydrates at trace concentrations. Further development in these (and other) methods will soon provide the analytical chemist with single molecule detection limits for carbohydrates.

10.4 STUDY QUESTIONS

1. Distinguish chemically between monosaccharides, oligosaccharides, and polysaccharides, and explain how solubility characteristics can be used in an extraction procedure to separate monosaccharides and oligosaccharides from polysaccharides.
2. Briefly explain one method that could be used for each of the following:
   a. to prevent hydrolysis of polysaccharides when free sugars are extracted from fruits with an alcohol extraction
   b. to remove pigments from a sugar-containing solution
   c. to remove proteins from solution for starch analysis
   d. to measure total carbohydrate
   e. to measure reducing sugars
   f. to measure the sucrose concentration of a pure sucrose solution by a physical method
   g. to measure glucose by an enzymatic method
   h. to measure simultaneously the concentration of individual free sugars
3. The Munson and Walker method, Lane-Eynon method, and Nelson-Somogyi method can all be used to measure reducing sugars. Explain the similarities and differences among these methods with regard to the principles involved and the procedures used.
4. Why are enzymatic methods so popular for the analyses of food carbohydrates?
5. What factors must be kept constant in enzymatic methods for the analyses of carbohydrates and other food components? Why?
6. The amount of NADH or NADPH generated is used in many enzymatic methods to quantitate the carbohydrate of interest. How are these generated, and how are they quantitated?
7. Explain how and why it is possible to measure the glucose content of a pure glucose solution using polarimetry.
8. What is a saccharometer? How and why can it be used to measure the percent sucrose in a water solution?
9. What types of stationary phases are commonly used for HPLC analysis of carbohydrate? Briefly describe the principle of each to achieve separation.
10. Why is the refractive index detector the most common detector used for HPLC analysis of carbohydrates? For each of the other common detectors, give one advantage over the use of refractive index.
11. Why is HPLC analysis of carbohydrates rapidly gaining favor over GC analysis?
12. What information concerning carbohydrates can be obtained by NMR analysis?
13. Differentiate supercritical fluid chromatography from traditional HPLC and GC.
14. Differentiate the solubility characteristics of pectic substances, hemicellulose, and cellulose, and explain how these characteristics can be used to help quantitate these structural polysaccharides.

10.5 REFERENCES

1. American Chemical Society, Nomenclature Committee, Division of Carbohydrate Chemistry. 1963. Rules of carbohydrate nomenclature. *J. Organic Chem.* 28:281.
2. Angyal, S.J. 1984. The composition of reducing sugars in solution. *Adv. Carbohydr. Chem. Biochem.* 42:15.
3. Helferich, B., Bohn, E., and Winkler, S. 1930. Ungesattigte derivative von gentiobiose und cellobiose. *Chem. Ber.* 63:989.
4. Hassid, W.Z., and Ballou, C.E. 1957. Oligosaccharides. Ch. 9, in *The Carbohydrates,* W. Pigman (Ed.), 478–535. Academic Press Inc., New York, NY.
5. Low, N.H., Nelson, D.L., and Sporns, P. 1988. Carbohydrate analysis of western Canadian honeys and their nectar sources to determine the origin of honey oligosaccharides. *J. Apic. Res.* 27:245.
6. Krol, B. 1978. Side reactions of acid hydrolysis of sucrose. *Acta Aliment. Pol.* 4:373.
7. Swallow, K.W., Petrus, D.R., and Low, N.H. 1991. Detection of orange juice adulteration with beet medium invert sugar using anion exchange liquid chromatography with pulsed amperometric detection. *J. Assoc. Off. Anal. Chem.* 74:341.
8. Adam, S. 1983. Recent developments in radiation chemistry of carbohydrates. Ch. 6, in *Recent Advances in Food Irradiation,* P.S. Elias and A.J. Cohen (Eds.), 149–170. Elsevier Press, Amsterdam, The Netherlands.
9. AOAC. 1990. *Official Methods of Analysis,* 15th ed. Association of Official Analytical Chemists, Washington, DC.
10. Southgate, D.A.T. 1969. Determination of carbohydrates in foods. I. Available carbohydrate. *J. Sci. Food Agric.* 20:331.
11. Friedemann, T.E., Witt, N.F., Neighbors, B.W., and Weber, C.W. 1967. Determination of available carbohydrates in plant and animal foods. *J. Nutr.,* 91, Part II, pp. 1–40.
12. Windholz, M. (Ed.) 1983. *Merck Index: An Encyclopedia of Chemicals, Drugs, and Biologicals,* 10th ed. Merck, Rahway, NJ.
13. Low, N.H., and Swallow, K.W. 1991. Nachweis des Zusatzes teilinvertierter Saccharose zu Orangensaft mit Hilfe der HPLC. *Flussiges Obst* 58:13.
14. Kilroe-Smith, T.A., and de Gier, J.F. 1955. Effects of cations on reducing-sugar determinations with Shaffer and Hartmann or Somogyi reagents. *Analysts* 80:627.

15. McDonald, E.J., and Turcotte, A.L. 1947. Further studies on Ofner's method for the determination of invert sugar. *J. Assoc. Off. Agric. Chem.* 48:124.
16. Nelson, N. 1944. A photometric adaptation of the Somogyi method for the determination of glucose. *J. Biol. Chem.* 153:375.
17. Somogyi, M. 1952. Notes on sugar determination. *J. Biol. Chem.* 195:19.
18. Dubois, M., Gilles, K.A., Hamilton, J.K., Rebers, P.A., and Smith, F. 1956. Colorimetric method for determination of sugars and related substances. *Anal. Chem.* 28:350.
19. Roe, J.H. 1955. The determination of sugar in blood and spinal fluid with anthrone reagent. *J. Biol. Chem.* 212:335.
20. Whistler, R.L., and Wolfrom, M.L. (Eds.). 1962. *Methods in Carbohydrate Chemistry,* Vol. I. Academic Press Inc., New York, NY.
21. Bottle, R.T., and Gilbert, G.A. 1958. The use of alkaline reagents to determine carbohydrate reducing groups. I. 3,5-Dinitrosalicylate ion and interference by air. *Analyst* 83:403.
22. Kulka, R.G. 1956. Colorimetric estimation of ketopentoses and ketohexoses. *Biochem. J.* 63:542.
23. Albaum, H.G., and Umbreit, W.W. 1947. Differentiation between ribose 3-phosphate and ribose 5-phosphate. *Biol. Chem.* 167:369.
24. Whitaker, J.R. 1985. Analytical uses of enzymes. In *Food Analysis Principles and Techniques,* Vol 3., D.W. Gruenwedel and J.R. Whitaker (Eds.), 297–377. Marcel Dekker, Inc., New York, NY.
25. Kearsley, M.W. 1985. Physical, chemical and biochemical methods of analysis of carbohydrates. Ch. 2, in *Analysis of Food Carbohydrate,* G.G. Birch (Ed.), 15–40. Elsevier Applied Science, London, UK.
26. Guilbault, G.G. 1976. *Handbook of Enzymatic Methods of Analysis.* Marcel Dekker, Inc., New York, NY.
27. Whitaker, J.R. 1974. Analytical application of enzymes. Ch. 2, in *Food Related Enzymes, a Symposium,* J.R. Whitaker (Ed.), Advances in Chemistry Series 136. American Chemical Society, Washington, DC.
28. Beutler, H.O. 1984. *Methods of Enzymatic Analysis,* Vol. VI, H.U. Bergmeyer (Ed.). Verlag Chemie, Berlin, Germany.
29. Taylor, P.J., Kmete, E., and Johnson, J.M. 1977. Design, construction and applications of a galactose selective electrode. *Anal. Chem.* 49:789.
30. Weast, R.C. (Ed.) 1978. *CRC Handbook of Chemistry and Physics,* 59th ed. CRC Press, Boca Raton, FL.
31. Plato, F. 1910. Tafel zur Umrechrung der Volumprozente in Gewichtsprozente und der Grewichspozente in Volumporzente bei Branntweinen. Berechnet nach der amtlichen Zahler der Kaisesrlichen Normal-Eichyungkommission.
32. Bates, J.F. 1942. Polarimetry, saccharimetry and the sugars. *Natl. Bur. Std. (U.S.) Circ.* 404:247.
33. Heyraud, A., and Rinaudo, M. 1981. Carbohydrate analysis by high-pressure liquid chromatography using water as the eluent. *J. Liq. Chromatog.* 4:175.
34. Kennedy, I.R., Mwandemele, O.D., and McWhirter, K.S. 1985. Estimation of sucrose, raffinose and stachyose in soybean seeds. *Food Chem.* 17:560.
35. Palla, G. 1981. C_{18} reversed-phased liquid chromatographic determination of invert sugar, sucrose, and raffinose. *Anal. Chem.* 53:1966.
36. Verzele, M., Simoens, G., and Van Damme, F. 1987. A critical review of some liquid chromatography systems for the separation of sugars. *Chromatographia* 23:292.
37. McGinnis, G.D., Prince, S., and Lowrimore, J. 1986. The use of reverse-phase columns for separation of unsubstituted carbohydrates. *J. Carbohyd. Chem.* 5:83.
38. Verhaar, L.A. Th., Kuster, B.F.M., and Claessens, H.A. 1984. Retention behaviour of carbohydrate oligomers in reverse-phase chromatography. *J. Chromatog.* 284:1.
39. Wight, A.W., and Datel, J.M. 1986. Evaluation of a reversed phase high performance liquid chromatographic column for estimation of legume seed oligosaccharides. *Food Chem.* 21:167.
40. Derler, H., Hormeyer, H.F., and Bonn, G. 1988. High-performance liquid chromatographic analysis of oligosaccharides. I. Separation on an ion-exchange stationary phase of low cross-linking. *J. Chromatog.* 440:281.
41. Ball, G.F.M. 1990. The application of HPLC to the determination of low molecular weight sugars and polyhydric alcohols in foods: A review. *Food Chem.* 35:117.
42. Dionex. 1988. IonPac Columns. Technical Bulletin LPN 034023. Dionex Corp., Sunnyvale, CA.
43. Swallow, K.W., and Low, N.H. 1990. Analysis and quantitation of the carbohydrates in honey using high-performance liquid chromatography. *J. Agric. Food Chem.* 38:1828.
44. Paik, J., Low, N.H., and Ingledew, W.M. 1991. Malt extract: relationship of chemical composition to fermentability. *Am. Soc. Brewing Chem. J.* 49:8.
45. Low, N.H., and Swallow, K.W. 1991. Detection of beet medium invert sugar addition to orange juice by HPLC. *Flussiges Obst.* 1:2.
46. Lobry de Bruyn, C.A., and van Ekenstein, W.A. 1895. *Trav. Chim. Pays-Bas* 14:203.
47. Wong, D.W.S. 1989. Carbohydrates. Ch. 3, in *Mechanism and Theory in Food Chemistry,* 105–146. Van Nostrand Reinhold, New York, NY.
48. Olechno, J.D., Carter, S.R., Edwards, W.T., and Gillen, D.G. 1987. Developments in the chromatographic determination of carbohydrates. *Am. Biotechnol. Lab* 5:38.
49. Bornhop, D.J., Nolan, T.G., and Dovichi, N.J. 1987. Subnanoliter laser-based refractive index detector for 0.25mm i.d. microbore liquid chromatography. *J. Chromatog.* 384:181.
50. Honda, S. 1984. Review: High-performance liquid chromatography of mono- and oligosaccharides. *Anal. Biochem.* 140:1.
51. Rocklin, R.D., and Pohl, C.A. 1983. Determination of carbohydrates by anion exchange chromatography with pulsed amperometric detection. *J. Liquid Chromatog.* 6:1577.
52. Turner, B. 1986. An evaporative analyzer for HPLC detection. *Laboratory Practice,* pp. 55–56, June.
53. Macrae, R., and Dick, J. 1981. Analysis of carbohydrates using the mass detector. *J. Chromatog.* 210:138.
54. Geisow, M. 1992. Shifting gear in carbohydrate analysis. *Bio/Technology* 10:277.
55. Tanaka, K., and Fritz, J.S. 1987. Ion-exclusion chroma-

tography of non-ionic substances with conductivity detection. *J. Chromatog.* 409:271.
56. Matsumoto, K., Hamada, O., Ukeda, H., and Osajima, Y. 1986. Amperometric flow injection determination of fructose with an immobilized fructose 5-dehydrogenase reactor. *Anal. Chem.* 58:2732.
57. Sakairi, M., and Kambara, H. 1988. Characteristics of a liquid chromatograph/atmospheric pressure ionization mass spectrometer. *Anal. Chem.* 60:774.
58. Low, N.H., and Sporns, P. 1988. Analysis and quantitation of minor di- and trisaccharides in honey, using capillary gas chromatography. *J. Food Sci.* 53:558.
59. Biermann, C.J. 1989. Introduction to analysis of carbohydrates by gas-liquid chromatography. Ch. 1, in *Analysis of Carbohydrates by GLC and MS,* C.J. Biermann and G.D. McGinnis (Eds.), 1–18. CRC Press, Boca Raton, FL.
60. Sweeley, C.C., Bentley, R., Makita, M., and Wells, W.W. 1963. Gas liquid chromatography of trimethylsilyl derivatives of sugars and related substances. *J. Am. Chem. Soc.* 85:2497.
61. Adam, S., and Jennings, W.G. 1975. Gas chromatographic separation of silylated derivatives of disaccharide mixtures on open tubular glass capillary columns. *J. Chromatog.* 115:218.
62. Biermann, C.J., and McGinnis, G.D. (Eds.) 1989. *Analysis of Carbohydrates by GLC and MS.* CRC Press, Boca Raton, FL.
63. Folkes, D.J. 1985. Gas-liquid chromatography. Ch. 5, in *Analysis of Food Carbohydrates,* G.G. Birch (Ed.), 91–123. Elsevier Applied Science, London, UK.
64. Barber, M., Bordoli, R.S., Sedgwick, R.D., and Tyler, A.N. 1981. Fast atom bombardment of solids (F.A.B.): A new ion source for mass spectrometry. *J. Chem. Soc. Chem. Comm.* 325.
65. Dell, A., and Thomas-Oates, J.E. 1989. Fast atom bombardment-mass spectrometry (FAB-MS): Sample preparation and analytical strategies. Ch. 10, in *Analysis of Carbohydrates by GLC and MS,* C.J. Biermann and G.D. McGinnis (Eds.), 217–236. CRC Press, Boca Raton, FL.
66. Derome, A.E. 1987. *Modern NMR Techniques for Chemistry Research.* Pergamon Press, Oxford, UK.
67. Rathbone, E.B. 1985. NMR spectroscopy in the analysis of carbohydrates. Ch. 7, in *Analysis of Food Carbohydrate,* G.G. Birch (Ed.), 149–223. Elsevier Applied Science, London, UK.
68. Lowman, D.W., and Maciel, G.E. 1979. Determination of sucrose in sugar beet juices by nuclear magnetic resonance spectrometry. *Anal. Chem.* 51:85.
69. Meredith, P., Dengate, H.N., Hennessy, W., Blunt, J.W., Hartshorn, M.P., and Munro, M.H.G. 1980. The developing starch granule. Part IX. Saccharides of developing wheat grain determined by carbon-13 NMR spectroscopy. *Starch* 32:198.
70. Blunt, J.W., and Munro, H.G. 1976. An automated procedure for qualitative and quantitative analysis of mixtures by means of carbon magnetic resonance spectroscopy: Applications to carbohydrate analysis. *Aust. J. Chem.* 29:975.
71. Low, N.H., Brisbane, T., Bigam, G., and Sporns, P. 1988. Carbon-13 nuclear magnetic resonance for the qualitative and quantitative analysis of structurally similar disaccharides. *J. Agric. Food Chem.* 36:953.
72. Tamate, J., and Bradbury, J.H. 1985. Determination of sugars in tropical root crops using ^{13}C-N.M.R. spectroscopy: Comparison with the h.p.l.c. method. *J. Sci. Food Agric.* 36:1291.
73. Altena, J.H., Van den Ouweland, G.A.M., Teunis, C.J., and Tjan, S.B. 1981. Analysis of the 220-MHz, PMR spectra of some products of the Amadori- and Heyns-rearrangements. *Carbohyd. Res.* 92:37.
74. Harvey, J.M., and Symons, M.C.R. 1978. The hydration of monosaccharides—an NMR study. *J. Soln. Chem.* 7:571.
75. Smith, K.M., Somanathan, R., Tabba, H.D., and Minch, M.J. 1982. High-field proton nuclear magnetic resonance study of hydration in concentrated solutions of monosaccharides. *Carbohyd. Res.* 106:160.
76. Sheth, H.B., Yaylayan, V.A., Low, N.H., Stiles, M.E., and Sporns, P. 1990. Reaction of reducing sugars with sulfathiazole and importance of this reaction to sulfonamide residue analysis using chromatographic, colorimetric, microbiological or ELISA methods. *J. Agric. Food Chem.* 38:1125.
77. Horvath, C.G., Preiss, B.A., and Lipsky, S.R. 1967. Fast liquid chromatography. Investigation of operating parameters and the separation of nucleotides on pellicular ion exchangers. *Anal. Chem.* 39:1422.
78. Ishii, D., Asai, K., Hibi, K., Jonokuchi, T., and Nagaya, M. 1977. A study of micro-high-performance liquid chromatography. I. Development of technique for miniaturization of high-performance liquid chromatography. *J. Chromatog.* 144:157.
79. Ishii, D. (Ed.) 1988. *Introduction to Microscale High-Performance Liquid Chromatography.* VCH Publishers, Inc., New York, NY.
80. Hannay, J.B., and Hogarth, J. 1879. On the solubility of solids in gases. *Proc. R. Soc. London* 29:324.
81. Klesper, E., Corwin, A.H., and Turner, D.A. 1962. High pressure gas chromatography above critical temperatures. *J. Org. Chem.* 27:700.
82. Li, S.F.Y. 1989. Experimental studies on supercritical fluid separation processes. *J. Chem. Tech. Biotechnol.* 46:1.
83. Latta, S. 1990. Supercritical fluids attracting new interest. *Inform* 1:810.
84. Honda, S., Iwase, S., Makino, A., and Fujiwara, S. 1989. Simultaneous determination of reducing monosaccharides by capillary zone electrophoresis as the borate complexes of N-2-pyridylglycamines. *Anal. Biochem.* 176:72.
85. Burolla, V.P., Pentoney, S.L., and Zare, R. 1989. High performance capillary electrophoresis. *Am. Biotech. Laboratory* November/December, 7:20–26.
86. Liu, J., Shirota, O., and Novotny, M. 1991. Capillary electrophoresis of amino sugars with laser-induced fluorescence detection. *Anal. Chem.* 63:413.
87. Albin, M., Weinberger, R., Sapp, E., and Moring, S. 1991. Fluorescence detection in capillary electrophoresis: Evaluation of derivatizing reagents and techniques. *Anal. Chem.* 63:417.
88. Low, N.H., and Swallow, K.W. 1992. Unpublished data. Dept. of Applied Microbiology and Food Science, University of Saskatchewan, Saskatoon, Saskatchewan, Canada.
89. Clendenning, K.A., and Wright, D.E. 1945. Detection of

starch in cereal products (III). Comparison and specific rotatory power of starches in relation to source and type. *Can J. Res.* 23B:131.

90. Rickard, J.E., and Behn, K.R. 1987. Evaluation of acid and enzyme hydrolytic methods for the determination of cassava starch. *J. Sci. Food Agric.* 41:373.
91. Griffith, L.S., and Sporns, P. 1990. Determination of dextrose equivalents in starch hydrolysates using Ce (IV). *Food Chem.* 38:1356.
92. McIntyre, D.D., Ho, C., and Vogel, H.J. 1990. One-dimensional nuclear magnetic resonance studies of starch and starch products. *Starch* 42:260.
93. Haissig, B., and Dickson, R. 1979. Starch measurement in plant tissue using enzymatic hydrolysis. *Physiol. Plant.* 47:151.
94. Mitchell, G.A. 1990. Methods of starch analysis. *Starch* 42:131.
95. Hassid, W., and Neufeld, E. 1964. Quantitative determination of starch in plant tissue. *Methods Carbohydr. Chem.* 4:33.
96. Rose, R., Rose, C.L., Omi, S.K., Forry, K.R., Durall, D.M., and Bigg, W.L. 1991. Starch determination by perchloric acid vs enzymes: Evaluating the accuracy and precision of six colorimetric methods. *J. Agric. Food Chem.* 39:2.
97. Bernetti, R., Kochan, D.A., Trost, V.W., and Young, S.N. 1990. Modern methods of analysis of food starches. *Cereal Foods World* 35:1100.
98. Southgate, D.A.T. 1991. The measurement of starch, its degradation products and modified starches. Ch. 4, in *Determination of Food Carbohydrates,* D.A.T. Southgate (Ed.), 48–60. Applied Science Ltd., New York, NY.
99. Zapsalis, C., and Beck, R.A. 1986. Carbohydrates: Chemistry, occurrence, and food applications. Ch. 6, in *Food Chemistry and Nutritional Biochemistry,* C. Zapsalis and R.A. Beck (Eds.), 315–414. Macmillan Publishing Company, New York, NY.
100. Graham, H.A. 1990. Semi-micro method for the quantitative determination of gellan gum in food products. *Food Hydrocolloids* 3:435.
101. Baird, J.K., and Smith, W.W. 1989. A simple colorimetric method for the specific analysis of food-grade galactomannans. *Food Hydrocolloids* 3:413.
102. Dickmann, R.S., Chism, G.W., Renoll, M.W., and Hansen, P.M.T. 1989. Detection of food gums on polyamide strips using horseradish peroxidase-benzidine staining and laser-beam densitometry. *Food Hydrocolloids* 3:33.
103. Patel, P.D., and Hawes, G.B. 1988. Estimation of food-grade galactomannans by enzyme-linked lectin assay. *Food Hydrocolloids* 2:107.
104. Joslyn, M.A. 1970. *Methods in Food Analysis.* Academic Press, New York, NY.
105. Rajakyla, E. 1986. Use of reversed-phase chromatography in carbohydrate analysis. *J. Chrom.* 353:1.
106. Pirisino, J.F. 1984. High-pressure liquid chromatography of carbohydrates in foods: Fixed-ion resin and amino-bonded silica columns. In *Food Constituents and Food Residues—Their Chromatographic Determinations.* J.F. Lawrence (Ed.), 159–193. Marcel Dekker, Inc., New York, NY.

CHAPTER

11

Fiber Analysis

Maurice R. Bennink

11.1 INTRODUCTION

11.1.1 Importance of Dietary Fiber

In the early 1970s, Burkitt and Trowel (1) postulated that the prevalence of heart disease and certain cancers in Western societies was related to inadequate consumption of dietary fiber. Their observations stimulated much interest, and a great deal of research has been done to test the fiber hypothesis. While the research has not always produced consistent results and while the great expectations inspired by Burkitt and Trowel have not materialized, it is clear that adequate consumption of dietary fiber is important for optimum health.

Liberal consumption of dietary fiber from a variety of foods will help protect against colon cancer and help normalize blood lipids and thereby reduce cardiovascular disease (2). Certain types of fiber can slow glucose absorption and reduce insulin secretion, which is of great importance for diabetics and probably for nondiabetics as well. Fiber helps prevent constipation and diverticular disease. With this wide range of beneficial effects attributable to dietary fiber, it is easy to lose perspective and consider dietary fiber a magic potion that will correct or prevent all diseases. A more correct view is that dietary fiber is an essential component of a well-balanced diet, and adequate intake of dietary fiber throughout one's lifetime will help minimize some of the most common health problems in the United States. The two review books edited by Leeds (3, 4) provide an extensive compilation of articles related to the physiological action of dietary fiber.

How much fiber is required for optimal health is still not precisely known. Certain fiber components will produce one physiologic response, while another fiber component produces a different physiologic response (2–5). For example, the pentose fraction of dietary fiber seems to be the most beneficial in preventing colon cancer and reducing cardiovascular disease. Pectin and the hydrocolloids are most beneficial in slowing glucose absorption and lowering insulin secretion. Pectins and hydrocolloids are of little value in preventing diverticulosis and constipation; however, a mixture of hemicellulose and cellulose will help prevent these gastrointestinal dysfunctions. Recognition of the importance of dietary fiber and recognition that certain physiologic effects can be related to specific fiber components has led to the emergence of a number of methodologies for determining dietary fiber. Several commonly used methods for analyzing dietary fiber are described in this chapter.

11.1.2 Definition of Fiber

Dietary fiber is generally defined as lignin plus plant polysaccharides that cannot be digested by human enzymes. Some starch is not digested in the small intestine and therefore fits the definition of dietary fiber. Small amounts of starch are measured as fiber in some of the fiber methodologies. But whether so-called **resistant starch** (see section 11.3.4.3.3) should be considered part of dietary fiber is controversial at this time.

11.1.3 Major Components of Dietary Fiber

The major components of dietary fiber are cellulose, hemicelluloses, pectins, hydrocolloids, and lignin. From a botanical view, fiber is categorized as cell wall polysaccharides, non-cell wall polysaccharides, and lignin.

11.1.3.1 Cell Wall Polysaccharides

11.1.3.1.1 Cellulose Cellulose is a long, practically linear polymer of β-1,4-linked glucose units. Some polymers may contain 10,000 glucose units. Hydrogen bonding between parallel polymers forms strong microfibrils. Cellulose microfibrils provide the strength and rigidity required in plant primary and secondary cell walls.

11.1.3.1.2 Hemicelluloses Hemicelluloses are a heterogenous group of substances containing a number of sugars in their backbone and side chains. Xylose, mannose, and galactose frequently form the backbone structure, while arabinose, galactose, and uronic acids are present in the side chains. Hemicelluloses by definition are soluble in dilute alkali but not in water. Molecular size and degree of branching are also highly variable. A typical hemicellulose molecule contains 50 to 200 sugar units. Hemicelluloses are matrix polysaccharides that tie together cellulose microfibrils and form covalent bonds with lignin.

11.1.3.1.3 Pectins Pectins are rich in uronic acids. They are soluble in hot water and form gels. The backbone structure consists of unbranched chains of 1,4-linked galacturonic acid. Side chains may contain rhamnose, arabinose, xylose, and fucose. Solubility is reduced by methylation of the free carboxyl group and by formation of calcium and magnesium complexes. Pectins like hemicelluloses are matrix polysaccharides in cell walls.

11.1.3.2 Non-Cell Wall Polysaccharides

Non-cell wall polysaccharides include hydrocolloids such as mucilages, gums, and algal polysaccharides. **Hydrocolloids** are hydrophilic polysaccharides that form viscous solutions or dispersions in cold or hot water. Typical **mucilages** are guar and locust bean gums. Oats and barley also contain mucilages. Plant exudate **gums** include arabic, ghatti, karaya, and tragacanth gums while **algal polysaccharides** include agar, alginates, and carrageenan. The non-cell wall polysaccharides contain a variety of neutral sugars and uronic acids.

11.1.3.3 Lignin

Lignin is a noncarbohydrate, three-dimensional polymer consisting of approximately 40 phenol units with strong intramolecular bonding. Lignin is often covalently linked to hemicellulose.

11.2 GENERAL CONSIDERATIONS

Fiber components or subfractions are methodology dependent and not distinct entities. These fractionations are somewhat arbitrary and frequently have little relationship to either plant or mammalian physiology. Although considerable progress has been made in relating fiber composition to physiological action during the past 15 years, much remains to be learned. Pectins and hydrocolloids have long been used as additives in food processing. However, utilizing complex sources of fiber to improve the nutritional value of foods remains a challenge for the food scientist.

11.3 METHODS

11.3.1 Overview

Dietary fiber is estimated by two basic approaches—**gravimetrically** or **chemically.** In the first approach, digestible carbohydrate, lipids, and proteins are selectively solubilized by chemicals and/or enzymes. Undigestible materials are then collected by filtration, and the fiber residue is quantitated gravimetrically. In the second approach, digestible carbohydrates are removed by enzymatic digestion, fiber components are hydrolyzed by acid, and monosaccharides are measured. The sum of monosaccharides in the acid hydrolysate represents fiber.

The food component that is most problematic in fiber analysis is starch. In both approaches, it is essential that all starch be removed for accurate estimates of fiber. With the gravimetric approach, incomplete removal of starch increases the residue weight and inflates the estimate of fiber. In the second approach, glucose in the acid hydrolysate is considered fiber. Therefore, glucose that is not removed in the early analytical steps causes an overestimation of dietary fiber. The starch hydrolases utilized in fiber methods include α-amylase, amyloglucosidase, and pullulanase. α-Amylase catalyzes the hydrolysis of internal α-1,4-linked D-glucose units, while pullulanase hydrolyzes internal α-1,6-linked glucose units. Amyloglucosidase hydrolyzes α-1,4- and α-1,6-glucosidic bonds from nonreducing ends of starch. Takadiastase® is a heat-stable fungal α-amylase, and Termamyl® is a heat-stable bacterial α-amylase.

All fiber methods include a heating step at 80 to 130°C for 10 min to 3 hrs to swell and disintegrate (gelatinize) starch granules. Even with gelatinization, some starch escapes digestion (solubilization) and inflates fiber values. As indicated earlier, whether or not resistant starch should be considered fiber is controversial.

In the gravimetric approach, it is essential that either all digestible materials be removed from the sample so that only undigestible polysaccharides remain or that the undigestible residue be corrected for remaining digestible contaminants. Lipids are easily removed from the sample with organic solvents and generally do not pose analytical problems for the fiber analyst. Protein and minerals that are not removed from the sample during solubilization steps should be corrected by Kjeldahl nitrogen analysis and by ashing portions of the fiber residue.

The descriptions of the various specific procedures in this chapter are meant to be overviews of the methods. The reader is referred to the referenced original articles for specifics regarding chemicals, reagents, apparatus, and step-by-step instructions.

11.3.2 Sample Preparation

Estimates of fiber are most consistent when the samples are low in fat (less than 5–10% fat), dry, and finely ground. If the sample contains more than 10 percent fat, extract fat by mixing the sample with 25 parts (v/w) petroleum ether or hexane. Centrifuge and decant the organic solvent. Repeat the lipid extraction two more times. Dry the sample overnight in a vacuum oven at 70°C and grind to pass through a 0.3–0.5 mm mesh screen. Record loss of weight due to fat and moisture removal and make appropriate correction to the final percentage dietary fiber value found in the analysis.

Nonsolid samples less than 10 percent fiber are best analyzed after lyophilization and treated as above. Nonsolid samples greater than or equal to 10 percent fiber can be analyzed without drying if the sample is homogeneous and low in fat and if particle size is sufficiently small to allow efficient removal of digestible carbohydrate and protein.

11.3.3 Gravimetric Methods

11.3.3.1 Crude Fiber

The **crude fiber** method was developed in the 1850s to estimate undigestible carbohydrate in animal feeds.

Since an easy alternative was not available, fiber in human foods was measured as crude fiber until the early 1970s (except for Southgate in England). Crude fiber is determined by sequential extraction of the sample with 1.25 percent H_2SO_4 and 1.25 percent NaOH. The insoluble residue is collected by filtration, and the residue is dried, weighed, and ashed to correct for mineral contamination of the fiber residue. Crude fiber measures variable amounts of the cellulose and lignin in the sample, but hemicelluloses, pectins, and the hydrocolloids are solubilized and not detected. Therefore, crude fiber determinations should be discontinued.

11.3.3.2 Detergent Methods

The **acid detergent fiber** and **neutral detergent fiber** methods were developed to more accurately estimate lignin, cellulose, and hemicellulose in animal feeds. Neutral detergent fiber is equal to acid detergent fiber plus hemicelluloses. Since pectins and hydrocolloids tend to be minor constituents for most feedstuffs, the detergent methods were quite adequate and well accepted for animal industries. The neutral detergent fiber method was the forerunner of the current American Association of Cereal Chemists (AACC) method (6) for determining insoluble fiber (AACC Method 32-20). Since pectins and hydrocolloids are important to human health, it is difficult to justify using these methods any longer to analyze foods.

11.3.3.3 Total, Insoluble and Soluble Fiber

With the recognition that insoluble and soluble fiber produced quite different physiological responses and that both types of fiber are important to human health, a number of methods with only minor differences were simultaneously proposed. A common method was developed from the earlier methods into what is now the widely accepted method of the Association of Official Analytical Chemists International (AOAC Method 991.43) (7). This method represents a slow evolution of methodologies that combined crude fiber, detergent fiber, and Southgate methodologies.

11.3.3.3.1 Principle Duplicate samples of dry, fat-extracted ground foods are enzymatically digested with α-amylase, amyloglucosidase, and protease to remove starch and protein. Insoluble fiber is collected by filtration. Soluble fiber is precipitated by bringing the filtrate to 78 percent ethanol and collected by filtration. The filtered fiber residues are washed with ethanol and acetone, oven dried, and weighed. One duplicate is analyzed for protein and the other is incinerated to determine ash content. Fiber = Residue weight − (Weight of protein + Ash).

11.3.3.3.2 Procedure A flow diagram outlining the general procedure for the AOAC method of determining total, insoluble, and soluble dietary fiber is shown in Fig. 11-1. Duplicate samples are mixed with buffer, a heat stable α-amylase is added, and the pH is adjusted. Starch is gelatinized and digested by heating the digestion mixture in a boiling water bath. After cooling, the pH is adjusted and a protease enzyme is added. Protein is digested and the digestion mixture is cooled. The pH is adjusted and starch digestion is completed with amyloglucosidase.

The next few steps differ depending on whether total, insoluble, or soluble fiber is being determined. If total fiber is to be determined without partitioning fiber into soluble and insoluble fractions, proceed as described in the note at the bottom of Fig. 11-1.

To determine insoluble and soluble fiber fractions, the digestion mixture following protease treatment is filtered through fritted crucibles containing Celite. The insoluble fiber retained by the filter is washed with water. The soluble fiber is in the filtrate. Four volumes (v/v) of 95 percent ethanol are added to the filtrate, plus water washes to precipitate the soluble fiber. The precipitate is allowed to form and the mixture is vacuum filtered through fritted crucibles containing Celite. The soluble fiber residue is washed successively three times with 78 percent ethanol.

The fiber residue (total, insoluble, or soluble fiber) in the crucibles is then washed with 95 percent ethanol and acetone. The crucibles are oven dried, cooled, and weighed. Since some protein and minerals are complexed with plant cell wall constituents, fiber values must be corrected for these contaminants. One duplicate is used to determine nitrogen content by the Kjeldahl procedure, and the other duplicate is incinerated to determine ash content.

Duplicate reagent blanks must be run through the entire procedure for each type of fiber determination. Figure 11-2 shows a sample and blank sheet used to calculate fiber percentages. Using the equation shown, percent dietary fiber is expressed on a dry weight basis if the sample weights are for a dried sample.

11.3.3.3.3 Applications The AOAC method for determining fiber has been extensively tested and has been found suitable for routine fiber analyses for research, legislation, and labeling purposes. This method can be used to determine fiber content of all foods. However, this method has been found to greatly overestimate the fiber content of foods with a high content of simple sugars.

11.3.4 Chemical Methods

11.3.4.1 Overview

In chemical methods for fiber determination, fiber is equal to the sum of all nonstarch monosaccharides plus lignin. Monosaccharides are measured either indirectly by colorimetric methods or by chromatographic methods [gas chromatography (GC) or high-performance liquid chromatography (HPLC)].

Carbohydrates in the presence of strong acids combine with a number of substances to produce chromagens that can be measured spectrophotometrically. Under specific, standardized conditions, hexoses can be measured with anthrone, pentoses with orcinol, and uronic acids with carbazole. There is mutual interference among groups of sugars that can and should be corrected mathematically (9). The sum of hexoses, pentoses, and uronic acids is taken as total polysaccharide content.

Uronic acid is technically difficult to measure by chromatography. Therefore, most procedures estimating fiber from monosaccharide analyses measure uronic acids colorimetrically by the carbazole method (10). The uronic acid values are then corrected for the presence of hexoses and pentoses as noted above.

Food sample (1 g).
⇓
Add 40 ml of buffer (pH 8.2).
Add heat stable α-amylase.
⇓
Incubate 15 min @ 95–100 °C.
⇓
Cool to 60°.
⇓
Add protease.
⇓
Incubate 30 min @ 60 °C.
⇓
Adjust pH to 4.0–4.7.
Add amyloglucosidase.
⇓
Incubate 30 min @ 60 °C.
⇓
Filter the digest.
⇓
Wash filtered residue with 10 ml water (2 times) then use for insoluble fiber determination.
⇓
Filtrate + water washes are used for soluble fiber determination (see note below).

Insoluble Fiber

Wash residue with 10 ml 95% ethanol (2 times).
⇓
Wash residue with 10 ml acetone (2 times).
⇓
Oven dry.
⇓
(1a) Weigh crucible.
⇓
(1b) Ash one of the duplicates and reweigh (525°C for at least 5 hr).

(1c) Determine residual protein on the other duplicate (Kjeldahl N × 6.25).

(1d) Calculate insoluble fiber content.

Soluble Fiber[1]

Bring filtrate and water washes to 80 g with water.
⇓
Add 320 ml of 95% ethanol preheated to 60°C.
⇓
Precipitate formation (1 hr @ room temp).
⇓
Filter the digest.
⇓
Wash filtered residue with 20 ml 78% ethanol (3 times).
⇓
Wash residue with 10 ml 95% ethanol (2 times).
⇓
Wash residue with 10 ml acetone (2 times).
⇓
Oven dry.
⇓
(2a) Weigh crucible.
⇓
(2b) Ash one of the duplicates and reweigh.

(2c) Determine residual protein on the other duplicate (Kjeldahl N × 6.25).

(2d) Calculate soluble fiber content.

Total Dietary Fiber = Insoluble Fiber + Soluble Fiber

[1]Total dietary fiber can be determined directly by weighing the digest following amyloglucosidase digestion and either 1) adding 4 volumes (v/v) of 95% ethanol preheated to 60°C, or 2) adjusting the volume to 80 g with water and then adding 320 ml of 95% ethanol preheated to 60°C. After (1) or (2), follow the procedure for determining soluble fiber starting at the precipitate formation step. Calculation of (2d) would produce total dietary fiber.

FIGURE 11-1.
Method to determine total dietary fiber content of foods. AOAC Method 991.43. Adapted from (7).

11.3.4.2 Southgate's Method

Southgate (11, 12) was the first to systematically quantitate dietary fiber in a wide range of foods. The carbohydrate chemistry used by Southgate has been improved and modernized, but his approach forms the foundation for many of the currently used gravimetric and chemical methods used in fiber determination.

	Sample				Blank			
	Insoluble fiber		Soluble fiber		Insoluble fiber		Soluble fiber	
Sample Weight (mg)	m_1	m_2						
Crucible + Celite weight (mg)								
Crucible + Celite + Residue Weight (mg)								
Residue Weight (mg)	R_1	R_2	R_1	R_2	R_1	R_2	R_1	R_2
Protein (mg) P								
Crucible + Celite + Ash Weight (mg)								
Ash Weight (mg) A								
Blank Weight (mg) B								
Fiber (%)								

$$\text{Blank (mg)} = \frac{R_1 + R_2}{2} - P - A$$

$$\text{Fiber (\%)} = \frac{\frac{R_1 + R_2}{2} - P - A - B}{\frac{m_1 + m_2}{2}} \times 100$$

FIGURE 11-2.
Dietary fiber sample and blank data sheet. AOAC Method 991.43. Adapted from (8).

11.3.4.2.1 Principle This method fractionates fiber into soluble and insoluble noncellulosic polysaccharides, cellulose, and lignin. Lignin is determined gravimetrically and polysaccharide content is determined from monosaccharide constituents that are measured colorimetrically.

11.3.4.2.2 Procedure The main steps in the Southgate procedure are shown in Fig. 11-3. This procedure will not be discussed in detail since the next two procedures (Englyst-Cummings procedure and Theander-Marlett approach) are modern versions of the Southgate procedure.

11.3.4.3 Englyst-Cummings Procedure

This procedure (13) is a modernized version of the Southgate procedure and is a reasonable alternative to the AOAC method.

11.3.4.3.1 Principle Starch is gelatinized and enzymatically digested. The remaining nonstarch polysaccharides are hydrolyzed by sulfuric acid to liberate free monosaccharides. Neutral sugars are determined by GC and uronic acids are determined colorimetrically. An alternate, rapid procedure measures all monosaccharides by a colorimetric method. Values for total, soluble, and insoluble fiber can be determined by both approaches. With the GC procedure, fiber can be divided into cellulose and noncellulosic polysaccharides with values for constituent sugars.

11.3.4.3.2 Procedure A flow diagram of the Englyst-Cummings procedure is shown in Fig. 11-4. Samples containing ≤200 mg of dry matter are mixed with dimethyl sulfoxide and heated in a boiling water bath to gelatinize and disperse starch. Starch and protein are digested with pancreatin and pullanase. Starch digestion is completed by incubating the samples with amyloglucosidase. Fiber is precipitated by adding 100

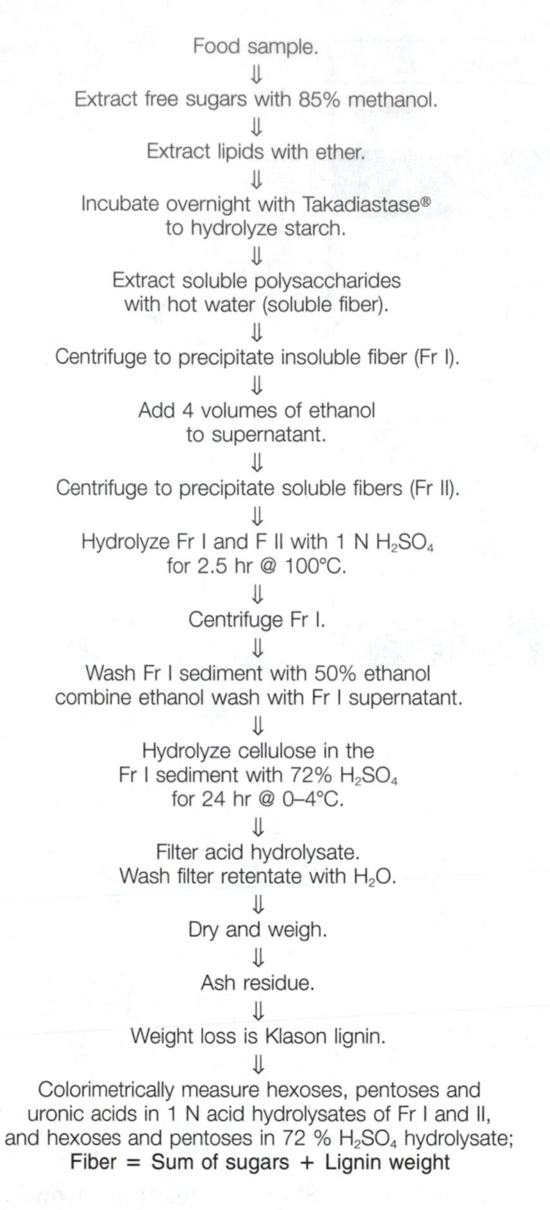

FIGURE 11-3.
Analytical scheme for determining fiber content of food by the Southgate method. Adapted from (11, 12).

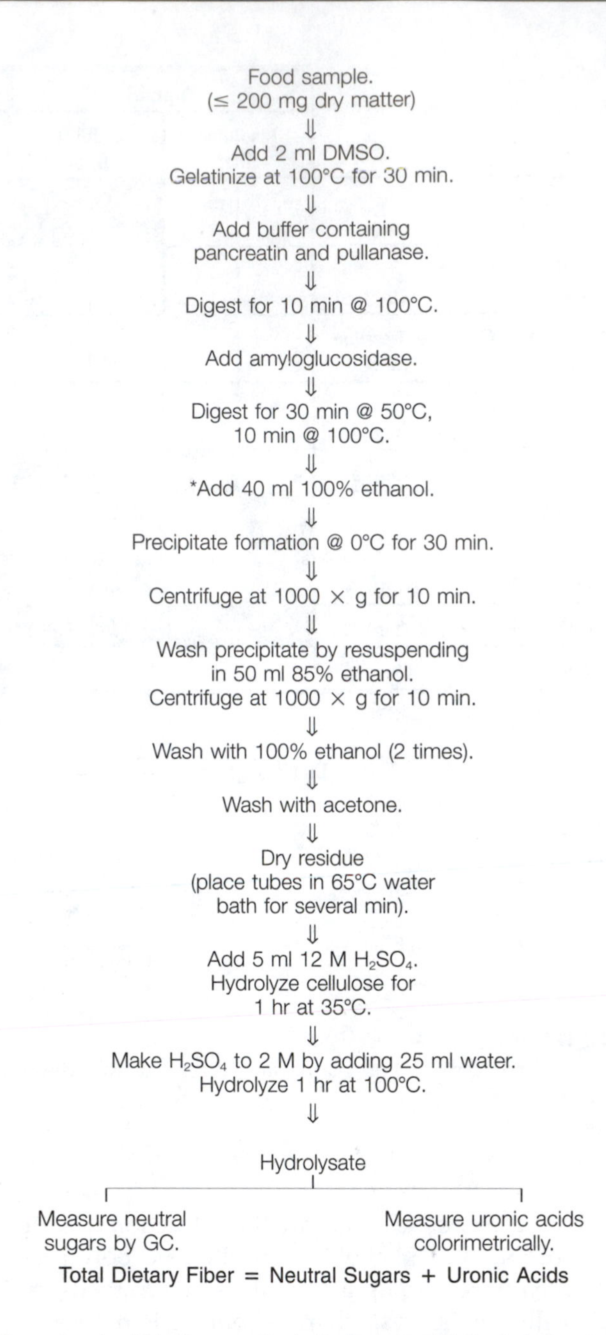

*To measure insoluble fiber, use 40 ml of buffer instead of 40 ml of ethanol, and extract soluble fiber at 100°C for 30 min. Continuation of the procedure yields insoluble, fiber.

Soluble Fiber = Total Fiber − Insoluble Fiber

FIGURE 11-4.
Englyst-Cummings procedure for determination of soluble, insoluble, and total dietary fiber. Adapted from (13).

percent ethanol and placing samples in a refrigerator. The fiber residue is collected by centrifugation and the supernatant is decanted as much as possible without disturbing the residue. The fiber is washed by resuspending the residue in 85 percent ethanol, and the residue is collected by centrifugation. Additional washes are done with 85 percent ethanol, 100 percent ethanol, and acetone. Then the fiber residue is dried. All supernatants from centrifugation steps are discarded.

Cellulose in the residue is hydrolyzed by mixing the dry residue with 12M H_2SO_4 and heating. Noncellulosic polysaccharides are hydrolyzed by rapidly adding water, mixing, and heating in a boiling water bath. The acid hydrolysate is used for sugar analysis. A portion of the hydrolysate is used to derivatize neutral sugars for GC analysis, and a second aliquot is used for uronic acid determination by a colorimetric procedure. The rapid version of the Englyst-Cummings method (14) estimates monosaccharide content of the acid hydrolysate by a single colorimetric method. Fiber weight

= monosaccharide weight in the rapid method *or* = neutral sugar + uronic acid weight in the GC method. No corrections are made for sugar losses during hydrolysis or for the addition of one molecule of water per glycoside bond since the corrections tend to offset each other.

The procedures described above yield total dietary fiber. To determine insoluble dietary fiber, the 40 ml of absolute ethanol is replaced with pH 7 buffer, and the water-soluble fiber is extracted by heating the digested mixture for 30 min in a boiling water bath. The insoluble fiber is collected by centrifugation, washed, dried, and hydrolyzed as described above. Soluble fiber = Total fiber − Insoluble fiber.

The cellulose content of the total fiber can be obtained by omitting the hydrolysis step with 12M H_2SO_4 and proceeding directly to hydrolysis of noncellulose polysaccharides with 2M H_2SO_4. Monosaccharide weight after hydrolysis with 2M H_2SO_4 = noncellulosic polysaccharides. Cellulose content = Total fiber − Noncellulosic polysaccharides.

11.3.4.3.3 Applications The rapid colorimetric procedure is essentially a single-tube assay and does not require special analytical skills or equipment other than a colorimeter. The GC-based procedure can be used to provide more detail regarding chemical composition of the fiber in addition to quantitating fiber content. The Englyst-Cummings procedures do not measure and therefore do not include lignin as a component of total dietary fiber. Since most foods do not contain significant amounts of lignin, this is a suitable method for determining fiber content of most foods. Foods containing a significant quantity of lignin should use the AOAC procedure or the procedures described in section 11.3.4.4.

A unique aspect of the Englyst-Cummings procedure is that it allows estimation of resistant starch. Resistant starch results from starch retrogradation, the Maillard reaction, crystalline starch that is not easily gelatinized, and the like. Cummings, Englyst, and Wood (15) provide details for measuring resistant starch.

11.3.4.4 Theander-Marlett Approach

Marlett (16, 17) has utilized and improved the constantly evolving procedures of Theander (18). Since the overall procedures of the two groups are so similar, they will simply be referred to as the Theander-Marlett approach. The unique aspects of the approach used by these two groups of researchers are: (1) extraction of free sugars from the sample in the initial analytical steps and (2) direct quantitation of lignin. Both aspects were part of the Southgate methodology but are not included in other current fiber methodologies.

11.3.4.4.1 Principle Free sugars and lipids are extracted with ethanol and hexane. Starch is removed by enzymatic digestion and insoluble fiber is separated from soluble fiber. Fiber fractions are hydrolyzed with sulfuric acid and the sugar content of the acid hydrolysates is determined. Lignin is determined gravimetrically. Fiber = Monosaccharides + Lignin.

11.3.4.3.2 Procedure A flow diagram for this approach is shown in Fig. 11-5. Duplicate samples of dry ground food are suspended in 80 percent ethanol and sonicated to extract and remove free sugars. The sonicated mixture is centrifuged, the supernatant is vacuum filtered, and the pellet is resuspended in 80 percent ethanol and extracted a second time. Lipids are extracted with hexane by sonication, centrifugation, and filtration. The extracted residues are dried and weighed to correct final percent fiber for sugar and lipid loss. The residues are pooled and aliquots are incubated with Termamyl® in a boiling water bath to digest starch. Starch digestion is completed by incubating the samples overnight with amyloglucosidase.

Insoluble residue is removed by centrifugation and filtration. The insoluble residue is resuspended in water and centrifuged a second time. The supernatant is filtered and the pellet is transferred to the funnel to yield the insoluble fraction. The original filtrate and the water wash filtrate are combined. Soluble polysaccharides are precipitated either by making the filtrate 80 percent ethanol or by dialyzing and then freeze-drying the filtrate. The soluble residue is hydrolyzed with H_2SO_4 in a boiling water bath. The resultant hydrolysate is analyzed for sugar content.

The insoluble residue in the funnel is washed with absolute ethanol and acetone, dried, weighed, and then mixed with 12N H_2SO_4 to hydrolyze cellulose. The acid is diluted to 1N, and noncellulose polysaccharides are hydrolyzed in a boiling water bath. The hydrolysate is centrifuged and the supernatant is saved for sugar analysis. The pellet is transferred to a funnel and washed several times with water and dried. The weight of the dried residue after acid hydrolysis is Klason lignin.

Aliquots of the acid-extracted soluble residue and insoluble residues are analyzed for neutral sugar content by HPLC or GC. Uronic acid content of the acid hydrolysates is measured colorimetrically. Fiber is equal to total monosaccharide content corrected for sugar losses during acid hydrolysis and corrected for the addition of one molecule of water for each glycoside bond hydrolyzed. Total fiber = Soluble fiber + Insoluble fiber (which includes lignin).

A dry, ground sample (10 g)
⇓
Sonicate and
extract with 80% ethanol (2 times)
and then hexane (2 times).
⇓
Filter between extractions
with Whatman No. 54 paper.
⇓
Determine sugar and lipid loss.
⇓
Dry sample, 4–5g.
⇓
Digest starch with Termamyl®
in 75 ml of 0.1 M acetate buffer,
pH 5.0, with 70 ppm Ca^{++};
96°C for 0.5 hr.
⇓
Digest starch with
amyloglucosidase (16 hr).
⇓
Centrifuge and filter to
separate soluble and insoluble fiber.

Filtrate.
⇓
Collect soluble polysaccharides by ppt with ethanol
or by dialyzing filtrate
and freeze-drying the
dialysate.
⇓
Hydrolyze soluble fiber
with 1 N H_2SO_4 (3 hr, 100°C)
⇓
*Sugar analysis
(*Soluble Fiber*)

Pellet and filter retentate.
⇓
Wash with 100% ethanol
then acetone.
⇓
Dry overnight under
vacuum @ 40°C.
⇓
Hydrolyze cellulose with
12 N H_2SO_4 (1 hr, room temp).
⇓
Dilute acid to 1 N and
hydrolyze non-cellulose,
insoluble polysaccharides
@ 100°C for 3 hr.
⇓
Centrifuge and vacuum filter.
Wash filter retentate with
water several times.

Dried filter retentate
is Klason lignin
(*Insoluble Fiber*)

Supernatant and filtrate
⇓
*Sugar analysis
(*Insoluble Fiber*)

*Sugars analysis: Uronic acids are measured colorimetrically;
Neutral sugars are measured by HPLC or GC

Soluble Fiber = sum of sugars**
Insoluble Fiber = sum of sugars** + lignin weight

**Corrected for incomplete recovery of sugars and multiplied by 0.9 to correct for the addition of one molecule of water per glycoside bond hydrolyzed

Total Fiber = Soluble Fiber + Insoluble Fiber

FIGURE 11-5.
Generalized analytical scheme for the Theander-Marlett approach for determining fiber.

11.3.4.4.3 Applications This approach to measuring dietary fiber probably provides the most accurate estimate of fiber over a wide range of foods when compared to the AOAC and Englyst-Cummings methods. Therefore, this method is suitable for research, legislation, and labeling purposes. However, the high degree of analytical skill, time commitment, and costly equipment required will probably prevent widespread adaptation for labeling purposes or routine analyses.

11.4 COMPARISON OF METHODS

The modified AOAC method, the Englyst-Cummings method, and the Theander-Marlett method are the most widely used methods for determining dietary fiber. These three methods and several other very similar methods all give quite comparable estimates of fiber content for a wide variety of foods. In general, the Englyst-Cummings method gives the lowest fiber values because lignin and resistant starch are not included as part of fiber in this method. Obviously, foods with a significant amount of resistant starch, such as corn flakes, and foods with significant lignin, such as cereal brans, will show the greatest deviation. The AOAC method overestimates fiber if the food is rich in simple sugars (glucose, fructose, and sucrose), such as in dry fruits and composite meals. It is hypothesized that some of the simple sugars are trapped and precipitated with ethanol if they are not extracted prior to fiber analysis. This does not appear to be a problem with the Englyst-Cummings procedure, possibly because of the small sample size relative to the large amount of ethanol used to precipitate soluble fiber. With the small sample size ($\leq$200 mg dry matter) in the Englyst-Cummings procedure, it is imperative that the food be completely homogeneous so that accurate subsamples can be taken for fiber analysis.

The AOAC and the Englyst-Cummings procedures incorporate a proteolytic enzyme to digest protein. Proteolysis allows some of the fiber to be solubilized, which in effect moves some of the insoluble fiber fraction into the soluble fiber fraction. In addition, proteolysis has the general effect of reducing the amount of material measured as lignin.

The AOAC procedure includes resistant starch as a dietary fiber component. Baked, flaked, and extruded products will have a significantly higher fiber value if determined by the AOAC procedure than if determined by the Englyst-Cummings methods. Which value is correct is debatable.

The rapid Englyst-Cummings procedure requires the least amount of time, technical skill, and specialized equipment compared to the other commonly used methods. Overall, the Englyst-Cummings and Theander-Marlett approaches are slightly more reproducible than the AOAC procedure.

Which fiber analysis method to choose is in part determined by (1) how much technical skill is available, (2) what the time constraints are, (3) availability of GC and/or HPLCs, and (4) the importance of knowledge of constituent sugar composition, cellulose, noncellulose, pectin, or lignin content. If only total, soluble, and insoluble fiber are needed, the AOAC or the rapid Englyst-Cummings methods are preferable. If the major components of fiber or constituent sugar composition are required, then the Englyst-Cummings GC procedure or the Theander-Marlett approach would be the method of choice.

11.5 SUMMARY

Dietary fiber is defined as plant polysaccharides that are undigestible by mammalian enzymes, plus lignin. The major components of dietary fiber are cellulose, hemicellulose, pectin, lignin, and hydrocolloids (gums). The crude fiber measurement drastically underestimates dietary fiber in foods since it measures only cellulose and lignin. All current methods use a combination of heat stable α-amylase and amyloglucosidase to digest and remove starch from the sample. Gravimetric procedures then digest and remove protein with a protease. The remaining undigestible material (fiber) is collected by filtration and weighed. The fiber residue is corrected for residual protein and ash contamination.

Chemical procedures collect macromolecules in the amylase-amyloglucosidase digest by filtration with or without ethanol precipitation. The polysaccharides in the precipitate are hydrolyzed with sulfuric acid and quantitated colorimetrically and/or chromatographically. The combined weight of sugars in the acid hydrolysate is equal to fiber weight.

Filtration and precipitation of polysaccharides with ethanol at specific steps in both the gravimetric and chemical methodologies allow differentiation of soluble, insoluble, and total fiber.

Current gravimetric and chemical methods produce similar estimates of total dietary fiber for most foods. However, there are method-dependent variabilities in estimates of soluble and insoluble fiber and in fiber estimates for foods that contain Maillard browning products.

11.6 STUDY QUESTIONS

1. Define the term *dietary fiber*, and give the constituents that compose dietary fiber.
2. Below are four values for the fiber content (percent dwb) of wheat bran. Using your knowledge of methods to measure crude fiber, acid detergent fiber, neutral detergent fiber (with amylase), and the AOAC method for total dietary fiber, indicate which method most likely fits with each of the four values below. Justify your answer by listing the constituents measured by each method.

Percent Fiber	*Method*	*Constituents Measured*	*Procedure*
46.0			
8.9			
40.2			
11.9			

3. You are teaching your technician to use the AOAC method for total dietary fiber to determine the dietary fiber content of a new line of high-fiber snack foods being developed by your company. Explain to your technician the purpose(s) of the steps in the total dietary fiber procedure listed below:
   a. heating sample and treating with amyloglucosidase
   b. treating sample with protease
   c. adding four volumes of 95 percent ethanol to sample after treatment with amyloglucosidase and protease
   d. after drying and weighing the filtered and washed residue, heating one duplicate final product to 525°C in a muffle furnace and analyzing the other duplicate sample for protein
4. Compare and contrast the AOAC method, the Englyst-Cummings method, and the Theander-Marlett method for determination of total dietary fiber. Consider the principles, procedures, applications, and advantages and disadvantages.
5. Explain how both gravimetric and chemical methods allow differentiation of soluble, insoluble, and total dietary fiber.

11.7 PRACTICE PROBLEMS

1. The following tabular data (see also pg. 180) were obtained when a high-fiber cookie was analyzed for fiber content by the AOAC Method 991.43.

	sample			
	insoluble		*soluble*	
Sample weight, mg	1,002.1	1,005.3		
Crucible + Celite weight, mg	31,637.2	32,173.9	32,377.5	33,216.4
Crucible + Celite + residue wt, mg	31,723.5	32,271.2	32,421.6	32,255.3
Protein, mg P	6.5		3.9	
Crucible + Celite + ash wt, mg		32,195.2		33,231.0

	blank			
	insoluble		*soluble*	
Crucible + Celite weight, mg	31,563.6	32,198.7	33,019.6	31,981.2
Crucible + Celite + residue wt, mg	31,578.2	32,213.1	33,033.4	31,995.6
Protein, mg P	3.2		3.3	
Crucible + Celite + ash wt, mg		32,206.8		31,989.1

What is the (a) total, (b) insoluble, and (c) soluble fiber content of the cookie?

2. The amount of dietary fiber in a navy bean cultivar was determined by the Englyst-Cummings procedure. The following data were obtained:

	sample a	*sample b*
Sample weight, mg	195.10	191.30
Neutral sugars, mg	26.36	25.74
Uronic acids, mg	8.74	8.64

How much dietary fiber is in the beans?

3. A sample of the same beans (question 2) was analyzed for fiber content by the Theander-Marlett approach. The following data were obtained:

	sample a	*sample b*
Sample weight, mg	4,503.10	4,635.70
Soluble fiber		
monosaccharide wt, mg	280.70	291.62
percent recovery	93.00	92.00
Insoluble fiber		
monosaccharide wt, mg	594.80	603.94
percent recovery	94.00	93.00
Crucible tare wt, mg	32,347.20	33,465.60
Crucible + retentate wt, mg	32,352.60	33,471.90

What is the (a) lignin, (b) soluble, (c) insoluble, and (d) total fiber content of the beans?

Answers

1. a = 8.06%; b = 6.06%; c = 2.00%. 2. 17.98%. 3. a = 0.13%; b = 6.09%; c = 12.76%; d = 18.85%.

11.8 REFERENCES

1. Burkitt, D.P., and Trowell, H.C. 1975. *Refined Carbohydrate Foods and Disease.* Academic Press, London, England.
2. Vahouny, G.V., and Kritchevsky, D. (Eds.) 1986. *Dietary Fiber: Basic and Clinical Aspects.* Plenum Press, New York, NY.
3. Leeds, A.R., and Avenell, A. (Eds.) 1985. *Dietary Fibre Perspectives: Reviews and Bibliography 1.* John Libby and Company Ltd., London, England.
4. Leeds, A.R., and Burley, V.J. (Eds.) 1990. *Dietary Fibre Perspectives: Reviews and Bibliography 2.* John Libby and Company Ltd., London, England.
5. Vahouny, G.V. 1982. Conclusions and recommendations of the symposium on "Dietary Fibers in Health and Disease," Washington, D.C., 1981. *Am. J. Clin. Nutr.* 35: 152–156.
6. AACC. 1983. *Approved Methods of the American Association of Cereal Chemists,* 8th ed. Sec. 32-20. American Association of Cereal Chemists, St. Paul, MN.
7. AOAC. 1990. *Official Methods of Analysis,* 15th ed. Method 991.43, 3rd Supplement 136–138, 1992. Association of Official Analytical Chemists, Washington, DC.
8. Prosky, L., Asp, N.-G., Schweizer, T.F., DeVries, J.W., and Furda, I. 1988. Determination of insoluble, soluble, and total dietary fiber in foods and food products: Interlaboratory study. *J. Assoc. Off. Anal. Chem.* 71:1017–1023.
9. Hudson, G.J., and Bailey, B.S. 1980. Mutual interference effects in the colorimetric methods used to determine the sugar composition of dietary fibre. *Food Chem.* 5:201–206.
10. Bitter, T., and Muir, H.M. 1962. A modified uronic acid carbazole reaction. *Anal. Biochem.* 4:330–334.
11. Southgate, D.A.T. 1969. Determination of carbohydrates in foods. II. Unavailable carbohydrates. *J. Sci. Fd. Agric.* 20:331–335.
12. Southgate, D.A.T. 1976. *Determination of Food Carbohydrates,* pp. 61–84, 107–112. Applied Science Publishers, London, UK.
13. Englyst, H.N., and Cummings, J.H. 1990. Non-starch polysaccharides (dietary fiber) and resistant starch. In *New Developments in Dietary Fiber,* I. Furda and C. J. Brine, (Eds.), pp. 205–255. Plenum Press, New York, NY.
14. Englyst, H.N., and Hudson, G.J. 1987. Colorimetric method for routine measurement of dietary fibre as non-starch polysaccharides. A comparison with gas-liquid chromatography. *Food Chem.* 24:63–76.
15. Cummings, J.H., Englyst, H.N., and Wood, R. 1985. Determination of dietary fibre in cereals and cereal products—collaborative trials. Part I: Initial trial. *J. Assoc. Publ. Analysts* 23:1–35.
16. Marlett, J.A. 1989. Measuring dietary fiber. *Anim. Feed Sci. Tech.* 23:1–13.
17. Marlett, J.A. 1990. Issues in dietary fiber. In *New Developments in Dietary Fiber,* I. Furda and C. J. Brine, (Eds.), pp. 183–192. Plenum Press, New York, NY.
18. Theander, O., and Westerlund, E. 1986. Studies on dietary fiber. 3. Improved procedures for analysis of dietary fiber. *J. Agric. Food Chem.* 34:330–336.

CHAPTER

12

Crude Fat Analysis

David B. Min

12.1 INTRODUCTION

12.1.1 Definitions

Lipids are a group of substances that, in general, are soluble in ether, chloroform, or other organic solvents but are sparingly soluble in water. They, with proteins and carbohydrates, constitute the principal structural components of foods. The solubility characteristic stated, rather than being a common structural feature, is unique to lipids (1). This definition describes a broad group of substances that have some common properties and compositional similarities (2). However, some lipids, such as triacylglycerols, are very hydrophobic. Other lipids, such as di- and monoacylglycerols, have both hydrophobic and hydrophilic moieties in their molecules and are soluble in relatively polar solvents (1). Triacylglycerols are fats and oils that represent the most prevalent category of a series of compounds known as lipids in foods. The terms **lipids, fats,** and **oils** are sometimes used interchangeably.

12.1.2 General Classification

The general classification of lipids that follows is useful to differentiate the lipids in foods (2).

12.1.2.1 Simple Lipids

Ester of fatty acids with alcohol:

- **Fats:** Esters of fatty acids with glycerol—triacylglycerols
- **Waxes:** Esters of fatty acids with long chain alcohols other than glycerols—for example, myricyl palmitate, cetyl palmitate, Vitamin A esters, and Vitamin D esters

12.1.2.2 Compound Lipids

Compounds containing groups in addition to an ester of a fatty acid with an alcohol:

- **Phospholipids:** Glycerol esters of fatty acids, phosphoric acids, and other groups containing nitrogen—for example, phosphatidyl choline, phosphatidyl serine, phosphatidyl ethanolamine, and phosphatidyl inositol
- **Cerebrosides:** Compounds containing fatty acids, a carbohydrate, and a nitrogen moiety—such as galactocerebroside and glucocerebroside
- **Sphingolipids:** Compounds containing fatty acids, a nitrogen moiety, and phosphoryl group—for example, sphingomyelins

12.1.2.3 Derived Lipids

Derived lipids are substances derived from neutral lipids or compound lipids. They have the general properties of lipids—examples are fatty acids, long chain alcohols, sterols, fat soluble vitamins, and hydrocarbons.

12.1.3 Content of Lipids in Foods

The lipid content in bovine milk is shown in Table 12-1. It shows the complexity of milk lipids that differ in polarity and concentrations.

Foods may contain any or all lipid compounds, but those of greatest importance are the triacylglycerols and the phospholipids. **Liquid triacylglycerols** at room temperature are referred to as **oils,** such as soybean oil and olive oil, and are generally of plant origin. **Solid triacylglycerols** at room temperature are termed as **fats.** Lard and tallow are examples of fats, which are generally from animals. The term *fat* is applicable to all triacylglycerols whether they are normally solid or liquid at ambient temperatures. Table 12-2 shows the wide range of lipid content in different foods.

12.1.4 Importance of Analysis

An accurate and precise quantitative analysis of lipids in foods is important for nutritional labeling, to determine whether the food meets the standard of identity and is uniform, and to understand the effects of fats and oils on the functional and nutritional properties of foods.

12.2 GENERAL CONSIDERATIONS

Lipids are soluble in organic solvents and insoluble in water. Therefore, water insolubility is the essential

TABLE 12-1. Lipids of Bovine Milk

Kinds of Lipids	*Percent of Total Lipids*
Triacylglycerols	97–99
Diacylglycerols	0.28–0.59
Monoacylglycerols	0.016–0.038
Phospholipids	0.2–1.0
Sterols	0.25–0.40
Squalene	Trace
Free fatty acids	0.10–0.44
Waxes	Trace
Vitamin A	7–8.5 μg/g
Carotenoids	8–10 μg/g
Vitamin D	Trace
Vitamin E	2–5 μg/g
Vitamin K	Trace

Adapted from (3), with permission of S. Patton, and (4) *Principles of Dairy Chemistry.* Jenness, R., and Patton, S. Copyright © 1959, John Wiley & Sons, Inc, with permission.

TABLE 12-2. Contents of Lipids in Foods

Foods	*Percent Lipid (wet weight basis)*
Lard, shortening, oils	close to 100
Butter and margarine	80
Salad dressings	40–70
Nuts	
Coconut	35
Almonds	54
Walnuts	64
Soybeans	18
Milk	3.5–4.3
Skim milk	0.1
Marine products	
Halibut	5.2
Cod	0.4
Cereals	
Grains	3–5
Germ	10
Bread	3–6
Raw meats	
Beef	11–28
Bacon	65
Pork	25–33
Eggs	12
Fruits and vegetables	
Apples	0.4
Oranges	0.2
Blackberries	1.0
Avocados	26.4
Asparagus	0.2
Lima beans	0.8
Sweet corn	1.2

Adapted from (5) (p. 149, 150), with permission.

analytical property used as the basis for the separation of lipids from proteins, water, and carbohydrates in foods. Glycolipids are soluble in alcohols and have a low solubility in hexane. In contrast, triacylglycerols are soluble in hexane and petroleum ether, which are nonpolar solvents. The wide range of relative hydrophobicity of different lipids makes the selection of a single universal solvent impossible for lipid extraction of foods. Some lipids in foods are in complex lipoproteins and liposaccharides; therefore, successful extraction requires that bonds between lipids and proteins or carbohydrates be broken so that the lipids can be freed and solubilized in the extracting organic solvents.

12.3 ANALYTICAL METHODS

The total lipid content of a food is commonly determined by organic solvent extraction methods. The accuracy of these methods greatly depends on the solubility of the lipids in the solvent used. The lipid content of a food determined by extraction with one solvent may be quite different from the content determined with another solvent of different polarity. In addition to solvent extraction methods, there are nonsolvent wet extraction methods and several instrumental methods that utilize the physical and chemical properties of lipids in foods for fat content determination.

Many of the methods cited in this chapter are official methods of the Association of Official Analytical Chemists (AOAC) International. Refer to these methods and other original references cited for detailed instructions of procedures.

12.3.1 Solvent Extraction Methods

12.3.1.1 Sample Preparation

The validity of the fat analysis of a food depends on **proper sampling** and **preservation of the sample** before the analysis (see also Chapter 3). Good sampling, sample preservation, and proper testing procedures are critical factors in food analyses. An ideal sample should be as close as possible in all of its intrinsic properties to the material from which it is taken. However, a sample is considered satisfactory if the properties under investigation correspond to those of the bulk material within the limits of the test (6).

The sample preparation for lipid analysis depends on the type of food and type and nature of lipids in the food (7). The extraction method for lipids in liquid milk is generally different from the extraction method for lipids in solid soybeans. In order to effectively analyze the lipids in foods, a knowledge of the structure, the chemistry, and the occurrence of the principal lipid classes and their constituents is necessary. Therefore, there is no single standard method for the extraction of all kinds of lipids in different foods. For the best results, sample preparation should be carried out under an inert atmosphere of nitrogen at low temperature to minimize chemical reactions such as lipid oxidation.

12.3.1.1.1 Predrying Sample Lipids cannot be effectively extracted with ethyl ether from moist food because the solvent cannot easily penetrate the moist food tissues. The ether, which is hygroscopic, becomes saturated with water and inefficient for lipid extraction. Drying the sample at elevated temperatures is undesirable because some lipids become bound to proteins and carbohydrates, and bound lipids are not easily extracted with organic solvents. Vacuum oven drying at low temperature or lyophilization increases the surface area of the sample for better lipid extraction. Predrying makes the sample easier to grind for better extraction, breaks fat-water emulsions to make fat dissolve easily in the organic solvent, and helps free fat from the tissues of foods (6).

12.3.1.1.2 Particle Size Reduction The extraction efficiency of lipids from dried foods depends on particle size; therefore, good grinding is very important. The classical method of determining fat in oilseeds involves the extraction of the ground seeds with selected solvent after repeated grinding at low temperature to minimize lipid oxidation. For better extraction, the sample and solvent are mixed in a high-speed comminuting device such as a blender. Lipids are extracted with difficulty from soybeans because of the limited porosity of the soybean hull and its sensitivity to dehydrating agents. The lipid extraction from soybeans is easily accomplished if the beans are broken mechanically by grinding.

12.3.1.1.3 Acid Hydrolysis A significant portion of the lipids in foods such as dairy, bread, flour, and animal products is bound to proteins and carbohydrates, and direct extraction with nonpolar solvents is inefficient. Such foods must be prepared for lipid extraction by acid hydrolysis (Table 12-3). Acid hydrolysis can break both covalently and ionically bound lipids into easily extractable lipid forms. The sample is predigested by refluxing for 1 hr with 3N hydrochloric acid, and ethanol and solid hexametaphosphate are added to facilitate separation of lipids from other components before foods are extracted with solvents (5, 6). For example, the acid hydrolysis of two eggs requires 10 ml HCl and heating in a water bath at 65°C for 15–25 min or until the solution is clear (5).

12.3.1.2 Solvent Selection

Ideal solvents for fat extraction should have a high solvent power for lipids and low or no solvent power for proteins, amino acids, and carbohydrates. They should evaporate readily and leave no residue, have a relatively low boiling point, and be nonflammable and nontoxic in both liquid and vapor states. The ideal solvent should penetrate sample particles readily, be single component to avoid fractionation, and be inexpensive and nonhygroscopic (5, 6). It is difficult to find an ideal fat solvent to meet all of these requirements. Ethyl ether and petroleum ether are the most commonly used solvents, but pentane and hexane are also used to extract oil from soybeans.

Ethyl ether has a boiling point of 34.6°C and is a better solvent for fat than petroleum ether. It is generally expensive compared to other solvents, has a greater danger of explosion and fire hazards, is hygroscopic, and forms peroxides (5). **Petroleum ether** is the low boiling point fraction of petroleum and is composed mainly of pentane and hexane. It has a boiling point of 35–38°C and is more hydrophobic than ethyl ether. It is selective for more hydrophobic lipids, cheaper, more nonhygroscopic, and less flammable than ethyl ether. The detailed properties of petroleum ether for fat extraction are described in AOAC Method 945.16 (9).

A combination of two or three solvents is frequently used. The solvents should be purified and peroxide free. The proper solvent-to-solute ratio must be used to obtain the best extraction of lipids from foods (6, 8).

TABLE 12-3. Effects of Acid Digestion on Fat Extraction from Foods

	Percent Fat Acid Hydrolysis	*Percent Fat No Acid Hydrolysis*
Dried egg	42.39	36.74
Yeast	6.35	3.74
Flour	1.73	1.20
Noodles	3.77–4.84	2.10–3.91
Semolina	1.86–1.93	1.10–1.37

Adapted from (5) (p. 154), with permission.

12.3.1.3 Continuous Solvent Extraction Methods

For continuous solvent extraction, sample is put in an extraction ceramic thimble and the solvent is added into the boiling flask. The continuous methods give faster, more efficient extraction than semicontinuous extraction methods. However, they may cause channeling, which results in incomplete extraction. The Wiley, Underwriters, and Goldfish tests are examples of continuous lipid extraction methods (5, 6), the last of which is described in more detail below.

12.3.1.3.1 Goldfish Method—Procedure

1. Weigh predried porous ceramic extraction thimble. Place vacuum oven dried sample in thimble and weigh again. (Sample could instead be combined with sand in thimble and then dried.)
2. Weigh predried extraction beaker.
3. Place ceramic extraction thimble into glass holding tube and then up into condenser of apparatus.
4. Place anhydrous ethyl ether (or petroleum ether) in extraction beaker and put beaker on heater of apparatus.
5. Extract for 4 hrs.
6. Lower heater and let sample cool.
7. Remove the extraction beaker and let air dry overnight, then at 100°C for 30 min. Cool beaker in desiccator and weigh.

12.3.1.3.2 Goldfish Method—Calculations

$$\text{weight of fat in sample} = (\text{beaker} + \text{fat}) - \text{beaker} \quad (1)$$

$$\% \text{ fat on dry weight basis} = (\text{g fat in sample/g dried sample}) \times 100 \quad (2)$$

12.3.1.4 Semicontinuous Solvent Extraction Methods

For semicontinuous solvent extraction, the solvent builds up in the extraction chamber for 5 to 10 minutes and completely surrounds the sample, then siphons back to the boiling flask. This method provides a soaking effect of sample and does not cause channeling. However, this method requires more time than the continuous method. The Soxhlet method (AOAC Method 920.39C for Cereal Fat; AOAC Method 960.39 for Meat Fat) is an example of the semicontinuous extraction method and is described below.

12.3.1.4.1 Soxhlet Method—Preparation of Sample If the sample contains more than 10 percent H_2O, dry the sample to constant weight at 95–100°C under pressure ≤ 100 mm Hg for about 5 hrs (AOAC Method 934.01).

12.3.1.4.2 Soxhlet Method—Procedure See Fig. 12-1.

1. Weigh, to the nearest mg, about 2 g predried sample into a predried extraction thimble, with porosity permitting a rapid flow of ethyl ether. Cover sample in thimble with glass wool.
2. Weigh predried boiling flask.

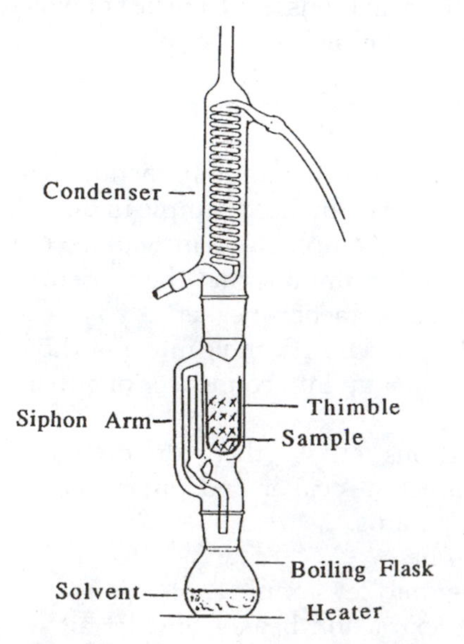

FIGURE 12-1.
Soxhlet extraction apparatus.

3. Put anhydrous ether in boiling flask.
   Note: The anhydrous ether is prepared by washing commercial ethyl ether with two or three portions of H_2O, adding NaOH or KOH, and letting stand until most of H_2O is absorbed from the ether. Add small pieces of metallic Na and let hydrogen evolution cease (AOAC Method 920.39B). Petroleum ether may be used instead of anhydrous ether (AOAC Method 960.39).
4. Assemble boiling flask, Soxhlet flask, and condenser.
5. Extract in a Soxhlet extractor at a rate of 5 or 6 drops per second condensation for about 4 hrs, or for 16 hrs at a rate of 2 or 3 drops per second by heating solvent in boiling flask.
6. Dry boiling flask with extracted fat in an air oven at 100°C for 30 min, cool in desiccator, and weigh.

12.3.1.4.3 Soxhlet Method—Calculation

$$\% \text{ fat on dry weight basis} = (\text{g fat in sample/g dried sample}) \times 100 \quad (3)$$

12.3.1.5 Discontinuous Solvent Extraction Methods

The Mojonnier test is an example of the discontinuous solvent extraction method. This extraction method does not require prior removal of moisture from the sample.

12.3.1.5.1 Modified Mojonnier Method for Milk Fat (AOAC Method 989.05)

1. Principle. Fat is extracted with a mixture of ethyl ether and petroleum ether, and the extracted fat is dried to a constant weight and expressed as percent fat by weight.

2. Preparation of Sample. Bring the sample to about 20°C; mix to prepare a homogeneous sample by pouring back and forth between clean beakers. Promptly weigh or measure the test portion. If lumps of cream do not disperse, warm the sample in a water bath to about 38°C and keep mixing until it is homogeneous, using a "policeman" if necessary to reincorporate the cream adhering to the container or stopper. When it can be done without interfering with dispersal of the fat, cool warmed samples to about 20°C before transferring the test portion.

3. First Extraction.

a. Weigh, to the nearest 0.1 mg, 10 g milk into a Mojonnier fat extraction flask (Fig. 12-2).
b. Add 1.5 mL NH_4OH and shake vigorously. Add 2 mL if the sample is sour. NH_4OH neutralizes acidic sample and dissolves protein.

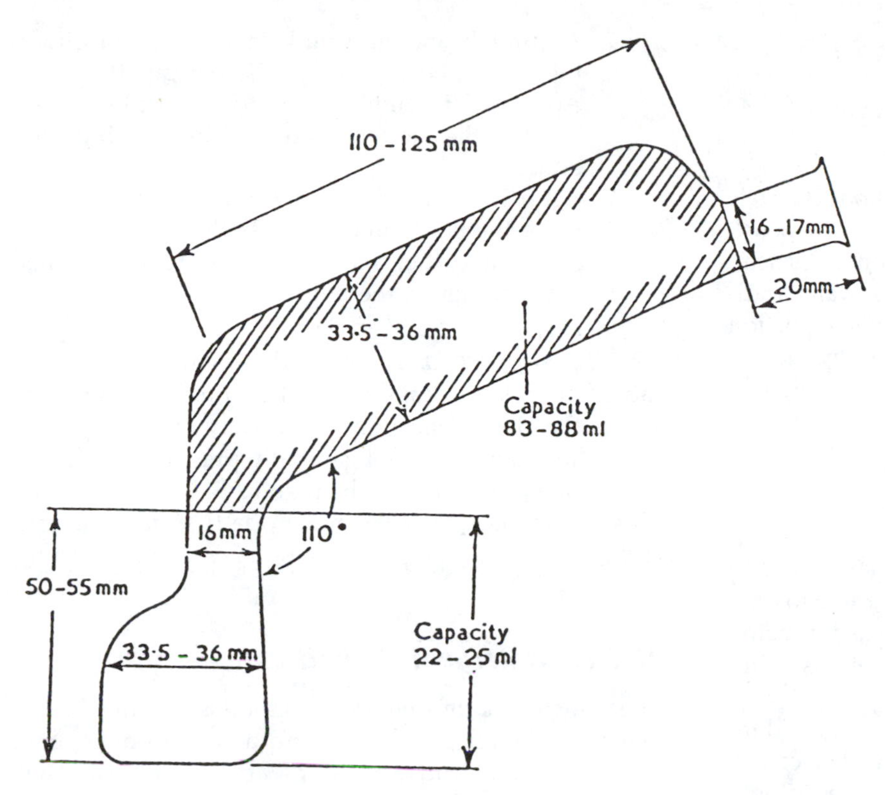

FIGURE 12-2.
Mojonnier fat extraction flask.

c. Add 10 mL 95 percent ethanol and shake 90 sec. The alcohol prevents possible gel formation.
d. Add 25 mL ethyl ether and shake 90 sec. The ether dissolves the lipid.
e. Cool if necessary, and add 25 mL petroleum ether and shake 90 sec. The petroleum ether removes moisture from the ethyl ether extract and dissolves more nonpolar lipid.
f. Centrifuge for 30 sec at 600 rpm.
g. Decant ether solution from the Mojonnier flask into the previously weighed Mojonnier fat dish.

4. Second Extraction.

a. Add 5 mL 95 percent ethanol and shake vigorously 15 sec.
b. Add 15 mL ethyl ether and shake 60 sec.
c. Add 15 mL petroleum ether and shake 60 sec.
d. Centrifuge for 30 sec at 600 rpm.
e. Decant solution into the same Mojonnier dish.

5. Third Extraction.

a. Add 15 mL ethyl ether and shake 60 sec.
b. Add 15 mL petroleum ether and shake 60 sec.
c. Centrifuge for 30 sec at 600 rpm.
d. Decant solution into the same Mojonnier dish.

e. Evaporate the solvent in the dish on the electric hot plate at ≤ 100°C in a hood.
f. Dry the dish and fat to a constant weight in a forced air oven at 100°C ± 1°C.
g. Cool the dish to room temperature and weigh.

6. Calculations.

$$\begin{aligned}\%\ \text{Fat} = 100 \times \{[&(\text{wt. dish} + \text{fat}) \\ &- (\text{wt. dish})] \\ &- (\text{av. wt. blank residue})\}/ \\ &\text{wt. sample}\end{aligned} \quad (4)$$

A pair of reagent blanks must be prepared every day. For reagent blank determination, use 10 mL distilled water instead of milk sample. The reagent blank should be < 0.002 g. Duplicate analyses should be < 0.03 percent fat.

12.3.1.5.2 Modified Mojonnier Method for Fat in Flour (AOAC Method 922.06) or Fat in Pet Food (AOAC Method 954.02) Mix 2 g sample and 2 mL ethyl alcohol well in 50 mL beaker. Add 10 mL HCl (25 + 11) and place the beaker in water bath held at 70–80°C with stirring for 30–40 min for hydrolysis. Add 10 mL alcohol and cool. The acid hydrolyzed flour is extracted by a combination of ethyl ether and petroleum ether as described in the Modified Mojonnier Method for Milk Fat (AOAC Method 989.05).

12.3.2 Nonsolvent Wet Extraction Methods

12.3.2.1 Babcock Method for Milk Fat (AOAC Method 989.04 and 989.10)

12.3.2.1.1 Principle In the Babcock method, H_2SO_4 is added to a known amount of milk in the Babcock bottle. The sulfuric acid digests protein, generates heat, and releases the fat. Centrifugation and hot water addition isolate fat for quantitation in the graduated portion of the test bottle. The fat is measured volumetrically, but the result is expressed as percent fat by weight.

12.3.2.1.2 Procedure

1. Accurately pipette milk sample (17.6 mL) into Babcock test bottle.
2. Add reagent grade (1.82 sp. gr.) sulfuric acid (17.5 mL) to the bottle, allowing the acid to flow gently down the neck of the bottle as it is being slowly rotated. The acid digests proteins to liberate the fat.
3. Centrifuge the mixture for 5 min, and liquid fat will rise into calibrated bottle neck. The centrifuge must be kept at 55–60°C during centrifugation.
4. Add hot water to bring liquid fat up into graduated neck of Babcock bottle.
5. The direct percentage of fat by weight is read to the nearest 0.05 percent from the graduation mark of the bottle.

12.3.2.1.3 Applications The Babcock method, which is a most common method for the determination of fat in milk, takes about 45 min and duplicate tests should agree within 0.1 percent. The Babcock method does not determine the phospholipids in the milk products (11). It is not applicable to products containing chocolate or added sugar without modification due to charring of chocolate and sugars by sulfuric acid. A modified Babcock method is used to determine essential oil in flavor extracts (AOAC Method 932.11) and fat in seafood (AOAC Method 964.12).

12.3.2.2 Gerber Method for Milk Fat

12.3.2.2.1 Principle The principle of the Gerber method (10) is similar to that of the Babcock method, but it uses sulfuric acid and amyl alcohol. The sulfuric acid digests proteins and carbohydrates, releases fat, and maintains the fat in a liquid state by generating heat.

12.3.2.2.2 Procedure

1. Transfer 10 mL of H_2SO_4 at 15–21°C into a Gerber milk bottle.
2. Accurately measure milk sample (11 mL) into the Gerber bottle, using a Gerber pipette.
3. Add 1 mL isoamyl alcohol to the bottle.
4. Tighten the stopper and mix by shaking the bottle.
5. Centrifuge the bottle for 4 min.
6. Place the bottle in a water bath at 60–63°C for 5 min, and then read the fat content from the graduations on the bottle neck.

12.3.2.2.3 Applications The Gerber method is comparable to the Babcock method but is simpler and faster and has wider application to a variety of dairy products. The isoamyl alcohol generally prevents the charring of sugar found with the regular Babcock method. This test is more popular in Europe than in America (12).

12.3.2.3 Detergent Method

The principle of the detergent method is that the detergents react with protein to form a protein-detergent complex to break up emulsions and release fat. The method was originally developed to determine fat in milk because of the corrosive properties of H_2SO_4 in the Babcock test (13). This method was later modified for use with other products. Milk is pipetted into a Babcock test bottle. An anionic detergent, dioctyl sodium phosphate, is added to disperse the protein layer that stabilizes the fat to liberate fat. Then, a strong hydrophilic nonionic polyoxyethylene detergent, sorbitan monolaurate, is added to separate fat from other food components. The percent fat is measured volumetrically and expressed as percent fat (6).

12.3.2.4 Refractive Index Method for Processed Meat

12.3.2.4.1 Principle The refractive index is characteristic of each kind of fat, and the values vary with degree and type of unsaturation, oxidation, heat treatment, temperatures of analyses, and the fat content. Fat is extracted with a solvent, and the refractive index of the solvent is compared to the refractive indices of the extracted fat solution and fat (5, 6).

12.3.2.4.2 Procedure

1. Weigh 2 g sample.
2. Transfer the sample to mortar and pestle.
3. Add 1.5 g dry sand.
4. Add 3 g anhydrous sodium sulfate.
5. Add 3 mL bromonaphthalene.
6. Grind the materials for 3 min.

7. Put filter paper in funnel in the rack above beaker.
8. Transfer contents of mortar to funnel.
9. Collect filtrate (i.e., solution with extracted fat).
10. Determine the refractive of index of filtrate.
11. Calculate percent fat in sample.

12.3.2.4.3 Calculations

$$\% \text{ Fat} = 100 \text{ V d } (n_1 - n_2) / \text{ W } (n_2 - n) \quad (5)$$

where: V = ml of bromonaphthalene
d = density of fat
n = refractive index of fat
n_1 = refractive index of bromonaphthalene
n_2 = refractive index of extracted solution
W = weight of sample

12.3.3. Instrumental Methods

12.3.3.1 Low Resolution NMR Method

The basic principles of **nuclear magnetic resonance** (NMR) are presented in Chapter 27. NMR can be used to measure fat and oil in food materials in a nondestructive way. Two kinds of NMR may be used: **time domain low resolution NMR** (sometimes called **pulsed NMR**), and **frequency domain NMR** (NMR spectra). The great advantages of NMR are that it is nondestructive and that it does not require that the sample be transparent.

In time domain NMR, signals from the hydrogen nuclei (^{1}H or protons) of different food components are distinguished by their different rates of decay or nuclear relaxation. Hydrogen nuclei in solid phases relax extremely fast (signal disappears), while protons in liquid phases relax very slowly. Furthermore, in samples such as oilseeds and some food products, water protons may relax faster than oil protons. The intensity of the signal is proportional to the number of hydrogen nuclei and therefore to hydrogen content. Thus, intensity can be converted to oil content using calibration methods (14, 15, 16). This method is used for water content, oil content, solid fat content, and solid-to-liquid ratio.

The equation for the so-called direct method to determine solid fat content follows:

$$S_{dir}(t) = \frac{fs'}{l + fs'} \times 100\% \quad (6)$$

where: S_{dir} (t) = solid fat content
f = correction factor
s′ = intensity at 10 μs
l = intensity at 70 μs (corresponding to liquid only)

The intensities s′ and l are obtained from the NMR signal as shown in Fig. 12-3.

Time domain NMR has been used to analyze fat-containing food materials including butter, margarine, shortening, chocolates, oilseeds, meats, milk powders, cheese, flours, and the like. An example is shown in Fig. 12-4. Here only the liquid signal intensity was measured. Since liquid oil dominates the liquidlike phases in seeds, a strong correlation of NMR signal with seed oil content (determined in an independent analysis) is obtained. Similar curves have been pre-

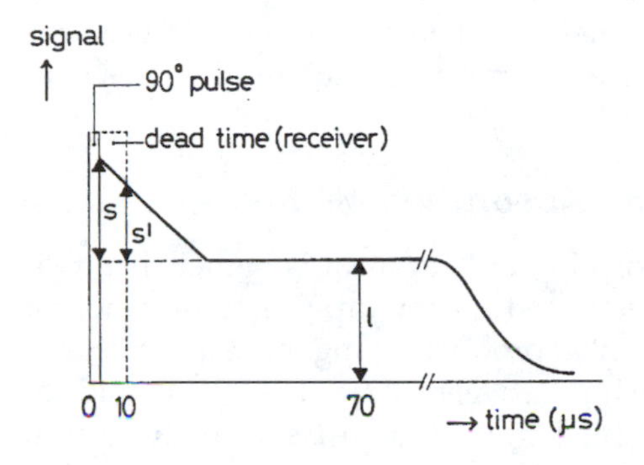

FIGURE 12-3.
Magnetization decay of a partly crystallized fat. From (17), used with permission.

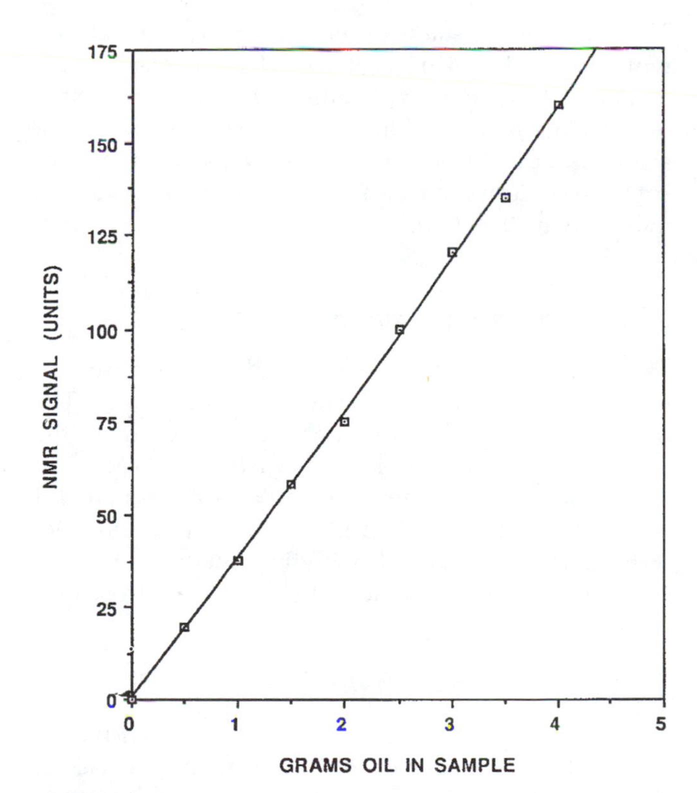

FIGURE 12-4.
The standard curve for the relationship between wide-line NMR signal and oil content of soybean determined by a solvent extraction method. From (14), used with permission.

pared for dairy foods, meats, and so on. Sometimes it is necessary to remove water either by drying the sample before analysis or by NMR tricks to isolate the oil signal.

Frequency domain NMR of foods is a new application in which food components are distinguished by the chemical shift (resonance frequency) of their peaks in an NMR spectrum. The pattern of oil resonances reflects degree of unsaturation and other chemical properties. This is useful for chemical analysis since intensities are proportional to amounts. Liquid glycerides have been detected this way in cheeses, fruits, meat, oilseeds, and other food materials. Frequency domain NMR analysis of fats and oils has been reviewed by Eads and Croasmun (18) and by Eads (19).

12.3.3.2 X-Ray Absorption Method

The X-ray absorption of lean meat is higher than that of fat. It has been used for the rapid determination of fat in meat and meat products using the standard curve of the relationship between X-ray absorption and fat content determined by a standard solvent extraction method (6).

12.3.3.3 Dielectric Method

The dielectric constants of foods change as the oil contents change. For example, the electrical current of lean meat is 20 times greater than that of fat. The coefficient of correlation for the linear regression between the amounts of induced current and the oil contents of soybeans determined by a standard solvent extraction method was 0.98 (20).

12.3.3.4 Infrared Method

This infrared (IR) method is based on absorption of infrared energy by fat at a wavelength of 5.73 μ. The more the energy absorption at 5.73 μ, the higher the fat content of the sample. This method was used to determine the fat content of milk using a standard curve of the infrared absorptions and fat contents determined by a standard analytical method (21). See Chapter 24 for a discussion of infrared spectroscopy.

12.3.3.5 Ultrasonic Method

The acoustic property of fat is different from that of solid-not-fat. The sound velocity in milk increases or decreases as the fat content increases or decreases above or below a certain critical temperature. The fat content of milk fat was determined by measuring the sound velocity (22).

12.3.3.6 Colorimetric Method

The fat content of milk was determined by measuring the color developed by the reaction between milk fat and hydroxamic acid. The color thus developed was compared to a standard curve of the color intensities and fat contents of samples determined by the Mojonnier method (23).

12.3.3.7 Density Measurement Method

The density and oil content of oilseeds have been found to be correlated with r = −0.96 (Fig. 12-5). The oil content of oilseeds can be determined by measuring the density of seeds using a linear regression line between the seed density and the fat content determined by a standard solvent extraction method (24).

12.4 COMPARISON OF METHODS

Soxhlet extraction or its modified method is the most common crude fat determination method in foods. However, this method requires a dried sample for the hygroscopic ethyl ether extraction. If the samples are moist or liquid foods, the Mojonnier method is gener-

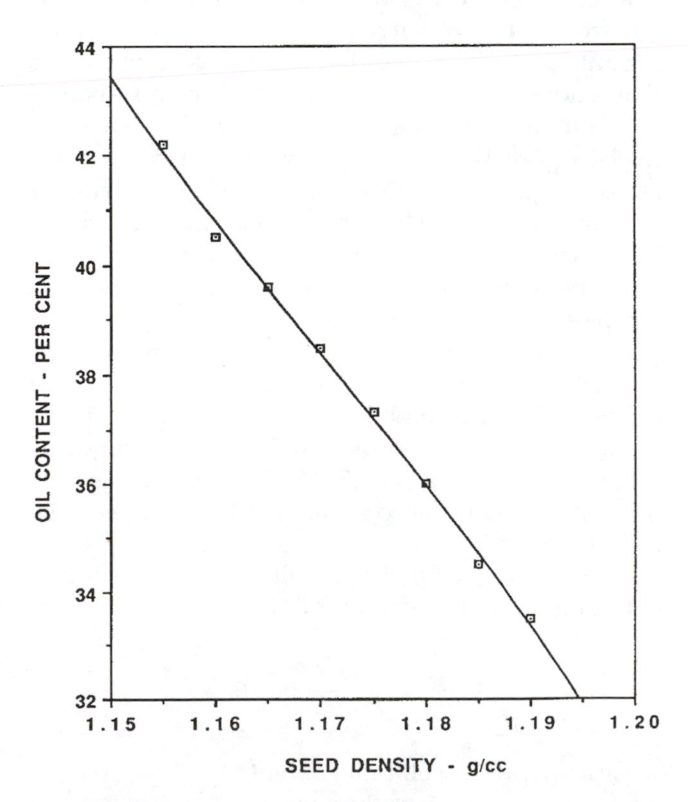

FIGURE 12-5.
The standard curve for the relationship between seed density and oil content of flax determined by a solvent extraction method. From (24), used with permission.

ally applicable to determine the fat content. The instrumental methods such as IR and NMR are very simple, reproducible, and fast but are only available for fat determination for specific foods. The application of instrumental methods for fat determination generally requires a standard curve between the signal of the instrument analysis and the fat content obtained by a standard solvent extraction method. However, a rapid instrumental method could be used as a quality control method for fat determination of a specific food.

12.5 SUMMARY

Lipids are generally defined by their solubility characteristics rather than by some common structural feature. Lipids in foods can be classified as simple, compound, or derived lipids. The lipid content of foods varies widely, but quantitation is important because of regulatory requirements, nutritive value, and functional properties. To analyze food for the fat contents accurately and precisely, it is essential to have a comprehensive knowledge of the general compositions of the lipids in the foods, the physical and chemical properties of the lipids as well as the foods, and the principles of fat determination. There is no single standard method for the determination of fats in different foods. The validity of any fat analysis depends on proper sampling and preservation of the sample prior to analysis. Predrying of the sample, particle size reduction, and acid hydrolysis prior to analysis may also be necessary. The total lipid content of foods is commonly determined by organic solvent extraction methods, which can be classified as continuous (e.g., Goldfish), semicontinuous (e.g., Soxhlet), or discontinuous (e.g., Mojonnier). Nonsolvent wet extraction methods, such as the Babcock, Gerber, detergent, and refractive index methods, are commonly used for certain types of food products. A variety of instrumental methods are also available for fat determination of specific foods. These methods are rapid and so may be useful for quality control but generally require correlation to a standard solvent extraction method.

12.6 STUDY QUESTIONS

1. Itemize the procedures that may be required to prepare a food sample for accurate fat determination by a solvent extraction method (e.g., Soxhlet method). Explain why each of these procedures may be necessary.
2. To extract the fat from a food sample, you have the choice of using ethyl ether or petroleum ether as the solvent, and you can use either a Soxhlet or a Goldfish apparatus. What combination of solvent and extraction would you choose? Give all the reasons for your choice.
3. Explain and contrast the principles (not procedures) involved in determining the fat content of a food product by the following methods. Indicate for each method the type of sample that would be appropriate for analysis.
   a. Soxhlet
   b. Babcock
   c. refractive index
   d. Mojonnier
   e. detergent
   f. low resolution NMR

12.7 PRACTICE PROBLEMS

1. To determine the fat content of beef by refractive index method, 5 mL of bromonaphthalene was used to extract fat from 20 grams of beef. The density of fat is 0.9 g/mL, and the refractive indices of beef fat, bromonaphthalene, and the bromonaphthalene beef fat extracted solution are 1.466, 1.658, and 1.529, respectively. Calculate the fat content of the beef.
2. To determine the fat content of a semimoist food by Soxhlet method, the food was first vacuum oven dried. The moisture content of the product was 25 percent. The fat in the dried food was determined by the Soxhlet method. The fat content of the dried food was 12.5 percent. Calculate the fat content of the original semimoist product.
3. The densities of milk fat and milk are 0.9 and 1.032, respectively. The fat content of the milk was 3.55 percent on a volume basis. Calculate the fat content of milk as percent, wet weight basis.
4. The fat content of 10 grams of commercial ice cream was determined by the Mojonnier method. The weights of extracted fat after the second extraction and the third extraction were 1.21 grams and 1.24 grams, respectively. How much of fat, as a percentage of the total, was extracted during the third extraction?

Answers

1. 46.1 percent; 2. 9.4 percent; 3. 3.09 percent; 4. 0.3 percent.

12.8 REFERENCES

1. Belitz, H.D., and Grosch, W. 1987. *Food Chemistry.* Springer-Verlag, Berlin, Germany.
2. Dugan, L., Jr. 1976. Lipids. In *Food Chemistry,* 1st ed. O.R. Fennema (Ed.). Marcel Dekker, Inc., New York, NY.
3. Patton, S., and Jensen, R.G. 1976. *Biomedical Aspects of Lactation,* p. 78. Pergamon Press, Oxford, England.
4. Jenness, R., and Patton, S. 1959. *Principles of Dairy Chemistry.* John Wiley & Sons, Inc., New York, NY.
5. Joslyn, M.A. 1970. *Methods in Food Analysis,* 2nd ed. Academic Press, Inc., New York, NY.
6. Pomeranz, Y., and Meloan, C.F. 1987. *Food Analysis: Theory and Practice,* 2nd ed. Van Nostrand Reinhold, New York, NY.
7. Marinetti, G.V. 1962. Chromatographic separation, identification, and analysis of phosphatides. *J. Lipid Res.* 3:1–20.

8. Entenmann, C. 1961. The preparation of tissue lipid extracts. *J. Am. Oil Chem. Soc.* 38:534–538.
9. AOAC. 1990. *Official Methods of Analysis*, 15th ed. Association of Official Analytical Chemists, Washington, DC.
10. Milk Industry Foundation. 1964. Laboratory Manual—Methods of Analysis of Milk and Its Products. Milk Industry Foundation, Washington, DC.
11. Levowitz, D. 1967. Determination of fats and total solids in dairy products. In *Laboratory Analysis of Milk and Milk Products*. U.S. Dept. of Health, Education, and Welfare, Cincinnati, OH.
12. Marshall, R.T. (Ed.) 1992. *Standard Methods for the Examination of Dairy Products*, 16th ed. American Public Health Association, Washington, DC.
13. Schain, P. 1949. The use of detergents for quantitative fat determination. Determination of milk fat. *Science* 110: 121–122.
14. Alexander, D.E., Silvela, L., Collins, I., and Rodgers, R.C. 1967. Analysis of oil content of maize by wide-line NMR. *J. Am. Oil Chem. Soc.* 44:555–558.
15. Collins, I., Alexander, D.E., Rodgers, R.C., and Silvela, L. 1967. Analysis of oil content of soybeans by wide-line NMR. *J. Am. Oil Chem. Soc.* 44:708–710.
16. Conway, T.F., and Earle, F.R. 1963. Nuclear magnetic resonance for determining oil content of seeds. *J. Am. Oil Chem. Soc.* 40:265–268.
17. van Putte, K. 1975. Pulsed NMR as a routine method in the fat and margarine industry, part 2. Bruker Minispec application note #5. Bruker Analytische Messtechnik GmbH, Karlsruhe, Germany.
18. Eads, T.M., and Croasmun, W.R. 1988. NMR applications to fats and oils. *J. Amer. Oil Chem. Soc.* 65:78–83.
19. Eads, T.M. 1991. Multinuclear high resolution and wide line NMR methods for analysis of lipids. Ch. 23, in *Analysis of Fats, Oils, and Lipoproteins*, E.G. Perkins (Ed.), 409–457. American Oil Chemists' Society, Champaign, IL.
20. Hunt, W.H., Neustadt, M.H., Hardt, J.R., and Zeleny, L. 1952. A rapid dielectric method for determining the oil content of soybeans. *J. Am. Oil Chem. Soc.* 29:258–261.
21. Biggs, D.A. 1967. Milk analysis with the infrared milk analyzer. *J. Dairy Sci.* 50:799–803.
22. Fitzgerald, J., Ringo, W.G.R., and Winder, W.C. 1961. An ultrasonic method for measurement of solid-not-fat and milk fat in fluid milk. *J. Dairy Sci.* 44:1165.
23. Katz, I., Keeney, M., and Bassette, R. 1959. Caloric determination of fat in milk and saponification number of a fat by the hydroxamic acid reaction. *J. Dairy Sci.* 42:903–906.
24. Zimmerman, D. C. 1962. The relationship between seed density and oil content in flax. *J. Am. Oil Chem. Soc.* 39:77–78.

CHAPTER

13

Fat Characterization

Oscar A. Pike

13.1 INTRODUCTION

Methods for characterizing edible lipids, fats, and oils can be separated into two categories: methods developed to analyze bulk fats and oils, and methods focusing on analysis of foodstuffs and their lipid extracts. For foodstuffs it is usually necessary to extract the lipids prior to analysis, and, if sufficient quantities of lipids are available, some of the methods developed for bulk fats and oils can be utilized in their characterization.

The methods in this chapter are divided into four subsections. Traditional analytical methods for bulk fats and oils are covered in the first subsection. Some of these have been supplemented by instrumental methods such as gas chromatography (GC), high-performance liquid chromatography (HPLC), and nuclear magnetic resonance (NMR). The second and third subsections give methods of measuring lipid oxidation. Some of these utilize intact foodstuffs and others require lipid extracts from the foodstuffs. The fourth subsection gives thin-layer chromatography (TLC) and GC methods for the analysis of lipid fractions, including fatty acid methyl esters and cholesterol.

Numerous methods exist for the characterization of lipids, fats, and oils (1–11). Those methods included here seem appropriate for use in an undergraduate food analysis laboratory. The understanding of basic concepts derived from traditional methods is valuable prior to learning more sophisticated instrumental methods. Many of the methods cited are official methods of the Association of Official Analytical Chemists (AOAC) International or the American Oil Chemists' Society (AOCS). The principles, general procedures, and applications are described for the methods. Refer to the specific method cited for detailed information on the procedure.

13.1.1 Definitions and Classifications

As explained in Chapter 12, the term **lipids** refers to a wide range of compounds soluble in organic solvents but only sparingly soluble in water. Chapter 12 also outlines the general classification scheme for lipids. The majority of lipids present in foodstuffs are of the following types: fatty acids, mono- di-, and triacylglycerols, phospholipids, sterols (including cholesterol), and lipid-soluble pigments and vitamins.

In contrast to **lipids, fats** and **oils** often refer to bulk products of commerce, crude or refined, that have already been extracted from animal products or oilseeds and other plants grown for their lipid content. The term **fat** signifies extracted lipids that are solid at room temperature, and **oil** refers to those that are liquid. However, the three terms, lipid, fat, and oil, are often used interchangeably.

In relation to the human diet and food labeling, the term **fat** (e.g., dietary fat, percent fat, or calories from fat) refers to the lipid components of the foodstuff, in contrast to the carbohydrate and protein components.

The term **triglyceride** is synonymous with triacylglycerol.

13.1.2 Importance of Analysis

Such issues as the effect of dietary fat on health and food labeling requirements necessitate that food scientists be able not only to measure the total lipid content of a foodstuff but also to characterize it. Health concerns require the measurement of such parameters as cholesterol content, amount of saturated and unsaturated fat, and perhaps even the type and amount of individual fatty acids. Measurements of lipid stability impact not only the shelf life of the product but also its safety, since some oxidation products (e.g., malonaldehyde, cholesterol oxides) have toxic properties. Another area of interest is the analysis of oils and fats used in deep-fat frying operations.

13.1.3 Content in Foods

Tables are available that indicate the total fat content of foods, as discussed in Chapter 12. Ongoing studies are reporting more accurately the quantities in food of saturated and unsaturated fat, cholesterol, and other specific parameters.

Commodities containing significant amounts of bulk fats and oils include confections, margarine and spreads, shortening, frying and cooking oils, salad oils, emulsified dressings, emulsifiers, peanut butter, and imitation dairy products.

Because of their usefulness as food ingredients, it is sometimes important to know the characteristics of bulk fats and oils. Definitions and specifications for bulk fats and oils (e.g., soybean oil, corn oil, coconut oil), including values for many of the tests described in this chapter, can be found in the Merck Index (12).

Foods containing even minor amounts of lipids (e.g., <1%) can have a shelf life limited by lipid oxidation and subsequent rancidity.

13.2 GENERAL CONSIDERATIONS

Various fat extraction solvents and methods are discussed in Chapter 12. For lipid characterization, extraction of fat or oil from foodstuffs can be accomplished by homogenizing with a solvent combination such as hexane-isopropanol (3:2, v/v). The solvent can then be removed using a rotary evaporator or by drying under a stream of nitrogen gas. The use of chloroform-methanol as a lipid solvent is common but is discouraged because of its toxicity. Lipid oxidation during extraction and testing can be minimized by adding antioxidants [e.g., 10–100 mg butylated hydroxytoluene (BHT) per liter] to solvents and by taking other precautions such as flushing containers with nitrogen gas and avoiding exposure to heat and light. Constituents present in lipid extractions that may present problems in lipid characterization include phosphatides, gossypol, carotenoids, chlorophyll, sterols, tocopherols, vitamin A, and metals.

After extraction from their parent source, bulk fats and oils typically undergo the following refinements: degumming, refining, bleaching, and deodorization. Modifications such as fractionation, winterization, interesterification, and hydrogenation may also be a part of the processing, depending on the commodity being produced. Methods discussed in this chapter can be used to monitor the refining process.

For the purpose of nutritional labeling, the lipid present in a foodstuff (or bulk fats and oils) can be analyzed for its **fatty acid profile.** This is done by determining the kind and amount of fatty acids that are present. Once the fatty acid profile has been determined, the following constituents can be **calculated:**

- percent saturated fatty acids
- percent unsaturated fatty acids
- percent monounsaturated fatty acids
- percent polyunsaturated fatty acids (PUFA)

Alternately, AOAC Method 979.19 gives a spectrophotometric determination for PUFA in oils.

Changes that lipids undergo during processing and storage include lipolysis, oxidation, and thermal degradation (such as during deep-fat frying operations). **Lipolysis** is the hydrolysis of fatty acids from the glyceride molecule. Because of their volatility, hydrolysis of short-chained fatty acids can result in off odors. The term **rancidity** refers to the off odors and flavors resulting from **lipolysis (hydrolytic rancidity)** or **lipid oxidation (oxidative rancidity).** Methods used to monitor the quality of the oil or fat used in deep-fat frying operations are based on physical and chemical changes occurring, which include an increase in viscosity, foaming, free fatty acids, degree of saturation, hydroxyl and carbonyl group formation, and saponification value.

Lipid oxidation (also called autoxidation) as it occurs in bulk fats and oils proceeds via a self-sustaining free radical mechanism that produces hydroperoxides (initial or primary products) that undergo scission to form various products including aldehydes, ketones, organic acids, and hydrocarbons (final or secondary products) (Fig. 13-1).

More recently, an appreciation has been gained for the mechanisms occurring in biological tissues, including foodstuffs, where abstraction reactions and rearrangements of alkoxyl and peroxyl radicals result in the production of endoperoxides and epoxide products as secondary products.

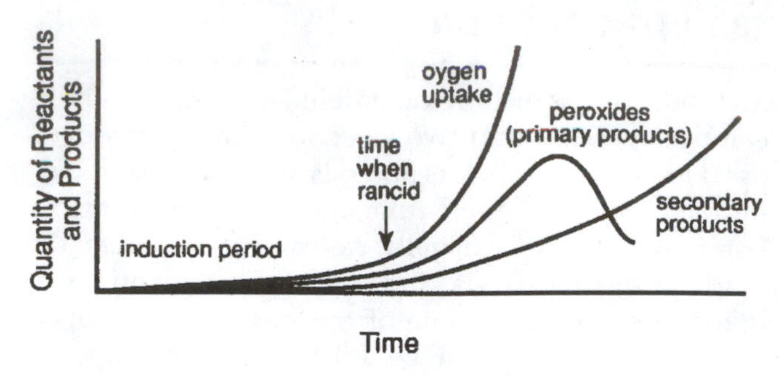

FIGURE 13-1.
Changes in quantities of lipid oxidation reactants and products over time. Adapted with permission from Labuza, T.P. 1971. Kinetics of lipid oxidation in foods. *CRC Crit. Rev. Food Technol.* 2:355–405. Copyright CRC Press, Inc., Boca Raton, FL.

Many methods have been developed measuring the different compounds as they form or degrade during lipid oxidation. Since the system is dynamic, it is recommended that two or more methods be used to obtain a more complete understanding of lipid oxidation.

13.3 METHODS

13.3.1 Bulk Oils and Fats

Numerous methods exist to measure the characteristics of fats and oils. Some methods (e.g., titer test) have limited use for edible oils (in contrast to soaps and industrial oils). Other methods (e.g., color measurement) require a special apparatus not commonly available or have been antiquated by common instrumental procedures [e.g., volatile acids (Reichert-Meissl, Polenske, and Kirschner values) have been largely replaced by determination of fatty acids using GC]. Finally, methods to determine impurities have not been listed, including moisture, insoluble impurities, and unsaponifiable matter.

13.3.1.1 Sample Preparation

Ensure that samples are visually clear and free of sediment. When required (e.g., iodine value), dry the samples prior to testing (AOAC Method 981.11). Because exposure to heat, light, or air promotes lipid oxidation, avoiding these conditions during sample storage will retard rancidity. For bulk oils and fats, sampling procedures are available (AOCS Method C 1–47; 7).

13.3.1.2 Refractive Index (R.I.)

13.3.1.2.1 Principle The **refractive index** of an oil is defined as the ratio of the speed of light in air (technically, a vacuum) to the speed of light in the oil.

13.3.1.2.2 Procedure Measure samples with a refractometer at 20 or 25°C for oils and 40°C for fats since most fats are liquid at this temperature. AOAC Method 921.08; AOCS Method Cc 7–25.

13.3.1.2.3 Applications R.I. is used to control hydrogenation; it decreases linearly as iodine value decreases. R.I. is also used as a measure of purity and means of identification, since each substance has a characteristic R.I. Tables of published data are available.

13.3.1.3 Melting Point

13.3.1.3.1 Principle Melting point may be defined in various ways, each corresponding to a different residual amount of solid fat. The **capillary tube method,** also known as the **clear point,** is the temperature at which fat heated at a given rate becomes completely clear and liquid in a one-end closed capillary. The **Slip melting point** is performed similarly to the capillary tube method and measures the temperature at which a column of fat moves in an open capillary when heated. The **Wiley melting point** measures the temperature at which a ⅛ × ⅜ in. disc of fat, suspended in an alcohol-water mixture of similar density, changes into a sphere.

13.3.1.3.2 Procedure

- Capillary tube melting point: AOAC Method 920.157; AOCS Method Cc 1–25.
- Slip melting point: AOCS Method Cc 3–25.
- Wiley melting point: AOAC Method 920.156; AOCS Method Cc 2–38.

13.3.1.3.3 Applications The capillary tube method is less useful for oils and fats (in comparison to pure compounds) since they lack a sharp melting point due to their array of various components. The Slip melting point is most often used in Europe, whereas the Wiley melting point is preferred in the United States. A disadvantage of the Wiley melting point is the subjective determination as to when the disc is spherical. A disadvantage of the Slip melting point is its 16 hour stabilization time.

13.3.1.4 Solid Fat Index (SFI)

13.3.1.4.1 Principle The **solid fat index** (SFI) is an empirical expression of the ratio of solids to liquids in fat at a given temperature. It is commonly measured using **dilatometry,** which determines the change in volume with change in temperature. As solid fat melts, it increases in volume. The SFI may also be measured using NMR where the actual percent solid fat can be determined.

Fat dilatometers are made of a bulb connected to a calibrated capillary tube. As fat in the bulb expands upon heating, it forces a liquid (i.e., colored water or mercury) into the capillary tube. Plotting volume against temperature gives a line where the fat is solid, a line where it is liquid, and a melting curve in between (Fig. 13-2). At a given temperature, the amount of solid and liquid can be estimated (9).

13.3.1.4.2 Procedure AOCS Method Cd 10–57.

13.3.1.4.3 Applications Amount of solid fat phase present in a plastic fat (e.g., margarine, shortening) depends on the type of fat, its history, and the temperature of measurement. The proportion of solids to liquids in the fat and how quickly the solids melt have an impact on functional properties, such as the mouthfeel of a food.

13.3.1.5 Cold Test

13.3.1.5.1 Principle The **cold test** is a measure of the resistance of an oil to crystallization.

13.3.1.5.2 Procedure Store oil in an ice bath (0°C) for 5.5 hrs. Observe for crystallization. Absence of crystals or turbidity indicates proper winterizing. AOAC Method 929.08; AOCS Method Cc 11–53.

13.3.1.5.3 Applications The cold test is a measure of success of the winterizing process. It ensures that oils remain clear even when stored at refrigerated temperatures.

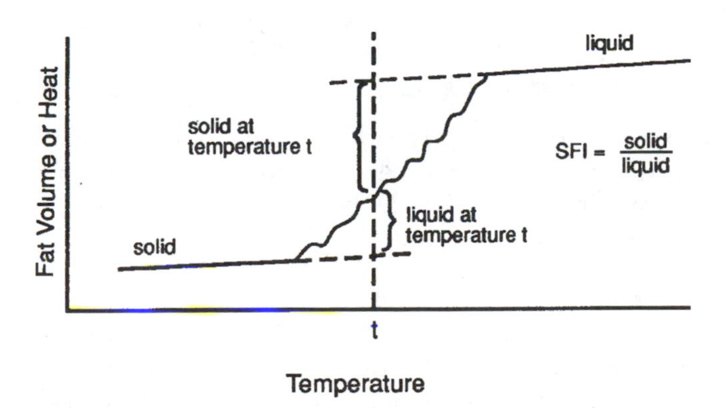

FIGURE 13-2.
Melting curve of a glyceride mixture. Adapted from (11), p. 160, by courtesy of Marcel Dekker, Inc.

13.3.1.6 Cloud Point

13.3.1.6.1 Principle The **cloud point** is the temperature at which a cloud is formed in a liquid fat due to the beginning of crystallization.

13.3.1.6.2 Procedure Heat the sample to 130°C. Cool the sample with agitation. Observe temperature of first crystallization (when a thermometer in the fat is no longer visible). AOCS Method Cc 6–25.

13.3.1.7 Smoke, Flash, and Fire Points

13.3.1.7.1 Principle The **smoke point** is the temperature at which the sample begins to smoke when tested under specified conditions. The **flash point** is the temperature at which a flash appears at any point on the surface of the sample; volatile gaseous products of combustion are produced rapidly enough to permit ignition. The **fire point** is the temperature at which evolution of volatiles (by decomposition of sample) proceeds with enough speed to support continuous combustion. These tests reflect the volatile organic material in oils and fats, especially free fatty acids and residual extraction solvents.

13.3.1.7.2 Procedure For smoke point, fill a Cleveland open cup with oil or melted fat, secure the thermometer, and place in cabinet. Heat the sample and note the temperature at which a thin, continuous stream of bluish smoke is given off. Flash and fire points are performed similarly, passing a test flame over the sample at 5°C intervals. AOCS Method Cc 9a–48.

For fats and oils that flash at temperatures below 149°C, use the flash point–closed cup method. AOCS Method Cc 9b-55.

13.3.1.7.3 Applications Frying oils and refined oils should have smoke points above 200 and 300°C, respectively.

13.3.1.8 Iodine Value

13.3.1.8.1 Principle The iodine value (or iodine number) is a measure of degree of unsaturation, the number of carbon-carbon double bonds in relation to the amount of fat or oil. **Iodine value** is defined as the g iodine absorbed per 100 g sample.

A quantity of fat or oil is reacted with a measured amount of iodine (or some other halogen). The amount of iodine left at the end of the reaction is then measured to calculate the amount of iodine absorbed. The higher the amount of unsaturation, the more iodine is absorbed; therefore, the higher the iodine value, the greater the degree of unsaturation.

13.3.1.8.2 Procedure The two most common procedures to measure iodine value are the **Wijs method** and the **Hanus method.** The Wijs method (AOAC Method 920.159; AOCS Method Cd 1–25) uses iodine monochloride (ICl), and the Hanus method (AOAC Method 920.158) uses iodine monobromide (IBr) as halogenating agents. AOCS Method Cd 1b–87 replaces carbon tetrachloride with cyclohexane.

The Wijs method:

1. Add a solution of iodine monochloride in acetic acid to a test portion of oil or fat dissolved in cyclohexane.
2. Allow the mixture to stand for a specific time period. Halogen addition to double bond takes place (eq. 1).
3. Add KI solution to reduce excess ICl to free iodine (eq. 2).
4. Titrate the liberated iodine with a standardized solution of sodium thiosulfate using a starch indicator (eq. 3).
5. Calculate the iodine value (eq. 4).

$$\begin{array}{c} \text{excess ICl} \\ + \text{R-CH} = \text{CH-R} \end{array} \rightarrow \begin{array}{c} \text{R-CHI-CHCl-R} \\ + \text{remaining ICl} \end{array} \quad (1)$$

$$ICl + 2\,KI \rightarrow KCl + KI + I_2 \quad (2)$$

$$\begin{array}{c} I_2 + \text{starch} \\ + 2Na_2S_2O_3 \\ \text{(blue)} \end{array} \rightarrow \begin{array}{c} 2NaI + \text{starch} \\ + Na_2S_4O_6 \\ \text{(colorless)} \end{array} \quad (3)$$

$$\text{iodine value} = \frac{(B - S) \times N \times 12.69}{\text{sample wt. (g)}} \quad (4)$$

where: B = blank titration
S = sample titration
N = normality of $Na_2S_2O_3$ solution
12.69 is used to convert from meq. thiosulfate to g iodine; M.W. of iodine is 126.9

13.3.1.8.3 Applications Iodine value is used to characterize oils, to follow the hydrogenation process in refining, and as an indication of lipid oxidation, since there is a decline in unsaturation during oxidation.

The Wijs method is probably more widely used and gives results closer to theoretical values; the Hanus method results are 2–5 percent below the Wijs method, but the Hanus method reagents are more stable (9).

13.3.1.9 Saponification Number

13.3.1.9.1 Principle Saponification is the process of breaking down or degrading a neutral fat into glycerol and fatty acids by treatment of the fat with alkali (eq. 5).

$$\begin{array}{l} H_2C-O-\overset{O}{\overset{\|}{C}}-R_1 \\ \quad | \\ H-C-O-\overset{O}{\overset{\|}{C}}-R_2 \\ \quad | \\ H_2C-O-\overset{O}{\overset{\|}{C}}-R_3 \end{array} + 3\,KOH \xrightarrow[\text{heat}]{}$$

Triacylglycerol

$$\begin{array}{l} H_2C-OH \\ \quad | \\ H-C-OH \\ \quad | \\ H_2C-OH \end{array} + \begin{array}{l} K^{+}\,{}^{-}O-\overset{O}{\overset{\|}{C}}-R_1 \\ K^{+}\,{}^{-}O-\overset{O}{\overset{\|}{C}}-R_2 \\ K^{+}\,{}^{-}O-\overset{O}{\overset{\|}{C}}-R_3 \end{array} \qquad (5)$$

Glycerol Fatty Acid Potassium Salt

The **saponification number** is defined as the mg of KOH required to saponify 1 g of fat.

13.3.1.9.2 Procedure The saponification number is determined as follows (AOAC Method 920.160; AOCS Method Cd 3–25):

1. Add excess alcoholic KOH to a known amount of fat.
2. Heat sample to saponify the fat.
3. Back-titrate the unreacted KOH with standardized HCl using phenolphthalein as the indicator.
4. Use weight of sample and titration values of the blank and samples to calculate the saponification number (eq. 6).

$$\text{saponification number} = \frac{(S - B) \times N \times 56.1}{\text{sample weight (g)}} \qquad (6)$$

where:
S = sample titration
B = blank titration
N = normality of the HCl
56.1 = the M.W. of KOH

13.3.1.9.3 Applications The saponification number is an index of the average molecular weight of the triacylglycerols in the sample (eq. 7). The molecular weight of the triacylglycerols may be divided by three to give an approximate average molecular weight for the fatty acids present. The smaller the saponification number, the longer the average fatty acid chain length.

$$\text{average M.W. of triacylglycerols} = \frac{56.1}{SN} \times \frac{1000\text{ mg}}{\text{g}} \times 3 \qquad (7)$$

where: 56.1 = the M.W. of KOH
SN = saponification number in mg/g

13.3.1.10 Acid Value and Free Fatty Acids (FFA)

13.3.1.10.1 Principle Measures of fat acidity normally reflect the amount of fatty acids hydrolyzed from triacylglycerols (eq. 8).

$$\begin{array}{l} H_2C-O-\overset{O}{\overset{\|}{C}}-R_1 \\ \quad | \\ H-C-O-\overset{O}{\overset{\|}{C}}-R_2 \\ \quad | \\ H_2C-O-\overset{O}{\overset{\|}{C}}-R_3 \end{array} + 3\,H_2O \longrightarrow$$

Triacylglycerol

$$\begin{array}{l} H_2C-OH \\ \quad | \\ H-C-OH \\ \quad | \\ H_2C-OH \end{array} + \begin{array}{l} HO-\overset{O}{\overset{\|}{C}}-R_1 \\ HO-\overset{O}{\overset{\|}{C}}-R_2 \\ HO-\overset{O}{\overset{\|}{C}}-R_3 \end{array} \qquad (8)$$

Glycerol Fatty Acids

In addition to free fatty acids, acid phosphates and amino acids can also contribute to acidity. **Acid value** is defined as the mg of KOH necessary to neutralize the free acids present in 1 g of fat or oil. **Free fatty acids** (FFA) is the percentage by weight of a specified fatty acid (e.g., percent oleic acid). In samples containing no acids other than fatty acids, acid value and FFA may be converted from one to the other using a conversion factor. Sometimes the acidity of edible oils and fats is expressed as ml N NaOH required to neutralize the fatty acids in 100 g fat or oil (9).

13.3.1.10.2 Procedure FFA are determined as follows (AOCS Method Ca 5a-40):

1. Accurately weigh a well-mixed liquid oil or melted fat sample.
2. Add the specified amount of neutralized 95 percent ethanol and phenolphthalein indicator.
3. Titrate with NaOH of specified normality, shaking constantly until pink color persists 30 sec.
4. Calculate percent FFA (eq. 9).

$$\%\ \text{FFA (as oleic)} = \frac{\text{ml alkali} \times \text{N of alkali} \times 28.2\ \text{mg}}{\text{sample wt.}} \qquad (9)$$

$$\%\ \text{FFA (as oleic)} \times 1.99 = \text{acid value} \qquad (10)$$

Acid value conversion factors for lauric and palmitic are 2.81 and 2.19, respectively. Acid value can be determined using AOCS Method Cd 3d-63.

13.3.1.10.3 Applications In crude fat, acid value or FFA estimates the amount of oil that will be lost during refining steps designed to remove fatty acids. In refined fats, a high acidity level means a poorly refined fat or fat breakdown after storage or use. If the fatty acids liberated are volatile, acid value or FFA may be a measure of **hydrolytic rancidity.**

13.3.1.11 Polar Components in Frying Fats

Standard tests used in the evaluation of frying fats include polar components, conjugated dienoic acids, and fatty acid analysis. In addition, several quick tests useful in day-to-day quality assurance of deep-fat frying operations have been outlined (14).

13.3.1.11.1 Principle Deterioration of used frying oils and fats can be monitored by measuring the polar components, which include monoacylglycerols, diacylglycerols, free fatty acids, and other products formed during heating of foodstuffs. Nonpolar compounds are primarily unaltered triacylglycerols.

13.3.1.11.2 Procedure Polar components are measured by dissolving 2.5 g fat in light petroleum ether: diethyl ether (87:13), then applying the solution to a silica gel column. Polar compounds are adsorbed onto the column. Nonpolar compounds are eluted, the solvent evaporated, the residue weighed, and the total polar components estimated by difference. Quality of the determination can be verified by eluting polar compounds and separating polar and nonpolar components using thin layer chromatography. [AOAC Method 982.27; International Union of Pure and Applied Chemistry (IUPAC) Method 2.507.]

13.3.1.11.3 Applications A suggested limit of 27 percent polar components in frying oil is a guide for when it should be discarded. A limitation of this method is the sample run time of 3.5 hours (14).

13.3.2 Lipid Oxidation—Measuring Present Status

Measuring the current status of a fat or oil in regards to lipid oxidation can be achieved using **peroxide value, thiobarbituric acid** (TBA) **test,** and **conjugated dienes.** These three procedures have been modified (especially with respect to sample size) for use in biological tissue assays (15).

In addition to the tests discussed below, other methods that monitor lipid oxidation (and that vary in usefulness) include the anisidine value, iodine value, acid value, Kreis test, and oxirane test, as well as the measurement of fluorescent compounds, total and volatile carbonyl compounds, polar compounds, and hydrocarbon gases (6, 11).

13.3.2.1 Sample Preparation

Most methods require lipid extraction prior to analysis. However, variations of some methods (e.g., TBA test) begin with the original foodstuff.

13.3.2.2 Peroxide Value

13.3.2.2.1 Principle Peroxide value measures the degree of lipid oxidation in fats and oils but not their stability. **Peroxide value** is defined as the milliequivalents (meq.) of peroxide per kg fat. It is a measure of the formation of **peroxide** or **hydroperoxide** groups that are the initial products of lipid oxidation.

13.3.2.2.2 Procedure Peroxide value is determined as follows (AOAC Method 965.33; AOCS Method Cd 8–53; AOCS Method 8b-90):

1. Dissolve 5.0 g of fat or oil in 30 ml of glacial acetic acid–chloroform (3:2, v/v).
2. Add 0.5 ml of saturated KI. I_2 is liberated by reaction with peroxides (eq. 11).

3. Titrate with standardized sodium thiosulfate using a starch indicator (eq. 12).
4. Calculate peroxide value (eq. 13).

$$\underset{\text{(excess)}}{ROOH + K^+I^-} \xrightarrow{H^+,\ \text{heat}} ROH + K^+OH^- + I_2 \quad (11)$$

$$\underset{\text{(blue)}}{I_2 + \text{starch}} + 2\,Na_2S_2O_3 \longrightarrow \underset{\text{(colorless)}}{Na_2S_4O_6 + \text{starch} + 2NaI} \quad (12)$$

$$\text{peroxide value (meq peroxide/kg fat)} = \frac{(S - B) \times N \times 1000}{\text{sample wt. (g)}} \quad (13)$$

where: S = sample titration
B = blank titration
N = normality of $Na_2S_2O_3$ solution

13.3.2.2.3 Applications Peroxide value measures a transient product of oxidation. A low value may represent either the beginning of oxidation or advanced oxidation (see Fig. 13-1); which one can be distinguished by measuring over time. For determination in foodstuffs, a disadvantage is the 5 g fat or oil sample size required; it is difficult to obtain sufficient quantities from foods low in fat. The method is highly empirical; modifications may change results. AOCS Method Cd 8b-90 replaces chloroform with isooctane. Despite drawbacks, peroxide value is one of the most common tests of lipid oxidation.

13.3.2.3 Thiobarbituric Acid (TBA) Test

13.3.2.3.1 Principle The **thiobarbituric acid** (TBA) **test** measures a secondary product of lipid oxidation, **malonaldehyde.** It involves reaction of malonaldehyde (or malonaldehyde-type products) with TBA to yield a colored end product. The sample may be reacted directly with TBA, but is often distilled to eliminate interfering substances, then the distillate is reacted with TBA. Many modifications of the test have been developed.

13.3.2.3.2 Procedure A common procedure for measuring TBA follows (16,17):

1. Combine weighed sample with distilled water and mix in blender.
2. Adjust to pH 1.1 to 1.2 with HCl.
3. Transfer quantitatively to distillation flask by washing with a set volume of distilled water.
4. Add an aliquot of BHT (optional), antifoam reagent, and boiling beads.
5. Distill sample rapidly, collecting the first 50 ml (12–14 min).
6. Combine aliquot of distillate with TBA reagent.
7. Cover tube and boil for 35 min.
8. Read absorbance at 530 nm against distilled water blank reacted with TBA reagent.
9. TBA number is expressed by converting absorbance readings to mg malonaldehyde per 1000 g sample.

13.3.2.3.3 Applications The TBA test correlates better with sensory evaluation of rancidity than does peroxide value, but it still measures a transient product of oxidation. Despite its limited specificity and the large sample sizes possibly required (depending on the method), the TBA test with minor modifications is frequently used.

13.3.2.4 Conjugated Dienes and Trienes

13.3.2.4.1 Principle Double bonds in lipids are changed from nonconjugated to conjugated upon oxidation. **Conjugated dienes** absorb light at 233 nm and **conjugated trienes** at 268 nm.

13.3.2.4.2 Procedure The lipid sample is dissolved in isooctane and diluted to about 0.01 mg of sample per ml solvent. Absorbance is measured using an ultraviolet (UV) spectrophotometer. (AOCS Method Ti 1a-64; see also AOAC Method 957.13.)

13.3.2.4.3 Applications The conjugated dienes and trienes method is useful for monitoring the early stages of oxidation. The magnitude of the changes in absorption are not easily related to extent of oxidation in advanced stages.

13.3.3 Lipid Oxidation—Evaluating Susceptibility

Accelerated tests have been developed for speeding up the determination of the susceptibility of fat to lipid oxidation. In such methods, lipid oxidation is artificially hastened by exposing lipid to heat, oxygen, metal catalysts, light, or enzymes. A major problem with accelerated tests is assuming reactions that occur at elevated temperatures or under other artificial conditions are the same as normal reactions occurring at the storage temperature of the product. An additional difficulty is ensuring that the apparatus is clean and completely free of metal contaminants and oxidation products from previous runs.

Measuring the induction period before the appearance of oxidation products and the onset of rancidity can give an indication of the effectiveness of antioxidants in preventing lipid oxidation. **Induction period** is defined as the length of time before detectable rancidity or rapid acceleration of lipid oxidation (see Fig. 13-1). Induction period can be calculated by such methods as calculating the maximum of the second derivative with respect to time or manually drawing tangents to the lines (Fig. 13-3).

13.3.3.1 Schaal Oven Test

13.3.3.1.1 Procedure A fat or oil of known weight is placed in the oven at a specified temperature (about 65°C). Results are reported as time elapsed until rancidity is detected. Detection is by sensory evaluation (odor or taste) or peroxide value.

13.3.3.1.2 Applications Automated systems based on this premise have been developed. Weaknesses of this method include lack of control of oxidation conditions and the subjective aspect of sensory evaluation. Nevertheless, results of the Schaal oven test correlate well with actual shelf life determinations (9).

13.3.3.2 Active Oxygen Method (AOM)

13.3.3.2.1 Procedure The lipid sample is held at 98°C and air is bubbled through it. Stability is expressed as hours of heating until detection of rancidity. The endpoint may be a rancid odor or a peroxide value of 100. (AOCS Method Cd 12–57.)

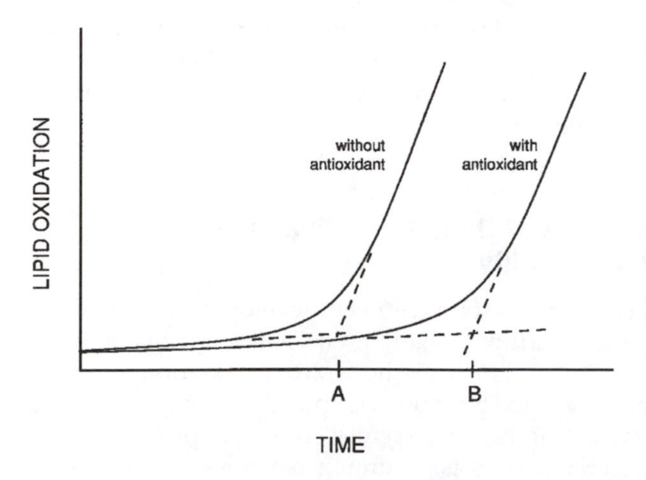

FIGURE 13-3.
A plot of lipid oxidation over time, showing the effect of an antioxidant on induction period. Time A is induction period of sample without antioxidant and Time B is induction period of sample with antioxidant.

13.3.3.2.2 Applications The AOM was originally designed to measure the effectiveness of antioxidants. The AOM is faster than the Schaal oven test, but does not correlate as well with actual shelf life (9). The Rancimat® and Oxidative Stability Instrument® automated systems have been developed from this method (see AOCS Method Cd 12b-92). These automated systems allow for continuous data monitoring by sweeping the acidic volatiles into deionized water, where conductivity is measured. Though expensive, these automated systems are less labor intensive.

13.3.4 Analysis of Lipid Fractions

Gas chromatography (GC) is ideal for the analysis of lipids. GC can be used for such determinations as total fatty acid composition, distribution and position of fatty acids in lipid, studies of fat stability and oxidation, assaying heat or irradiation damage to lipids, and detection of adulterants and antioxidants (9). Methods exist detailing the analysis of various lipid fractions using GC (5). GC combined with **mass spectroscopy** (MS) is a powerful tool used in identification of compounds. GC methodology is covered in detail in Chapter 30. **High-performance liquid chromatography** (HPLC), discussed in Chapter 29, is also useful in lipid analyses (18).

Thin-layer chromatography (TLC) has been used extensively in the past by lipid chemists. Partly due to low cost and ease, TLC is still useful, though many assays may be better performed using GC or HPLC (better resolution, more quantitative). Chapter 28 presents the principles of TLC (section 28.3.4.2).

13.3.4.1 Separation of Lipid Fractions by TLC

13.3.4.1.1 Procedure TLC is performed using silica gel G as the adsorbent and hexane-diethyl ether-formic acid (80:20:2, v/v/v) as the eluting solvent system (Fig. 13-4). Plates are sprayed with 2′,7′-dichlorofluorescein in 95 percent methanol and placed under UV light to view yellow bands against dark background (8).

13.3.4.1.2 Applications This procedure permits rapid analysis of the presence of lipid fractions in a food lipid extract. For small-scale preparative purposes, TLC plates can be scraped to remove various bands for further analysis using GC or other means. Many variations in TLC parameters are available which will separate various lipids.

13.3.4.2 Fatty Acid Methyl Esters by GC

13.3.4.2.1 Principle Short columns and high temperatures are needed for analysis of intact triacylglycerols

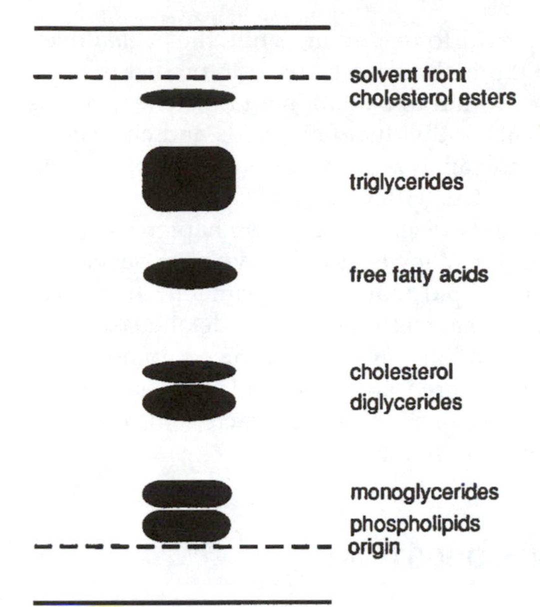

FIGURE 13-4.
Schematic thin-layer chromatography (TLC) separation of lipid fractions on silica gel G. Adapted with permission from (5).

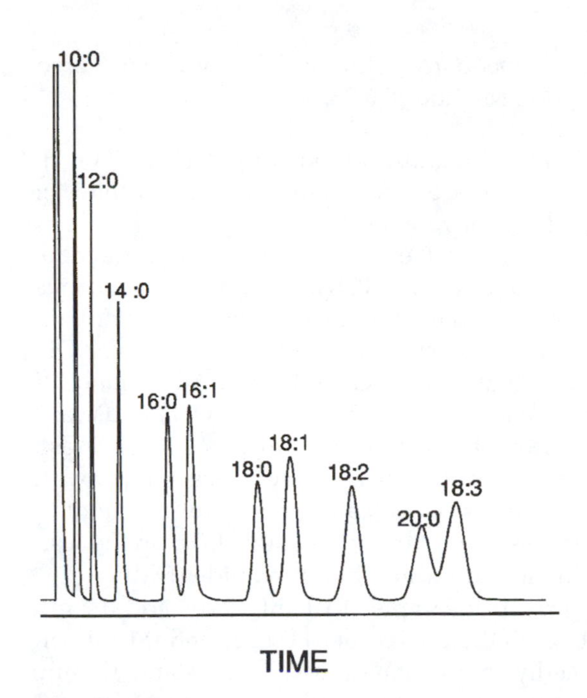

FIGURE 13-5.
Gas chromatogram of separation of fatty acid methyl esters.

IUPAC Method 2.323). To increase volatility before GC analysis, triacylglycerols and phospholipids are typically saponified and fatty acids thus liberated are esterified to form **fatty acid methyl esters** (FAME) (eq 14). Figure 13-5 is a chromatograph showing separation of FAME of varying length and unsaturation.

$$\begin{array}{l} H_2C-O-C(=O)-R_1 \\ H-C-O-C(=O)-R_2 \\ H_2C-O-C(=O)-R_3 \\ \text{Triacylglycerol} \end{array} \xrightarrow[\text{NaOH, BF}_3]{CH_3OH} \begin{array}{l} H_2C-OCH_3 \\ H-C-OCH_3 \\ H_2C-OCH_3 \end{array} + \begin{array}{l} H_3C-O-C(=O)-R_1 \\ H_3C-O-C(=O)-R_2 \\ H_3C-O-C(=O)-R_3 \\ \text{Fatty Acid Methyl Esters (FAME)} \end{array} \quad (14)$$

13.3.4.2.2 Procedure

1. Extract fat or oil from the food [e.g., by homogenizing with hexane-isopropanol (3:2, v/v) and then evaporating the solvent].
2. Prepare FAME (AOAC method 969.33; AOCS Method Ce 2-66): extracted lipid + NaOH + CH_3OH + BF_3 + heptane; reflux. [*Note:* An alternative to the use of boron trifluoride (BF_3) is sulfuric acid.]
3. Remove aliquot of upper heptane solution, dry with anhydrous Na_2SO_4, and dilute to concentration of 5–10 percent for injection on GC (AOAC Method 963.22; AOCS Method Ce 1-62).

13.3.4.2.3 Applications Determination of the fatty acid profile permits the **calculation** of the following categories of fats that pertain to health issues and food labeling: percent saturated fatty acids, percent unsaturated fatty acids, percent monounsaturated fatty acids, and percent polyunsaturated fatty acids.

13.3.4.3 Cholesterol Content by GC

13.3.4.3.1 Principle The lipid extracted from food is saponified. Cholesterol (in nonsaponifiable fraction) is extracted with benzene (some methods use hexane) and derivatized to form **trimethylsilyl** (TMS) ethers. Quantitation is achieved using GC.

13.3.4.3.2 Procedure Cholesterol may be measured using AOAC Method 976.26:

1. Extract lipids from food using $CHCl_3$:CH_3OH:H_2O (procedure is different for dried whole egg and for mayonnaise).
2. Saponify and extract unsaponifiable fraction: Filter aliquot of $CHCl_3$ layer through anhydrous Na_2SO_4 and evaporate to dryness in water bath using nitrogen stream. Add concentrated KOH and ethanol and reflux 1 hour. Add aliquot of benzene. Shake with 1 N KOH and discard aqueous layer. Repeat with 0.5 N KOH. Wash four times with water. Dry benzene layer with anhydrous Na_2SO_4 and evaporate an aliquot to dryness on rotary evaporator. Take up the residue in 3 mL dimethylformamide (DMF).
3. Derivatize sample: To 1 mL DMF, add 0.2 mL hexamethyldisilazane (HMDS) and 0.1 mL trimethylchlorosilane (TMCS). Let stand 15 min, then add 5α-cholestane in heptane (internal standard) and 10 mL H_2O and centrifuge.
4. Inject 3 μL of heptane layer into GC.

13.3.4.3.3 Applications GC quantitation of cholesterol is recommended since many spectrophotometric methods have not been specific for cholesterol. Other GC, HPLC, and enzymatic methods are available. For example, cholesterol methods developed for frozen foods (19) and for meat products (20) eliminate the fat extraction step, directly saponifying the sample; in comparison to the AOAC method outlined above, they are more rapid and avoid exposure to toxic solvents.

Cholesterol oxidation products can be measured using GC, HPLC, and TLC.

13.4 SUMMARY

The importance of fat characterization is evident in many aspects of the food industry, including ingredient technology, product development, quality assurance, product shelf life, and regulatory aspects. The effort to reduce the amount of calories consumed as fat in the United States accentuates the significance of understanding the lipid components of food.

The methods discussed above help to characterize bulk oils and fats and the lipids in foodstuffs. Methods described for bulk oils and fats can be used to determine characteristics such as purity, melting rate and melting point, crystallization temperature, flash and fire points, degree of unsaturation, average fatty acid chain length, and amount of polar components. The peroxide value, TBA, and conjuncted diene and triene tests can be used to measure the present status of a lipid with regard to oxidation, while the Schaal oven test and AOM method can be used to predict the susceptibility of a lipid to oxidation and the effectiveness of antioxidants. Individual fatty acids and cholesterol are commonly analyzed by chromatographic techniques such as GC, HPLC, and TLC.

The methods discussed in this chapter represent only a few of the many tests that have been developed to characterize lipid material. Consult the references cited for additional methods or more detailed explanations. Time, funding, availability of equipment and instruments, required accuracy, and purpose will all dictate the choice of method in characterizing oils, fats, and foodstuffs containing lipids.

13.5 STUDY QUESTIONS

1. You want to compare several fat/oil samples for the chemical characteristics listed below. For each characteristic, name one test (give full name, not abbreviation) that could be used to obtain the information desired:
   a. degree of unsaturation
   b. predicted susceptibility to oxidative rancidity
   c. present status with regard to oxidative rancidity
   d. average fatty acid molecular weight
   e. amount of solid fat at various temperatures
   f. hydrolytic rancidity
2. Your analysis of an oil sample gives the following results. What do each of these results tell you about the characteristics of the sample? Briefly describe the principle for each method used.
   a. large (vs. small) saponification number
   b. low (vs. high) iodine value
   c. high (vs. low) TBA number
   d. high (vs. low) free fatty acid content
   e. long (vs. short) time in active oxygen method
3. Describe how and why refractive index can be useful in fat characterization.
4. Define solid fat index, describe how it is measured using dilatometry, and explain the usefulness of this measurement.
5. Peroxide value, TBA number, and the value from the test for conjugated dienes and trienes all can be used to help characterize a fat sample.
   a. What do the results of these tests tell you about a fat sample?
   b. Differentiate these three tests as to what chemical is being measured.
6. What methods would be useful in determining the effectiveness of various antioxidants added to an oil?
7. The Nutrition Education and Labeling Act of 1990 (see Chapter 2) requires that the nutritional label on food products contain information related to lipid constituents. In addition to the amount of *total fat* (see Chapter 12), the label must state the amount of *saturated fat* and the *cholesterol* content. Explain an appropriate method for the analysis of each lipid constituent.

13.6 PRACTICE PROBLEMS

1. A 5.00 g sample of oil was saponified with excess KOH. The unreacted KOH was then titrated with 0.500 N HCl (standardized). The difference between the blank and the sample was 25.8 ml of titrant. Calculate the saponification number.
2. A sample of oil containing only triacylglycerols has a saponification number of 192. What is the approximate average molecular weight of the fatty acids in the oil?
3. A sample (5.0 g) of food-grade oil was reacted with excess KI to determine peroxide value. The free iodine was titrated with a standardized solution of 0.10 N $Na_2S_2O_3$. The amount of titrant required was 0.60 ml (blank corrected). Calculate the peroxide value of the oil.

Answers

1. 145, 2. 292, 3. 12.

13.7 REFERENCES

1. AOAC. 1990. *Official Methods of Analysis*, 15th ed. Association of Official Analytical Chemists, Washington, DC.
2. AOCS. 1990. *Official Methods and Recommended Practices of the American Oil Chemists' Society*, 4th ed. 2nd printing (additions and revisions through 1993.) American Oil Chemists' Society, Champaign, IL.
3. IUPAC. 1987. *Standard Methods for Analysis of Oils, Fats and Derivatives*, 7th ed. International Union of Pure and Applied Chemistry, Commission on Oils, Fats and Derivatives, C. Paquot and A. Hautfenne (Eds.). Blackwell Scientific Publications, Oxford, England.
4. Christie, W. W. 1982. *Lipid Analysis. Isolation, Separation, Identification, and Structural Analysis of Lipids*, 2nd ed. Pergamon Press Ltd., Oxford, England.
5. Christie, W. W. 1989. *Gas Chromatography and Lipids. A Practical Guide.* The Oily Press Ltd., Ayr, Scotland.
6. Gray, J. I. 1978. Measurement of lipid oxidation: A review. *J. Am. Oil Chem. Soc.* 55:539–546.
7. Hamilton, R. J., and Rossell, J. B. 1986. *Analysis of Oils and Fats.* Elsevier Applied Science Publishers, London, England.
8. Melton, S. L. 1983. Methodology for following lipid oxidation in muscle foods. *Food Technol.* 37(7):105–111, 116.
9. Pomeranz, Y., and Meloan, C. E. 1987. *Food Analysis: Theory and Practice*, 2nd ed. Van Nostrand Reinhold, New York, NY.
10. Sonntag, N. O. V. 1982. Analytical methods. Ch. 7 in *Bailey's Industrial Oil and Fat Products*, Vol. 2., 4th ed., D. Swern (Ed.), 407–525. John Wiley & Sons, New York, NY.
11. Nawar, W. W. 1986. Lipids. Ch. 4 in *Food Chemistry*, 2nd ed. O. R. Fennema (Ed.), 139–244. Marcel Dekker, Inc., New York, NY.
12. Windholz, M. (Ed.). 1989. *The Merck Index*, 11th ed. Merck & Co., Inc., Rahway, NJ.
13. Labuza, T. P. 1971. Kinetics of lipid oxidation in foods. *CRC Crit. Rev. Food Technol.* 2:355–405.
14. White, P. J. 1991. Methods for measuring changes in deep-fat frying oils. *Food Technol.* 45(2):75–80.
15. Buege, J. A., and Aust, S. D. 1978. Microsomal lipid peroxidation. *Methods Enzymol* 52:302–310.
16. Tarladgis, B. G., Watts, B. M., Younathan, M. T., and Dugan, L. R. 1960. A distillation method for the quantitative determination of malonaldehyde in rancid foods. *J. Am. Oil Chem. Soc.* 37:1.
17. Rhee, K. S., and Watts, B. M. 1966. Evaluation of lipid oxidation in plant tissues. *J. Food Sci.* 31:664–668.
18. Perkins, E.G. 1991. *Analyses of Fats, Oils, and Lipoproteins.* American Oil Chemists' Society, Champaign, IL.
19. Al-Hasani, S.M., Shabany, H., and Hlavac, J. 1990. Rapid determination of cholesterol in selected frozen foods. *J. Assoc. Off. Anal. Chem.* 73:817–820.
20. Adams, M.L., Sullivan, D.M., Smith, R.L., and Richter, E.F. 1986. Evaluation of direct saponification method for determination of cholesterol in meats. *J. Assoc. Off. Anal. Chem.* 69:844–846.

CHAPTER

14

Protein Analysis

Sam K. C. Chang

14.1 INTRODUCTION

14.1.1 Classification and General Considerations

Proteins are an abundant component in all cells, and almost all except storage proteins are important for biological functions and/or cell structure. Food proteins are very complex. Many food proteins have been purified and characterized. Proteins vary in molecular mass, ranging from approximately 5,000 daltons to more than a million daltons. They are composed of elements including hydrogen, carbon, nitrogen, oxygen, and sulfur. Twenty α-amino acids are the building blocks of proteins. The linkages between amino acid residues in a protein are peptide bonds. Nitrogen is the most distinguishing element present in proteins. However, nitrogen content in various food proteins ranges from 13.4 to 19.1 percent (1) due to the variation in the specific amino acid composition of proteins. Generally, proteins rich in basic amino acids contain more nitrogen.

Proteins can be classified by their composition, structure, biological function, or solubility properties. For example, simple proteins contain only amino acids upon hydrolysis, but conjugated proteins also contain non-amino acid components.

Proteins have unique conformations that could be altered by denaturants such as heat, acid, alkali, 8M urea, 6M guanidine-HCl, organic solvents, and detergents. The solubility as well as functionality of proteins could be altered by denaturants.

The analysis of proteins is complicated by the fact that some food components possess similar physicochemical properties. Nonprotein nitrogen could come from free amino acids, small peptides, nucleic acids, phospholipids, amino sugars, porphyrin, and some vitamins, alkaloids, uric acid, urea, and ammonium ions. Therefore, the total organic nitrogen in foods would represent nitrogen primarily from proteins and to a lesser extent from all organic nitrogen-containing nonprotein substances. Depending upon methodology, other macro food components, including lipids and carbohydrates, may interfere physically with analysis of food proteins.

Numerous methods have been developed to measure protein content. The basic principles of these methods include the determinations of nitrogen, peptide bonds, aromatic acids, UV absorptivity of proteins, free amino groups, light scattering properties, and dye binding capacity. In addition to factors such as sensitivity, accuracy, precision, speed, and cost of analysis, what is actually being measured must be considered in the selection of an appropriate method for a particular application.

14.1.2 Importance of Analysis

Protein analysis is important for:

1. **Biological activity determination.** Some proteins, including enzymes or enzyme inhibitors, are relevant to food science and nutrition: for instance, the proteolytic enzymes in the tenderization of meats, pectinases in the ripening of fruits, and trypsin inhibitors in legume seeds are proteins.
2. **Functional property investigation.** Proteins in various types of food have unique food functional properties: for example, gliadin and glutenins in wheat flour for breadmaking, casein in milk for coagulation into cheese products, and egg albumen for foaming.
3. **Nutritional labeling.**

Protein analysis is required when you want to know:

1. Total protein content
2. Amino acid composition
3. Content of a particular protein in a mixture
4. Protein content during isolation and purification of a protein
5. Nonprotein nitrogen
6. Nutritive value (digestibility, protein efficiency ratio, or nitrogen balance) of a protein

14.1.3 Content in Foods

Protein content in food varies widely. Foods of animal origin and legumes are excellent sources of proteins. The protein contents of selected food items are listed in Table 14-1.

14.2 METHODS

Principles, general procedures, and applications are described below for various protein determination methods. Refer to the referenced methods for detailed instructions of the procedures. Several of the methods cited are from the *Official Methods of Analysis* of the Association of Official Analytical Chemists (AOAC) International (3).

14.2.1 Kjeldahl Method

14.2.1.1 Principle

In the Kjeldahl procedure, proteins and other organic food components in a sample are digested with sulfuric acid in the presence of catalysts. The **total organic**

TABLE 14-1. Protein Content of Selected Foods

Food Item	*Percent Protein (wet weight basis)*
Dairy foods	
Whole milk	3.5
Skim milk	3.6
Nonfat dry milk	35.9
Cheddar cheese	25.0
Meats	
Beef, boneless chuck, lean with fat	18.7
Beef, round steak, lean with fat	28.6
Beef, dried, chipped	34.3
Pork, loin, lean with fat, without bone	17.1
Chicken, breast	27.3
Egg, raw, whole	12.9
Fish	
Cod, cooked	28.5
Tuna, canned in oil	24.2
Cereals	
Rice, brown, raw	7.5
Rice, polished, raw	6.7
Flour, wheat, whole, hard varieties	13.3
Flour, corn	7.8
Spaghetti, dry	12.5
Starch, corn	0.3
Potato, whole, raw, round	1.6
Tapioca	0.6
Legumes	
Bean, dry, all varieties	22.3
Soybean, dry	34.1
Soybean, curd	7.8
Fruits and vegetables	
Apple, raw, whole	0.2
Asparagus, green	2.5
Almonds, whole	18.6

From (2).

nitrogen is converted to ammonium sulfate. The digest is neutralized with alkali and distilled into a boric acid solution. The borate anions formed are titrated with standardized acid, which is converted to nitrogen in the sample. The result of the analysis represents the **crude protein** content of the food since nitrogen also comes from nonprotein components.

14.2.1.2 Historical Background

14.2.1.2.1 Original Method In 1883, Johann Kjeldahl developed the basic process of today's Kjeldahl method to analyze organic nitrogen. General steps in the original method include:

1. **Digestion** with sulfuric acid, with the addition of powdered potassium permanganate to complete oxidation and conversion of nitrogen to ammonium sulfate.
2. **Neutralization** of the diluted digest, followed by **distillation** into a known volume of standard acid, which contains potassium iodide and iodate.
3. **Titration** of the liberated iodine with standard sodium thiosulfate.

14.2.1.2.2 Improvements Several important modifications have improved the original Kjeldahl process:

1. Metallic catalysts such as mercury, copper, and selenium are added to sulfuric acid for complete digestion. Mercury has been found to be the most satisfactory. Selenium dioxide and copper sulfate in the ratio of 3:1 have been reported to be effective for digestion. Copper and titanium dioxide also have been used as a mixed catalyst for digestion (AOAC Method 945.01) (3). The use of titanium dioxide and copper poses less **safety concern** than mercury in the post-analysis disposal of the waste.
2. Potassium sulfate is used to increase the boiling point of the sulfuric acid to accelerate digestion.
3. Sulfide or sodium thiosulfate are added to the diluted digest to help release nitrogen from mercury, which tends to bind ammonium.
4. The ammonia is distilled directly into a boric acid solution and followed by titration with standard acid.
5. Colorimetric, Nesslerization, and ion chromatography are used to determine nitrogen content after digestion.

An excellent book to review the Kjeldahl method for total organic nitrogen was written by Bradstreet (4). The basic AOAC Kjeldahl procedure is Method 955.04. Semiautomation, automation, and modification for microgram nitrogen determination (micro Kjeldahl method) have been established by AOAC in Methods 976.06, 976.05 and 960.52, respectively.

14.2.1.3 General Procedures and Reactions

14.2.1.3.1 Sample Preparation Solid foods are ground to pass a 20-mesh screen. Samples for analysis should be homogeneous. No other special preparations are required.

14.2.1.3.2 Digestion Place sample (accurately weighed) in Kjeldahl flask. Add acid and catalyst; digest until clear to get complete breakdown of all organic matter. Nonvolatile ammonium sulfate is formed from the reaction of nitrogen and sulfuric acid.

$$\text{Protein} \xrightarrow[\text{Heat, catalyst}]{\text{Sulfuric acid}} (NH_4)_2SO_4 \quad (1)$$

During digestion, protein nitrogen is liberated to form ammonium ions; sulfuric acid oxidizes organic matter and combines with ammonium formed; carbon and hydrogen elements are converted to carbon dioxide and water.

14.2.1.3.3 Neutralization and Distillation The digest is diluted with water. Alkali containing sodium thiosulfate is added to neutralize the sulfuric acid. The ammonia formed is distilled into a boric acid solution containing the indicators methylene blue and methyl red.

$$(NH)_2SO_4 + 2\ NaOH \longrightarrow 2\ NH_3 + Na_2SO_4 + 2\ H_2O \quad (2)$$

$$NH_3 + H_3BO_3\ \text{(boric acid)} \longrightarrow NH_4 + \underset{\text{(borate ion)}}{H_2BO_3^-} \quad (3)$$

14.2.1.3.4 Titration Borate anion (proportional to the amount of nitrogen) is titrated with standardized HCl.

$$H_2BO_3^- + H^+ \longrightarrow H_3BO_3 \quad (4)$$

14.2.1.3.5 Calculations

$$\text{Moles HCl} = \text{Moles } NH_3 = \text{Moles N in the sample} \quad (5)$$

A reagent blank should be run to subtract reagent nitrogen from the sample nitrogen.

$$\%\ N = N\ HCl \times \frac{\text{Corrected acid volume}}{\text{g of sample}} \times \frac{14\ \text{g N}}{\text{mole}} \times 100 \quad (6)$$

where: N HCl = Normality of HCL, in moles/1000 ml
Corrected acid vol. = (mL std. acid for sample) − (mL std. acid for blank)
14 = atomic weight of nitrogen

A factor is used to convert percent N to percent crude protein. Most proteins contain 16 percent N, so the conversion factor is 6.25 (100/16 = 6.25).

$$\frac{\%\ N}{0.16} = \%\ \text{protein} \quad \text{or} \quad \%\ N \times 6.25 = \%\ \text{protein} \quad (7)$$

Conversion factors for various foods are given in Table 14-2.

14.2.1.3.6 Alternate Procedures In place of distillation and titration with acid, ammonia or nitrogen can be quantitated by:

1. Nesslerization

$$NH_4OH + \underset{\text{mercuric iodide}}{2HgI_2} + 2KI + 3KOH \longrightarrow \underset{\text{ammonium dimercuric iodide, red-orange, 440 nm}}{NH_4Hg_2I} + 7KI + 4H_2O \quad (8)$$

   This method is rapid and sensitive, but the ammonium dimercuric iodide is colloidal and color is not stable.

2. $$NH_3 + \text{phenol} + \text{hypochlorite} \xrightarrow[OH^-]{} \text{indophenol (blue, 630 nm)} \quad (9)$$

3. pH measurement after distillation into known volume of boric acid.
4. Direct measurement of ammonia, using ion chromatographic method.

14.2.1.4 Applications

Advantages:

1. Applicable to all types of foods
2. Relatively simple
3. Inexpensive
4. Accurate; the official method for crude protein content

TABLE 14-2. Nitrogen to Protein Conversion Factors for Various Foods

	Percent N in Protein	*Factor*
Egg or meat	16.0	6.25
Milk	15.7	6.38
Wheat	18.0	5.70
Corn	16.0	6.25
Oat	17.15	5.83
Soybean	17.51	5.71

From (1).

5. Has been modified (micro Kjeldahl method) to measure microgram quantities of proteins

Disadvantages:

1. Measures total organic nitrogen, not just protein nitrogen
2. Time consuming (at least 2 hrs to complete)
3. Poorer precision than the biuret method
4. Corrosive reagent

14.2.2 Biuret Method

14.2.2.1 Principle

A violet-purplish color is produced when **cupric ions** are complexed with **peptide bonds** (substances containing at least two peptide bonds, i.e., biuret, large peptides, and all proteins) under **alkaline conditions.** The absorbance of the color produced is read at 540 nm. The color intensity (absorbance) is proportional to the protein content of the sample (5).

14.2.2.2 Procedure

1. A 5 mL biuret reagent is mixed with a 1 mL portion of protein solution (1 to 10 mg protein/mL). The reagent includes copper sulfate, NaOH, and potassium sodium tartrate, which is used to stabilize the cupric ion in the alkaline solution.
2. After standing at room temperature for 15 or 30 min, the absorbance is read at 540 nm against a reagent blank.
3. Filtration or centrifugation before reading absorbance is required if the reaction mixture is not clear.
4. A standard curve of concentration versus absorbance is constructed using **bovine serum albumin** (BSA).

14.2.2.3 Applications

The biuret method has been used for determining proteins in cereal (6, 7), meat (8), soybean proteins (9), and as a qualitative test for animal feed [AOAC Method 935.11 (refers to Methods 22.012–22.013, AOAC, 10th ed., 1965)] (10). The biuret method also is used widely to measure the protein content of isolated proteins.
Advantages:

1. Less expensive than the Kjeldahl method; rapid (can be completed in less than 30 min); simplest method for analysis of proteins
2. Color derivations encountered less frequently than with Lowry, ultraviolet (UV) absorption, or turbidimetric methods (described below)
3. Very few substances other than proteins in foods interfere with the biuret reaction
4. Does not detect nitrogen from nonpeptide or nonprotein sources

Disadvantages:

1. Not very sensitive as compared to the Lowry method; requires at least 2 to 4 mg protein for assay
2. Absorbance could be contributed from bile pigments if present
3. High concentration of ammonium salts interfere with the reaction
4. Color varies with different proteins; gelatin gives a pinkish-purple color
5. Opalescence could occur in the final solution if high levels of lipid or carbohydrate are present
6. Not an absolute method: color must be standardized against known protein (e.g., BSA) or against the Kjeldahl nitrogen method

14.2.3 Lowry Method

14.2.3.1 Principle

The Lowry method (11, 12) combines the **biuret reaction** with the reduction of the **Folin-Ciocalteau phenol reagent** (phosphomolybdic-phosphotungstic acid) by **tyrosine and tryptophan** residues in the proteins. The bluish color developed is read at 750 nm (high sensitivity for low protein concentration) or 500 nm (low sensitivity for high protein concentration). The original procedure has been modified by Miller (13) and Hartree (14) to improve the linearity of the color response to protein concentration.

14.2.3.2 Procedure

The following procedure is based on the modified procedure of Hartree (14):

1. Proteins to be analyzed are diluted to an appropriate range (20 to 100 mg).
2. K Na Tartrate-Na_2CO_3 solution is added after cooling and incubated at room temperature for 10 min.
3. $CuSO_4$-K Na Tartrate-NaOH solution is added after cooling and incubated at room temperature for 10 min.

4. Freshly prepared Folin reagent is added, then mixed and incubated at 50°C for 10 min.
5. Absorbance is read at 650 nm.
6. A standard curve of BSA is carefully constructed for estimating protein concentration of the unknown.

14.2.3.3 Applications

Because of its simplicity and sensitivity, the Lowry method has been widely used in protein biochemistry. However, it has not been widely used to determine proteins in food systems without first extracting the proteins from the food mixture.
Advantages:

1. Very sensitive
   a. 50 to 100 times more sensitive than biuret method
   b. 10 to 20 times more sensitive than 280 nm UV absorption method (described below)
   c. Several times more sensitive than ninhydrin method (described below)
   d. Similar sensitivity as Nesslerization; however, more convenient than Nesslerization
2. Less affected by turbidity of the sample
3. More specific than most other methods
4. Relatively simple; can be done in 1 to 1.5 hrs.

Disadvantages:
For the following reasons, the Lowry procedure requires careful standardization for particular applications:

1. Color varies with different proteins to a greater extent than the biuret method.
2. Color is not strictly proportional to protein concentration.
3. The reaction is interfered with to varying degrees by sucrose, lipids, phosphate buffers, monosaccharides, and hexoamines.
4. High concentrations of reducing sugars, ammonium sulfate, and sulfhydryl compounds interfere with the reaction.

14.2.4 Bicinchoninic Acid (BCA) Method

14.2.4.1 Principle

Smith et al. (15) proposed that proteins reduce **cupric ions** to **cuprous ions** under **alkaline conditions.** The cuprous ion complexes with apple-greenish **BCA reagent** to form a purplish color. The color formed is proportional to protein concentration.

14.2.4.2 Procedure

1. Mix (one step) the protein solution with the BCA reagent, which contains BCA sodium salt, sodium carbonate, NaOH, and copper sulfate, pH 11.25.
2. Incubate at 37°C for 30 min, or room temperature for 2 hrs, or 60°C for 30 min. The selection of the temperature depends upon sensitivity desired. A higher temperature gives a greater color response.
3. Read the solution at 562 nm against a reagent blank.
4. Construct a standard curve using BSA.

14.2.4.3 Applications

The BCA method has been used in protein isolation and purification. The suitability of this procedure for measuring protein in complex food systems has not been reported.
Advantages:

1. Sensitivity is comparable to the Lowry method; sensitivity of the micro BCA method (0.5 mg to 10 mg) is slightly better than that of the Lowry method.
2. One-step mixing is easier than the Lowry method.
3. The reagent is more stable than for the Lowry reagent.
4. Nonionic detergent and buffer salts do not interfere with the reaction.
5. Medium concentrations of denaturing reagents (4M guanidine-HCl or 3M urea) do not interfere.

Disadvantages:

1. Color is not stable with time. The analyst needs to carefully control the time for reading absorbance.
2. Reducing sugars interfere to a greater extent than in the Lowry method. High concentrations of ammonium sulfate also interfere.
3. Color variations among proteins are similar to the Lowry method.
4. Response of absorbance to concentration is not linear.

14.2.5 Ultraviolet (UV) 280 nm Absorption Method

14.2.5.1 Principle

Proteins show strong absorption at **UV 280 nm,** primarily due to **tryptophan** and **tyrosine residues** in the proteins. Since the content of tryptophan and tyrosine in each protein is fairly constant, the absorbance at 280 nm could be used to estimate the concentration of proteins, using **Beer's law.**

Since each protein has a unique aromatic amino acid composition, the extinction coefficient (E_{280}) or molar absorptivity (E_m) must be determined for individual proteins for protein content estimation.

14.2.5.2 Procedure

1. Proteins are solubilized in buffer or alkali.
2. Absorbance of protein solution is read at 280 nm against a reagent blank.
3. Protein concentration is calculated according to the equation

$$A = a\,b\,c \tag{10}$$

where: A = absorbance
a = absorptivity
b = cell or cuvette path length
c = concentration

14.2.5.3 Applications

The UV 280 nm method has been used to determine the protein contents of milk (16) and meat products (17). This method has not been used widely in food systems. This technique applies better in a purified protein system or to proteins that have been extracted in alkali or denaturing agents such as 8 M urea.

Although peptide bonds in proteins absorb more strongly at 190–210 nm than at 280 nm, the low UV region is more difficult to measure.

Advantages:

1. Rapid and relatively sensitive (several times more sensitive than the biuret method)
2. No interference from ammonium sulfate and other buffer salts
3. Nondestructive; samples can be used for other analyses after protein determination; used very widely in post-column detection of proteins.

Disadvantages:

1. Nucleic acids also absorb at 280 nm. The absorption 280 nm/260 nm ratios for pure protein and nucleic acids are 1.75 and 0.5, respectively. One can correct the absorption of nucleic acids at 280 nm if the ratio of the absorption of 280 nm/260 nm is known. Nucleic acids also can be corrected using a method based on the absorption difference between 235 nm and 280 nm (18).
2. Aromatic amino acid contents in various proteins differ considerably.
3. Solution must be clear and colorless. Turbidity due to particulates in the solution will increase absorbance falsely.
4. A relatively pure system is required to use this method.

14.2.6 Dye Binding Method

14.2.6.1 Anionic Dye Binding

14.2.6.1.1 Principle The protein-containing sample is mixed with a known excess amount of **anionic dye** in a buffered solution. **Proteins bind the dye** to form an **insoluble complex.** The **unbound soluble dye** is measured after equilibration of the reaction and the removal of insoluble complex by centrifugation or filtration.

$$\text{protein} + \text{excess dye} \longrightarrow \begin{matrix}\text{protein-dye}\\ \text{insoluble complex}\end{matrix} + \text{unbound soluble dye} \tag{11}$$

The anionic sulfonic acid dye, including acid orange 12, orange G, and Amido black 10B, binds cationic groups of the **basic amino acid residues** (imidazole of histidine, guanidine of arginine, and ϵ-amino group of lysine) and **free amino terminal group** of the proteins. The unbound dye is inversely related to the protein content of the sample (19).

14.2.6.1.2 Procedure

1. The sample is finely ground (60-mesh or smaller sizes) and added to an excess dye solution.
2. The content is shaken vigorously to equilibrate the dye binding reactions and filtered or centrifuged to remove insoluble substances.
3. Absorbance of the unbound dye solution in the filtrate is measured and dye concentration estimated from a dye standard curve.
4. A straight calibration curve can be obtained by plotting the unbound dye concentration against total nitrogen (as determined by Kjeldahl method) of a given food covering a wide range of protein content.
5. Protein content of the unknown sample of the same food type can be estimated from the calibration curve or from a regression equation calculated by the least squares method.

14.2.6.1.3 Applications Anionic dye binding has been used in estimating proteins in milk (20, 21), wheat flour (22), soy products (9), and meats (8). The AOAC includes two dye-binding methods (Method 967.12

using Acid Orange 12 and Method 967.13 using Amido Black 10B) for analyzing proteins in milk.
Advantages:

1. Rapid (15 min or less), inexpensive, and relatively accurate for analyzing protein content in food commodities
2. May be used to estimate the changes in available lysine content of cereal products during processing since the dye does not bind altered, unavailable lysine. Since lysine is the limiting amino acid in cereal products, the available lysine content represents protein nutritive value of the cereal products (23).
3. No corrosive reagents
4. Does not measure nonprotein nitrogen
5. More precise than the Kjeldahl method

Disadvantages:

1. Not sensitive; mg quantities of protein are required.
2. Proteins differ in basic amino acid content and so differ in dye-binding capacity. Therefore, a calibration curve for a given food commodity is required.
3. Many nonprotein components bind dye (i.e., starch) and/or protein (i.e., calcium or phosphate) and cause errors in final results. The problem with calcium and heavy metal ions can be eliminated using properly buffered reagent containing oxalic acid.

14.2.6.2 Bradford Method

14.2.6.2.1 Principle When Coomassie Brilliant Blue G-250 binds to protein, the **dye changes** from reddish to bluish color, and the absorption maximum of the dye is shifted from 465 to 595 nm. The change in the absorbance at 595 nm is proportional to the protein concentration of the sample (24).

14.2.6.2.2 Procedure

1. Coomassie Brilliant Blue G-250 is dissolved in 95 percent ethanol and acidified with 85 percent phosphoric acid.
2. Samples containing proteins (1–100 μg per mL) and standard BSA solutions are mixed with the Bradford reagent.
3. Absorbance at 595 nm is read against a reagent blank.
4. Protein concentration in the sample is estimated from the BSA standard curve.

14.2.6.2.3 Applications The Bradford method has been successfully used to determine protein content in worts and beer products (25) and in potato tubers (26). This procedure has been improved to measure microgram quantities of proteins (27). Due to its rapidity, sensitivity, and fewer interferences than the Lowry method, the Bradford method has been widely used in protein purification.
Advantages:

1. Rapid; reaction can be completed in 2 min
2. Reproducible
3. Sensitive; several-fold more sensitive than the Lowry method
4. No interference from cations such as K^+, Na^+, and Mg^{+2}
5. No interference from ammonium sulfate
6. No interference from polyphenols and carbohydrate such as sucrose
7. Measures protein or peptides with molecular mass approximately equal to or greater than 4,000 daltons

Disadvantages:

1. Interfered with by both nonionic and ionic detergents, such as Triton X-100 and sodium dodecyl sulfate. However, errors due to small amounts (0.1%) of these detergents can be corrected using proper controls.
2. The protein-dye complex can bind to quartz cuvettes. The analyst must use glass or plastic cuvettes.
3. Color varies with different types of proteins. The standard protein must be selected carefully.

14.2.7 Ninhydrin Method

14.2.7.1 Principle

Amino acids, ammonia, and **primary amino groups** in a protein, when boiled in a pH 5.5 buffer in the presence of ninhydrin and hydrindantin, form a **Ruhemann purple color** (28, 29).

14.2.7.2 Procedure

1. Mix 1 mL sample solution with 1 mL ninhydrin solution in a test tube.
2. Heat in a boiling bath for 15 min.
3. Add 5 mL ethanol or propanol diluent, shake, and cool.
4. Read absorbance at 570 nm against a water blank.

14.2.7.3 Applications

The Ninhydrin method has not been used widely for the determination of protein quantity in foods. However, it can be used to determine the hydrolysis of peptide bonds during food processing.
Advantage:

1. Relatively rapid as compared to the Kjeldahl method

Disadvantages:

1. The presence of a small quantity of amino acids, peptides, primary amines, and ammonia causes an overestimation of the protein content.
2. Low precision.
3. Color varies with different amino acid compositions. Proline absorbs maximum at 440 nm; other amino acids at 570 nm.
4. A standard calibration curve must be prepared on each occasion.

14.2.8 Turbidimetric Method

14.2.8.1 Principle

Low concentrations (3–10%) of trichloroacetic acid, sulfosalicylic acid (30, 31), and potassium ferricyanide in acetic acid (32) can be used to **precipitate extracted proteins** to form a turbid suspension of protein particles. The **turbidity** can be measured from the reduction in the transmission of radiation. The reduction in radiation transmission is due to radiation scattering by the protein particles. The intensity of the radiation reduction can be related to protein concentration in the solution.

14.2.8.2 Procedure

The general procedure for measuring wheat proteins by the sulfosalicylic acid method (31) is as follows:

1. Wheat flour is extracted with 0.05 N sodium hydroxide.
2. Protein solubilized in alkali is separated from the nonsoluble materials by centrifugation.
3. Sulfosalicylic acid is mixed with a portion of the protein solution.
4. The degree of turbidity is measured by reading the light transmittance at 540 nm against a reagent blank.
5. The protein content can be estimated from a calibration curve, which is established using the Kjeldahl nitrogen method.

14.2.8.3 Applications

The turbidimetric method has been used to measure protein content of wheat flour (31) and corn (33).
Advantages:

1. Rapid; can be completed in 15 min
2. Does not measure nonprotein nitrogen other than nucleic acids

Disadvantages:

1. Different proteins precipitate at different rates.
2. Turbidity varies with different concentrations of acid reagents.
3. Nucleic acids also are precipitated by acid reagents.

14.3 COMPARISON OF METHODS

- **Sample preparation:** The Kjeldahl method requires little preparation. Sample particle size of 20-mesh or smaller is satisfactory; other methods require fine particles for extraction of proteins from the complex food systems.
- **Principle:** The Kjeldahl method measures directly the total amount of organic nitrogen element in the foods; other methods measure the various physiocochemical properties of proteins. For instance, the biuret method measures peptide bonds, and the Lowry method measures a combination of peptide bonds and the amino acids tryptophan and tyrosine.
- **Sensitivity:** Kjeldahl, biuret, and anionic dye binding are less sensitive than UV, Lowry, BCA, or Bradford methods.
- **Speed:** Methods involving spectrophotometric (colorimetric) measurements usually must separate proteins from the interfering insoluble materials before mixing with the color reagents or must remove the insoluble materials from the colored protein-reagent complex after mixing. However, the speed of determination in the colorimetric methods is faster than with the Kjeldahl method.

14.4 SPECIAL CONSIDERATIONS

1. To select a particular method for a specific application, sensitivity, accuracy, and reproducibility as well as physicochemical properties of food materials must be considered. The data should be interpreted carefully to reflect what actually is being measured.

2. Food processing methods, such as heating, may reduce the extractability of proteins for analysis and cause an underestimation of the protein content measured by methods involving an extraction step (8).
3. All methods, except for the Kjeldahl and the UV method for purified proteins, require the use of a standard or reference protein or a calibration with the Kjeldahl method. In the methods using a standard protein, proteins in the samples are assumed to have similar composition and behavior compared to the standard protein. The selection of an appropriate stand ard for a specific type of food is important.
4. **Nonprotein nitrogen** is present in practically all foods. To determine **protein nitrogen,** the samples usually are extracted with alkaline and followed by trichloroacetic acid or sulfosalicylic acid precipitation. The concentration of the acid used affects the precipitation yield. Therefore, nonprotein nitrogen content may vary with the type and concentration of the reagent used. Heating could be used to aid protein precipitation by acid, alcohol, or other organic solvents. In addition to acid precipitation methods used for nonprotein nitrogen determination, less empirical methods such as dialysis and ultrafiltration and column chromatography could be used to separate proteins from small nonprotein substances.
5. In the determination of the nutritive value of food proteins, including **protein digestibility** and **protein efficiency ratio** (PER), the Kjeldahl method with a 6.25 conversion factor is usually used to determine crude protein content. The PER could be underestimated if a substantial amount of nonprotein nitrogen is present in foods. A food sample with a higher nonprotein nitrogen content (particularly if the nonprotein nitrogen does not have many amino acids or small peptides) may have a lower PER than a food sample containing similar protein structure/composition and yet with a lower amount of nonprotein nitrogen.
6. Other methods, including the nitrogen combustion method and the **near infrared** (NIR) **method,** also have been used to quantitate protein content in foods. The NIR method is described in Chapter 24.

14.5 SUMMARY

Methods based on the unique characteristics of proteins and amino acids have been described to determine the protein content of foods. The Kjeldahl method measures organic nitrogen. Copper-peptide bond structure interactions contribute to the analysis by the biuret and Lowry methods. Amino acids are involved in the UV 280 nm, dye-binding, Ninhydrin, and Lowry methods. The BCA method utilizes the reducing power of proteins in an alkaline solution. The various methods differ in their speed and sensitivity.

In addition to the commonly used methods discussed, there are other methods available for protein quantification. Because of the complex nature of various food systems, problems may be encountered to different degrees in protein analysis by available methods. Rapid methods may be suitable for quality control purposes, while a sensitive method is required in working with a minute amount of protein. Indirect colorimetric methods usually require the use of a carefully selected protein standard or a calibration with the official Kjeldahl method.

14.6 STUDY QUESTIONS

1. What factors should one consider when choosing a method for protein determination?
2. The Kjeldahl method of protein analysis consists of three major steps. List these steps in the order they are done, and describe in words what occurs in each step. Make it clear why ml HCl can be used as an indirect measure of the protein content of a sample.
3. Why is the conversion factor from Kjeldahl nitrogen to protein different for various foods, and how is the factor of 6.25 obtained?
4. How can Nesslerization or the procedure that uses phenol and hypochlorite be used as part of the Kjeldahl procedure, and why might they best be put to use?
5. Differentiate and explain the chemical basis of the following techniques that can be used to quantitate proteins in quality control/research:
   a. Kjeldahl method
   b. turbidimetric method
   c. Ninhydrin method
   d. absorbance at 280 nm
   e. absorbance at 220 nm
   f. biuret method
   g. Lowry method
   h. Bradford method
   i. bicinchoninic acid method
6. Differentiate the principles of protein determination by dye binding with an anionic dye such as Amido black versus with the Bradford method, which uses the dye Coomassie Brilliant Blue G-250.
7. With the anionic dye binding method, would a sample with a higher protein content have a higher or a lower absorbance reading than a sample with a low protein content? Explain your answer.

14.7 PRACTICE PROBLEMS

1. A dehydrated precooked pinto bean was analyzed for crude protein content in duplicate using the Kjeldahl

method. The following data were recorded:

moisture content = 8.0%
weight of sample no. 1 = 1.015 g
weight of sample no. 2 = 1.025 g
normality of HCL used for titration = 0.1142
mL HCl used for sample no. 1 = 22.0 mL
mL HCl used for sample no. 2. = 22.5 mL
mL HCl used for reagent blank = 0.2 mL

Calculate crude protein content on both wet and dry weight basis of the pinto bean, assuming pinto bean protein contains 17.5 percent nitrogen.

2. A 10 mL protein fraction recovered from a column chromatography was analyzed for protein using the BCA method. The following data were obtained from a duplicate analysis using BSA as a standard:

BSA mg/mL	*Mean Absorbance at 562 nm*
0.2	0.25
0.4	0.53
0.6	0.74
0.8	0.92
1.0	1.20

The average absorbance of a 0.5 mL sample was 0.44. Calculate protein content (mg/mL) and total protein quantity of this column fraction.

Answers:

1. Protein content = 19.8 percent on a wet weight basis; 21.4 percent on a dry weight basis. 2. Protein content = 0.68 mg/mL. Total protein quantity = 6.8 mg.

14.8 REFERENCES

1. Jones, D.B. 1931. Factors for converting percentages of nitrogen in foods and feeds into percentages of proteins. U.S. Dept. Agric. Circular No. 183. August. USDA, Washington, DC.
2. Adams, C.F. 1975. Nutritive value of American foods in common units. USDA Agricultural Handbook No. 456. USDA Agricultural Research Service, Washington, DC.
3. AOAC. 1990. *Official Methods of Analysis*, 15th edition. Association of Official Analytical Chemists, Washington, DC.
4. Bradstreet, R.B. 1965. *The Kjeldahl Method for Organic Nitrogen.* Academic Press, New York, NY.
5. Robinson, H.W., and Hodgen, C.G. 1940. The biuret reaction in the determination of serum protein. I. A study of the conditions necessary for the production of the stable color which bears a quantitative relationship to the protein concentration. *J. Biol. Chem.* 135:707–725.
6. Jennings, A.C. 1961. Determination of the nitrogen content of cereal grain by colorimetric methods. *Cereal Chem.* 38:467–479.
7. Pinckney, A.J. 1961. The biuret test as applied to the estimation of wheat protein. *Cereal Chem.* 38:501–506.
8. Torten, J., and Whitaker, J.R. 1964. Evaluation of the biuret and dye-binding methods for protein determination in meats. *J. Food Sci.* 29:168–174.
9. Pomeranz, Y. 1965. Evaluation of factors affecting the determination of nitrogen in soya products by the biuret and orange-G dye-binding methods. *J. Food Sci.* 30: 307–311.
10. AOAC. 1965. *Official Methods of Analysis*, 10th edition. Association of Official Analytical Chemists, Washington, DC.
11. Lowry, O.H., Rosebrough, N.J., Farr, A.L., and Randall, R.J. 1951. Protein measurement with the Folin phenol reagent. *J. Biol. Chem.* 193:265–275.
12. Peterson, G.L. 1979. Review of the Folin phenol protein quantitation method of Lowry, Rosebrough, Farr, and Randall. *Anal. Biochem.* 100:201–220.
13. Miller, G.L. 1959. Protein determination for large numbers of samples. *Anal. Chem.* 31:964.
14. Hartree, E.F. 1972. Determination of protein: A modification of the Lowry method that gives a linear photometric response. *Anal. Biochem.* 48:422–427.
15. Smith, P.K., Krohn, R.I., Hermanson, G.T., Mallia, A.K., Gartner, F.H., Provensano, M.D., Fujimoto, E.K., Goeke, N.M., Olson, B.J., and Klenk, D.C. 1985. Measurement of protein using bicinchoninic acid. *Anal. Biochem.* 150:76–85.
16. Nakai, S., Wilson, H.K., and Herreid, E.O. 1964. Spectrophotometric determination of protein in milk. *J. Dairy Sci.* 47:356–358.
17. Gabor, E. 1979. Determination of the protein content of certain meat products by ultraviolet absorption spectrophotometry. *Acta Alimentaria* 8(2):157–167.
18. Whitaker, J.R., and Granum, P.E. 1980. An absolute method for protein determination based on difference in absorbance at 235 and 280 nm. *Anal. Biochem.* 109: 156–159.
19. Fraenkel-Conrat, H., and Cooper, M. 1944. The use of dye for the determination of acid and basic groups in proteins. *J. Biol. Chem.* 154:239–246.
20. Udy, D.C. 1956. A rapid method for estimating total protein in milk. *Nature* 178:314–315.
21. Tarassuk, N.P., Abe, N., and Moats, W.A. 1966. The dye binding of milk proteins. Technical Bulletin No. 1369. USDA Agricultural Research Service in cooperation with California Agricultural Experiment Station, Washington, D.C.
22. Udy, D.C. 1954. Dye-binding capacities of wheat flour protein fractions. *Cereal Chem.* 31:389–395.
23. Hurrel, R.F., Lerman, P., and Carpenter, K.J. 1979. Reactive lysine in foodstuffs as measured by a rapid dye-binding procedure. *J. Food Sci.* 44:1221–1227.
24. Bradford, M. 1976. A rapid and sensitive method for the quantitation of microgram quantities of protein utilizing the principle of protein-dye binding. *Anal. Biochem.* 72:248–254.
25. Lewis, M.J., Krumland, S.C., and Muhleman, D.J. 1980. Dye-binding method for measurement of protein in wort and beer. *J. Am. Soc. Brewing Chemists* 38:37–41.
26. Snyder, J., and Desborou, S. 1978. Rapid estimation of potato tuber total protein content with Coomassie Brilliant Blue G-250. *Theor. Appl. Genet.* 52:135–139.
27. Bearden, Jr., J.C. 1978. Quantitation of submicrogram

quantities of protein by an improved protein-dye binding assay. *Biochim. Biophys. Acta.* 533:525–529.
28. Duggan, E.C. 1957. Measurement of amino acids by column chromatography. *Methods Enzymol.* 3:492–495.
29. Spackman, D.H., Stein, W.H., and Moore, S. 1958. Automatic recording apparatus for use in the chromatography of amino acids. *Anal. Chem.* 30:1191–1206.
30. Layne, E. 1957. Spectrophotometric and turbidimetric methods for measuring proteins. *Methods Enzymol.* 3:447–454.
31. Feinstein, L., and Hart, J.R. 1959. A simple method for determination of the protein content of wheat and flour samples. *Cereal Chem.* 36:191–193.
32. Tappan, D.V. 1966. A light scattering technique for measuring protein concentration. *Anal. Biochem.* 14:171–182.
33. Paulis, J.W., Wall, J.S., and Kwolek, W.F. 1974. A rapid turbidimetric analysis for zein in corn and its correlation with lysine content. *J. Agric. Food Chem.* 22:313–317.

CHAPTER

15

Protein Separation and Characterization Procedures

Denise M. Smith

15.1 INTRODUCTION

Many protein separation techniques are available to food scientists. Several of the separation techniques described in this chapter are used commercially for the production of food or food ingredients, whereas other separation methods are used to purify a protein from a food for further study in the laboratory. In general, separation techniques exploit the biochemical differences in protein solubility, size, charge, adsorption characteristics, and biological affinities for other molecules. These physical characteristics are then used to purify individual proteins from complex mixtures. Some techniques to characterize the biochemical properties of a protein are also presented in this chapter.

15.2 INITIAL CONSIDERATIONS

Usually, several separation techniques are used in sequence to purify a protein from a food. In general, the more separation steps used, the higher the purity of the resulting preparation. Food ingredients such as protein concentrates may be prepared using only one separation step because high purity is not necessary. To prepare a pure protein for laboratory study it is often necessary to use three or more separation steps in sequence to achieve a highly purified protein preparation.

Before starting a separation sequence, it is necessary to learn as much as possible about the biochemical properties of a protein, such as molecular weight, isoelectric point (pI), solubility properties, and denaturation temperature, to determine any unusual physical characteristics that will make separation easier. The first separation step should be one that can easily be used with large quantities of material. This is often a technique that utilizes the differential solubility properties of a protein. Each succeeding step in a purification sequence will use a different mode of separation. The most common methods of purification include precipitation, ion-exchange chromatography, affinity chromatography, and size-exclusion chromatography (1).

15.3 METHODS OF PROTEIN SEPARATION

15.3.1 Separation by Differential Solubility Characteristics

15.3.1.1 Principle

Separation by precipitation exploits the differential solubility properties of proteins in solution. Proteins are polyelectrolytes; thus, solubility characteristics are determined by the type and charge of amino acids in the molecule. Proteins can be selectively precipitated or solubilized by changing **buffer pH, ionic strength, dielectric constant,** or **temperature.** These separation techniques are advantageous when working with large quantities of material, are relatively quick, and are not usually influenced by other food components. Precipitation techniques are used most commonly during early stages of a purification sequence.

15.3.1.2 Procedures

15.3.1.2.1 Salting Out Proteins have unique solubility profiles in neutral salt solutions. Low concentrations of neutral salts usually increase the solubility of proteins; however, proteins are precipitated from solution as ionic strength is increased. This property can be used to precipitate a protein from a complex mixture. **Ammonium sulfate** [$(NH_4)_2SO_4$] is commonly used because it is highly soluble, although other neutral salts such as NaCl or KCl may be used to salt out proteins. Generally a two-step procedure is used to maximize separation efficiency. In the first step, $(NH_4)_2SO_4$ is added at a concentration just below that necessary to precipitate the protein of interest. When the solution is centrifuged, less soluble proteins are precipitated while the protein of interest remains in solution. The second step is performed at an $(NH_4)_2SO_4$ concentration just above that necessary to precipitate the protein of interest. When the solution is centrifuged, the protein is precipitated, while more soluble proteins remain in the supernatant. One disadvantage of this method is that large quantities of salt contaminate the precipitated protein and often must be removed before the protein is resolubilized in buffer. Tables and formulas are available in many biochemistry textbooks for calculating the proper amount of $(NH_4)_2SO_4$ to achieve a specific concentration (2).

15.3.1.2.2 Isoelectric Precipitation The **isoelectric point** (pI) is defined as the pH at which a protein has no net charge in solution. Proteins aggregate and precipitate at their pI because there is no electrostatic repulsion between molecules. Proteins have different pIs; thus, they can be separated from each other by adjusting solution pH. When the pH of a solution is adjusted to the pI of a protein, the protein precipitates while proteins with different pIs remain in solution. The precipitated protein can be resolubilized in another solution of different pH.

15.3.1.2.3 Solvent Fractionation Protein solubility at a fixed pH and ionic strength is a function of the dielectric constant of a solution. Thus, proteins can be separated based on solubility differences in organic solvent-water mixtures. The addition of water miscible organic solvents, such as ethanol or acetone, decreases the dielectric constant of an aqueous solution and decreases the solubility of most proteins. Organic solvents decrease ionization of charged amino acids, resulting in protein aggregation and precipitation. The optimum quantity of organic solvent to precipitate a protein varies from 5 to 60 percent. Solvent fractionation is usually performed at 0°C or below to prevent protein denaturation caused by temperature increases that occur when organic solvents are mixed with water.

15.3.1.2.4 Denaturation of Contaminating Proteins Many proteins are denatured and precipitated from solution when heated above a certain temperature or by adjusting a solution to highly acid or basic pHs. Proteins that are stable at high temperatures or at extremes of pH are most easily separated by this technique because many contaminating proteins can be precipitated while the protein of interest remains in solution.

15.3.1.3 Applications

All of the above techniques are commonly used to fractionate proteins. The differential solubility of water-soluble muscle proteins in $(NH_4)_2SO_4$ and acetone and temperature stability at 55°C are illustrated in Table 15-1. These three techniques can be combined in sequence to prepare muscle proteins of high purity.

One of the best examples of the commercial use of differential solubility to separate proteins is in production of protein concentrates. Soy protein concentrate can be prepared from defatted soybean flakes or flour using several methods described above. Soy proteins can be precipitated from other soluble constituents in the flakes or flour using a 60–80 percent aqueous alcohol solution, by isoelectric precipitation at pH 4.5 (which is the pI of many soy proteins), or by denaturation with moist heat. These methods have been used to produce concentrates containing greater than 65 percent protein. Two or three separation techniques can be combined in sequence to produce soy protein isolates with protein concentrations above 90 percent.

15.3.2 Separation by Adsorption

15.3.2.1 Principle

Adsorption chromatography is defined as the separation of compounds by adsorption to, or desorption from, the surface of a solid support by an eluting solvent. Separation is based on differential affinity of the protein for the adsorbent or eluting buffer. Affinity chromatography and ion-exchange chromatography are two types of adsorption chromatography that will be described briefly below (see Chapter 28 for a more detailed description).

15.3.2.2 Procedures

15.3.2.2.1 Ion-Exchange Chromatography Ion-exchange chromatography is defined as the reversible adsorption between charged molecules and ions in solution and a charged solid support matrix. Ion-exchange chromatography is the most commonly used protein separation technique and results in an average eightfold purification (1). A positively charged matrix is called an **anion-exchanger** because it binds negatively charged ions or molecules in solution. A negatively charged matrix is called a **cation-exchanger** because it binds positively charged ions or molecules. The most commonly used exchangers for protein purification are anionic diethylaminoethyl derivatized

TABLE 15-1. Conditions for Fractionating Water-Soluble Muscle Proteins Using Differential Solubility Techniques

	PRECIPITATION RANGE		
Enzyme	*$(NH_4)_2SO_4$ pH 5.5, 10°C (Percent Saturation)*	*Acetone pH 6.5, −5°C (Percent v/v)*	*Stability[1] pH 5.5, 55°C*
Phosphorylase	30–40	18–30	U
Pyruvate kinase	55–65	25–40	S
Aldolase	45–55	30–40	S
Lactate dehydrogenase	50–60	25–35	S
Enolase	60–75	35–45	U
Creatine kinase	60–80	35–45	U
Phosphoglycerate kinase	60–75	45–60	S
Myoglobin	70–90	45–60	U

Adapted from (3) with permission.
[1]Indicated by U = unstable and S = stable at heating temperature.

supports, followed by carboxymethyl and phospho cation-exchangers (see Chapter 28).

The protein of interest is first adsorbed to the ion-exchanger under buffer conditions (ionic strength and pH) that maximize the affinity of the protein for the matrix. Contaminating proteins of different charges pass through the exchanger unabsorbed. Proteins bound to the exchangers are selectively eluted from the column by gradually changing the ionic strength or pH of the eluting solution (Fig. 15-1). As the composition of the eluting buffer changes, the charges of the proteins change and their affinity for the ion-exchange matrix is decreased.

15.3.2.2.2 Affinity Chromatography Affinity chromatography is a type of adsorption chromatography in which a protein is separated in a chromatographic matrix containing a ligand covalently bound to a solid support. A **ligand** is defined as a molecule with a reversible, specific, and unique binding affinity for a protein. Ligands include enzyme inhibitors, enzyme substrates, coenzymes, antibodies, and certain dyes. Covalently bound ligands can be purchased commercially or prepared in the laboratory.

The protein is passed through a column containing the ligand bound to a solid support, under buffer conditions (pH, ionic strength, temperature, and protein concentration) that maximize binding of the protein to the ligand. Contaminating proteins and molecules that do not bind the ligand are eluted. The bound protein is then desorbed or eluted from the column under conditions that decrease the affinity of the protein for the bound ligand, by changing the pH, temperature, or concentration of salt or ligand in the eluting buffer.

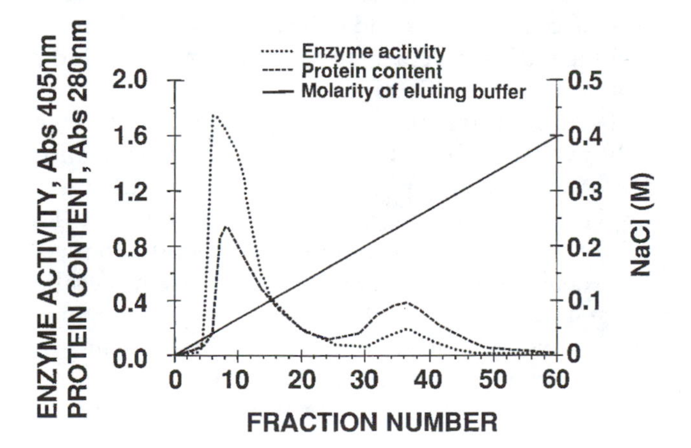

FIGURE 15-1.
Elution profile of two α-galactosidase isozymes from a carboxymethyl-cellulose ion-exchange column. Adapted from *Photochemistry,* 28, S.R. Alani, D.M. Smith, and P. Markakis, α-Galactosidases of *Vigua unguiculata,* p. 2048, copyright 1989, with kind permission from Pergamon Press Ltd., Headington Hill Hall, Oxford OX3 OBW, U.K.

Affinity chromatography is a very powerful technique and is the second most commonly used protein purification procedure (1). The average purification achieved by affinity chromatography is approximately 100-fold, although 1000-fold increases in purification have been reported. This technique is more powerful than size-exclusion, ion-exchange, and other separation methods that usually achieve less than a twelve-fold purification. The development of new affinity chromatography procedures can be very time consuming because many variables must be optimized, which is a major disadvantage to the method. Also, the affinity materials are often more expensive than other separation media.

15.3.2.2.3 High-Performance Liquid Chromatography Many chromatographic methods have been adapted for use with **high-performance liquid chromatography systems** (HPLC). The use of HPLC to separate proteins was made possible by development of macroporous, microparticulate packing materials that withstand high pressures. This technique is discussed in more detail in Chapter 29.

15.3.2.3 Applications

Ion-exchange chromatography is commonly used to separate proteins in the laboratory and can be used for quantification of amino acids in a protein as described in section 15.3.5. Affinity chromatography has many uses in the analytical lab, and may be used for commercial preparation of protein reagents by chemical suppliers, but is not generally used for commercial production of food protein ingredients due to the high costs involved.

Affinity chromatography is used to purify many glycoproteins. Glycoproteins can be separated from other proteins in a complex mixture by utilization of the high carbohydrate binding affinity of lectins. Lectins, such as concanavalin A, are carbohydrate-binding proteins that can be bound to a solid support and used to bind the carbohydrate moiety of glycoproteins that are applied to the column. Once the glycoproteins are bound to the column, they can be desorbed using an eluting buffer containing an excess of lectin. The glycoproteins bind preferentially to the free lectins and elute from the column.

15.3.3 Separation by Size

15.3.3.1 Principle

Protein molecular weights range from about 10,000 to over 1,000,000; thus, size is a logical parameter to exploit for separations. Actual separation occurs based on the **Stokes radius** of the protein, not on the molecular weight. Stokes radius is the average radius of the protein in solution and is determined by protein conformation. For example, a globular protein may have an actual radius very similar to its Stokes radius, whereas a fibrous or rod-shaped protein of the same molecular weight may have a Stokes radius that is much larger than that of the globular protein. As a result, the two proteins may separate as if they had different molecular weights.

15.3.3.2 Procedures

15.3.3.2.1 Dialysis Dialysis is used to separate molecules in solution by the use of semipermeable membranes that permit passage of small molecules but not larger molecules. To perform dialysis, a protein solution is placed into dialysis tubing that has been tied or clamped at one end. The other end of the tubing is sealed, and the bag is placed in a large volume of water or buffer (usually 500–1000 times greater than the sample volume inside the dialysis tubing) which is slowly stirred. Low-molecular-weight solutes diffuse from the bag, while buffer diffuses into the bag. Dialysis is simple; however, it is a relatively slow method, usually requiring at least 12 hr and one change of buffer. The protein solution inside the bag is often diluted during dialysis, due to osmotic strength differences between the solution and dialysis buffer.

This technique can be used to concentrate protein by coating the dialysis bag containing a protein solution with polyethylene glycol. **Polyethylene glycol** absorbs water and concentrates the solution within the dialysis bag. Equilibrium dialysis can be used for determining the stoichiometry of protein-ligand binding.

15.3.3.2.2 Ultrafiltration Ultrafiltration is a technique using a semipermeable membrane for the separation of solutes on the basis of size under an applied pressure. This method is similar to dialysis but is much faster. Semipermeable membranes with **molecular weight cut-offs** from 500 to 300,000 are available. Molecules larger than the membrane cut-off are retained and become part of the retentate, while smaller molecules pass through the membrane and become part of the ultrafiltrate. Ultrafiltration can be used to concentrate a protein solution, remove salts, exchange buffer, or fractionate proteins on the basis of size.

Several types of laboratory and production scale ultrafilters are commercially available. A stirred cell ultrafiltration unit is illustrated in Fig. 15-2. The protein solution in the stirred cell is filtered through the semipermeable membrane by gas pressure, leaving a concentrated solution of proteins larger than the membrane cut-off point inside the cell. Some ultrafiltration devices are designed for use in a centrifuge.

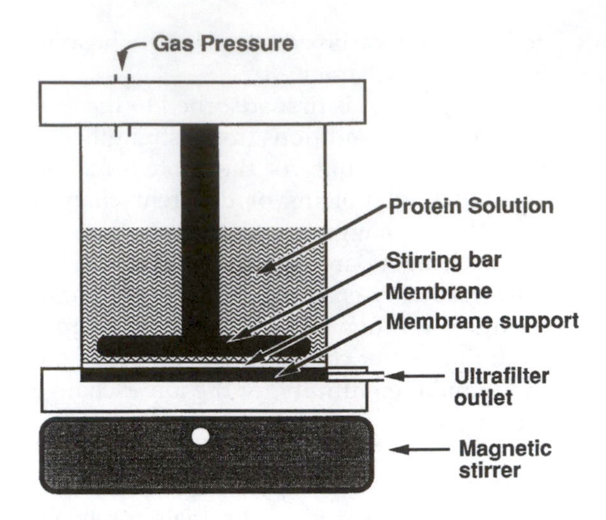

FIGURE 15-2.
Schematic diagram of a stirred cell ultrafiltration unit.

15.3.3.2.3 Size-Exclusion Chromatography Size-exclusion chromatography, also known as gel filtration or gel permeation chromatography, is a column technique that can be used to separate proteins on the basis of size. A protein solution is allowed to flow down a column packed with a solid support of porous beads made of a cross-linked polymeric material such as agarose or dextran. Molecules larger than the pores in the beads are excluded, moving quickly through the column and eluting from the column in the shortest times. Small molecules enter the pores of the beads and are retarded, thus moving very slowly through the column. Molecules of intermediate sizes partially interact with the porous beads and elute at intermediate times. Consequently, molecules are eluted from the column in order of decreasing size.

Beads of different average pore sizes are commercially available that allow for efficient fractionation of proteins of different molecular weights. Chemical suppliers list working molecular weight ranges for each of their gel permeation solid support products. Size-exclusion chromatography is used to remove salts, change buffers, fractionate proteins, and estimate molecular weight. Molecular weight can be calculated by chromatographing the unknown protein and several proteins of known molecular weights. Standards of known molecular weight are commercially available and can be used to prepare a standard curve. A plot of the **elution volume** (V_e) of each protein versus **log of the molecular weight** yields a straight line. Size-exclusion techniques generally can be used to estimate molecular weights within ±10 percent; however, errors can occur if the Stokes radius of the unknown protein and standards are quite different. More information on size-exclusion chromatography is available in Chapter 28.

15.3.3.3 Applications

Dialysis and size-exclusion chromatography are primarily used in the analytical laboratory in a protein separation sequence. Dialysis is commonly used to change the buffer to one of the appropriate pH and ionic strength prior to electrophoresis of a protein sample. Dialysis is usually performed after $(NH_4)_2SO_4$ precipitation of a protein to remove excess salt and other small molecules and to solubilize protein in a new buffer.

Ultrafiltration is used both in the laboratory and for commercial applications. Ultrafiltration is commonly used for the preparation of protein concentrates from whey, which is a by-product of the cheese-making industry. In this process, a semipermeable ultrafiltration membrane with a molecular weight cut-off of 10,000 to 20,000 is used to partially remove lactose, salts, and water from whey and concentrate proteins in the retentate (5).

15.3.4 Separation by Electrophoresis

15.3.4.1 Polyacrylamide Gel Electrophoresis

15.3.4.1.1 Principle Electrophoresis is defined as the migration of charged molecules in a solution through an electrical field. The most common type of electrophoresis performed with proteins is zonal electrophoresis in which proteins are separated from a complex mixture into bands by migration in aqueous buffers through a solid polymer matrix called a gel. **Polyacrylamide gels** are the most common matrix for protein electrophoresis, although other matrices such as starch and agarose may be used. Gel matrices can be formed in glass tubes or as slabs between two glass plates.

Separation depends on the friction of the protein within the matrix and the charge of the protein molecule as described by the following equation:

$$\text{mobility} = \frac{(\text{applied voltage})(\text{net charge on molecule})}{\text{friction of the molecule}} \quad (1)$$

Proteins are positively or negatively charged, depending on solution pH and their pI. A protein is negatively charged if solution pH is above its pI, whereas a protein is positively charged if solution pH is below its pI. The magnitude of the charge and applied voltage will determine how far a protein will migrate in an electrical field. The higher the voltage and stronger the charge on the protein, the greater the migration within the electrical field. Molecular size and shape, which determine the Stokes radius of a protein, also determine migration distance within the gel matrix. Mobility of proteins decreases as molecular friction increases due to an increase in Stokes radius; thus, smaller proteins tend to migrate faster through the gel matrix. Similarly, a decrease in pore size of the gel matrix will decrease mobility.

In nondenaturing or **native electrophoresis,** proteins are separated in their native form based on charge, size, and shape of the molecule. Another form of electrophoresis commonly used for separating proteins is denaturing electrophoresis (6). **Polyacrylamide gel electrophoresis** (PAGE) with an anionic detergent, **sodium dodecyl sulfate** (SDS), is used to separate protein subunits by size. Proteins are solubilized in a buffer containing SDS and a **reducing agent, mercaptoethanol** or **dithiothreitol,** to dissociate the protein into subunits and reduce disulfide bonds. Proteins bind SDS, become negatively charged, and are separated based on size alone.

15.3.4.1.2 Procedures A power supply and electrophoresis apparatus containing the polyacrylamide gel matrix and two buffer reservoirs are necessary to perform a separation. A representative slab gel and electrophoresis unit is shown in Fig. 15-3. The power supply is used to make the electric field by providing a source of constant current, voltage, or power. The

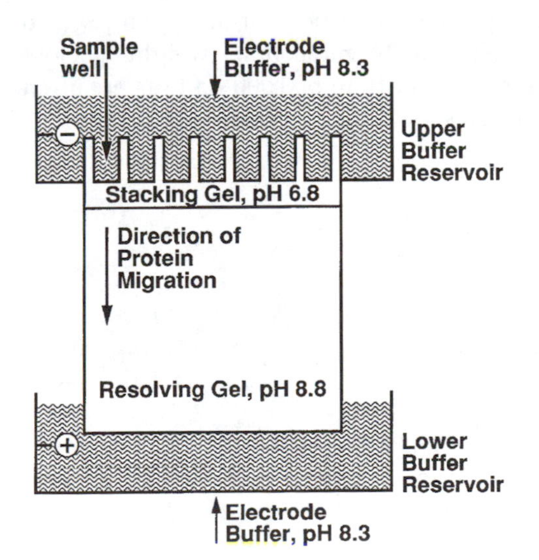

FIGURE 15-3.
Schematic diagram of a slab gel electrophoresis unit indicating the pHs of the stacking and resolving gels and the electrode buffer in an anionic discontinuous buffer system.

electrode buffer controls the pH to maintain the proper charge on the protein and conducts the current through the polyacrylamide gel. Commonly used buffer systems include an anionic tris-(hydroxymethyl)aminomethane buffer with a resolving gel at pH 8.8 (7) and a cationic acetate buffer at pH 4.3 (8).

The polyacrylamide gel matrix is formed by polymerizing **acrylamide** and a small quantity (usually 5% or less) of the cross-linking reagent, **N,N′-methylene-bisacrylamide,** in the presence of a **catalyst, tetramethylethylenediamine** (TEMED), and **source of free radicals, ammonium persulfate,** as illustrated Fig. 15-4. Gels can be made in the laboratory or purchased pre-cast.

A discontinuous gel matrix is usually used to improve resolution of proteins within a complex mixture (9, 10). The discontinuous matrix consists of a **stacking gel** with a large pore size (usually 3–4% acrylamide) and a resolving gel of a smaller pore size. The stacking gel, as its name implies, is used to stack or concentrate the proteins into very narrow bands prior to their entry into the resolving gel. At pH 6.8, a voltage gradient is formed between the chloride (high negative charge) and glycine ions (low negative charge) in the electrode buffer that serves to stack the proteins into narrow bands between the ions. Migration into the **running gel** of a different pH disrupts this voltage gradient and allows separation of the proteins into discrete bands.

The pore size of the running gel is selected based on the molecular weight of the proteins of interest and is varied by altering the concentration of acrylamide in solution. Proteins are usually separated on running gels that contain 4 to 15 percent acrylamide. Acrylamide concentrations of 15 percent are often used to separate proteins with molecular weights below 50,000. Proteins greater than 500,000 daltons are often separated on gels with acrylamide concentrations below 7 percent. A **gradient gel** in which the acrylamide concentration increases from the top to the bottom of the gel is often used to separate a mixture of proteins with a large molecular weight range.

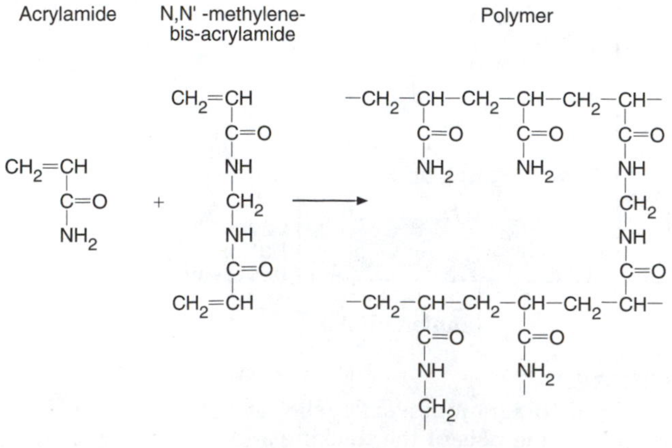

FIGURE 15-4.
Free radical polymerization reaction of polyacrylamide.

To perform a separation, proteins in a buffer of the appropriate pH are loaded on top of the stacking gel. **Bromophenol blue** tracking dye is added to the protein solution. This dye is a small molecule that migrates ahead of the proteins and is used to monitor the progress of a separation. After an electrophoresis run, the bands on the gels are generally visualized using a protein stain such as **Coomassie Brilliant Blue** or **silver stain.** Specific enzyme stains or antibodies can be used to detect a protein.

The electrophoretic or **relative mobility** (R_m) of each protein band is calculated as

$$R_m = \frac{\text{distance protein migrated from start of resolving gel}}{\text{distance between start of running gel and tracking dye}} \quad (2)$$

15.3.4.1.3 Applications

Electrophoresis is often used to determine the protein composition of a food product. For example, differences in the protein composition of soy protein concentrates and whey protein concentrates produced by different separation techniques can be detected. Electrophoresis can also be used to determine the purity of a protein extract.

SDS-PAGE is used to determine subunit composition of a protein and to estimate subunit molecular weight within an error of ± 5 percent, although highly charged proteins or glycoproteins may be subject to a larger error. Molecular weight is determined by comparing R_m of the protein subunit with R_m of protein standards of known molecular weight (Fig. 15-5.) Commercially prepared protein standards are available in several molecular weight ranges. To prepare a standard curve, logarithms of protein standard molecular weights are plotted against their corresponding R_m. The molecular weight of the unknown protein is determined from its R_m using the standard curve.

15.3.4.2 Isoelectric Focusing

15.3.4.2.1 Principle Isoelectric focusing is a modification of electrophoresis, in which proteins are separated by charge in an electric field on a gel matrix in which a pH gradient has been generated using ampholytes.

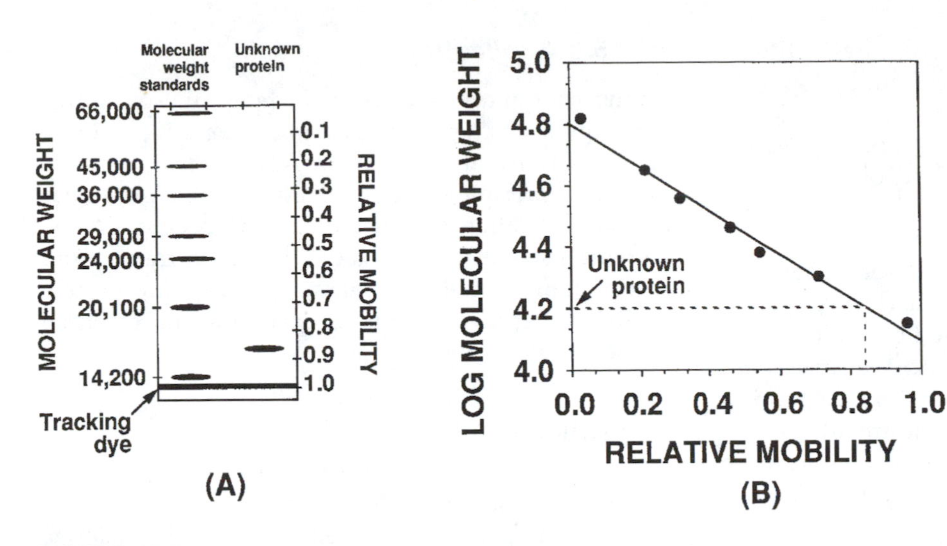

FIGURE 15-5.
Use of sodium dodecyl sulfate polyacrylamide gel electrophoresis to determine the molecular weight of a protein. A) Separation of molecular weight standards and the unknown protein. B) Standard curve for estimating protein molecular weight.

Proteins are focused or migrate to the location in the gradient at which pH equals the pI of the protein. Resolution is among the highest of any protein separation technique and can be used to separate proteins with pIs that vary less than 0.02 of a pH unit.

15.3.4.2.2 Procedure A pH gradient is formed using **ampholytes,** which are small polymers (molecular weight of about 5000 daltons) containing both positively and negatively charged groups. An ampholyte mixture is composed of thousands of polymers that exhibit a range of pK values. Ampholytes are added to the gel solution prior to polymerization. After the gel is formed and a current applied, the ampholytes migrate to produce the pH gradient; negatively charged ampholytes migrate toward the anode while positively charged ampholytes migrate toward the cathode. Ampholyte mixtures are available that cover a narrow pH range (2–3 units) or a broad range (pH 3–10) and should be selected for use based on properties of the proteins to be separated.

15.3.4.2.3 Applications Isoelectric focusing is the method of choice for determining the isoelectric point of a protein and is an excellent method for determining the purity of a protein preparation. For example, isozymes of polyphenol oxidase and other plant and animal proteins are identified using isoelectric focusing. Isoelectric focusing is used to differentiate closely related fish species based on protein patterns.

Isoelectric focusing and SDS-PAGE can be combined to produce a two-dimensional electrophoretogram that is extremely useful for separating very complex mixtures of proteins. This technique is called **two-dimensional electrophoresis** (11). Proteins are first separated in tube gels by isoelectric focusing. The tube gel containing the separated proteins is then placed on top of an SDS-PAGE slab gel, and proteins are separated. Thus, proteins are separated first on the basis of charge and then on the basis of size and shape. Over 1000 proteins in a complex mixture have been resolved using this technique.

15.3.5 Amino Acid Analysis

15.3.5.1 Principle

Amino acid analysis is used to quantitatively determine the amino acid composition of a protein. The protein sample is first hydrolyzed to release the amino acids. Amino acids are then separated using chromatographic techniques and quantified. Ion-exchange chromatography, reversed-phase liquid chromatography, and gas-liquid chromatography are three separation techniques used. This section will describe the use of ion-exchange and reversed-phase liquid chromatography.

15.3.5.2 Procedures

In general, a protein sample is hydrolyzed in constant boiling 6N HCl for 24 hr to release amino acids prior to chromatography. Accurate quantification of some amino acids is difficult because they react differently during hydrolysis. Consequently, special hydrolysis procedures must be used to prevent errors.

Tryptophan is completely destroyed by acid hydrolysis. Methionine, cysteine, threonine, and serine are progressively destroyed during hydrolysis; thus, the time of hydrolysis will influence results. Asparagine and glutamine are quantitatively converted to aspartic and glutamic acid, respectively, and cannot be measured. Isoleucine and valine are hydrolyzed more slowly in 6N HCl than other amino acids, while tyrosine may be oxidized.

In general, losses of threonine and serine can be estimated by hydrolysis of samples for three periods of time (i.e., 24, 48, and 72 hr) followed by amino acid analysis. Compensation for amino acid destruction may be made by calculation to zero time assuming first order kinetics. Valine and isoleucine are often estimated from a 72 hr hydrolysate. Cysteine and cystine can be converted to the more stable compound, cysteic acid, by hydrolysis in performic acid and then hydrolyzed in 6M HCl and chromatographed. Tryptophan can be separated chromatographically after a basic hydrolysis or analyzed using a method other than amino acid analysis.

In the original method developed by Moore and colleagues (12) and later revised by Stein et al. (13), amino acids were separated by ion-exchange chromatography using a stepwise elution with buffers of increasing pH and ionic strength. Amino acids eluting from the column were quantified by reaction with ninhydrin to produce a colored product that was measured spectrophotometrically. This method was automated in the late 1970s and is the basis of many amino acid analysis systems in use today. This methodology was adapted for use with high-performance liquid chromatographs in the 1980s. This adaptation was made possible because new ion-exchange resins were developed that could withstand high pressures and extremes of pH, ionic strength, and temperature.

Other new methods were also developed in the 1980s using HPLC and a reversed-phase column. The hydrolyzed amino acids are derivatized prior to chromatography with phenylthiocarbamyl or other compound, separated by reversed-phase HPLC, and quantitated by ultraviolet (UV) spectroscopy. Methods using HPLC techniques can detect picomole quantities of amino acids. Chromatographic runs usually take 30 min or less. A chromatogram showing the separation of amino acids in an infant formula is shown in Fig. 15-6.

The quantity of each amino acid in a peak is usually determined by spiking the sample with a known quantity of internal standard. The internal standard is usually an amino acid, such as norleucine, not commonly found in a food product. Results are usually expressed as mole percent. This quantity is calculated by dividing the mass of each amino acid (determined from the chromatogram) by its molecular weight, summing the values for all amino acids, dividing each by the total moles, and multiplying the result by 100.

15.3.5.3 Applications

Amino acid analysis is used to determine the nutritional quality of a protein and to characterize or identify newly isolated protein. Amino acid analysis provides information for calculating the molecular weight of a protein as well as its partial specific volume. Proteins used in animal diets, infant formulas, and special human diets are often analyzed for protein quality to ensure adequate quantities of essential amino acids.

15.4 SUMMARY

This chapter has provided a brief introduction into a few techniques used to separate and characterize proteins that rely on the differences between protein molecules in their solubility, size, charge, adsorption characteristics, and biological affinity for other molecules. More detailed information on these techniques can be found in other publications (2, 15–18).

15.5 STUDY QUESTIONS

1. For each of the techniques listed below, identify the basis by which it can be used to separate proteins within a protein solution (e.g., precipitation, adsorption, size, charge) and give a brief explanation of how/why it works in that way.
   a. dialysis
   b. adjustment of pH to pI
   c. addition of ammonium sulfate
   d. ultrafiltration
   e. heating to high temperature
   f. addition of ethanol
   g. affinity chromatography
   h. size-exclusion chromatography
2. Compare and contrast the principles and procedures in using SDS-PAGE versus IEF to separate proteins. Include in your explanation how and why it is possible to separate proteins by each method and what you can learn about the protein by running it on each type of system.
3. You are submitting a soy protein sample to a testing laboratory with an amino acid analyzer (ion-exchange chromatography) so that you can obtain the amino acid composition. Explain how (a) the sample will be treated initially and (b) the amino acids will be quantitated as they elute from the ion-exchange column. Describe the procedures. Note: You want to quantitate *all* the amino acids.
4. In amino acid analysis, a protein sample hydrolyzed to individual amino acids is applied to a cation-exchange column. The amino acids are eluted by gradually increasing the pH of the mobile phase.
   a. Describe the principles of ion-exchange chromatography.
   b. Differentiate anion- versus cation-exchangers.
   c. Explain why changing the pH allows different amino acids to elute from the column at different times.

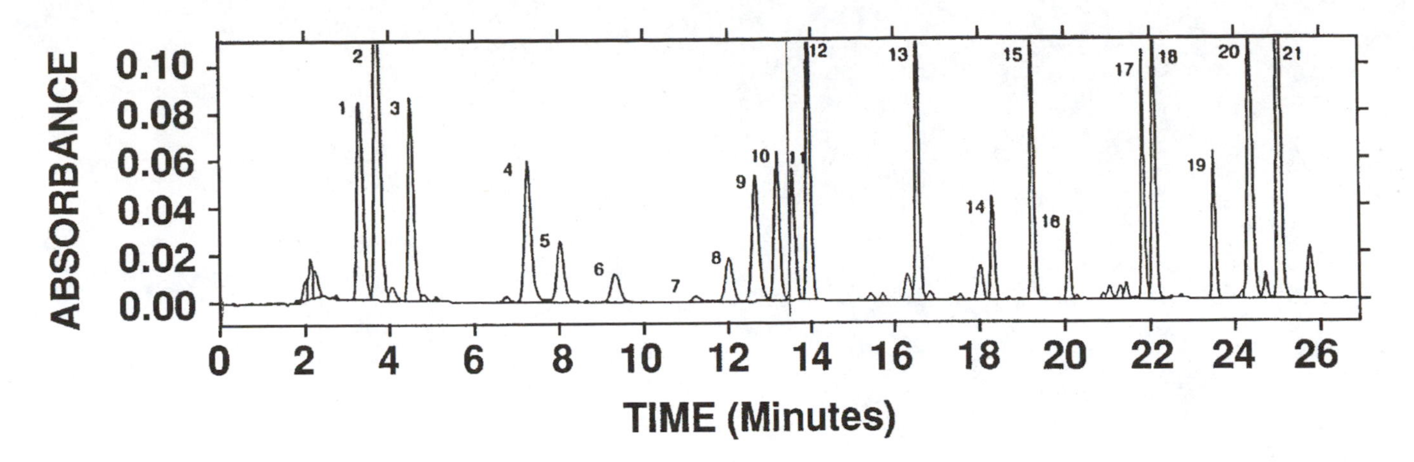

FIGURE 15-6.
High-performance liquid chromatographic analysis of phenylthiocarbamyl-derived amino acids from infant formula separated on a reversed-phase column. Sample was spiked with taurine (1. ASP; 2. GLU; 3. internal standard; 4. SER; 5. GLY; 6. HIS; 7. taurine; 8. ARG; 9. THR; 10. ALA; 11. NH_3; 12. PRO; 13. internal standard; 14. TYR; 15. VAL; 16. MET; 17. ILE; 18. LEU; 19. PHE; 20. reagent; 21. LYS). Adapted from (14).

15.6 REFERENCES

1. Bonnerjea, J., Oh, S., Hoare, M., and Dunnell, P. 1986. Protein purification: The right step at the right time. *Biotech.* 4:954–958.
2. Suelter, C.H. 1985. Purification of an enzyme. Ch. 3, in *A Practical Guide to Enzymology,* 63–132. John Wiley & Sons, Inc., New York, NY.
3. Scopes, R.K. 1970. Characterization and study of sarcoplasmic proteins. Ch. 22, in *Physiology and Biochemistry of Muscle as a Food,* Vol. 2. E.J. Briskey, R.G. Cassens, and B.B. Marsh (Eds.), 471–492. Univ. Wisconsin Press, Madison, WI.
4. Alani, S., Smith, D.M., and Markakis, P. 1989. α-Galactosidases of *Vigna unguiculata. Phytochem.* 28:2047–2051.
5. Kosikowski, R.V. 1986. Membrane separations in food processing. Ch. 9, in *Membrane Separations in Biotechnology,* W.C. McGregor (Ed.), 201–254. Marcel Dekker, Inc., New York, NY.
6. Weber, K., and Osborn, M. 1969. The reliability of molecular weight determinations by dodecyl sulfate-polyacrylamide gel electrophoresis. *J. Biol. Chem.* 244:4406–4412.
7. Davis, B.J. 1964. Disc electrophoresis. II. Method and application to human serum proteins. *Ann. N.Y. Acad. Sci.* 121:404–427.
8. Reisfeld, R.A., Lewis, U.J., and Williams, D.E. 1962. Disk electrophoresis of basic proteins and peptides on polyacrylamide gels. *Nature* 195:281–282.
9. Ornstein, L. 1964. Disk electrophoresis. I. Background and theory. *Ann. N.Y. Acad. Sci.* 121:321–349.
10. Laemmli, U.K. 1970. Cleavage of structural proteins during the assembly of the head of bacteriophage T4. *Nature* 227:680–685.
11. O'Farrall, P.H. 1974. High resolution of two-dimensional electrophoresis of proteins. *J. Biol. Chem.* 250:4007–4021.
12. Moore, S., and Stein, W.H. 1951. Chromatography of amino acids on sulfonated polystyrene resins. *J. Biol. Chem.* 192:663–681.
13. Moore, S., Spackman, D.H., and Stein, W.H. 1958. Chromatography of amino acids on sulfonated polystyrene resins: An improved system. *Anal. Chem.* 30:1185–1190.
14. Millipore Corp. 1990. Liquid chromatography analysis of amino acids in feeds and foods using a modification of the Pico-Tag method. Technical Bulletin, Millipore Corp., Milford, MA.
15. Scopes, R.K. 1987. *Protein Purification,* 2nd ed. Springer-Verlag, New York, NY.
16. Robyt, J.F., and White, B.J. 1987. *Biochemical Techniques, Theory and Practice,* Waveland Press, Prospect Heights, IL.
17. Cherry, J.P., and Barford, R.A. 1988. *Methods for Protein Analysis.* American Oil Chemists' Society, Champaign, IL.
18. Deutscher, M.P. (Ed.) 1980. *Guide to Protein Purification,* Methods in Enzymology, Vol. 182. Academic Press, New York, NY.

CHAPTER

16

Protein Quality Tests

Barbara A. Rasco

16.1 INTRODUCTION

Protein quality measurements are used to predict the nutritional quality of food proteins for growth and maintenance (1–3). Biological *(in vivo)* assays measure growth or N balance as indicators of protein utilization and metabolism. These tests are designed to reflect the essential amino acid content, bioavailability of the amino acids present in the protein, and protein digestibility of the food or food ingredient being tested. Instead of animal feeding studies, microbiological assays can also be used to estimate protein quality. Because of the large amount of time (30–50 days) and expense of biological assays, chemical or biochemical *(in vitro)* assays also have been developed to predict how a protein will meet nutritional and growth requirements. *In vitro* assays for protein quality include enzyme assays that model mammalian digestion for estimation of protein digestibility. Amino acid composition data is commonly compared to reference proteins, then corrected for digestibility (*in vivo* or *in vitro*) to obtain a protein quality estimate. Chemical methods to determine the availability of amino acids can also provide the basis for protein quality measurements when coupled with other tests.

16.2 GENERAL CONSIDERATIONS

16.2.1 Estimates of Protein Needs

Estimates of protein needs by humans have been established by national and international expert groups. Recommended intake levels have been set by the **Food and Agricultural Organization/World Health Organization** (FAO/WHO) and by the **Food and Nutrition Board of the National Academy of Sciences** (FNB/NAS). For example, in 1985 the FAO/WHO set its recommended intake level for the human adult at 0.75 g/kg body weight, or 52.5 g for a 70 kg adult, for protein with the digestibility of milk or eggs (4). Recommended levels for infants and children are higher and are progressively lowered during childhood until reaching the adult level.

The FAO/WHO and the **National Research Council** (NRC) have set estimated requirements for essential amino acids (see Table 16-1 for FAO/WHO values). The methods to measure protein quality are designed to determine how well a food or food ingredient meets these requirements for essential amino acids. This is determined by the amino acid content of the food and the digestibility of the protein, which in turn determine

TABLE 16-1. Comparison of Suggested Patterns of Amino Acid Requirements for Humans with that of the Rat and with the Composition of Casein

	SUGGESTED PATTERN OF REQUIREMENT[1]					
Amino Acid (mg/g crude protein)	*Infant Mean (range)*[2]	*Preschool Child (2–5 years)*[3]	*School Child (10–12 years)*	*Adult*	*Laboratory Rat*[4]	*Reported Composition Casein*[5]
Histidine	26 (18–36)	(19)[6]	(19)	16	25	32
Isoleucine	46 (41–53)	28	28	13	42	54
Leucine	93 (83–107)	66	44	19	62	95
Lysine	66 (53–76)	58	44	16	58	85
Methionine + cystine	42 (29–60)	25	22	17	50[7]	35
Phenylalanine + tyrosine	72 (68–118)	63	22	19	66	111
Threonine	43 (40–45)	34	28	9	42	42
Tryptophan	17 (16–17)	11	(9)	5	12.5	14
Valine	55 (44–77)	35	25	13	50	63
Total						
including histidine	460 (408–588)	339	241	127	407.5	531
minus histidine	434 (390–552)	320	222	111	382.5	499

Adapted from (11) with permission.
[1]Values for humans from FAO/WHO/UNU (4).
[2]Amino acid composition of human milk (5–7).
[3]Amino acid requirement/kg divided by safe level of reference protein/kg. For adults, safe taken as 0.75 g/kg; children (10–12 years), 0.99 g/kg; children (2–5 years), 1.10 g/kg. (This age range is chosen because it coincides with the age range of the subjects from whom the amino acid data were derived. The pattern of amino acid requirements of children between 1 and 2 years may be taken as intermediate between that of infants and preschool children.)
[4]National Research Council (9), based on a protein requirement of 12 percent plus an ideal protein (100% true digestibility and 100% biological value).
[5]Steinke et al., 1980 (8).
[6]Values in parentheses interpolated from smoothed curves of requirement versus age.
[7]A lower rat requirement of 40 mg/g protein for methionine and cystine has been reported (10).

the bioavailability of the amino acids contained within the food.

16.2.2 Regulatory Actions Related to Protein Quality Tests

The selection and approval of methods to assess protein quality are important from a regulatory perspective with regard to both public health needs and economic impact (11). Human clinical studies that measure growth and/or other metabolic indicators (e.g., N balance) provide the most accurate assessment of protein quality. However, use of such techniques is inappropriate and impractical to routinely measure protein quality, for reasons primarily of time and cost and sometimes of ethics. Therefore, animals are often used in protein quality evaluation tests.

One of the first widely used methods for assessing protein quality was the **protein efficiency ratio** (PER) test, developed in 1919 (described in section 16.3.1.1 below). This method, which measures the ability of a protein to support growth in young, rapidly growing rats, has been used to predict protein quality for humans. Until recently, the PER procedure with casein as the reference protein was the only method specified by the FDA for nutritional labeling purposes. However, it has been recognized for some time that the PER is an inadequate method for assessing the nutritional quality of food proteins in human nutrition (11–13). It overestimates the value of some animal proteins for human nutrition and underestimates the value for some vegetable proteins. The PER method is also critized because it does not account for the protein used for maintenance. From 1981 through 1989, the Codex Committee on Vegetable Proteins (CCVP) evaluated procedures for assessing the nutritional quality of vegetable proteins for human nutrition. This eventually led to a recommendation by the Joint FAO/WHO Expert Consultation on Protein Quality Evaluation that a **protein digestibility-corrected amino acid score** (PDCAAS) method (described in section 16.3.2.2 below) be adopted internationally as the official method for routine evaluation of protein quality for humans. The PDCAAS method requires an accurate measurement of amino acid composition and a careful assessment of protein digestibility.

In 1990, the FDA proposed to continue the use of the PER method to evaluate protein quality as part of the Nutritional Labeling and Education Act. However, arguments raised by Young and Pellett (12) and a citizen petition, both in support of the PDCAAS method, caused the FDA to reconsider their proposal. As a result, the FDA decided in 1991 to adopt the PDCAAS method in the new nutritional labeling regulations for all foods other than those intended for infants (13). Details of the regulations relating to protein quality evaluation methods in the Nutritional Labeling and Education Act are published in the *Federal Register* of January 6, 1993 (14).

16.3 METHODS

16.3.1 Growth and Nitrogen Balance Techniques

The quality of proteins has historically been based largely on rat nutrition studies that measure N balance or growth. Of the **growth methods,** the PER test has been most widely used. An improvement upon the PER method is the **net protein ratio** (NPR) method. For **N balance,** the **biological value** (BV) method and the **net protein utilization** (NPU) method are utilized, with NPU being a modification and improvement of the BV method.

16.3.1.1 Protein Efficiency Ratio

The PER is a biological assay approved by the Association of Official Analytical Chemists International (AOAC Method 960.48) (15) to estimate the protein quality of different foods or food ingredients. For this and other assays described below, details of the procedures can be found in the methods cited.

16.3.1.1.1 Principle The PER method is based upon the weight gain of a group of male weanling rats fed a test protein, compared to those fed a control diet that contains casein as the sole source of protein. The better the nutritional quality of the test protein, the more rapidly the animals will grow. The quality of the test protein is reported relative to the casein control.

Since PER is an *in vivo* test, protein digestibility and amino acid bioavailability are encompassed to some extent within the assay. However, it is difficult from a PER assay to determine the individual contribution of each of these factors toward the overall protein quality of any given food ingredient.

16.3.1.1.2 Procedure Two sets of male weanling rats from the same colony (21–28 days old) are fed 10 percent protein diets. There should be at least 10 animals per assay group. One set of animals is fed a reference casein (control) diet, and the other set, a diet containing the test protein. The test protein must contain a minimum of 1.80 percent nitrogen if it is to be incorporated into the test diet at the proper level. Diets are isocaloric and contain added carbohydrate in the form of corn starch, crude lipid as cottonseed oil, crude fiber as cellulose, and a balance of minerals and vitamins. To account for differences in the protein content of test materials, corn starch content of the diet is adjusted. Animals are housed in individual cages and are provided with water *ad libitum.*

The weight of each animal is recorded at the beginning of the assay, and body weight and food intake are measured at regular intervals during the course of the 28-day feeding trial. PER is calculated as the weight gain per g protein (%N × 6.25) fed. The quality of the test protein is reported as the ratio of the test protein PER (× 100) to the PER for reference casein. Casein is assigned a PER of 2.5, and results of the test protein are normalized to this value in an attempt to reduce the interlaboratory variation that has been observed in collaborative experiments. PER is calculated using the average weight gain and protein intake per each diet group (at 28 d):

$$\text{PER} = \frac{\text{weight gain (g)}}{\text{protein intake (g)}} \qquad (1)$$

Relative PER (RPER) gives a value for a test protein relative to a casein control run as part of the same experiment:

$$\text{RPER} = \frac{\text{PER of the test protein}}{\text{PER for casein}} \times 2.5 \qquad (2)$$

16.3.1.1.3 Applications PER can discriminate between proteins based upon their nutritional quality even though the test has a tendency to overestimate the protein quality of certain animal protein sources for the human diet and underestimate the value of some vegetable protein sources, as mentioned above. This underestimation problem is due to the relatively higher need of the rapidly growing weanling rat for certain dietary essential amino acids compared to humans. From a public health standpoint, underestimating PER is not necessarily detrimental. However, there is a tendency of the PER assay to overestimate the nutritional requirements for histidine, isoleucine, threonine, valine, and sulfur-containing amino acids (methionine and cystine). Casein is a less than ideal reference protein and is deficient by 15–30 percent in meeting the sulphur amino acid requirements of the rat (Table 16-1).

The major flaw with the PER method is that it is a growth assay and, as such, does not adequately account for the protein used for maintenance. A protein that does not support growth has a PER of zero, even though it may be suitable for meeting the protein requirements of adults. Problems such as this one have led to recommendations to replace the PER with other methods.

16.3.1.2 Net Protein Ratio

The **net protein ratio** (NPR) is a method to correct the PER value for maintenance. This procedure is often run in conjunction with a PER. One group of animals is fed the test protein and a control group is fed a basal diet that contains no protein. The average weight loss of each group of animals is recorded at 10 and 14 d. The NPR value calculated accounts for the protein requirement for maintenance and represents the weight gain for animals on the test diet plus the average weight loss of animals on the zero protein diet per g protein consumed.

$$\text{NPR} = \frac{\text{avg. wt. gain of test group (g)} + \text{avg. wt. loss of basal diet group (g)}}{\text{protein consumed by test group (g)}} \qquad (3)$$

16.3.1.3 Biological Value and Net Protein Utilization

Unlike PER and NPR, which are growth techniques, **biological value** (BV) and **net protein utilization** (NPU) are determined from N balance (i.e., N intake − N losses) assays. The BV for a test protein is the proportion of absorbed nitrogen that is retained for maintenance and/or growth, corrected for metabolic and endogenous losses of nitrogen (BV = N retained/N absorbed). The NPU is the proportion of nitrogen intake that is retained and is equivalent to BV × true digestibility. The NPU can be determined by comparing the carcass nitrogen content of a group of animals fed the test protein diet to that of animals fed a zero protein diet.

$$\text{NPU} = \frac{\text{N retained}}{\text{N intake}} = \text{BV} \times \text{true digestibility} \qquad (4)$$

$$\text{NPU} = \frac{\text{body N of test group} - \text{body N of zero protein group} + \text{N consumed by zero protein group}}{\text{N consumed by test group}} \qquad (5)$$

16.3.2 Amino Acid Scoring Patterns

16.3.2.1 Overview

Several protein quality testing methods utilize amino acid content data. Amino acid content of a test protein is compared to that of a reference protein. The estimate of protein quality is based upon either the first limiting amino acid or all essential amino acids. This estimate may be corrected for digestibility as determined by an *in vivo* or *in vitro* assay. Some of the methods that utilize amino acid content data are suitable as quality control tests. When the methods include a correction for digestibility, they may be more appropriate than the use of the PER assay for estimating protein quality.

Tests that utilize amino acid content data compare the amino acid composition of the test protein with that of egg or milk protein, human milk, or a reference pattern that is based on human amino acid requirements. Because the requirements of amino acids for growth and maintenance vary as people age, different reference patterns are used when estimating the nutritional quality of a given protein for infants, older children, and adults (Table 16-1). For the amino acid score method described below (section 16.3.2.2), the reference pattern for preschool age children is recommended for evaluating protein quality for all groups except infants, even though this may overestimate protein requirements and underestimate protein quality for adults and older children (11). For infants, the recommended scoring pattern is the amino acid composition of human milk.

The protein quality evaluation methods based upon amino acid data require the accurate analysis of a food for amino acid composition. This is generally determined by hydrolyzing the protein to its constituent amino acids and separating the amino acids chromatographically (see Chapter 15). Separate assays are required for cystine, methionine, and tryptophan. The FAO/WHO has reviewed the methods for amino acid analysis and has made recommendations regarding procedures and further studies needed to improve analytical methods (11).

16.3.2.2 Amino Acid Score

16.3.2.2.1 Procedures An amino acid score is calculated for each essential amino acid by dividing the amount of each essential amino acid in a given test protein by the amount in the reference protein. The lowest ratio indicates which amino acid is first limiting, and that is the **amino acid score** for the test protein. The amino acid score is expressed either as a ratio compared to the standard, or as a percentage.

$$\text{amino acid score} = \frac{\text{mg of amino acid in 1 g of test protein}}{\text{mg of amino acid in 1 g reference protein}} \quad (6)$$

In eq. 6, the amino acid in the test protein is the first limiting amino acid, compared to the amino acid value in the reference pattern based on human amino acid requirements. The amino acid score is in practice equivalent to **chemical score** and **protein score** methods. However, chemical score was originally defined relative to the amino acid composition of egg protein.

The amino acid score combined with a determination of protein digestibility (see section 16.3.3.1) is the basis of the **protein digestibility-corrected amino acid score** (PDCAAS) method.

$$\text{PDCAAS} = \text{amino acid score} \times \%\ \text{true digestibility} \quad (7)$$

16.3.2.2.2 Applications Unless the amino acid score is corrected for digestibility, it may not be a good reflection of protein quality. An amino acid score may not be valid for some mixtures of food proteins, even though a value could be readily calculated knowing the amino acid composition of each protein in the composite and the relative amounts of each. Likewise, a calculated value of protein quality for various foods may not provide a good indication of the overall protein quality of a diet containing a wide variety of foods. This is because the body's utilization of dietary protein is affected by a number of different factors that are not reflected in an amino acid score.

There is a general consensus that the PDCAAS method, which uses amino acid score corrected for protein digestibility, can provide a better estimate of protein quality for humans than the PER rat growth assay. The PDCAAS method is recommended by the FAO/WHO for measuring protein quality (11) and has been adopted by the FDA in the new nutritional labeling regulations for measuring the protein quality of all foods other than those intended for infants (14).

16.3.2.3 Calculated PER and Discriminate Calculated PER

16.3.2.3.1 Procedures Amino acid composition data for all essential amino acids in a food are compared to the FAO/WHO standard for use in calculating a **calculated PER** (C-PER) (AOAC Method 982.30). The C-PER is a PER calculated from the amino acid composition of the test protein and *in vitro* protein digestibility (see section 16.3.3.2). A related method for estimating protein quality, the **discriminate calculated PER** (DC-PER) (AOAC Method 982.30I), is calculated only from the essential amino acid composition of the food, compared to the FAO/WHO standard. The calculations for both C-PER and DC-PER involve complicated algorithms that are given in the AOAC procedures.

16.3.2.3.2 Applications Unlike the amino acid score method, C-PER and DC-PER consider the content of all essential amino acids. Such consideration of all dietary essential amino acids may be useful, especially for foods with more than one essential amino acid present in relatively low amounts.

The C-PER and DC-PER methods are intended as

alternative methods for routine quality control screening of foods. Used together, they provide a reliable estimate of protein quality for the majority of foods and food ingredients. Most of the concerns about the C-PER method relate to the *in vitro* digestibility portion of the assay, as later described in section 16.3.3.2.2.

16.3.2.4 Essential Amino Acid Index

16.3.2.4.1 Procedure The **essential amino acid index** is calculated by using the ratio of test protein to the reference protein for each of the eight essential amino acid plus histidine in the equation that follows:

$$\text{essential amino acid index} = \sqrt[9]{\frac{\text{mg Lysine in 1 g test protein}}{\text{mg Lysine in 1 g reference protein}} \times \begin{array}{c}\text{etc. for all}\\ \text{8 essential}\\ \text{amino acids}\\ \text{+ His}\end{array}} \qquad (8)$$

Methionine and cystine are counted as a single amino acid value in this equation, as are phenylalanine and tyrosine (see example in practice problem 3 in section 16.6).

16.3.2.4.2 Applications The essential amino acid index method can be useful as a rapid tool to evaluate food formulations for protein quality. Like the C-PER and DC-PER methods, it is calculated using the content of all dietary essential amino acids. However, unlike the PDCAAS and C-PER methods, it does not include an estimate of protein digestibility. Therefore, it would not account for differences in protein quality due to various processing methods or certain chemical reactions, as further described in the next section.

16.3.3 Protein Digestibility Assays

All proteins are digested, absorbed, and utilized to different extents. Differences in protein digestibility arise from the susceptibility of a protein to enzymatic hydrolysis in the digestive system. This is directly related to the primary, secondary, and tertiary structure of the protein. The presence of nonprotein dietary constitutents consumed at the same time as the protein can also affect protein digestion. Some of these components include phytate, dietary fiber, and various toxigenic agents. How a protein has been treated is also important because processing or storage conditions can alter the three-dimensional structure of the protein, either improving or lessening its susceptibility to digestive enzymes. In addition, chemical reactions involving amino acids such as Maillard browning can significantly reduce the biological availability of dietary essential amino acids, particularly lysine.

16.3.3.1 *In Vivo* Assays

Digestibility of a protein is the proportion of protein nitrogen that is absorbed. Common *in vivo* digestibility assays that provide the best indication of protein digestibility in humans measure nitrogen balance in animals.

16.3.3.1.1 Procedure (AOAC Method 991.29) Male weanling rats (50–70 g) are initially fed a protein-free diet for 4 days prior to the study to obtain an estimate of metabolic fecal nitrogen. The test diet (10% protein) is then fed at a rate of 15 g (dry matter)/day for 5 days. Food intake is monitored. Feces are collected daily and analyzed for nitrogen by the Kjeldahl method (AOAC Method 955.04C, 976.05; see also Chapter 14). Nitrogen analysis by the Kjeldahl method is also needed for diet formulation, as are analyses for moisture (AOAC Method 927.05 or 934.01), fat (AOAC Method 920.38A, 983.23, or an equivalent method), and dietary fiber (AOAC Method 985.29).

The **true digestibility** is calculated based on the amount of nitrogen ingested and the feed intake and is corrected for metabolic losses in the feces.

$$\text{true digestibility (\%)} = \frac{\text{Ni} - (\text{Fn} - \text{Mn})}{\text{Ni}} \times 100 \qquad (9)$$

where: Ni = N intake
Fn = Fecal N
Mn = Fecal metabolic N loss

If the correction for metabolic losses in the feces is not made, the value is termed **apparent digestibility:**

$$\text{apparent digestibility (\%)} = \frac{\text{Ni} - \text{Fn}}{\text{Ni}} \times 100 \qquad (10)$$

16.3.3.1.2 Applications Protein digestibility data from rat studies must be used with caution when estimating protein quality for humans. When at all possible, protein quality evaluation should be obtained from nitrogen balance studies with humans. Obviously, this is not always possible due to safety, ethical, or monetary restraints. Fortunately, where there are comparative data available from nitrogen balance studies for the same food proteins, the results from rat and human studies are similar (11).

16.3.3.2 In Vitro Assays

16.3.3.2.1 Overview Various *in vitro* enzymatic hydrolysis methods have been proposed to evaluate the digestibility and availability of proteins, usually by a one- or two-step process, using mammalian gastric and/or pancreatic and intestinal enzymes:

- pepsin
- pepsin-pancreatin
- papain
- papain-trypsin
- trypsin
- trypsin-chymotrypsin-peptidase
- trypsin-chymotrypsin-peptidase-bacterial protease

The *in vitro* digestibility assays can be generally classified as methods that measure either the extent of hydrolysis or the initial rate of hydroysis. Further subclassifications can be based upon the enzymes used, the method of digest fractionation (if used), and the method of analysis. In the *in vitro* digestibility assays, the constituents, pH, and temperature of the incubation medium are generally fixed according to requirements of the enzyme reaction. The enzyme to substrate ratio often varies among various *in vitro* digestion methods. This can influence the reaction rate, the size of the peptides produced, and the enzymatic release of different amino acids. The digestibility and availability of protein constituents may be determined directly using the enzymatic hydrolyzate or after the hydrolyzate has been further treated. Since the accumulation of digestion products during proteolysis may inhibit the enzyme reaction, some methods include procedures to remove these digestion products.

Values from *in vitro* digestibility assays generally do not take into account the fermentation of food proteins in the lower bowel or the amino acid balance of the protein being tested. *In vivo* tests, such as rat nitrogen balance or amino acid balance methods, generally provide a better assessment of protein digestibility for processed or complex food mixtures than an *in vitro* assay because protein may be altered in susceptibility to enzymatic hydrolysis as a result of processing.

16.3.3.2.2 pH-Shift Method An *in vitro* digestibility assay is conducted as part of the C-PER assay as described above (section 16.3.2.3). The digestibility measurement is suitable for correcting the amino acid composition data for protein digestibility. The degree of digestibility by enzymes is calculated relative to casein and is based upon a drop in pH resulting from protein hydrolysis. As the proteolytic enzymes attack and break the peptide bonds in the protein, the freed carboxyl groups liberate a H^+ ion, which causes a decrease in pH. This is the reason such methods are sometimes call **pH-shift** or **pH-drop** procedures.

The AOAC Method 982.30 for *in vitro* digestibility is based on studies by Hsu et al. (16) and Satterlee et al. (17) and is summarized in Fig. 16-1. The percent digestibility and essential amino acid values (see section 16.3.2.3) are used to calculate C-PER. The digestibility portion of the C-PER assay is rapid and is unaffected by food lipids and buffering salts commonly found in foods. It is sensitive enough to detect the presence of soybean trypsin inhibitors and to detect changes in protein digestibility that may occur during processing. However, there are problems with the C-PER method for certain foods (e.g., connective tissue protein, low protein foodstuffs). The C-PER and DC-PER values may not be similar and may not correlate well with biological methods for assessing protein quality for foods that contain large amounts of cell wall materials (e.g., yeast, bran), partially digested protein, or protease inhibitors.

Although *in vitro* digestibility values for pH-shift methods generally correlate well with data for *in vivo* digestibility assays when ranking protein sources, they do not accurately estimate quantitative differences between samples with low and high protein digestibility. The major limitation of the pH shift methods is that pH is not constant during the course of the reaction. Buffering capacity of peptides, proteins, and other substances in the food may influence how the pH changes during this type of assay.

16.3.3.2.3 pH-Stat Methods To overcome the problem that enzyme hydrolysis of the test protein in the pH-shift assay is not conducted at a constant pH, Pedersen and Eggum (19) developed an enzyme hydrolysis

Put specified quantity of protein (test protein or casein) (based on Kjeldahl N) in solution at 37°C and adjust to pH 8.0.

⇓

Add freshly prepared trypsin, chymotrypsin, and peptidase solutions (type, activity and amounts of these enzymes are specified, as are preparation procedures).

⇓

After 10 min., add bacterial protease (specified amount and activity).

⇓

Transfer to 55°C water bath.

⇓

Transfer back to 37°C bath at 19 minutes.

⇓

Read pH at 20 minutes.

⇓

Calculate digestibility as:

$$\% \text{ Digestibility} = 234.84 - 22.56(x)$$

x = pH at 20 minutes

FIGURE 16-1.
Procedure for *in vitro* digestibility assay for C-PER assay (AOAC Method 982.30).

method in which the pH is kept constant during the incubation period (Fig. 16-2). This is the **pH-stat** method. The pH-stat assay is similar to that of Hsu et al. (16) and Satterlee et al. (17) (summarized in Fig. 16-1) in that the same enzymes are used for protein hydrolysis. Protein digestibility in the pH-stat method is estimated from the volume of standard alkali (0.1 N NaOH) used to maintain a constant pH of 8.0 during the hydrolysis period. The pH-stat assay is generally more accurate than a pH-shift assay in estimating protein digestibility and gives better correlation with *in vivo* digestibility values. Pedersen and Eggum (19) used a single equation to calculate *in vitro* digestibility by the pH-stat method for 31 plant and animal proteins and obtained a correlation coefficient of >0.90 compared to *in vivo* digestibilities.

Dimes et al. (20) have used a pH-stat procedure to measure the degree of protein hydrolysis (%DH) as enzymic digestion proceeds. Protein hydrolysis is calculated from hydrolysis equivalent (h) based upon the volume of standard alkali required to maintain the pH of the reaction mixture at pH 8.0. Correlation coefficients of this pH-stat method compared with *in vitro* methods for protein digestibility are approximately 0.86. Comparable results for pH-shift assays are lower at 0.71 to 0.78.

$$DH\ (\%) = (h/h_{tot}) \times 100 \quad (11)$$

$$h = (B \times 1/a \times N_b)/M \times (S\%/100) \quad (12)$$

where:
- h = hydrolysis equivalent
- $a = 10^{pH - pK/1} + 10^{pH - pK}$
- N_b = normality of titrant
- M = mass (g) or reaction mixture
- S = protein concentration in reaction mixture
- B = volume of titrant (ml)
- h_{tot} = the total number of peptide bonds in the protein

Put specified quantity of protein (test protein or casein) (based on Kjeldahl N) in solution at 37°C and adjust to pH 8.0.

⇓

Add suspension of trypsin, chymotrypsin and peptidase (type, activity and amounts of these enzymes are specified, as is the preparation procedure).

⇓

Record the amount of 0.1 N NaOH required to maintain pH at 7.98 for exactly 10 min.

⇓

Calculate true digestibility as:
% True Digestibility = 76.14 + 47.77B
B = ml of 0.1 N NaOH added

FIGURE 16-2.
Procedure for pH-stat method for *in vitro* digestibility. Adapted from (19).

For a well defined protein, the h_{tot} is calculated from the mean average molecular weight of amino acid residue. For ill-defined systems, such as mixed food systems, the average amino acid molecular residue weight is assumed to be 120.

16.3.3.2.4 Immobilized Digestive Enzyme Assay (IDEA) In a recently developed *in vitro* digestibility assay, digestive tract enzymes are covalently immobilized on large-pore diameter (2000 A) glass beads via an amide linkage (21). The large-pore diameter beads allow access of the protein substrate to the enzymes. The enzymes immobilized using this technique have been shown to have structural and kinetic behavior quite similar to that of the soluble enzymes. The method uses a biodigester containing immobilized pepsin and a second biodigester containing immobilized trypsin, chymotrypsin, and intestinal mucosal peptidases. The procedure is done as summarized in Fig. 16-3.

Advantages of using immobilized forms of the enzymes in the protein digestibility assay include:

1. Autolysis of the enzyme is prevented.
2. Structural stability of enzyme can be increased.
3. Digest is not contaminated with the enzyme or its autolysis products.
4. Digest is readily separated from the enzymes.
5. A number of assays can be done with the same immobilized enzyme biodigester.

The IDEA has been shown to correlate well with rat bioassays for a wide variety of food proteins. It can apparently be used to accurately assess the reduced digestibility of proteins modified by cross-linking, racemization, or the Maillard reaction. Different correla-

Sample is dissolved in 0.01 M HCl, adjusted to pH 2, at 1–3 mg protein/ml.

⇓

Digested in pepsin biodigester (digester #1), 18–20 hr., 37°C.

⇓

Pepsin digest is adjusted to pH 7.5 with Na_2HPO_4.

⇓

Digested in intestinal enzyme biodigester (digester #2, with trypsin, chymotrypsin, and intestinal mucosal peptidases), 20–24 hr., 37°C.

⇓

Digest is reacted with o-phthalaldehyde (OPA) in the presence of β-mercaptoethanol to measure primary amines.

⇓

Read Abs. at 340 nm.

⇓

Digestibility is defined as the fraction of total peptide bonds that are hydrolyzed.

FIGURE 16-3.
Procedure for immobilized digestive enzyme assay. Adapted from (18, 21).

tion equations are not needed for plant versus animal protein.

16.3.4 Amino Acid Availability

16.3.4.1 Overview

Amino acid availability methods for proteins measure the relative digestibilities of the individual amino acids. Amino acids in proteins may be digested and absorbed at different rates, for various reasons, to affect utilization of the protein. For example, the rate of amino acids uptake from a protein mixture compared with a protein supplemented with free amino acids differs even if the amino acid composition is the same for both. Free amino acids are absorbed more quickly than amino acids in protein.

Amino acid scoring pattern methods for proteins are based on the assumption that there is a direct linear relationship between the concentration of a limiting amino acid and utilization of the limiting amino acid in a protein. A second assumption is that the amino acid balance in a protein has no effect on the utilization of dietary essential amino acids, particularly the limiting amino acid. Additionally, the amino acid balance plays an important role in the overall quality of food proteins, but this is not generally reflected in the amino acid scoring pattern assay corrected for digestibility, particularly if an *in vitro* method for protein digestibility was used.

Because an amino acid analysis is conducted on a protein that has been acid hydrolyzed, it often does not provide a good indication of the protein bioavailability. A number of factors can affect how a protein is digested. For example, if the secondary or tertiary structure of a protein has been altered during heating, or by some other processing treatment, proteolytic enzymes that usually hydrolyze it may not be effective because the enzyme may no longer recognize its active site.

Also, the amino acid composition does not provide an indication of how well an amino acid will be utilized, so other assays are needed to determine availability of individual essential amino acids. For example, foods that undergo Maillard browning are particularly susceptible to loss of lysine. Sulfur-containing amino acids (methionine and cystine) can also be lost during processing. Severe heat treatment can damage protein to such a degree that digestibility is reduced and bioavailability of all dietary essential amino acids is affected.

16.3.4.2 In Vivo Amino Acid Availability

An *in vivo* assay for amino acid availability referred to as **amino acid balance** is similar to that for apparent digestibility. However, instead of simply measuring the nitrogen content of the diet and the feces, the amino acid profile of each is determined. The amino acid balance can be calculated for all amino acids, but generally it is restricted to the first, or first and second, limiting amino acids:

$$\begin{array}{c}\text{amino}\\ \text{acid}\\ \text{balance}\end{array} = \begin{array}{c}\text{amino}\\ \text{acid}\\ \text{intake (g)}\end{array} - \begin{array}{c}\text{amino acid}\\ \text{excreted}\\ \text{(fecal content) (g)}\end{array} \qquad (12)$$

In vivo amino acid digestibility can overestimate protein quality because a significant fraction of certain limiting dietary essential amino acids (lysine, methionine, cystine, threonine, and tryptophan) are lost through microbial fermentation in the large intestine.

16.3.4.3 Microbiological Assay for Amino Acid Availability

Microbiological assays using the bacteria *Streptococcus zymogenes* or *Pediococcus cerevisia* or the protozoan *Tetrahymena pyriformis W* can also be used to measure amino acid availability. The protein is treated with an enzyme preparation (e.g., papain) prior to microbial incubation to reduce the time needed for the assay. The organism is incubated in media containing the hydrolyzed food protein at various concentrations and allowed to grow and multiply for several days before the analyst counts the number of microorganisms. Growth can also be determined using a Coulter counter.

S. zymogenes can be used to assay available arginine, histidine, leucine, isoleucine, valine, methionine, and tryptophan; the organism does not require lysine and cannot be used to assay for this amino acid. Microbiological assays for lysine often use the organism *P. cerevisiae*. The protozoan *T. pyriformis* assay can be used to assay for 10 amino acids: arginine (required for the rat), histidine, isoleucine, leucine, lysine, methionine + cystine, phenylalanine + tyrosine, theonine, tryptophan, or valine. Data from the *T. pyriformis* correlate well with those from rat bioassays. However, several common food additives including propionates, benzoates, sorbates, nitrate, erythorbate, ascorbate, and certain spices interfere with the protozoan assay.

16.3.4.4 In Vitro Amino Acid Availability

Amino acid availability can be measured by *in vitro* enzymatic digestion utilizing assay systems that mimic the mammalian digestion. Similar to *in vitro* tests for protein digestibility, these assays are useful for ranking proteins, particularly in studies in which the effect of processing treatments on a given type of food protein is of interest. One such test measures the amino acid composition of a filtrate recovered after a test protein has been hyrolyzed with a mixture of trypsin, pepsin, and pancreatin.

16.3.5 Available Lysine

16.3.5.1 Overview

The free amino group on the side chain of lysine can react chemically with many constituents during food processing and storage to give biologically unavailable lysine complexes. These lysine complexes can result from reactions with reducing sugars (producing Maillard products), oxidized polyphenols (such as caffeic acid present in oilseeds), oxidized lipids, glutaminyl and asparaginyl residues (during severe toasting), and alkaline solutions (to destroy or racemize lysine or to produce lysinoalanine). Utilizable lysine is not equal to lysine content. Essentially, the only source of utilizable lysine in a food is the lysine residue with its ϵ-amino group free (referred to as **reactive lysine**).

Reactive lysine can be measured directly by using reagents such as 1-fluoro-2, 4-dinitrobenzene (DNFB), trinitrobenzenesulfonic acid (TNBS), ω-methylisourea, or *o*-phthalaldehyde. It can be measured indirectly by the DNFB difference method, dye-binding procedure, furosine method, or reduction by $NaBH_4$. Several microbiological assays have been developed for available lysine, as have *in vitro* enzymatic hydrolysis methods. *In vitro* digestibility assays assume that protein digestibility gives a good approximation of the digestibility of each amino acid, including reactive lysine. Enzymatic hydrolysis with pepsin plus pancreatin or with pronase makes use of the fact that these proteolytic enzymes release only reactive lysine.

16.3.5.2 Assays with 1-Fluoro-2, 4-Dinitrobenzene (DNFB)

Available lysine is measured by difference in the AOAC procedure (Method 975.44) using 1-fluoro-2, 4-dinitrobenzene (DNFB) (also referred to as FDNB). Protein is reacted with DNFB which reacts with free ϵ-amino groups in lysine. An amino acid profile is run to obtain the amount of unreacted (unavailable) lysine. Total lysine of the protein is determined from an amino acid profile conducted on a suspension of the same protein that had not been treated with DNFB. The percent available lysine is the percent lysine in the untreated sample minus that in the DNFB treated sample.

Spectrophotometric assays for available lysine are also used and involve the reaction of DNFB with protein, hydrolysis of the sample with acid, and comparison of the amount of DNFB-reactive lysine (g/16 g nitrogen) with a standard, mono-ϵ-N-dinitrophenyllysine hydrochloride monohydrate (DNP-lysine). The DNFB-reactive lysine generally provides a good indication of bioavailable lysine in oilseeds, milk powder, and fish flour. The method is less suitable for proteins that have undergone partial hydrolysis (hydrolyzed vegetable and meat proteins and certain fish meals) or protein foods that contain high concentrations of reducing sugars, such as cereals. Sugars released during acid hydrolysis can reduce up to 30 percent of the DNP-lysine derivatives so that they are no longer measurable. Adding excess DNFB to the reaction mixture can have a protective effect when these foods are assayed.

16.3.5.3 Assay with Trinitrobenzenesulphonic Acid (TNBS)

A water-soluble reagent, trinitrobenzenesulphonic acid (TNBS) can be used in place of DNFB for free lysine measurement. However, TNBS-lysine derivatives are more susceptible to loss during acid hydrolysis than DNFB derivatives. Like DNFB, TNBS reacts with lysine derivatives formed early in Maillard browning which may or may not be bioavailable. The TNB derivatives formed during Maillard browning will break down to yield labeled lysine complexes, whereas the DNP derivatives will not.

16.3.5.4 Enzymatic Methods

Enzymatic methods for bioavailable lysine have been particularly useful for carbohydrate-containing foods. Lysine decarboxylase has a high degree of specificity for L-lysine, yielding carbon dioxide and the biogenic amine, cadaverine, either of which can be easily measured by gas chromatography or other methods. Unfortunately, there is little comparative data for this method with bioassay procedures for amino acid availability. A comparison of results from different assays for available lysine is given in Table 16-2.

16.3.5.5 Dye-Binding Methods

The binding capacity of **azo dyes** such as Orange 12 (Acrilane Orange G, 1-phenylazo-2-naphthol-6-sulphonic acid) to the free basic amino groups of lysine, histidine, or arginine often correlates well with bioassays. These assays are relatively rapid and are particularly useful for monitoring heat damage to oilseed and cereal proteins. Results are less reliable for fish and meat proteins. Azo dyes can bind to the basic reaction products formed early in Maillard browning and because of this, cannot be used to detect heat damage to

TABLE 16-2. Lysine Levels (mg/g N) in Four Samples of Milk Powder

Sample of Milk Powder	*Total Value after Acid Hydrolysis*[1]	*Lysine Reacting with DNFB*[2]	*Lysine Released by in vitro Enzymzatic Digestion*	*Value Obtained by Growth Assay with the Rat*
Good quality	500	513	519	506
Slightly damaged	475	400	388	381
Scorched	425	238	281	250
Severely scorched	380	119	144	125

From (1), used with permission.
[1]By amino acid analysis
[2]DNFB = 1-fluoro-2, 4-dinitrobenzene

milk proteins. For dried milk and other susceptible products, a dye-binding assay utilizing Remazol Brilliant Blue R can be helpful in detecting the initial products of Maillard browning. This dye reacts with the free amino acid group of lysine and also the thiol group in cystine.

16.3.6 Availability of Sulfur-Containing Amino Acids

The sulfur-containing amino acids, methionine and cysteine/cystine, are often the limiting amino acids in foods. Because these amino acids can be readily oxidized to forms that are no longer bioavailable during drying, bleaching, and other processing operations, it is important to have suitable assay methods for nutritionally available forms. Available cysteine/cystine can be measured by converting cystine to cysteine with dithiothreitol, reacting cysteine with 5,5′-dithiobis-2-nitrobenzoic acid (DTNB) and measuring the quantity of the derivatized form. Methionine can reduce dimethyl sulphoxide (Me_2SO) to dimethyl sulphide (Me_2S), which can be quantified by headspace gas chromatography. Results from this method correlate well with biological assays for methionine. Methionine can also be reacted with cyanogen bromide (CNBr) and the reaction product methylthiocyanate (MeSCN) measured by gas chromatography as an indication of available methionine. The oxidized forms of the amino acid, methionine sulphone and methionine sulphoxide, are not measured by either of these methods.

16.4 SUMMARY

Biological assays, principally rat feeding studies, have been commonly used to assess protein quality, by measuring either nitrogen balance (BV and NPU methods) or growth (PER or NPR methods). The PER method involves feeding a set of male weanling rats a test diet that contains a single source of dietary protein at a fixed level for a set period, monitoring food intake and weight gain. The PER procedure with casein as the reference protein was until recently the only protein quality testing method specified by the FDA for nutritional labeling purposes. However, problems associated with the PER method and improvements in the estimates for human essential amino acid requirements have led to adoption of the PDCAAS method by the FDA for most nutritional labeling purposes. This involves calculation of the amino acid score and the determination of protein digestibility. Other methods also utilize amino acid composition data compared to a reference protein, with (C-PER) or without (DC-PER, essential amino acid index) correction for digestibility. While the PDCAAS method uses an *in vivo* (rats) determination of digestibility, the C-PER procedure uses an *in vitro* assay that measures the drop in pH due to enzymatic digestion. Improvements in this *in vitro* assay for digestibility have been made in the pH-stat method and the immobilized digestive enzyme assay. Like the determination of protein digestibility, the bioavailability of amino acids can be tested by *in vivo* and *in vitro* methods and by a microbiological assay. Chemical or microbiological tests can be used to predict the availability of limiting amino acids, particularly lysine and the sulfur-containing amino acids. These assays for bioavailability of particular amino acids are helpful when evaluating the effect of various processing treatments on the limiting amino acids in a given food, without having to resort to more extensive testing.

16.5 STUDY QUESTIONS

1. You want to provide the amount of protein per serving, expressed as a percent of the Daily Value, on a nutritional label for a new food product (not an infant food).
   a. What method must be used to estimate the protein quality?
   b. List the procedures that must be performed and calculations required to obtain this estimate of protein quality.
   c. What method has been *previously* the official method required by the FDA for protein quality measurement for nutritional labeling of all foods?

d. Give two reasons why the method you identified in part (1a) is better than the method you identified in part (1c) (for all foods except infant foods).

2. Describe the basic procedure used for a PER study.
3. Define and briefly describe the differences between the following assay procedures used in assessing protein quality:
   a. PER versus RPER
   b. PER versus NPR
   c. PER versus BV
   d. BV versus NPU
   e. amino acid score versus essential amino acid index
   f. PER versus C-PER
   g. C-PER versus DC-PER
   h. PDCAAS versus amino acid score
   i. true digestibility versus apparent digestibility
   j. pH-shift versus pH-stat method for *in vitro* digestibility
4. How are certain microorganisms used to measure amino acid availability?
5. Explain how *in vitro* assays can be used to assess protein digestibility and amino acid availability. What are the advantages and disadvantages of an *in vitro* assay?
6. You are helping to develop a new process for making high-protein snack foods from cereal grains and soy. You want to determine the protein quality of snack foods generated under various processing conditions. Considering the number of samples to be tested, you cannot afford an expensive *in vivo* assay, and you cannot wait more than a few days to get the answer.
   a. Briefly describe the procedure you would use to estimate and compare the protein nutritional quality of snack foods made under different processing conditions. Include an explanation of the principles involved.
   b. You suspect that the time-temperature combinations lead to overprocessed products. Your testing from (6a) shows that these samples have a lower nutritional quality. What amino acid(s) in the snack food would you suspect to be most adversely affected by thermal abuse from overprocessing? What test(s) could you use to confirm that these amino acid(s) have become nutritionally unavailable by the overprocessing? How do these test(s) work?

16.6 PRACTICE PROBLEMS

1. PER, RPER, NPR, and apparent digestibility.
   Based on the information below, calculate the (a) PER, (b) RPER, (c) NPR, and (d) apparent digestibility for casein, ingredient X (Ing. X), and ingredient X with supplemental amino acids (AA).

Weight Gain and Food Intake for Protein Ingredient X and Casein

	NONCUMULATIVE WEIGHT GAIN (g)			
	wk1	*wk2*	*wk3*	*wk4*
Casein	31.2	30.0	31.2	27.
Ing. X + AA	26.1	28.7	24.3	30.6
Ing. X	3.0	4.8	6.4	4.9

	FOOD INTAKE (g)			
	wk1	*wk2*	*wk3*	*wk4*
Casein	90.3	134.5	112.1	147.9
Ing. X + AA	91.3	131.6	101.0	129.6
Ing. X	80.7	74.2	52.3	64.6

Casein was added at 12 percent of diet and ingredient X or ingredient X with supplemental amino acids at 24.7 percent of diet by weight. The protein concentration of casein (%N × 6.25) is 83.3 percent and of ingredient X is 40.4 percent protein.

To determine NPR, a group of rats was fed a basal diet (no protein) in conjunction with the study described above. By day 14, these animals had lost an average of 14.8 g each. The following data were collected in a study to determine apparent digestibility, with each diet containing 9.1 percent protein.

	Food Intake (g)	*Fecal Nitrogen (g)*
casein	260	0.27
Ing. X + AA	231	0.40
Ing. X	117	0.27

2. Protein digestibility.
   Calculate the *in vitro* digestibility for the following proteins using the pH-shift method (AOAC Method 982.30; see section 16.3.3.2.2). For soy, final pH was 6.7; for whey, final pH was 6.3.
3. Essential amino acid index, amino acid score, and PDCAAS.
   Using the data provided in the table below, (a) calculate the essential amino acid index for defatted soy flour, (b) determine the amino acid score for the soy flour, and (c) calculate the PDCAAS, using the true digestibility value of 87 percent for defatted soy flour (14).

Amino Acid	*Soy*[1]	*Reference Pattern*[2]
	(mg/g protein)	
isoleucine	46	28
leucine	78	66
lysine	64	58
methionine/cystine	26	25
phenylalanine/tyrosine	88	63
threonine	39	34
tryptophan	14	11
valine	46	35
histidine	26	19

[1]From (22).
[2]From (4).

Answers

1a. PER preliminary calculations:
Total weight casein group = 120 g
Total food intake casein group = 485 g

Amount protein consumed by casein group = 485 g × 12/100 × 83.3/100 = 48.5g

Ing. X + AA:
wt gain = 110 g
food intake = 453 g
protein consumed = 45.2 g

Ing. X
wt gain = 19 g
food intake = 272 g
protein consumed = 27.1 g

PER calculations:

PER = wt gain/protein intake

Casein: 120/48.5 = 2.47
Ing. X + AA: 110/45.2 = 2.43
Ing. X 19/27.1 = 0.70

1b. Calculate the RPER for Ing. X and Ing. X + AA:

RPER = (PER of ingredients/PER of casein) × 2.5

Ing. X + AA: (2.43/2.47) × 2.5 = 2.45
Ing. X: (0.70/2.47) × 2.5 = 0.71

1c. $NPR = \frac{\text{wt gain} - \text{wt loss (g) (0 protein)}}{\text{protein intake (g)}}$

Casein: (120 − 14.8)/48.5 = 2.2
Ing. X + AA: (110 − 14.8)/45.2 =2.1
Ing. X: (19 − 14.8)/27.1 = 0.15

1d.

$$\text{Apparent digestibility} = \frac{\text{g nitrogen ingested} - \text{g nitrogen in feces}}{\text{g nitrogen ingested}} \times 100$$

Percent nitrogen in diet, e.g., for casein: 260 g diet × (0.091/6.25) = 3.79 g nitrogen

Apparent digestibility for casein: (3.79 − 0.27)/3.79 × 100 = 93%
for Ing. X + AA: (3.36 − 0.4)/3.36 × 100 = 88%
for Ing. X: (1.70 − 0.27)/1.70 × 100 = 84%

2. Percent protein digestibility (*in vitro*) = 238.84 − 22.56x

For soy: 238.84 − 22.56(6.7) = 88%
For whey: 238.84 − 22.56(6.3) = 97%

3a.

$$\sqrt[9]{\frac{(46/28)(78/66)(64/58)(26/25)(88/63)}{(39/34)(14/11)(46/35)(26/19)}} =$$

$$\sqrt[9]{\frac{(1.64)(1.18)(1.10)(1.04)(1.39)}{(1.15)(1.27)(1.31)(1.37)}} = 1.26$$

3b. Amino acid score = 1.04

3c. PDCAAS = amino acid score × true digestibility
1.04 (0.87) = .905

16.7 REFERENCES

1. Pellett, P. L., and Young, V. R. (Eds.) 1980. *Nutritional Evaluation of Protein Foods*. The United Nations University, Tokyo, Japan.
2. Bodwell, C. E., Adkins, J. S., and Hopkins, D. T. (Eds.) 1981. *Protein Quality in Humans: Assessment and In Vitro Estimation*. AVI Publishing Co., Westport, CT.
3. Walker, A. F. 1983. The estimation of protein quality. Ch. 8, in *Developments in Food Proteins—2*, B.J.F. Hudson (Ed.). Applied Science Publishers, New York, NY.
4. World Health Organization. 1985. *Energy and Protein Requirements*. Joint FAO/WHO/UNU Expert Consultation. WHO Tech. Rept. Ser. No. 724. World Health Organization, Geneva, Switzerland.
5. Finley, J. W. 1985. Reducing variability in amino acid analysis. In *Digestibility and Amino Acid Availability in Cereals and Oilseeds*, J. W. Finley and D. T. Hopkins (Eds.), 15–30. Amer. Assoc. Cereal Chem., St. Paul, MN.
6. Williams, A. P. 1988. Determination of amino acids. In *HPLC in Food Analysis*, 2nd ed. R. MacRae (Ed.), 441–470. Academic Press, New York, NY.
7. Zumwalt, R. W., Absheer, J. S., Kaiser, F. E., and Gehrke, C. W. 1987. Acid hydrolysis of proteins for chromatographic analysis of amino acids. *J. Assoc. Off. Anal. Chem.* 70:147–151.
8. Steinke, F. H., Prescher, E. E., and Hopkins, D. T. 1980. Nutritional evaluation (PER) of isolated soybean protein and combinations of food proteins. *J. Food Sci.* 45:323–327.
9. National Research Council. 1978. *Nutrient Requirements for Laboratory Animals*, No. 10. National Academy of Sciences, Washington, DC.
10. Sawar, G., Peace, R. W., and Botting, H. G. 1985. Corrected relative net protein ratio (CRNPR) method based on differences in rat and human requirements for sulfur amino acids. *J. Assoc. Off. Anal. Chem.* 68:689–693.
11. FAO/WHO. 1990. *Protein Quality Evaluation Report of the Joint FAO/WHO Expert Consultation on Protein Quality Evaluation*. Held in Bethesda, MD, Dec. 4–8, 1989. Food and Agriculture Organization of the United Nations. Rome, Italy.
12. Young, V. R., and Pellett, P. L. 1991. Protein evaluation, amino acid scoring and the Food and Drug Administration's proposed Food Labeling Regulations. *J. Nutr.* 121:145–150.
13. Henley, E. C. 1992. Food and Drug Administration's proposed labeling rules for protein. *J. Am. Diet. Assoc.* 92:293,294,296.
14. *Federal Register*. 1993. 21 CFR Part 1, et al. Food Labeling; General Provisions; Nutrition Labeling; Label Format; Nutrient Content Claims; Health Claims; Ingredient Labeling; State and Local Requirements; and Exemptions; Final Rules. January 6, 1993. 58(3). Superintendent of Documents. U.S. Government Printing Office, Washington, DC.
15. AOAC. 1990. *Official Methods of Analysis*, 15th ed. Association of Official Analytical Chemists, Washington, DC.
16. Hsu, H. W., Satterlee, L. D., and Miller, G. A. 1977. A multi-enzyme technique for estimating protein digestibility. *J. Food Sci.* 42:1269–1273.
17. Satterlee, L. D., Marshall, H. F., and Tennyson, J. M. 1979. Measuring protein quality. *J. Am. Oil Chem. Soc.* 56:103–109.
18. Swaisgood, H. E., and Catagnani, G. L. 1991. Protein digestibility: *In vitro* methods of assessment. *Adv. Food Nutr. Res.* 35:185–236.
19. Pedersen, B., and Eggum, B. O. 1983. Prediction of protein digestibility by an *in vitro* enzymatic pH-stat procedure. *Z. Tierphysiol., Tierneehr. Fullermittelkde.* 49:265–277.
20. Dimes, L. E., Garcia-Carreno, F., Dong, F., Rasco, B., Fair-

grieve, W., Hardy, R., Barrows, R., Higgs, D. A., and Haard, N. F. 1994. Estimation of protein digestibility in salmonids by an *in vitro* method. *Comp. Biochem. Physiol. B*. In press.

21. Chang, H. I., Catagnani, G. L., and Swaisgood, H. E. 1990. Protein digestibility of alkali- and fructose-treated protein by rat true digestibility and by the immobilized digestive enzyme assay system. *J. Agric. Food Chem.* 38:1016–1018.
22. Cavins, J. F., Kwolek, D. F., Inglett, G. E., and Cowen, J. C. 1972. Amino acid analysis of soybean meal: Interlaboratory study. *J. Assoc. Off. Anal. Chem.* 55:686–694.

CHAPTER

17

Vitamin Analysis

Jörg Augustin

17.1 INTRODUCTION

17.1.1 Definition and Importance

Vitamins are defined as relatively low molecular weight compounds that the human body, and for that matter, any type of living organism that depends on organic matter as a source of nutrients, requires in small quantities for normal metabolism. With few exceptions, humans cannot sythesize most vitamins and so depend on their presence through intake from outside sources such as food. Insufficient supply of these vitamins in the diet has resulted in some deficiency diseases, such as scurvy and pellagra, which are due to the lack of ascorbic acid and niacin, respectively.

17.1.2 Importance of Analysis

The analysis of vitamins plays an essential role in assuring their adequate supply from the existing food regimen. It entails not only the quantitative determination concentrations in food but also their concentrations in food as affected by post-harvest and post-slaughter handling. Furthermore, its use extends to the assessment of vitamin bioavailability for its users.

17.2 METHODS

17.2.1 Overview

Vitamin assays basically fall into three categories:

1. **Bioassays** involving humans and animals.
2. **Microbiological assays** making use of protozoan organisms, bacteria, and yeast.
3. **Physicochemical assays** that include spectrophotometric, fluorometric, chromatographic, enzymatic, and radiometric methods.

In terms of ease of performance, but not necessarily with regard to accuracy and precision, the three systems follow the reverse order. It is for this reason that bioassays, on a routine basis at least, are limited in their use to those instances in which no satisfactory alternative methods are available.

The selection criteria for a particular assay depend on a number of factors, including accuracy and precision, but also economic factors and the sample load to be handled. Applicability of certain methods for a particular matrix also needs to be considered. It is important to bear in mind that many official methods are limited in their applicability to certain matrices, such as vitamin concentrates, milk, or cereals, and thus cannot be translated to other matrices without some procedural modifications, if at all.

Because of the sensitivity of some vitamins to adverse conditions such as light, oxygen, pH, and heat, proper precautions need to be taken to prevent any deterioration throughout the analytical process, regardless of the type of assay used. Such precautionary steps need to be followed with the test material in bioassays throughout the feeding period. They are required with microbiological and physicochemical methods during extraction as well as during the analytical procedure.

Just as with any type of analysis, proper sampling and subsampling as well as the preparation of a homogenous sample are critical aspects of vitamin analysis. General guidelines regarding this matter are provided in Chapter 3 of this text.

The principles, critical points, procedures, and calculations are described in this chapter for various vitamin analysis methods. Many of the methods cited are official methods of the Association of Official Analytical Chemists (AOAC) International. Refer to these methods and other original references cited for detailed instructions of procedures.

17.2.2 Extraction Methods

With the exception of some biological feeding studies, vitamin assays in most instances involve the extraction of a vitamin from its biological matrix prior to analysis. This generally includes one or several of the following treatments: heat, acid, alkali, solvents, and enzymes.

In general, extraction procedures are specific for each vitamin. However, in some instances some procedures are applicable to the combined extraction of more than one vitamin, for example, for thiamin and riboflavin and some of the fat-soluble vitamins (1–4). Typical extraction procedures are:

- Ascorbic acid: Cold extraction with metaphosphoric acid/acetic acid.
- Vitamin B-1 and B-2: Boiling or autoclaving in acid plus enzyme treatment.
- Niacin: Autoclaving in acid (noncereal products) or alkali (cereal products).
- Vitamin A, E, or D: Organic solvent extraction, saponification and reextraction with organic solvents.

17.2.3 Bioassay Methods

17.2.3.1 Applications

Outside of vitamin bioavailability studies, bioassays at the present are used only for the analysis of **Vitamins B-12 and D.** For the latter it is the reference standard method of analysis in food materials (AOAC Method 936.14), known as the **line test.** Since the determination of Vitamin D involves deficiency studies as well as sacrificing the test organisms, it is limited to animals rather than humans as test organisms.

17.2.3.2 Vitamin D Line Test—Principle

The measurement of **bone calcification** uses rats as test organisms.

17.2.3.3 Vitamin D Line Test—Procedures

Figure 17-1 outlines the procedural protocol for this assay and shows how time consuming this procedure is—a minimum of 25 days including the production of Vitamin D-deficient rats and the feeding studies that follow with the food to be tested.

Calcification is measured following staining of the proximal end of the tibia or the distal end of the radius or ulna by rating the discolored areas against those established from results obtained with a standard reference group.

Extract sample.

⇓

Incorporate into rat diet.

⇓

Use rat fed vitamin D deficient diet for 19–25 days.

⇓

Feed rats known and unknown amounts of vitamin D in diets.

⇓

Measure calcification of bones after killing rats: Immerse specified bone in 1.5% $AgNO_3$; get Ag_3P ppt; Ag^+ complexes w/PO_4 in calcified regions; dark areas (lines) are calcified regions; give score based on darkness of line.

⇓

Compare unknown to standard curve.

FIGURE 17-1.
Vitamin D, bioassay method, line test. AOAC Method 936.14. Adapted from (4).

17.2.4 Microbiological Assays

17.2.4.1 Applications

Microbiological assays are limited to the analysis of water-soluble vitamins. The methods are very sensitive and specific for each vitamin. With certain biological matrices, they are the only feasible methods for some vitamins. The methods are somewhat time consuming, and strict adherence to the analytical protocol is critical for accurate results.

17.2.4.2 Principle

The growth of microorganisms is proportional to their requirement for a specific vitamin. Thus, in microbiological assays the growth of a certain microorganism in an extract of a vitamin-containing sample is compared with the growth of this microorganism in the presence of known amounts of that vitamin. Bacteria, yeast, or protozoan organisms are feasible as test organisms. **Growth** can be measured in terms of **turbidity, acid production,** or **gravimetry.** With bacteria and yeast, turbidimetry is the most commonly used system. If turbidity measurements are involved, clear sample and standard extracts, versus turbid ones, are essential. In terms of incubation time, turbidity measurement is also a less time consuming method. The microorganism are specified by ATCC numbers and are obtainable from the American *Type Culture* Collection (12301 Parkway Drive, Rockville, MD 20852).

17.2.4.3 Procedure

The procedural sequence for the analysis of **niacin** is outlined in Fig. 17-2 [AOAC Method 944.13 and (5)]. *Lactobacillus plantarum* (ATCC 8014) is the test organism. A stock culture needs to be prepared and maintained by inoculating the freeze-dried culture on bacto-lactobacilli agar, incubating at 37°C for 24 hr, and storing it under refrigeration until use, but no more than 2 to 4 weeks prior to retransfer or inoculum preparation.

The inoculum is prepared by transferring the stock culture to a tube containing bacto-lactobacillus broth and incubating it at 37°C for 24 hr prior to sample and standard inoculation. A second transfer may be advisable in the case of poor growth of the inoculum culture.

Growth measurement in general is done by tur-

Inoculum: Prepare using L. plantarum, ATCC 8014 culture using Bacto Lactobacilli Broth AOAC.

⇓

Sample preparation: Weigh out enough sample to contain ca. 0.1 mg niacin, add 1N H_2SO_4, macerate, autoclave 1 hr at 121°C and cool. Adjust pH to 6.8, dilute to volume (ca. 0.1 g niacin/ml), mix and filter.

⇓

Assay tube preparation: In at least duplicate use 0.0, 0.5, 1.0, 2.0, 3.0, 4.0, and 5.0 ml sample filtrate and make up the difference to to 5.0 ml with H_2O, then add 5.0 ml Difco Basal Medium for Niacin Assay broth to each tube, autoclave 10 min at 121°C and cool.

⇓

Standard preparation: Prepare assay tube in at least duplicate using 0.0, 0.5, 1.0, 1.5, 2.0, 2.5, 3.0, 4.0, and 5.0 ml standard solution (0.1 μl/ml niacin), make up difference to 5.0 ml with H_2O, then add 5.0 ml of assay broth and treat identical as sample tubes.

⇓

Inoculation and incubation: Add one drop of inoculum to each tube, cover tubes and then inoculate at 37°C for 16–18 hr, i.e. until max. turbidity is reached in tubes containing highest concentration of niacin.

⇓

Determination: Measure %T at any wavelength between 540 and 660 nm.

FIGURE 17-2.
Niacin, microbiological assay. AOAC Method 944.13 (4), and adapted from (5).

bidity. If lactobacilli are used as the test organism, acidimetric measurements can be used as well. The latter may be necessary in the case where a clear sample extract prior to inoculation and incubation (which is a prerequisite for turbidimetry) cannot be obtained. In making a choice between the two methods of measurement, one needs to bear in mind the prolonged incubation period of 72 hr that is required with acidimetry.

17.2.5 Physicochemical Methods

17.2.5.1 Vitamin A

Vitamin A is sensitive to ultraviolet (UV) light, air (and any prooxidants, for that matter), high temperatures, and moisture. Therefore, steps need to be taken to avoid any adverse changes in this vitamin due to such effects, using low actinic glassware, nitrogen and/or vacuum, as well as avoiding excessively high temperatures. The addition of an antioxidant at the onset of the procedure is highly recommended.

17.2.5.1.1 Colorimetric (Carr-Price) Method [AOAC Method 974.29 and (6)] **1. Principle.** This method measures the unstable color at A_{620} nm that results from the reaction between vitamin A and $\mathbf{SbCl_3}$. The color reaction does not differentiate between retinol isomers or retinyl esters. Colors developed with retinaldehyde, dehydroretinol, and similar compounds exhibit maximum absorption at slightly higher wavelengths.

2. Critical Points. $SbCl_3$ needs to be kept free of moisture, and **carotenoids** must be removed from the final extract or appropriate corrections must be made for their presence.

3. Procedure. Figure 17-3 summarizes the procedural steps of this analysis. The antioxidant pyrogallol is added prior to saponification. Corrections for the presence of carotenoids are accomplished either by the removal of these compounds by column chromatography or, in the case of low carotenoid concentration relative to the vitamin A content, by taking readings of a chloroform extract at 440 to 460 nm without prior chromatographic separation. These corrective steps need not be taken in the absence carotenoids from the sample to be tested.

4. Calculations.

$$\mu\text{g Vitamin A/g} = A_{620} \times SL \times (V/WT) \quad (1)$$

where: A_{620} = corrected absorbance at 620 nm, which equals A_{440} − CF (correction factor for carotenoids)
SL = slope of standard curve (Vitamin A concentration/A_{620} reading)
V = final volume in colorimeter tube
WT = sample weight, g

Weigh out appropriate amount of sample and homogenize.
⇓
Saponify with ethanolic KOH (add antioxidant) for 30 min.
⇓
Transfer to separatory funnel, add H_2O, then extract with 1–1.5 volume of hexane, repeat extraction and combine extracts. Wash extract repeatedly with equal volumes of H_2O.
⇓
Filter through paper containing 5 g anhydrous Na_2SO_4 into volumetric flask, rinse filter with hexane and make up to volume.
⇓
Prepare standard curve: Using USP Vitamin A reference standard, prepare a standard dilutions yielding A_{620} values ranging from 0.07 to 0.7, and plot against μg Vitamin A.
⇓
Determine vitamin A content in sample at 620 nm as follows: Remove hexane by evaporation from standard solution in colorimetry tube, add 1 ml chloroform, plus a measured amount of $SbCl_3$ solution, and place into colorimeter and adjust to 0 Abs. (or 100 % T). Evaporate hexane from sample and standard solutions, add 1 ml chloroform followed by an amount of $SbCl_3$ solution equal to that added to the above standard blank and read immediately.

FIGURE 17-3.
Vitamin A, Carr-Price method. AOAC Method 974.29 (4), and adapted from (6).

17.2.5.1.2 High-Performance Liquid Chromatography (HPLC) Method (6) **1. Principle.** This method involves chromatographic separation and quantitative determination at 325 nm.

2. Critical Points. As with any HPLC method, evidence not only of peak identity but also of peak purity is essential to provide good quantitative results.

3. Procedure. Figure 17-4 outlines the procedural sequence of this method. If carotenoids are to be determined simultaneously, a second UV detector using 435 nm wavelength is required. Such a situation may also require modification of the extraction procedures as well as the HPLC separation protocol.

4. Calculations.

$$\text{Vitamin A in } \mu\text{g/g} = C \times (DV/WT) \quad (2)$$

Weigh sample and saponify as outlined in Fig. 17-3.
⇓
Transfer to separatory funnel using H_2O/ETOH and repeatedly extract with hexane. Combine extracts and evaporate to dryness under reduced pressure.
⇓
Dissolve residue in mobile phase (see below).
⇓
Separate and determine by HPLC under the following conditions: Use either MEOH/H_2O, 95/5 (v/v) or acetonitrile/methylene chloride/MEOH, 70/20/10 as the mobile phase; a 20–25 cm, 5-10 particle μ size C_{18} column; a UV detector set at 325 nm; if carotene is also to be determined simultaneously, a second UV detector, set at 436 nm, is needed.

FIGURE 17-4.
Vitamin A, HPLC method. Adapted from (6).

where: C = Vitamin A concentration resulting from sample and standard peak height or area determinations
DV = final dilution volume of sample
WT = sample weight, g

17.2.5.2 Vitamin C

This vitamin (**L-ascorbic acid** and **L-dehydroascorbic acid**) is very susceptible to oxidative deterioration, which is enhanced by high pH and by the presence of ferric and cupric ions. For these reasons, the entire analytical procedure needs to be performed at low pH and, if necessary, in the presence of some chelating agent.

Mild oxidation of ascorbic acid results in the formation of dehydroascorbic acid that is also biologically active and that is reconvertible to ascorbic acid by treatment with **reducing agents.**

17.2.5.2.1 2,6-Dichloroindophenol Titrimetric Method [AOAC Method 967.21 and (7)] **1. Principle.** L-ascorbic acid is oxidized to L-dehydroascorbic acid by the indicator dye. At the endpoint, excess unreduced dye is rose-pink in acid solution. L-dehydroascorbic acid can be determined by first converting it to L-ascorbic acid with a suitable reducing agent.

2. Procedure. Figure 17-5 outlines the the protocol followed with this method. In the presence of significant amounts of iron or copper in the biological matrix to be analyzed, it is advisable to include a **chelating agent** such as EDTA (ethylenediaminetetraacetic acid) with the extraction.

The red-color endpoint should last at least 10 seconds in order to be valid. With colored samples such as red beets or heavily browned products, the endpoint is impossible to detect by human eyes. Therefore, in such cases it needs to be determined by observing the change of transmittance using a spectrophotometer with the wavelength set at 545 nm.

Weigh and extract by homogenizing sample in metaphosphoric acid/acetic acid solution (15 g HPO_4 and 40 ml HOAc in 500 ml H_2O).
⇓
Prepare standard solution: 50 mg L-ascorbic acid diluted to 100 ml H_2O.
⇓
Filter (and/or centrifuge) sample extract and dilute appropriately to a final concentration of 10–100 mg ascorbic acid/100 ml.
⇓
Titrate three replicates each of standard and sample with dichloroindophenol solution to a pink endpoint lasting at least 10 sec.

FIGURE 17-5.
Ascorbic acid, dichloroindophenol method. AOAC Method 967.21 (4), and adapted from (7).

3. Calculations.

$$\text{mg ascorbic acid/g or ml sample} = C \times V \times (DF/WT) \qquad (3)$$

where: C = mg ascorbic acid/ml dye
V = ml dye used for titration of diluted sample
DF = dilution factor
WT = sample weight, g

17.2.5.2.2 Fluorometric Method [AOAC Method 984.26 and (8)] **1. Principle.** This method measures both ascorbic acid and dehydroascorbic acid. Ascorbic acid following oxidation to dehydroascorbic acid, upon reaction with **o-phenylenediamine,** forms a **fluorescent quinoxaline compound.**

2. Procedure. The procedural sequences for this method are outlined in Fig. 17-6. In order to compensate for the presence of interfering extraneous material, blanks need to be run using boric acid prior to the addition of the diamine solution.

3. Calculations.

$$\text{mg ascorbic acid/g or ml} = [(A - B)/(C - D)] \times S \times DF/WT \qquad (4)$$

where: A and C = fluorescence of sample and standard, respectively
B and D = fluorescence of sample and standard blanks, respectively

Prepare sample and extract as outlined in Fig. 17-5.
⇓
To 100 ml each of standard and sample solution add 2 g of acid washed Norit, shake vigorously and filter.
⇓
Transfer 5 ml of each filtrate to separate 100 ml volumetric flask, each containing 5 ml boric acid/NaOAc solution, and allow to stand for 15 min with occasional swirling, then dilute to 100 ml with H_2O.
⇓
Transfer 2 ml of each solution to each of 3 fluorescence tubes.
⇓
Transfer 5 ml of each standard and sample solution to 100 ml volumetric flasks, each containing 5 ml 50% NaOAc trihydride and ca. 27 ml H_2O, and dilute to volume with H_2O, then transfer 2 ml of standard and sample solution to each of 2 sets of 3 fluorescence tubes.
⇓
Add 5 ml of 20 mg% aq. o-phenylene diamine·HCl solution to each tube, swirl using a Vortex mixer and allow to stand for 35 min at room temperature.
⇓
Measure fluorescence at 356 nm/440 nm ex/em.

FIGURE 17-6.
Ascorbic acid, microfluorometric method. AOAC Method 984.26 (4), and adapted from (8).

S = concentration of standard in mg/ml
DF = dilution factor
WT = sample weight, g

17.2.5.3 Thiamin, Thiochrome Fluorometric Method (AOAC Method 942.23, 953.17, or 957.17)

17.2.5.3.1 Principle Following extraction and enzymatic hydrolysis of thiamin's phosphate esters and chromatographic clean-up, this method is based on the fluorescence measurement of its oxidized form **thiochrome.**

17.2.5.3.2 Critical Points Thiochrome is light sensitive. Therefore, the analytical steps following the oxidation must be performed under subdued light. Thiamin is sensitive to heat, especially at alkaline pH. The analytical steps starting with the oxidation of thiamin through the fluorescence measurement (Fig. 17-7) need to be carried out rapidly and precisely according to the instructions.

17.2.5.3.3 Procedure Figure 17-7 outlines the procedural sequence of the thiamin analysis. The enzymatic treatment or the chromatographic clean-up may not be necessary with certain matrices, such as vitamin concentrates that contain nonphosphorylated thiamin and no significant amounts of substances that could interfere with the determination.

17.2.5.3.4. Calculations

$$\mu g \text{ thiamin/g} = (S - S_b/Std - Std_b) \times C/A \times 25/V_p \times V_o/WT \quad (5)$$

where:

S and S_b = fluorescence of sample and sample blank, respectively
Std and Std_b = fluorescence of standard and standard blank, respectively
C = concentration of standard
A = aliquot taken
25 = final volume of column eluate
V_p = volume passed through chromatography column
V_o = dilution volume of original sample
WT = sample weight, g

Weigh out sufficient composite and finely ground sample to contain ca. 10–20 μg thiamin, add 0.14 N HCl, mix and autoclave 15 min at 121°C, then cool.

⇓

Adjust pH to 4.5–5.0 with HCl, then add 5 ml 6% enzyme solution (Mylase 100, Miles Laboratories or Takadiastase, Pfalz and Bauer), and incubate 3 hr at 45–50°C, cool, adjust pH to 3.5, dilute to volume with H_2O, mix and filter.

⇓

Identically as sample, but separately, subject standard solution to the same enzyme treatment.

⇓

Purify by passing 5–25 ml sample extract through a 6–8 mm × 15 cm column containing 50–100 mesh Bio-Rex 70 (BioRad Laboratories), Na form using gravity flow. Wash with 3 × 10 ml hot H_2O, then elute thiamin with 5 ml portions of hot acid KCl into a 25 ml volumetric flask. Treat standard identically.

⇓

Under subdued light, convert thiamin to thiochrome as follows: To each of 4 test tubes, each containing 1.5 g NaCl add 5 ml sample or standard eluate. To two tubes rapidly add 3 ml $K_3Fe(CN)_6$ solution, swirl gently and immediately add 15 ml isobutylalcohol, shake vigorously for 90 sec, centrifuge, decant isobutyl alcohol fraction into fluorescence reading tube and read at 365nm/435nm ex/em. Treat the two remaining tubes identically but instead of the ferricyanide solution add 3 ml 15 % NaOH. Treat standard solution identically.

FIGURE 17-7.
Thiamin, thiochrome fluorometric method. AOAC Method 942.23, 953.17, or 957.17. Adapted from (4).

17.2.5.4 Riboflavin, Fluorometric Method (AOAC Method 970.65)

17.2.5.4.1 Principle Following extraction, clean-up, and compensation for the presence of interfering substances, riboflavin is determined fluorometrically.

17.2.5.4.2 Critical Points Because of the extreme sensitivity of the vitamin to UV light, all operations need to be conducted under subdued light. The analyst also needs to be aware that exact adherence to the permanganate oxidation process is essential for reliable results.

17.2.5.4.3 Procedure An outline of the procedural protocol for this analysis is shown in Fig. 17-8. In spite

Weigh out homogenized sample, add 0.1 N HCl, mix, then autoclave 30 min at 121°C and cool.

⇓

Precipitate interfering substances by adjusting pH to 6.0 immediately followed by a pH readjustment to 4.5, dilute to volume with H_2O and filter.

⇓

Oxidize as follows: Place 10 ml filtrate into each of 4 tubes. To 2 of the tubes add 1.0 ml H_2O and add 1.0 ml standard solution (0.5 μg / ml riboflavin) to each of the 2 remaining tubes. To each tube, one at the time, add 1.0 ml glacial HOAc, followed by 0.5 ml 3 % $KMnO_4$, allow to stand for 2 min, then add 0.5 ml 3 % H_2O_2. Shake well.

⇓

Measure fluorescence at 440 nm/565 nm ex/em. Note: First read sample extracts containing H_2O, then add 20 mg $Na_2S_2O_4$ and mix and reread. Next read standard samples.

FIGURE 17-8.
Riboflavin, fluorometric method. AOAC Method 970.65. Adapted from (4).

of the fact that riboflavin is classified as a water-soluble vitamin, it does not easily dissolve in water. When preparing the standard solution, the analyst needs to pay special attention that the riboflavin is completely dissolved.

17.2.5.4.4 Calculations

$$\mu g \text{ riboflavin/g sample} = [(A - C)/(B - A)] \times (CS/V) \times (DF/WT) \quad (6)$$

where: A and C = fluorescence of sample containing water and sodium hydrosulfite, respectively
B = fluorescence of sample containing riboflavin standard
CS = concentration of standard, μg/ml
V = volume of sample for fluorescence measurement
DF = dilution factor
WT = weight of sample, g

17.2.5.5 Thiamin and Riboflavin, HPLC Method (1, 2)

17.2.5.5.1 Principle The basic principles are the same as outlined above for the fluorometric methods used for the separate determination of thiamin and riboflavin. However, in this method a common extract of the vitamins is concentrated in a precolumn, and they are separated by HPLC followed by fluorescence determination.

17.2.5.5.2 Critical Points Because of the light sensitivity of thiochrome and riboflavin, all operations must be conducted under subdued light. In order to further assure the stability of the two vitamins during the actual HPLC separation, is is necessary to keep the pH above 7.0.

17.2.5.5.3 Procedure The procedural sequences for this method are outlined in Fig. 17-9, and an example chromatogram is given in Fig. 17-10. The precolumn SepPak C-18 (Waters, Division of Millipore, Melford, MA) cartridge can be reused repeatedly for the concentration of these vitamins.

17.2.5.5.4 Calculations

$$\text{thiamin or riboflavin, } \mu g/g = C \times V \times (DF/WT) \quad (7)$$

Weigh sample and extract as outlined in Fig 17-7 including the enzyme treatment.

⇓

Oxidize 10 ml of the diluted extract by passing it through a SepPak C_{18} (Waters Associates) cartridge followed by 5 ml 0.01 M phosphate buffer, pH 7.0, then elute thiamin and riboflavin with 5 ml 50% MEOH into a 5 ml volumetric flask.

⇓

Separate the vitamins by HPLC using the following conditions: a 4.6 mm × 25 cm Ultrasphere ODS, 5 column or equivalent, a 20/80 MEOH/H_2O mobile phase containing 0.005 M tetrabutylammoniuphosphate, fluorescence detection at 360 nm/415 nm ex/em.

FIGURE 17-9.
Thiamin and riboflavin, HPLC method. Adapted from (1) and (2).

where: C = concentration of vitamin in μg/ml obtained from peak height or area of sample and standard
V = precolumn eluate sample volume, ml
DF = dilution factor
WT = sample weight, g

17.2.5.6 Niacin, Colorimetric Method (6)

17.2.5.6.1 Principle This method involves a reaction between niacin (nicotinic acid) and **cyanogen bromide,** which under proper conditions forms a colored compound with an intensity proportional to the niacin concentration.

17.2.5.6.2 Critical Points The color development phase must be executed exactly according to protocol. Because of the extreme cyanogen bromide toxicity, it needs to carried out under an efficient hood.

17.2.5.6.3 Procedure The procedural steps for the analysis of cereal and noncereal products are outlined in Fig. 17-11. The main difference in the analysis of the two products is in the extraction. Aqueous calcium hydroxide and sulphuric acid are the respective extraction media.

17.2.5.6.4 Calculations

$$\mu g \text{ niacin/g sample} = C \times DF/WT \quad (8)$$

where: C = concentration of niacin, μg/ml
DF = dilution factor
WT = sample weight, g

17.2.5.7 Vitamin D, HPLC Method (for Fortified Milk and Milk Powder) (AOAC Method 981.17)

17.2.5.7.1 Principle Following extraction and saponification, interfering substances are removed by passage through an HPLC clean-up column. Vitamin D and its

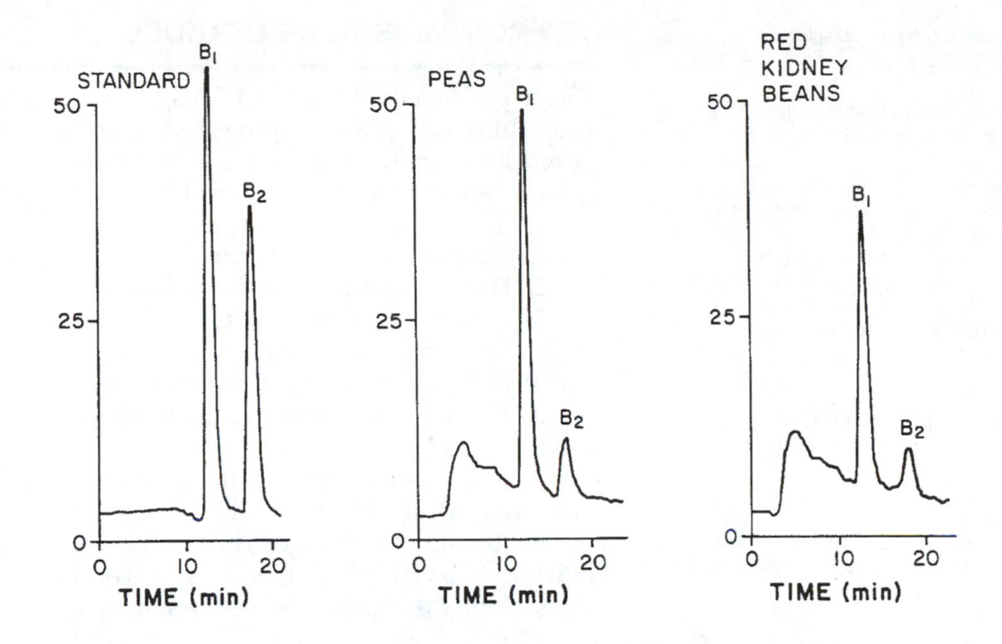

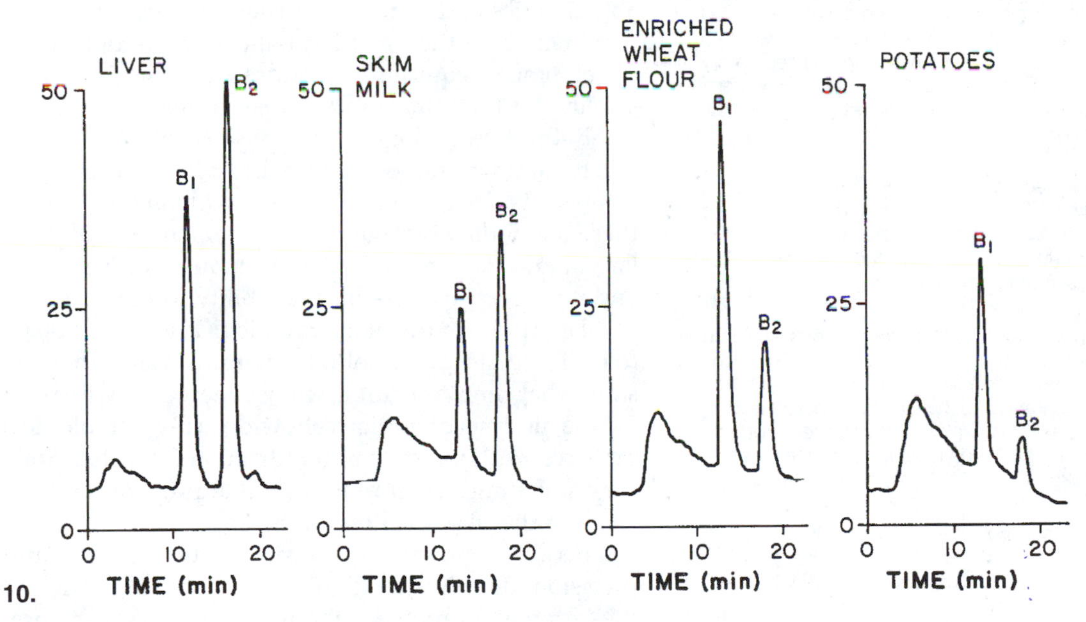

FIGURE 17-10.
Chromatogram from analysis of thiamin and riboflavin by HPLC (15). Reprinted with permission from the *Journal of the AOAC*, Volume 67, Number 5, pages 1012–1015, 1984. Copyright 1984 by AOAC International.

isomers are then separated by HPLC and determined at 254 nm.

17.2.5.7.2 Critical Points Vitamin D is sensitive to air and high temperatures. Thus, the use of antioxidants and not too high temperatures during the analytical process is essential.

17.2.5.7.3 Procedure Figure 17-12 outlines the procedural sequence of the analysis.

17.2.5.7.4 Calculations

$$\text{IU Vitamin D/g sample} = \frac{(1.25 \times P \times WS \times V \times 40{,}000)}{(PS \times W \times VS)} \quad (9)$$

where:

1.25 = correction factor for previtamin D formed during saponification

Extract niacin in 1N H_2SO_4 and autoclaving for 30 min at 121°C, then cool and adjust pH to 4.5, dilute to volume and filter.

⇓

Purify by pipetting 40 ml filtrate into a 50 ml volumetric flask containing 17 g $(NH_4)_2SO_4$ followed by vigorous shaking. Then filter.

⇓

Develop color and determine as follows: For standard and sample blanks pipet 1.0 ml of standard and sample extract into each of one tube, add 5.0 ml H_2O, mix and add 0.5 ml 2 % aq. NH_4OH, 2.0 ml 10 % sulfanilic acid and 0.5 ml diluted HCl, allow to stand for 1.5 min, then read at 430 nm. To measure standard and sample solutions treat standard and sample extract as above, except substitute the 5.0 ml H_2O by 5.0 ml 10% CNBr solution.

FIGURE 17-11.
Niacin, colorimetric method. Adapted from (5).

Weigh out and extract 50 g milk powder of 200 ml milk and saponify for 45 min in ethanolic KOH solution containing sodium ascorbate, cool rapidly.

⇓

Transfer to separatory funnel using with sequential use of H_2O, ETOH and ether, shake vigorously. Transfer aq. phase to second separatory funnel and reextract with 25/100 ml mixture of ETOH /pentane. Place pentane fraction into first separatory funnel and aq. phase into a third separatory funnel. Wash second unit with 10 ml pentane portions and add these portions to first separatory funnel. Wash aq. phase with 100 ml ETOH/pentane mixture and add pentane fraction to first separatory funnel, then wash pentane fraction with 3% KOH in 10 % ETOH portions followed by H_2O portions until neutral pH is obtained. Remove H2O traces by adding filter paper to separatory funnel and shaking.

⇓

Transfer to round bottom flask add 1 ml 0.1 % BHT in hexane, and evaporate to dryness. Dissolve residue in a few ml 5/95 toluene/hexane with the latter containing 0.35% n-amylalcohol and reevaporate to dryness at room temperature under a N stream, then dissolve in 2.0 ml of the same solvent mixture.

⇓

Clean up mixture by passing 200 µl through a 250 x 4.6mm column packed with 10 µm Sil-60D-10CN, using n-hexane containing 0.35% n-amylalcohol as the mobile phase and 254 nm UV detection at a flow rate of 1.0 ml/min. Collect the fraction between 2 min before and 2 min after the the vitamin D peak in a 10 ml volumetric flask. Add I ml 0.1% BHT in hexane and evaporate to dryness at room temperature under a N stream, then take up in 2.0 ml 5/95 toluene/hexane.

⇓

Chromatograph 500 µl extract as follows: a 250 x 4.6 ID mm column containing 5 Partisil, a n-hexane containing 0.35 % n-amylalcohol mobile phase, a UV detector set at 254 nm and 0.008 AUFS, a 2.5 ml/min flow rate.

FIGURE 17-12.
Vitamin D, HPLC method. AOAC Method 981.17. Adapted from (4).

P and PS = peak height of sample and standard, respectively
W and WS = weight of sample in g and standard in mg, respectively
V and VS = total volume of sample and standard in ml, respectively
40,000 IU = International Units of Vitamin D/mg

17.3 COMPARISON OF METHODS

Each type of method has its advantages and disadvantages. In selecting a certain method of analysis for a particular vitamin or vitamins, a number of factors need to be considered, some of which are listed below:

1. Method accuracy and precision
2. The need for bioavailability information
3. Time and instrumentation requirements
4. Personnel requirements
5. The type of biological matrix to be analyzed
6. The number of samples to be analyzed

Bioassays are extremely time consuming. Their uses are generally limited to those instances in which no suitable alternate methods are available, or in cases in which bioavailability of the analyte is desired, especially if other methods have not been demonstrated to provide this information. Bioassays have the advantage that they sometimes do not require the preparation of an extract, thus eliminating the potential of undesirable changes of the analyte during the extract preparation. On the other hand, in the case of deficiency development requirements prior to analysis, bioassays are limited to animals rather than humans.

Both microbiological and physicochemical methods require vitamin extraction, i.e. solubilization prior to analysis. In general, the results obtained through these methods represent the total content of a particular vitamin in a certain biological matrix, such as food, and not necessarily its bioavailability to humans.

The applicability of microbiological assays appears to be limited to water-soluble vitamins. While they are somewhat time consuming they generally can be used for the analysis of a relatively wide array of biological matrices without major modifications. Furthermore, they sometimes require less sample preparation than some of the physicochemical assays.

Because of their relative simplicity, accuracy, and precision, the physicochemical methods, in particular the chromatographic methods using HPLC, are preferred. While HPLC involves a high capital outlay, it is applicable to several water- and fat-soluble vitamins and lends itself in some instances to simultaneous analysis of several vitamins and/or vitamers (isomers of vitamins). However, although its applicability has been demonstrated in some cases to a wide variety of biological matrices with no or only minor modifications, one must always bear in mind that all chromatographic methods, including HPLC, are separation and not identification methods. Therefore, during adaptation of an existing HPLC method to a new matrix, establishing evidence of peak identity and purity is an essential step of the method adaptation or development.

When selecting a system for analysis, at least initially it is wise to consider the use of official methods that have been tested through interlaboratory studies and that are published by such organizations as the AOAC International (4) or the American Association of Cereal Chemists (AACC) (9). Again, one must realize that these methods are limited to certain biological matrices.

Recent developments regarding new methods involve mostly HPLC systems (10–14). In order to keep current with new developments in the area of vitamin analysis, the journal *Food Chemistry* is an excellent reference.

17.4 SUMMARY

The three most used types of methods currently used for the analysis of vitamins—bioassays and microbiological and physicochemical assays—have been outlined in this chapter. They are in general applicable to the analysis of more than one vitamin and several food matrices. However, the analytical procedures need to be properly tailored to the analyte and the biological matrix to be analyzed, including sample preparation, extraction, and quantitative measurements. It is essential with any new application to validate it appropriately by assessing its accuracy and precision. Method validation is especially important with the chromatographic methods such as HPLC, since basically these methods accent separations rather than identification of compounds. For this reason, it is essential to ensure not only identity of these compounds but, just as importantly, their purity.

17.5 STUDY QUESTIONS

1. What factors should be considered in selecting the assay for a particular vitamin?
2. To be quantitated by most methods, vitamins must be extracted from foods. What treatments are commonly used to extract the vitamins? For one fat-soluble vitamin and one water-soluble vitamin, give an appropriate extraction procedure.
3. What two vitamins must be listed on the new nutritional label? What vitamins were listed on the old nutritional label but are no longer required on the new nutritional label? (See Chapter 2.)
4. The standard by which all chemical methods to measure Vitamin D content are compared is a bioassay method. Describe this bioassay method. (Be sure to include in your description the appropriate extraction procedure for fat-soluble vitamins and the chemical basis of the bioassay method.)
5. Explain why it is possible to use microorganisms to quantitate a particular vitamin in a food product, and describe such a procedure.
6. Compare and contrast the principles and procedures for the Carr-Price method versus the HPLC method to quantitate Vitamin A.
7. There are two commonly used AOAC methods to measure the Vitamin C content of foods. Identify these two methods; then compare and contrast them with regard to the principles involved.
8. During processing and storage of foods, L-ascorbic acid can be oxidized to L-dehydroascorbic acid. Using the 2,6-dichloroindophenol titrimetric for Vitamin C, how could you quantitate total Vitamin C and each form individually?
9. What are the advantages and disadvantages of using HPLC for vitamin analysis?

17.6 PRACTICE PROBLEMS

Calculate the concentration of vitamin in the original sample for each of the vitamin and assay conditions described below.

1. Niacin, microbiological assay
   - Sample weight: 0.1120 g
   - Dilutions: 1 to 100 and 0.1 to 500
   - Niacin concentration in sample: 0.86 μg/ml
2. Vitamin A, HPLC method
   - Sample weight: 5.0340 g
   - Final dilution volume: 5 ml
   - Vitamin A concentration: 1.25 μg/μl
3. Vitamin C, dichloroindophenol method
   - Sample weight: 100 g, diluted to 500 ml with extracting solution
   - Amount of filtrate titrated: 25 ml
   - Amount of dye used in titration: 9.1
   - Ascorbic acid concentration in dye: 0.175 mg/ml
4. Thiamin, fluorometric method
   - Sample weight: 2.0050 g
   - Dilutions: Dilute to 100 ml, take 25 ml for chromatography, then dilute eluate to 25 ml and use 5 ml for fluorometry
   - Standard concentration: 0.1μg/ml
   - Fluorometry reading ratio: 0.850
5. Riboflavin, fluorometric method
   - Sample weight: 1.0050 g
   - Dilutions: to 50 ml; use 10 ml for fluorometry
   - Fluorometry readings: A60/B85/C10
   - Riboflavin concentration: 0.1 μg/ml
6. Thiamin/riboflavin, HPLC method
   - Sample weight: 2.1050 g
   - Dilution factor: 10
   - Precolumn elution volume: 5 ml
   - Concentrations in extract injected: 0.160 μg B-1 and 0.750 μg B-2
7. Vitamin D, HPLC method
   - Weight of sample: 50.00 g
   - Peak heights: sample 79, standard 82
   - Sample volume: 16
   - Standard volume: 40,000

Answers:

1. 4.3 mg/g; 2. 12.39 μg/g; 3. 0.228 mg/g; 4. 0.8479 μg/g; 5. 0.9950 μg/g; 6. 3.8 μg/g, B-1 and 18.0 μg/g, B-2; 7. 6.021 IU/g

17.7 References

1. Fellman, J.K., Artz, W.E., Tassinari, P.D., and Augustin, J. 1982. Simultaneous determination of thiamin and riboflavin in selected foods by high-performance liquid chromatography. *J. Food Sci.* 47:2048–2051.
2. Augustin, J. 1984. Simultaneous determination of thiamine and riboflavin in foods by liquid chromatography. *J. Assoc. Off. Anal. Chem.* 67:1012–1015.
3. Beaulieu, N., Curran, N.M., Gragney, C., Gravelle, M., and Lovering, E.G. 1989. Liquid chromatographic methods for vitamin A and D in multivitamin formulations. *J. Assoc. Off. Anal. Chem.* 72:247–254.
4. AOAC. 1990. *Official Methods of Analysis*, 15th ed. Association of Official Analytical Chemists, Washington, DC.
5. Eitenmiller, R. R., and DeSouza, S. 1985. Niacin, in *Methods of Vitamin Assay*, J. Augustin, B.P. Klein, D.A. Becker, and P.B. Venugopal (Eds.), 4th ed., 389–392 and 393–397. John Wiley & Sons, Inc., New York, NY.
6. Parrish, D.B., Moffitt, R.A., Noel, R.J., and Thompson, J.N. 1985. Vitamin A, in *Methods of Vitamin Assay.*, J. Augustin, B.P. Klein, D.A. Becker, and P.B. Venugopal (Eds.), 4th ed., 175–179. John Wiley & Sons, Inc., New York, NY.
7. Pelletier, O. 1985. Vitamin C (L-ascorbic and dehydro-L-ascorbic acids), in *Methods of Vitamin Assay.*, J. Augustin, B.P. Klein, D.A. Becker, and P.B. Venugopal (Eds.), 4th ed., 334–336. John Wiley & Sons, Inc., New York, NY.
8. Ibid. pp. 338–341.
9. AACC. 1983. *Approved Methods of the American Association of Cereal Chemists.*, 8th ed. American Association of Cereal Chemists, St. Paul, MN.
10. Ponsello, A., and Rizzolo, A. 1986. Application of HPLC to the determination of water-soluble vitamins in foods. A review 1981 to 1985. *J. Micronutr. Anal.* 2:153–187.
11. Ibid. 1990. A review 1985 to 1990. *J. Micronutr. Anal.* 8:105–158.
12. Brubacher, G., Muller-Munot, W., and Southgate, D.A.T. (Eds.) 1985. *Methods for the Determination of Vitamins in Foods.* Elsevier Science Publishing Co., New York, NY.
13. Macrae, R., Beacher, G.R., and Crow, R. (Eds.) 1985–1991. Bibliography of Micronutrient Analysis. 4. Vitamins and Cofactors. *J. Micronutr. Anal.* Vols. 1–7.
14. Grary, J.I., Macrae, R., and Rogen, J.P. (Eds.) 1992 to date. Bibliography on Analytical Food Chemistry. 6. Vitamins and Cofactors. *Food Chemistry,* Vols. 45 to date.
15. Augustin, J. 1984. Simultaneous determination of thiamine and riboflavin in foods by liquid chromatography. *J. Assoc. Off. Anal. Chem.* 67:1012–1015.

CHAPTER

18

Pigment Analysis

Steven J. Schwartz

18.1 INTRODUCTION

18.1.1 Importance of Color and Food Quality

Color is one of the most important quality attributes of foods. The first impression of the quality and acceptability of a particular food is judged upon its appearance. Therefore, the pigments, which are the prime contributors to coloration, are important quality constituents to analyze in foods.

Measurement of both natural and synthetic pigments in foods is an analytical challenge to food chemists. The diversity of naturally occurring pigments, their derivatives, and the formation of degradation components that contribute to the color of foods complicate both qualitative and quantitative measurements. Furthermore, compartmentalization of natural pigments within foods and interactions that may occur between pigments and other food components, especially during thermal processing, may lead to difficulties in liberating or extracting the pigments for analysis. Despite these obstacles, a number of excellent methods have been developed specifically for the extraction, separation, and measurement of pigments in foods. This chapter will summarize some of the current methods used for this analysis with applications aimed toward methodology appropriate for use in a food analysis course.

18.1.2 Presence and Distribution of Pigments in Foods

Basically, there are five major classes of naturally occurring pigments in foods. Specifically, four pigment classes are distributed throughout the plant kingdom and the fifth in animal tissues. The lipid-soluble **chlorophylls** and **carotenoids** and the water-soluble **anthocyanins** and **betalains** are found in plants. In animal tissues, meat color is due to a heme protein **myoglobin.** In some fish tissue, such as salmon and trout and in crustaceans, the orange red coloration arises because of the presence of carotenoid pigments. However, carotenoid pigments found in the animal kingdom are not biosynthesized but derived from plant sources.

18.1.3 Basic Principles in Handling and Storage of Pigments

It is well known that naturally occurring pigments and synthetic dyes can be sensitive to oxygen, heat, light, metal ions, and catalysts that enhance the rate of oxidative and reductive reactions. Therefore, care should always be taken to minimize these reactions during extraction, handling, and storage of pigments. Plant tissue pigments can be extracted fresh from the raw state; however, some losses may occur because of enzymatic activity. Plant pigments may be extracted from the frozen tissue; however, prior to freezing, the tissue samples should be blanched to inactivate the native enzymes that can decompose the pigments. Once extracted, the analyst should ensure that contaminant-free glassware is always used and pigment extracts are stored under a nitrogen headspace.

Extracts under nitrogen should be stored under reduced temperature conditions in amber glass vials or clear glass vials wrapped in aluminum foil to prevent exposure to light. Aqueous extracts should preferably not be frozen since freeze concentration can enhance pigment-pigment interactions. However, some pigments' storage stabilities are enhanced at frozen temperatures, and standard control solutions should be checked regularly for the extent of degradation. Optimally, freshly prepared extracts should be analyzed immediately to minimize chemical alterations and pigment decomposition.

If a quantitative measurement of the pigments is required, all tissues should be subjected to a moisture content analysis. This will provide quantitative data on pigments on both a wet-weight and dry-weight basis. Since the moisture content of tissues may differ from sample to sample and changes during processing, it may be necessary to monitor the changes in solids content to determine if losses in pigment content have occurred.

In the following sections, the chemical properties and techniques involved in the analysis of specific food pigments are discussed. All the methods covered involve spectrophotometric measurements, and therefore, the reader is advised to refer to Chapter 22 on Basic Principles of Spectroscopy as well as Chapter 23 for a thorough review of ultraviolet and visible spectroscopy.

18.2 CHLOROPHYLLS

In higher plants, the **chlorophylls,** both **a** and **b,** are found ubiquitously throughout photosynthetic tissues. In vegetable tissues, a number of different chlorophyll derivatives can be present, especially after thermal processing, that contribute to the overall color of vegetable products. The structures of the chlorophylls and their derivatives found in food products are depicted in Fig. 18-1.

Several analytical methods have been developed for the analysis of chlorophylls in foods. Many of these methods have been compiled in a recent review by Schwartz and Lorenzo (1). Early spectrophotometric methods allowed for the determination of chlorophylls by measuring absorbance at the absorption maxima of the two chlorophylls. These methods are suitable

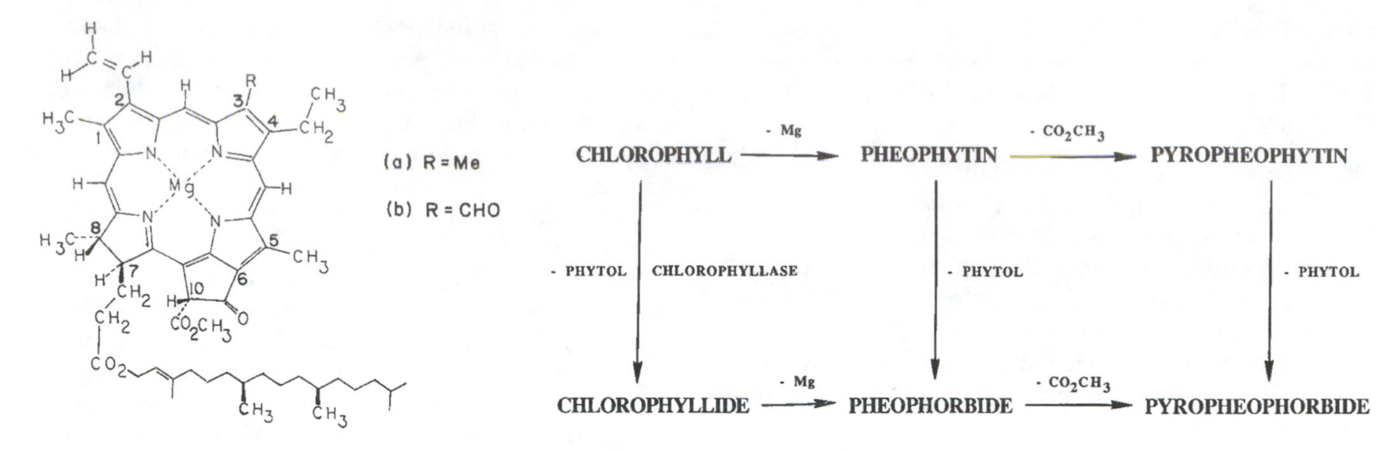

FIGURE 18-1.
The structure of chlorophyll and its derivatives. Reprinted with permission from *Critical Reviews in Food Science and Nutrition.* 1990. Vol. 29, No. 1, pp. 1–17. Copyright CRC Press, Inc., Boca Raton, Fl.

only for fresh plant materials where no pheophytin degradation components are present. This is the basis for the Association of Official Analytical Chemists (AOAC) International spectrophotometric procedure (2) (Method 942.04), which provides results for total chlorophyll, chlorophyll a, and chlorophyll b content. Vernon (3) developed a quantitative spectrophotometric method for the analysis of both chlorophylls a and b as well as pheophytin a and b. The method utilizes specific absorptivities of the four components to derive a set of equations to calculate the quantities of each individual pigment in 80 percent acetone solutions. The method is generally applicable to vegetable tissue in which only chlorophyll and pheophytin derivatives are present. High-performance liquid chromatography (HPLC) methods are preferred when other chlorophyll components such as chlorophyllides, pheophorbides, and pyropheophytins are expected.

Acetone extracts are generally used for quantitative extraction of chlorophylls from plant tissue. Tissue samples are usually blended in acetone in the presence of a small quantity of $CaCO_3$. The $CaCO_3$ base neutralizes any acids that may be liberated from the tissue and prevents the formation of pheophytins during extraction. Since water is always present in the tissue cells, a final concentration of 20 percent water and 80 percent acetone is often used, and therefore, drying of the sample is not necessary. Furthermore, specific absorption coefficients and molar absorptivity values are widely published for 80 percent acetone solutions, allowing for simple concentration calculations of pure pigment extracts. Acetone extracts of pigments are also generally compatible with reversed-phase HPLC methods, which use some water in the mobile phase. Alternatively, chlorophyll pigments can be easily transferred to other organic solvents, such as diethyl ether, by repeated washings of the acetone-ether mixture with water in a separatory funnel.

A simple reversed-phase HPLC method for the analysis of chlorophylls and their derivatives in fresh and processed plant tissues has been described by Schwartz et al. (4). The method is applicable to determine chemical alterations in chlorophyll composition during the processing of foods and can separate and quantitate the major chlorophyll derivatives including the pheophytins and pyropheophytins. The method involves a gradient elution technique that can be eliminated if chlorophyllides and pheophorbides are not present or found in small quantities (5). Typical HPLC chromatograms of fresh, blanched, frozen, and canned spinach are shown in Fig. 18-2. An advantage of the method is that chromatograms are monitored at 654 nm, which selectively screens for chlorophyll components while the yellow-colored carotenoids are excluded. More sophisticated methods have been published for the separation of the water-soluble chlorophyll components (6, 7). Identification of the purified chlorophyll pigments present in extracts can be performed by comparison of ultraviolet-visible (UV-Vis) spectra and coelution with authentic standards. Mass spectrometry has also been used to confirm the identity of individual chlorophylls (8, 9).

18.3 CAROTENOIDS

The carotenoid pigments consist of two major classes: the **hydrocarbon carotenes** and the **oxygenated xanthophylls.** Not only do the carotenoids provide yellow to red coloration in foods, but some also serve as precursors to Vitamin A. For this reason, many analytical methods for carotenoids are aimed at measurement of

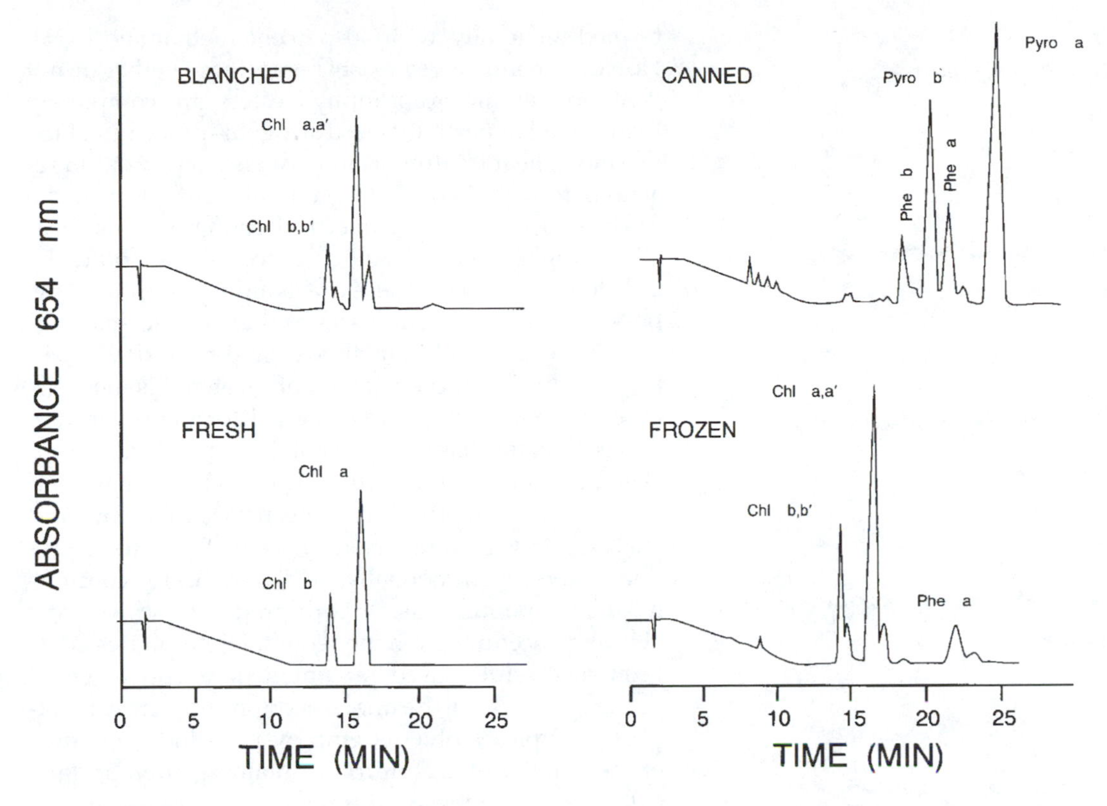

FIGURE 18-2.
HPLC chromatograms of chlorophylls in fresh, blanched, frozen, and canned spinach. Reprinted with permission from (4). Copyright 1981 American Chemical Society.

the **provitamin A carotenoids** for determination of their nutritional value. In addition to over 500 naturally occurring carotenoids, β-carotene, β-apo-8′-carotenal, and canthaxanthin have been synthesized and are used to enhance the appearance of a variety of manufactured food products. The carotenoids present in annatto, saffron, tumeric, and paprika, to mention a few, are also used to color foods. The structures of some prominent carotenoids are shown in Fig. 18-3 as examples.

The complex nature and diversity of carotenoid compounds present in plant foods necessitates chromatographic separation. The AOAC Method 941.15 (2) recommends extraction with acetone/hexane, filtration, and removal of the acetone by repeated washings with water. Hexane extracts are then applied to a MgO_2 (activated) diatomaceous earth column and eluted with a mixture of acetone and hexanes. Because most carotenoid extracts consist of a mixture of nonpolar carotenes and more polar **xanthophylls,** a carotene fraction elutes early from the column in a chromatographic elution. As the acetone concentration is increased, the more polar xanthophylls elute separately as the monohydroxy and dihydroxy pigments. Alternatively, total xanthophylls can be eluted and isolated by adding methanol to the eluting solvent. After dilution to volume, the concentrations of carotenes and xanthophylls are measured by absorbance readings based on standard solutions. The major concern about using the AOAC procedure (Method 941.15) is its failure to distinguish between co-eluting carotenes: no separation occurs between **α- and β-carotene** as well as other hydrocarbon carotenoids. The method also calculates all carotenes as β-carotene, which possesses the highest provitamin A activity of all the carotenes. Thus, if the provitamin A activity is to be determined, the results for Vitamin A content can be severely overestimated.

Extraction procedures for quantitative removal of carotenoids from plant tissues involve the use of organic solvents that must penetrate through a hydrophilic matrix. Polar solvents such as acetone and tetrahydrofuran can be used for this purpose (10). More polar solvents are best for extraction of xanthophylls, while mixtures of hexane-acetone have been successful for extraction of total carotenoids. Recommended procedures involve first blending the tissue with water, followed by precipitation of carbohydrates and proteins with methanol/ethanol, which also serves to dehydrate the tissue. This allows for easy penetration and subsequent extraction of tissue with organic solvents. Homogenization of the tissue in the presence of the extracting solvents is preferred for complete quantitative extractions.

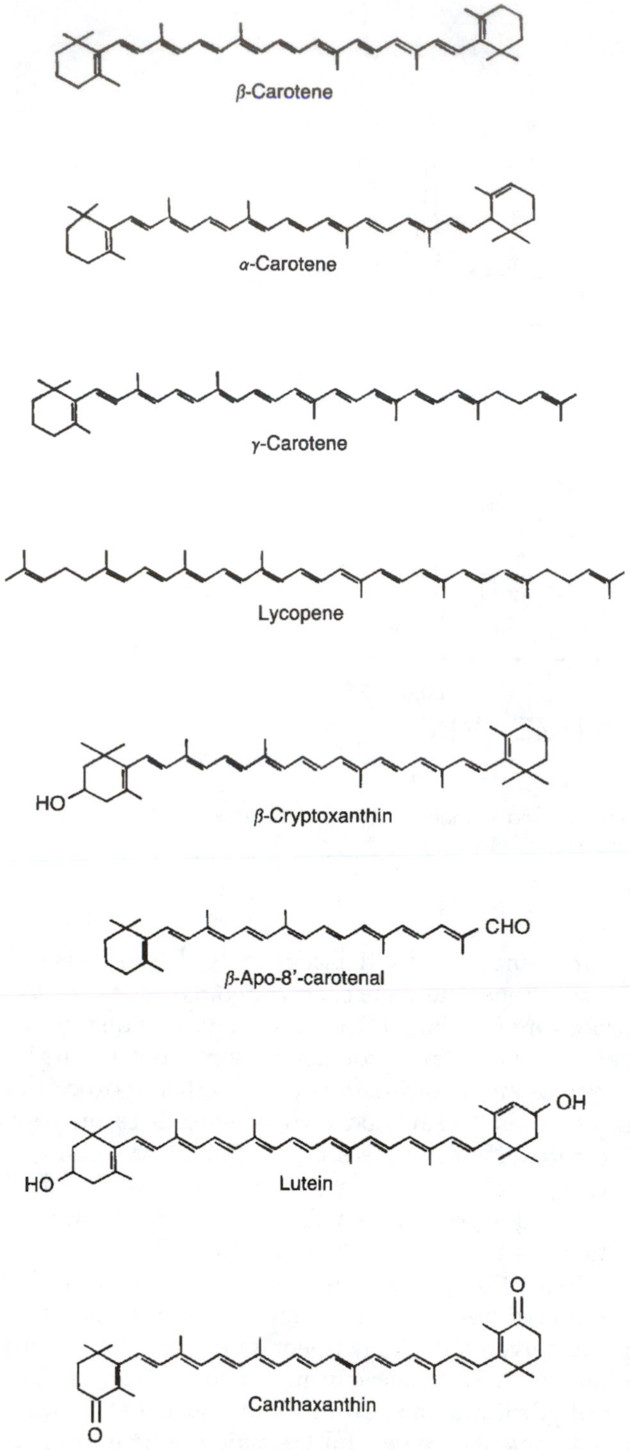

β-Carotene

α-Carotene

γ-Carotene

Lycopene

HO β-Cryptoxanthin

CHO β-Apo-8'-carotenal

OH HO Lutein

O O Canthaxanthin

FIGURE 18-3.
Structures of prominent carotenoids.

For samples high in fats and oils, saponification may be necessary to remove the lipids (11). Also, if chlorophylls are present and interfere with the chromatographic separation of carotenoids, saponification will rapidly degrade the chlorophyll constituents. Either hot or cold saponification procedures can be performed, generally with 40 percent methanolic KOH. However, controversy exists regarding whether or not xanthophylls and xanthophyll esters are completely hydrolyzed or partially lost during this procedure. Furthermore, heat treatments may cause isomerization reactions to occur, converting the all-trans form of the carotenoid to its cis geometrical configurations. If at all possible, saponification procedures should be avoided to alleviate the extra handling step and the possibility for pigment losses and artifact formation.

Numerous HPLC methods have been developed and optimized for the analysis of carotenoids in foods. Often the procedures are designed for the separation of specific carotenoids present in a particular fruit, vegetable, or food product (12). Both normal- and reversed-phase methods have been used for carotenoid analyses; however, reversed-phase methods have predominated. Commercially available C-18 columns have been popular, using both gradient and isocratic elution procedures. Nonaqueous mobile phases have been successfully used in conjunction with reversed phase columns for the analysis of provitamin A carotenoids. Typical solvents employed include mixtures of methanol and acetonitrile containing ethyl acetate, chloroform, or tetrahydrofuran.

Detection wavelengths for monitoring carotenoids range from approximately 430 to 480 nm. The higher wavelengths are usually used for some xanthophylls as well as to prevent detection from interfering chlorophylls. Since β-carotene absorption maximum in hexane is 453 nm, many methods have detected the carotenoids near this region. When fixed wavelength detectors are available, 436 or 440 nm is popular but less sensitive. Detection sensitivity can be optimized by monitoring eluants at the absorption maxima of the carotenoid in question.

Often the identity of a particular carotenoid is found by co-elution with a standard. It is also advisable to compare UV-Vis spectra of unknowns with authentic compounds. Standards can be purified on TLC plates using absorbants such as MgO_2, $Ca(OH)_2$, and diatomaceous earth. Many carotenoids exhibit spectral shifts upon reactions with various reagents, and these spectral changes are useful to assist in identification. Freshly prepared purified standards are recommended for instrument calibration since many carotenoids are labile and subject to decomposition. Further, the identity of the carotenoid can be confirmed by mass spectral and NMR data. However, this still excludes the determination of the structural configuration of optical isomers, which requires much more detailed analyses.

A typical chromatogram of the carotenoids present in a carrot extract is shown in Fig. 18-4. The HPLC conditions used for the analysis are designed for the separation of provitamin A carotenoids (i.e., α, β, γ carotenes and cryptoxanthin). Most fruits and vegeta-

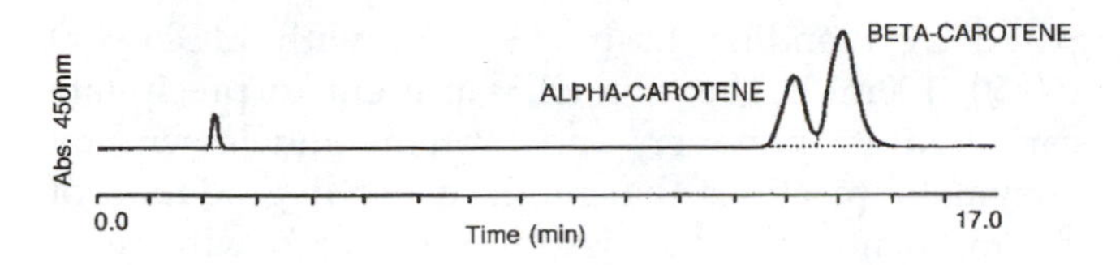

FIGURE 18-4.
Typical HPLC chromatogram of a carrot extract. From (13).

bles that possess provitamin A activity contain significant quantities of the mentioned carotenoids; however, the provitamin A activity varies considerably (32). Other methods are more appropriately designed for the analysis of the more polar xanthophylls. To resolve and separate the xanthophylls, more polar solvent systems are often used. However, it is possible to resolve a complex mixture of xanthophylls in the presence of the less polar carotenes by using a selective solvent mixture or eluting the carotenoids with a gradient (14). Further challenges are present and commonly encountered in fruits when carotenoid esters must be separated for analysis.

18.4 ANTHOCYANINS

The **anthocyanins** are important pigments responsible for the red, blue, and purple colors of some flowers, fruits, vegetables, juices, wines, and jams. Many different anthocyanins are found in commonly consumed plant products. A unique aspect of most anthocyanins is their ability to reversibly change color as a function of pH. The basic structure of the anthocyanins and the structural transformations that occur with pH changes are illustrated in Fig. 18-5.

The water solubility of anthocyanins provides for their relative ease of extraction from plant tissues. Timberlake and Bridle (15) have compiled an extensive list of anthocyanins found in commonly consumed plants. These pigments are glycosides, commonly glucosylated, but other sugar moieties may be present along with a variety of esters. Methanol or ethanol containing 1 percent or less HCl are best for extraction of finely ground tissue. Lower concentrations of HCl (0.01 to 0.05%) may be necessary to prevent hydrolysis if anthocyanins are present.

Measurements of anthocyanins present in extracts, juices, or wines has been explained by Wrolstad (16) and can be estimated by determining the absorbance of a diluted sample at pH 1.0 at the wavelength maximum (510 to 540 nm). The samples must be centrifuged if they are not clear. Concentrations can be estimated by using an appropriate known absorptivity value for a pure pigment solution. Selection of the absorptivity value is based upon the predominant anthocyanin present in the plant material and the extracting solvent used. Interferences are subtracted out by using a pH differential method. Both pH 1.0 and pH 4.5 buffers are used to dilute the extracts. At pH 4.5, the anthocyanins are colorless and thus absorbance measurements attributed to only the anthocyanins are obtained by the difference. If any haze is apparent in the solutions, this can be corrected by absorbance readings at 700 nm as follows:

$$\text{Absorbance of Anthocyanins} = (A/\lambda\text{max pH } 1.0 - A/700\text{nm pH } 1.0) - (A/\lambda\text{max pH } 4.5 - A/700\text{nm pH } 4.5) \qquad (1)$$

Absorbance measurements for anthocyanin content provide estimates for total quantity; however, this determination does not account for the complexity and

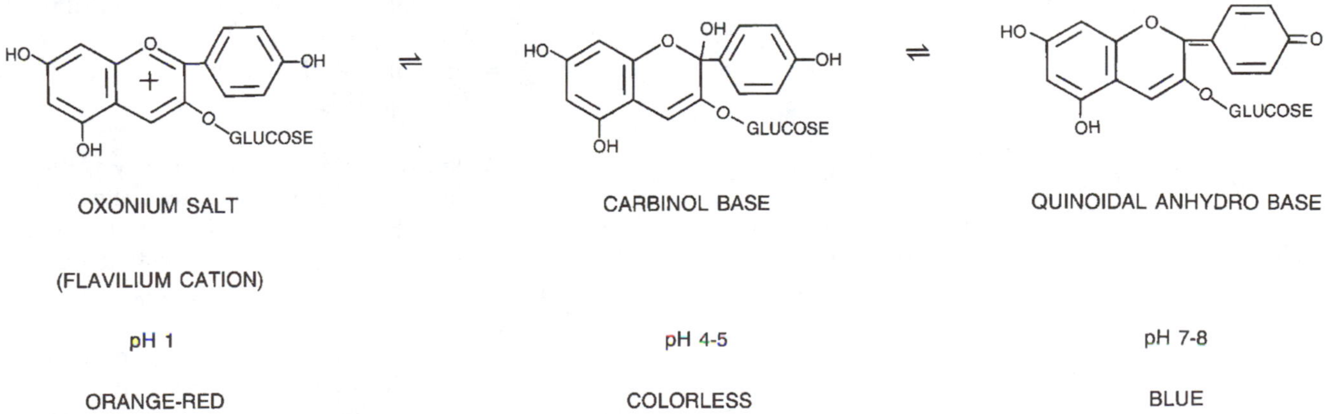

HO OH OH O + O GLUCOSE ⇌ HO OH OH O O GLUCOSE OH ⇌ HO O O O GLUCOSE OH

OXONIUM SALT (FLAVILIUM CATION) pH 1 ORANGE-RED

CARBINOL BASE pH 4-5 COLORLESS

QUINOIDAL ANHYDRO BASE pH 7-8 BLUE

FIGURE 18-5.
Structural transformation of anthocyanins with change in pH. From (16), used with permission.

identity of the various anthocyanin pigments that are present in plant tissues. Early chromatographic methods utilized both paper and column chromatography to separate the various components. Paper chromatography is still popular for isolation and characterization where mg quantities of pigments are needed. Many modern HPLC methods are now available for this analysis. The water solubility of anthocyanins makes this class of compounds ideal for reversed-phase methods employing C-18 columns. Solvent systems usually consist of a mixture of water acetic, formic, or phosphoric acids and either methanol or acetonitrile. Recently, characterization of the anthocyanins has been enhanced with the use of photodiode array detectors where full UV-Vis spectral information can be obtained as the compounds elute from the column (17).

18.5 BETALAINS

The betalain pigments are not widely distributed throughout the plant kingdom. The purple red beet root contains a high concentration of betalain pigments, which consist of the predominant purple-red **betacyanins** and lower concentration of the yellow **betaxanthins.** Figure 18-6 shows the structures for these compounds.

Betalain pigments are ionic, exhibit high water solubilities, and are therefore extracted from plant tissues readily with water (18). Initial homogenates are prepared by blending tissue (~100g) with EtOH:H_2O (50:50, 100mL). The ethanol is present to precipitate carbohydrate polymers and proteins and lessen any enzymatic reactions that might cause degradation of the pigments. Blended tissues are filtered with celite or filter aid and are then quantitatively washed with water until completely extracted.

BETANIN R = GLUCOSE

BETANIDINE R = H

PREBETANINE

R = GLUCOSE-6-SULFATE

VULGAXANTHIN I R = NH_2

VULGAXANTHIN II R = OH

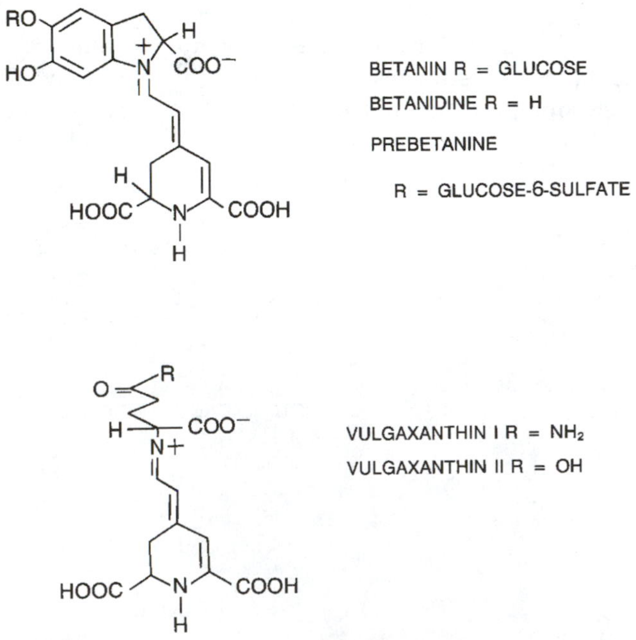

FIGURE 18-6.
Structures of betalains in red beets. Reprinted from (18), p. 65, by courtesy of Marcel Dekker, Inc.

The fact that the betalains consist of both the yellow betaxanthines and the red betacyanins precludes the use of direct spectophotometric measurements of either species. However, quantitative measurements can be performed prior to separation by accounting for the absorbances of each pigment as well as some impurities (19). The method is based on the fact that **betanin** and **vulgaxanthin,** the major betacyanin and betaxanthin, maximum absorbance wavelength regions are at 535 to 540 nm and 476 to 478 nm, respectively. Secondly, betanin absorbs light at the absorption maximum of vulgaxanthin; however, vulgaxanthin does not interfere at the maximum absorption of betanin. Therefore, the ratios of the absorbances at 538 nm and 436 nm with a correction for impurities lead to a set of equations for the determination of the pigments. The equations Nilsson (19) developed were:

$$x = a - z \tag{2}$$

$$y = b - z - x/A538/A476 \tag{3}$$

$$z = c - x/A538/A600 \tag{4}$$

where:
- a = the light absorbance values of the extract at 538 nm
- b = the light absorbance values of the extract at 476 nm
- c = the light absorbance values of the extract at 600 nm
- x = the calculated absorbance contributions for betanin
- y = the calculated absorbance contribution for vulgaxanthin
- z = the calculated absorbance contribution for impurities

Using the known 1 percent absorptivity values of 1,120 for betanin and 750 for vulgaxanthin, the concentrations of each can be calculated. Alternatively, a nonlinear curve fitting procedure can be used to determine individual pigment contents from the mixture (20).

Because of the coexistence of the light-absorbing pigments present in extracts, chromatographic procedures have been developed to separate and analyze the individual components. Separation of the charged pigments can be achieved by electrophoresis, but they are more rapidly resolved and analyzed by HPLC. Individual betacyanins can be separated on reversed-phase columns by using an ion-pairing or ion-suppression technique (21, 22). These methods enhance resolution by minimizing complete ionization of the carboxylic acid groups present on the pigment structure. This allows for greater interaction of the molecule with the stationary phase and better separation of the ionic pigments, as shown in Fig. 18-7.

18.6 MYOGLOBINS

Myoglobins (metmyoglobin, oxymyoglobin, and deoxymyoglobin) responsible for the color of meats are usually determined by reflectance spectrophotometry of the meat surface (23, 24). These consist of three forms, **metmyoglobin, oxymyoglobin,** and **deoxymyoglobin.** The oxidative state of the iron located within the **porphyrin structure** dictates whether the color is red (oxymyoglobin, Fe^{+2}) or brown (metmyoglobin, Fe^{+3}). Deoxymyoglobin (Fe^{+2}) is purplish red in color but can be oxidized to the undesirable brown metmyoglobin. The chemistry of myoglobin systems has been recently reviewed by Faustman and Cassens (25).

The pigments present in meat can also be extracted and measured (26). Extraction is carried out by using phosphate buffer, pH 6.8, at an ionic strength of 0.04. The absorbance of the extract is due to a multicomponent system consisting of oxidized, oxygenated, and reduced pigment forms. By measuring the absorbances at 572, 565, 545, and 525 nm, the concentration of each pigment species can be calculated using derived equations (27).

18.7 SYNTHETIC FOOD DYES

There are seven synthetic food dyes that are approved for use in the United States under the Food, Drug, and Cosmetic (FD&C) Act of 1938, amended by the 1960 Color Additive Amendment (28). These dyes are generally much more stable to heat treatments, pH changes, and extended storage conditions relative to the natural colorants (29). For these reasons, they are used in small quantities in a variety of food products. The trivial names and FD&C numbers of these compounds are shown in Table 18-1.

Extraction of synthetic colorants from foods is usually not a difficult process for qualitative measurements. However, for quantitative measurements, complete removal of the colorants is problematic because of an affinity of the dyes to bind to protein. Acidic solutions or various buffers may be used to extract the dyes. If acid dyes are present, ammonical alcohol is suitable as an extracting solvent. In the AOAC Method 34.008 (2) procedure, a liquid anion exchange resin is employed to trap and purify the pigments from other food components. Digestion of the proteins and lipids with papain and lipase has also been proposed, but recoveries are not completely quantitative.

Once extracted, identification and quantification of the pigment is obtained by measuring the UV-Vis spectrum. If more than one pigment is present, it is possible to derive a set of equations for the determination of the individual components. However, modern HPLC methods have been developed to rapidly separate and identify the dyes by co-elution and comparison of spectra to readily available authentic standards. Most successful methods utilize an ion-pairing technique that forms reversible ion-pairs with the dye molecule anions (30). The ion-pairs are then resolved on a C-18 reversed-phase column using polar eluant mixtures of water, methanol, and acid (31).

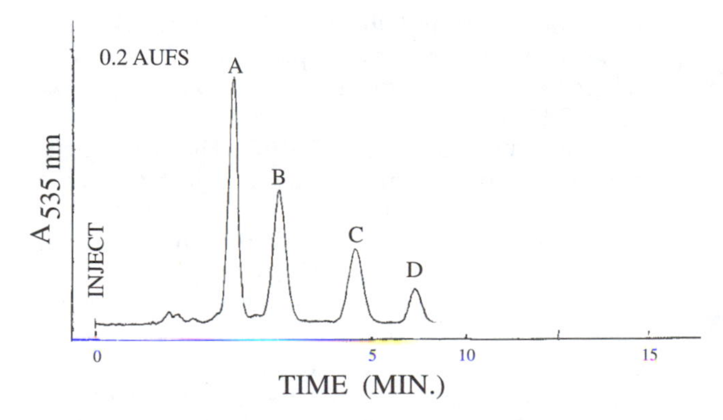

FIGURE 18-7.
HPLC chromatogram of betacyanin pigments. A = betanin, B = isobetanin, C = betanidin, D = isobetanidin. Reprinted from (18), p. 67, by courtesy of Marcel Dekker, Inc.

18.8 SUMMARY

This chapter has attempted to summarize some of the current methods for pigment analysis applicable to a food analysis course. Basic principles of extracting, handling, and storing pigments are emphasized since many naturally occurring pigments are relatively labile

TABLE 18-1. Certified Food Colorants

FD&C Color	Common Name
FD&C Red No. 40	Allura Red
FD&C Blue No. 1	Brilliant Blue
FD&C Blue No. 2	Indigotine
FD&C Green No. 3	Fast Green
FD&C Yellow No. 5	Tartrazine
FD&C Yellow No. 6	Sunset Yellow
FD&C Red No. 3	Erythrosine

and susceptible to degradation. In addition, specific procedures for the analysis of plant and animal pigments including the chlorophylls, carotenoids, anthocyanins, betalaines, and myoglobins are provided. Relatively little information on the analysis of synthetic dyes present in food products is published. However, a brief section on the measurement of FD&C dyes is included for the reader's reference.

Although the advent of HPLC techniques has markedly improved the analyst's ability to detect and quantitate the food pigments, many complex analytical challenges exist. Further understanding of the chemistry and interactions of pigments in food systems will be necessary to enhance our capabilities in extraction, separation, and detection techniques. In addition, advances in instrumentation and rapid detection methods will improve and provide the food chemist with additional tools for rapid measurements of pigments and colorants in food products.

18.9 STUDY QUESTIONS

1. Given that chlorophyll and carotenoid pigments are both lipid soluble, what chromatographic assays (i.e., stationary phase and solvent composition) could be considered for their measurement in sample extracts? Consider specific procedures that measure for both classes of compounds in a single analysis.
2. Describe appropriate extraction methods for the lipid-soluble (chlorophylls, carotenoids) and water-soluble (anthocyanins, betalains, and FD&C dyes) pigments. For the water-soluble myoglobins, what changes in extraction procedures would be required?
3. Explain the limitations of the AOAC procedure for analysis of carotenoids in comparison to HPLC techniques. How do these limitations influence provitamin A measurements?
4. In high-fat-containing foods, lipids may interfere with the analysis of lipophilic pigments. Discuss what procedures can be used to minimize the interference.
5. Discuss why pH is an important factor when trying to analyze and separate anthocyanin pigments present in fruit juice extracts.
6. Once FD&C dyes are added to formulated food products, explain why it may be difficult to quantitatively extract and measure the dyes that have been added.
7. Describe the techniques used to measure the quantity of a single pigment present in an extract containing a mixture of pigments. Consider procedures that do and do not require chromatographic separation steps.

18.10 REFERENCES

1. Schwartz, S.J., and Lorenzo, T.V. 1990. Chlorophylls in foods. *Critical Reviews in Food Sci. and Nutr.* 29(1):1–17.
2. AOAC. 1990. *Official Methods of Analysis*, 15th ed. Association of Official Analytical Chemists, Washington, DC.
3. Vernon, L.P. 1960. Specrophotometric determination of chlorophylls and pheophytins in plant extracts. *Anal. Chem.* 32:1144–1150.
4. Schwartz, S.J., Woo, S.L., and von Elbe, J.H. 1981. High-performance liquid chromatography of chlorophylls and their derivatives in fresh and processed spinach. *J. Agric. Food Chem.* 29:533–535.
5. Schwartz, S.J., and von Elbe. J.H. 1983. Kinetics of chlorophyll degradation to pyropheophytin in vegetables. *J. Food Sci.* 48:1303–1306.
6. Minguez-Mosquera, M.I., Garrido-Fernandez, J., and Gandul-Rojas, B. 1990. Quantification of pigments in fermented manzanilla and hojiblanca olives. *J. Agric. Food Chem.* 38:1662–1666.
7. Canjura, F.L., and Schwartz, S.J. 1991. Separation of chlorophyll compounds and their polar derivatives by high-performance liquid chromatography. *J. Agric. Food Chem.* 39:1102–1105.
8. Grese, R.P., Cerny, R.L., Gross, M.L., and Senge, M. 1990. Determination of structure and properties of modified chlorophylls by using fast atom bombardment combined with tandem mass spectrometry. *J. Am. Soc. Mass Spectrom.* 1:72–84.
9. van Breeman, R.B., Canjura, F.L., and Schwartz, S.J. 1991. Identification of chlorophyll derivatives by mass spectrometry. *J. Agric. Food Chem.* 39:1452–1456.
10. Bushway, R.J., and Wilson, A.M. 1982. Determination of a- and b- carotene in fruit and vegetables by high performance liquid chromatography. *Can. Inst. Food Sci. Tech. J.* 15(3):165–169.
11. Khachik, F., Beecher, G.R., and Whittaker, N.F. 1986. Separation, identification and quantification of the major carotenoid and chlorophyll constituents in extracts of several green vegetables by liquid chromatography. *J. Agric. Food Chem.* 34:603–616.
12. Bureau, J.L., and Bushway, R.J. 1986. HPLC determination of carotenoids in fruits and vegetables in the United States. *J. Food Sci.* 51:128–130.
13. Corbet, A., and Schwartz, S.J. 1991. Unpublished data. Dept. of Food Science, North Carolina State Univ., Raleigh, NC.
14. Khachik, F., Beecher, G.R., and Lusby, W.R. 1989. Separation, identification and quantification of the major carotenoids in extracts of apricots, peaches, cantaloupe, and pink grapefruit by liquid chromatography. *J. Agric. Food Chem.* 37:1465–1473.
15. Timberlake, C.F., and Bridle, P. 1971. The anthocyanins of apples and pears: The occurrence of acyl derivatives. *J. Sci. Food Agric.* 22:509–513.
16. Wrolstad, R.E. 1976. Color and pigment analyses in fruit products. Station Bulletin 624, October 1976, pp. 1–17. Agric. Exper. Station, Oregon State Univ., Corvallis, OR.
17. Hong, V., and Wrolstad, R.E. 1990. Use of HPLC separation/photodiode array detection for characterization of anthocyanins. *J. Agric. Food Chem.* 38:708–715.
18. Schwartz, S.J., and von Elbe, J.H. 1982. High performance liquid chromatography of plant pigments—A review. *J. Liq. Chromatog.* 5:43–73.

19. Nilsson, T. 1970. Studies into the pigment in beetroot. *Lantbrukshoegskolans Annaler.* 36:179–219.
20. Saguy, I., Kopelman, I.J., and Mizrahi, S. 1978. Computer-aided determination of beet pigments. *J. Food Sci.* 43:124–127.
21. Schwartz, S.J., and von Elbe, J.H. 1980. Quantitative determination of individual betacyanin pigments by high-performance liquid chromatography. *J. Agric. Food Chem.* 28:540–543.
22. Huang, A.S., and von Elbe, J.H. 1985. Kinetics of the degradation and regeneration of betanine. *J. Food Sci.* 50:1115–1120, 1129.
23. Krzywicki, K. 1979. Assessment of relative content of myoglobin, oxymyoglobin and metmyoglobin at the surface of beef. *Meat Sci.* 3:1–10.
24. Hunt, M.C. 1980. Meat color measurements. Reciprocal Meat Conference Proceedings. 33:41–46.
25. Faustman, C., and Cassens, R.G. 1990. The biochemical basis for discoloration in fresh meat: A review. *J. Muscle Foods.* 1:217–243.
26. Warriss, P.D. 1979. The extraction of haem pigments from fresh meat. *J. Food Tech.* 14:75–80.
27. Krzywicki, K. 1982. The determination of haem pigments in meats. *Meat Sci.* 7:29–36.
28. Hallagan, J.B., and Thompson, D.R. 1991. The use of certified food color additives in the United States. *Cereal Foods World.* 36:945–948.
29. Coulson, J. 1980. Synthetic organic colours for food. Ch. 3, in *Developments in Food Colours-1,* John Walford (Ed.), pp. 71–74, 83–87. Applied Science Publishers, LTD., London, England.
30. Lawrence, J.F., Lancaster, F.E., and Conacher, H.B.S. 1981. Separation and detection of synthetic food colors by ion-pair high-performance liquid chromatography. *J. Chromatog.* 210:168–173.
31. Chudy, J., Crosby, N.T., and Patel, I. 1978. Separation of synthetic food dyes using high-performance liquid chromatography. *J. Chromatog.* 154:306–312.
32. Bauernfeind, J.C. 1972. Carotenoid vitamin A precursors and analogs in foods and feeds. *J. Agric. Food Chem.* 20:456–473.

CHAPTER

19

Analysis of Pesticide, Mycotoxin, and Drug Residues in Foods

William D. Marshall

19.1 PESTICIDE RESIDUES

19.1.1 Introduction

19.1.1.1 Regulations Governing Pesticides Residues in Foods

Pesticides continue to be used, on a massive scale, to mitigate or limit the economic losses associated with decreases in crop yields (and/or quality) that are caused by noxious insects, fungi, weeds, or other pests. When applied improperly, residues of some of these pesticides can remain on foods and, as such, can pose a significant hazard to human health. Thus, in most countries of the world, the sale, distribution, and ultimately the application and end use of these chemical and biological poisons are strictly controlled by law (1). To be offered for sale/distribution within the United States, a candidate control agent must be registered with the **Environmental Protection Agency** (EPA) (2). Other countries have adopted analogous procedures. The process of **registration** is an administrative procedure in which the agency reviews a detailed compilation of the chemical, biochemical, and environmental fate/behavior of the active agent as well as an extensive review of the toxicology of the pesticide to both target and nontarget organisms. Included in this data package must be analytical methods for the determination of the terminal residues of the ingredient on each crop that might be treated with the pesticide (and for more recent registrations, a requirement for information about the behavior of the active ingredient in standard multiresidue analytical methods has also been included). Thus, the process of registration involves an application to use the biological or chemical control agent at specified levels on specific crops.

The agency is charged with balancing the risks and benefits associated with the proposed use(s) of the candidate pesticide and to assure itself that neither humans nor the environment will be placed at undue risk. Registration status can be reviewed, altered, or revoked based on new toxicological, environmental, and/or residue data. In the latter case, the agency would issue a **Rebuttable Presumption Against Registration** (RPAR), which would be printed in the *Federal Register* and would outline the agency's reasons for the proposed revocation-alteration of registration and would provide interested parties a set time to respond and rebut the agency's arguments.

Based on the intended use of the candidate pesticide, EPA establishes a legal limit of pesticide residue at harvest (which will include the active ingredient as well as toxic metabolites and transformation products). Conventionally, **tolerance levels** are expressed in units of concentration and are in the range of low to sub mg/kg of fresh weight of edible produce [hence the term *ppm*, **part per million** (parts)]. These tolerances must be established prior to registration. Crop-specific tolerances cannot be legally exceeded, and residues are legally prohibited in or on food crops for which a tolerance has not been established. If experiments demonstrate that processing the raw agricultural commodity will concentrate residues of the control agent, a separate tolerance for processed products is issued.

Generic names (common names) for the active ingredients of products offered for sale have been developed by pesticide science societies so as to avoid reference to either different formulations (which may be sold by different companies) containing the same active ingredient or to their trade names. Wherever possible, the **International Union of Pure and Applied Chemistry** (IUPAC) common name is used when more than one exists.

19.1.1.2 The Enforcement of Tolerances

Whereas registration in the United States is the prerogative of EPA, the responsibility for enforcing the tolerance limits is the responsibility of the Food and Drug Administration (FDA), which oversees all foods and feeds moving within interstate commerce, with the exception of meat and poultry products, which are the responsibility of the U.S. Department of Agriculture (USDA) (3).

19.1.1.3 Changing Pesticide Usage Patterns

In the vast majority, the pesticides in current use are synthetic organic chemicals that are foreign to the environment (hence the term xenobiotic). This can be expected to change slowly with time as accepted agricultural practice evolves from a chemical based to a combination of chemical and biological control practices. For example, the annual Western world market for microbial pesticides is anticipated to increase some 200-fold, to $8 billion, by the turn of the century (4). It can also be anticipated that the market share will continue to shift in favor of control agents that are less persistent, that are more selective in their biochemical mode of action, and/or that can be used efficaciously at lower rates of application.

Pesticides are not biocides but rather are selectively toxic to target organisms. These chemical control agents have the ability to selectively disrupt certain specific biological processes. It is this high degree of selectivity that makes them useful as agrochemicals. Given the vast array of crop pathogens, crop predators, and other plants (weeds) that compete with the crop plants for limited nutrients, it is not surprising that no one control agent can protect even a single crop from all predators. Since these control agents selectively interfere with different biochemical processes, it is not surprising that they have different chemical structures and therefore that they have very different physical and chemical properties.

Currently, there are over 315 pesticides (active ingredients) for which tolerances have been set (5). Additionally, there are other pesticides whose registration status has been revoked but that persist in the environment, pesticides for which no tolerance has been established, pesticides used in food production in other countries of the world, and transformation products and/or impurities that can be formed during the manufacture of the active ingredient. Table 19-1 provides an estimate of the numbers of control agents within these different classes, which collectively exceed 450. Regardless of the exact number, it is small relative to the more than 8,000 tolerances that have been established. Given the number of different possible residues and possible food types and the current pesticide residue monitoring technologies, it is not physically possible to test every food commodity for every pesticide residue.

19.1.1.4 Are Foods Safe?

Public concern for pesticide residues in foods has increased dramatically in the last decade (6). Increasingly articulated fears concerning food chemical safety have placed added pressures on government regulatory and monitoring agencies as well as on producers and processors to demonstrate that foods offered for sale are free from toxic chemical contaminants. These concerns are heightened by the possible deleterious health effect(s) of residues of pesticides in the diet and a lack of information on the effect (if any) of continued exposure to combinations of pesticide residues. Finally, there is a sense that exposure to these residues is beyond the control of the consumer. The fact that it is currently not physically possible to test all foods for all possible residues is often interpreted as a lack of proof of the chemical safety of the food. Frequent surveys of our food supplies indicate, repeatedly, that the majority of the samples do not contain any detectable pesticide residues.

19.1.2 Contemporary Analytical Techniques for Pesticide Residues in Foods

19.1.2.1 General Considerations

Given that the objectives of an analysis for residues can be quite different, the methodological approach must be matched to the problem. On the one hand, the problem might be to detect suspected residues of a particular active ingredient within a shipment of fresh produce that is known to have been field treated with the pesticide. Alternately, a considerably more complicated problem would be to determine the levels of residues of any pesticide(s) within that same shipment. As with all analytical procedures, the results are generated from a small quantity of an extract that is actually presented to the instrument. The results are of no use unless they can be used to make predictions about the levels of pesticide(s) that are present within the rest of the shipment. This can only be done if two conditions are met: (1) The analytical sample must be homogeneous and (2) the sample must be a miniature replica of the shipment itself. The problem of obtaining a sample that is truly representative of the levels of trace substances in a heterogeneous matrix is difficult to solve. The problems associated with obtaining representative samples for the determination of trace contaminants in foods have been reviewed by Horwitz

TABLE 19-1. Numbers of Pesticides That Are Determined or Identified by the Principal FDA Multiresidue Methods[1]

Type of Pesticide	Total in Database	Total[2] for All 5 Methods	Methods in PAM I[3]				
			4	5	6	7	8
Pesticide with:							
tolerance	316	163	68	85	55	140	20
temporary or pending tolerance	74	10	4	3	4	9	4
no EPA tolerance	56	25	17	21	7	10	0
metabolites, impurities, etc.[9]	297	92	20	32	31	61	18

[1]Data as of May 1988 (reference 5)
[2]Entries in this column are not cumulative because more than one method can detect the same pesticide.
[3]Standard multiresidue methods as outlined in *Pesticide Analytical Manual (Volume 1) (PAM I).*
[4]GC method for relatively nonpolar (organochlorine and organophosphate) pesticides in fatty foods.
[5]GC method for relatively nonpolar (organochlorine and organophosphate) pesticides in nonfatty foods.
[6]GC method for organophosphate pesticides and metabolites.
[7]GC method for polar and nonpolar pesticides, using a variety of detectors.
[8]HPLC method for N-methyl carbamates.
[9]Only some of these pesticide-related compounds occur in foods and/or are of toxicological concern.

and Howard (7) and by Kratochvil and Peak (8). It is highly likely that the residues are unevenly distributed on exposed surfaces. Moreover, the food matrix itself is highly heterogeneous. For certain commodities, recommended sampling protocols have been established by both the Association of Official Analytical Chemists (AOAC) International (9) and by FDA (10).

Even when the samples are truly representative of the shipment, we often fail to recognize or to communicate to our customers that all of our analytical results are only estimates. The numbers we generate are not exact; rather, they are subject to uncertainties. These uncertainties can be appreciable (11)—typically, ± 10 percent if the analyte is present at the 1 to 10 mg/kg level and ± 30 to 60 percent if that same analyte is present at the 1 to 10 μg/kg level. It is not routinely possible to reduce these uncertainties appreciably.

19.1.2.2 Types of Analytical Methods for Pesticide Residues

There are several approaches to pesticide residue analysis. These methodological approaches vary in their degree of complexity; in the time, effort, and analytical instrumentation required to complete them; and in the degree of confidence that can be placed in the final results. Procedures may be quantitative multi- or single-residue methods, or only semiquantitative or even qualitative. Typically, one would use the least demanding procedure that will provide a level of confidence in the final results sufficient to answer the questions being posed.

19.1.2.2.1 Multiresidue Methods **Multiresidue methods** (MRMs) come closest to meeting the analytical needs of monitoring agencies charged with regulatory roles (12). They have been designed to detect and measure a multiplicity of residues in a range of foods. They are sufficiently precise to provide reliable estimates of residue levels for many pesticides at or below the established tolerance levels. These multistep methods typically contain the steps of sample preparation, extraction, and cleanup, followed by chromatographic separation with on-line detection using a highly selective detector and automated quantitation. Of the 10 MRMs currently used by FDA and USDA, eight are based on gas chromatography and the remaining two on high-performance liquid chromatography. The MRMs that are used currently have evolved over a period of many years, yet they continue to be modified, expanded, and optimized. However, none of these MRM procedures can detect all residues on all crop types. In practice, they represent a compromise among the number of residues that can be detected, the range of food types that can be handled, and the levels of residues that can be measured. Their principal advantage resides in the numbers of different residues that they can detect and determine. A detailed description of the MRMs currently used by FDA is available in ***Volume I of the Pesticide Analytical Manual (PAM I)*** (10).

19.1.2.2.2 Single-Residue Methods In contrast to MRMs, **single-residue methods** (SRMs) have been designed to measure a single analyte and, often, its principal metabolites and transformation products of toxicological importance. The majority of SRMs have been developed in support of applications for registration (including tolerance setting) and/or research into the metabolism and environmental fate of the analyte. The majority of SRMs are based on the same sequential steps as MRMs; however, each step has been optimized for the analyte(s) of interest. Generally, they are less time consuming to perform and often provide lower limits of detection than MRMs. However, they do vary in the level of validation to which they have been subjected. ***Volume II of the Pesticide Analytical Manual (PAM II)*** (10) consists solely of SRMs. Included in this volume are methods that have undergone EPA review (and possibly EPA laboratory evaluation) as well as certain methods that have been published in peer-reviewed scientific journals of high quality. In the latter case, these methods are similar to methods that have been approved by EPA but have been optimized for other commodities. In PAM II, those methods that have received EPA review are listed with roman numerals, whereas methods that have not been reviewed are lettered.

19.1.2.2.3 Semiquantitative and Qualitative Methods Semiquantitative and qualitative methods range widely in their abilities to estimate the level of a particular pesticide residue in a sample. In general, they are capable of detecting a limited number of somewhat similar pesticide residues. These methods are often referred to as **screening methods** in the sense that they are capable of assaying a large number of samples for the presence of a limited number of pesticide residues in a relatively short time. Additionally, they are generally robust in character (they are less sensitive to small changes in the purity of reagents, quantities of reagents, time factors, reaction temperatures, and/or environmental conditions) and are not limited to a highly controlled lab environment. Whereas semiquantitative methods provide an estimate of the concentration range for a detected residue, qualitative methods will detect the pesticide if present above some predetermined level. The principal benefits of these methods are their low cost, relative speed, and simplicity. These methods make use of such techniques as thin-layer chromatography, enzyme inhibition, and immunoassay.

19.1.3 Quantitative Methods

19.1.3.1 Overview

The basic steps of a quantitative analytical method for pesticide residues include the following:

1. **Sample preparation**—The plant parts are separated into edible and nonedible fractions followed by chopping, grinding, or macerating of the sample.
2. **Extraction**—Pesticide residues are removed from most of the sample's other constituents by solubilizing them in a suitable solvent. This step often involves blending the chopped sample with solvent in a homogenizer, followed by a filtration.
3. **Cleanup (isolation)**—The crude extract is purified further by removing those coextractives that can interfere in the subsequent determination step(s).
4. **Separation**—The components of the purified extract are further separated by a differential partitioning between a mobile phase (liquid or gas) and a stationary phase.
5. **Detection** and **quantitation**—A physical parameter of the separated components in the mobile phase is measured as they pass through a detector; this signal is then related to the quantity of analyte via a quantitation step.

There is a voluminous technical literature on pesticide residue analyses. Several methods have been used widely and have undergone extensive testing and validation for many different residues and different food types. There are several manuals (which are updated periodically) that provide detailed descriptions of the individual steps of these methods. Included among these compendia of methods are *Pesticide Analytical Manual (Volumes I and II)* (10); *Analytical Methods for Pesticide Residues in Foods* (13), published by the Health Protection Branch, Health and Welfare Canada; and *Analytical Methods for Residues of Pesticides* (14), published by the Government Publishing Office for the Ministry of Welfare, Health, and Cultural Affairs of the Netherlands.

The choice of a method can be simplified somewhat by recognizing that the following criteria dictate which procedure(s) will be most efficient: the properties of the food matrix, the properties of the analyte(s), and the detectors that are available to the analyst. It is convenient to group the food types into one of four broad categories based on moisture and fat content: (1) high moisture, low fat (fruits and vegetables), (2) high moisture, high fat (meats), (3) low moisture, low fat (dried fruits), and (4) low moisture, high fat (cocoa beans). Although not exact, the high-low boundary is roughly 2 percent for fat and 75 percent for water. A detailed compilation of the moisture, fat, and sugar contents of a wide variety of fresh and processed foods has been compiled by Luke and Masumoto (15). The polarity, volatility, chemical reactivity, and thermal stability of the analyte(s) will determine the efficiency of recovery from the food matrix, the behavior on cleanup columns, the choice of an appropriate analytical column, and the selection of an analytical instrument.

19.1.3.2 Sample Handling

Samples have a habit of arriving at the most inopportune times; often it is not possible to perform an analysis immediately. Sample handling procedures are designed to prevent any change of that sample in a way that would affect either the determination of the concentration of the analyte(s) or the nature of the analyte(s). If samples must be stored, it is essential that the conditions of storage be chosen so that the sample [both the analyte(s) and the matrix] deteriorates as little as possible. As with all biological materials, sample decomposition is usually retarded by storage in sealed containers at low temperature. Freezing the samples is recommended for certain protocols but otherwise avoided. Often it is preferable to perform part of the analysis and then to store a partially purified sample extract rather than store the whole laboratory sample itself.

19.1.3.3 Sample Preparation

Assuming that a laboratory sample has been provided that is a miniature replica of the food commodity for which residue data is required and that a working method that will provide the information required is available, the laboratory sample is prepared for analysis. The sample, as received, is divided into edible and nonedible portions; then a composite sample (often 1.5 kg) is prepared by chopping or grinding, followed by blending and mixing. A Hobart food chopper is very suitable for this purpose. Other food matrices can be best handled with a meat grinder, a hammer mill, or a larger-capacity food blender. These steps have two objectives: to reduce the structural features of the sample and facilitate the subsequent extraction and to produce a homogeneous composite sample from which subsamples can be taken. Care should be taken to avoid contaminating the sample and/or exposing it to unnecessary heat, which can cause loss of volatile analyte(s) and/or accelerate decomposition. Most pesticides are not systemic. They are not translocated within plants, nor do they traverse plant membranes. They can be expected to occur as surface residues on fresh produce and to be unequally distributed on those surfaces.

Without special care, residues on the outer damaged leaves of leafy greens that typically are not eaten can easily contaminate inner leaves. Fresh fruits and vegetables as offered for sale are not washed prior to analysis (this has usually been performed by the producer), and only damaged outer leaves are removed. Pesticide tolerances have been established for **raw agricultural commodities** (RAC). Thus, the RAC is analyzed as received. If the outer skins are not usually consumed (onions, melons, or kiwi fruit), only the outer 2 to 3 mm are removed. Similarly, only stems of grapes and strawberries and the stems and cores of apples are removed. Although it would appear to be somewhat rare, the degradation of certain pesticides' surface residues can be accelerated by the process of maceration with the food matrix. The degradation of the fungicide chlorothalonil on peas and captan and folpet on green beans has been reported to be accelerated by maceration. For field crops, when it is known that there has been no pretreatment, samples can be rubbed lightly to remove soil particles and other visible adhering contaminants.

19.1.3.4 Extraction

The objective of this stage of the analysis is to recover as much of the analytes as possible by solubilizing them. Often, a subsample (250 g or less) of the chopped or macerated composite sample is blended with a suitable **organic solvent,** often acetonitrile (CH_3CN) or acetone ($CH_3C(O)CH_3$). **Anhydrous salt** (NaCl or Na_2SO_4) can be added to absorb water, or water can be intentionally added so that the crude extract can be purified with a subsequent partitioning step with a second water-immiscible solvent. The solvent is separated from insoluble solids by filtration.

A not infrequent problem at this stage of the analysis is the formation of an **emulsion** (a suspension of one solvent in a second immiscible solvent that masks the interface between them). Emulsion formation often can be minimized by adding a salt to the predominantly aqueous phase. Once formed, it can sometimes be broken down by adding a small quantity of saturated salt solution, or with a few drops of alcohol or a commercial antifoaming agent, or by centrifuging the mixture if feasible. Less desirably, most emulsions will break down when left undisturbed for a few hours.

An alternate procedure (16) can be used to simplify somewhat the number of operations. Variations among different food types can be reduced appreciably by adding water to the sample to obtain a suspension that is ~70 percent water by weight. For samples that are > 70 percent water (fruits, vegetables, wines, milk), a 100 g aliquot is taken. Dry samples (less than 40 percent water), 10 to 50 g, are presoaked (up to 2 hrs) with sufficient added water to bring the water content to 100 g. For matrices that contain appreciable quantities of both water and fat (butter, animal tissues), sufficient water is added to the sample (10–30 g) to obtain a total of 100 g of water. Acetonitrile (200 mL) and dichloromethane (150 mL) or acetone (200 mL) and petroleum ether (150 mL) are added to the water-amended sample together with sodium chloride (30 g). The mixture is blended at high speed for 1–2 min. The organic phase is dried over sodium sulfate, reduced in volume to 3–5 mL, and diluted with 5 mL of an appropriate solvent and reconcentrated. The dilution-reconcentration steps are repeated to ensure the complete removal of those extraction solvents that can disrupt the operation of the detector. The resulting concentrate can be used for analysis by gas chromatography without further purification unless an electron capture detection is to be used. For fruits and vegetables (100 g), the coextractives amount to a small fraction of 1 g. The more widely used procedures for extraction and cleanup have been reviewed recently by Walters (17).

19.1.3.5 Cleanup

Often, the crude sample extract is partially purified before the separation/determination steps. The necessity for and the degree of cleanup required depends, to a large extent, on the instrumental detector to be used and to a lesser extent on the type of chromatography in the automated separation stage of the analysis. In general, it is the sample cleanup that is the most time-consuming, labor-intensive, and error-prone step of standard analytical methods. There is no universal cleanup procedure; instead, a variety of techniques have been used successfully. In this stage of the analysis, the analytes are separated from coextractives that can interfere with the detection of the analyte(s). Often, the preliminary partitioning step is followed by a preparative chromatography step. (See Chapter 28 for basic information on chromatography.) The crude water-acetone or water-acetonitrile extract is partitioned with a relatively nonpolar organic solvent. The organic phase is dried and reduced in volume. The residues are then sometimes further purified by a column chromatographic procedure (using either adsorption or size-exclusion chromatography). Typically, for adsorption chromatography, a 10 to 20 cm by 2.5 cm column packed with Florisil, silica gel, or less often alumina is used. The choice of packing material is both analyte and matrix dependent (Table 19-2). The activity of the adsorbent must be standardized (heated in an oven overnight, then deactivated by equilibrating with a prescribed quantity of water). It is sometimes convenient to make selective separations of the residues in the crude extract into groups based on their order of

TABLE 19-2. Stationary Phases Used for the Preparative Chromatographic Cleanup of Pesticide Residues

Florisil[R] (usually 60–100-mesh)
1. A diatomaceous earth adsorbent that is well suited to the cleanup of nonpolar pesticides in fatty foods. Efficiently removes interferents when eluted with nonpolar solvents.
2. Less effective for the cleanup of more polar pesticides in fruits and vegetables.
3. Prone to batch-to-batch variations in activity.
4. Can accelerate the oxidation of certain organophosphate (OP) esters containing thioether linkages (thiolo compounds) and can adsorb, irreversibly, certain other OPs (oxons).
5. Thioethers (R-S-R) can also be oxidized to sulfones (R-S(O)-R) and to sulfones (R-S(O)(O)-R) and strongly retained by this material.

Silica Gel (a highly gelatinous form of silica)
1. Useful for the isolation of more polar pesticides and for the partial cleanup of nonpolar pesticides (organochlorines, OCls) from animal fats.
2. Will not adequately separate plant coextractives from certain pesticides.

Alumina
1. This material tends to be more alkaline and can decompose certain OPs.
2. Can be substituted for Florisil in the cleanup of fatty foods.
3. Does not adequately separate plant coextracts from certain pesticides.

Carbon
1. Preferentially absorbs nonpolar and high molecular weight pesticides.
2. Efficiently removes chlorophylls but not waxes from vegetable extracts.
3. Pretreatment of this material strongly affects its adsorption behavior. Flow rates with open tubular columns are difficult to maintain.

Size Exclusion Packings
1. These materials separate mixtures principally according to size (not exclusively). High molecular weight materials are eluted first.
2. They must be pre-equilibrated (swollen) with the solvent or solvent mixture.
3. Losses of analyte residues are minimal (less than 10%) and are effective for the separation of animal fats and plant waxes from nonpolar pesticides.
4. Procedures often couple size exclusion with a short alumina column to remove residual carotenoids.

elution from the preparative column. Separate fractions of column eluate can be analyzed. The behavior of more than 200 pesticide residues on Florisil has been compiled in *PAM I* (10). Either isocratic elution or a stepped gradient of increasing solvent polarity can be used. Preparative size-exclusion chromatography represents an attractive alternative because it represents a different chromatographic mode to adsorption, it provides excellent recoveries (usually >80%), and it can be automated readily. Typically, the extract is added to 50 to 60 g of 200- to 400-mesh Biobeads SX3 (that have been preswollen by equilibration with the mobile phase) and eluted with methylene chloride or with acetone-cyclohexane mixtures (typically 25:75, v/v).

19.1.3.6 Derivatization

It is sometimes advantageous to alter the chemical structure of analyte residues to make them more suitable for detection by chromatographic techniques. This process of structural modification by chemical reaction, referred to as **derivatization,** can be used to enhance the thermal stability and/or volatility of the analyte(s), to modify chromatographic behavior, to increase the selectivity of the detection, and/or to increase the sensitivity of detection. Most often, derivatization is used to overcome limitations of sensitivity. Typically, substitution or addition reactions are used to introduce a chromaphore, fluorophore, or other functional group into the analyte(s) to augment the detector response to the resulting product(s). Typically, the derivatization reaction is performed precolumn, but postcolumn reactions can also be used provided that the reaction can be performed *in situ* and automated. The one requirement is that the identity of the original analyte(s) not be compromized by the reaction. The advantages of the derivatization must outweigh the disadvantages of increased sample handling, analysis time, and, typically, a decrease in precision. A wide variety of derivatization reactions have been used for selected pesticide residues (18, 19). Not surprisingly, this approach is limited to either single-residue methods or methods that determine a limited group of structurally similar pesticide residues.

19.1.3.7 Automated Chromatographic Separation

19.1.3.7.1 Gas Chromatography (GC) GC has been used routinely for pesticide residue determinations for some 30 years; it is the instrumental approach of choice for most analysts. (See Chapter 30 for general information on gas chromatography.) An extensive technical literature on the behavior of residues with this technique has been built up. In practice, the purified extract dissolved in a volatile solvent is added to the head of a glass column containing an adsorbing material (the stationary phase), which is maintained at elevated temperature inside an oven. Components of the sample mixture are volatilized at the normal operating temperatures of the instrument, and they partition between the stationary phase and a carrier gas (the mobile phase). Since components of the sample can only advance through the column when dissolved in the mobile phase, separations are achieved because different components have different relative affinities for the two phases. Typically, for routine MRMs, the column is maintained under isothermal conditions; however, a temperature program that increases the column temperature with time can also be used.

A variety of column formats are available commercially. **Packed columns** that provide low resolution chromatography are typically glass coils 2 to 4 mm in inner diameter and some 1.5 to 2 m in length. These columns are filled with a solid support (usually 100- to 120-mesh), often a fire brick or a diatomaceous earth that has been extensively washed, neutralized, and sized and then precoated (typically at 4–6% by weight) with an active polymer (the liquid phase). It is this thin liquid coating that is the active part of the stationary phase. There is a bewildering array (several hundred) of liquid phases available for packed column GC, and they can be purchased in bulk or precoated (and pretested) on the inert support of your choice. For the majority of residue laboratories, there is no advantage to coating inert supports in house. Stationary phases with higher batch-to-batch reproducibilities are inexpensive and are available from numerous chromatographic supply houses.

For the vast majority of pesticide residue problems, it is enormously advantageous to follow a standard method. To choose a stationary phase for a custom analysis or for method development studies, a simple rule of thumb is helpful: Like dissolves like (nonpolar analytes will interact most strongly with nonpolar liquid phases and vice versa). Analytes will behave quite differently on different liquid phases even if all other analytical conditions are identical. An extensive compilation of the retention times of many pesticides relative to an internal standard (chlorpyrifos, which contains the heteroatoms P, S, N, and Cl in addition to C, H, and O) on OV-17, on OV-101, and on DEGS has been presented by Froberg and Doose (20). These authors also provide many helpful suggestions on packed column preparation, conditioning, care, and rejuvenating. With time, there will be a gradual but inevitable buildup of sample coextractives at the head of the packed column (signaled by peak tailing, reduced response and/or retention times, or even analyte degradation). To restore performance, the initial 4 to 8 cm of packing can be removed, the exposed glass cleaned, and the packing replaced with fresh preconditioned stationary phase. Relative to other stationary phases, the DEGS material has a greater tendency to bleed (slowly volatilize), which can foul detectors.

A second format of GC column, **capillary columns** are fabricated from fused silica and are jacketed with an impervious polyamide coating. Interestingly, it is the coating that renders these columns quite flexible. Typically, capillary columns are 0.25 to 0.50 mm in diameter and can range in length up to 50 m (although 25 m is often recommended). The active stationary phase is usually permanently bonded to the inner surface of the columns so that they can be used at higher operating temperatures than packed columns. Their principal advantage is their increased resolving power (roughly twelvefold over packed columns of similar lengths), which is achieved at the expense of sample capacity. In addition to providing increased resolution, the increased chromatographic efficiency of these columns (sharper peaks) results in better limits of detection. Despite these appreciable advantages, capillary columns are used only infrequently for routine pesticide residue analysis because of fouling problems due to coextractives. The use of true capillary columns requires that older chromatographs be modified to optimize their resolving capabilities (21, 22).

Columns intermediary between packed and true capillary (0.75–1.0 mm i.d.) are also available and present many of the best characteristics of both formats. Older chromatographs need not be modified; **megabore columns** retain much of the resolving power of true capillary columns even at higher flow rates, yet they retain sample capacities that are midway between packed and capillary columns.

19.1.3.7.2 Gas Chromatographic Detection of Residues The successful application of GC to pesticide residue analysis is critically dependent on the use of sensitive and highly selective detectors. The great majority of pesticides contain one or more heteroatoms (atoms other than H, C, or O) within their molecular framework. The presence of heteroatoms is exploited advantageously by using detectors that provide a greatly enhanced response to these heteroatoms. These include flame photometric (for S or P detection), electron capture (halogens, S and N), electrolytic conductivity (halogens or N and S), and thermionic detectors (N or P).

Typically, a **flame photometric detector** (FPD) is operated with a hydrogen-rich flame under conditions that favor the formation of S_2 or HPO decomposition products when S- or P-containing pesticides are combusted. By interposing the appropriate narrow bandpass optical filter between the flame and the photomultiplier tube, operating conditions can be chosen so as to detect either of these species with a high degree of selectivity (several thousandfold for heteroatom-containing analyte over carbon-containing coextractive). However, the response to S is only 22 times greater at 394 nm than at 526 nm (i.e., some response to S-compounds can be anticipated under operation in the phosphorous mode). This detector is more robust (more forgiving of dirty samples) yet highly sensitive and highly selective. The response of this detector to sulfur-containing analytes is not linear (it varies as the quantity of analyte raised to the power of approximately 1.8), hence calibration requires several standard concentrations to accurately define a calibration curve. Additionally, as the oxidation state of the sulfur increases, the detector response decreases, and it has also been reported that high concentrations of co-eluting hydrocarbon(s) also decrease the response.

Electron capture detection is characterized by only moderate selectivity (enhanced response to halogen, nitrogen, phosphorous, sulfur, metals, and sites of unsaturation within the analyte) but it can be made to be very sensitive (picogram amounts). Relative to other detectors, sample extracts must be purified more extensively when using this detector. When run in a constant current mode, the detector becomes saturated, easily resulting in a narrow linear range. This linear range is increased appreciably by operating the detector in a pulsed mode. Halogenated and aromatic solvents must be avoided, and, when possible, alkanes are used. However, for certain applications, acetone must be used to avoid the loss of certain OPs that can become adsorbed on the inner surfaces of the syringe.

Within a Hall-type **electrolytic conductivity detector** analyte is converted to an ionic species, interfering ions are scrubbed from the effluent, and the ionic analyte-transformation product is detected within an electrolytic conductivity cell. When operated in the nitrogen mode, analyte is mixed with H_2 gas and hydrogenated over a nickel catalyst at 850°C in a quartz tube. Acidic hydrogenation products are removed from the effluent by passage over a $Sr(OH)_2$ trap, and the NH_3 from the analyte passes to the conductivity cell. Detection limits can be as little as 0.1 ng (as nitrogen). The solution in the conductivity cell is cleaned by continuous circulation through an anion exchanger. When operated in the chlorine mode, this detector also provides some response to bromine, phosphates, and sulfates.

A **thermionic emission detector** is a modified flame ionization detector in which a nonvolatile ceramic bead is used to suppress the ionization of hydrocarbons as they pass through the low temperature fuel-poor hydrogen plasma. The bead (often rubidium silicate fused into silica) is heated to 600–800°C and controlled independently of the plasma temperature. This detector responds to both nitrogen and phosphorous at low He flow rates to the plasma. A highly selective response to phosphorus containing analytes is achieved with higher flow rates of H_2 and inverting the polarity between the plasma tip and the collector.

Mass spectrometry (MS) (see Chapter 26) offers enormous possibilities as a highly selective means of detection and quantitation of pesticide residues. This technique offers unparalleled performance in terms of selectivity, in confirmatory power, and in universality of analytes that can be detected. Dedicated systems for gas chromatography (mass selective and ion trap types) continue to decrease in cost and to improve in performance. These devices provide maximum response when only a few masses are monitored selectively (**selected ion monitoring**, SIM, as opposed to recording complete mass spectra). For multiresidue monitoring, there is a requirement that the mass detector be capable of rapidly switching between preselected masses as the chromatogram develops. It is in this area where there has been continued improvement in recent years. The design and operation of GC detectors has been reviewed in detail by Holland and Greenhalgh (23).

19.1.3.7.3 High-Performance Liquid Chromatography (HPLC) The application of HPLC to pesticide multiresidue analysis has been largely restricted to those analytes that do not possess either the volatility or the thermal stability required for GC. Typical among these analytes are N-methyl carbamates [R-C(O)NH-CH_3], which are decomposed thermally at normal GC operating temperatures. (See Chapter 29 for general information concerning HPLC.) In HPLC, samples are separated, normally at ambient temperatures, by partitioning between a solid stationary and a liquid mobile phase. The sample is introduced into a flowing solvent stream with an injection device and transferred to the head of the analytical column, typically a 4 mm i.d. by 15 to 25 cm stainless steel tube packed with 3, 5, or 10 μm particles of silica (or polymeric resin). One benefit of the smaller particle size for the stationary phase is that faster mobile phase flow rates can be used without appreciably degrading the resolving power of the column. As with GC, separation is achieved because components of the mixture possess different affinities for the two phases. Optimal chromatography is normally achieved with mobile phase flow rates of 1 mL/min and requires a highly reproducible pumping system to deliver the solvent against a back pressure of 1,200 to 2,000 psig. As with classical gravity flow liquid chromatography (Chapter 28), HPLC columns containing different packing materials (stationary phases) are commercially available to provide a variety of separation modes. The most popular approach, **reversed-phase** HPLC, employs a modified silica stationary phase in which the accessible surface silanol functional groups have been substituted with C_{18}, C_8, or even C_2 normal alkyl chains. The treated packing material preferentially adsorbs nonpolar analytes. A very polar mobile phase, typically a mixture of water with either methanol or acetonitrile forces the solutes to interact with the nonpolar stationary phase. The most polar components of the sample mixture elute first, and the order of elution is reversed relative to the order observed for untreated silica. In addition, some solid supports are designed to interact with specific functional groups [possibly thiols (-SH) or diols (-CH(OH)-CH(OH)-)] that may be present in the mixture to be separated. However, virtually no separation occurs by only one chromatographic mode; there are always other interactions in addition to the predominant type. As a consequence, the optimal chromatographic conditions cannot be predicted with certainty and must be confirmed by experiment. One advantage of the HPLC approach to pesticide residue analysis is that sample cleanup is usually less extensive. An extensive compendium of references for the analysis of pesticides by HPLC has been compiled by Muszkat and Aharonson (24); other reviews by Lawrence (25) and Moye (26) also provide an overview of the application of the technique to pesticide residues.

19.1.3.7.4 HPLC Detection of Pesticide Residues The two detection techniques that have been most popular for HPLC determination of pesticide residues are ultraviolet (UV) and fluorescence spectrometry. Typically, the column eluate is transferred directly to a low volume sample cell (8 to 20 μL), which has been designed to withstand moderate pressures. For **UV detection,** two designs have been popular. Fixed wavelength detectors can be operated at 254 (or 280 nm) and have the advantages of high sensitivity and low cost. A mercury lamp with a narrow bandpass filter provides the high energy incident radiation in this device. A phosphor can be inserted into the light path to shift the incident radiation from 254 to 280 nm. Variable wavelength detectors have the appreciable advantage that they can be operated at wavelengths between 200 and 650 nm but are less sensitive because only a small fraction of the radiation from their polychromatic light sources is actually used. However, the operator can select the wavelength to be monitored to increase the response and/or the selectivity of detection. **Fluorescence detectors** exploit the fact that the fluorescence emitted by an analyte is at longer wavelength (lower energy) than the wavelength of light required to induce this phenomenon. Typically, a filter or monochromator (interposed between the light source and sample cell) is used to select the wavelength(s) of radiation incident on the sample, and a second and separate filter or monochromator (interposed between the sample cell and the detector) is used to select the wavelength(s) to monitor for the light that is re-emitted from the analyte (emission radiation). Fluorophores (fluorescent tags) can readily be added to reactive functional groups of those analytes that are not naturally fluorescent. This process of derivatization, which increases the number of analytes and/or the response to these analytes, can be performed either pre- or postcolumn.

19.1.3.8 Quantitation

19.1.3.8.1 Overview Since some physical parameter of the sample is actually measured (ability to capture electrons, or to absorb light, etc.) by the detector, the analog signal from this device must be related to the quantity of analyte via a separate process of calibration. It is assumed that the change in the detector signal as the component analyte passes through the detector is caused only by that component. A recording device provides a record of the changes in the detector signal with time (a **chromatogram**). The result is a series of approximately Gaussian peaks corresponding to the separated components of the mixture. The time (or volume of mobile phase) required to reach the maximum of a particular peak is referred to as its retention time, and the peak height (vertical displacement from the base line) or peak area is used to determine the quantity of analyte that must have caused that displacement. Components of the mixture are identified, tentatively, based solely on their retention times (i.e., the retention time of the component is identical to the retention time of an authentic standard that has been chromatographed under identical conditions). Frequently, the retention time for an analyte is expressed relative to the retention time for an **internal standard** that has been added intentionally to the sample. The pesticide chlorpyrifos has been used often as an internal standard because it chromatographs well on many columns and is detected by all selective GC detectors.

Quantitation is performed, typically, by a method of **external standards.** A series of standard solutions is prepared by dissolving known quantities of authentic standard pesticide in a suitable solvent. These standard solutions are separately chromatographed to establish the response of the detector (in terms of either peak area or peak height) for the quantity of analyte injected into the instrument. A relationship between response and quantity of analyte is then established by regression analysis (the former on the latter).

19.1.3.8.2 Ancillary Devices for Automated Chromatography It was inferred earlier that the major source of variation and uncertainty in most residue methods resulted from the cleanup step, which is laborious and time consuming as well as being error prone. This is especially true if ancillary devices are used to increase the degree of automation of the separation and detection/quantitation steps of the analysis. A **recording integrator** will not only determine the retention time of each component of the mixture, but will also determine the peak area and/or peak height of each component as well as the corresponding area/height ratio. These robust, low-cost dedicated units make the successful outcome of the determination somewhat less critically dependent on the volume of sample injected into the chromatograph. The determination of peak height or peak area is achieved using sophisticated mathematical algorithms that provide more precise and more accurate measurements of these parameters than is available using a strip chart recorder. Moreover, these measurements are independent of the visual presentation of the chromatogram. Thus, accurate quantitation can be achieved even if attenuation of the detector signal results in very small peaks for some trace components of the mixture. Since real chromatograms rarely contain truly Gaussian peaks, operator-selected variables (peak width, threshold, and area reject) are optimized to achieve reliable results.

Other microcomputer-based **chromatography software packages** can increase the ease of the quantitation process. Retention times of peaks, which have been detected, can be calculated relative to an internal standard; then the resulting relative retention times can by compared with a data base for standards to assign a probable identity. Additionally, two or more chromatograms (or regions of chromatograms) can be visualized on the screen for comparison. Typical applications might include the comparison of two chromatograms generated from the same extract with different selective detectors or the comparison of a test chromatogram with a control (pesticide-free) chromatogram of the same food matrix. A review by Stan (27) provides a concise overview of the capabilities of personal-computer-based software for the evaluation of pesticide residue chromatographic data.

An **autoinjector** can be used to deliver a preset volume of sample to the chromatograph. This increase in automation frees the operator for other tasks (and, in theory at least, permits 24 hr operation) but, more importantly, it increases the precision associated with the injection step. Other automation can involve the use of column switching valves so that samples can be directed to different columns.

19.1.3.9 Chemical Confirmation of Pesticide Residues

The degree of confidence that can be placed in a peak assignment may not be sufficiently high; it is often preferable to corroborate the identity of the residue that has been detected. It is generally agreed that the most reliable way to increase the level of confidence concerning the identity and amount of an analyte present in a sample is to obtain concordant results using two independent methods that are based on entirely different analytical principles. This is not always feasible—multiresidue methods that are not based on a chromatographic separation are simply unavailable. For pesticide residues, it is customary to make use of one or more of the following less than ideal approaches. GC-MS can be used to (1) record the mass spectrum of the analyte (corroborate identity) and (2) provide a separate estimate (quantitation) of residue levels. Alternately, a different selective detector can be used to replace the one used for the original analysis, and/or a second column with an appreciably different stationary phase can be employed. Finally, the analyte can be chemically altered and then rechromatographed to demonstrate that the signal for the analyte has disappeared and has been replaced by a second signal at the predicted retention time for the anticipated transformation product. Two reviews (19, 28) explore these approaches in depth.

19.1.3.10 Immunoassays

Immunoassays are a group of related analytical techniques whose basis of commonality is the use of antibodies that have a high affinity for, and only for, the pesticide analyte. Immunoreactions are highly selective (virtually specific) addition reactions between the antibody (a high molecular weight glycoprotein that exhibits the properties of immunoglobulins) and the analyte of interest. This interaction can be exploited analytically if there is a means of detecting and, preferably, of quantifying the reaction, typically by competitive inhibition or by displacement in which the binding of the pesticide to the antibody competes with or displaces a tracer molecule. (See Chapter 33 for a detailed description of immunoassay techniques). A major concern is the degree of **cross-reactivity** (affinity) the antibodies show for related chemical structures. Thus, antibodies that were developed to one pesticide may react with other related chemical structures. Usually, this cross-reactivity is characterized by a lower affinity for the related structures than for the analyte itself. The merits of immunoassays include their high selectivities and the simplicity, speed, and moderate cost of the procedure relative to other methods for pesticide quantitation. For certain assays, such as for carbendazim, successful analyses could be performed on crude ethyl acetate extracts without cleanup. By contrast, other procedures (for polychlorinated biphenyls) were only successful when the cleanup was as extensive as was required for gas chromatography. The principal disadvantage of the immunochemical approach is the extensive effort (and time) required to elicit the antibodies in a vertebrate host. Recent reviews (29, 30, 31) on the application of immunoassays to pesticide residues are illustrative of the current status and the potential of this approach.

19.1.4 Thin-Layer Chromatography as Semiquantitative and Qualitative Method for Pesticide Residues

Relative to automated chromatographic techniques, **Thin-layer chromatography** (TLC) generally provides roughly one tenth of the resolving capabilities of a packed column for GC. In addition, the reduced precision associated with quantitation (relative to either GC or HPLC) and the limits of detection have caused this approach to fall somewhat into disfavor. (Refer to Chapter 28 for a discussion of TLC.) However, TLC has certain features that can be exploited advantageously, particularly when used as a semiquantitative screen for a limited group of pesticide residues. Quantitation need not be performed on line; thus, a wide variety of static visualization techniques have been developed. In addition to appreciable improvements in resolving power that can be obtained with a smaller particle size stationary phase (**high-performance thin-layer chromatography,** HPTLC), devices for reproducibly spotting the sample onto the plate and for quantitation by *in situ* densitometry appreciably improve the reproducibility of the technique. Extensive reviews on the detection and determination of pesticide residues by TLC are available (32, 33, 34, 35).

One successful application of TLC has been the detection and estimation of pesticides that inhibit cholinesterases (36). Many organophosphate (OPs) and carbamate insecticides are capable of inhibiting this group of enzymes. A crude extract containing the suspect residues is separated by TLC. The developed plate is dried and sprayed with a solution containing one (or more) of these enzymes (often a partially purified tissue extract) and then with a substrate that, when hydrolyzed by the enzyme, liberates a colored product. The lack of color development indicates enzyme inhibition, so that the pesticide residues are visualized as colorless areas within a colored background. The zone of inhibition is proportional to the quantity of inhibitor present. However, many OPs (thiophosphate and dithiophosphate esters) must be activated by oxidation to their oxon analogs to serve as potent inhibitors. The stability of oxons is such that it is often preferable to perform the oxidation after the plate has been developed. Care must be taken that traces of oxidant do not interfere with the subsequent color development. More importantly, there are a number of other inhibitors of cholinesterases that occur naturally in foods. A kit based on cholinesterase inhibition has been developed by EnzyTech Inc. (Kansas City, MO). It is claimed to detect, at low ng/g levels, all cholinesterase-inhibiting insecticides (37).

A somewhat similar approach is the chromogenic detection of photosynthesis-inhibiting herbicides. Phenylureas, phenyl carbamates, 13 uracils, acyl anilides, and triazines can be detected on TLC separations of crude extracts that have been partitioned into dichloromethane (38). The eluted plate is sprayed with a suspension of chloroplasts, then with a redox indicator (2,6-dichloroindophenol), and exposed to sunlight. Photosynthetic inhibitors are visible as blue spots of unreacted dye.

19.2 MYCOTOXIN RESIDUES

19.2.1 Introduction

Mycotoxins are a broad group of chemically diverse toxins (natural products that are toxic to some other organism) that are produced as secondary metabolites, principally by the terrestrial filamentous fungi commonly known as molds. The filamentous nature of molds is ideally suited to growth over surfaces and through solid substrates; the mycelium forms an efficient system for utilizing the available nutrients and transporting them to the growing hyphal tips. A consequence is that secondary metabolic products, including mycotoxins, will remain localized and highly concentrated in the food. The description as **secondary metabolites** suggests only that these chemicals are not required by the producing organism for growth. Mycotoxins can be produced as a result of fungal infection either in the field or in storage. Additionally, there is not a good correlation between the level of fungal infection and the levels of the mycotoxin(s) in the contaminated produce. Failure to detect viable inocula of a particular toxigenic species in fresh or stored produce is not a certain indicator that the mycotoxin is absent, nor does the presence of that species guarantee that it will have produced toxin(s). Thus, features of this class of toxins are that individual members are frequently produced only by a specific species, and levels of production vary greatly not only among different strains of that species but also in response to environmental and nutritional conditions. Thus, particular strains of *Aspergillus flavus* are used in the manufacture of koji, whereas other strains can produce aflatoxins. Analytical methods for the detection and quantitation for mycotoxin residues in foods and feeds are necessary to ensure that these commodities are safe for human and/or animal consumption. Although the toxic effects vary greatly among different members of this class, they are generally relatively small molecules (MW <1,000) and, typically, are not antigenic. They do not appear to accumulate in the body, and their toxicological effects, which can be acute but rarely fatal, vary widely. The principal concern, at least for humans, remains the deleterious health effects associated with chronic exposure. Hesseltine (39) has suggested the following order of relative importance, on a worldwide basis, for the common mycotoxins: aflatoxins (hepatotoxins), ochratoxin (nephrotoxin), trichothecenes (dermatoxins), zearalenone (estrogen), deoxynivalenol (dermatoxin), and citrinin (nephrotoxin). To limit human exposure to mycotoxins, legal limits (typically 5 to 25 μg/kg for aflatoxins) in foods and animal feeds have been established in many countries (40, 41). As with pesticide residues, a number of analytical methods for screening, survey, and regulatory control have been developed and validated by interlaboratory collaborative studies. Organizations such as the AOAC, American Oil Chemists' Society (AOCS), American Association of Cereal Chemists (AACC), and the IUPAC have mycotoxin method validation programs.

19.2.2 Sampling

For mycotoxins, the problem of obtaining a sample of the produce that is representative with respect to burdens of mycotoxins in the shipment is especially severe. (See Chapter 3 for a general discussion of sampling.) It is very informative to consider the findings of a study (42) in which the levels of aflatoxin were determined in 20 2.27 kg samples from each of 10 contaminated lots of cottonseed (Table 19-3). Each 2.27 kg sample was comminuted in a sampling mill and a 100 g subsample was then removed and analyzed following a standard method (extracted and the levels of aflatoxin determined densitometrically after separation on a minicolumn). In Table 19-3, the 20 analytical results for each of the 10 lots are ranked from low to high to facilitate comparisons. Rather than being symmetrically distributed about the mean (the "best" estimate of the aflatoxin concentration in the lot), the distribution of test results, for each of the 10 lots, is positively skewed (there are more values below the mean than there are above the mean). If a single 2.27 kg sample had been tested, there would have more than a 50 percent chance that the result would have been less than the true lot concentration. In general, the degree of skewness is greatest for small sample sizes and decreases as the size of the sample is increased. However, it can also be seen from the results of Table 19-3 that, even for the same sample size (2.27 kg), as the level of contaminant increases (lower to higher lot number), the distribution among replicate determinations becomes somewhat more symmetrical. This is reflected in the decreasing coefficient of variation (CV) with increasing mean level of analyte. Research has demonstrated a similar distribution of aflatoxin in contaminated corn, Brazil nuts (field contamination), and peanuts (post-harvest contamination) (43).

As with all analytical procedures, the final result is obtained from a series of sequential steps. The uncertainty associated with the estimates that are generated (final results) is cumulative and contains contributions from each of the steps. Variance can be used as a parameter of this uncertainty. The **total variance** (V_T) for the overall testing procedure is equal to the sum of the variances from several sources:

$$V_T = V_S + V_{SS} + V_A \qquad (1)$$

where: V_S = variance associated with the sampling procedure

V_{SS} = variance associated with the subsampling process

V_A = variance associated with the analytical method

Since variance is equal to the square of the standard deviation, it represents a measure of the precision (repeatability) associated with a process or a step; the most efficient way to make our estimates of the levels of toxins more precise is to improve the repeatability of the step with the greatest variance. Improving the repeatability of steps that are not major contributors

TABLE 19-3. Results of Aflatoxin Analysis for 20 Replicate 2.27 Kg Samples from Each of 10 Contaminated Lots of Cottonseed

Lot Number	*Aflatoxin Test Results (μg/kg)*																				*Mean (μg/kg)*	*Variance*	*CV[1] (%)*
1	0	0	0	0	0	0	0	0	0	1	1	1	1	1	3	5	7	9	10	14	2.7	17.1	156.0
2	0	0	0	1	1	1	1	1	1	1	4	6	10	10	12	13	16	27	40	44	9.5	174	139.6
3	0	0	0	0	1	1	1	1	1	1	8	9	12	23	24	24	25	30	40	50	12.6	234	122.0
4	0	1	1	1	1	1	1	4	4	11	12	14	19	20	21	22	24	28	38	56	14.0	223	107.0
5	0	4	9	14	16	16	25	27	30	30	31	32	32	32	34	37	40	42	42	100	30.3	819	94.5
6	6	6	7	14	20	22	24	24	31	33	38	40	42	45	54	60	67	68	68	165	41.7	1260	85.1
7	10	10	14	20	25	31	32	34	37	37	55	61	61	65	70	74	83	86	101	117	51.1	959	60.5
8	15	16	20	21	27	30	48	52	57	67	70	80	80	90	111	118	133	136	144	160	73.8	2183	63.4
9	1	16	29	40	53	73	85	89	100	104	113	118	120	121	121	143	157	175	260	266	109.9	4990	64.3
10	70	80	91	110	114	116	127	130	133	150	178	178	192	196	200	206	237	252	269	281	169.8	4741	40.6

From (62).

[1]CV = coefficient of variation = 100(Variance)$^{1/2}$/mean

to V_T will not impact greatly on the magnitude of V_T. Studies of aflatoxin levels on granular (peanuts and cottonseed) produce have indicated that the sampling step, especially for smaller sample sizes, is the major source of error (uncertainty) in the overall analytical process (44). Sampling error is especially large because the aflatoxin is present on only a very small percentage of the kernels within the lot (<< 0.1%), but the concentration within that kernel can be extremely high (up to 1×10^6 μg/kg). The one way to reduce V_S is to take a larger sample size. Table 19-4 presents estimates of the range of aflatoxin test results (at the 95% level of confidence) that can be expected for different composite sample sizes taken from a cottonseed shipment that is contaminated with 100 μm g/kg of toxin. The predicted range of results does not decrease at a constant rate with increasing sample size, indicating that increasing the sample size beyond a certain point may not be the best use of resources. It is assumed that the sample is taken in such a way that all parts of the shipment have an equal chance of being included in the sample. The contaminated particles occur in isolated pockets that are unevenly distributed; a composite sample must be accumulated from many different locations throughout the shipment. One of the characteristics of sampling statistics is that the reliability of a random sample does not depend so much on the size of the shipment as on the size of the sample. Thus, as a first approximation, the size of the shipment can be ignored.

Once a composite sample has been obtained, the aflatoxins must be recovered, usually by a process of extraction. Not surprisingly, it is not feasible to extract the mycotoxins from the total sample, which is much too large to be handled conveniently. In practice a subsample is prepared. As with the sampling step, the reproducibility of the subsampling is also dependent on the mycotoxin concentration. One way to reduce V_{SS} is to take a larger subsample. However, there is a practical limit to the size of subsample that can be handled. If the commodity is granular, it is essential that the sample be comminuted (ground to a smaller particle size and mixed thoroughly) in a suitable mill. This comminution process not only reduces the variance by the particle size (more particles per unit mass), it also increases the homogeneity of the product. The final particle size to which the sample can be reduced is limited by the screen mesh of the mill.

The variance associated with a particular analytical method for aflatoxins is also concentration dependent. The variance V_A can only be reduced by performing more replicate analyses (V_A is inversely proportional to the number of replicate analyses). Detailed sampling plans for separate commodities have been developed by the AOAC and AOCS and by FDA in collaboration with USDA.

For raw shelled peanuts where marketability is certified if the lot is found to contain less than the 25 μg/kg (sum of aflatoxins B_1, B_2, G_1, G_2) action level, a sequential sampling plan is recommended. It is considered that this action level will permit processors to meet the FDA's 20 μg/kg practical action guideline in finished products. Three 48 lb (21.8 kg) samples are taken. The first sample is ground in a special subsampling mill; a subsample (1,100 g) is taken and analyzed in duplicate. If the average of the two determinations is less than 16 μg/kg, the shipment is passed; if the average is >75 μg/kg, it is rejected. For averages between these two extremes, a second 48 lb sample is analyzed in duplicate, and the average of the four results is used to decide whether to accept (<22 μg/kg) or reject (>48 μg/kg). If the average is between this second set of extremes, the third 48 lb sample is also analyzed, and the average of the six determinations is used to decide whether the level of contamination is more or less than the actionable level (43). By contrast, an actionable level of 3 μg/kg (for B_1 alone) in shelled peanuts has been legislated in the Netherlands.

Aflatoxin contamination of foods of animal origin is not considered likely because livestock and poultry

TABLE 19-4. Predicted Range of Test Results (at the 95% Level of Confidence) When Testing a Cottonseed Lot That is Contaminated with 100 μg/kg Aflatoxin

Sample Size (kg)	*Low[1] Value (μg/kg)*	*High[2] Value (μg/kg)*	*Standard Deviation[3]*	*Range (High-Low)*
1	0	271	87	271
2	0	222	62	222
4	13	187	45	174
8	37	163	32	126
16	53	147	24	94
32	65	136	19	72

From (62).
[1]Low = 100 − 1.96 (std deviation). If low was <0, a 0 was recorded.
[2]High = 100 + 1.96 (std deviation).
[3]Standard deviation is based on a single 100 g subsample taken from the sample and one analysis.

have the ability to dilute and detoxify these chemicals. The feed to tissue ratio for aflatoxin B_1 has been estimated to be 14,000 for beef cattle liver and 2,200 for chicken eggs. By contrast, this ratio drops to between 100 to 200 for milk (from dairy cattle) where the predominant aflatoxin is M_1 (a ring hydroxylated metabolite of B_1). FDA has established an actionable level of 0.5 μg/kg for fluid milk.

19.2.3 Chemical Screening Procedures

Nonquantitative procedures include **minicolumn screening tests** and **single dimensional TLC** procedures that are intended to provide a preliminary indication of the presence of analyte mycotoxins. Since the majority of samples are anticipated to contain no detectable residues, a rapid yet sensitive screening procedure can identify those samples that will require a more time-consuming quantitative analysis. Minicolumn procedures exploit the native fluorescence of aflatoxins, zearalenone, or ochratoxin A analytes. Glass chromatography tubes (4–6 mm i.d.) containing different adsorbants are available commercially in a prepacked format and are used in simplified procedures to detect either the aflatoxins or zearalenone and ochratoxin A. After partial cleanup, the extract is added to the head of the column and then eluted with various solvents. In the final step, the analyte(s) are eluted from upper absorbing layers into a Florisil lower layer. When viewed under long wavelength UV light (365 nm), the analyte(s) appear as a blue fluorescent band. The method limit of detection is 10 to 15 ng/g. The precolumn sample treatment is somewhat matrix dependent (10, 45).

TLC has been used frequently both for screening assays and separately for quantitation. The advantages of using a TLC-based preliminary screen of samples include the fact that more samples can be assayed on the same plate because fewer standards are required. Quantitative assessments require a series of standard concentrations for each analyte. The screening procedure can also be used to provide a crude estimate of the levels of toxin so that the volume of extract can be adjusted to obtain a response that is within the linear range of a quantitative method subsequently applied to the same sample. The principal limitation of unidirectional TLC is the presence of coextractives that may interfere with the detection and quantitation. This limitation is often overcome by resorting to two-dimensional TLC; however, the number of samples or standards that can be run on the same plate is severely limited. There are numerous environmental factors that influence the relative mobility of an analyte (R_f) on a TLC plate, including temperature, the activity of the stationary phase, and the degree of solvent undersaturation of the chromatography tank. In consequence, many of the official methods do not report R_f values from their procedures. Authentic standards must be run frequently (if not with every sample) to ensure that the chromatographic conditions remain unchanged. Despite the increased variability associated with TLC relative to other analytical techniques, it is important to remember that, for mycotoxin analyses, the sampling error is probably the predominant contributor to the overall uncertainty associated with the final results. Multitoxin screening TLC procedures have been reviewed by Steyn et al. (46) and by Romer (47).

19.2.4 Quantitative Chemical Procedures

Quantitative chemical methods for mycotoxins are inevitably multistage methods that follow the same general procedures as for pesticide residues. These procedures typically involve separate steps of **extraction, filtration, cleanup, concentration, chromatographic separation, detection/quantitation, and confirmation.** Extraction procedures tend to be similar to those for pesticide residues. Aqueous methanol, acetone, and/or acetonitrile are added to the subsample and either blended at high speed or vigorously shaken for 0.5 hr. Defatting of lipid-rich food matrices is frequently required; typically, extractions with hexane or isooctane can be performed prior to, during, or after the mycotoxin solubilization step. The use of diethyl ether for defatting operations is avoided because of analyte (aflatoxin) losses.

After filtration to remove suspended solids, cleanup procedures are used to further purify the extract. In addition to preparative chromatographic columns and/or solvent partitioning, using procedures described in the pesticide residues section, an aqueous anionic precipitation procedure is sometimes used to remove plant pigments and proteinaceous substances. The precipitation can be induced with a variety of additives including lead, zinc, and ammonium salts and phosphotungstic acid. This procedure is limited to those extracts in polar organic solvents containing more than 55 percent water.

After the partially purified extract is concentrated, it is subjected to chromatographic separation by two-dimensional TLC, HPLC, or, for certain classes of mycotoxins, by GC. An excellent overview of two-dimensional TLC procedures is provided by van Egmond (48). The most popular automated chromatographic technique for mycotoxins has been HPLC. Many of these toxins (trichothecenes are the exception) can be detected by ultraviolet or fluorescence with sufficient sensitivity to provide for quantitation at the low μg/kg levels. Reversed-phase chromatographic separation with precolumn treatment with trifluoroacetic acid (to convert aflatoxins B_1 and G_1 to their corresponding hemiacetals) or postcolumn derivatization with iodine has been used to increase the fluorescence response of these analytes (48) in aqueous mobile phases. For the trichothecenes, procedures are often optimized for the recovery of either the relatively less polar class A subgroup (diacetoxyscirpenol, T-2 toxin, HT-2 toxin, and neosolaniol) or the more polar B subgroup [deoxynivalenol (DON), nivalenol, and fusarenon-X]. The purified extracts are treated to convert the analytes to either their trimethylsilyl or their heptafluorobutyryl derivatives, and the derivatives are detected by electron capture GC (or less often by flame ionization) following separation on packed or capillary columns.

19.2.5 Biochemical Methods

19.2.5.1 Immunoassays

Although a variety of biological assays (acute toxicity to the larvae of brine shrimp or fish or to chick embryos) have been described that are useful in tracing sources of mycotoxins, their use in the surveillance of foods and feeds is only of minor importance. By contrast, the potential for the determination of mycotoxins by immunoassay techniques (49) has been amply demonstrated by the development of both **radioimmunoassay** (RIA) and **enzyme-linked immunosorbent assay** (ELISA) procedures for several mycotoxins. These include a number of aflatoxins (B_1, G_1, Q_1, and M_1), trichothecenes (DON, DON triacetate, T-2 toxin, diacetoxyscirpenol), ochratoxin A, zearalenone, and rubratoxin in a variety of food matrices. Limits of detection are in the picogram to nanogram range. The speed (10 minutes for the competitive binding step), simplicity (directly applicable to liquid samples and to water methanol extracts of solid samples), and reliability have been improved steadily. Estimates have suggested that, excluding research and development, ELISA procedures cost only 6 and 3 percent as much as GC and GC-MS based procedures, respectively.

Recently, kits for the detection and quantitation of aflatoxins have become available from several commercial sources. As suggested by Pestka (49), answers to a number of questions should be considered in choosing a source of an assay kit:

1. Are the limits of detection and dynamic range of the assay relevant to the needs of the laboratory? If the anticipated levels of mycotoxin residues are high, extensive dilution of the extract might be necessary for quantitation. Since the absorbance recorded in an ELISA procedure is inversely related to the logarithm of analyte concentration, measurements made near the middle of the standard s-shaped curve will be more precise than measurements at either extreme. It would be advantageous to have the concentration corresponding to the actionable level at the midpoint of the calibration curve.
2. In view of the extreme heterogeneity of contamination, are realistic sampling protocols described in the literature accompanying the product?
3. Since the antibodies might also react with closely related chemical structures, what is the cross-reactivity of each analog relative to the target analyte? Antibodies developed against aflatoxin B_1 will cross-react with aflatoxin B_2 [however, the strength of the binding (avidity) will be different for the two aflatoxins]. Typically, analog competition curves do not overlap and might not have similar slopes. Moreover, the specificity of the antibodies might vary from batch to batch. Ideally, antibody lot characteristics should be defined in terms of limit of detection and sensitivity range for the analytical procedure, resistance to organic solvents, and the variability of the antibody-coated solid support.
4. Since certain kits come in a modular format with assay units that can be physically separated, can the kit be used to analyze a single sample? By contrast, other kits are designed for large numbers of samples that are analyzed simultaneously.

As an example, an enzyme-linked immunosorbent screening method for aflatoxin B_1 in cottonseed and mixed feed that is applicable to screening B_1 at >15 μg/kg has been adopted first action as a joint AOAC-IUPAC method (49). Antibodies (to aflatoxins) coated onto plastic microliter wells, lyophilized horseradish peroxidase–conjugated aflatoxin B_1, an enzyme substrate (2,2′-azino-bis(3-ethylbenzthiazoline-6-sulfonate) (ATBS) in pH 4 citrate buffer, hydrogen peroxide (30%) in pH 4 citrate buffer, and a color stopping solution are available as an Agri-Screen for Aflatoxin kit (Neogen Corp., Lansing, MI). Other similar kits include the Immunodot Screen Cup and the Alfa-20 Cup (International Diagnostic Systems Corp., St. Joseph, MO) and the Cite Probe (IDEXX Corp., Portland, ME). Other kits that employ a card containing a glass fiber filter impregnated with antibodies include the EZ-Screen (Environmental Diagnostics Inc., Burlington, NC). Several companies in Great Britain have also developed and distribute kits.

19.2.5.2 Immunoaffinity Separations

A further very promising recent development is the use of antibodies to isolate aflatoxins from a biological matrix. This potentially simple approach can result in a degree of cleanup that is superior to the more elaborate standard methods. Commercial single-use columns consisting of anti-aflatoxin antibodies covalently bound to a gel matrix (a beaded agarose) in a plastic cartridge have been introduced. The column packing material selectively adsorbs aflatoxins from coextractives in a crude extract serving as a concentration technique. The analytes are then readily released from the packing with a polar organic solvent. One application involves the use of monoclonal antibody affinity chromatography as a one-step column cleanup prior to the fluorometric determination of aflatoxin M_1 in milk (50). Raw milk samples (40 mL) were mixed with NaCl (1 g), centrifuged at 2,000 × g for 5 min, and then filtered. A 25 mL aliquot of the filtrate was passed through a column (Aflatest™, Vicam, Sommerville, MA) under a slight positive pressure. The column was washed with two successive 10 mL portions of 10 percent (v/v) methanol; then the analyte was eluted with 1 mL 80 percent methanol. The eluate was diluted with 1 mL aqueous bromine, and the fluorescence measured at 450 nm (λ_{EX} = 360 nm). Recoveries at the 0.05 to 2 µg/l levels were excellent. A method detection limit was estimated to be 0.05 µg/L (tenfold less than the FDA actionable level). One appreciable advantage of this general approach is that it circumvents the rather narrow linear dynamic range of analyte concentrations associated with ELISA and RIA methods. Other substrates that have been used in a collaborative study (51) using the same immunoaffinity column isolation technique include corn, peanuts, and peanut butter. A concise overview of the techniques for preparing immunoaffinity columns has been published recently (52).

19.3 DRUG RESIDUES

19.3.1 Introduction

In addition to the drugs administered at therapeutic levels to combat diseases in food-producing animals, the subtherapeutic use of antimicrobial drugs has also played an important role in animal husbandry. FDA approved the addition of subtherapeutic levels of antibiotics to animal feeds almost 40 years ago. These levels (1) reduce the incidences of infectious diseases caused by bacteria and protozoa (prophylactic effect), (2) increase the rate of weight gain of treated animals, and (3) decrease the amount of feed needed to achieve these weight gains. The continued use of subtherapeutic levels of antibiotics can pose a serious but indirect hazard to humans. Abuse of this practice has resulted in the dominance of antimicrobial-resistant enteric bacteria in some food animals and can result in the transmission to the human reservoir of bacterial resistance to antimicrobial agents.

The use of antimicrobial agents in feeds is strictly regulated—monitoring of compliance is the responsibility of the **Center for Veterinary Medicine** (CVM) of FDA. A tolerance in edible tissues has been set for each antibiotic and sulfonamide approved for use in animal feeds. By analogy to pesticide registration, a tolerance, based on uncooked edible tissues, is established after an extensive review of the toxicology, chemistry, and biochemistry of the active product and the development (by the sponsors of the application) of an analytical method for determining residues of the drug in tissues. As part of the approval process, restrictions on the dosage level and duration, species that may be treated, and the withdrawal period (the time between the last availability of the drug to the animal and slaughter or the use of milk or eggs by humans) are established. The withdrawal time can vary between 0 and 30 days. Tolerances, if established, represent total residues (parent compound plus all compounds derived from it, including metabolites, conjugates, and residues bound to macromolecules). In an effort to reduce the number of methods required for monitoring, the concepts of a marker residue and a target tissue have been advanced. The marker residue is a selected residue (possibly the parent compound) that has a known relationship with the level of total residues in each of the edible tissues. The traditional method of establishing this relationship is by performing a feeding trial (under conditions of the proposed use of the candidate drug) with a radiolabeled parent compound and monitoring the depletion and fate of the label with time. Based on the pharmacokinetics of the drug and its subsequent depletion from the different tissues with time, an estimate of the levels in tissues at sacrifice can be obtained.

Estimates are that, at some time during their lives, nearly 80 percent of poultry, 75 percent of swine, 60 percent of feedlot cattle, and 75 percent of dairy calves have been fed with antibiotics (53). The **Food, Safety, and Inspection Service** (FSIS) of USDA monitors edible tissues destined for commerce for residues of drugs, pesticides, industrial chemicals, and heavy metals. The U.S. Code of Federal Regulations lists tolerances for some 80 animal drugs in foods, of which some 30 (mostly antibiotics) are readily detected by microbiological screening assays. By contrast to pesticides multiresidue procedures, chemically-based multiresidue procedures for drug residues tend to be more limited in scope in that they are directed at specific classes of drug residues.

19.3.2 Screening Assays for Antibiotic Residues

Many preliminary screening assays rely on the inhibition of growth of microorganisms by antibiotic residues present in the test sample. These traditional assay procedures are based on either **diffusion processes** or on **turbidity.** The growth of an indicator organism, in a transparent liquid culture, can be followed by monitoring the increased turbidity with time. In diffusion-dependent assays, the material to be assayed diffuses through an agar-based nutrient medium that has been uniformly seeded with spores of a susceptible organism. Upon incubation, a zone of inhibition of germination and growth develops, indicating the presence of inhibitor(s). There are several factors that affect the size and appearance of zones of inhibition. The number of viable organisms used to inoculate the medium is critical because the density of growth (and therefore the visualization of zones) is dependent on the initial numbers of organisms. The temperature of incubation must also be rigidly controlled because both the rate of growth of the organism and the rate(s) of diffusion of inhibitor(s) are temperature-dependent phenomena. Porosity of the medium also influences the rate of diffusion. In general, lower proportions of agar result in larger zones of inhibition. Other factors include the depth of the agar layer, the age of the inoculum, the technique of adding the sample to the plate, and the presence of coextractives from the sample. Both turbidity and diffusion-based techniques can be carried out manually, or many of the steps can be automated. A detailed monograph on the theory and application of microbial assays has been prepared by Hewitt and Vincent (54).

Two procedures, STOP **(swab test on premises)** and CAST **(calf antibiotic and sulfa test),** are typical. Cotton swabs that have been used to sample the suspect tissue are placed in contact with gelled growth media that has been amended with *Bacillus subtilis* (ATCC 6633) spores and incubated at 32°C for 16 to 18 hr (STOP) or with *Bacillus magterium* (ATCC 9885) and incubated at 44°C (CAST). If positive, these screen are usually followed by a thin-layer bioautography assay.

Milk represents another commodity that is routinely screened for antibiotic residues; antibiotic-contaminated milk (and milk products) are considered as adulterated by FDA. In addition to the hazard these residues can pose for certain ultrasensitive consumers, antibiotic residues can interfere with the acid and flavor production during the manufacture of buttermilk and similar products, the acid production of starter cultures used in processed milk products and cheeses, and starter culture growth when propagated in reconstituted skim milk. Disc assay procedures in which the test sample is impregnated on filter paper disks (7 mm–1.27 cm diam.) and placed on solidified agar that has been seeded with *Bacillus subtilis* (ATCC 6633) spores, *Bacillus stearothermophilus* (ATCC 10149) spores, or *Bacillus megaterium* (ATCC 9855) spores have been used extensively to screen milk samples for residues of antibiotics (55). The Delvotest-P-Ampule kit (Gist Brocades USA Inc., Charlotte, NC) is based on a similar screening procedure in which the rapid growth and acid production of *B. stearothermophilus* causes the whole of the agar medium to turn yellow in the absence of antibiotic inhibitors. In total, these screening assays are efficient monitoring procedures in that they are simple and rapid (permitting numerous samples to be screened). However, they are nonspecific and respond only to biologically active residues that inhibit the growth of the indicator organism.

Within the last decade, a number of commercial assay materials have become available. The **Charm II** (Penicillin Assays, Inc., Malden, MA) test (56) is a rapid radioisotopic assay procedure that can detect the following antibiotic groups: β-lactams, tetracyclines, macrolides, aminoglycosides, novobiocin, sulfonamides, and chloramphenicols. The assay is based on a competition between labeled drug and residues in the milk sample for a limited number of specific binding sites on the surfaces of bacteria that are added to the sample. In brief, the procedure involves the addition of the radiolabeled tracer antibiotic and the binding microorganism to the milk sample, a short incubation, centrifugation, fat removal, resuspension of the microbial plug in a scintillation fluid, and counting. The greater the concentration of antibiotic residue, the less radiolabeled tracer will become bound to the microorganism. Two antibiotic groups can be assayed in each tube by using a combination of ^{14}C and ^{3}H tracer antibiotics. The various antibiotic types can be assayed in approximately 12 minutes, and with limits of detection that are very much lower than the more traditional diffusion assays. For β-lactam antibiotic residues, Angenics Inc. (Cambridge, MA) has developed a rapid (6-minute) antibody-based assay, in kit form, that is performed on a glass slide and evaluated in a monitoring device. Penzyme On-Farm and Laboratory III tests (Smithkline Animal Health Products, West Chester, PA) are enzyme-based colorimetric assays for β-lactam antibiotics.

19.3.3 Chemically-Based Approaches to Quantitative Determinations

19.3.3.1 Overview

The approaches to the quantitative determination of drug residues in tissues, in feeds, or in other food products follow the same general steps as for other trace analytes. After an optional sample pretreatment (to aid in the release of bound residues), drug residues are solubilized with an appropriate solvent, the crude extract is purified by partitioning and/or column chromatography, and then the purified extract is analyzed using an automated chromatographic technique. However, for drug residues, cleanup procedures often involve acid-base partitioning against organic solvents to take advantage of the acidic character (phenolic compounds) or the basic character (benzimidazoles, sulfonamides, tetracyclines) of the analytes. For ionic analytes, ion exchange cleanup columns have been used extensively. Alternately, the acid/base character of analytes can be exploited by performing the initial extraction with mineral acids (HCl, $HClO_4$) or with aqueous buffers to recover tetracyclines from meats, fish, and blood (57). Buffers can sometimes be combined advantageously with a water-miscible organic solvent to improve the selectivity of the solubilization step. The addition of the water-miscible organic solvent appreciably reduces the solubility of proteinaceous materials and avoids a separate deproteination step. Other advantages of this approach (57) are that the procedure is simple and rapid, appears to be widely applicable, and results in consistently high recoveries. However, the resulting extracts cannot be injected directly onto a reversed-phase HPLC column for lack of analyte retention. The organic solvent is usually removed (in part or totally). One approach that can be effective for polar analytes (tetracyclines) is to add a nonpolar water-immiscible solvent (CH_2Cl_2 or hexane). In contrast to most pesticides, the polar drug residues are retained in the aqueous phase. Analytes in the crude aqueous phase can be concentrated on a solid phase extraction (SPE) cartridge or directly on the head of an HPLC column.

19.3.3.2 Automated Chromatographic Separations

As is the case for mycotoxins, the most popular automated approach to separation/quantitation of drug residues has been HPLC. The major applications of this approach have been for confirmatory rather than for screening purposes. One of the advantages of HPLC relative to GC is that frequently little sample preparation is required. A multiresidue procedure for eight benzimidazole residues in liver and muscle is typical of recent developments in drug residue methodology (58). Briefly, previously blended and frozen tissue sample (bovine, ovine, or swine liver or muscle), 10 g; Na_2SO_4, 5g; 4M K_2CO_3, 1 mL; and ethyl acetate, 30 mL are blended, and the filtrate is evaporated. The residue, in hexane, is partitioned against ethanol -0.2 M HCl. An aliquot of the aqueous phase is basified with K_2CO_3 and then purified on a C_2 minicolumn. The analytes are recovered from the minicolumn with ethyl acetate. The organic solvent is evaporated, the residues are resuspended in ethanol-ammonium phosphate mobile phase and separated by reversed-phase HPLC, and detected at $\lambda = 298$ nm. The identity of residues can be corroborated by GC-MS after hydrolyzing a second aliquot of the extract with HCl and derivitizing the liberated amine(s) with N-methyl-N-(t-butyldimethylsilyl)-trifluoroacetamide. Recoveries, at the 100 μg/kg spiking level, averaged 88 percent $\pm$ 6 percent. A somewhat similar procedure, when subjected to interlaboratory evaluation (59), illustrated the value of the method validation process.

An HPLC screen for 10 sulfonamides in raw bovine milk is also illustrative (60). Sample preparation is minimal and involves partitioning the residues into chloroform-acetone, evaporating the solvent, resuspending the residues in 0.1M potassium dihydrogen phosphate, and subsequently removing fatlike coextractives with hexane. After filtration, the aqueous phase is analyzed by reversed-phase HPLC (using either of two isocratic mobile phases) with detection at $\lambda = 265$ nm. The coefficients of variation were 3 to 13 percent at the 10 μg/kg level of spiking. A recent overview of the application of HPLC to antibiotic residue determinations has been provided by Moats (57).

19.4 SUMMARY

As a society, we expend enormous effort and considerable resources to identify and control "synthetic" chemicals that are considered to pose a carcinogenic risk to society of greater than one in a million. The detection and determination of traces of pesticide, mycotoxin, and antibiotic residues in foods represents a formidable challenge for the analyst. In support of existing legislation, detailed sampling protocols and analytical procedures have been developed both to screen for and to measure toxic residues at trace and ultratrace levels. Yet, despite a generation of dedicated research and hundreds of thousands of analyses per year, the level of consumer confidence in our food supplies appears to be declining somewhat. In response to concerns regarding the safety of our food supplies, newer methodologies will have to be developed and optimized to screen even greater numbers of samples and to analyze for greater numbers of toxicants. Above all, we have to continue to ensure the rapid publication of the results of monitoring programs in the open literature to help allay consumers' concerns.

However, we must also not lose sight of the concepts associated with relative risk. It has been estimated that we consume about 10,000 times more natural than synthetic pesticides (61). It seems probable that virtually every plant food that is available in supermarkets contains natural plant toxins, many of which are probable carcinogens. It has been suggested that "the carcinogenic hazard from current levels of pesticide residues or water pollution is likely to be minimal relative to the background levels of natural substances" (61).

19.5 STUDY QUESTIONS

1. What is meant by the "tolerance level" for a pesticide on a fresh agricultural product?
2. Why is it that all analytical methods provide only estimates of the level(s) of an analyte within the sample? In your opinion, how exact are these estimates?
3. Briefly outline the five major steps involved in a quantitative single-residue method for residues of a pesticide in a fresh plant food.
4. What strategies can be followed in an attempt to corroborate the presence of a pesticide, a mycotoxin, or a drug residue in a sample? What is the value, if any, of these approaches?
5. For immunoassay-based analytical methods, what is meant by the term *cross reactivity?* How is this parameter measured?
6. Why haven't microbiological assays been developed to detect the presence of toxicogenic fungi in fresh or stored produce and used as an indicator of possible mycotoxin contamination of that product?
7. Why are sampling procedures for pesticide residues appreciably different from sampling procedures for mycotoxins even when dealing with the same sample matrix?
8. What are the advantages and disadvantages of analytical screening procedures for pesticide, for mycotoxin, and/or for drug residues?
9. Having suffered through this chapter, are you any closer to deciding whether foods are safe?

19.6 REFERENCES

1. Codex Alimentarius Commission. 1985. *Recommended National Regulatory Practices to Facilitate Acceptance and Use of Codex Maximum Limits for Pesticide Residues in Foods. Guide to Codex Recommendations Concerning Pesticide Residues.* Part 9. Food and Agriculture Organization, World Health Organization, Rome, Italy.
2. Kovacs, M.F., and Trichilo, C.L. 1987. Regulatory perspective of pesticide analytical enforcement methodology in the United States. *J. Assoc. Off. Anal. Chem.* 70:937–940.
3. Wessel, J.R., and Yess, N.J. 1991. Pesticide residues in foods imported into the United States. *Revs. Environ. Contam. Toxicol.* 120:83–103.
4. Forsyth, S.F. 1990. Regulatory issues for plant disease biocontrol. *Can. J. Plant Pathol.* 12:318–321.
5. Office of Technology Assessment. 1988. *Pesticide Residues in Food—Technologies for Detection,* OTA-F-398. U.S. Government Printing Office, Washington, DC.
6. Ott, S. L. 1990. Supermarket shoppers' pesticide concerns and willingness to purchase certified pesticide residue-free fresh produce. *Agribusiness* 6:593–602.
7. Horwitz, W., and Howard, J.W. 1979. National Bureau of Standards (US - NIST) Spec. Publ. 519, 231–242.
8. Kratochvil, B., and Peak, J. 1989. Sampling techniques for pesticide analysis, in *Advanced Analytical Techniques, Analytical Methods for Pesticides and Plant Growth Regulators,* Vol. XVII, J. Sherma (Ed.), 1–29. Academic Press, Inc., New York, NY.
9. AOAC. 1990. *Official Methods of Analysis,* 15th. ed., Vol. 1, Chapter 10, 274–310. Association of Official Analytical Chemists, Washington, DC.
10. FDA. 1985. *Pesticide Analytical Manual,* Vol. 1 *(Methods Which Detect Multiple Residues)* and Vol. 2 *(Methods for Individual Pesticide Residues).* National Technical Information Service, Springfield, VA. Also available from Public Records and Documents Center, Food and Drug Administration, HFI-35, Rockville, MD.
11. Horowitz, W. 1980. Analytical aspects: An introduction, in *The Pesticide Chemist and Modern Toxicology,* S.K. Benbal, J.J. Marco, L. Goldberg, and M.L. Leng (Eds.), 331–334. ACS Symposium Series No. 160. American Chemical Society, Washington, DC.
12. Ambrus, A., and Thier, H.P. 1986. Applications of multiresidue procedures in pesticide residue analysis. *Pure Appl. Chem.* 58:1035–1062.
13. McLeod, H.A., and Graham, R.A. (Eds.) 1986. *Analytical Methods for Pesticide Residues in Foods.* Canadian Government Publishing Center, Supply and Services Canada, Ottawa, Ont., Canada, K1A 0S9.
14. Greve, P.A. (Ed.) 1988. *Analytical Methods for Residues of Pesticides,* 5th ed. Government Publishing Office, The Hague, Netherlands.
15. Luke, M.A., and Masumoto, H.T. 1986. Pesticide residue analysis in foods, in *Principles, Statistics and Applications, Analytical Methods for Pesticides and Plant Growth Regulators,* Vol XV, G. Zweig and J. Sherma (Eds.), 161–200. Academic Press, Inc., New York, NY.
16. Steinwandter, H. 1989. Universal extraction and cleanup methods, in *Advanced Analytical Techniques, Analytical Methods for Pesticides and Plant Growth Regulators,* Vol XVII, J. Sherma (Ed.), 35–71. Academic Press, Inc., New York, NY.
17. Walters, S.M. 1986. Cleanup of samples, in *Principles, Statistics and Applications, Analytical Methods for Pesticides and Plant Growth Regulators,* Vol XV, G. Zweig and J. Sherma (Eds.), 67–110. Academic Press, Inc., New York, NY.
18. Cochrane, W.P. 1981. Chemical derivatization in pesticide analysis, in *Chemical Derivatization in Analytical Chemistry,* R.W. Frei and J.F. Lawrence (Eds.), 1–97. Plenum Press, New York, NY.

19. McMahon, D. (Guest Ed.) 1987. Derivatization techniques in chromatography. *J. Chromatogr. Sci.* 25:434–478.
20. Froberg, J.E., and Doose, G.M. 1986. Practical aspects of gas chromatography, in *Modern Analytical Techniques, Analytical Methods for Pesticides and Plant Growth Regulators,* Vol XIV, G. Zweig and J. Sherma (Eds.), 41–74. Academic Press, Inc., New York, NY.
21. Zweig, G. 1986. Capillary column chromatography of pesticides, in *Modern Analytical Techniques, Analytical Methods for Pesticides and Plant Growth Regulators,* Vol XIV, G. Zweig and J. Sherma (Eds.), 75–93. Academic Press, Inc., New York, NY.
22. Roseboom, H. 1984. Recent advances in pesticide residue analysis, in *Food Constituents and Food Residues—Their Chromatographic Determination, Food Science and Technology,* Vol 11, J.F. Lawrence (Ed.), 489–532. Marcel Dekker, Inc., New York, NY.
23. Holland, P.T., and Greenhalgh, R. 1981. Selection of gas chromatographic detectors for pesticide residue analysis, in *Analysis of Pesticide Residues, Chemical Analysis,* Vol 58, H.A. Moye (Ed.), 51–136. John Wiley & Sons, New York, NY.
24. Muszkat, L., and Aharonson, N. 1986. High performance liquid chromatography, in *Modern Analytical Techniques, Analytical Methods for Pesticides and Plant Growth Regulators,* Vol XIV, G. Zweig and J. Sherma (Eds.), 95–131. Academic Press, Inc., New York, NY.
25. Lawrence, J.F. 1982. *High Performance Liquid Chromatography of Pesticides—Analytical Methods for Pesticides and Plant Growth Regulators,* Vol XII, G. Zweig and J. Sherma (Eds.). Academic Press, Inc., New York, NY.
26. Moye, H.A. 1981. High performance liquid chromatographic analysis of pesticide residues, in *Analysis of Pesticide Residues, Chemical Analysis,* Vol 58, H.A. Moye (Ed.), 52–136. John Wiley & Sons, New York, NY.
27. Stan, H-J. 1989. Application of computers for the evaluation of gas chromatographic data, in *Advanced Analytical Techniques, Analytical Methods for Pesticides and Plant Growth Regulators,* Vol XVII, J. Sherma (Ed.), 167–215. Academic Press, Inc., New York, NY.
28. Lawrence, J.F. 1981. Confirmatory tests, in *Pesticide Analysis,* K.G. Das (Ed.), 425–460. Marcel Dekker, Inc., New York, NY.
29. Mumma, R.O., and Brady, J.F. 1987. Immunological assays for agrochemicals, in *Pesticide Science and Technology,* Sixth IUPAC Congress of Pesticide Chemistry, R. Greenhalgh and T.R. Roberts (Eds.), 341–348. Blackwell Scientific, Oxford, UK.
30. Van Emon, J.M., Seiber, J.N., and Hammock, B.D. 1989. Immunoassay techniques for pesticide analysis, in *Advanced Analytical Techniques, Analytical Methods for Pesticides and Plant Growth Regulators,* Vol XVII, J. Sherma (Ed.), 217–263. Academic Press, Inc., New York, NY.
31. Mumma, R.O., and Hunter, K.W. 1988. Potential of immunoassays in monitoring pesticide residues in foods, in *Office of Technology Assessment Pesticide Residues in Foods, Technologies for Detection,* 171–181. OTA-F-398. U.S. Government Printing Office, Washington, DC. Also available from Technomic Publishing Co., Inc., Lancaster, PA.
32. Sherma, J. 1986. Thin layer chromatography, in *Modern Analytical Techniques, Analytical Methods for Pesticides and Plant Growth Regulators,* Vol XIV, G. Zweig and J. Sherma (Eds.), 1–39. Academic Press, Inc., New York, NY.
33. Getz, M.E. 1980. *Paper and Thin Layer Chromatography of Environmental Toxicants.* Heyden, London, UK.
34. Freid, B., and Sherma, J. 1986. *Thin Layer Chromatography: Techniques and Applications,* 2nd ed. Chromatographic Science Series, Vol. 36. Marcel Dekker, Inc., New York, NY.
35. Armbrus, A., Hargitai, E., Karoly, G., Fulop, A., and Lantos, J. 1981. General method for determination of pesticide residues in samples of plant origin, soil, and water. II. Thin layer chromatographic determination. *J. Assoc. Off. Anal. Chem.* 64:743–768.
36. Mendoza, C.E. 1972. Analysis of pesticides by the thin-layer chromatographic-enzyme inhibition technique. *Residue Revs.* 43:105–142.
37. EnzyTec, Inc. Pesticide Detector Ticket. Product Bulletin. Kansas City, MO.
38. Lawrence, J.F. 1980. Simple, sensitive and selective thin layer chromatographic technique for detecting some photosynthesis inhibiting herbicides. *J. Assoc. Off. Anal. Chem.* 63:758–761.
39. Hesseltine, C.W. 1986. Global significance of mycotoxins, in *Bioactive Molecules, Vol 1: Mycotoxins and Phycotoxins,* P.S. Steyn and R. Vleggaar (Eds.), 1–18. Elsevier, Amsterdam, The Netherlands.
40. van Egmond, H.P. 1989. Current situation on regulations for mycotoxins. Overview of tolerances and status of standard methods of sampling and analysis. *Food Addit. Cont.* 6:139–188.
41. van Egmond, H.P. 1991. Limits and regulations for mycotoxins in raw materials and animal feeds, in *Mycotoxins and Animal Foods,* J.E. Smith and R.S. Henderson (Eds.), 423–436. CRC Press, Inc., Boca Raton, FL.
42. Whitaker, T.B., Dickens, J.W., and Giesbrecht, F.G. 1991. Testing animal feedstuffs for mycotoxins: Sampling, subsampling, and analysis, in *Mycotoxins and Animal Foods,* J.E. Smith and R.S. Henderson (Eds.), 153–164. CRC Press, Inc., Boca Raton, FL.
43. Horwitz, W. 1988. Sampling and preparation of sample for chemical examination. *J. Assoc. Off. Anal. Chem.* 71:241–245.
44. Keith, L.H. 1990. *Environmental Sampling and Analysis: A Practical Guide.* Lewis Publishers, Boca Raton, FL.
45. Holaday, C.E. 1981. Minicolumn chromatography: State of the art. *J. Amer. Oil Chem. Soc.* 58:931A–934A.
46. Steyn, P.S., Thiel, P.G., and Tinder, D.W. 1991. Detection and quantification of mycotoxins by chemical analysis, in *Mycotoxins and Animal Foods,* J.E. Smith and R.S. Henderson (Eds.), 165–221. CRC Press, Inc., Boca Raton, FL.
47. Romer, T. 1984. Chromatographic techniques for mycotoxins, in *Food Constituents and Food Residues—Their Chromatographic Determination, Food Science and Technology,* Vol 11, J.F. Lawrence (Ed.), 355–393. Marcel Dekker, Inc., New York, NY.
48. van Egmond, H.P. 1984. Determination of mycotoxins, in *Developments in Food Analysis Techniques,* Vol 3, R.D. King (Ed.), 99–144. Elsevier Applied Science Publishers, New York, NY.
49. Pestka, J.J. 1988. Enhanced surveillance of foodborne mycotoxins by immunochemical assay. *J. Assoc. Off. Anal. Chem.* 71:1075–1081.

50. Hansen, T.J. 1990. Affinity column cleanup and direct fluorescence measurement of aflatoxin M1 in raw milk. *J. Food Prot.* 53:75–77.
51. Trucksess, M.W., Stack, M.E., Nesheim, S., Page, S.W., Albert, R.H., Hansen, T.J., and Donahue, K.F. 1991. Immunoaffinity column coupled with solution fluorometry or liquid chromatography postcolumn derivatization for determination of aflatoxins in corn, peanuts, and peanut butter: Collaborative study. *J. Assoc. Off. Anal. Chem.* 74:81–88.
52. Katz, S.E., and Brady, M.S. 1990. High-performance immunoaffinity chromatography for drug residue analysis. *J. Assoc. Off. Anal. Chem.* 73:557–560.
53. Franco, D.A., Webb, J., and Taylor, C.E. 1990. Antibiotic and sulfonamide residues in meat: Implications for human health. *J. Food Prot.* 53:178–185.
54. Hewitt, W., and Vincent, S. 1989. *Theory and Application of Microbiological Assay.* Academic Press, Inc., New York, NY.
55. Bruhn, J.C., Ginn, R.E., Messer, J.W., and Mikolajcik, E.M. 1985. Detection of antibiotic residues in milk and dairy products, in *Standard Methods for the Examination of Dairy Products*, 15th ed., G.H. Richardson (Ed.), 265–288. American Public Health Association, Washington, DC.
56. Charm, S.E., and Chi, R. 1988. Microbial receptor assay for rapid detection and identification of seven families of antimicrobial drugs in milk: Collaborative study. *J. Assoc. Off. Anal. Chem.* 71:304–316.
57. Moats, W.A. 1990. Liquid chromatographic approaches to antibiotic residue analysis. *J. Assoc. Off. Anal. Chem.* 73:343–346.
58. Wilson, R.T., Groneck, J.M., Henry, C., Rowe, L.D. 1991. Multiresidue assay for benzimidazole anthelmintics by liquid chromatography and confirmation by gas chromatography / Selected-ion monitoring electron impact mass spectrometry. *J. Assoc. Off. Anal. Chem.* 74: 56–67.
59. LeVan, L.W., and Barnes, C.J. 1991. Liquid chromatographic method for multiresidue determination of benzimidazoles in beef liver and muscle: Collaborative study. *J. Assoc. Off. Anal. Chem.* 74:487–494.
60. Smedley, M.D., and Weber, J.D. 1990. Liquid chromatographic determination of multiple sulfonamide residues in bovine milk. *J. Assoc. Off. Anal. Chem.* 73:875–879.
61. Ames, B.N., and Gold, L.S. 1989. Pesticides, risk and applesauce. *Science* 244:755–757.
62. Marshall, W.D. 1991. Unpublished data. Department of Food Science and Agricultural Chemistry, Macdonald Campus of McGill University, Ste Anne de Bellevue, Québec, Canada.

CHAPTER

20

Analysis for Extraneous Matter

John R. Pedersen

20.1 INTRODUCTION

Analysis for extraneous matter is an important element both in the selection of raw materials for food manufacturing and for monitoring the quality of processed foods.

20.1.1 Food, Drug, and Cosmetic Act

The Federal Food, Drug, and Cosmetic Act (FD&C Act) of 1938 with Amendments (1) defines a food as **adulterated** "if it consists in whole or in part of any filthy, putrid, or decomposed substance, or if it is otherwise unfit for food [Section 402 (a)(3)]; or if it has been prepared, packed, or held under insanitary conditions whereby it may have become contaminated with filth, or whereby it may have been rendered injurious to health" [Section 402 (a)(4)]. The filthy, putrid, or decomposed substances referred to in the law include the extraneous matter addressed in this chapter. In addition, extraneous matter includes adulterants that may be encountered in processing systems, such as lubricants, metal particles, or other contaminants (animate or inanimate) that may be introduced into a food because of a poorly operated food processing system.

20.1.2 Good Manufacturing Practices

The **Current Good Manufacturing Practice in Manufacturing, Packing, or Holding Human Food** (GMPs) was published by the Food and Drug Administration (FDA) (2) to provide guidance for compliance with the FD&C Act. That regulation provides guidelines for operating a food processing facility in compliance with Section 402 (a)(4). Paramount to complying with the FD&C Act and GMPs is the thorough inspection of raw materials and routine monitoring of food processing operations to ensure protection of the consuming public from harmful or filthy food products.

20.1.3 Defect Action Levels

Most of our foods are made from or consist in part of ingredients that are obtained from plants or animals and are mechanically stored, handled, and transported in large quantities. It would be virtually impossible to keep those materials completely free of various forms of contaminants. In recognition of that, the FDA (2) has established **defect action levels** (DALs) that reflect current maximum levels for natural or unavoidable defects in food for human use that present no health hazard. They reflect the maximum levels that are considered unavoidable under good manufacturing practices and apply mainly to contaminants that are unavoidably carried over from raw agricultural commodities into the food processing system. The manner in which foods are manufactured may lead to their contamination with extraneous materials if strict controls in processing are not maintained. This latter kind of contamination is not excused by the DALs.

20.1.4 Purposes of Analyses

The major purposes for conducting analyses for extraneous matter in foods are to ensure the protection of the consuming public from harmful or filthy food products, to meet regulatory requirements of the FD&C Act Sections 402 (a)(3) and 402 (a)(4), and to comply with defect action levels.

20.2 GENERAL CONSIDERATIONS

20.2.1 Methodologies

There are various methods for separating (isolating) extraneous materials from foods and for identifying and enumerating them. The most authoritative source, and that generally considered official by the FDA, is the *Official Methods of Analysis* of the Association of Official Analytical Chemists (AOAC) International. Chapter 16, Extraneous Materials: Isolation (3) includes methods for extraneous matter isolation in 16 food categories, including dairy products; grains and their products; poultry, meat, fish, and other marine products; snack food products; spices and other condiments; animal excretions; and mold and rot fragments.

The American Association of Cereal Chemists (AACC) (4) has established methods for isolating and identifying extraneous matter in cereal grains and their products (AACC Method 28-00). In most instances, the AACC methods are based on FDA or AOAC methods, but the format is slightly different. The AACC presents each procedure in an outline form that includes the scope, apparatus, and reagents required and the procedure in itemized steps; the AOAC uses a narrative paragraph form.

20.2.2 Definition of Terms

Terms used to classify and/or characterize various types of extraneous materials are defined as follows:

20.2.2.1 Extraneous Materials

Any foreign matter associated with objectionable conditions or practices in production, storage, or distribution; included are various classes of filth, decomposed material (decayed tissues due to parasitic or nonparasitic causes), and miscellaneous matter such as sand and soil, glass, rust, or other foreign substances. Bacterial counts are not included.

20.2.2.2 Filth

Any objectionable matter contributed by animal contamination such as rodent, insect, or bird matter; or any other objectionable matter contributed by insanitary conditions.

20.2.2.3 Heavy Filth

Heavier material separated from products by sedimentation based on different densities of filth, food particles, and immersion liquids. Examples of such filth are sand, soil, and some animal excreta pellets.

20.2.2.4 Light Filth

Lighter filth particles that are oleophilic and are separated from products by floating them in an oil-aqueous liquid mixture. Examples are insect fragments, whole insects, rodent hairs and fragments, and feather barbules.

20.2.2.5 Sieved Filth

Filth separated from food products on the basis of particle size using selected sieve mesh sizes.

20.2.2.6 Diagnostic Characteristics of Filth

Examples include specific diagnostic characteristics of **molds** (i.e., parallel hyphal walls, septation, granular appearance of cell contents, branching of hyphae, blunt ends of hyphal filaments, nonrefracted appearance of hyphae); diagnostic characteristics of **insect fragments** (i.e., recognizable shape, form, or surface sculpture, an articulation or joint, setae or setal pits, sutures), **rodent hairs** (i.e., pigment patterns and structural features), **feather barbules** (i.e., structural features); diagnostic characteristics of **insect-damaged** grains and packaging materials; and chemical identification of **animal urine and excrement.**

20.2.3 Objectivity of Methods

Insect parts, rodent hairs, and feather barbules in food products are generally reported as the total number of filth elements counted of each kind encountered per sample unit. They are identified on the basis of objective criteria (see section 20.2.2.6 above). Identifying insect fragments is not a simple task. Training and supervised practice are required to achieve competence and consistency. Some fragments are easily identified on the basis of structural shape and form. Mandibles, for example, are quite distinctive in their shape and configuration; certain species of insects can be determined on the basis of this one structure. In other instances, fragments may be mere chips of insect cuticle that have neither distinctive shape nor form but can be identified as being of insect origin if they have one or more of the characteristics given in section 20.2.2.6. Experienced analysts should rarely misinterpret fragments.

Isolation of extraneous material from a food product so that it can be identified and enumerated can be a very simple procedure or one that requires a series of several rather involved steps. In the process of isolating fragments from flour by the acid hydrolysis method, the sample is transferred from the digestion container to the separatory container and then to the filter paper for identification and enumeration. At each of those transfers there is an opportunity for loss of fragments. Although the analyst may have made every effort to maintain the isolation "quantitative," there are opportunities for error. Both fragment loss and analyst variation are minimized by common use of standard methods and procedures and proper training and supervised practice.

Another concern involves the significance of insect fragment counts (as well as particles of sand, pieces of rodent excreta, rodent hairs, etc.) in relation to fragment size. Fragment counts are reported on a numerical basis; they do not reflect the total contaminant biomass that is present. A small fragment is counted the same as a large fragment. The size of the fragment may be a reflection of the process to which a common raw material, such as wheat, has been subjected, a more vigorous process producing more and smaller fragments than a less vigorous process. These factors have been of concern to food processors for some time and have prompted the search for more objective means of determining insect contamination.

20.3 METHODS

Various basic methods for isolation of extraneous matter were suggested in section 20.2.2, which defined different types of filth: separation on the basis of differences in density, affinity for oleophilic solvents, particle size; diagnostic characteristics for identification of filth; and chemical identification of contaminants. Not all methods of analysis for extraneous matter for all categories of food can be discussed in this chapter. However, a few representative methods have been selected to illustrate the various principles of separation and isolation. As indicated, some isolations are effected very simply and others are more complex, requiring a series of various steps. AOAC methods (3) are used as reference sources, but they will be presented in a manner similar to that used for AACC methods (4). Refer to the specific AOAC methods cited for detailed instructions of the procedures.

20.3.1 Foreign Matter in Spices and Condiments—Sieving Method (AOAC Method 960.51)

20.3.1.1 Scope

Applicable to ground allspice, anise, curry powder, dill seed, fennel, fenugreek, poppyseed, savory, and condiments; heavy filth only: caraway seed, cardamon, celery seed, cloves, coriander, cumin, ginger, mace, marjoram, mustard, oregano, rosemary, sage, and thyme.

20.3.1.2 Apparatus

1. No. 20 sieve (850-micron openings) and pan
2. Widefield stereoscopic microscope

20.3.1.3 Reagents

There are no reagents in this procedure.

20.3.1.4 Procedure

1. Sift 200–400 g ground spice through No. 20 sieve.
2. Transfer insects or other filth retained on sieve to suitable dish.

20.3.1.5 Explanation

This method illustrates a very simple technique based on particle size separation. The No. 20 sieve refers to the U.S. National Bureau of Standards (now called the National Institute of Standards and Technology) sieve with 20-mesh per inch (plain weave), which provides a sieve opening of 850 microns. That sieve is coarse enough to let the ground spices pass through and yet retain the adult stages of insects and other contaminants of spices and condiments. The analyst must then be able to identify the insect species, stage, and whether live or dead, as well as any other foreign objects.

20.3.2 Filth in Shelled Nuts—Heavy Filth by Sedimentation (AOAC Method 968.33A)

20.3.2.1 Scope

Applicable to shelled nuts except pecans.

20.3.2.2 Apparatus

1. 600 mL beaker
2. Ashless filter paper
3. Stereoscopic microscope

20.3.2.3 Reagents

Note: Use effective fume removal device to remove vapors of flammable solvents (petroleum ether) and toxic solvents (carbon tetrachloride and chloroform).

1. Petroleum ether (pet ether)
2. Chloroform ($CHCl_3$)
3. Carbon tetrachloride (CCl_4)

20.3.2.4 Procedure

1. Weigh 100 g sample into 600 mL beaker.
2. Add ca. 350 mL pet ether and boil gently 30 minutes, adding pet ether to maintain original volume. (*Note:* petroleum ether is extremely flammable.)
3. Decant solvent, taking care not to lose any coarse nut tissue, and discard.
4. Add 300 mL $CHCl_3$ to beaker and let settle 10–15 min.
5. Pour off floating nut meats and about two-thirds of the $CHCl_3$ and discard.
6. Repeat separation with smaller volumes of mixture of $CHCl_3$ and CCl_4 (1 + 1) until residue in beaker is relatively free of nutmeat particles.
7. Transfer residue in beaker to ashless paper and examine for heavy filth.
8. Count number of particles of shell, sand, and soil.
9. If appreciable amount of sand and/or soil is present, ignite paper in weighed crucible at ca. 500°C and weigh. Report residue as percent of 100 g sample.

20.3.2.5 Explanation

The petroleum ether is used as a means for defatting the nut meats in preparation for continued analysis to determine light filth by an additional procedure. The chloroform and chloroform:carbon tetrachloride solvents allow pieces of shell, sand, and soil to settle to the bottom of the beaker on the basis of specific gravity and cause the defatted nut meats to float and be decanted. Although the AOAC method does not specify microscopic examination, that is desirable to identify the extraneous material.

Essentially the same procedure is used to isolate pieces of rodent excreta from corn grits, rye and wheat meal, whole wheat flour, farina, and semolina in AOAC Method 941.16A. It should be noted that the use of the more toxic solvents such as carbon tetrachloride, chloroform, and petroleum ether is avoided in most contemporary analytical methods.

20.3.3 Insect Infestation (Internal) of Wheat—Cracking Flotation Method (AOAC Method 982.31)

20.3.3.1 Scope

Applicable to wheat. [Also available are specific cracking flotation methods for oats (AOAC Method 985.36) and grains and seeds (except wheat and oats) (AOAC Method 955.42)]

20.3.3.2 Apparatus

1. Jones sampler (riffle-type divider)
2. No. 12 sieve (1680-micron openings), No. 100 sieve (150-micron openings)
3. Cutting-type mill
4. 2 L glass beaker
5. Magnetic stirring hotplate
6. 2 L trap flask (Wildman)
7. Wide-stem funnel
8. Filter flask
9. Ruled filter paper–smooth, high wet strength, rapid acting filter paper ruled with oil-, alcohol-, and water-proof lines 5 mm apart
10. Widefield stereoscopic microscope, 15×

20.3.3.3 Reagents

1. Hydrochloric acid (HCl)
2. 40 percent isopropanol
3. Tween 80:40 percent isopropanol mixture—to 40 mL polysorbate 80 (ICI Americas, Inc.) add 210 mL 40 percent isopropanol, mix, and filter
4. Na_4EDTA:40 percent isopropanol mixture—dissolve 5 g Na_4EDTA in 100 mL H_2O, add 100 mL isopropanol, mix, and filter
5. Mineral oil—paraffin oil, white, light, 125/135 Saybolt Universal Viscosity (38°), sp. gr. 0.840–0.860 (24°)

20.3.3.4 Procedure

1. Sample preparation
   a. Mix grain by passing six times through Jones divider (Fig. 20-1).
   b. Separate out slightly more than 50 g and weigh 50 g.
   c. Transfer weighed sample to No. 12 sieve and work any visible insects thru sieve using stiff brush (Fig. 20-2).
   d. Grind sieved sample in cutting-type mill set at 0.061 in.
2. Isolation
   a. Transfer cracked wheat to 2 L glass beaker containing magnetic stirring bar and mixture of 600 mL H_2O + 50 mL HCl. Stir *gently* while boiling 15 min on hotplate.
   b. Transfer sample to No. 100 sieve with a gentle stream of hot tap water. Wash material on sieve with *very gentle* stream of hot (55–70°C) tap water until washings show no acidity when tested with blue litmus paper.
   c. Place a wide-stem funnel in the neck of a 2 L **Wildman trap flask** (Fig. 20-3) containing a magnetic stirring bar. Quantitatively transfer the residue from the sieve to the flask with 40 percent isopropanol. Add 40 percent isopropanol to make 800 mL total volume.
   d. Clamp stirring rod of trap flask so stopper is

FIGURE 20-1.
Jones sampler (also called a Jones divider or riffle divider). Alternate channels cause grain and other flowable solids to be systematically divided into representative fractions.

FIGURE 20-2.
Sieve and pan. Standard mesh sizes are available for various particle size separations.

FIGURE 20-3.
Wildman trap flask. Stopper on shaft is lifted up to neck of flask to trap off floating layer.

above liquid in flask. Stir *gently* while boiling 7 min on magnetic stirring hotplate.

e. Remove flask from hotplate and wash down sides with minimal 40 percent isopropanol.
f. Immediately add 100 mL 1 + 1 mixture of Tween 80:40 percent isopropanol solution and Na_4EDTA:40 percent isopropanol solution slowly down stirring rod. Hand stir *gently* 1 min and let stand 3 min.
g. Add 50 mL mineral oil by pouring down stirring rod. Stir magnetically 5 min on cool stirrer and let stand 3 min.
h. Fill flask with 40 percent isopropanol, added slowly down stirring rod to avoid agitation of flask contents, and let stand 20 min *undisturbed*.
i. Trap off oil layer, rinsing neck of flask with 40 percent isopropanol. Add rinse to trappings in beaker.
j. Add 35 mL mineral oil to flask and hand stir 1 min. Clamp stirring rod with stopper at mid-point of flask and let stand for 5 min; then spin stirring rod to free settlings from top of stopper and adjust oil level to ca. 1 cm above fully raised stopper. Let stand *undisturbed* 15 min.
k. Trap off oil layer, combining trappings with those in beaker. Rinse neck of flask well with 40 percent isopropanol and add rinsings to beaker.
l. Filter trappings/rinsings on ruled filter paper, rinsing beaker well with isopropanol (Fig. 20-4).
m. Examine filter paper at 15x, counting only whole or equivalent insects and cast skins (Fig. 20-5).

20.3.3.5 Explanation

This method was developed to replace a previous simpler method that resulted in excessive amounts of plant

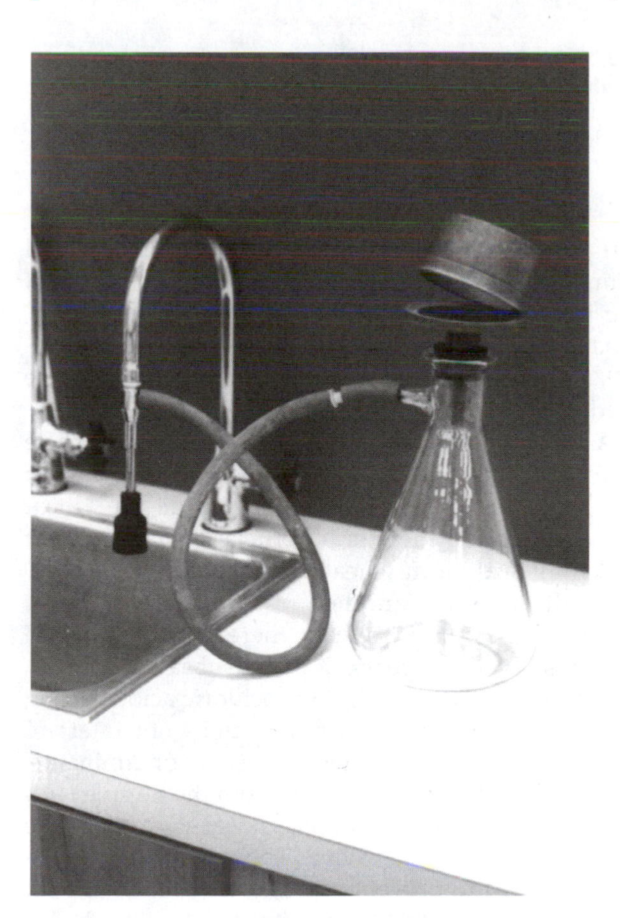

FIGURE 20-4.
Filter flask and funnel. A specially designed funnel has a collar (partially raised) that holds ruled filter paper in place on the funnel base for trapping filth for examination. Suction is applied with a water aspirator.

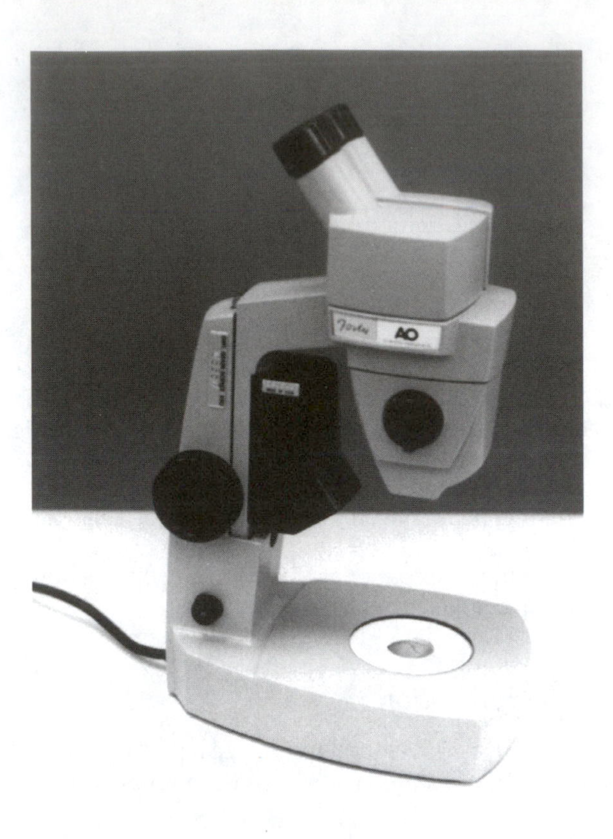

FIGURE 20-5.
Widefield stereoscopic microscope. Used to examine ruled filter papers for insect fragments, rodent hairs, feather barbules, and other microscopic contaminants.

debris trapped on filter papers. The previous method required as many as 5 to 10 filter papers to collect material trapped from a 100 g sample of wheat.

The initial screening with the No. 12 sieve (Fig. 20-2) is to remove any forms of insect that may be outside the sample wheat kernels.

A 50 g sieved wheat sample is ground to a particle size roughly that of coarsely ground coffee to expose and release insect forms that exist inside the wheat kernels. Although there is no official action level for internal infestation in wheat, the insects that develop inside wheat kernels are the primary source of insect fragments in flour, for which there is a DAL of 75 fragments (average) per 50 g in six subsamples (5).

The sample is boiled in hydrochloric acid solution to hydrolyze the starch so that it does not interfere with the flotation separation of insect contaminants. The hydrolyzed starch is washed from the sample with hot tap water while the sample and contaminants are retained on the No. 100 sieve. The tap water also removes the acid to prevent any chemical reaction with the Na_4EDTA used later in the procedure.

After the sample has been washed and neutralized, it is transferred to a Wildman trap flask (Fig. 20-3) to separate any insect material present from the remaining plant material. The sample is boiled in 40 percent isopropanol to deaerate the remaining plant material to prevent particles from being trapped off with the contaminants. The mixture of Tween 80 and Na_4EDTA solutions serves to prevent lighter plant materials, such as particles of bran, from being trapped off with the contaminants.

Tween 80 (polyoxyethylene sorbitan mono-oleate) is a nonionic agent used in many microscopic filth procedures. It appears to have certain surface active properties that make it a useful adjunct to Na_4EDTA. **Na_4EDTA** in the presence of Tween 80 appears to be a depressor for food materials (such as bran and other light plant matter) that otherwise tend to float. It has been suggested that the chelating properties of Na_4EDTA may result in its adsorption onto the surfaces of food particles along with the surfactant Tween 80, thereby preventing an attraction of the food particles to oils used to separate out insect contaminants. By preventing plant material from being collected in the mineral oil layer that is trapped off, contaminants such as oleophilic insect parts (exoskeleton) that are contained in the separating oil are much easier to distinguish and identify.

20.3.4 Light Filth (Pre- and Post-Milling) in Flour (White)—Flotation Method (AOAC Method 972.32)

20.3.4.1 Scope

Applicable to wheat flour (white).

20.3.4.2 Apparatus

1. 2–2.5 L beaker and 1 L beaker
2. Autoclave
3. Magnetic stirrer hotplate
4. Kilborn funnel or percolator
5. Ruled filter paper

20.3.4.3 Reagents

1. HCl solution
2. Mineral oil—paraffin, light 125/135 Saybolt
3. 5 percent detergent solution (aqueous Na lauryl sulfate)

20.3.4.4 Procedure

1. Digest (hydrolyze) 50 g flour in 2–2.5 L beaker with 600 mL HCl solution (3 + 97) in autoclave 5 min at 121°C.

2. Immediately transfer digest to 1 L beaker using HCl (3 + 97) at room temperature to assist transfer.
3. Add 50 mL mineral oil and stir magnetically 5 min.
4. Quantitatively transfer sample to Kilborn funnel or percolator (Fig. 20-6). Retain beaker.
5. Let stand 30 min, stirring gently with long glass rod several times during first 10 min.
6. Drain lower layer to ca. 3 cm of interface, wash sides of funnel with cold tap water, and let layers separate ca. 2 to 3 min. Repeat drain and H_2O wash until lower phase is clear.
7. After final wash, drain oil layer into retained beaker, rinsing sides of funnel with H_2O and alcohol.
8. Add HCl to ca. 3 percent (v/v) and boil 3 to 4 min on hot plate.
9. Filter hot solution through ruled filter paper (Fig. 20-4) and thoroughly rinse beaker and funnel with H_2O, alcohol, and 5 percent detergent solution onto ruled filter paper.
10. Examine microscopically (Fig. 20-5) and record light filth (insect fragments, rodent hairs and fragments, and feather barbules).

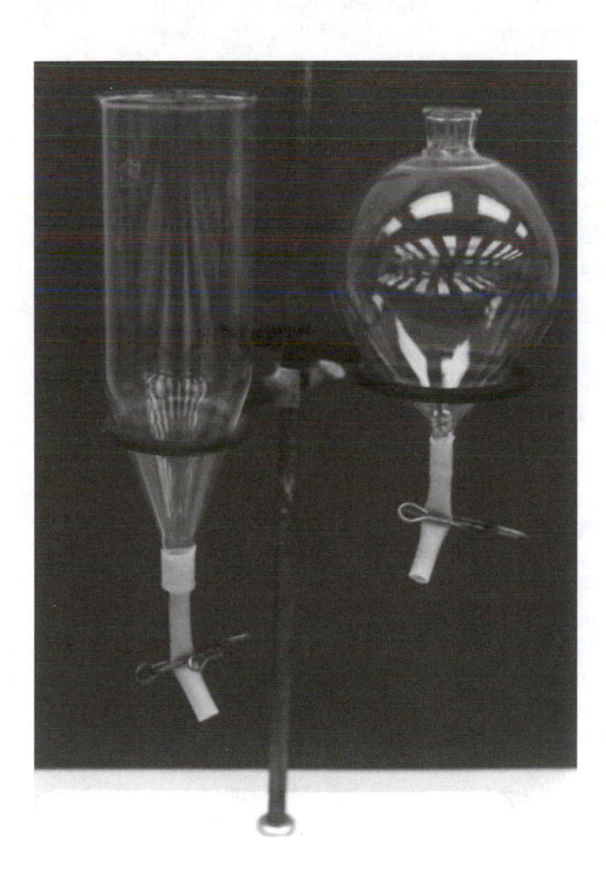

FIGURE 20-6.
Separatory funnel(s). Straight-sided, open-top funnel (left); leveling-bulb type with narrow opening at top (right). These types may be used to substitute for the Kilborn funnel or percolator.

20.3.4.5 Explanation

This is a common method used for insect fragments, rodent hairs, and other forms of light filth determination. The acid digestion breaks down the starch in the flour and allows the other flour constituents to more cleanly separate from the dilute acid solution. Although the AOAC method calls for digestion by autoclaving, AACC Method 28-41A provides for an alternative hotplate digestion, which might be more convenient for some laboratories. The oleophilic property of insect fragments, rodent hairs, and feather barbules allows them to be coated by the mineral oil and trapped in the oil layer for separation and collection on ruled filter paper. The heavier sediments of the digestion are washed and drained from the funnel. Fragments and rodent hairs are reported on the basis of 50 g of flour. The current FDA DAL for insect fragments in flour is an average of 75 or more per 50 g in six subsamples, and an average of one or more rodent hairs per 50 g in six subsamples (6).

20.4 COMPARISON OF METHODS

There are limited sources of methods for isolation of extraneous materials from foods. AOAC methods (3) are generally considered official and are often the basis for methods recommended by other organizations, such as the AACC (4). The difference in format used by the AACC has already been mentioned. The AACC has included colored pictures of representative insect fragments commonly found in cereal products, pictorial examples of rodent hair structure, and radiographic examples of grain kernels that contain internal insect infestation. Kurtz and Harris (6) provide a virtual parts catalog of insect fragments with a series of micrographs. Gentry et al. (7) is an updated version of the Kurtz and Harris publication and includes colored micrographs of common insect fragments.

Although **X-ray radiography** is not an official method for isolation of extraneous material, it is used by some processors as a means of inspecting wheat for internal insect infestation, the source of insect fragments in processed cereal products (8).

The most complete resource on analysis for extraneous matter is *Principles of Food Analysis for Filth, Decomposition and Foreign Matter,* FDA Technical Bulletin No. 1, edited by J. R. Gorham (9). The FDA *Training Manual for Analytical Entomology in the Food Industry* (10) is a looseleaf publication prepared to facilitate the orientation of food analysts to the basic techniques they will need for filth analysis.

20.5 SPECIAL CONSIDERATIONS

Mention was made earlier of the subjectivity involved in the isolation and especially identification of some biological contaminants commonly assessed in food products. Results have been based on numbers of particular contaminants rather than on mass determined in an objective manner. Some research has been directed at measuring **uric acid** content of samples as a basis for measuring insect and/or bird contamination by excrement (11). The AOAC has three methods for uric acid detection (Methods 962.20, 986.29, 969.46). More recently, an ELISA **(enzyme-linked immunosorbent assay)** method has been developed to objectively measure quantitatively the amount of insect material in a sample (12). It involves the measurement of the insect protein myosin. The advantage of uric acid and the myosin ELISA techniques is their adaptability to automation and objective measurement. Until these methods are accepted, one of the major needs for satisfactory filth isolation is properly trained laboratory analysts.

20.6 SUMMARY

Extraneous matter in raw ingredients and in processed foods might be unavoidable in the array of foods that are stored, handled, processed, and transported. Defect action levels may be set for amounts considered unavoidable and no health hazard. A variety of methods are available to isolate extraneous matter from foods. Those methods largely prescribed by the AOAC employ a series of physical and chemical means to separate the extraneous material for identification and enumeration. Major concerns in the analysis of food products for extraneous matter are the objectivity of methods and the availability of adequately trained analysts.

20.7 STUDY QUESTIONS

1. Indicate why the FDA has established defect action levels.
2. List three major reasons for conducting analysis for extraneous matter in foods.
3. What two resources provide methods for separating extraneous matter from cereal grains and their products?
4. There are several basic principles involved in separating (isolating) extraneous matter from foods. List five of these principles and give an example of each principle.
5. Briefly describe the major constraint(s) to currently accepted methods for analyses of extraneous matter in foods.

20.7 REFERENCES

1. FDA. 1986. Federal Food Drug and Cosmetic Act, as Amended, and Related Laws (though August 1985). U.S. Department of Health and Human Services, HHS Publication No. (FDA) 86-1051. U.S. Government Printing Office, Washington, DC.
2. FDA. 1991. Current Good Manufacturing Practice in Manufacturing, Packing, or Holding Human Food. Part 110, Title 21 Food and Drugs, in *Code of Federal Regulations*. Office of the Federal Register National Archives and Records Administration, Washington, DC.
3. Boese, J.L., and Bandler, R. (Assoc. Chap. Eds.) 1990. Extraneous materials: isolation. Ch. 16 (pp. 369–424), in *Official Methods of Analysis,* 15th ed. Association of Official Analytical Chemists, Arlington, VA.
4. AACC. 1983. AACC Method 28-00 Extraneous Matter. In *Approved Methods of the American Association of Cereal Chemists,* 8th ed. American Association of Cereal Chemists, Inc., St. Paul, MN.
5. FDA. 1988. FDA Compliance Policy Guide 7104.06. Office of Enforcement, Division of Compliance Policy, FDA. January 20, 1988. U.S. Government Printing Office, Washington, DC.
6. Kurtz, O. L., and Harris, K. L. 1962. *Micro-Analytical Entomology for Food Sanitation Control*. Association of Official Analytical Chemists, Washington, DC.
7. Gentry, J. W., Harris, K. L., and Gentry, J. W., Jr. 1991. *Microanalytical Entomology for Food Sanitation Control.* Vols. 1 and 2. J.W. Gentry and K.L. Harris, Melbourne, FL.
8. Milner, M., Lee, M. R., and Katz, R. 1950. Application of X-ray technique to the detection of internal insect infestation of grain. *J. Econ. Entomol.* 43:933–935.
9. FDA. 1981. *Principles of Food Analysis for Filth, Decomposition, and Foreign Matter.* FDA Technical Bulletin No. 1, J.R. Gorham (Ed.). Association of Official Analytical Chemists, Arlington, VA.
10. FDA. 1977. *Training Manual for Analytical Entomology in the Food Industry.* FDA Technical Bulletin No. 2, J.R. Gorham (Ed.). Association of Official Analytical Chemists, Arlington, VA.
11. Wehling, R. L., Wetzel, D. L., and Pedersen, J. R. 1984. Stored wheat insect infestation related to uric acid as determined by liquid chromatography. *J. Assoc. Off. Anal. Chem.* 67:644–647.
12. Quinn, F. A., Burkholder, W. E., and Kitto, G. B. 1992. Immunological technique for measuring insect contamination of grain. *J. Econ. Entomol.* 85:1463–1470.

Contribution No. 92-443-B Kansas Agricultural Experiment Station, Kansas State University, Manhattan, Kansas.

CHAPTER

21

Determination of Oxygen Demand

Yong D. Hang

21.1 INTRODUCTION

Oxygen demand is a commonly used parameter to evaluate the potential effect of organic pollutants on either a wastewater treatment process or a receiving water body. Because microorganisms utilize these organic materials, the concentration of dissolved oxygen is greatly depleted from the water. The oxygen depletion in the environment can have a detrimental effect on fish and plant life.

The three main methods used to measure the oxygen demand of water and wastewater are **biochemical oxygen demand** (BOD), **chemical oxygen demand** (COD), and **total organic carbon** (TOC). This chapter briefly describes the principles, procedures, applications, and limitations of each method. Methods described are from *Standard Methods for the Examination of Water and Wastewater,* published by the American Public Health Association (APHA). Refer to methods cited for detailed procedures.

21.2 METHODS

21.2.1 Biochemical Oxygen Demand (BOD)

21.2.1.1 Principle

The BOD determination is a measure of the amount of oxygen required by microorganisms to oxidize the biodegradable organic constituents present in water and wastewater. The method is based on the direct relationship between the concentration of organic matter and the amount of oxygen used to oxidize the pollutants to water, carbon dioxide, and inorganic nitrogenous compounds. The **oxygen demand** of water and wastewater is proportional to the amount of **organic matter** present. The BOD method measures the biodegradable carbon (carbonaceous demand) and, under certain circumstances, the biodegradable nitrogen (nitrogeneous demand).

21.2.1.2 Procedure

Place a known amount of a water or wastewater sample that has been seeded with an effluent from a biological waste treatment plant in an airtight BOD bottle and measure the initial dissolved oxygen immediately. Incubate the sample at 20°C and, after 5 days, measure the dissolved oxygen content again (APHA Method 5210 B). The dissolved oxygen content can be determined by the membrane electrode method (APHA Method 4500-O G) or the azide modification of the iodometric method (APHA Method 4500-O C). The BOD value, which is expressed as mg/L, can be calculated from the difference in the initial dissolved oxygen and the content of dissolved oxygen after the incubation period according to the following equation (APHA Method 5210B):

$$\text{BOD (mg/L)} = 100/\text{P} \times (\text{DOB} - \text{DOD}) \quad (1)$$

where: DOB = initial oxygen in diluted sample, mL
DOD = oxygen in diluted sample after 5-day incubation, mg/L
P = mL sample × 100 / capacity of bottle

21.2.1.3 Applications and Limitations

The BOD test is used most widely to measure the organic loading of waste treatment processes, to determine the efficiency of treatment systems, and to assess the effect of wastewater on the quality of receiving waters. The 5-day BOD test has some drawbacks because:

1. The procedure requires an incubation time of at least 5 days.
2. The BOD method does not measure all the organic materials that are biodegradable.
3. The test is not accurate without a proper seeding material.
4. Toxic substances such as chlorine present in water and wastewater may inhibit microbial growth.

21.2.2 Chemical Oxygen Demand (COD)

21.2.2.1 Principle

The COD determination is a rapid way to measure the quantity of oxygen used to oxidize the organic matter present in water and wastewater by a strong oxidizing agent. Most organic compounds are destroyed by refluxing in a strong acid solution with a known quantity of an oxidizing agent such as **potassium dichromate.** The excessive amount of potassium dichromate left after digestion of the organic matter is measured. The amount of organic matter that is chemically oxidizable is directly proportional to the potassium dichromate consumed.

21.2.2.2 Procedure

A known quantity of sample of water or wastewater is refluxed at elevated temperatures for up to two hours with a known quantity of potassium dichromate and sulfuric acid. The amount of potassium dichromate left after digestion of the organic matter is titrated with a standard ferrous ammonium sulfate solution using orthophenanthroline ferrous complex as an indicator. The amount of oxidizable organic matter, determined as oxygen equivalent, is proportional to the potassium dichromate used in the oxidative reaction. The COD value can be calculated from the following equation (APHA Method 5220 B):

$$\text{COD, mg/L} = (A - B) \times M \times 8{,}000/D \quad (2)$$

where: A = mL thiosulfate used for blank
B = mL thiosulfate used for sample
M = molarity of thiosulfate
D = mL sample used

21.2.2.3 Applications and Limitations

Potassium dichromate is widely used for the COD method because of its advantages over other oxidizing compounds in oxidizability, applicability to a wide variety of waste samples, and ease of manipulation. The dichromate reflux method can be used to measure the samples with COD values of greater than 50 mg/L.

The COD test measures carbon and hydrogen in organic constituents but not nitrogenous compounds. Furthermore, the method does not differentiate between biologically stable and unstable compounds present in water and wastewater. The COD test is a very important procedure for routinely monitoring industrial wastewater discharges and for the control of waste treatment processes. The test is faster and more reproducible than the BOD method. The obvious disadvantages of the COD method are:

1. Aromatic hydrocarbons, pyridine, and straight-chain aliphatic compounds are not readily oxidized.
2. The method is very susceptible to interference by **chloride,** and thus the COD of certain food processing waste effluents such as pickle and sauerkraut brines cannot be readily determined without modification. This difficulty may be overcome by adding **mercuric sulfate** to the sample prior to refluxing. Chloride concentrations greater than 500 to 1,000 mg/L may not be corrected by the addition of mercuric sulfate. A chloride correction factor can be developed for a particular waste by the use of proper blanks.

21.2.3 Total Organic Carbon (TOC)

21.2.3.1 Principle

The TOC method is another rapid and convenient means for determining the amount of organic matter present in water and wastewater. The test uses high temperatures and a strong oxidizing agent to oxidize the organic carbon in water and wastewater to carbon dioxide and water. The carbon dioxide that is produced in the oxidative reaction is measured.

21.2.3.2 Procedure

Place a known quantity of a properly homogenized water or wastewater sample in an elevated-temperature reactor (900°C) containing an oxidizing agent such as cobalt oxide as a catalyst. Under controlled conditions, the carbon atoms of organic matter present in water and wastewater are converted to carbon dioxide and water. The amount of carbon dioxide produced in the oxidative reaction is determined quantitatively by means of an infrared total organic carbon analyzer (APHA Method 5310 B).

21.2.3.3 Applications and Limitations

The TOC test is especially useful for determining the amount of total organic carbon in water and wastewater. It is a rapid method and is more precise than COD and BOD. The major disadvantages of the TOC test are:

1. The test requires a well-trained technician and expensive equipment not normally found in a laboratory.
2. The TOC method does not completely oxidize all the organic carbon compounds present in water and wastewater.
3. TOC is generally not as reliable as BOD or COD in predicting the oxygen demand potential of water and wastewater because oxygen demands differ between organic compounds with the same number of organic carbons in their structure. For example, the oxygen demand of ethanol is six times greater than that of oxalic acid.

21.3 COMPARISON OF BOD, COD, AND TOC METHODS

The BOD and COD analyses of water and wastewater can result in different values because the two methods measure different materials. As shown in Table 21-1, the COD value of a waste sample is usually higher than its BOD and TOC values because:

TABLE 21-1. Oxygen Demand of Tomato Processing Wastes

Item	*1973*	*1974*	*1975*
BOD, mg/L	2,400	1,300	1,200
COD, mg/L	5,500	3,000	2,800
TOC, mg/L	2,000	1,100	1,000

From EPA. 1977. Pollution Abatement in the Fruit and Vegetable Industry. EPA-625/3-77-0007. Environmental Protection Agency, Washington, DC.

1. Many organic compounds that can be chemically oxidized cannot be biochemically oxidized. For example, cellulose cannot be determined by the BOD method but can be measured by the COD test.
2. Certain inorganic compounds such as ferrous iron, nitrites, sulfides, and thiosulfates are readily oxidized by potassium dichromate. This inorganic COD introduces an error when computing the organic matter of water and wastewater.
3. The BOD test can give low values because of a poor seeding material. The COD test does not requires an inoculum.
4. Some aromatics and nitrogenous (ammonium) compounds are not oxidized by the COD method. Other organic constituents such as cellulose or lignin, which are readily oxidized by potassium dichromate, are not biologically degraded by the BOD method.
5. Toxic materials present in water and wastewater that do not interfere with the COD test can affect the BOD results.

The COD, however, has value for specific wastes since it is possible to obtain a direct correlation between COD and BOD values. The **COD/BOD ratio** can be a useful tool for rapid determination of the biodegradability of organic matter present in the wastes. A low COD/BOD ratio indicates the presence of a small amount of nonbiodegradable organic matter. Samples of wastewater with high COD/BOD ratios have a large amount of organic matter that is nonbiodegradable.

The TOC test is a rapid and convenient means to estimate BOD or COD, once an empirical relationship between TOC and BOD or TOC and COD has been established. The TOC method may give a lower value than the actual amount of organic matter present in water and wastewater because certain organic matter may not be oxidized completely (Table 21-1). Furthermore, samples of water and wastewater that contain certain inorganic components, such as carbonates and bicarbonates, may give slightly higher TOC values.

21.4 SAMPLING AND HANDLING REQUIREMENTS

Samples of water and wastewater collected for oxygen demand determinations must be analyzed as soon as possible or stored under properly controlled conditions until analyses can be made.

Samples for the BOD test can be kept at low temperatures (4°C or below) for up to 48 hrs. Chemical preservatives should not be added to water and wastewater because they can interfere with BOD analysis. Untreated wastewater samples for the COD test must be collected in glass containers and analyzed promptly. The COD samples can be stored at 4°C or below for up to 28 days if they are acidified with a concentrated mineral acid (sulfuric acid) to a pH value of 2.0 or below.

Samples of water and wastewater for the TOC test must be collected in amber glass containers. If the test cannot be made immediately, the TOC samples must be kept in the dark at low temperatures (4°C or below) or they can be preserved by acidification to pH 2.0 or below using concentrated sulfuric or phosphoric acid.

21.5 SUMMARY

Oxygen demand is most widely used to determine the effect of organic pollutants present in water and wastewater on receiving streams and rivers. The three important methods used to measure oxygen demand are BOD, COD, and TOC.

The BOD test measures the amount of oxygen required by microorganisms to oxidize the biodegradable organic matter present in water and wastewater.

The COD method determines the quantity of oxygen consumed during the oxidation of organic matter in water and wastewater by potassium dichromate. The TOC determination is a measure of the amount of total organic carbon present in water and wastewater that can be converted to carbon dioxide and water by means of a strong oxidizing agent such as potassium dichromate at elevated temperatures.

Of the three methods used to measure oxygen demand, the BOD test has the widest application in measuring waste loading to treatment systems, in determining the efficiency of treatment processes, and in evaluating the quality of receiving streams and rivers because it most closely approximates the natural conditions of the environment. The COD or TOC test can be used to monitor routinely the biodegradability of organic matter in water and wastewater if a relationship between COD and BOD or TOC and BOD has been established.

21.6 STUDY QUESTIONS

1. Differentiate the principles of the BOD, COD, and TOC methods to measure oxygen demand.
2. In your new job as supervisor of a lab that has previously been using the BOD method to determine oxygen demand of wastewater, you have decided to change to the COD method.
   a. Describe the basic principle and procedure of the COD method to your lab technicians.
   b. In what case would they be instructed to use mercuric sulfate in the COD assay?

c. You realize there are advantages and disadvantages of all three potential methods—BOD, COD, and TOC. Give two advantages and two disadvantages for each of the BOD and TOC methods as compared to the COD method.

3. In each case described below, indicate if you would expect the COD value to be higher or lower than results from a BOD test. Explain your answer.
   a. poor seed material in BOD test
   b. sample contains toxic materials
   c. sample high in aromatics and nitrogenous compounds
   d. sample high in nitrites and ferrous iron
   e. sample high in cellulose and lignin

21.7 PRACTICE PROBLEMS

1. Determine the BOD value of a sample given the following data:
   DOB = 9.0 mL
   DOD = 6.6 mL
   P = 15 mL
   Capacity of bottle = 300 mL

2. Determine the COD value of a sample given the following data:
   thiosulfate for blank = 37.8 mL
   thiosulfate for sample = 34.4 mL
   molarity of thiosulfate = 0.025 M
   sample = 5 mL

Answers

1. BOD = 48; 2. COD = 136.

21.8 RESOURCE MATERIALS

EPA. 1977. Pollution Abatement in the Fruit and Vegetable Industry. EPA-625/3-77-0007. Environmental Protection Agency, Washington, DC.

Greenberg, A.E., Clesceri, L.S., and Eaton, A.D. (Eds.) 1992. *Standard Methods for the Examination of Water and Wastewater,* 18th ed. American Public Health Association, Washington, DC.

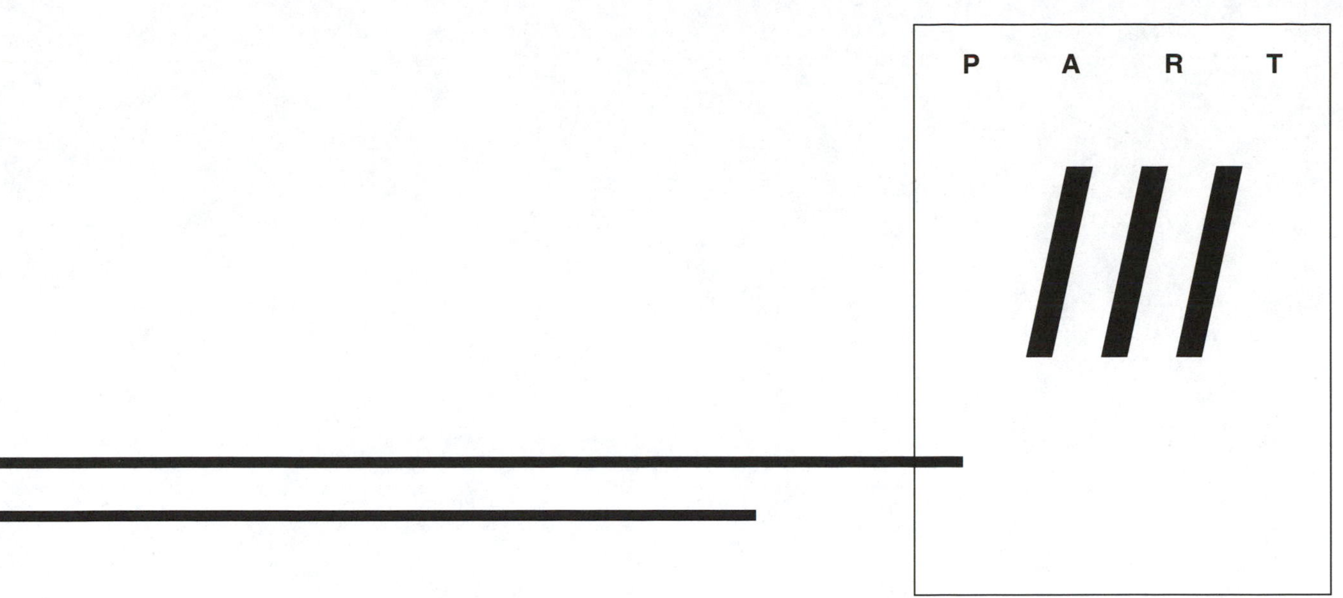

PART III

Spectroscopy

CHAPTER

22

Basic Principles of Spectroscopy

Michael H. Penner

22.1 INTRODUCTION

Spectroscopy deals with the production, measurement, and interpretation of spectra arising from the interaction of electromagnetic radiation with matter. There are many different spectroscopic methods available for solving a wide range of analytical problems. The methods differ with respect to the species to be analyzed (such as molecular or atomic spectroscopy), the type of radiation-matter interaction to be monitored (such as absorption, emission, or diffraction), and the region of the electromagnetic spectrum used in the analysis. Spectroscopic methods are very informative and widely used for both quantitative and qualitative analyses. Spectroscopic methods based on the absorption or emission of radiation in the **ultraviolet** (UV), **visible** (Vis), **infrared** (IR), and radio (**nuclear magnetic resonance,** NMR) frequency ranges are most commonly encountered in traditional food analysis laboratories. Each of these methods is distinct in that they monitor different types of molecular and/or atomic transitions. The basis of these transitions will be rationalized in the following sections.

22.2 LIGHT

22.2.1 Properties

Light may be thought of as particles of energy that move through space with wavelike properties. This image of light suggests that the energy associated with a ray of light is not distributed continuously through space along the wave's associated electric and magnetic fields but rather that it is concentrated in discrete packets. Light is therefore said to have a dual nature: **particulate** and **wavelike.** Phenomena associated with light propagation, such as interference, diffraction, and refraction, are most easily explained using the wave theory of electromagnetic radiation. However, the interaction of light with matter, which is the basis of absorption and emission spectroscopy, may be best understood in terms of the particulate nature of light. Light is not unique in possessing both wavelike and particulate properties. For example, fundamental particles of matter, such as electrons, protons, and neutrons, are known to exhibit wavelike behavior.

The wave properties of electromagnetic radiation are described in terms of the wave's frequency, wavelength, and amplitude. A graphical representation of a plane-polarized electromagnetic wave is given in Fig. 22-1. The wave is plane polarized in that the oscillating electric and magnetic fields making up the wave are each limited to a single plane. The **frequency** *(v)* of a wave is defined as the number of oscillations the wave will make at a given point per second. This is the reciprocal of the **period** (p) of a wave, which is the time in seconds required for successive maxima of the wave to pass a fixed point. The **wavelength** (λ) represents the distance between successive maxima on any given wave. The units used in reporting wavelengths will depend on the region of electromagnetic radiation used in the analysis. Spectroscopic data are sometimes reported with respect to **wavenumbers** ($\overline{v}$), which are reciprocal wavelengths in units of cm^{-1}. Wavenumbers are most often encountered in infrared spectroscopy. The **velocity of propagation** (v_i) of an electromagnetic wave, in units of distance per second, in any given medium "i" can be calculated by taking the product of the frequency of the wave, in cycles per second, and its wavelength in that particular medium.

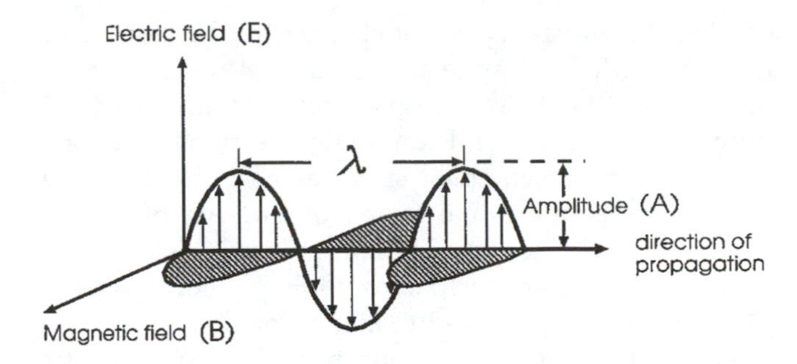

FIGURE 22-1.
Representation of plane-polarized electromagnetic radiation propagating along the x-axis. The electric and magnetic fields are in phase, perpendicular to each other and to the direction of propagation.

$$v_i = v\lambda_i \tag{1}$$

where: v_i = velocity of propagation in medium "i"
v = wave frequency
λ_i = wavelength in medium "i"

The frequency of an electromagnetic wave is determined by the source of the radiation, and it remains constant as the wave traverses different media. However, the velocity of propagation of a wave will vary slightly depending on the medium through which the light is propagated. The wavelength of the radiation will change in proportion to changes in wave velocity as defined by equation 1. The **amplitude of the wave** (A) represents the magnitude of the electric vector at the wave maxima. The **radiant power** (P) and **radiant intensity** (I) of a beam of radiation are proportional to the square of the amplitude of the associated waves making up that radiation. Figure 22-1 indicates that electromagnetic waves are composed of oscillating magnetic and electric fields, the two of which are mutually perpendicular, in phase with each other, and perpendicular to the direction of wave propagation. As drawn, the waves represent changes in the respective field strengths with time at a fixed location or changes in the respective field strengths over distance at a fixed time. The electrical and magnetic components of the waves are represented as a series of vectors whose lengths are proportional to the magnitude of the respective field. It is the oscillating electric field that is of most significance to spectroscopic phenomena such as absorption, transmission, and refraction. However, a purely electric field, without its associated magnetic field, is impossible.

22.2.2 Terminology

The propagation of electromagnetic waves is often described in terms of wavefronts and/or trains of waves (Fig. 22-2). A **wavefront** represents the locus of a set of points all of which are in phase. For a point source of light, a concentric ring that passes through the maxima of adjacent light rays will represent a wavefront. The entire ring need not be drawn in all cases, such that wavefronts may represent planes of light in cases where the observation is sufficiently removed from the point source that the curved surface appears planer. Wavefronts are most typically drawn by connecting maxima, minima, or both for adjacent rays. If maxima are used for depicting wavefronts, then each of the wavefronts will be separated by one wavelength. A **train of waves,** or **wave-train,** refers to a series of wavefronts all of which are in phase, that is, each individual wave will have a maximum amplitude at the same location in space. A wave-train may also be represented by a series of light rays. Rays of light are generally used with reference to the corpuscular nature of light, representing the path of photons. A wave-train would indicate that a series of photons, all in phase, followed the same path.

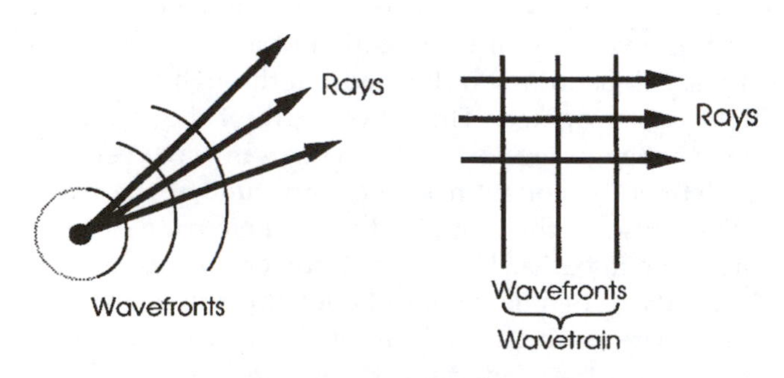

FIGURE 22-2.
Wavefronts, wave-trains, and rays. From Hugh D. Young, *University Physics,* Eighth Edition (p. 947), © 1992 by Addison-Wesley Publishing Company, Inc. Courtesy of the publisher.

22.2.3 Interference

Interference is the term used to describe the observation that when two or more wave-trains cross one another, they result in an instantaneous wave, at the point of intersection, whose amplitude is the algebraic sum of the amplitudes of the individual waves at the point of intersection. The law describing this wave behavior is known as the **principle of superposition.** Superposition of sinusoidal waves is illustrated in Fig. 22-3. Note that in all cases, the effective amplitude of the perceived wave at the point in question is the combined effect of each of the waves that crosses that point at any given instant. In spectroscopy, the amplitude of most general interest is that corresponding to the magnitude of the resulting electric field intensity. **Maximum constructive interference** of two waves occurs when the waves are completely in phase (i.e., the maxima of one wave aligns with the maxima of the other wave), while **maximum destructive interference** occurs when waves are 180° out of phase (the maxima of one wave aligns with the minima of the other wave). This concept of interference is fundamental to the interpretation of diffraction data, which represents a specialized segment of qualitative spectroscopy. Interference phenomena are also widely used in the design of spectroscopic instruments that require the dispersion and/or selection of radiation, such as those instruments employing grating monochromators or interference filters.

Interference phenomena are best rationalized by considering the wavelike nature of light. However, phenomena like the absorption and emission of radiation are more easily understood by considering the particulate nature of light. The particles of energy that move through space with wavelike properties are

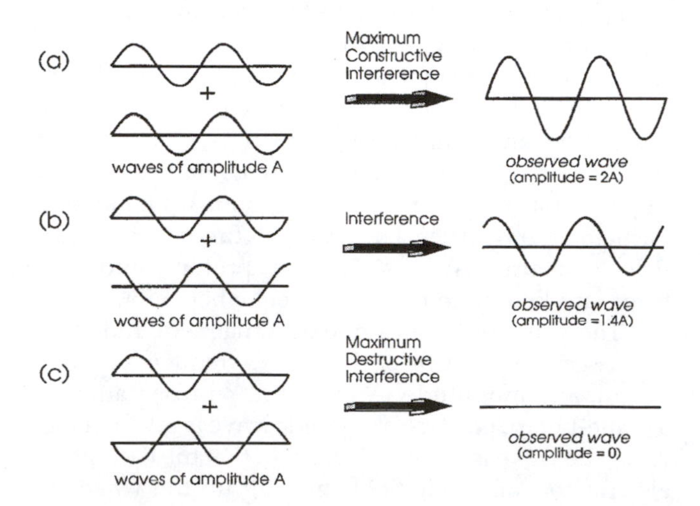

FIGURE 22-3.
Interference of identical waves that are (a) in phase, (b) 90° out of phase, and (c) 180° out of phase.

called **photons.** The energy of a photon can be defined in terms of the frequency of the wave with which it is associated (eq. 2).

$$E = h\upsilon \qquad (2)$$

where: E = energy of a photon
h = Planck's constant
υ = frequency of associated wave

This relationship indicates that the photons making up monochromatic light, which is electromagnetic radiation composed of waves having a single frequency and wavelength, are all of equivalent energy. Furthermore, just as the frequency of a wave is a constant determined by the radiation source, the energy of associated photons will also be unchanging. The brightness of a beam of monochromatic light, when expressed in terms of the particulate nature of light, will be the product of the photon flux and the energy per photon. The **photon flux** refers to the number of photons flowing across a unit area perpendicular to the beam per unit time. It follows that to change the brightness of a beam of monochromatic light will require a change in the photon flux. In spectroscopy, the term *brightness* is generally not used, but rather one refers to the **radiant power** (P) or the **radiant intensity** (I) of a beam of light. Radiant power and radiant intensity are often used synonymously when referring to the amount of radiant energy striking a given area per unit time. In terms of SI Units (time, seconds; area, meters^2; energy, joules), radiant power equals the number of joules of radiant energy impinging on a 1 meter^2 area of detector per second. The basic interrelationships of light-related properties and a general scheme of the electromagnetic spectrum are presented in Table 22-1 and Fig. 22-4, respectively.

22.3 ENERGY STATES OF MATTER

22.3.1 Quantum Nature of Matter

The energy content of matter is quantized. Consequently, the potential or internal energy content of an atom or molecule does not vary in a continuous manner but rather in a series of discrete steps. Atoms and molecules, under normal conditions, exist predominantly in the ground state, which is the state of lowest energy. Ground state atoms and molecules can gain energy, in which case they will be elevated to one of their higher energy states, referred to as excited states. The quantum nature of atoms and molecules puts limitations on the energy levels that are available to these species. Consequently, there will be specific "allowed" internal energy levels for each atomic or molecular species. Internal energy levels not corresponding to an allowed value for that particular species are unattainable. The set of available energy levels for any given atom or molecule will be distinct for that species. Similarly, the potential energy spacings between allowed internal energy levels will be characteristic of a species. Therefore, the set of potential energy spacings for a species may be used qualitatively as a distinct fingerprint. Qualitative absorption and emission spectroscopy make use of this phenomenon in that these techniques attempt to determine an unknown compound's relative energy spacings by measuring transitions between allowed energy levels.

TABLE 22-1. Properties of Light

Symbols/Terms	*Relationship*	*Frequently Used Units*
λ = wavelength	$\lambda\nu = c$	nm (nanometers, 10^{-9} m) Å (Ångstrom units, 10^{-10} m) μ (microns, 10^{-6} m) mμ (millimicrons, 10^{-9} m)
ν = frequency		Hz (hertz, 1 Hz = 1 oscillation per second) s^{-1}
c = speed of light		2.9979×10^8 m s^{-1} in vacuum
$\bar{\nu}$ = wavenumber	$\bar{\nu} = 1/\lambda$	cm^{-1} kK (kilokayser, 1 kK = 1000 cm^{-1})
p = period	$p = 1/\nu$	seconds
E = energy	$E = h\nu$ $= hc/\lambda$ $= hc/\bar{\nu}$	J (1 joule = 1 kg m^2 s^{-2}) cal (calorie, 1 cal = 4.184 J) erg (1 erg = 10^{-7} J) eV (1 electron volt = 1.6022×10^{-19} J)
h = Planck's constant		6.6262×10^{-34} J s
P = radiant power	the amount of energy striking a given unit area per unit time	(Joules)(meter^2)$^{-1}$(sec)$^{-1}$

Adapted from (4).

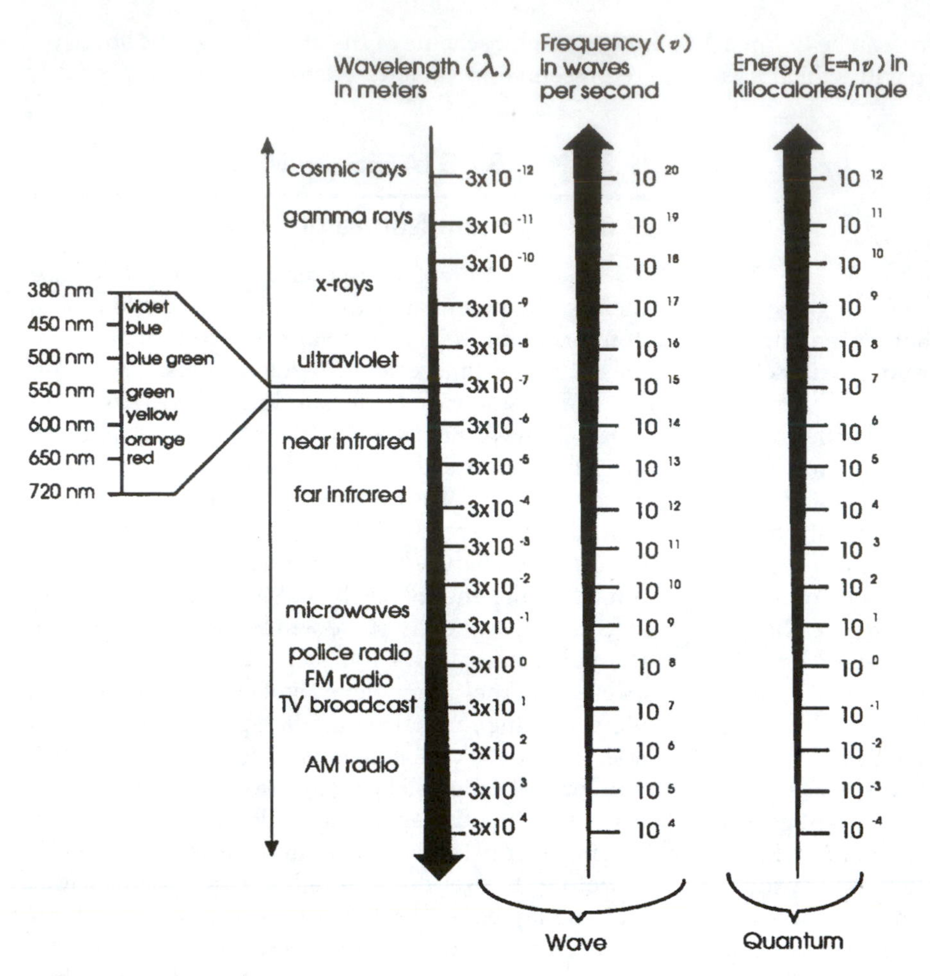

FIGURE 22-4.
The electromagnetic spectrum. *Milton Roy Educational Manual*, 1989, p. 3. Figure courtesy of Milton Roy Company, Rochester, NY, a subsidiary of Sundstrand Corporation.

22.3.2 Electronic, Vibrational, and Rotational Energy Levels

The relative **potential energy** of an atom or molecule corresponds to the energy difference between the energy state in which the species exists and that of the ground state. Figure 22-5 is a partial molecular energy level diagram depicting potential energy levels for an organic molecule. The lowest energy state in the figure, bottom line in bold, represents the ground state. There are three **electronic energy states** depicted, each with its corresponding vibrational and rotational energy levels. Each of the electronic states corresponds to a given electron orbital. Electrons in different orbitals are of different potential energy. When an electron changes orbitals, such as when absorbing or emitting a photon of appropriate energy, it is termed an **electronic transition** since it is the electron that is changing energy levels. However, any change in the potential energy of an electron will, by necessity, result in a corresponding change in the potential energy of the atom or molecule that the electron is associated with. Atoms are like molecules in that only specific energy levels are allowed for atomic electrons. Consequently, an energy level diagram of an atom would consist of a series of electronic energy levels. In contrast to molecules, the electronic energy levels of atoms have no corresponding vibrational and rotational levels and, hence, may appear less complicated. Atomic energy levels correspond to allowed electron shells (orbits) and corresponding subshells (i.e., 1s, 2s, 2p, etc.). The magnitude of the energy difference between the ground state and first excited states for valence electrons of atoms and bonding electrons of molecules is generally of the same range as the energy content of photons associated with UV and Vis radiation.

The wider lines within each electronic state of Fig. 22-5 depict the species **vibrational energy levels.** The atoms that comprise a molecule are in constant motion, vibrating in many ways. However, in all cases the energy associated with this vibrational motion corresponds to defined quantized energy levels. The energy

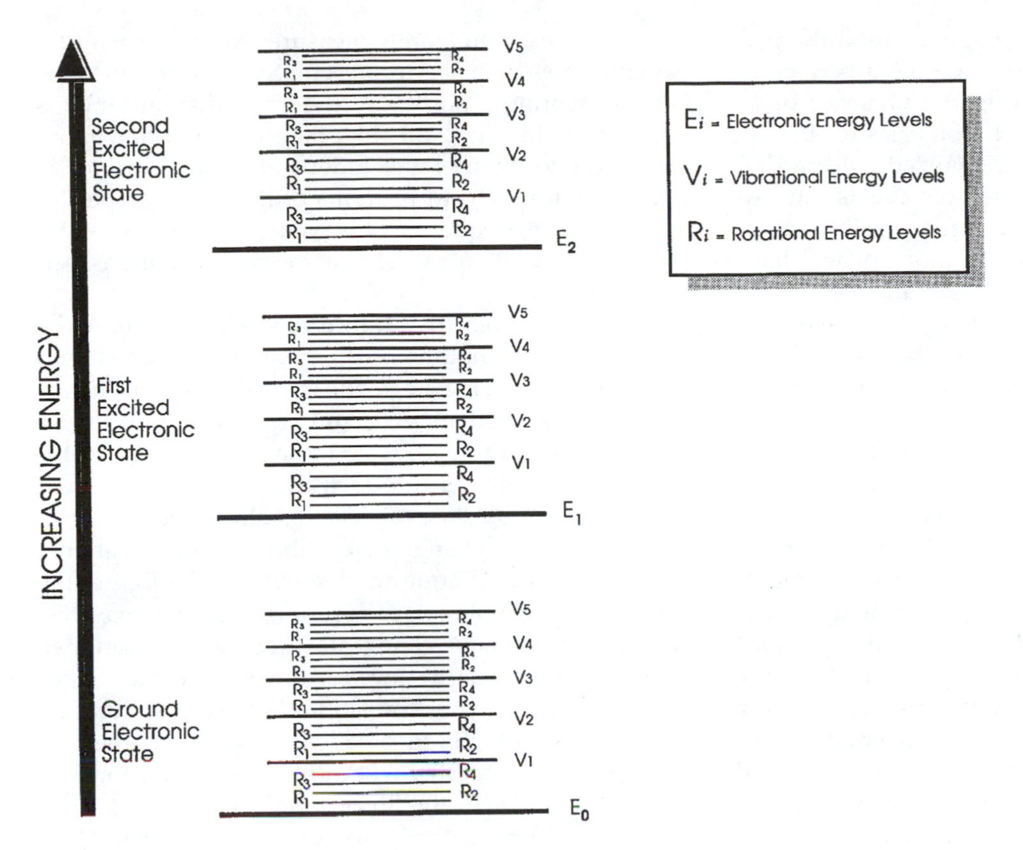

FIGURE 22-5.
Partial molecular energy level diagram depicting three electronic states.

differences between neighboring vibrational energy levels are much smaller than those between adjacent electronic energy levels. Therefore, it is common to consider that several vibrational energy levels are superimposed on each of the molecular electronic energy levels. Energy differences between allowed vibrational energy levels are of the same magnitude as the energy of photons associated with radiation in the infrared region. Vibrational energy levels would not be superimposed on an atomic potential energy level diagram since this vibrational motion does not exist in a single atom. In this respect, the potential energy diagram for an atom is less complex than that for a molecule, the atomic energy level diagram having fewer energy levels. The potential energy of a molecule is also quantized in terms of the energy associated with the rotation of the molecule about its center of gravity. These **rotational energy levels** are yet more closely spaced than the corresponding vibrational levels, as depicted by the narrow lines within each electronic state shown in Fig. 22-5. Hence it is customary to consider several rotational energy levels superimposed on each of the permitted vibrational energy levels. The energy spacings between rotational energy levels are of the same magnitude as the energy associated with photons of microwave radiation. Microwave spectroscopy is not commonly used in food analysis laboratories; however, the presence of these different energy levels will impact the spectrum observed in other forms of spectroscopy, as will be discussed later. Similar to the situation of vibrational energy levels, rotational energy levels are not of consequence to atomic spectroscopy. In summation, the internal energy of an atom is described in terms of its electronic energy levels, while the internal energy of a molecule is dependent on its **electronic, vibrational,** and **rotational** energies. The algebraic form of these statements follows.

$$E_{atom} = E_{electronic} \tag{3}$$

$$E_{molecule} = E_{electronic} + E_{vibrational} + E_{rotational} \tag{4}$$

The spectroscopist makes use of the fact that each of these associated energies is quantized and that different species, atoms, or molecules, will have somewhat different energy spacings.

22.3.3 Nuclear Energy Levels in Applied Magnetic Fields

Nuclear magnetic resonance (NMR) spectroscopy makes use of yet another type of quantized energy level. The energy levels of importance to NMR spectroscopy differ with respect to those described above in that they are relevant only in the presence of an applied external **magnetic field.** The basis for the observed energy levels may be rationalized by considering that the nuclei of some atoms behave as tiny bar magnets. Hence, when the atoms are placed in a magnetic field, their nuclear magnetic moment will have a preferred orientation, just as a bar magnet would behave. The NMR-sensitive nuclei of general relevance to the food analyst have two permissible orientations. The energy difference between these allowed orientations depends on the effective magnetic field strength that the nuclei experience. The effective magnetic field strength will itself depend on the strength of the applied magnetic field and the chemical environment surrounding the nuclei in question. The applied magnetic field strength will be set by the spectroscopist, and it is essentially equivalent for each of the nuclei in the applied field. Hence, differences in energy spacings of NMR-sensitive nuclei will solely depend on the identity of the nucleus and its environment. In general, the energy spacings between permissible nuclear orientations, under usable external magnetic field strengths, are of the same magnitude as the energy associated with radiation in the radio-frequency range.

22.4 ENERGY LEVEL TRANSITIONS IN SPECTROSCOPY

22.4.1 Absorption of Radiation

The **absorption of radiation** by an atom or molecule is that process in which energy from a photon of electromagnetic radiation is transferred to the absorbing species. When an atom or molecule absorbs a photon of light, its internal energy increases by an amount equivalent to the amount of energy in that particular photon. Therefore, in the process of absorption, the species goes from a lower energy state to a more **excited state.** In most cases, the species is in the **ground state** prior to absorption. Since the absorption process may be considered quantitative—all of the photon's energy is transferred to the absorbing species—the photon being absorbed must have an energy content that exactly matches the energy difference between the energy levels across which the transition occurs. This must be the case due to the quantized energy levels of matter, as discussed above. Consequently, if one plots photon energy versus the relative absorbance of radiation uniquely composed of photons of that energy, one observes a characteristic **absorption spectrum,** the shape of which is determined by the relative absorptivity of photons of different energy. The **absorbtivity** of a compound is a wavelength-dependent proportionality constant that relates the absorbing species concentration to its experimentally measured absorbance under defined conditions. A representative absorption spectrum covering a portion of the UV radiation range is presented in Fig. 22-6. In practice, it is often more practical to express the independent variable of an absorption spectrum in terms of the wave properties (wavelength, frequency, or wavenumbers) of the radiation, as in Fig. 22-6, rather than the energy of the associated photons.

Various molecular transitions resulting from the absorption of photons of different energy are shown schematically in Fig. 22-7. The transitions depicted represent those that may be induced by absorption of UV, Vis, IR, and microwave radiation. The figure also includes transitions in which the molecule is excited from the ground state to an exited electronic state with a simultaneous change in its vibrational or rotational energy levels. Although not shown in the figure, the absorption of a photon of appropriate energy may also cause simultaneous changes in electronic, vibrational, and rotational energy levels. The ability of molecules to have simultaneous transitions between the different energy levels tends to broaden the peaks in the UV-Vis absorption spectrum of molecules relative to those peaks observed in the absorption spectrum of atoms. This would be expected when one considers that vibrational and rotational energy levels are absent in an atomic energy level diagram. The depicted transitions between vibrational energy levels, without associated electronic transitions, are induced by radiation in the infrared region. Independent transitions between al-

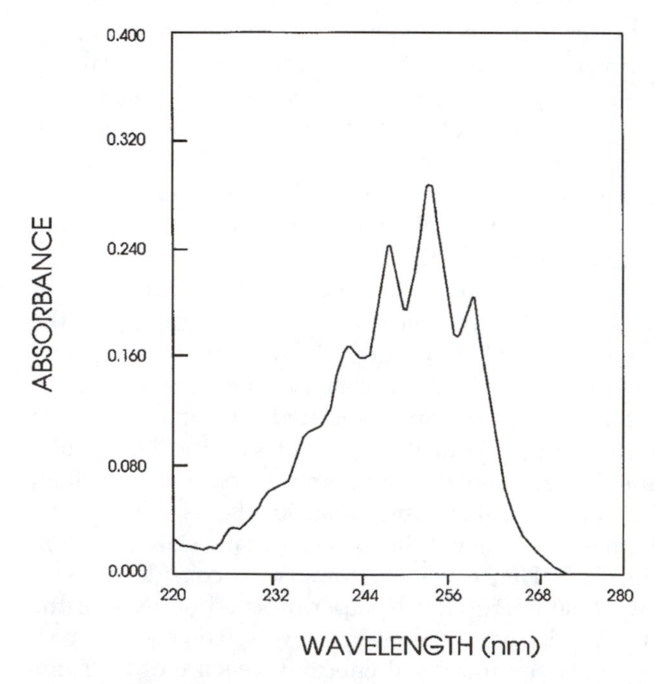

FIGURE 22-6.
Absorption spectrum of a 0.005 M benzene in water solution.

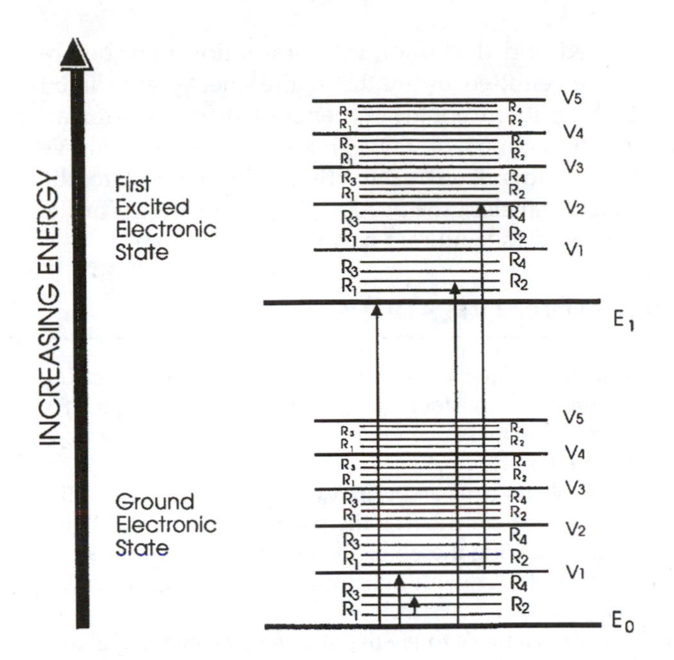

FIGURE 22-7.
Partial molecular energy level diagram including electronic, vibrational, and rotational transitions.

lowed rotational energy levels are also depicted, these resulting from the absorption of photons of microwave radiation. A summary of transitions relevant to atomic and molecular absorption spectroscopy, including corresponding wavelength regions, is presented in Table 22-2.

22.4.2 Emission of Radiation

Emission is essentially the reverse of the absorption process, occurring when energy from an atom or molecule is released in the form of a photon of radiation. A molecule raised to an excited state will typically remain in the excited state for a very short period of time before relaxing back to the ground state. There are several **relaxation processes** through which an excited molecule may dissipate energy. The most common relaxation process is for the excited molecule to dissipate its energy through a series of small steps brought on by collisions with other molecules. The energy is thus converted to kinetic energy, the net result being the dissipation of the energy as heat. Under normal conditions, the dissipated heat is not enough to measurably affect the system. In some cases, molecules excited by the absorption of UV or Vis light will lose a portion of their excess energy through the emission of a photon. This emission process is referred to as either **fluorescence** or **phosphorescence,** depending on the nature of the excited state. In molecular fluorescence spectroscopy, the photons emitted from the excited species will generally be of lower energy, longer wavelength, than the corresponding photons that were absorbed in the excitation process. This is because, in most cases, only a fraction of the energy difference between the excited and ground states is lost in the emission process. The other fraction of the excess energy is dissipated as heat during vibrational relaxation. This process is depicted in Fig. 22-8, which illustrates that the excited species undergoes vibrational relaxation down to the lowest vibrational energy level within the excited electronic state, and then undergoes a transition to the ground electronic state through the emission of a photon. The photon emitted will have an energy that equals the energy difference between the lowest vibrational level of the excited electronic state and the ground electronic state level it descends to. The fluorescing molecule may descend to any of the vibrational levels within the ground electronic state. If the fluorescence transition

TABLE 22-2. Wavelength Regions, Spectroscopic Methods, and Associated Transitions

Wavelength Region	*Wavelength Limits*	*Type Spectroscopy*	*Usual Wavelength Range*	*Types of Transitions in Chemical Systems with Similar Energies*
Gamma ray	0.01–1 Å	Gamma ray emission	<0.1 Å	Nuclear proton/neutron arrangements
X-ray	0.1–10 nm	X-ray; absorption, emission, fluorescence, and diffraction	0.1–100 Å	Inner-shell electrons
Ultraviolet	10–380 nm	Ultraviolet; absorption, emission, and fluorescence	180–380	Outer-shell electrons in atoms, bonding electrons in molecules
Visible	380–750 nm	Visible; absorption, emission, and fluorescence	380–750 nm	Same as ultraviolet
Infrared	0.75–1000 μm	Infrared absorption	0.78–300 μm	Vibrational position of atoms in molecular bonds
Microwave	0.1–100 cm	Microwave absorption Electron spin resonance	0.75–3.75 mm 3 cm	Rotational position in molecules Orientation of unpaired electrons in an applied magnetic field
Radiowave	1–1000 m	Nuclear magnetic resonance	0.6–10 m	Orientation of nuclei in an applied magnetic field

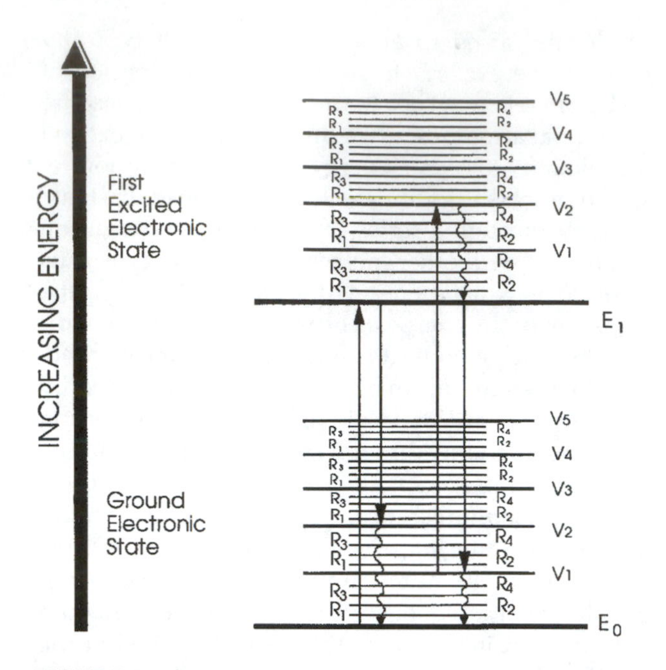

FIGURE 22-8.
Partial molecular energy level diagram including absorption, vibrational relaxation, and fluorescence relaxation.

is to an excited vibrational level within the ground electronic state, then it will quickly return to the ground state (lowest energy level) via vibrational relaxation. In yet other cases, an excited species may be of sufficient energy to initiate some type of photochemistry that ultimately leads to a decrease in the system's potential energy. In all cases, the relaxation process is driven by the tendency for a species to exist at its lowest permissible internal energy level. The relaxation process that dominates a system will be the one that minimizes the lifetime of the excited state. Under normal conditions, the relaxation process is so rapid that the population of molecules in the ground state is essentially unchanged.

22.5 SUMMARY

Spectroscopy deals with the interaction of electromagnetic radiation with matter. Spectrochemical analysis, a branch of spectroscopy, encompasses a wide range of techniques used in analytical laboratories for the qualitative and quantitative analysis of the chemical composition of foods. Common spectrochemical analysis methods include ultraviolet, visible, and infrared absorption spectroscopy, molecular fluorescence spectroscopy, and nuclear magnetic resonance spectroscopy. In each of these methods, the analyst attempts to measure the amount of radiation either absorbed or emitted by the analyte. All of these methods make use of the facts that the energy content of matter is quantized and that photons of radiation may be absorbed or emitted by matter if the energy associated with the photon equals the energy difference for allowed transitions of that given species. The above methods differ from each other with respect to the radiation wavelengths used in the analysis and/or the molecular versus atomic nature of the analyte.

22.6 STUDY QUESTIONS

1. Which phenomena associated with light are most readily explained by considering the wave nature of light? Try rationalizing these phenomena based on your understanding of interference.
2. Which phenomena associated with light are most readily explained by considering the particulate nature of light? Try rationalizing these phenomena based on your understanding of the quantum nature of electromagnetic radiation.
3. What does it mean to say that the energy content of matter is quantized?
4. Molecular absorption of radiation in the UV-Vis range results in transitions between what types of energy levels?
5. Molecular absorption of radiation in the IR range results in transitions between what types of energy levels?
6. Why is an applied magnetic field necessary for NMR spectroscopy?
7. How do the allowed energy levels of molecules differ from those of atoms? Answer with respect to the energy level diagram depicted in Fig. 22-5.
8. In fluorescence spectroscopy, why is the wavelength of the emitted radiation longer than the wavelength of the radiation used for excitation of the analyte?

22.7 RESOURCE MATERIALS

Christian, G.D., and O'Reilly, J.E. (Eds.) 1986. *Instrumental Analysis*, p. 164. Allyn and Bacon, Inc., Newton, MA.

Hargis, L.G. 1988. *Analytical Chemistry—Principles and Techniques*. Prentice Hall, Inc., Englewood Cliffs, NJ.

Harris, D.C. 1991. *Quantitative Chemical Analysis*, 3rd ed. W.H. Freeman and Co., New York, NY.

Harris, D.C., and Bertolucci, M.D. 1978. *Symmetry and Spectroscopy*. Oxford University Press, New York, NY.

Ingle, J.D. Jr., and Crouch, S.R. 1988. *Spectrochemical Analysis*. Prentice Hall, Inc., Englewood Cliffs, NJ.

Milton Roy Educational Manual for the SPECTRONIC® 20 & 20D Spectrophotometers. 1989. Milton Roy Co., Rochester, NY.

Ramette, R.W. 1981. *Chemical Equilibrium and Analysis*. Addison-Wesley Publishing Co., Reading, MA.

Sears, F.W., and Zemansky, M.W. 1964. *University Physics*, 3rd ed. Addison-Wesley Publishing Co., Reading, MA.

Skoog, D.A., and Leary, J.J. 1992. *Principles of Instrumental Analysis*, 4th ed. Harcourt Brace Jovanovich, Orlando, FL.

Tinoco, I. Jr., Sauer, K., and Wang, J.C. 1978. *Physical Chemistry—Principles and Applications in Biological Sciences*. Prentice Hall, Inc., Englewood Cliffs, NJ.

CHAPTER

23

Ultraviolet, Visible, and Fluorescence Spectroscopy

Michael H. Penner

23.1 INTRODUCTION

Spectroscopy in the ultraviolet-visible (UV-Vis) range is one of the most commonly encountered laboratory techniques in food analysis. Electromagnetic radiation in the UV-Vis portion of the spectrum ranges in wavelength from approximately 200 to 700 nm. The **UV range** runs from 200 to 350 nm and the **Vis range** from 350 to 700 nm (Table 23-1). The UV range is colorless to the human eye, while different wavelengths in the visible range each have a characteristic color, ranging from violet at the short wavelength end of the spectrum to red at the long wavelength end of the spectrum. Spectroscopy utilizing radiation in the UV-Vis range may be divided into two general categories, **absorbance** and **fluorescence** spectroscopy, based on the type of radiation-matter interaction that is being monitored. Each of these two types of spectroscopy may be further subdivided into **qualitative** and **quantitative** techniques. In general, quantitative absorption spectroscopy is the most common of the subdivisions within UV-Vis spectroscopy.

23.2 ULTRAVIOLET AND VISIBLE ABSORPTION SPECTROSCOPY

23.2.1 Basis of Quantitative Absorption Spectroscopy

The objective of quantitative absorption spectroscopy is to determine the concentration of analyte in a given sample solution. The determination is based on the measurement of the amount of light absorbed from a reference beam as it passes through the sample solution. In some cases, the analyte may naturally absorb radiation in the UV-Vis range, such that the chemical nature of the analyte is not modified during the analysis. In other cases, analytes that do not absorb radiation in the UV-Vis range are chemically modified during the analysis, converting them to a species that absorbs radiation of the appropriate wavelength. In either case, the presence of analyte in the solution will affect the amount of radiation transmitted through the solution and, hence, the relative transmittance or absorbance of the solution may be used as an index of analyte concentration.

In actual practice, the solution to be analyzed is contained in an absorption cell and placed in the path of radiation of a selected wavelength(s). The amount of radiation passing through the sample is then measured relative to a reference sample. The relative amount of light passing through the sample is then used to estimate the analyte concentration. The process of absorption may be depicted as in Fig. 23-1.

The radiation incident on the absorption cell, P_o, will have significantly greater radiant power than the radiation exiting the opposite side of the cell, P. The decrease in radiant power as the beam passes through the solution is due to the capture (absorption) of photons by the absorbing species. The relationship between the power of the incident and exiting beams is typically expressed in terms of either the transmittance or the absorbance of the solution. The **transmittance** (T) of a solution is defined as the ratio of P to P_o as given in eq. 1. Transmittance may also be expressed as a percentage as given in eq. 2.

$$T = \frac{P}{P_o} \quad (1)$$

$$\%T = \left(\frac{P}{P_o}\right) \times 100 \quad (2)$$

where:
T = transmittance
P_o = radiant power at beam incident on absorption cell
P = radiant power of beam exiting the absorption cell
%T = percent transmittance

TABLE 23-1. Spectrum of Visible Radiation

Wavelength	*Color*	*Complementary Hue*[1]
< 380	ultraviolet	
380–420	violet	yellow-green
420–440	violet-blue	yellow
440–470	blue	orange
470–500	blue-green	red
500–520	green	purple
520–550	yellow-green	violet
550–580	yellow	violet-blue
580–620	orange	blue
620–680	red	blue-green
680–780	purple	green
> 780	near-infrared	

[1]Complementary hue refers to the color observed for a solution that shows maximum absorbance at the designated wavelength assuming a continuous spectrum "white" light source.

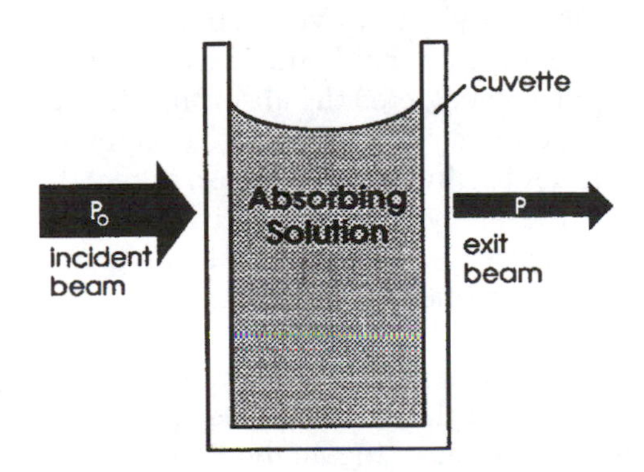

FIGURE 23-1.
Attenuation of a beam of radiation as it passes through a cuvette containing an absorbing solution.

The terms T and %T are intuitively appealing, as they are expressing the fraction of the incident light absorbed by the solution. However, T and %T are not directly proportional to the concentration of the absorbing analyte in the sample solution. The nonlinear relationship between transmittance and concentration is an inconvenience since analysts are generally interested in analyte concentrations. A second term used to describe the relationship between P and P_o is **absorbance** (A). Absorbance is defined with respect to T as shown in eq. 3.

$$A = \log \frac{P_o}{P} = -\log T = 2 - \log \%T \qquad (3)$$

where: A = absorbance
T, %T as in eq. 1 and 2, respectively

Absorbance is a convenient expression in that, under appropriate conditions, it is directly proportional to the concentration of the absorbing species in the solution. Note that, based on these definitions for A and T, the absorbance of a solution *is not* simply unity minus the transmittance. In quantitative spectroscopy, the fraction of the incident beam that is not transmitted does not equal the solution's absorbance (A).

The relationship between the absorbance of a solution and the concentration of the absorbing species is known as **Beer's law** (eq. 4).

$$A = abc \qquad (4)$$

where: A = absorbance
c = concentration of absorbing species
b = path length through solution (cm)
a = absorptivity

There are no units associated with absorbance, A, since it is the log of a ratio of beam powers. The concentration term, c, may be expressed in any appropriate units (M, mM, mg/ml, %). The path length, b, is in units of cm. The **absorptivity,** a, of a given species is a proportionality constant dependent on the molecular properties of the species. The absorptivity is wavelength dependent and may vary depending on the chemical environment (pH, ionic strength, solvent, etc.) the absorbing species is experiencing. The units of the absorptivity term are $(cm)^{-1}$ $(concentration)^{-1}$. In the special case where the concentration of the analyte is reported in units of molarity, the absorptivity term has units of $(cm)^{-1}$ $(M)^{-1}$. Under these conditions, it is designated by the symbol ϵ, which is referred to as the **molar absorptivity.** Beer's law expressed in terms of the molar absorptivity is given in eq. 5. In this case, c refers specifically to the molar concentration of the analyte.

$$A = \epsilon bc \qquad (5)$$

where: A, b as in eq. 4
ϵ = molar absorptivity
c = concentration in units of molarity

Quantitative spectroscopy is dependent on the analyst being able to accurately measure the fraction of an incident light beam that is absorbed by the analyte in a given solution. This apparently simple task is somewhat complicated in actual practice due to processes other than analyte absorption that also result in significant decreases in the power of the incident beam. A pictorial summary of reflection and scattering processes that will decrease the power of an incident beam is given in Fig. 23-2. It is clear that these processes must be accounted for if a truly quantitative estimate of analyte absorption is necessary. In practice, a reference cell is used to correct for these processes. A **reference cell** is one that, in theory, exactly matches the sample absorption cell with the exception that it contains no analyte. A reference cell is often prepared by adding distilled water to an absorption cell. This reference cell is then placed in the path of the light beam, and the power of the radiation exiting the reference cell is measured and taken as P_o for the sample cell. This procedure assumes that all processes except the selective absorption of radiation by the analyte are equivalent for the sample and reference cells. The absorbance actually measured in the laboratory approximates eq. 6.

$$A = \log \frac{P_{solvent}}{P_{analyte\ solution}} \approx \log \frac{P_o}{P} \qquad (6)$$

where: $P_{solvent}$ = radiant power of beam exiting cell containing solvent (blank)
$P_{analyte\ solution}$ = radiant power of beam exiting cell containing analyte solution
P_o, P as in eq. 1
A as in eq. 3

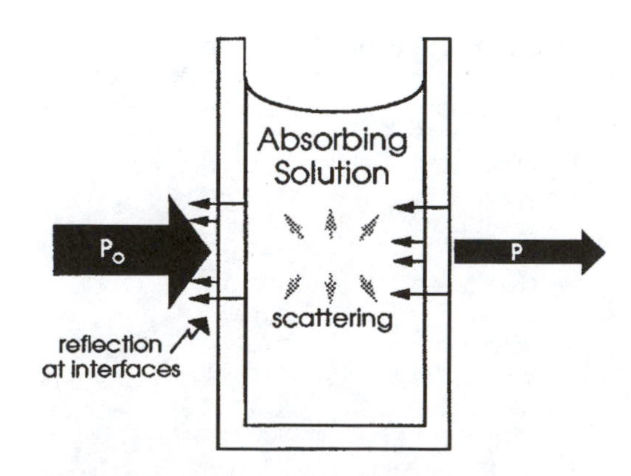

FIGURE 23-2.
Factors contributing to the attenuation of a beam of radiation as it passes through a cuvette containing an absorbing solution.

23.2.2 Deviations from Beer's Law

It should never be assumed that Beer's law is strictly obeyed. Indeed, there are several reasons for which the predicted linear relationship between absorbance and concentration may not be observed. In general, Beer's law is only applicable to dilute solutions, up to approximately 10 mM for most analytes. The actual concentration at which the law becomes limiting will depend on the chemistry of the analyte. As analyte concentrations increase, the intermolecular distances in a given sample solution will decrease, eventually reaching a point at which neighboring molecules mutually affect the charge distribution of the other. This perturbation may significantly affect the ability of the analyte to capture photons of a given wavelength; that is, it may alter the analyte's absorptivity (a). This causes the linear relationship between concentration and absorption to break down since the absorptivity term is the constant of proportionality in Beer's law (assuming a constant path length, b). Other chemical processes may also result in deviations from Beer's law, such as the reversible association-dissociation of analyte molecules or the ionization of a weak acid in an unbuffered solvent. In each of these cases, the predominant form of the analyte may change as the concentration is varied. If the different forms of the analyte, for example ionized versus neutral, have different absorptivities (a), then a linear relationship between concentration and absorbance will not be observed.

A further source of deviation from Beer's law may arise from limitations in the instrumentation used for absorbance measurements. Beer's law strictly applies to situations in which the radiation passing through the sample is monochromatic, since under these conditions a single absorptivity value describes the interaction of the analyte with all the radiation passing through the sample. If the radiation passing through a sample is polychromatic and there is variability in the absorptivity constants for the different constituent wavelengths, then Beer's law will not be obeyed. An extreme example of this behavior occurs when radiation of the ideal wavelength and stray radiation of a wavelength that is not absorbed at all by the analyte simultaneously pass through the sample to the detector. In this case, the observed transmittance will be defined as in eq. 7. Note that a limiting absorbance value will be reached as $P_s >> P$, which will occur at relatively high concentrations of the analyte.

$$A = \log \frac{P_o + P_s}{P + P_s} \tag{7}$$

where: P_s = radiant power of stray light
A as in eq. 3
P, P_o as in eq. 1

23.2.3 Procedural Considerations

In general, the aim of quantitative measurements is to determine the concentration of an analyte with optimum precision and accuracy, in a minimal amount of time and at minimal cost. To accomplish this, it is essential that the analyst consider potential errors associated with each step in the methodology of a particular assay. Potential sources of error for spectroscopic assays include inappropriate sample preparation techniques, inappropriate controls, instrumental noise, and errors associated with inappropriate conditions for absorbance measurements (such as extreme absorbance/transmittance readings).

Sample preparation schemes for absorbance measurements vary considerably. In the simplest case, the analyte-containing solution may be measured directly following homogenization and clarification. Except for special cases, homogenization is required prior to any analysis in order to ensure a representative sample. Clarification of samples is essential prior to taking absorbance readings in order to avoid the apparent absorption due to scattering of light by turbid solutions. The **reference solution** for samples in this simplest case will be the sample solvent, the solvent being water or an aqueous buffer in many cases. In more complex situations, the analyte to be quantified may need to be chemically modified prior to making absorbance measurements. In these cases, the analyte that does not absorb radiation in an appropriate spectral range is specifically modified, resulting in a species with absorption characteristics compatible with a given spectrophotometric measurement. Specific reactions such as these form the basis of many colorimetric assays that measure the absorption of radiation in the Vis range. The reference solution for these assays is prepared by treating the sample solvent in a manner identical with that of the sample. The reference solution will therefore help to correct for any absorbance due to the modifying reagents themselves and not the modified analyte.

A **sample-holding cell** or **cuvette** should be chosen after the general spectral region to be used in a spectrophotometric measurement has been determined. Sample holding cells vary in composition and dimensions. The sample holding cell should be composed of a material that does not absorb radiation in the spectral region being used. Cells meeting this requirement for measurements in the UV range may be composed of quartz or fused silica. Cells made of silicate glass are appropriate for measurements in the Vis range. Inexpensive plastic cells are also available for some applications in the Vis range. The dimensions of the cell will be important with respect to the amount of solution required for a measurement and with regard to the path length term used in Beer's law. A typical absorption cell is 1 cm^2 and approximately 4.5 cm long. The path length for this traditional cell is 1 cm, and the minimum volume of solution needed for standard absorption measurements is approximately 1.5 ml. Absorption cells with path lengths ranging from 1–100 mm are commercially available. Narrow cells, approximately 4 mm in width, with optical path lengths of 1 cm, are also available. These narrow cells are convenient for absorbance measurements when limiting amounts of solution, less than 1 ml, are available.

In many cases, the analyst will need to **choose an appropriate wavelength** at which to make absorbance measurements. If possible, it is best to choose the wavelength at which the analyte demonstrates maximum absorbance and where the absorbance does not change rapidly with changes in wavelength (Fig. 23-3). This position usually corresponds to the apex of the highest absorption peak. Taking measurements at this apex has two advantages: (1) maximum sensitivity, defined as the absorbance change per unit change in analyte concentration and (2) greater adherence to Beer's law since the spectral region making up the radiation beam is composed of wavelengths with relatively small differences in their molar absorptivities for the analyte being measured (Fig. 23-3). The latter point is important in that the radiation beam used in the analysis will be composed of a small continuous band of wavelengths centered about the wavelength indicated on the instrument's wavelength selector.

The actual **absorbance measurement** is made by first calibrating the instrument for 0 and then 100 percent transmittance. The 0 percent transmittance adjustment is made while the photodetector is screened from the incident radiation by means of an occluding shutter, mimicking infinite absorption. This adjustment sets the base level current or "dark current" to the appropriate level, such that the readout indicates zero. The 100 percent transmittance adjustment is then made with the occluding shutter open and an appropriate reference cell/solution in the light path. The reference cell itself should be equivalent to the cell containing sample, a matched set of cells. In many cases, the same cell is used for both the sample and reference solutions. The reference cell is generally filled with solvent, that often being distilled/deionized water for aqueous systems. The 100 percent T adjustment effectively sets T = 1 for the reference cell, which is equivalent to defining P_o in eq. 1 as equivalent to the radiant power of the beam exiting the reference cell. The 0 percent T and 100 percent T settings should be confirmed as necessary throughout the assay. The sample cell containing analyte is then measured without changing the adjustments. The adjustments made with the reference cell will effectively set the instrument to give a sample readout in terms of eq. 6. The readout for the sample solution will be between 0 and 100 percent T. Most modern spectrophotometers allow the analyst to make readout measurements in either absorbance units or as percent transmittance. It is generally most convenient to make readings in absorbance units since, under optimum conditions, absorbance is directly proportional to concentration. When making measurements with an instrument that employs an analog swinging-needle type of readout, it may be preferable to use the linear percent transmittance scale and then calculate the corresponding absorbance using eq. 3. This is particularly true for measurements in which the percent transmittance is less than 20.

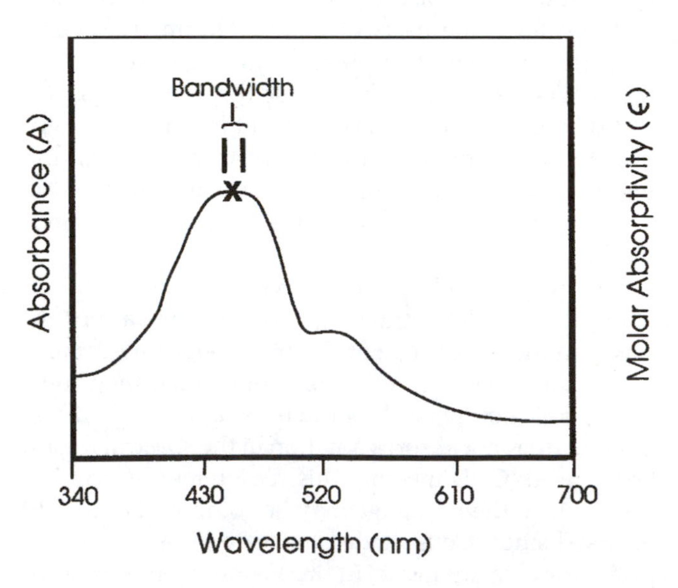

FIGURE 23-3.
Hypothetical absorption spectrum between 340 and 700 nm. The effective bandwidth of the radiation used in obtaining the spectrum is assumed to be approximately 20 nm. Note that at the point indicated there is essentially no change in molar absorptivity over this wavelength range.

23.2.4 Calibration Curves

In most instances, it is advisable to use calibration curves for quantitative measurements. In food analysis, there are a large number of empirical assays for which calibration curves are essential. The calibration curve is used to establish the relationship between analyte concentration and absorbance. This relationship is established experimentally through the analysis of a series of samples of known analyte concentration. The standard solutions are best prepared with the same reagents and at the same time as the unknown. The concentration range covered by the standard solutions must include that expected for the unknown. Typical calibration curves are depicted in Fig. 23-4. **Linear calibration curves** are expected for those systems that obey

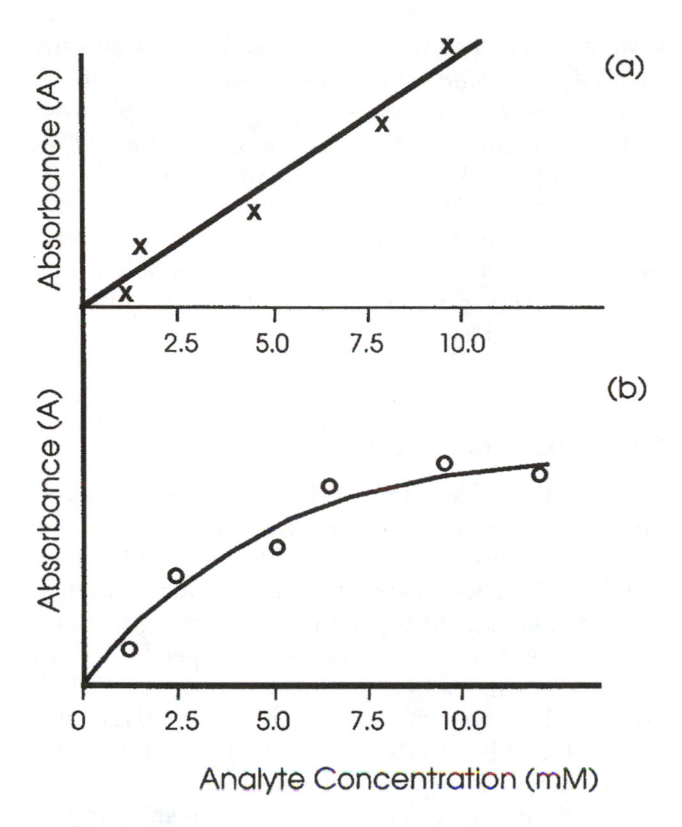

FIGURE 23-4.
Linear (a) and nonlinear (b) calibration curves typically encountered in quantitative absorption spectroscopy.

Beer's law. **Nonlinear calibration curves** are used for some assays, but linear relationships are generally preferred due to the ease of processing the data. Nonlinear calibration curves may be due to concentration-dependent changes in the chemistry of the system or to assay limitations associated with instrumentation not being used under optimal conditions. The nonlinear calibration curve in Fig. 23-4b reflects the fact that the **calibration sensitivity,** defined as change in absorbance per unit change in analyte concentration, is not constant. For the case depicted in Fig. 23-4b, the assay's concentration-dependent decrease in sensitivity obviously begins to limit its usefulness at analyte concentrations above 10 mM.

In many cases, truly representative calibration standards cannot be prepared due to the complexity of the unknown sample. This scenario must be assumed when insufficient information is available on the extent of interfering compounds in the unknown. **Interfering compounds** include those that absorb radiation in the same spectral region as the analyte, those that influence the absorbance of the analyte, and those compounds that react with modifying reagents that are supposedly specific for the analyte. This means that calibration curves are potentially in error if the unknown and the standards differ with respect to pH, ionic strength, viscosity, types of impurities, and the like. In these cases, it is advisable to calibrate the assay system by using a **standard addition protocol.** One such protocol goes as follows: To a series of flasks add a constant volume of the unknown (V_u) for which you are trying to determine the analyte concentration (C_u). Next, to each individual flask add a known volume (V_s) of a standard analyte solution of concentration C_s, such that each flask receives a unique volume of standard. The resulting series of flasks will contain identical volumes of the unknown and different volumes of the standard solution. Next, dilute all flasks to the same total volume, V_t. Each of the flasks is then assayed, treating each flask identically. If Beer's law is obeyed, then the measured absorbance of each flask will be proportional to the total analyte concentration as defined in eq. 8.

$$A = k\left(\frac{V_sC_s + V_uC_u}{V_T}\right) \tag{8}$$

where: V_s = volume of standard
V_u = volume of unknown
V_T = total volume
C_s = concentration of standard
C_u = concentration of unknown
k = proportionality constant (path length × absorptivity)

The results from the assays are then plotted with the volume of standard added to each flask (V_s) as the independent variable and the resulting absorbance (A) as the dependent variable (Fig. 23-5). Assuming Beer's law, the line describing the relationship will be as in eq. 9, in which all terms other than V_s and A are constants. Taking the ratio of the slope of the plotted line (eq. 10) to the line's intercept (eq. 11) and rearranging gives eq. 12, from which the concentration of the unknown,

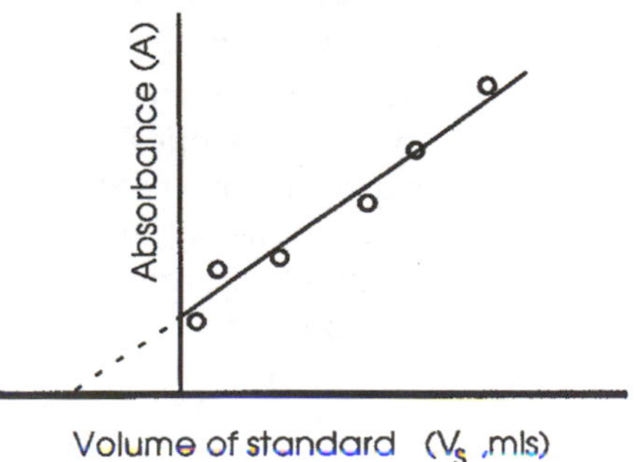

FIGURE 23-5.
Calibration curve for the determination of the analyte concentration in an unknown using a standard addition protocol. A, absorbance; V_s, volume of standard analyte solution; as discussed in text.

C_u, can be calculated since C_s and V_u are experimentally defined constants.

$$A = \frac{kC_sV_s}{V_T} + \frac{V_uC_uk}{V_T} \quad (9)$$

$$\text{slope} = \frac{kC_s}{V_T} \quad (10)$$

$$\text{intercept} = \frac{V_uC_uk}{V_T} \quad (11)$$

$$C_u = \left(\frac{\text{measured intercept}}{\text{measured slope}}\right)\left(\frac{C_s}{V_u}\right) \quad (12)$$

where: V_s, V_u, V_T, C_s, C_u, and K as in eq. 8

23.2.5 Effect of Indiscriminate Instrumental Error on the Precision of Absorption Measurements

All spectrophotometric assays will have some level of **indiscriminant error** associated with the absorbance/transmittance measurement itself. Indiscriminant error of this type is often referred to as **instrument noise.** It is important that the assay be designed such that this source of error is minimized, the objective being to keep this source of error low relative to the variability associated with other aspects of the assay, such as sample preparation, subsampling, reagent handling, and so on. Indiscriminant instrumental error is observed with repeated measurements of a single homogeneous sample. The relative concentration uncertainty resulting from this error is not constant over the entire percent transmittance range (0 to 100%). Measurements at intermediate transmittance values tend to have lower relative errors, thus greater relative precision, than measurements made at either very high or very low transmittance. **Relative concentration uncertainty** or relative error may be defined as S_c/C, where S_c = sample standard deviation and C = measured concentration. Relative concentration uncertainties of from 0.5 to 1.5 percent are to be expected for absorbance/transmittance measurement taken in the optimal range. The optimal range for absorbance measurements on simple, less expensive spectrophotometers is from approximately 0.2 to 0.8 absorbance units, or from 15 to 65 percent transmittance. On more sophisticated instrumentation, the range for optimum absorbance readings may be extended up to 1.5 or greater. To be safe, it is prudent to always make absorbance readings under conditions where the absorbance of the analyte solution is less than 1.0. If there is an anticipated need to make measurements at absorbance readings greater than 1.0, then the relative precision of the spectrophotometer should be established experimentally by repetitive measurements of appropriate samples. Absorbance readings outside the optimal range of the instrument may be used, but the analyst must be prepared to account for the higher relative error associated with these extreme readings. When absorbance readings approach the limits of the instrumentation, then relatively large differences in analyte concentrations may not be detected.

23.2.6 Instrumentation

There are many variations of spectrophotometers available for UV-Vis spectrophotometry. Some instruments are designed for operation in only the visible range, while others encompass both the UV and Vis range. Instruments may differ with respect to design, quality of components, and versatility. A basic spectrophotometer is composed of five essential components: the light source, the monochromator, the sample/reference holder, the radiation detector, and a readout device. A power supply is required for instrument operation. A schematic depicting component interrelationships is shown in Fig. 23-6.

Light sources used in spectrophotometers must continuously emit a strong band of radiation encompassing the entire wavelength range for which the instrument is designed. The power of the emitted radiation must be sufficient for adequate detector response, and it should not vary sharply with changes in wavelength or drift significantly over the experimental time scale. The most common radiation source for Vis spectrophotometers is the **tungsten filament lamp.** These lamps emit adequate radiation covering the wavelength region from 350 to 2500 nm. Consequently, tungsten filament lamps are also employed in near-infrared spectroscopy. The most common radiation sources for measurements in the UV range are **deuterium electrical-discharge lamps.** These sources provide a continuous radiation spectrum from approximately 160 nm through 375 nm. These lamps employ quartz windows and should be used in conjunction with quartz sample holders since glass significantly absorbs radiation below 350 nm.

The component that functions to isolate the specific, narrow, continuous group of wavelengths to be used in the spectroscopic assay is the **monochromator.** The monochromator is so named because light of a single wavelength is termed **monochromatic.** Theoretically, **polychromatic radiation** from the source enters the monochromator and is dispersed according to wavelength, and **monochromatic radiation** of a selected single wavelength exits the monochromator. In practice, light exiting the monochromator is not of a

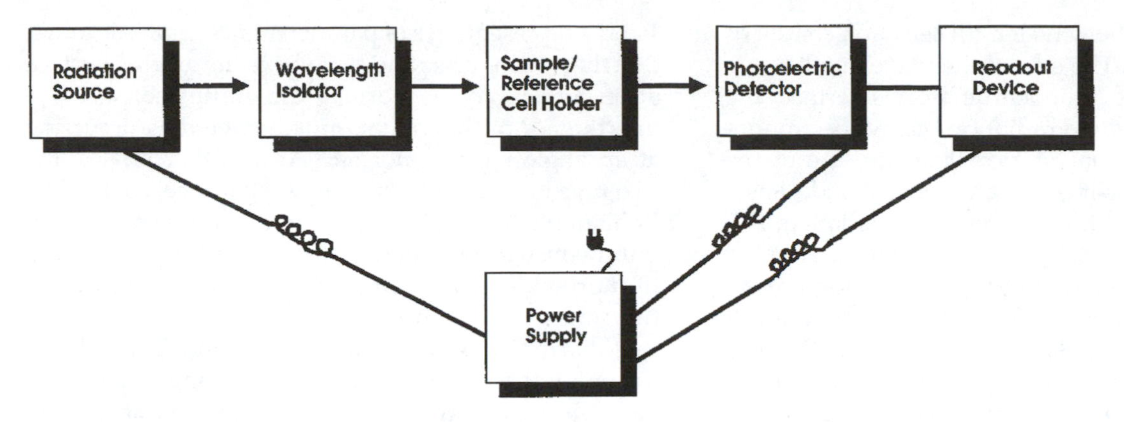

FIGURE 23-6.
Arrangement of components in a simple single-beam, UV-Vis absorption spectrophotometer.

single wavelength, but rather it consists of a narrow continuous band of wavelengths. A representative monochromator is depicted in Fig. 23-7. As illustrated, a typical monochromator is composed of **entrance and exit slits, concave mirror(s),** and a **dispersing element** (the grating in this particular example). Polychromatic light enters the monochromator through the entrance slit and is then culminated by a concave mirror. The culminated polychromatic radiation is then dispersed, dispersion being the physical separation in space of radiation of different wavelengths. The radiation of different wavelengths is then reflected from a concave mirror that focuses the different wavelengths of light sequentially along the focal plane. The radiation that aligns with the exit slit in the focal plane is thus emitted from the monochromator. The radiation emanating from the monochromator will consist of a narrow range of wavelengths presumably centered around the wave-

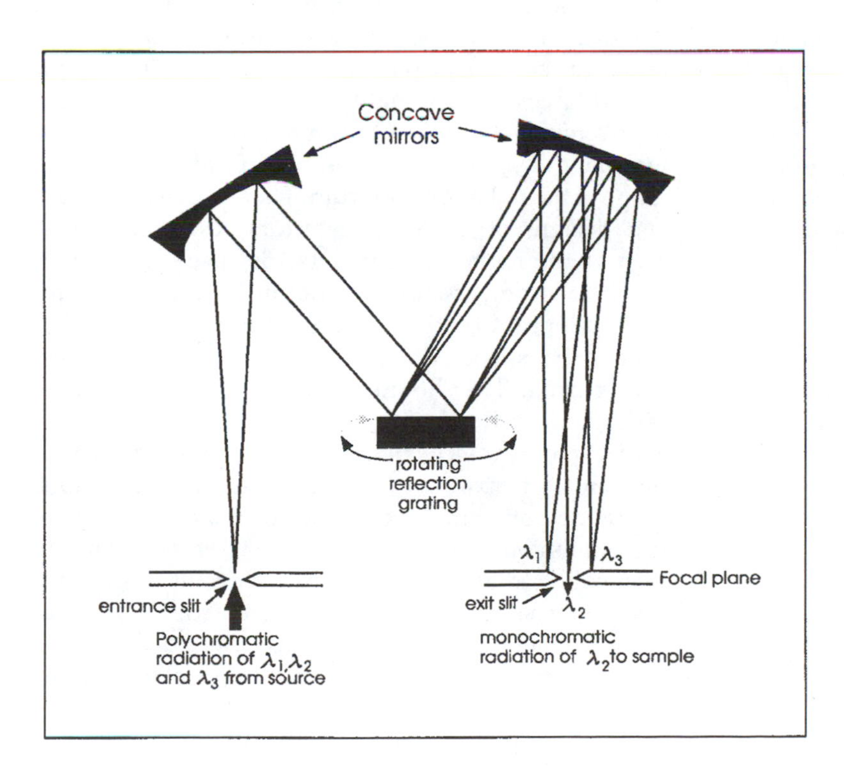

FIGURE 23-7.
Schematic of a monochromator employing a reflection grating as the dispersing element. The concave mirrors, collimating mirrors, serve to collimate the radiation into a beam of parallel rays.

length specified on the wavelength selection control of the instrumentation. The size of the wavelength range passing through the monochromator is termed the **bandwidth** of the emitted radiation. Many spectrophotometers allow the analyst to adjust the size of the monochromator exit slit (and entrance slit) and, consequently, the bandwidth of the emitted radiation. Decreasing the exit slit width will decrease the associated bandwidth and the radiant power of the emitted beam. Conversely, further opening of the exit slit will result in a beam of greater radiant power, but one that has a larger bandwidth. In some cases where resolution is critical, such as some qualitative work, the narrower slit width may be advised. However, in most quantitative work a relatively open slit may be used since adsorption peaks in the UV-Vis range are generally broad relative to spectral bandwidths and because the signal-to-noise ratio associated with transmittance measurements is improved due to the higher radiant power of the measured beam.

The effective bandwidth of a monochromator is determined not only by the slit width but also by the quality of its dispersing element. The dispersing element functions to spread out the radiation according to wavelength. **Reflection gratings,** as depicted in Fig. 23-8, are the most commonly used dispersing elements in modern spectrophotometers. Gratings are sometimes referred to as **diffraction gratings** since the separation of component wavelengths is dependent on the different wavelengths being diffracted at different angles relative to the grating normal. A reflection grating incorporates a reflective surface in which a series of closely spaced grooves has been etched, typically between 1200 and 1400 grooves per mm. The grooves themselves serve to break up the reflective surface such that each point of reflection behaves as an independent point source of radiation. Referring to Fig. 23-8, lines 1 and 2 represent rays of parallel monochromatic radiation that are in phase and that strike the grating surface at an angle i to the normal. Maximum constructive interference of this radiation is depicted as occurring at an angle r to the normal. At all other angles, the two rays will partially or completely cancel each other. Radiation of a different wavelength would show maximum constructive interference at a different angle to the normal. The wavelength dependence of the diffraction angle can be rationalized by considering the relative distance the photons of rays 1 and 2 travel and assuming that maximum constructive interference occurs when the waves associated with the photons are completely in phase. Referring to Fig. 23-8, prior to reflection, photon 2 travels a distance CD greater than photon 1. After reflection, photon 1 travels a distance AB greater than photon 2. Hence, the waves associated with photons 1 and 2 will remain in phase after reflection only if the net difference in the distance traveled is an integral multiple of their wavelength. Note that for a different angle r the distance AB would change and, consequently, the net distance CD − AB would be an integral multiple of a different wavelength. The net result is that the component wavelengths are each diffracted at their own unique angles r.

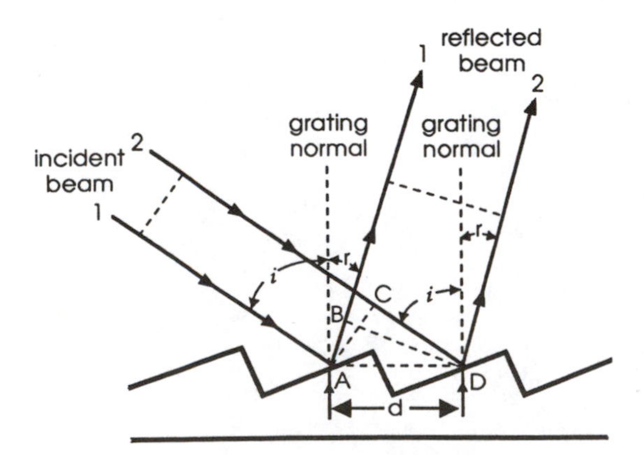

FIGURE 23-8.
Schematic illustrating the property of diffraction from a reflection grating. Each reflected point source of radiation is separated by a distance d.

In a spectroscopic measurement, the light transmitted through the reference or sample cell is quantified by means of a **detector.** The detector is designed to produce an electric signal when it is struck by photons. An ideal detector would give a signal directly proportional to the radiant power of the beam striking it; it would have a high signal-to-noise ratio; and it would have a relatively constant response to light of different wavelengths, such that it was applicable to a wide range of the radiation spectrum. There are several types and designs of radiation detectors currently in use. Two of the more popular detectors used in modern spectrophotometers are the **phototube** and the **photomultiplier tube.** Both detectors function by converting the energy associated with incoming photons into electrical current. The phototube consists of a semicylindrical cathode covered with a photoemissive surface and a wire anode, the electrodes being housed under vacuum in a transparent tube (Fig. 23-9a). When photons strike the pothoemissive surface of the cathode, there is an emission of electrons, and the freed electrons are then collected at the anode. The net result of this process is that a measurable current is created. The number of electrons emitted from the cathode and the subsequent current through the system are directly proportional to the number of photons, or radiant power of the beam, impinging on the photoemissive surface. The photomultiplier tube is of similar design. However, in the photomultiplier tube there is an amplification of the number of electrons collected at the anode per photon striking the photoemissive surface of the cath-

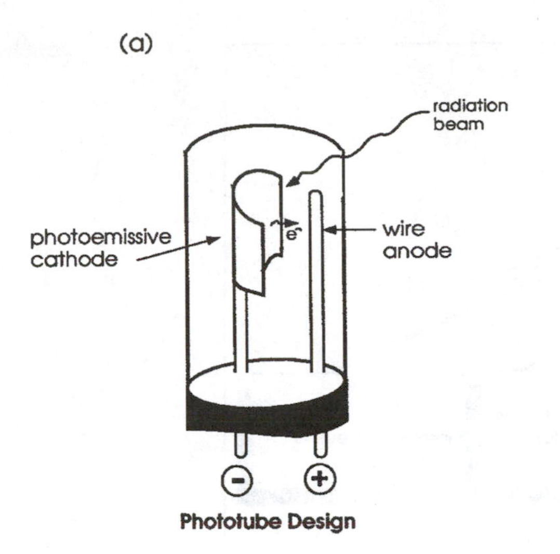

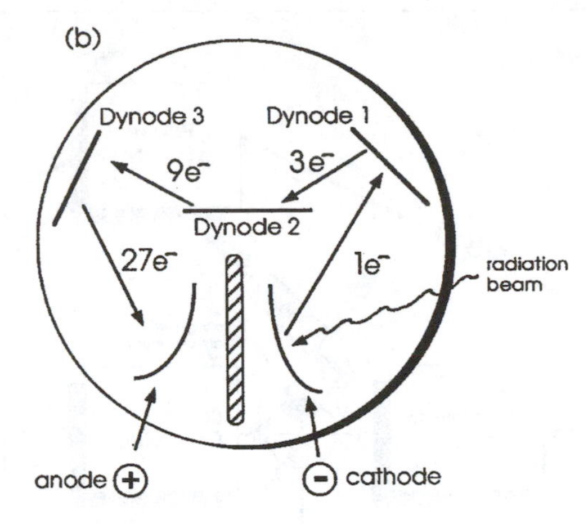

FIGURE 23-9.
Schematic diagram of a typical phototube design (a) and the cathode-dynode-anode arrangement of a representative photomultiplier tube (b).

ode (Fig. 23-9b). The electrons originally emitted from the cathode surface are attracted to a dynode with a relative positive charge. At the dynode, the electrons strike the surface, causing the emission of several more electrons per original electron, resulting in an amplification of the signal. Signal amplification continues in this manner, as photomultiplier tubes generally contain a series of such dynodes, with electron amplification occurring at each dynode. The cascade continues until the electrons emitted from the final dynode are collected at the anode of the photomultiplier tube. The final gain may be as many as 10^6 to 10^9 electrons collected per photon.

The signal from the detector is generally amplified and then displayed in a usable form to the analyst. The final form in which the signal is displayed will depend on the complexity of the system. In the simplest case, the analog signal from the detector is displayed on an **analog meter** through the position of a needle on a meter face calibrated in percent transmission and/or absorbance. Analog readouts are adequate for most routine analytical purposes; however, analog meters are somewhat more difficult to read and, hence, the resulting data is expected to have somewhat lower precision than that obtained on a digital readout (assuming the digital readout is given to enough places). **Digital readouts** express the signal as numbers on the face of a meter. In these cases, there is an obvious requirement for signal processing between the analog output of the detector and the final digital display. In virtually all cases, the signal processor is capable of presenting the final readout in terms of either absorbance or transmittance. Many of the newer instruments include microprocessors capable of more extensive data manipulations on the digitized signal. For example, the readouts of some spectrophotometers may be in concentration units, provided the instrument has been correctly calibrated with appropriate reference standards.

23.2.7 Instrument Design

The optical systems of spectrophotometers fall into one of two general categories: They are either single-beam or double-beam instruments. In a **single-beam instrument,** the radiant beam follows only one path, that going from the source through the sample to the detector (Fig. 23-6). When using a single-beam instrument, the analyst generally measures the transmittance of a sample after first establishing 100 percent T, or P_o, with a reference sample or blank. The blank and the sample are read sequentially since there is but a single light path going through a single cell-holding compartment. In a **double-beam instrument,** the beam is split such that one half of the beam goes through one cell-holding compartment and the other half of the beam passes through a second. The schematic of Fig. 23-10 illustrates a double-beam optical system in which the beam is split in time between the sample and reference cell. In this design, the beam is alternately passed through the sample and reference cells by means of a rotating sector mirror with alternating reflective and transparent sectors. The double-beam design allows the analyst to simultaneously measure and compare the relative absorbance of a sample and a reference cell. The advantage of this design is that it will compensate for deviations or drifts in the radiant output of the source since

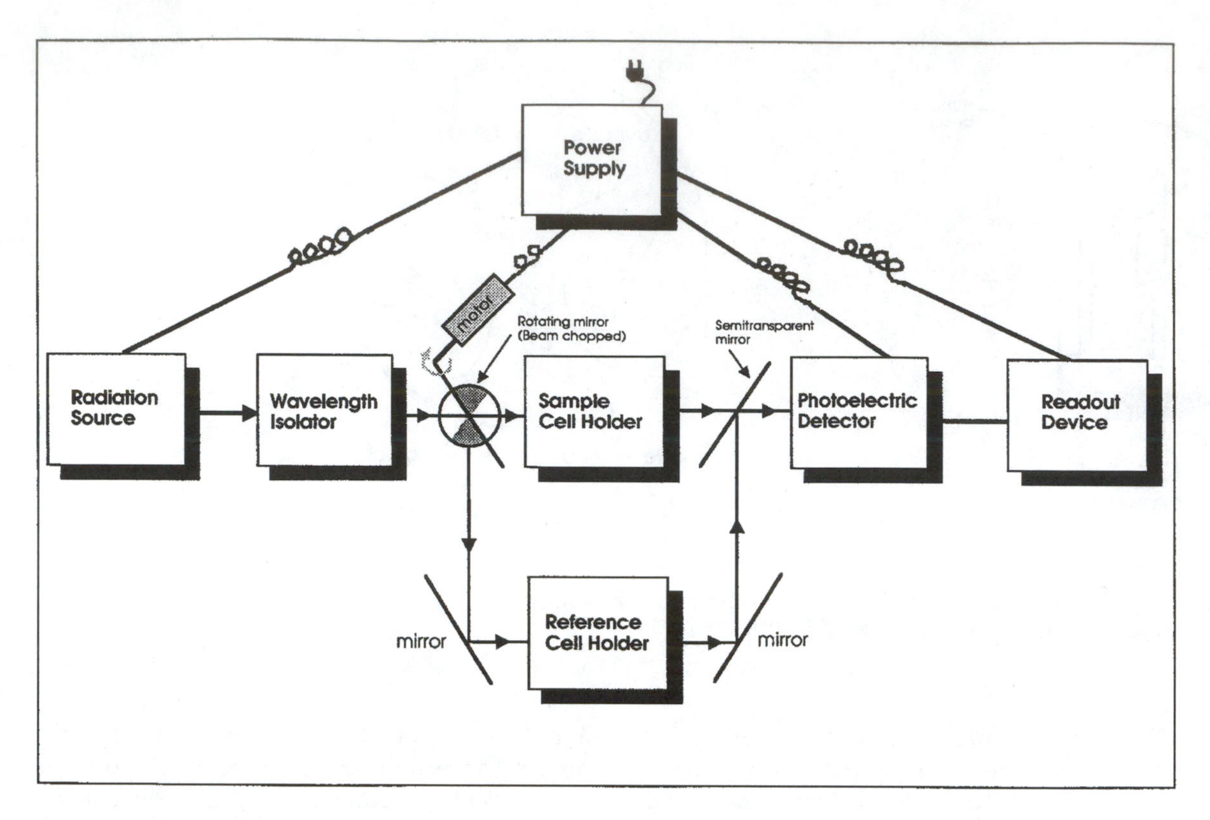

FIGURE 23-10.
Arrangement of components in a representative double-beam UV-Vis absorption spectrophotometer. The incident beam is alternatively passed through the sample and reference cells by means of a rotating beam chopper.

the sample and reference cells are compared many times per second. The disadvantage of the double-beam design is that the radiant power of the incident beam is diminished because the beam is split. The lower energy throughput of the double-beam design is generally associated with inferior signal-to-noise ratios. Computerized single-beam spectrophotometers are now available that claim to have the benefits of both the single- and double-beam designs. Their manufacturers report that previously troublesome source and detector drift and noise problems have been stabilized such that simultaneous reading of the reference and sample cell is not necessary. With these instruments, the reference and sample cells are sequentially read, and the data is stored, then processed, by the associated computer.

The Spectronic 20 is a classic example of a simple single-beam visible spectrophotometer (Fig. 23-11). The white light emitted from the source passes into the monochromator via its entrance slit; the light is then dispersed into a spectrum by a diffraction grating; and a portion of the resulting spectrum then leaves the monochromator via the exit slit. The radiation emitted from the monochromator passes through a sample compartment and strikes the measuring phototube, resulting in an induced photocurrent that is proportional to the intensity of impinging light. The lenses depicted in Fig. 23-11 function in series to focus the light image on the focal plane that contains the exit slit. To change the portion of the spectrum exiting the monochromator, one rotates the reflecting grating by means of the wavelength cam. A shutter automatically blocks light from exiting the monochromator when no sample/reference cell is in the instrument; the zero percent T adjustment is made under these conditions. The **light control occluder** is used to adjust the radiant power of the beam exiting the monochromator. The occluder consists of an opaque strip with a V-shaped opening that can be physically moved in or out of the beam path. The occluder is used to make the 100 percent T adjustment when an appropriate reference cell is in the instrument.

23.3 FLUORESCENCE SPECTROSCOPY

The technique of fluorescence spectroscopy is generally one to three orders of magnitude more sensitive

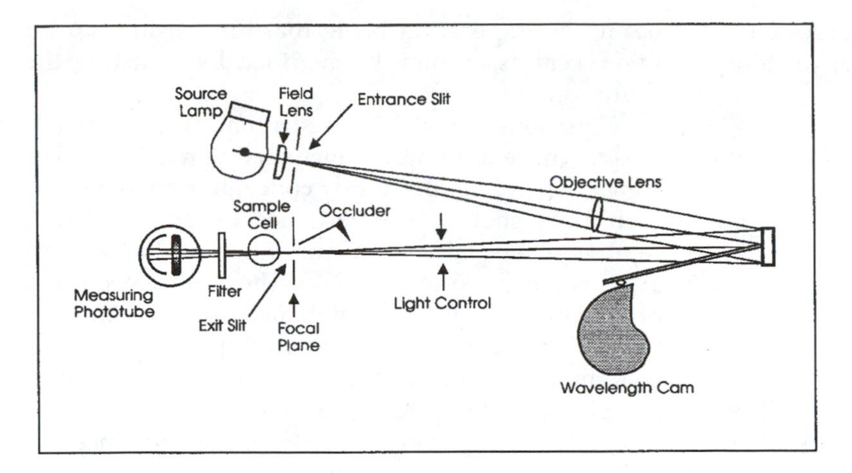

FIGURE 23-11.
Optical system for the SPECTRONIC® 20 spectrophotometer. Figure 3, *Bausch & Lomb SPECTRONIC® 20 Spectrophotometer Operator's Manual.* Figure courtesy of Milton Roy Company, Rochester, NY, a subsidiary of Sundstrand Corporation.

than corresponding absorption spectroscopy methods. In **fluorescence spectroscopy,** the signal being measured is the electromagnetic radiation that is emitted from the analyte as it relaxes from an excited electronic energy level to its corresponding ground state. The analyte is originally activated to the higher energy level by the absorption of radiation in the UV or Vis range. The processes of activation and deactivation occur simultaneously during a fluorescence measurement. For each unique molecular system, there will be an optimum radiation wavelength for sample excitation and another, of longer wavelength, for monitoring fluorescence emission. The respective wavelengths for excitation and emission will depend on the chemistry of the system under study.

The instrumentation used in fluorescence spectroscopy is composed of essentially the same components as the corresponding instrumentation used in UV-Vis absorption spectroscopy. However, there are definite differences in the arrangement of the optical systems used for the two types of spectroscopy (compare Figs. 23-6 and 23-12). In fluorometers and spectrofluorometers, there is a need for two wavelength selectors, one for the **excitation beam** and one for the **emission beam.** In some simple fluoroimeters, both wavelength selectors are filters such that the excitation and emission wavelengths are fixed. In more sophisticated spectrofluorometers, the excitation and emission wavelengths are selected by means of grating monochromators. The photon detector of fluorescence instrumentation is generally arranged such that the emitted radiation that strikes the detector is traveling at an angle of 90° relative to the axis of the excitation beam. This detector placement minimizes signal interference due to transmitted source radiation and radiation scattered from the sample.

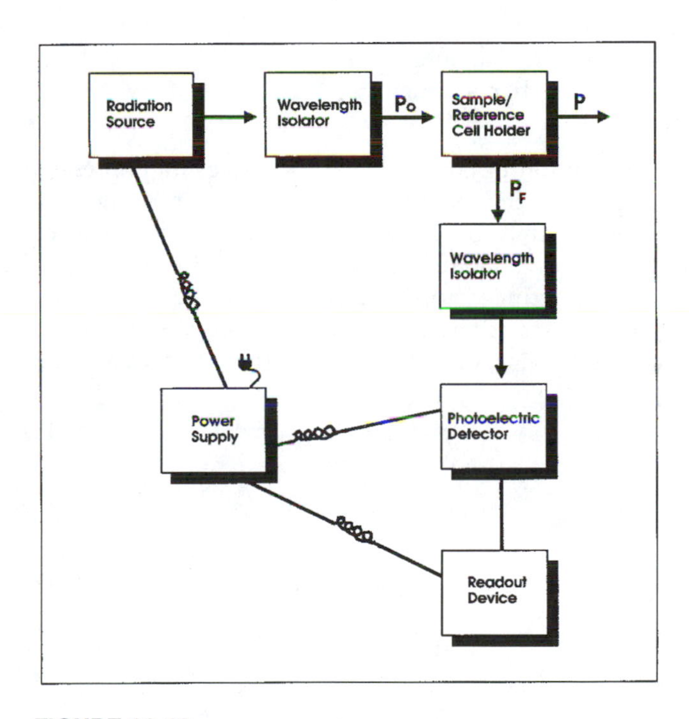

FIGURE 23-12.
Schematic diagram depicting the arrangement of the source, excitation and emission wavelength selectors, sample cell, photoelectric detector, and readout device for a representative fluorometer or spectrofluorometer.

The **radiant power** of the fluorescence beam (P_F) emitted from a fluorescent sample is proportional to the change in the radiant power of the source beam as it passes through the sample cell (eq. 13). Expressing this another way, the radiant power of the fluorescence beam will be proportional to the number of photons absorbed by the sample.

$$P_F = \phi(P_o - P) \qquad (13)$$

where: P_F = radiant power of beam emitted from fluorescent cell
ϕ = constant of proportionality
P_o, P as in eq. 1

The constant of proportionality used in eq. 13 is termed the **quantum efficiency** (ϕ), which is specific for any given system. The quantum efficiency equals the ratio of the total number of photons emitted to the total number of photons absorbed. Combining eq. 3 and 5 allows one to define P in terms of the analyte concentration and P_o, as given in eq. 14.

$$P = P_o\, 10^{-\epsilon bc} \qquad (14)$$

where: P_o, P as in eq. 1
ϵ, b, c as in eq. 5

Substitution of eq. 14 into eq. 13 gives an expression that relates the radiant power of the fluorescent beam to the analyte concentration and P_o, as shown in eq. 15. At low analyte concentrations, $\epsilon bc < 0.01$, eq. 15 may be reduced to the expression of eq. 16. Further grouping of terms leads to the expression of eq. 17, where k incorporates all terms other than P_o and c.

$$P_F = \phi P_o(1 - 10^{-\epsilon bc}) \qquad (15)$$

$$P_F = \phi P_o\, 2.303\, \epsilon bc \qquad (16)$$

$$P_F = kP_oc \qquad (17)$$

where: k = constant of proportionality
P_F as in eq. 13
c as in eq. 5

Equation 17 is particularly useful since it emphasizes two important points that are valid for the conditions assumed when deriving the equation, particularly the assumption that analyte concentrations are kept relatively low. First, the fluorescent signal will be directly proportional to the analyte concentration, assuming other parameters are kept constant. This is very useful since a linear relationship between signal and analyte concentration simplifies data processing and assay troubleshooting. Second, the sensitivity of a fluorescent assay is proportional to P_o, the power of the incident beam, the implication being that the sensitivity of a fluorescent assay may be modified by adjusting the source output.

Equations 16 and 17 will eventually break down if analyte concentrations are increased to relatively high values. Therefore, the **linear concentration range** for each assay should be determined experimentally. A representative calibration curve for a fluorescence assay is presented in Fig. 23-13. The nonlinear portion of the curve at relatively high analyte concentrations results from decreases in the fluorescence yield per unit concentration. The fluorescence yield for any given sample is also dependent on its environment. Temperature, solvent, impurities, and pH may all influence this parameter. Consequently, it is imperative that these environmental parameters be accounted for in the experimental design of fluorescence assays. This may be particularly important in the preparation of appropriate reference standards for quantitative work.

23.4 SUMMARY

Absorption and fluorescence spectroscopy utilizing electromagnetic radiation in the ultraviolet and visible regions are widely used in food analysis. These techniques may be used for either qualitative or quantitative measurements. Qualitative measurements are based on the premise that each analyte has a unique set of energy spacings that will dictate its absorption/emission spectrum. Hence, qualitative assays are generally based on the analysis of the absorption and/or emission spectrum of the analyte.

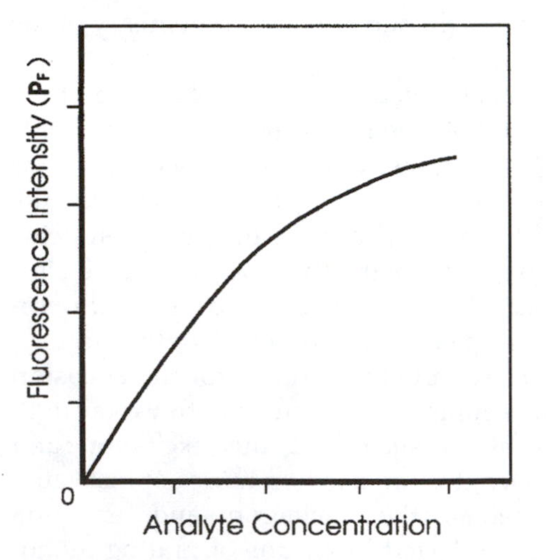

FIGURE 23-13.
Relationship between the solution concentration of a fluorescent analyte and that solution's fluorescence intensity. Note that there is a linear relationship at relatively low analyte concentrations that eventually goes nonlinear as the analyte concentration increases.

In contrast, quantitative assays are most often based on measuring the absorbance and/or fluorescence of the analyte solution at one wavelength. Quantitative absorption assays are based on the premise that the absorbance of the test solution will be a function of the solution's analyte concentration.

Under optimum conditions, there is a direct linear relationship between a solution's absorbance and its analyte concentration. The equation describing this linear relationship is known as Beer's law. The applicability of Beer's law to any given assay should always be verified experimentally by means of a calibration curve. The calibration curve should be established at the same time and under the same conditions that are used to measure the test solution. The analyte concentration of the test solution should then be estimated from the established calibration curve.

Molecular fluorescence methods are based on the measurement of radiation emitted from excited analyte molecules as they relax to lower energy levels. The analytes are raised to the excited state as a result of photon absorption. The processes of photon absorption and fluorescence emission are occurring simultaneously during the assay. Quantitative fluorescence assays are generally one to three orders of magnitude more sensitive than corresponding absorption assays. Like absorption assays, under optimal conditions there will be a direct linear relationship between the fluorescence intensity and the concentration of the analyte in the unknown solution. Most molecules do not fluoresce and, hence, cannot be assayed by fluorescence methods.

The instrumentation used for absorption and fluorescence methods have similar components, including a radiation source, wavelength selector(s), sample holding cells, radiation detector(s), and a readout device.

23.5 STUDY QUESTIONS

1. Why do we commonly obtain and utilize absorbance values rather than transmittance values to quantitate compounds by UV-Vis spectroscopy?
2. For a particular assay, your plot of absorbance versus concentration is not linear. Explain the possible reasons for this.
3. What criteria should be used to choose an appropriate wavelength at which to make absorbance measurements, and why is that choice so important?
4. In a particular assay, the absorbance reading on the spectrophotometer for one sample is 2.033 and for another sample is 0.032. Would you trust these values? Why or why not?
5. A technician in your lab has never used your UV-Vis spectrophotometer for an assay that requires reading the absorbance in the UV range. How would you explain to the technician what the UV range is versus the visible range? What special instructions would you give your technician on use of the spectrophotometer for the UV range?
6. When you set the wavelength on a spectrophotometer, what inside the spectrophotometer are you really adjusting? What does this thing inside the spectrophotometer do, and why is it necessary to utilize absorbance readings to determine concentration?
7. If the width of the exit slit between the monochromator and the sample on a spectrophotometer is decreased, what effects will this have on the light incident to the sample?
8. Describe the similarities and differences between a phototube and a photomultiplier tube as the detector for a spectrophotometer. What is the advantage of one over the other?
9. Your lab has been using an old single-beam spectrophotometer that must now be replaced by a new spectrophotometer. You obtain sales literature that describes single-beam and double-beam instruments. What are the basic differences between a single-beam and a double-beam spectrophotometer, and what are the advantages and disadvantages of each?
10. Explain the similarities and differences between UV-Vis spectroscopy and fluorescence spectroscopy with regard to instrumentation and principles involved. What is the advantage of using fluorescence spectroscopy?

23.6 PRACTICE PROBLEMS

1. A particular food coloring has a molar absorptivity of 3.8 $\times$ 10^3 cm^{-1} M^{-1} at 510 nm.
   a. What will be the absorbance of a 2 $\times$ 10^{-4}M solution in a 1 cm cuvette at 510 nm?
   b. What will be the percent transmittance of the solution in (a)?
2. a. You measure the percent transmittance of a solution containing chromophore X at 400 nm in a 1 cm pathlength cuvette and find it to be 50 percent. What is the absorbance of this solution?
   b. What is the molar absorptivity of chromophore X if the concentration of X in the solution measured in question 2a is 0.5 mM?
   c. What is the concentration range of chromophore X that can be assayed if, when using a sample cell of pathlength 1, you are required to keep the absorbance between 0.2 and 0.8?
3. What is the concentration of compound Y in an unknown solution if the solution has an absorbance of 0.846 in a glass cuvette with a pathlength of 0.2 cm? The absorptivity of compound Y is 54.2 cm^{-1} $(mg/ml)^{-1}$ under the conditions used for the absorption measurement.
4. a. What is the molar absorptivity of compound Z at 295 nm and 348 nm, given the absorption spectrum shown in Fig. 23-14 (which was obtained using a UV-Vis spectrophotometer and a 1 mM solution of compound Z in a sample cell with a pathlength of 1 cm)?
   b. Now you have decided to make quantitative measure-

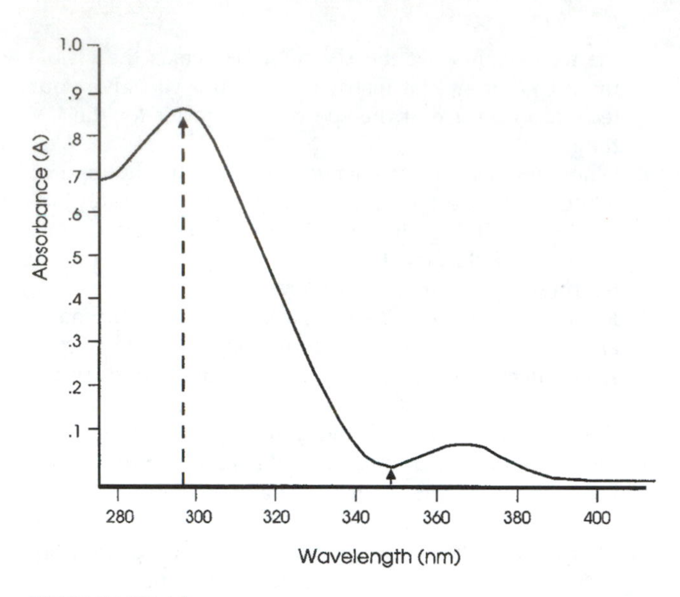

FIGURE 23-14.
Absorption spectrum of compound Z, to be used in conjunction with problems 4a and 4b.

ments of the level of compound Z in different solutions. Based on the above spectrum, which wavelength will you use for your measurements? Give two reasons why this is the optimum wavelength.

Answers

1. a = .76, b = 17.4; 2. a = .301, b = 602 $cm^{-1}M^{-1}$, c = .33 $\times 10^{-3}$M to 1.33 $\times 10^{-3}$M; 3. 0.078 mg/ml; 4. a = 860 at 295 nm, 60 at 348 nm; b = 295 nm; optimum sensitivity and more likely to adhere to Beer's law.

23.7 RESOURCE MATERIALS

Christian, G.D., and O'Reilly, J.E. (Eds.) 1986. *Instrumental Analysis*, p. 164. Allyn and Bacon, Inc., Newton, MA.

Hargis, L.G. 1988. *Analytical Chemistry—Principles and Techniques*. Prentice Hall, Inc., Englewood Cliffs, NJ.

Harris, D.C. 1991. *Quantitative Chemical Analysis*, 3rd ed. W.H. Freeman and Co., New York, NY.

Harris, D.C., and Bertolucci, M.D. 1978. *Symmetry and Spectroscopy*. Oxford University Press, New York, NY.

Ingle, J.D. Jr., and Crouch, S.R. 1988. *Spectrochemical Analysis*. Prentice Hall, Inc., Englewood Cliffs, NJ.

Milton Roy Educational Manual for the SPECTRONIC® 20 & 20D Spectrophotometers. 1989. Milton Roy Co., Rochester, NY.

Ramette, R.W. 1981. *Chemical Equilibrium and Analysis*. Addison-Wesley Publishing Co., Reading, MA.

Skoog, D.A., and Leary, J.J. 1992. *Principles of Instrumental Analysis*, 4th ed. Harcourt Brace Jovanovich, Orlando, FL.

Tinoco, I. Jr., Sauer, K., and Wang, J.C. 1978. *Physical Chemistry—Principles and Applications in Biological Sciences*. Prentice Hall, Inc., Englewood Cliffs, NJ.

CHAPTER

24

Infrared Spectroscopy

Randy L. Wehling

24.1 INTRODUCTION

Infrared (IR) spectroscopy refers to measurement of the absorption of different frequencies of infrared radiation by foods or other solids, liquids, or gases. Infrared spectroscopy began in 1800 with an experiment by Herschel. When he used a prism to create a spectrum from white light and placed a thermometer at a point just beyond the red region of the spectrum, he noted an increase in temperature. This was the first observation of the effects of infrared radiation. By the 1940s, infrared spectroscopy had become an important tool used by chemists to identify functional groups in organic compounds. In the 1970s, commercial near-infrared reflectance instruments were introduced that provided rapid quantitative determinations of moisture, protein, and fat in cereal grains and other foods. Today, infrared spectroscopy is widely used in the food industry for both qualitative and quantitative analysis of ingredients and finished foods.

In this chapter, we will describe the techniques of mid- and near-infrared spectroscopy, including the principles by which molecules absorb infrared radiation, the components and configuration of commercial infrared spectrometers, sampling methods for infrared spectroscopy, and qualitative and quantitative applications of these techniques to food analysis.

24.2 PRINCIPLES OF INFRARED (IR) SPECTROSCOPY

24.2.1 The Infrared Region of the Electromagnetic Spectrum

Infrared radiation is electromagnetic energy with **wavelengths** (λ) longer than visible light but shorter than microwaves. Generally, wavelengths from 0.8 to 100 micrometers (μm) can be used for IR spectroscopy and are divided into the near-IR (0.8–2.5 μm), the mid-IR (2.5–15 μm), and the far-IR (15–100 μm) regions. One μm is equal to 1×10^{-6} m. The near- and mid-IR regions of the spectrum are most useful for quantitative and qualitative analysis of foods.

Infrared radiation can also be measured in terms of its **frequency,** which is useful because frequency is directly related to the energy of the radiation by the following relationship:

$$E = hv \quad (1)$$

where: E = the energy of the system
h = Planck's constant
v = the frequency in hertz

Frequencies are also commonly expressed as **wavenumbers** ($\bar{v}$, reciprocal centimeters = cm^{-1}). Wavenumbers are calculated as follows:

$$\bar{v} = 1/(\lambda \text{ in cm}) = 10^4/(\lambda \text{ in } \mu\text{m}) \quad (2)$$

24.2.2 Molecular Vibrations

A molecule can absorb IR radiation if it vibrates in such a way that its charge distribution, and therefore its electric dipole moment, changes during the vibration. Although there are many possible vibrations in a polyatomic molecule, the most important vibrations that produce a change in dipole moment are stretching and bending (scissoring) motions. Examples of these vibrations for the water molecule are shown in Fig. 24-1. Note that the stretching motions vibrate at a higher frequency than the scissoring motions, indicating that more energy is required.

24.2.3 Factors Affecting the Frequency of Vibration

The energy level for any molecular vibration is given by the following equation:

$$E = \left(v + \frac{1}{2}\right)\frac{h}{2\pi}\sqrt{k/\left(\frac{m_1 m_2}{m_1 + m_2}\right)} \quad (3)$$

where:
v = the vibrational quantum number (positive integer values, including zero, only)
h = Planck's constant
k = the force constant of the bond
m_1 and m_2 = the masses of the individual atoms involved in the vibration

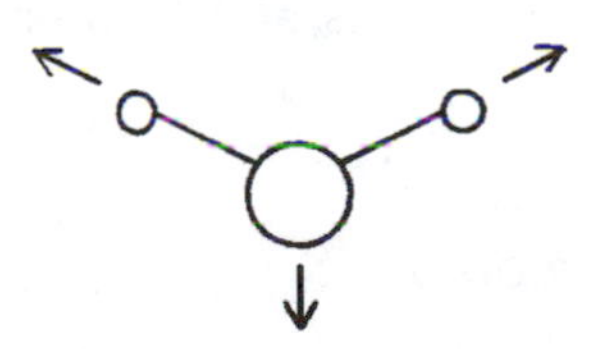

FIGURE 24-1.
Vibrational modes of the water molecule. Frequencies of the fundamental vibration for the symmetrical stretch, asymmetric stretch, and scissoring motion are 3,652 cm^{-1}, 3,756 cm^{-1}, and 1,596 cm^{-1}, respectively.

Note that the vibrational energy, and therefore the frequency of vibration, is directly proportional to the strength of the bond and inversely proportional to the mass of the molecular system. The vibrating molecular functional group can absorb radiant energy to move from the lowest (v = 0) vibrational state to the first excited (v = 1) state, and the frequency of radiation that will make this occur is identical to the initial frequency of vibration of the bond. This frequency is referred to as the **fundamental absorption.** Molecules can also absorb radiation to move to a higher (v = 2 or 3) excited state, such that the frequency of the radiation absorbed is two or three times that of the fundamental frequency. These absorptions are referred to as **overtones,** and the intensity of these absorptions is much lower than the fundamental since these transitions are less favored. The overall result is that each functional group within the molecule absorbs IR radiation in distinct wavelength bands rather than as a continuum.

24.3 MID-INFRARED SPECTROSCOPY

Mid-infrared spectroscopy measures a sample's ability to absorb light in the 2.5–15 μm (4,000–650 cm^{-1}) region. Fundamental absorptions are primarily observed in this spectral region.

24.3.1 Instrumentation

Two types of spectrometers are routinely used for mid-IR spectroscopy: dispersive instruments and Fourier transform instruments.

24.3.1.1 Dispersive Instruments

Dispersive instruments use a **monochromator** to disperse the individual frequencies of radiation and sequentially pass them through the sample so that the absorption of each frequency can be measured. Infrared spectrometers have components similar to ultraviolet-visible (UV/Vis) spectrometers, including a radiation source, a monochromator, a sample holder, and a detector connected to an amplifier system for recording the spectra. Most IR spectrometers are double-beam instruments.

The most common IR source is a coil of **Nichrome wire** wrapped around a ceramic core, which glows when an electrical current is passed through it. A **Globar,** which is a silicon carbide rod across which a voltage is applied, can also be used as a more intense source. Older spectrometers used NaCl prisms to disperse the radiation into monochromatic components, but modern instruments use a diffraction grating to achieve this effect. The most common detector is a **thermocouple,** whose output voltage varies with temperature changes caused by varying levels of radiation striking the detector. More sensitive detectors include the **Golay detector,** in which radiation striking a sealed tube of xenon gas warms the gas and causes pressure changes within the tube, and newer **semiconductor detectors** whose conductivities vary according to the amount of radiation striking the detector surface.

24.3.1.2 Fourier Transform Instruments

In Fourier transform (FT) instruments, the radiation is not dispersed, but rather all wavelengths arrive at the detector simultaneously and a mathematical treatment, called a Fourier transform, is used to convert the results into a typical IR spectrum. Instead of a monochromator, the instrument uses an **interferometer,** which splits and then recombines a light beam (Fig. 24-2). As the pathlength of one beam is varied, the two beams can interfere either constructively or destructively depending on their phase difference. When the sample is placed in the recombined beam ahead of the detector, an interferogram, which is a measure of energies as a function of optical path difference, is obtained. This interferogram is then converted by FT mathematics into an infrared absorption spectrum. A computer allows the mathematical transformation to be completed rapidly. Because all wavelengths are measured at once, FT instruments can acquire spectra more rapidly, with a greatly improved signal-to-noise ratio, as compared to dispersive instruments.

24.3.1.3 Sample Handling Techniques

Liquids are most commonly measured by **transmission IR spectroscopy,** using cells with a pathlength of

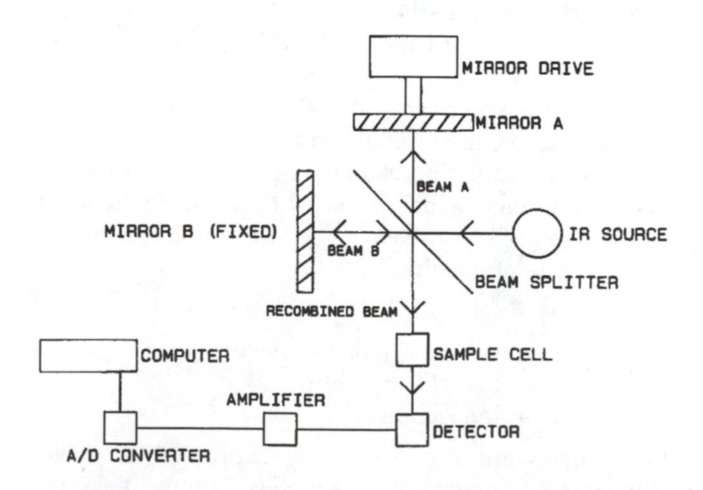

FIGURE 24-2.
Block diagram of an interferometer and associated electronics typically used in an FTIR instrument.

0.01 to 1.0 mm. Because quartz and glass absorb in the mid-IR region, cell windows of halide or sulfide salts are most commonly used. Since many of these materials are soluble in water, care must be taken when selecting cells for use with aqueous samples. Transmission spectra of solids can be obtained by finely grinding a small amount of the sample with KBr, pressing the mixture into a pellet under high pressure, and inserting the pellet into the IR beam. An alternative technique is to disperse a finely divided solid in Nujol mineral oil to form a mull. Also, **attenuated total reflectance** (ATR) cells are available for obtaining spectra from solid samples. ATR measures the total amount of energy reflected from the surface of a sample in contact with an IR transmitting crystal. The radiation penetrates a short distance into the sample before it is reflected back into the transmitting medium; therefore, the itensity of the reflected radiation is decreased at wavelengths where the sample absorbs radiation, allowing an absorption spectrum to be obtained. Similarly, internal reflectance cells are also available for use with liquid samples, where the IR radiation penetrates a few micrometers into the liquid before being reflected back into an IR transmitting crystal in contact with the liquid. These types of cells are especially useful for samples such as aqueous liquids that absorb strongly in the mid-IR region.

Transmission spectra can be obtained from gas samples using a sealed 2 to 10 cm glass cell with IR transparent windows. For trace analysis, multiple-pass cells are available that reflect the IR beam back and forth through the cell many times to obtain pathlengths as long as several meters. FTIR instruments, because of their speed of data acquisition, can also be interfaced to a gas chromatograph in order to obtain spectra of compounds eluting from the chromatography column.

24.3.2 Applications of Mid-IR Spectroscopy

24.3.2.1 Absorption Bands of Organic Functional Groups

The wavelength bands where a number of important organic functional groups absorb radiation in the mid-IR region are shown in Table 24-1.

24.3.2.2 Presentation of Mid-IR Spectra

Spectra are normally presented with either wavenumbers or wavelengths plotted on the x-axis and either percent transmittance or absorbance plotted on the y-axis. The mid-IR spectrum of polystyrene is shown in Fig. 24-3 and is typical of the common method of presentation of IR spectra.

24.3.2.3 Qualitative Applications

The center frequencies and relative intensities of the absorption bands can be used to identify specific functional groups present in an unknown substance. A

TABLE 24-1. Mid-Infrared Absorption Frequencies of Various Organic Functional Groups

Group	*Absorbing Feature*	*Frequency (cm^{-1})*
Alkanes	-CH stretch and bend	3,000–2,800
	$-CH_2$ and $-CH_3$ bend	1,470–1,420 and 1,380–1,340
Alkenes	Olefinic -CH stretch	3,100–3,000
Alkynes	Acetylenic -CH stretch	3,300
Aromatics	Aromatic -CH stretch	3,100–3,000
	-C=C- stretch	1,600
Alcohols	-OH stretch	3,600–3,200
	-OH bend	1,500–1,300
	C-O stretch	1,220–1,000
Ethers	C-O asymmetric stretch	1,220–1,000
Amines	Primary and secondary -NH stretch	3,500–3,300
Aldehydes and ketones	-C=O stretch	1,735–1,700
	-CH (doublet)	2,850–2,700
Carboxylic acids	-C=O stretch	1,740–1,720
Amides	-C=O stretch	1,670–1,640
	-NH stretch	3,500–3,100
	-NH bend	1,640–1,550

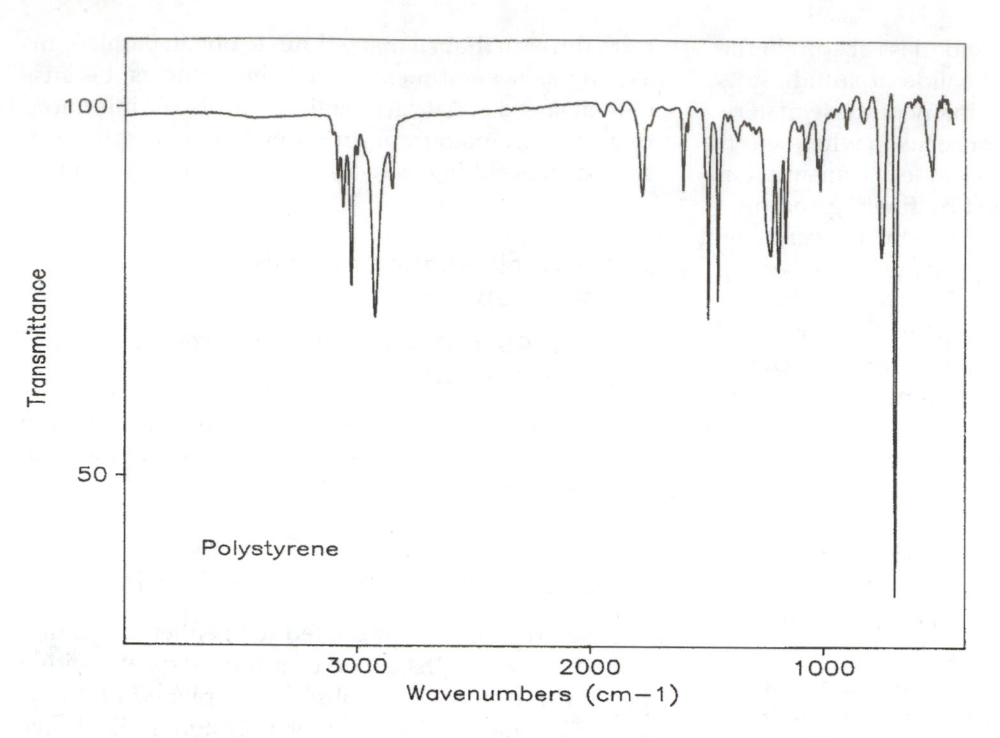

FIGURE 24-3.
Mid-IR spectrum of polystyrene, showing percent transmittance versus frequency in wavenumbers. The absorption bands just above 3,000 cm^{-1} and at 1,600 cm^{-1} indicate the presence of aromatic ring structures in the molecule, while the -CH bands just below 3,000 cm^{-1} indicate that saturated hydrocarbon regions are also present.

substance can also be identified by comparing its mid-IR spectrum to a set of **standard spectra** and determining the closest match. Probably the largest collection of standard spectra are the Sadtler Standard Spectra (Samuel P. Sadtler and Sons, Inc., Philadelphia, PA). Standard spectra are available in books for visual comparison or on magnetic media for search by computer algorithm. Common food applications include the identification of flavor and aroma compounds, particularly when FTIR measurements are coupled with gas chromatography. IR spectra are also useful in identifying packaging films.

24.3.2.4 Quantitative Applications

Infrared spectroscopic measurements obey **Beer's law,** although deviations may be greater than in UV-Vis spectroscopy due to the low intensities of IR sources, the low sensitivities of IR detectors, and the relative narrowness of mid-IR absorption bands. However, quantitative measurements can be successfully made. Perhaps the most extensive use of this technique is in the Infrared Milk Analyzers. The fat, protein, and lactose contents of milk can be determined simultaneously with one of these instruments. The wavelengths used are 5.73 μm for the carbonyl groups of lipid, 6.46 μm for the amide groups of protein, and 9.60 μm for the hydroxyl groups of lactose. Automated versions of these instruments can homogenize and analyze several hundred milk samples per hour. Commercial instruments are also available for measuring the fat content of emulsified meat samples by IR spectroscopy. Other quantitative applications include measurement of the degree of unsaturation in fats and oils.

24.4 NEAR-INFRARED (NIR) SPECTROSCOPY

Measurements in the **near-IR** (NIR) spectral region (0.7–2.5 μm, equal to 700–2,500 nm) are more widely used for quantitative analysis of foods than are mid-IR measurements. Several commercial instruments are available for compositional analysis of foods using NIR spectroscopy. A major advantage of NIR spectroscopy is its ability to directly measure the composition of solid food products by use of diffuse reflectance techniques.

24.4.1 Principles

24.4.1.1 Principles of Diffuse Reflectance Measurements

When radiation strikes a solid or granular material, part of the radiation is reflected from the sample surface. This mirrorlike reflectance is called **specular re-**

flectance, and has little useful information about the sample. Most of the specularly reflected radiation is directed back toward the energy source. Another portion of the radiation will penetrate through the surface of the sample and be reflected off several sample particles before it exits the sample. This is referred to as **diffuse reflectance,** and this diffusely reflected radiation emerges from the surface at random angles through 180°. Each time the radiation interacts with a sample particle, the chemical constituents in the sample can absorb a portion of the radiation. Therefore, the diffusely reflected radiation contains information about the chemical composition of the sample, as indicated by the amount of energy absorbed at specific wavelengths.

The amount of radiation penetrating and leaving the sample surface is affected by the size and shape of the sample particles. Therefore, in order to make reproducible reflectance measurements, granular materials should be ground to a fine particle size with a uniform size distribution. Sample preparation mills such as the Udy Cyclotec (Boulder, CO), equipped with a 1 mm screen, are useful for preparing granular materials, such as cereal grains, for analysis by NIR reflectance techniques.

24.4.1.2 Absorption Bands in the NIR Region

The absorption bands observed in the near-IR region are primarily overtones. Therefore, the absorptions tend to be weak in intensity. However, this is actually an advantage, since absorption bands that have sufficient intensity to be observed in the NIR region arise primarily from functional groups that have a hydrogen atom attached to a carbon, nitrogen, or oxygen, which are common groups in the major constituents of food such as water, proteins, lipids, and carbohydrates. Table 24-2 lists the absorption bands associated with a number of important food constituents.

The absorption bands in the NIR region tend to be broad and frequently overlap, yielding spectra that are quite complex. While the overlapping bands limit the value of NIR spectra for qualitative work, the broad bands are useful for quantitative analysis. Typical NIR spectra of wheat, dried egg white, and cheese are shown in Fig. 24-4. Note that strong absorption bands associated with the -OH groups of water are centered at ca. 1,450 and 1,940 nm. These bands are the dominant features in the spectrum of cheese, which contains 30–40 percent moisture, and they are still prominent even in the lower-moisture wheat and egg white samples. Bands arising from the -NH groups in protein can be observed at 2,060 nm and 2,180 nm in the egg white spectrum but are partially obscured by a starch absorption band, centered at 2,100 nm, in the wheat sample. Relatively sharp absorption bands arising from -CH groups in lipid can be observed at 2,310 and 2,350 nm, and another band from these groups is seen around 1,730 nm. The 1,730 nm band overlaps a weak protein absorption. The lipid bands are distinctly observable in the cheese spectrum.

24.4.2 Instrumentation

A commercial NIR spectrometer is shown in Fig. 24-5. The radiation source in most NIR instruments is a tungsten-halogen lamp with a quartz envelope, similar to a projector lamp. These lamps emit significant amounts of radiation in both the visible and NIR spectral regions. Semiconductor detectors are most commonly used in NIR instruments, with Si detectors used in the 700 to 1,100 nm range, and PbS used in the 1,100 to 2,500 nm region. Most commercial NIR instruments use monochromators rather than interferometers. Some instruments have monochromators based on diffraction gratings that allow measurements to be taken at each wavelength over the entire NIR spectral region. Other instruments are dedicated to specific applications and use optical interference filters to select 6 to 20 discrete wavelengths that can be impinged on the sample. The filters are selected to transmit wavelengths that are known to be absorbed by the sample constituents. The instrument inserts filters one at a time into the light beam to direct individual wavelengths of radiation onto the sample.

TABLE 24-2. Near-Infrared Absorption Bands of Various Food Constituents

Constituent	*Absorber*	*Wavelength (nm)*
Water	-OH stretch/deformation combination	1,920–1,950
	-OH stretch	1,400–1,450
Protein—peptides	-NH deformation	2,080–2,220 and 1,560–1,670
Lipid	methylene -CH stretch	2,300–2,350
	$-CH_2$ and $-CH_3$ stretch	1,680–1,760
Carbohydrate	C-O, O-H stretching combination	2,060–2,150

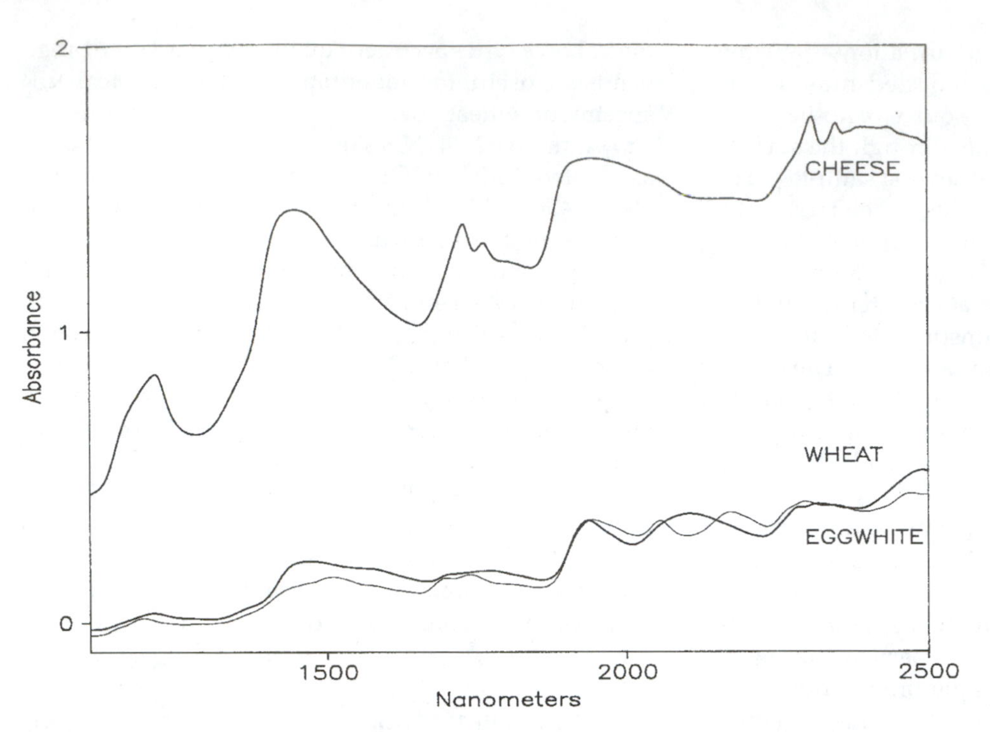

FIGURE 24-4.
Near-IR spectra of cheese, wheat, and dried egg white plotted as log (1/R) versus wavelength in nm.

FIGURE 24-5.
A modern commercial near-infrared spectrometer. Photograph courtesy of NIRSystems, Inc., Division of Perstorp Analytical, Silver Springs, MD.

Either **reflectance** or **transmittance** measurements may be made in NIR spectroscopy, depending on the type of sample. In the reflectance mode, used primarily for solid or granular samples, it is desirable to measure only the diffuse reflectance that contains information about the sample. In some instruments, this is accomplished by positioning the detectors at a 45° angle with respect to the incoming infrared beam, so that the specularly reflected radiation is not measured (Fig. 24-6a). Other instruments use an integrating sphere, which is a gold-coated metallic sphere with the detectors mounted inside (Fig. 24-6b). The sphere collects the diffusely reflected radiation coming at various angles from the sample and focuses it onto the detectors. The specular component escapes from the sphere through the same port by which the incident beam enters and strikes the sample.

Most samples are prepared by packing the food tightly into a cell against a quartz window, thereby providing a smooth, uniform surface from which reflection can occur. Quartz does not absorb in the near-IR region. At each wavelength, the intensity of light reflecting from the sample is compared to the intensity reflected from a nonabsorbing reference, such as a ceramic or fluorocarbon material or the interior of the integrating sphere. Reflectance (R) is calculated by the following formula:

$$R = I/I_0 \tag{4}$$

where: I = the intensity of radiation reflected from the sample at a given wavelength
I_0 = the intensity of radiation reflected from the reference at the same wavelength

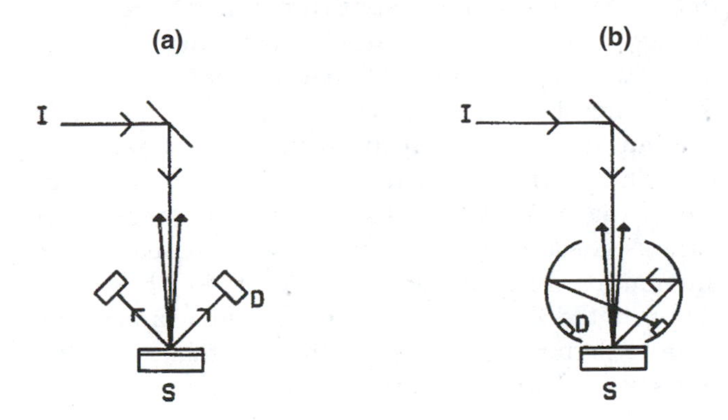

FIGURE 24-6.
Typical instrument geometries for measuring diffuse reflectance from solid food samples. Radiation from the monochromator (I) is directed by a mirror onto the sample (S). Diffusely reflected radiation is measured directly by detectors (D) placed at a 45° angle to the incident beam (panel a) or is collected by an integrating sphere and focused onto the detectors (panel b). In both cases, the specularly reflected radiation is not measured.

Reflectance data are most commonly expressed as log (1/R), an expression analogous to absorbance in transmission spectroscopy. Reflectance measurements are also sometimes expressed as differences, or derivatives, of the reflectance values obtained from adjacent wavelengths:

$$(\log R_2 - \log R_1) \tag{5}$$

or

$$(2\log R_2 - \log R_1 - \log R_3) \tag{6}$$

These derivative values are measures of the changes in slope of the spectrum.

Transmission measurements can also be made in the NIR region, and this is usually the method of choice for liquid samples. A liquid is placed in a quartz cuvette and the absorbance measured at the wavelengths of interest. Transmission measurements can also be taken from solid samples, but generally only in the 700 to 1,100 nm range. In this wavelength region, the absorption bands are higher overtones that are very weak, allowing the radiation to penetrate through several millimeters of a solid sample.

24.4.3 Quantitative Methods Using NIR Spectroscopy

NIR instruments can be calibrated to measure various constituents in food and agricultural commodities. Because of the overlapping nature of the NIR absorption bands, it is usually necessary to take measurements at two or more wavelengths in order to reliably quantitate a food component. The instrument uses an equation of the following form to predict the amount of a constituent present in the food from the spectral measurements:

$$\% \text{ constituent} = z + a \log(1/R_1) + b \log(1/R_2) + c \log(1/R_3) + \ldots\ldots \tag{7}$$

where each term represents the spectral measurement at a different wavelength multiplied by a corresponding coefficient. Each coefficient and the intercept (z) are determined by multivariate regression analysis. Absorbance or derivatized reflectance data can also be used in lieu of the log(1/R) format. Use of derivatized reflectance data has been found to provide improved results in some instances, particularly with samples that may not have uniform particle sizes.

24.4.3.1 Calibration Methods Using Multivariate Statistics

The first step in calibrating an NIR instrument is to select a set of calibration, or training, samples. The samples should be representative of the products that will be analyzed, contain the constituent of interest at levels covering the range that is expected to be encountered, and have a relatively uniform distribution of concentrations across that range. The calibration samples are analyzed by the classical analytical method normally used for that constituent, and spectral data are also obtained on each sample with the NIR instrument at all available wavelengths. All data are stored into computer memory. Multiple linear regression is then most commonly used to select the optimum wavelengths for measurement and the associated coefficients for each wavelength. Wavelengths are selected based on statistical significance by using a step forward or reverse stepwise regression procedure or by using a computer algorithm that tests regressions using all possible combinations of two, three, or four wavelengths in order to determine the combination that provides the best results. Most calibrations will use between two and six wavelengths, and one should always check to make certain that the wavelengths chosen on the basis of statistical significance also make sense from a spectroscopic standpoint. Calibration results are evaluated by comparing the multiple correlation coefficients, Fs of regression, and standard errors for the various equations developed. It is desirable to maximize the correlation coefficient (generally R should be > 0.9) and minimize the standard error. A calibration should always be tested by using the instrument to predict the composition of a set of test samples that are completely independent of the calibration set and comparing the results obtained to the classical method.

Recently, calibration techniques such as Partial Least Squares Regression and Principal Components Regression have been developed that use all the wavelengths in the entire NIR spectrum rather than a few selected wavelengths. These methods are reported to yield improved results for some samples (1).

24.4.4 Applications of NIR Spectroscopy to Food Analysis

Theory and applications of NIR spectroscopy to food analysis have been discussed in several publications (2, 3, 4). The technique is widely used for measuring moisture, protein, lipid, and starch in grains, cereal products, and oilseeds (5, 6). NIR spectroscopy can also be used for applications such as moisture, protein, and fat in meats (7, 8); lactose, moisture, and protein in milk powders (9); moisture and fat in cheese (10, 11); moisture, fat, and protein in egg products (12); total sugars and soluble solids in fruits and vegetables (13, 14); and sugar content of corn sweeteners. These are some examples of current applications, but if a substance absorbs in the NIR region and is present at a level of a few tenths of a percent or greater, it has potential for being measured by this technique.

The primary advantage of NIR spectroscopy is that once the instrument has been calibrated, several constituents in a sample can be measured rapidly (from 30 sec to one or two min) and simultaneously. No sample weighing is required, and the technique can be used by employees without extensive training. It is also applicable for on-line measurement systems. Disadvantages include the high initial cost of the instrumentation, which may require a large sample load to justify the expenditure, and the fact that specific calibrations must be developed and maintained for each product to be measured. Also, the results produced by the instrument can be no better than the data used to calibrate it, which makes careful analysis of the calibration samples of highest importance.

24.5 SUMMARY

Infrared spectroscopy measures the absorption of radiation in the near (λ = 0.8–2.5 μm) or mid (λ = 2.5–15 μm) infrared regions by molecules in food or other substances. Infrared radiation is absorbed as molecules change their vibrational energy levels. Mid-infrared spectroscopy is especially useful for qualitative analysis, such as identifying specific functional groups present in a substance. Different functional groups absorb different frequencies of radiation, allowing the groups to be identified from the spectrum of a sample. Near-infrared spectroscopy is used most extensively for quantitative applications. Diffuse reflectance measurements are often used in NIR spectroscopy to obtain spectral information directly from solid samples. By using multivariate statistical techniques, NIR instruments can be calibrated to measure the amounts of various constituents in a food sample based on the amount of infrared radiation absorbed at specific wavelengths. NIR spectroscopy requires much less time to perform quantitative analysis than do many conventional wet chemical or chromatographic techniques.

24.6 STUDY QUESTIONS

1. Describe the factors that affect the frequency of vibration of a molecular functional group and thus the frequencies of radiation that it absorbs. Also, explain how the funda-

mental absorption and overtone absorptions of a molecule are related.

2. Describe the essential components of a Fourier transform (FT) mid-infrared spectrophotometer and their function, and compare the operation of the FT instrument to a dispersive instrument. What advantages do Fourier transform instruments have over dispersive infrared spectrophotometers?
3. Of the three antioxidants BHT (butylated hydroxytoluene), BHA (butylated hydroxyanisole), and propyl gallate, which would you expect to have a strong infrared absorption band in the 1700–1750 cm^{-1} spectral region? Look up these compounds in a reference book if you are uncertain of their structure.
4. Describe the two ways in which radiation is reflected from a solid or granular material. Which type of reflected radiation is useful for making quantitative measurements on solid samples by near-infrared (NIR) reflectance spectroscopy? How are NIR reflectance instruments designed to select for the desired component of reflected radiation?
5. Describe the steps involved in calibrating a near-infrared reflectance instrument to measure the protein content of wheat flour. Why is it usually necessary to make measurements at more than one wavelength?

24.7 REFERENCES

1. Martens, H., and Naes, T. 1987. Mutivariate calibration by data compression. Ch. 4, in *Near-Infrared Technology in the Agricultural and Food Industries*, P. C. Williams and K. H. Norris (Eds.), 57–87. American Association of Cereal Chemists, St. Paul, MN.
2. Wetzel, D. L. 1983. Near-infrared reflectance analysis: Sleeper among spectroscopic techniques. *Anal. Chem.* 55:1165A.
3. Osborne, B. G., and Fearn, T. 1986. *Near-Infrared Spectroscopy in Food Analysis*. Longman, Essex, England.
4. Williams, P. C., and Norris, K. H. (Eds.) 1987. *Near-Infrared Technology in the Agricultural and Food Industries*. American Association of Cereal Chemists, St. Paul, MN.
5. Hymowitz, T., Dudley, J. W., Collins, F. I., and Brown, C. M. 1974. Estimation of protein and oil concentration in corn, soybean, and oat seed by near-infrared light reflectance. *Crop Sci.* 14:713.
6. Orman, B. A., and Schumann, R. A. 1991. Comparison of near-infrared spectroscopy calibration methods for the prediction of protein, oil, and starch in maize grain. *J. Agric. Food Chem.* 39:883.
7. Lanza, E. 1983. Determination of moisture, protein, fat and calories in raw pork and beef by near-infrared spectroscopy. *J. Food Sci.* 48:471.
8. Bartholomew, D. T., and Osuala, C. I. 1988. Use of the InfraAlyzer in proximate analysis of mutton. *J. Food Sci.* 53:379.
9. Baer, R. J., Frank, J. F., Loewenstein, M., and Birth, G. S. 1983. Compositional analysis of whey powders using near-infrared diffuse reflectance spectroscopy. *J. Food Sci.* 48:959.
10. Wehling, R. L., and Pierce, M. M. 1988. Determination of moisture in Cheddar cheese by near-infrared reflectance spectroscopy. *J. Assoc. Off. Anal. Chem.* 71:571.
11. Wehling, R. L., and Pierce, M. M. 1989. Application of near-infrared reflectance spectroscopy to determination of fat in Cheddar cheese. *J. Assoc. Off. Anal. Chem.* 72:56.
12. Wehling, R. L., Pierce, M. M., and Froning, G. W. 1988. Determination of moisture, fat and protein in spray-dried whole egg by near-infrared reflectance spectroscopy. *J. Food Sci.* 53:1356.
13. Birth, G. S., Dull, G. G., Renfroe, W. T., and Kays, S. J. 1985. Nondestructive spectrophotometric determination of dry matter in onions. *J. Am. Soc. Hort. Sci.* 110:297.
14. Dull, G. G., Birth, G. S., Smittle, D. A., and Lefler, R. G. 1989. Near-infrared analysis of soluble solids in intact cantaloupe. *J. Food Sci.* 54:393.

CHAPTER

25

Atomic Absorption and Emission Spectroscopy

Dennis D. Miller

25.1 INTRODUCTION

Atomic spectroscopy has played a major role in the development of our current database for mineral nutrients and toxicants in foods. When atomic absorption spectrometers became widely available in the sixties and seventies, the development of atomic absorption methods for accurately measuring trace amounts of mineral elements in biological samples paved the way for unprecedented advances in fields as diverse as food analysis, nutrition, biochemistry, and toxicology. The application of plasmas as excitation sources for atomic emission spectroscopy led to the commercial availability of the inductively coupled plasma emission spectrometer beginning in the late seventies. This instrument has further enhanced our ability to measure the mineral composition of foods and other materials rapidly, accurately, and precisely. These two instrumental methods have largely replaced traditional wet chemistry methods for mineral analysis of foods, although traditional methods for iron and phosphorous remain in wide use today (see Chapter 9).

In theory, virtually all of the elements in the periodic chart may be determined by atomic absorption and/or atomic emission spectroscopy. In practice, atomic spectroscopy is used primarily for the determination of mineral elements. Table 25-1 lists mineral elements of concern in foods. The database for Ca, Fe, Na, and K in foods is reasonably good. The database for the trace elements and toxic heavy metals is incomplete and should be expanded.

TABLE 25-1. Mineral Elements in Foods Classified According to Nutritional Essentiality, Potential Toxic Risk, and Inclusion in USDA Handbook 8

Essential Nutrient	*Toxicity Concern*	*USDA Handbook 8 Database*
Calcium	Lead	Calcium
Phosphorus	Mercury	Iron
Sodium	Cadmium	Magnesium
Potassium	Nickel	Phosphorus
Chlorine	Arsenic	Potassium
Magnesium		Sodium
Iron		Zinc
Iodine		Copper
Zinc		Manganese
Copper		
Selenium		
Chromium		
Manganese		
Arsenic		
Boron		
Molybdenum		
Nickel		
Silicon		

From (1).

This chapter deals with the basic principles that underlie analytical atomic spectroscopy and provides an overview of the instrumentation available for measuring atomic absorption and emission. In addition, some practical problems associated with the use of the technology are addressed. Readers interested in a more thorough treatment of the topic are referred to two excellent monographs available from the Perkin-Elmer Corporation. One is by Beaty (2); the other is by Boss and Fredeen (3).

The following abbreviations will be used throughout the chapter:

AAS: atomic absorption spectroscopy

AES: atomic emission spectroscopy

ICP: inductively coupled plasma

25.2 GENERAL PRINCIPLES

Atomic **absorption** spectroscopy quantifies the absorption of electromagnetic radiation by well-separated atoms or ions in the gaseous state, while atomic **emission** spectroscopy measures emission of radiation from atoms excited by heat or other means. Atomic spectroscopy is particularly well suited for analytical measurements because atomic spectra consist of discrete lines, and every element has a unique spectrum. Therefore, individual elements can be identified and quantified with accuracy and precision even in the presence of atoms or ions of other elements.

25.2.1 Energy Transitions in Atoms

Atomic absorption spectra are produced when **ground state atoms** (or ions) **absorb energy** from a radiation source. Atomic emission spectra are produced when **excited atoms emit energy** on returning to the ground state. Absorption of a photon of radiation causes an outer shell electron to jump to a higher energy level, moving the atom into an **excited state.** The excited atom may fall back to a lower energy state, releasing a photon in the process. Atoms absorb or emit radiation of discrete wavelengths because the allowed energy levels of electrons in atoms are fixed (not random). The energy change associated with a transition between two energy levels is directly related to the frequency of the absorbed radiation:

$$E_e - E_g = hv \quad (1)$$

where: E_e = energy in excited state

E_g = energy in ground state

h = Planck's constant

v = frequency of the radiation

Rearranging, we have:

$$v = (E_e - E_g)/h \quad (2)$$

or, since $c = \lambda v$

$$\lambda = hc/(E_e - E_g) \quad (3)$$

where: c = speed of light
λ = wavelength of the absorbed light

The above relationships clearly show that for a given electronic transition, radiation of a discrete wavelength is either absorbed or emitted. Each element has a unique set of allowed transitions and therefore a unique spectrum. The absorption and emission spectra for sodium are shown in Fig. 25-1. For **absorption,** transitions involve primarily the excitation of electrons in the ground state, so the number of transitions is relatively small. **Emission,** on the other hand, occurs when electrons in various excited states fall to lower energy levels including, but not limited to, the ground state. Therefore, the emission spectrum has more lines than the absorption spectrum. An energy level diagram for an electron in the 3s orbital of sodium is shown in Fig. 25-2. When a transition is from or to the ground state, it is termed a **resonance transition,** and the resulting spectral line is called a **resonance line.**

See Chapter 22 for a more detailed discussion of atomic and molecular energy transitions.

25.2.2 Atomization

Atomic spectroscopy requires that atoms of the element of interest be in the **atomic state** (not combined with other elements in a compound) and that they be well separated in space. In foods, virtually all elements are present as compounds or complexes and must therefore be atomized before atomic absorption or emission measurements can be made. **Atomization** involves separating particles into individual molecules (vaporization) and breaking molecules into atoms. It is accomplished by exposing the analyte to high temperatures in a flame or plasma. A solution containing the analyte is introduced into the flame or plasma as a fine mist. The solvent quickly evaporates, leaving solid particles of the analyte that vaporize and decompose to atoms that may absorb radiation (atomic absorption) or become excited and subsequently emit radiation (atomic emission). This process is shown schematically in Fig. 25-3.

Three methods for atomizing samples are summarized in Table 25-2.

25.3 ATOMIC ABSORPTION SPECTROSCOPY (AAS)

Atomic absorption spectroscopy is an analytical method based on the absorption of ultraviolet or visible radiation by free atoms in the gaseous state. It is a relatively simple method and may be accomplished

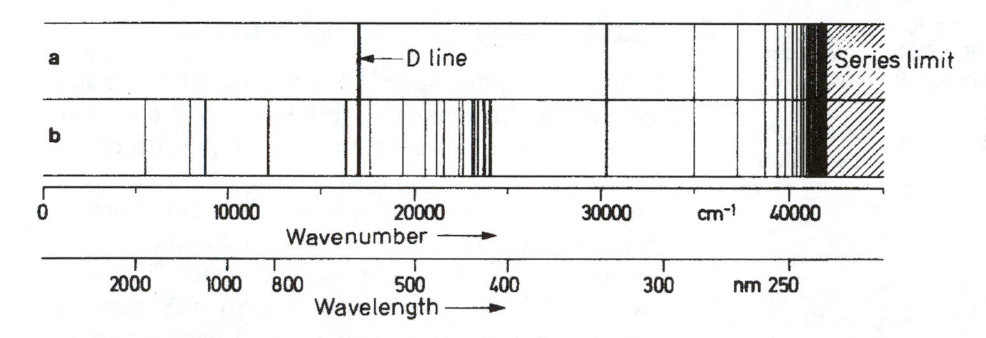

FIGURE 25-1.
Spectra for sodium. The upper spectrum (a) is the absorption spectrum and the lower (b) is the emission spectrum. From (4). Reprinted with permission by VCH Publishers (1985).

TABLE 25-2. Methods for Atomization of Analytes

Source of Energy for Atomization	*Approximate Atomization Temperature, °C*	*Analytical Method*
Flame	1,700–3,150	AAS, AES
Electrothermal	1,200–3,000	AAS (graphite furnace)
Inductively coupled argon plasma	6,000–8,000	ICP-AES

Adapted from (5).

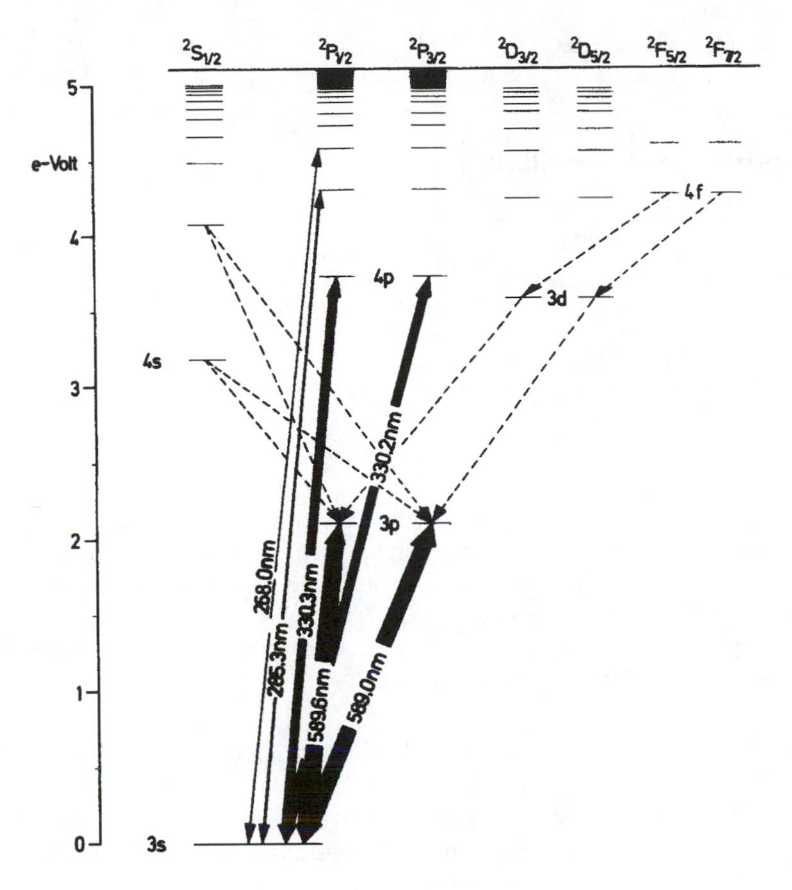

FIGURE 25-2.
An energy level diagram for sodium, showing transitions between allowed energy levels. The width of the lines is proportional to the intensity of the absorbed or emitted radiation. Solid lines represent transitions allowed in either absorption or emission. Broken lines represent transitions that occur only during emission. From (4). Reprinted with permission of VCH Publishers (1985).

with instruments ranging in price from $10,000 to $50,000. It is the most widely used form of atomic spectroscopy in food analysis. Two types of atomization are commonly used in atomic absorption spectroscopy: **flame atomization** and **electrothermal (graphite furnace) atomization.**

25.3.1 Principles of Flame Atomic Absorption Spectroscopy

Figure 25-4 shows a simplified diagram of a flame atomic absorption spectrometer.

In flame atomic absorption spectroscopy, a nebulizer-burner system is used to convert a solution of the sample into an atomic vapor. It is important to note that the sample must be in solution (usually an aqueous solution) before it can be analyzed by flame atomic absorption spectroscopy. The sample solution is nebulized (dispersed into tiny droplets), mixed with fuel and an oxidant, and burned in a flame produced by oxidation of the fuel by the oxidant. Atoms and ions are formed within the hottest portion of the flame as

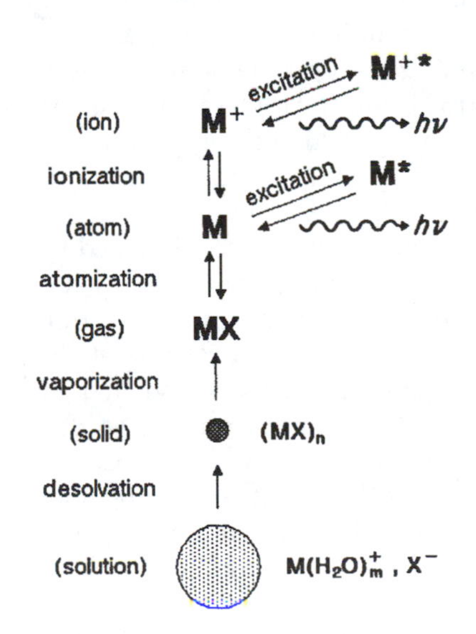

FIGURE 25-3.
A schematic representation of the atomization of an element in a flame or plasma. The large circle at the bottom represents a tiny droplet of a solution containing the element (M) as part of a compound. From (3), used with permission. Figure courtesy of the Perkin-Elmer Corporation, Norwalk, CT.

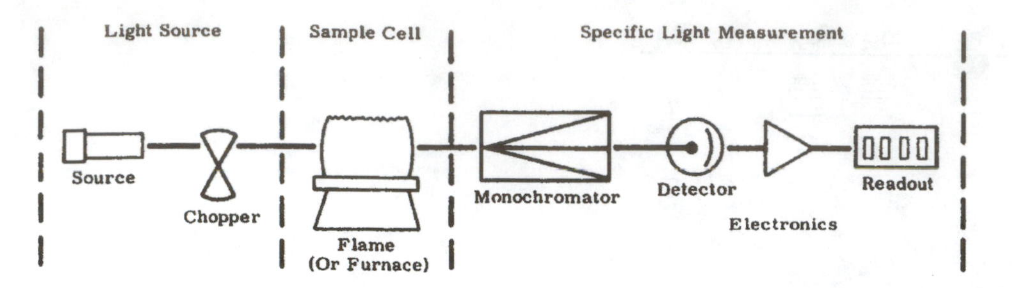

FIGURE 25-4.
A simplified diagram of a single-beam atomic absorption spectrometer. The sample enters the flame following dispersal in a nebulizer (not shown). From (2), used with permission. Figure courtesy of the Perkin-Elmer Corporation, Norwalk, CT.

analyte compounds are decomposed by the high temperatures. The flame itself serves as the sample compartment. The temperature of the flame is important because it will affect the efficiency of converting compounds to atoms and ions and because it influences the distribution between atoms and ions in the flame. Atoms and ions of the same element produce different spectra, and it is desirable to choose a flame temperature that will maximize atomization and minimize ionization. Both atomization efficiency and ionization increase with increasing flame temperature, so choice of the optimal flame is not a simple matter. Flame characteristics may be manipulated by choice of oxidant and fuel and by adjustment of the oxidant-fuel ratio. The most common oxidant-fuel combinations are air-acetylene and nitrous oxide-acetylene. The instrument instruction manual and/or the literature should be consulted for recommended flame characteristics.

Once the sample is atomized in the flame, its quantity is measured by determining the attenuation of a beam of radiation passing through the flame. In order for the measurement to be specific for a given element, the radiation source is chosen so that the emitted radiation contains an emission line that corresponds to one of the most intense lines in the atomic spectrum of the element being measured. This is accomplished by fabricating lamps in which the element to be determined serves as the cathode. Thus, the radiation emitted from the lamp is the emission spectrum of the element. The emission line of interest is isolated by passing the beam through a monochromator so that only radiation of a very narrow band width reaches the detector. Usually, one of the strongest spectral lines is chosen; for example, for sodium the monochromator is set to pass radiation with a wavelength of 589.0 nm (see Fig. 25-2). The principle of this process is illustrated in Fig. 25-5. Note that the intensity of the radiation leaving the flame is less than the intensity of radiation coming from the source. This is because sample atoms in the flame absorb some of the radiation. Notice also that the line width of the radiation from the source is narrower than the corresponding line width in the absorption spectrum. This is because the higher temperature of the flame causes a broadening of the line width.

The amount of radiation absorbed by the sample is given by **Beer's law:**

$$A = \log(I_o/I) = abc \quad (4)$$

where: A = absorbance
I_o = intensity of radiation incident on the flame
I = intensity of radiation exiting the flame
a = molar absorptivity
b = pathlength through the flame
c = concentration of atoms in the flame

Clearly, absorbance is directly related to the concentration of atoms in the flame.

25.3.2 Principles of Electrothermal Atomic Absorption Spectroscopy (Graphite Furnace AAS)

Electrothermal atomic absorption spectroscopy is identical to flame atomic absorption spectroscopy except for the atomization process. **Electrothermal atomization** involves heating the sample to a temperature (2,000–3,000°C) that produces volatilization and atomization. This is accomplished in a tube or cup positioned in the light path of the instrument so that absorbance is determined in the space directly above the surface where the sample is heated. The advantages of electrothermal atomization are that it can accommodate smaller samples than are required for flame atomic absorption and that detection limits are lower. Disadvantages are the added expense of the electrothermal furnace, lower sample throughput, more difficult operation, and lower precision.

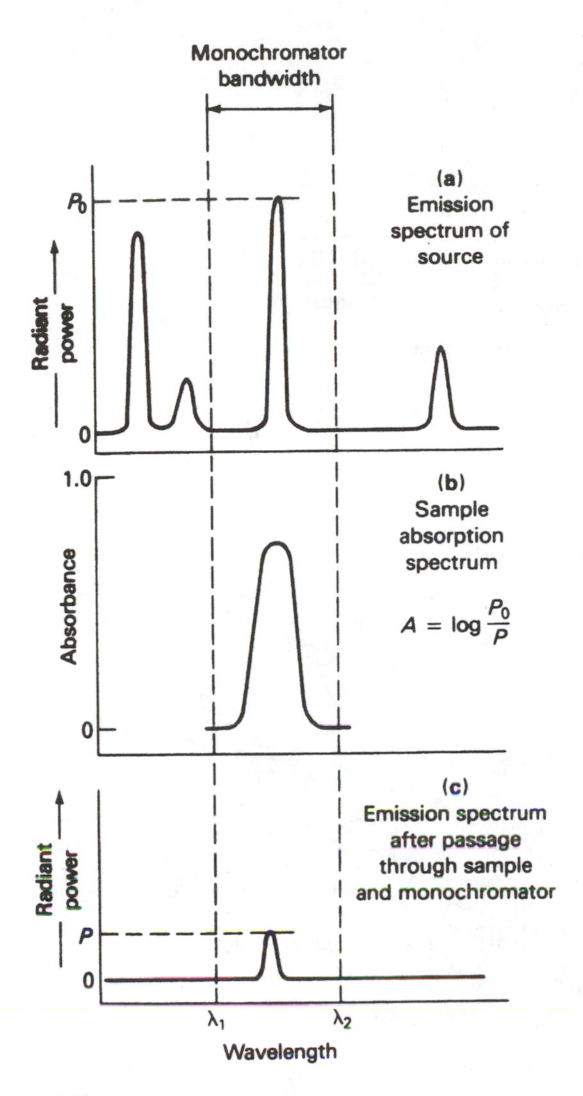

FIGURE 25-5.
Schematic representation of the absorption of radiation by a sample during an atomic absorption measurement. The spectrum of the radiation source is shown in (a). As the radiation passes through the sample (b), it is partially absorbed by the element of interest. Absorbance is proportional to the concentration of the element in the flame. The radiant power of the radiation leaving the sample is reduced because of absorption by the sample (c). From (5), used with permission. Illustration from PRINCIPLES OF INSTRUMENTAL ANALYSIS, Third Edition, Solutions Manual by Douglas A. Skoog, copyright © 1985 by Holt, Rinehart & Winston, Inc., reprinted by permission of the publisher.

25.3.3 Instrumentation for Atomic Absorption Spectroscopy

Atomic absorption spectrometers consist of the following components:

1. **Radiation source,** usually a hollow cathode lamp
2. **Atomizer,** usually a nebulizer-burner system or an electrothermal furnace
3. **Monochromator,** usually an ultraviolet-visible (UV-Vis) grating monochromator
4. **Detector,** usually a photomultiplier tube
5. **Readout device,** analog or digital

The configuration of a double-beam atomic absorption spectrometer is illustrated in Fig. 25-6. (See Fig. 25-4 for a diagram of a **single-beam instrument.**) In **double-beam instruments,** the beam from the light source (hollow cathode lamp) is split by a rotating mirrored chopper into a reference beam and a sample beam. The reference beam is diverted around the sample compartment (flame or furnace) and recombined before passing into the monochromator. The electronics are designed to produce a ratio of the reference and sample beams. This way, fluctuations in the radiation source and the detector are canceled out, yielding a more stable signal.

25.3.3.1 Radiation Source

The radiation source in atomic absorption spectrometers is called a **hollow cathode lamp.** Hollow cathode lamps consist of a hollow tube filled with argon or neon, an anode made of tungsten, and a cathode made of the metallic form of the element being measured (Fig. 25-7). When voltage is applied across the electrodes, the lamp emits radiation characteristic of the metal in the cathode; if the cathode is made of iron, an iron spectrum is emitted. When this radiation passes through a flame containing the sample, iron atoms in the flame will absorb some of it because it contains radiation of exactly the right energy for exciting iron atoms. This makes sense when we remember that for a given electronic transition, either up or down in energy, the energy of an emitted photon is exactly the same as the energy of an absorbed photon. Of course, this means that it is necessary to use a different lamp for each element analyzed (there are a limited number of multi-element lamps available that contain cathodes made of more than one element). Hollow cathode lamps for about 40 metallic elements may be purchased from commercial sources, which means atomic absorption may be used for the analysis of up to 40 elements.

Radiation reaching the monochromator comes from two sources, the attenuated beam from the hollow cathode lamp and excited atoms in the flame. Instruments are designed to discriminate between these two sources either by modulating the lamp so that the output fluctuates at a constant frequency or by positioning a **chopper** perpendicular to the light path between the lamp and the flame (Fig. 25-6). A chopper is a disk with segments removed. The disk is rotated at a constant speed so that the light beam reaching the flame is either on or off at regular intervals. The radiation

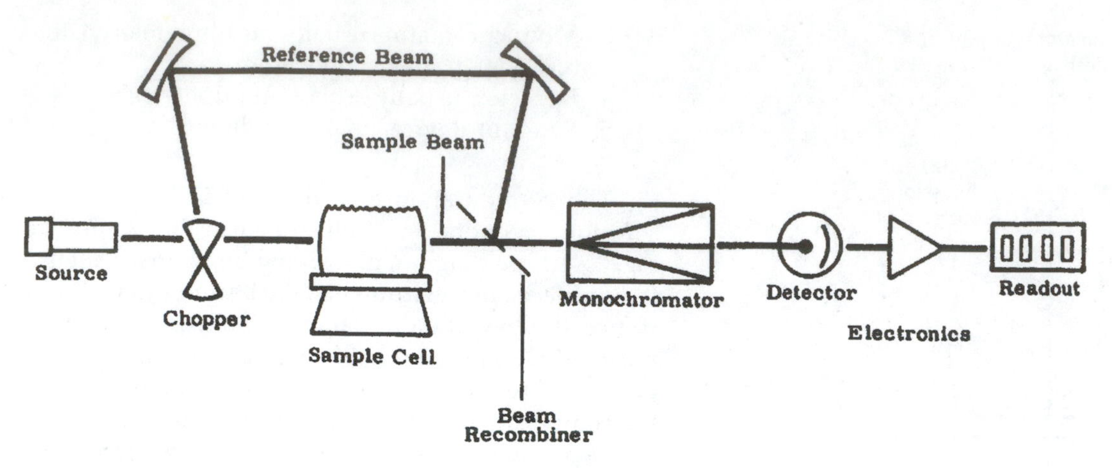

FIGURE 25-6.
Schematic representation of a double-beam atomic absorption spectrophotometer. From (2), used with permission. Figure courtesy of the Perkin-Elmer Corporation, Norwalk, CT.

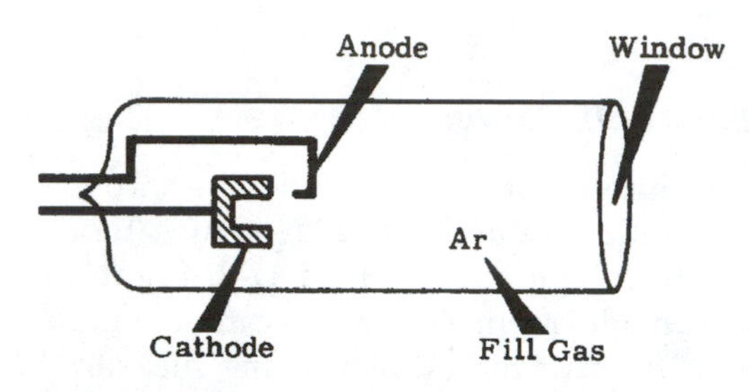

FIGURE 25-7.
Schematic representation of a hollow cathode lamp. From (2), used with permission. Figure courtesy of the Perkin-Elmer Corporation, Norwalk, CT.

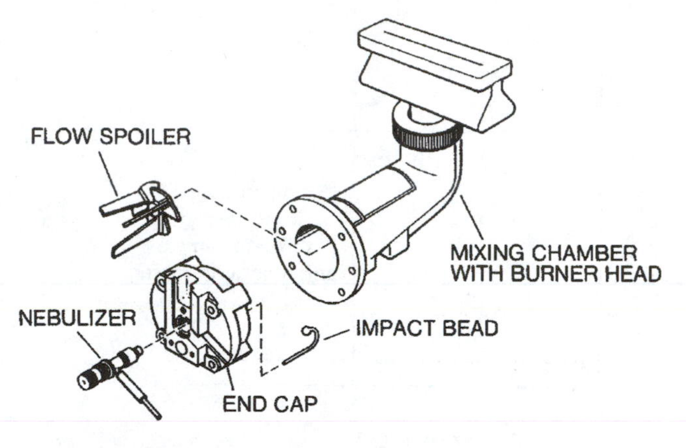

FIGURE 25-8.
Schematic representation of a nebulizer-burner assembly for an atomic absorption spectrophotometer. From (2), used with permission. Figure courtesy of the Perkin-Elmer Corporation, Norwalk, CT.

from the flame is continuous. Therefore, the radiation reaching the detector consists of the sum of an alternating and a direct signal. Instrument electronics subtract the direct signal and send only the alternating signal to the readout. This effectively eliminates the contribution of emissions from elements in the flame to the final signal.

25.3.3.2 Atomizers

Two types of atomizers are used in atomic absorption spectroscopy: **flame** and **electrothermal.** Many manufacturers now offer instruments that will accept either flame atomizers or electrothermal atomizers.

The **flame atomizer** consists of a **nebulizer** and a **burner** (Fig. 25-8). The nebulizer is designed to convert the sample solution into a fine mist or aerosol. This is accomplished by aspirating the sample through a capillary into a chamber through which oxidant and fuel are flowing. The chamber contains baffles which remove larger droplets, leaving a very fine mist that is carried into the flame by the oxidant-fuel mixture. The larger droplets fall to the bottom of the mixing chamber and are collected as waste. The burner head contains a long, narrow slot which produces a flame that may be 5 to 10 centimeters in length. This gives a long pathlength that increases the sensitivity of the measurement.

Flame characteristics may be manipulated by adjusting oxidant-fuel ratios and by choice of oxidant and fuel. **Air-acetylene** and **nitrous oxide-acetylene** are the most commonly used oxidant-fuel mixtures although other oxidants and fuels may be used for some elements. There are three types of flames: (1) **Stoichiometric.** This flame is produced from stoichiometric amounts of oxidant and fuel so the fuel is completely burned and the oxidant is completely consumed. It is characterized by yellow fringes. (2) **Oxidizing.** This flame is produced from a fuel-lean mixture. It is the hottest flame and has a clear blue appearance. (3) **Reducing.** This flame is produced from a fuel-rich mixture. It is a relatively cool flame and has a yellow color. Analysts should follow guidelines for the proper type of flame for each element.

Flame atomizers have the advantage of being stable and easy to use. However, sensitivity is relatively low because much of the sample never reaches the flame and the residence time of the sample in the flame is short.

Electrothermal atomizers are typically cylindrical graphite tubes connected to an electrical power supply. They are commonly referred to as **graphite furnaces.** The sample is introduced into the tube through a small hole using a microliter syringe (sample volumes normally range from 0.5 to 10 μ). During operation, the system is flushed with an inert gas to prevent the tube from burning and to exclude air from the sample compartment. The tube is heated electrically. Through a step-wise increase in temperature, first the sample solvent is evaporated, then the sample is ashed, and finally the temperature is rapidly increased to 2,000 to 3,000°C to rapidly vaporize and atomize the sample.

25.3.3.3 Monochromator

The monochromator is positioned in the optical path between the flame or furnace and the detector (Fig. 25-6). Its purpose is to isolate the resonance line of interest from the rest of the radiation coming from the flame or furnace and the lamp so that only radiation of the desired wavelength reaches the detector. Typically, monochromators of the grating type are used. (See Chapter 22.)

25.3.3.4 Detector/Readout

The detector is a **photomultiplier tube** that converts the radiant energy reaching it into an electrical signal. This signal is processed to produce either an analog or a digital readout. Modern instruments may be interfaced with computers for data collection, manipulation, and storage. (See Chapter 22.)

25.4 ATOMIC EMISSION SPECTROSCOPY (AES)

In contrast to atomic absorption spectroscopy, the source of radiation in atomic emission spectrometers is the excited atoms or ions in the sample rather than an external source. Figure 25-9 shows a simplified diagram of an atomic emission spectrometer. As with atomic absorption spectroscopy, the sample must be atomized in order to produce usable spectra for quantitative analysis. The difference is that in emission spectroscopy, sufficient heat is applied to the sample to excite atoms to higher energy levels. Aside from the external radiation source required for atomic absorption spectroscopy, instrumentation for atomic emission spectroscopy is similar. In fact, many instruments may be operated in either the absorption or emission mode.

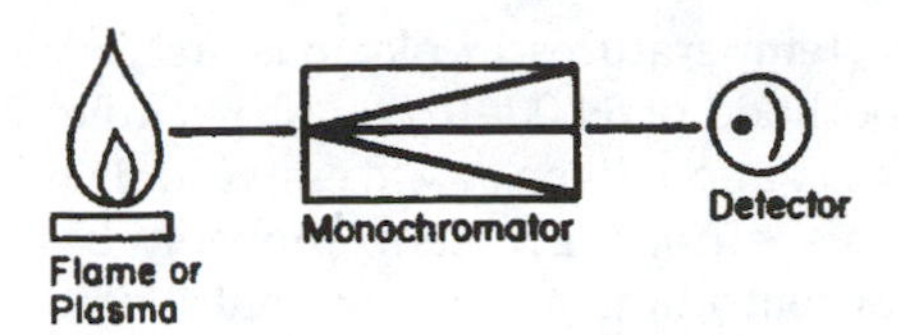

FIGURE 25-9.
A simplified diagram of an atomic emission spectrometer. From (3), used with permission. Figure courtesy of the Perkin-Elmer Corporation, Norwalk, CT.

Emissions are produced when electrons in excited atoms fall back to lower energy states. Emissions have wavelengths characteristic of individual elements since, as discussed above, the allowed energy levels for electrons are unique for each element. Energy for excitation may be produced by several methods, including **heat** (usually from a flame), **light** (from a laser), **electricity** (arcs or sparks), or **radio waves** (inductively coupled plasma) (6). Emissions are passed through monochromators or filters prior to detection by photomultiplier tubes.

The two most common forms of atomic emission spectroscopy used in food analysis are **flame emission spectroscopy** and **inductively coupled plasma** (ICP) **atomic emission spectroscopy.**

25.4.1 Principles of Flame Emission Spectroscopy

Flame emission spectroscopy employs a nebulizer-burner system to atomize and excite the sample. The instrument may be either a spectrophotometer (which uses a monochromator to isolate desired emission line) or a photometer (which uses a filter to isolate emission lines). Flame emission is most useful for elements with relatively low excitation energies. These include sodium, potassium, and calcium.

25.4.2 Principles of Inductively Coupled Plasma Atomic Emission Spectroscopy (ICP-AES)

Inductively coupled plasma emission spectroscopy has become widely available only in the last decade or so. In ICP spectroscopy, a plasma is used as the atomization-excitation source. A **plasma** is defined as gaseous mixture containing significant concentrations of cations and electrons. Temperatures in plasmas are very high (in the neighborhood of 8,000–10,000°K) resulting in very effective atomization. Even so, excessive ionization of sample atoms is not a problem, probably because of the high concentration of electrons contributed by the ionization of the argon.

25.4.3 Instrumentation for Flame Emission Spectroscopy

Flame emission spectrometers consist of the following components:

1. **Atomization-excitation source,** usually a nebulizer-laminar flow burner
2. **Monochromator or filter.** Instruments with monochromators are more versatile because any wavelength in the UV-Vis spectrum can be selected. Instruments designed for routine analysis of alkali and alkaline earth metals may employ interference filters to isolate the desired emission line.
3. **Detector**
4. **Readout device**

A comparison of the components of atomic absorption and flame emission spectrometers quickly reveals their similarities. In emission spectrometers, the flame is the radiation source, so the hollow cathode lamp and the chopper are not required. Many modern atomic absorption instruments can also be operated as flame emission spectrometers. Specialized instruments specifically designed for the analysis of sodium, potassium, lithium, and calcium in biological samples are made by some manufacturers. These instruments are called flame photometers. They employ interference filters to isolate the spectral region of interest. Low flame temperatures are used so that only easily excited elements like the alkali and alkaline earth metals produce emissions. This results in a simpler spectrum and reduces interference from other elements that may be present.

25.4.4 Instrumentation for ICP-AES

There are two basic types of ICP-AES instruments available today: the **simultaneous spectrometer** (or polychromator) and the **sequential spectrometer** (or monochromator). Both instruments are capable of determining multiple elements in the same sample, but the simultaneous spectrometer determines a limited number simultaneously, while the sequential spectrometer determines multiple elements sequentially in rapid succession.

Inductively coupled plasma atomic emission spectrometers consist of the following components (see Fig. 25-10):

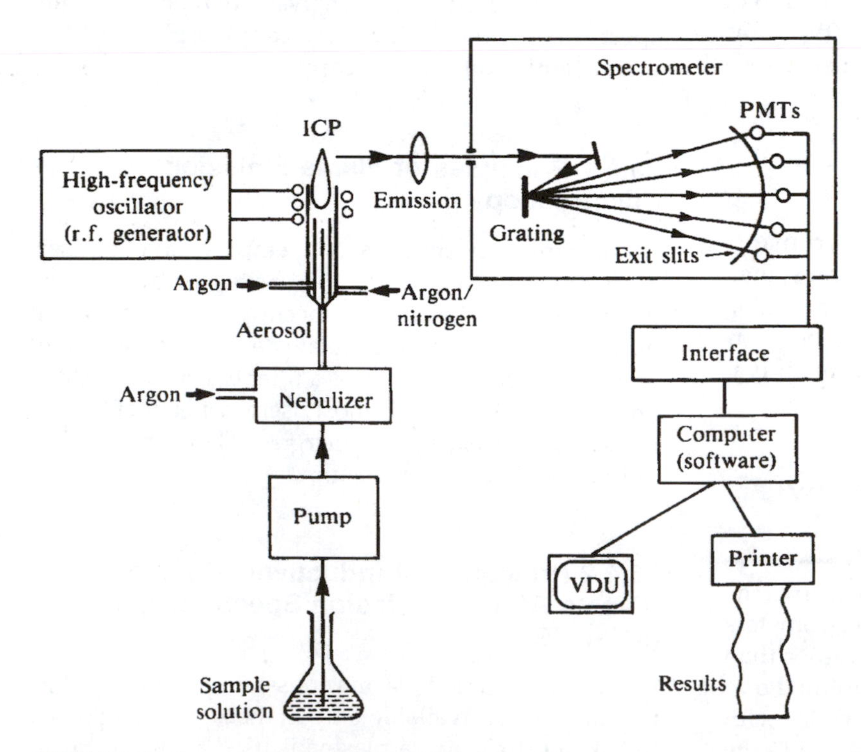

FIGURE 25-10.
Schematic of an inductively coupled plasma-atomic emission simultaneous spectrometer. PMT stands for photomultiplier tube. From (6), used with permission.

1. **Argon plasma torch**
2. **Monochromator or polychromator**
3. **Detector(s):** one photomultiplier tube for sequential spectrometers, multiple photomultiplier tubes for simultaneous spectrometers
4. **Readout device**

25.4.4.1 Argon Plasma Torch

The development of the inductively coupled argon plasma torch was a major advance in the field of quantitative atomic emission spectroscopy. The **ICP torch** is shown schematically in Fig. 25-11. It consists of three concentric quartz tubes. The open (top) end of the outermost tube is encased in the induction coil, which is connected to a radio frequency generator. Argon gas flows through the three tubes during operation. The torch is started by ionizing the argon gas with a spark from a Telas discharge. The oscillating magnetic field generated by the radio frequency induction coil couples with the argon ions and electrons inside the torch, causing them to accelerate in an annular path. The ions and electrons collide with argon atoms, causing heat generation and further ionization of the argon. The resulting plasma reaches temperatures of 6,000–10,000°K. In order to prevent the quartz tube from melting, a stream of argon (or nitrogen) is directed tangentially upward along the inside of the outer quartz tube. This cools the tube and isolates the plasma in the center of the tube. The innermost tube serves as a sample injection port. The sample is aspirated and nebulized in a fashion similar to that in a flame instrument and is injected into the base of the plasma by the innermost tube.

The extremely high temperatures and the inert atmosphere of argon plasmas are ideal for atomizing and exciting analytes. The absence of oxygen prevents the formation of oxides, which is sometimes a problem with flame methods. The relatively uniform temperature in the plasma (compared to nonuniform temperatures in flames) and the relatively long residence time in the plasma give good linear responses over a several orders of magnitude concentration range.

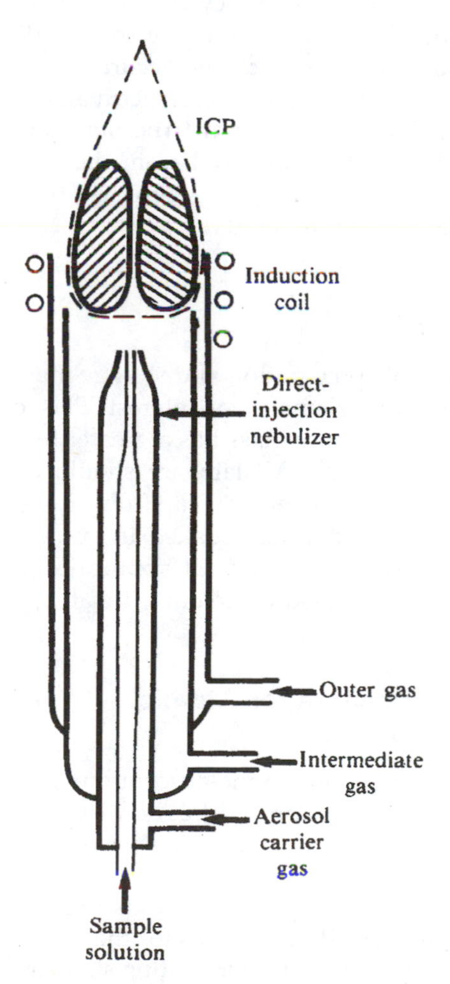

FIGURE 25-11.
Schematic of an ICP torch. From (6), used with permission.

25.4.4.2 Monochromator or Polychromator

Sequential ICP-AES instruments are equipped with **monochromators** that are capable of scanning over a wavelength range, so that readings at several preselected wavelengths can be made rapidly but not simultaneously. In this way, several elements in a single sample can be determined during a single aspiration.

Simultaneous ICP-AES instruments are capable of monitoring several wavelengths simultaneously. These instruments are equipped with **polychromators** that are preset to separate and focus several spectral lines on a series of photomultiplier tubes arranged around a semicircle inside the instrument (Fig. 25-12). These instruments have the advantage of being able to analyze several elements very rapidly and with excellent precision. However, they are expensive to buy, and the wavelengths are preset at the factory.

25.5 APPLICATIONS OF ATOMIC ABSORPTION AND EMISSION SPECTROSCOPY

25.5.1 Uses

Atomic absorption and emission spectroscopy are widely used for the quantitative measurement of mineral elements in foods. In principle, any food may be analyzed with any of the atomic spectroscopy methods discussed. In most cases, it is necessary to **ash** the food to destroy organic matter and to dissolve the ash in a

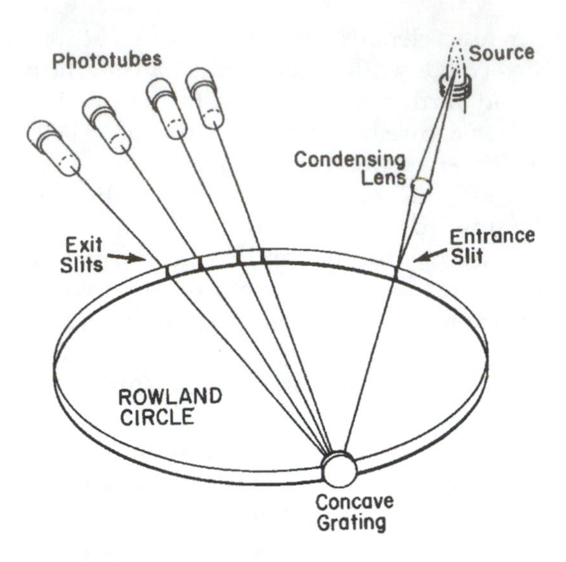

FIGURE 25-12.
Schematic of a polychromator for a simultaneous ICP-AES instrument. From (3), used with permission. Figure courtesy of the Perkin-Elmer Corporation, Norwalk, CT.

suitable solvent (usually water or dilute HCl) prior to analysis (see Chapter 8 for details on ashing methodology). Proper ashing is critical to accuracy. Some elements may be volatile at temperatures used in **dry ashing** procedures. Volatilization is less of a problem in **wet ashing,** but ashing reagents may be contaminated with the analyte. It is therefore wise to carry blanks through the ashing procedure.

Some liquid products may be analyzed without ashing, provided appropriate precautions are taken to avoid interferences. For example, vegetable oils may be analyzed by dissolving the oil in an organic solvent such as acetone or ethanol and aspirating the solution directly into a flame atomic absorption spectrometer. Milk samples may be treated with trichloroacetic acid to precipitate the protein; the resulting supernatant is analyzed directly. A disadvantage of this approach is that the sample is diluted in the process. This may be a problem when analytes are present in low concentrations. An alternative approach is to use a graphite furnace for atomization. For example, an aliquot of an oil may be introduced directly into a graphite furnace for atomization. The choice of method will depend on several factors, including instrument availability, cost, precision/sensitivity, and operator skill.

25.5.2 Practical Considerations

25.5.2.1 Reagents

Since concentrations of many mineral elements in foods are at the trace level, it is essential to use highly pure chemical reagents and water for preparation of samples and standard solutions. Only reagent grade chemicals should be used. Water may be purified by distillation, deionization, or a combination of the two. Reagent blanks should always be carried through the analysis.

25.5.2.2 Standards

Quantitative atomic spectroscopy depends on comparison of the sample measurement with appropriate standards. Ideally, standards should contain the analyte metal in known concentrations in a solution that closely approximates the sample solution in composition and physical properties. A series of standards of varying concentrations should be run to generate a calibration curve. Since many factors can affect the measurement, such as flame temperature, aspiration rate, and the like, it is essential to run standards frequently, preferably right before and/or right after running the sample. Standard solutions may be purchased from commercial sources, or they may be prepared by the analyst. Obviously, standards must be prepared with extreme care since the accuracy of the analyte determination depends on the accuracy of the standard. Perhaps the best way to check the accuracy of a given assay procedure is to analyze a reference material of known composition and similar matrix. Standard reference materials may be purchased from the U.S. National Institute of Standards and Technology (formerly the National Bureau of Standards).

25.5.2.3 Labware

Vessels used for sample preparation and storage must be clean and free of the elements of interest. Plastic containers are preferable since glass has a greater tendency to adsorb metal ions. All labware should be thoroughly washed with a detergent, carefully rinsed with distilled or deionized water, soaked in an acid solution (1N HCl is sufficient for most applications), and rinsed again with distilled or deionized water.

25.5.3 General Procedure for Atomic Absorption Analysis

While the basic design of all atomic absorption spectrometers is similar, operation procedures do vary from one instrument to another. Therefore, it is always good practice to carefully review operating procedures provided by the manufacturer before using the instrument. Most manuals have detailed procedures for the operation of the instrument as well as tables listing standard conditions (wavelength and slit width requirements, interferences and steps for avoiding them, flame characteristics, linear range, and suggestions for preparing standards) for each element. Be certain to pay close attention to safety precautions recommended by the manufacturer. *Acetylene is an explosive gas,* and great care must be taken to avoid dangerous and damaging explosions.

25.5.3.1 Operation of a Flame Atomic Absorption Instrument

The following is a *generalized procedure* that will be *similar but not identical* to procedures found in instrument operating manuals:

1. Turn the lamp current control knob to the off position.
2. Install the required hollow cathode lamp in the lamp compartment.
3. Turn on main power and power to lamp. Set lamp current to current shown on the lamp label.
4. Select required slit width and wavelength and align light beam with optical system.
5. Ignite flame and adjust oxidant and fuel flow rates.
6. Aspirate distilled water. Aspirate blank and zero instrument.
7. Aspirate standards and sample.
8. Aspirate distilled water.
9. Shut down instrument.

25.5.3.2 Calibration

According to Beer's law, absorbance is directly related to concentration. However, a plot of absorbance versus concentration will deviate from linearity when concentration exceeds a certain level (Fig. 25-13). Therefore, it is always necessary to calibrate the instrument using appropriate standards. This may be done by running a series of standards and plotting absorbance versus concentration or, in the case of most modern instruments, programming the instrument to read in units of concentration.

25.5.3.2.1 Selection of Standards The first step in calibration is to select the number and concentrations of standards to use. When operating in the linear range, only one standard is needed. The linear range may be determined by running a series of standards of increasing concentration and plotting absorbance versus concentration. Operating manuals should contain values for linear ranges. The concentration of the standard should be higher than the most concentrated sample. If the range of concentration exceeds the linear range, multiple standards must be used. Again, the concentration of the most concentrated standard should exceed the concentration of the most concentrated sample.

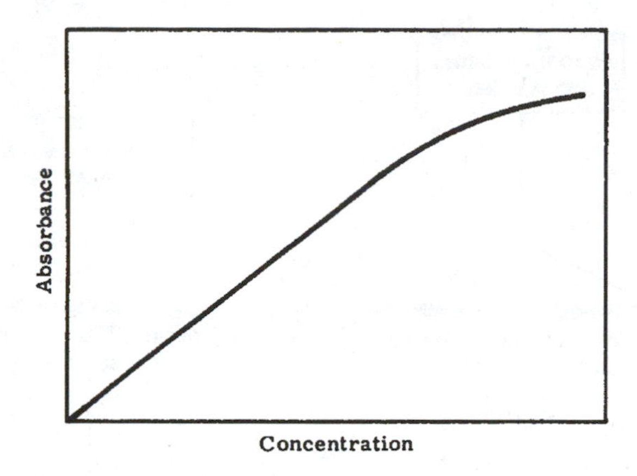

FIGURE 25-13.
A plot of absorbance versus concentration showing nonlinearity above a certain concentration. From (2), used with permission. Figure courtesy of the Perkin-Elmer Corporation, Norwalk, CT.

25.5.3.2.2 Sensitivity Check Since many factors can influence the operating efficiency of an instrument, it is a good idea to check instrument output using a standard of known concentration. Operating manuals should have values for characteristic concentrations for each element. For example, manuals for Perkin-Elmer atomic absorption spectrophotometers state that a 5.0 mg/L aqueous solution of iron "will give a reading of *approximately* 0.2 absorbance units." If the measured absorbance reading deviates significantly from this value, appropriate adjustments (e.g., flame characteristics, lamp alignment, etc.) should be made.

25.5.4 General Procedure for ICP-AES

As is the case with atomic absorption spectrometers, operating procedures for atomic emission spectrometers vary somewhat from instrument to instrument. Boss and Fredeen (3) have designed a flow chart that leads the operator through a series of steps to produce a final readout (Fig. 25-14). ICP atomic emission spectrometers are controlled by computers. The software contains **methods** that specify instrument operating conditions. The computer may be programmed by the operator, or, in some cases, default conditions may be used. Once the method is established, operation is highly automated.

25.6 INTERFERENCES

With any analytical technique, it is important to be on the lookout for possible interferences. Atomic spectros-

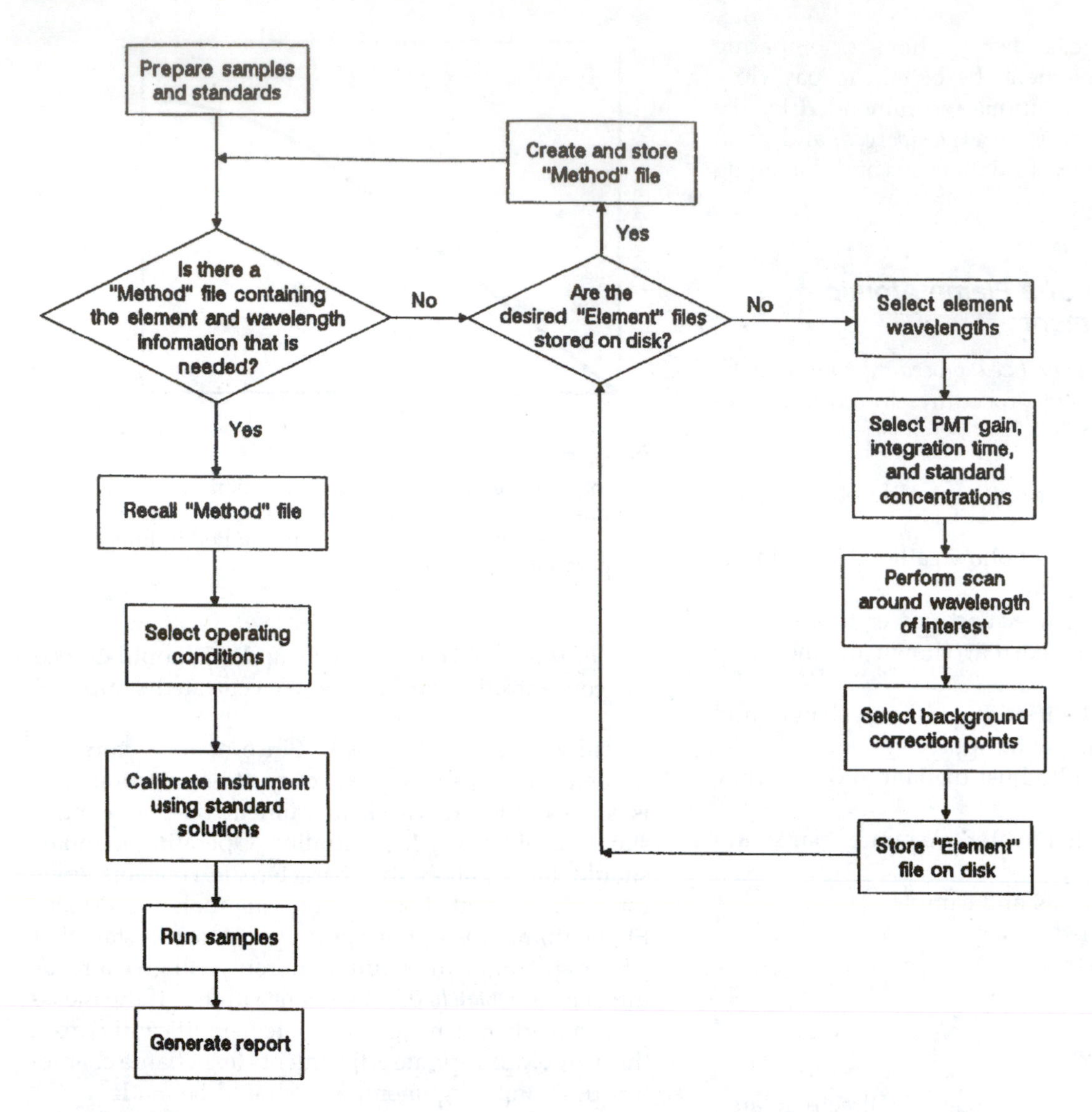

FIGURE 25-14.
Steps for operation of an ICP-AES instrument. Once the computer is programmed for a given set of elements, operation is highly automated. From (3), used with permission. Figure courtesy of the Perkin-Elmer Corporation, Norwalk, CT.

copy techniques are powerful partly because measurements of individual elements can usually be made without laborious separations. There are two main reasons for this. First, as mentioned above, a single narrow emission line is used for the measurement. Second, these are relative techniques; that is, quantitative results for an unknown sample are possible only through comparison with a standard of known concentration. If there are matrix effect problems, they can often be overcome by using the same matrix for the standard or by employing the method of additions approach.

25.6.1 Interferences in Atomic Absorption Spectroscopy

The following is a brief discussion of common interference problems in atomic absorption spectroscopy. See references 4 and 6 or your instrument manual for a thorough discussion of interference problems in atomic absorption spectroscopy and reference 5 for a list of interferences for each element. Two types of interferences are encountered in atomic absorption spectroscopy: **spectral** and **nonspectral interference.**

25.6.1.1 Spectral Interference

25.6.1.1.1 Absorption of Source Radiation An element in the sample other than the element of interest may absorb at the wavelength of the spectral band being used. Such interference is rare because emission lines from hollow cathode lamps are so narrow that only the element of interest is capable of absorbing the radiation. One example where this problem does occur is with the interference of iron in zinc determinations. Zinc has an emission line at 213.856, which overlaps the iron line at 213.859. The problem may be solved by choosing an alternative emission line for measuring zinc or by narrowing the monochrometer slit width. See reference 4 for a listing of interferences caused by overlapping spectral lines.

25.6.1.1.2 Background Absorption of Source Radiation Particulates present as a result of incomplete atomization may scatter source radiation, thereby attenuating the radiation reaching the detector. This problem may be overcome by going to a higher flame temperature to ensure complete atomization of the sample. Some instruments are equipped with automatic background correction devices. See reference 2 for a description of these devices.

25.6.1.2 Nonspectral Interferences

25.6.1.2.1. Transport Interferences This results when something in the sample solution affects the rate of aspiration, nebulization, or transport into the flame. Transport interferences are rarely a problem with graphite furnace instruments but may cause substantial errors in flame atomic absorption spectroscopy. Such factors as viscosity, surface tension, vapor pressure, and density of the sample solution can influence the rate of transport of sample into the flame. Acid concentration, organic solvents, or dissolved solids may affect the absorbance of the analyte. Transport interferences can often be overcome by matching as closely as possible the physical properties of the sample and the standards. For example, use the same solvent for the sample and the reference, or add the interferant in the sample (e.g., sugar) to the standard. The method of additions may also be used to overcome transport interferences.

25.6.1.2.2 Solute Volatilization Interferences This occurs when an interferant combines with the element of interest to form a compound of low volatility. The result is a falsely low result because some of the element remains unatomized in the flame. A common example of this is the decrease in calcium absorbance caused by the presence of phosphate in the sample. One approach for overcoming this type of interference is to **add another element,** such as lanthanum (as lanthanum oxide), to the sample and reference that will compete with the analyte for compound formation, thereby reducing or eliminating the problem. Another strategy is to use a **higher temperature flame;** for example, use nitrous oxide-acetylene instead of air-acetylene. A third approach is to **add a ligand** such as EDTA, which will complex the analyte and prevent it from reacting with the interferant.

25.6.1.2.3 Ionization Interference Ionization of analyte atoms in the flame may cause a significant interference. (Remember that absorption and emission lines of atoms and ions of the same element are different and that atomic absorption spectrometers are tuned to measure *atomic* absorption, not *ionic* absorption. Therefore, any factor that reduces the concentration of atoms in the flame will lower the absorbance reading.) The ionization of atoms results in an equilibrium situation:

$$M \rightleftharpoons M^+ + e^- \quad (5)$$

Ionization increases with increasing flame temperature and normally is not a problem in air-acetylene flames because the temperature is not high enough. It can be a problem in nitrous oxide-acetylene flames with elements that have ionization potentials of 7.5 eV or less. Ionization is suppressed by the presence of easily ionized elements, such as potassium, through mass action. When potassium ionizes, it increases the concentration of electrons in the flame and shifts the above equilibrium to the left. Reagents added to reduce ionization are called **ionization suppressors.**

25.6.2 Interferences in ICP-AES

Generally, interferences in ICP-AES analyses are less of a problem than with AAS, but they do exist and must be taken into account. **Spectral interferences** are the most common. Samples containing high concentrations of certain ions may cause an increase (shift) in background emissions at some wavelengths. This will cause a positive error in the measurement, referred to as **backround shift interference** (see Fig. 25-15). Correction is relatively simple. An emission measurement is made at a wavelength above or below the emission line of the analyte. This emission is then sub-

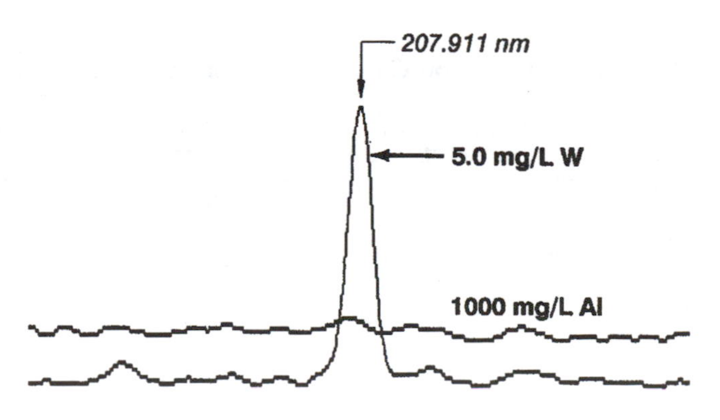

FIGURE 25-15.
An example of background shift caused by aluminum in a sample containing tungsten (w). Note that aluminum increases the baseline over a fairly wide wavelength range. From (3), used with permission. Figure courtesy of the Perkin-Elmer Corporation, Norwalk, CT.

tracted from the emission of the analyte. Alternatively, another emission line for tungsten in a region where there is no background shift could be chosen.

25.7 COMPARISON OF AAS AND ICP-AES

AAS and ICP-AES have many advantages in common. Both are capable of measuring trace metal concentrations in complex matrices with excellent precision and accuracy. Sample preparation is relatively simple. For most applications, sample preparation for both techniques involves destruction of organic matter by ashing, followed by dissolution of the ash in an aqueous solvent, usually a dilute acid. In comparison with traditional wet chemistry methods, measurements with AAS and ICP-AES are extremely rapid.

AAS has the advantage of being a more mature technique. There are literally thousands of papers in the literature describing methods for measuring trace element concentrations in hundreds of different matrices. This can be of great value to an analyst because time-consuming methods development can be avoided. Moreover, interferences are well established and relatively easily overcome. Another advantage of AAS is the wide availability of instruments.

ICP-AES instruments are capable of determining concentrations of multiple elements in a single sample with a single aspiration. This offers a significant speed advantage over AAS when the objective is to quantify several elements in a given sample. ICP-AES may also offer an advantage over AAS when analyzing for elements in refractory compounds. Refractory compounds are compounds that are unusually stable at high temperatures and may not be fully atomized in the flame of an AAS. Most refractory compounds are readily atomized in the much higher temperatures of a plasma torch.

One way of comparing analytical methods is to compare detection limits. **Detection limit** has been defined qualitatively as the lowest concentration of the element that can be distinguished from the blank at a given level of confidence. Skoog (reference 5) defines detection limit as follows:

$$\text{Detection limit} = \frac{S_m - S_{bl}}{m} \quad (6)$$

where: S_{bl} = the blank signal

S_m = the minimum distinguishable analytical signal; according to Skoog (5), $S_m = S_{bl} + ks_{bl}$ where s_{bl} is the standard deviation of the blank and k is a constant

m = the slope of the calibration curve at the concentration of interest

Note that the detection limit is a function of the standard deviation of the blank signal and the slope of the calibration curve. The smaller the standard deviation of the blank signal and the steeper the calibration curve, the lower the detection limit. Therefore, detection limit tells us something about the sensitivity (which is a function of m) and the precision of the method.

Table 25-3 lists detection limits for the various methods for elements that may be of interest to food analysts. It should be noted that these are approximate values, and detection limits will vary depending on the sample matrix, the stability of the instrument, and other factors. A two- or threefold difference in detection limit is probably not meaningful, but an order of magnitude difference probably is. Nevertheless, detection limits are useful for choosing among various methods. For example, if you had an atomic spectrometer that could be operated in either the flame atomic absorption mode or the flame emission mode and you wanted to analyze for iron and calcium in spinach, you would probably choose atomic absorption for the iron and atomic emission for the calcium.

TABLE 25-3. Detection Limits (ng/ml) for Selected Elements

Element	*Flame Atomic Absorption*	*Electrothermal Atomic Absorption*	*Flame Emission*	*ICP Emission*
Al	30	0.005	5	2
Ca	1	0.02	0.1	0.02
Cd	1	0.001	800	2
Cu	2	0.002	10	0.01
Fe	5	0.005	30	0.3
Hg	500	0.01	0.0004	1
Mg	0.01	0.00002	5	0.05
Mn	2	0.0002	5	0.06
Na	2	0.0002	0.1	0.2
Pb	10	0.002	100	2
Zn	2	0.00005	0.0005	2

Adapted from (5).

25.8 SUMMARY

Atomic absorption spectroscopy quantifies the absorption of electromagnetic radiation by well-separated atoms or ions in the gaseous state, while atomic emission spectroscopy measures emission of radiation from atoms exited by heat or other means. An atomic absorption spectrophotometer uses a hollow cathode lamp as the radiation source and a flame or graphite furnace to atomize the sample. In emission spectroscopy, a flame or plasma (ICP) serves as both the atomizer and excitation source. Development of ICP-AES has revived the use of emission spectroscopy because of its advantages with regard to sensitivity, interferences, and multi-element analysis.

Atomic spectroscopy is a powerful tool for the quantitative measurement of elements in foods. The development of this technology over the past 35 to 40 years has had a major impact on several fields, including food science and technology, food safety and toxicology, nutrition, biochemistry, and biology. Today, accurate and precise measurements of a large number of mineral nutrients and nonnutrients in foods can be made rapidly and with minimal sample preparation using commercially available instrumentation.

25.9 STUDY QUESTIONS

1. Explain the significance of energy transitions in atoms and of atomization for the techniques of atomic absorption and atomic emission spectroscopy.
2. Describe the similarities and differences between AAS and AES for mineral analysis.
3. A new employee in your laboratory is somewhat familiar with the application, principles involved, instrumental components, and quantitation procedure for UV-Vis spectroscopy. The employee must now learn to do analyses using AAS. Explain to the new employee the (a) applications, (b) principles, (c) instrumental components and their arrangement (use diagrams and explain differences), and (d) quantitation procedure for AAS by comparing and contrasting these same items for UV-Vis spectroscopy. (Assume you are talking about double-beam systems.)
4. What would be the advantages of having an atomic absorption unit that had a graphite furnace?
5. Your company has just purchased an inductively coupled plasma-atomic emission spectrometer.
   a. Explain the instrumentation and principle of its operation to analyze foods for specific minerals.
   b. Explain how AAS differs in instrumentation and principle of operation from what you described above for ICP-AES.
   c. What are the advantages of ICP-AES over AAS?
   d. For most types of food samples other than clear liquids, what type of sample preparation and treatment is generally required before using ICP-AES, AES, or AAS for analysis?
6. In your preparation of an ashed milk sample for calcium determination by atomic absorption, you forgot to add either EDTA or $LaCl_3$. Would you likely over- or underestimate the true Ca content? Why would it likely be necessary to add one of these to obtain accurate results? Briefly explain how each of these works.
7. In the quantitation of Na by atomic absorption, KCl or LiCl was not added to the sample. Would you likely over- or underestimate the true Na content? Explain why either KCl or LiCl is necessary to obtain accurate results.
8. Give five potential sources of error in sample preparation prior to atomic absorption analysis.
9. The detection limit for calcium is lower for ICP emission than it is for flame atomic absorption. How is the detection limit determined, and what does it mean?

25.10 PRACTICE PROBLEMS

1. The following data were recorded during a procedure for determining the iron content of enriched flour. Calculate the iron concentration in the flour. Express your answer as mg Fe/lb flour. The protocol was as follows:
   Weigh out 10.00 g of the flour. Transfer to a 800 mL Kjeldahl flask. Add 20 mL H_2O, 5 mL H_2SO_4, and 25 mL HNO_3. Heat to SO_3 fumes. Cool, add 25 mL H_2O, filter quantitatively into a 100 mL vol. flask. Dilute to volume. Prepare iron standards with concentrations of 0, 2, 5 mg Fe/L. Read absorbances of standards and sample on an atomic absorption spectrophotometer.

Sample	*Fe Concentration*	*Absorbance*	*Corrected Absorbance*
	(mg/L)		
Std 1	0.00	0.01	0.00
Std 2	2.0	0.21	0.20
Std 3	5.0	0.51	0.50
Flour Fe	?	0.38	0.37

2. Describe a procedure for determining calcium, potassium, and sodium in infant formula using an ICP-AES. *Note:* Concentrations of Ca, K, and Na in infant formula are around 700, 730, and 300 mg, respectively.

Answers

1. 18.16 mg Fe/lb. flour
2. Consult AOAC. 1990. *Official Methods of Analysis,* 15th ed., pp. 1106–1107. Association of Official Analytical Chemists Method 984.27. The following approach may be used:
   1. Shake can vigorously.
   2. Transfer 15.0 ml of formula to a 100 ml Kjeldahl flask. (Carry 2 reagent blanks through with sample.)
   3. Add 30 ml HNO_3-$HClO_4$ (2:1).
   4. Leave samples overnight.
   5. Heat until ashing is complete (follow AOAC procedure carefully—mixture is potentially explosive).
   6. Transfer quantitatively to a 50 mL vol. flask. Dilute to volume.

7. Calibrate instrument. Choose wavelengths of 317.9 nm, 766.5 nm, and 589.0 nm for Ca, K, and Na, respectively. Prepare calibration standards containing 200, 200, and 100 ng/mL for Ca, K, and Na, respectively.
8. The ICP-AES computer will calculate concentrations in the samples as analyzed. To convert to concentrations in the formula, use the following equation:

$$\text{Concentration in formula} = \text{concentration measured by ICP} \times \frac{50 \text{ mL}}{15 \text{ mL}}$$

25.11 REFERENCES

1. Annomyous. 1976–1989. Composition of Foods: Raw, Processed, Prepared. Agriculture Handbook No. 8-1 through 8-20. Agriculture Research Service, U.S. Dept. of Agriculture, Washington, DC.
2. Beaty, R.D. 1988. *Concepts, Instrumentation and Techniques in Atomic Absorption Spectrophotometry.* Perkin-Elmer Corporation, Norwalk, CT.
3. Boss, C.B., and Fredeen, K.J. 1989. *Concepts, Instrumentation and Techniques in Inductively Coupled Plasma Atomic Emission Spectrometry.* Perkin-Elmer Corporation, Norwalk, CT.
4. Welz, G. 1985. *Atomic Absorption Spectrometry.* VCH, Weinheim, Federal Republic of Germany.
5. Skoog, D.A. 1985. *Principles of Instrumental Analysis,* 3rd. ed. Saunders College Publishing, Philadelphia, PA.
6. Moore, G.L. 1989. *Introduction to Inductively Coupled Plasma Atomic Emission Spectrometry.* Elsevier Science Publishing Co., New York, NY.
7. Varma, A. 1984. *CRC Handbook of Atomic Absorption Analysis,* Vols. I and II. CRC Press, Inc., Boca Raton, FL.

CHAPTER

26

Mass Spectrometry

J. Scott Smith

26.1 INTRODUCTION

Mass spectrometry is unique among the various spectroscopy techniques in both theory and instrumentation. As you may recall, spectroscopy involves the interaction of electromagnetic radiation or some form of energy with molecules. The molecules absorb the radiation and produce a spectrum either during the absorption process or as the excited molecules return to the ground state. With mass spectrometry, the molecule is exposed to high energy electrons and, through a sequence of steps, is broken down into unique charged molecular fragments. The uniqueness of this process allows the method to be used for detection and identification of an unknown compound.

Because of recent advances in instrument design, electronics, and computers, mass spectrometry has become routine in many analytical labs. Probably the most common application is the interfacing of a mass spectrometer with **gas chromatography** (GC), in which the mass spectrometer is used to confirm the identity of compounds as they elute off the GC column. The use of **high-performance liquid chromatography** (HPLC) with the mass spectrometer also will become more routine as advances are made in the interconnecting interface.

26.2 INSTRUMENTATION—THE MASS SPECTROMETER

The **mass spectrometer** (MS) performs three basic functions. There must be a way to ionize the molecules and **produce the charged molecular fragments.** The fragments must then be **separated** according to size, and the separated charged fragments must be **monitored** by a detector. The block diagram below presents the various components of a mass spectrometer (Fig. 26-1).

The diagram of the mass spectrometer contains five major components; the three most important are a way to introduce the sample into the ion source, a mass analyzer to separate the fragments, and a detector to measure the amount. A data computer is not necessary (a chart recorder would work), though it does make the task of collecting and comparing mass spectra much easier. Even lower-quality mass spectrometers are now equipped with a moderately priced personal computer.

Fig. 26-2 depicts the inside of a typical **quadrupole mass spectrometer.** The entire inside of the instrument is closed to the atmosphere and maintained under a strong vacuum. This prevents corrosion of the internal parts by atmospheric oxygen and excludes other gases and compounds in the air that would otherwise contaminate the instrument. The ability to maintain a strong vacuum is an important aspect of obtaining good MS data.

The initial step in operating the MS is to get the sample into the ion source chamber. If the material is a gas or a volatile liquid, it is directly injected into the source region. This requires no special equipment or apparatus and is much the same as injecting a sample into a GC. Thus, this method of introducing the sample to the source is called **direct injection.** With solids that are at least somewhat volatile, the sample is placed in a small cup at the end of a stainless steel rod or **probe.** The probe is inserted into the ion source through one of the sample inlets, and the source is heated until the solid vaporizes. The mass spectrum is then obtained on the vaporized solid material as with the direct injection method.

Both direct injection and direct probe work well

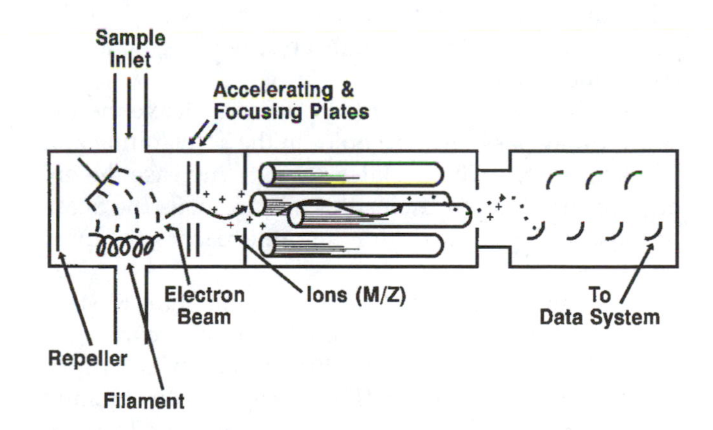

FIGURE 26-2.
Schematic of a typical mass spectrometer. The sample inlets (interfaces) at the top and bottom can be used for direct injection or interfacing to a GC or HPLC.

FIGURE 26-1.
A block diagram of the major components of a mass spectrometer.

with pure samples, but their use is limited when analyzing complex mixtures of several compounds. For mixtures, the sample must be separated into the individual compounds and then analyzed by the mass spectrometer. This is typically done by gas chromatography or high-performance liquid chromatography connected to a mass spectrometer by an interface. The interface removes excess GC carrier gas or HPLC solvent that would otherwise overwhelm the vacuum pumps of the MS. These interfaces will be discussed in more detail in the sections on GC-MS and HPLC-MS. Once in the ion source, the compound is exposed to a beam of electrons emitted from a filament composed of rhenium or tungsten metal. When a direct current is applied to the filament (usually 70 electron volts), it heats and emits electrons that move across the ion chamber toward a positive electrode on the other side. As the electrons pass through the source region, they come in close proximity to the sample molecule and extract an electron forming an ionized molecule. Once ionized, the molecules are unstable and, through a series of reactions, break into smaller molecular fragments. This entire process is called **electron impact (EI) ionization,** although the emitted electrons rarely hit a molecule.

The eventual outcome of the ionization process is both negatively and positively charged molecules of various sizes unique to each compound. When the repeller plate at the back of the source is positively charged, it repels the positive fragments toward the quadrupole mass analyzer. Thus, we only look at the **positive fragments,** although negative fragments are sometimes analyzed.

As the positively charged fragments leave the ion source, they pass through holes in the accelerating and focusing plates. These plates serve to increase the energy of the charged molecules and to focus the beam of ions, so that a maximum amount reaches the mass analyzer.

The **mass analyzer** is the component of the MS that separates or filters the ions from the source according to size and charge. Historically, three major types of filters have been used: **time of flight, magnetic sector,** and **quadrupoles.** Both magnetic sector and quadrupoles are used quite extensively in current instruments, although quadrupoles are becoming increasingly popular because of their small size and because they are easily computer controlled. Some references presented at the end of this chapter discuss the magnetic sector analyzer in detail.

A diagram of the **quadrupole analyzer** is presented in Fig. 26-2. As can be seen, it gets its name from the array of four parallel stainless steel rods. As ions traverse the length of the passageway between the rods, they are exposed to an oscillating electrostatic field caused by the application of both a **radio frequency** (RF) and a **direct current (DC) voltage** to the quadrupoles. The ions acquire an oscillation with a certain amplitude unique for each fragment. If the amplitude of the oscillation is too large, the ions eventually strike one of the rods and do not reach the detector at the end. However, if the amplitude is stable, the ions reach the detector and a mass fragment is recorded. For any given RF and DC voltage applied to the rods, only one ion of a particular mass-to-charge ratio will be detected. Scanning very rapidly through the various combinations of RF and DC voltage allows any ions that are present (up to about 1000) to be detected. This occurs almost instantaneously, which explains why a fast computer is very helpful in collecting data from the detector. The result of a scan is a table or plot of intensity versus the various RF and DC voltage combinations. All mass spectrometers automatically convert the RF/DC voltage combinations to more useful numbers, the mass-to-charge ratios.

26.3 INTERPRETATION OF MASS SPECTRA

As previously indicated, a **mass spectrum** is a plot (or table) of the intensity of various mass fragments produced when a molecule is bombarded with an electron beam. In most cases, the electron beam used to ionize the molecules is kept at a constant potential of 70 electron volts. This produces sufficient ions without too much fragmentation, which would result in a loss of the higher molecular weight ions. Typical mass spectra include only positive fragments that usually have a charge of $+1$. Thus, the mass-to-charge ratio is the molecular mass of the fragment divided by $+1$, which equals the mass of the fragment. As yet, the mass-to-charge ratio unit has no name and is currently abbreviated by the symbol m/z (older books use m/e). The American Society for Mass Spectrometry is considering naming the unit of mass-to-charge ratio the **thomson** after the late J.J. Thomson, who constructed the first instrument for the determination of the m/z of ions.

A mass spectrum for butane is illustrated in Fig. 26-3. The relative abundance is plotted on the y axis and the m/z is plotted on the x axis. Each line on the bar graph represents an m/z fragment with the abundance unique to a specific compound. The spectrum always contains what is called the **base peak** or **base ion.** This is the fragment (m/z) that has the highest abundance or intensity. When the signal detector is processed by the computer, the m/z with the highest intensity is taken to be 100 percent, and the abundance of all the other m/z ions is adjusted relative to the base peak. The base peak will always be presented as 100 percent relative abundance. Butane has the base peak at an m/z of 43.

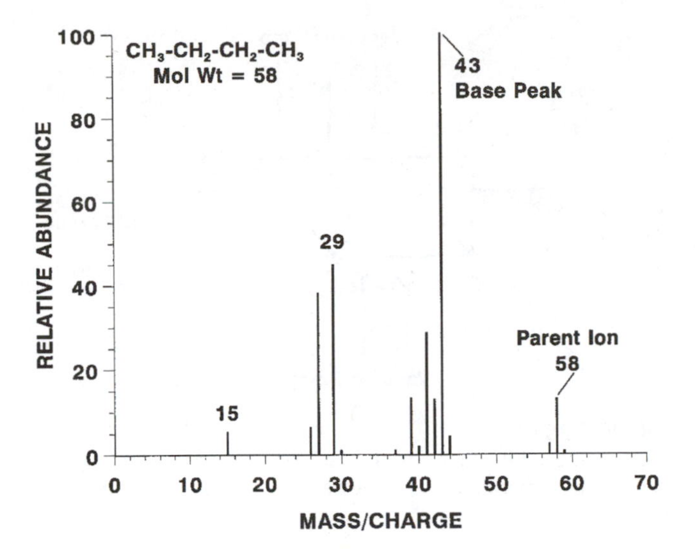

FIGURE 26-3.
Mass spectrum of butane obtained by electron impact ionization.

Another important fragment is the **parent ion** or **molecular ion,** designated by the symbol M^+. This peak has the highest mass number and represents the positively charged intact molecule with an m/z equal to the molecular mass. Because all other molecular fragments originate from this charged species, it is easy to see why it is called the parent ion. It is not always present since sometimes the parent ion decomposes before it has a chance to traverse the mass analyzer. However, a mass spectrum is still obtained, and this only becomes a problem when determining the molecular mass of an unknown. The remainder of the mass spectrum is a consequence of the stepwise cleavage of large fragments to yield smaller ones. The process is relatively straightforward for alkanes, such as butane, making identification of many of the fragments possible.

As indicated before, the initial step in electron impact ionization is the abstraction of an electron from the molecule as electrons from the beam pass in close proximity. The equation below illustrates the first reaction that produces the positively charged parent ion.

$$\text{M} + \text{e (from electron beam)} \rightarrow \text{M}^{\dot{+}} \text{ (parent ion)} + 2\text{e (one electron from the beam and one from the parent ion, M)} \quad (1)$$

The M symbolizes the un-ionized molecule as it reacts with the electron beam and forms a radical cation. The cation will have an m/z equal to the molecular weight. The parent ion then sequentially fragments in a unimolecular fashion. (Note that the parent ion is often written as M^+ where the free electron, symbolized by the dot, is assumed. Regardless, the molecule has lost one electron and still retains all the protons; thus, the net charge must be positive.)

The reactions of butane as it forms several of the predominant fragments are shown below.

$$CH_3\text{-}CH_2\text{-}CH_2\text{-}CH_3 + e \rightarrow CH_3\text{-}CH_2\text{-}CH_2\text{-}CH_3^{\dot{+}} \; (m/z = 58) + 2e \quad (2)$$

$$CH_3\text{-}CH_2\text{-}CH_2\text{-}CH_3^{\dot{+}} \rightarrow CH_3\text{-}CH_2\text{-}CH_2\text{-}CH_2^{+} \; (m/z = 57) + \cdot H \quad (3)$$

$$CH_3\text{-}CH_2\text{-}CH_2\text{-}CH_3^{\dot{+}} \rightarrow CH_3\text{-}CH_2\text{-}CH_2^{+} \; (m/z = 43) + \cdot CH_3 \quad (4)$$

$$CH_3\text{-}CH_2\text{-}CH_2\text{-}CH_3^{\dot{+}} \rightarrow CH_3\text{-}CH_2^{\dot{+}} \; (m/z = 29) + \cdot CH_3\text{-}CH_3 \quad (5)$$

$$CH_3\text{-}CH_2 \rightarrow CH_3^{\dot{+}} \; (m/z = 15) + \cdot CH_2 \quad (6)$$

Many of the fragments for butane result from direct cleavage of the methylene groups. With alkanes, you will always see fragments in the mass spectrum that are produced by the sequential loss of CH_2 or CH_3 groups.

Close examination of the butane mass spectrum in Fig. 26-3 reveals a peak that is 1 m/z unit larger than the parent ion at m/z = 58. This peak is designated by the symbol M + 1 and is due to the naturally occurring isotopes. The most abundant isotope of carbon has a mass of 12; however, a small amount of ^{13}C is also present (1.11%). Any parent ions that contained a ^{13}C or a deuterium isotope would be 1 m/z unit larger, although the relative abundance would be low.

Another example of MS fragmentation patterns is shown for methanol in Fig. 26-4. Again, the fragmenta-

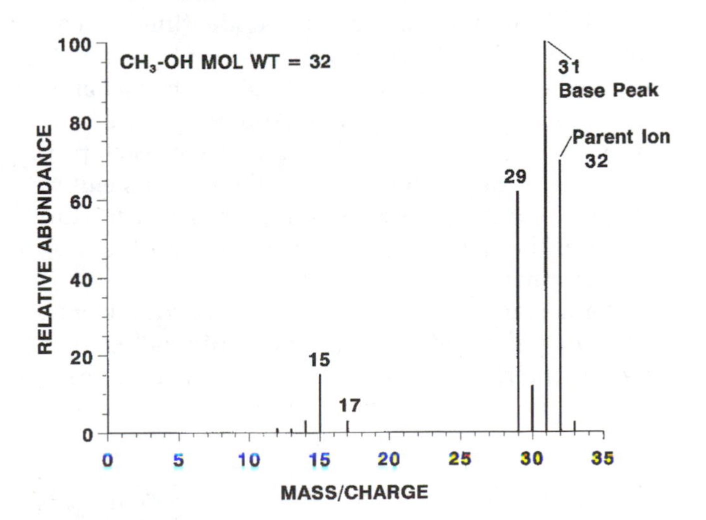

FIGURE 26-4.
Mass spectrum of methanol obtained by electron impact ionization.

tion pattern is straightforward. The parent ion (CH_3-$OH^{+\cdot}$) is at an m/z of 32, which is the molecular weight. Other fragments include the base peak at an m/z of 31 due to CH_2-OH^+, the CHO^+ fragment at an m/z of 29, and the CH_3^+ fragment at an m/z of 15.

So far, only EI types of ionization have been discussed. Another common fragmentation method is **chemical ionization** (CI). In this technique, a gas is ionized, such as methane (CH_4), which then directly ionizes the molecule. This method is classified as a **soft ionization** because only a few fragments are produced. The most important use of CI is in the determination of the parent ion since there is usually a fragment that is 1 m/z unit larger than that obtained with EI. Thus, a mass spectrum of butane taken by the CI method would have a parent ion at m/z = 59 (M + H). As can be seen, the reactions of the cleavage process can be quite involved. Many of the reactions are covered in detail in the books by McLafferty and Davis and Frearson listed in the resource materials section.

26.4 GAS CHROMATOGRAPHY–MASS SPECTROMETRY

Although samples can be introduced directly into the mass spectrometer ion source, many applications require separation before analysis. The rapid development of **gas chromatography–mass spectrometry** (GS-MS) has allowed for the coupling of the two methods for routine separation problems. As discussed in Chapter 30, gas chromatography is a powerful separation method applicable to many different types of food compounds. However, one of the problems with GC analysis is identifying the many compounds that elute off the column. In some cases, there may be GC peaks present that are unknown. A mass spectrometer coupled to GC allows the peaks to be identified or confirmed, and, if an unknown is present, it can be identified using a computer-assisted search of a library containing known MS spectra. Another critical function of GC-MS is to ascertain the purity of each peak as it elutes from the column. Does the material eluting in a peak contain one compound, or is it a mixture of several that just happen to co-elute with the same retention time?

Connecting a GC to a mass spectrometer is straightforward and requires one additional component, the interface. Since a GC operates with a carrier gas, there must be some way to remove or minimize the large amounts of gas going into the MS.

Figure 26-5 shows a diagram of a jet-separator type of interface. As the carrier gas and compounds pass through the interface, the small gas molecules are pumped away, while the compounds of interest move in a straight path toward the ion source.

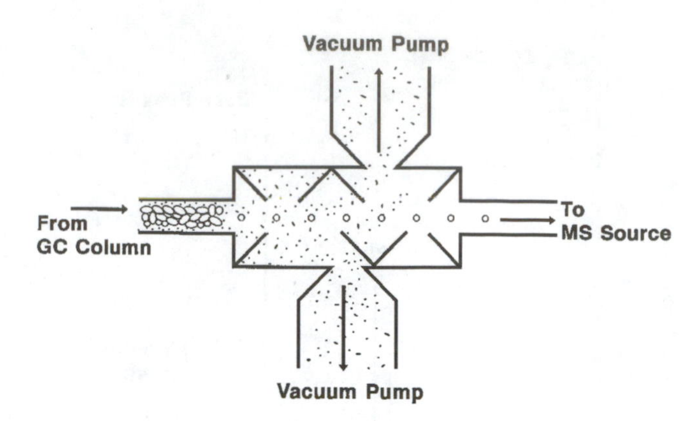

FIGURE 26-5.
Jet separator interface for the GC-MS. The small dots represent carrier gas, and the open circles are the compounds eluting off the columns.

Since the advent of **capillary GC columns,** the interfaces have become much simpler, with many consisting of only a heated region where both the gas and compounds go directly into the MS source. The direct interface is possible since capillary columns require considerably less gas flow. A cutaway view of the inside of a GC-MS is shown in Fig. 26-6. As you can see, the sample flows through the GC column into the interface and then on to be processed by the MS. A computer is used to store and process the data from the MS.

One other modification required for GC-MS is a detector for the ions as they leave the source region. A small ion collector plate is placed at the end of the source and monitors the total ions or total ion current as they go into the quadrupoles. This allows a total-ion GC chromatogram to be recorded in addition to the mass spectra.

An example of the power of GC-MS is shown below in the separation of the methyl esters of several long-chain fatty acids (Fig. 26-7). Long-chain fatty acids must have the acid group converted or blocked with a methyl group to make them volatile. Methyl esters of palmitic (16:0), oleic (18:1), linoleic (18:2), linolenic (18:3), stearic (18:0), and arachidic (20:0) acids were injected onto a column that was supposed to be able to separate all the naturally occurring fatty acids. However, the GC tracing only showed four peaks, when it was known that six different methyl esters were in the sample. The logical explanation is that several of the peaks contain a mixture of methyl esters resulting from poor resolution on the GC column.

The purity of the peaks is determined by running the GC-MS and taking mass spectra at very short increments of time (1 sec or less). If a peak is pure, then the mass spectra taken throughout the peak should be the same. In addition, the mass spectrum can be compared with the library of spectra stored in the computer.

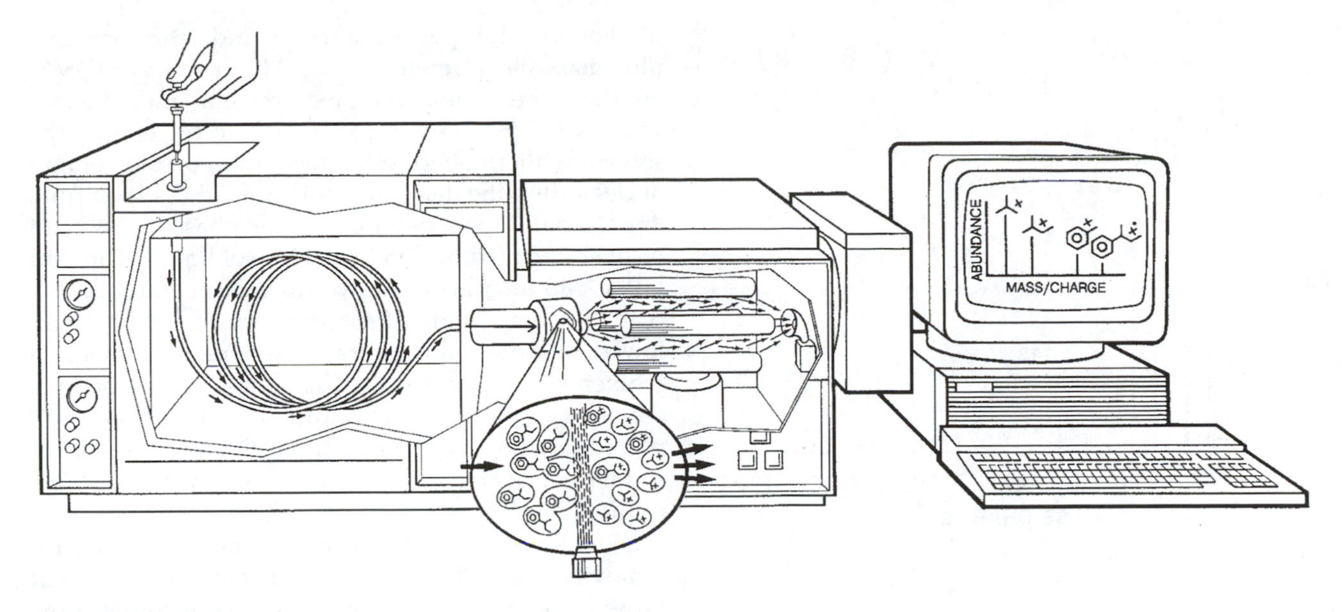

FIGURE 26-6.
Diagram of a gas chromatograph connected to a small tabletop mass selective detector.
Figure courtesy of Hewlett-Packard Co., Avondale, PA.

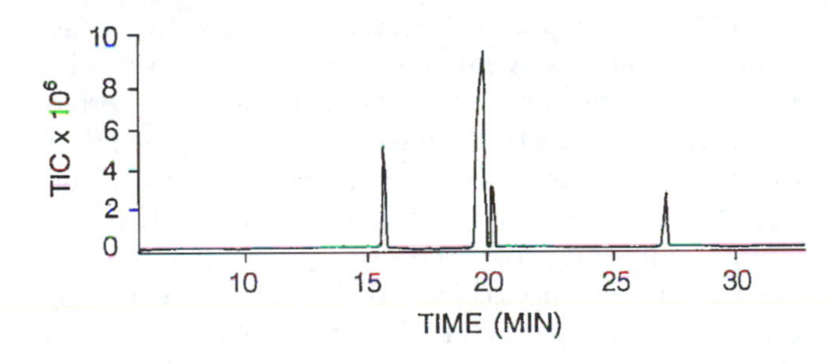

FIGURE 26-7.
Total ion current (TIC) GC chromatogram of the separation of the methyl esters of six fatty acids. Detection is by electron impact (EI) ionization using a direct capillary interface.

The **total ion current** (TIC) **chromatogram** of the separation of the fatty acid methyl esters is shown in Fig. 26-7. There are four peaks eluting off the column between 15.5 min and 28 min. The first peak at 15.5 min has the same mass spectrum throughout, indicating that only one compound is eluting. A computer search of the MS library gives an identification of the peak to the methyl ester of palmitic acid. The mass spectra shown in Figure 26-8 compare the material eluting from the column to the library mass spectrum.

Most of the fragments match, although the GC-MS scan does have a lot of small fragments not present on the library mass spectrum. This is common background noise and usually does not present a problem. Looking at the rest of the chromatogram, the data indicate that the peaks at both 20 min and 27 min contain only one component. The computer match identifies the peak at 20 min as stearic acid, methyl ester and the peak at 27 min as arachidic acid, methyl ester. However, the peak located at 19.5 min is shown to have several different mass spectra, indicating impurity or co-eluting compounds.

In Fig. 26-9, the region around 19 min has been enlarged. The arrows indicate where different mass spectra were obtained. The computer identified the material in the peak at 19.5 min as linoleic acid, methyl ester; the material at 19.7 min as oleic acid, methyl ester; and the material at 19.8 min as linolenic acid, methyl ester. Thus, as we originally suspected, several of the methyl esters were co-eluting off the GC column. This example illustrates the tremendous power of GC-MS used in both a quantitative and a qualitative manner.

26.5 LIQUID CHROMATOGRAPHY–MASS SPECTROMETRY

Until about 10 years ago, the only way to obtain a decent mass spectrum of material from high-performance liquid chromatography (HPLC) separations was to collect fractions, evaporate off the solvent, and introduce the sample into a conventional MS by direct injection or direct probe. Although this method was adequate, the direct on-line coupling of the two instruments was a tremendous advantage in terms of time and ease of operation. As with GC-MS, the major component necessary for coupling an HPLC to a mass spectrometer is the interface. However, more demands are placed on the interface with HPLC because it must remove liquid solvents, which is not an easy task.

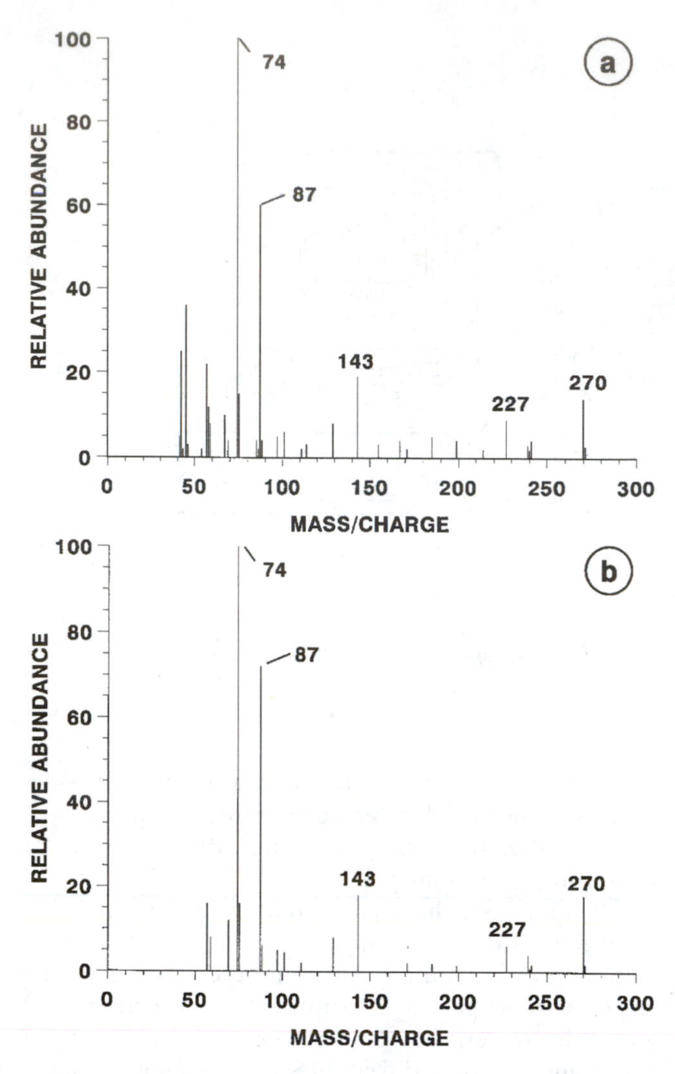

FIGURE 26-8.
Mass spectra of (a) the peak at 15.5 min in the TIC chromatogram shown in Fig. 26-7 and (b) the methyl ester of palmitic acid from a computerized MS library.

For a **high-performance liquid chromatography–mass spectrometry** (HPLC-MS, or simply LC-MS) interface, the same overall requirements must be met as for GC-MS. There must be a way to remove the solvent without altering the separation on the column or degrading the determination of the mass spectrum. An MS only operates in the vapor phase and would be overwhelmed by a large volume of liquid. Thus, all HPLC-MS interfaces remove the solvent carefully, so that the compounds of interest are retained and allowed to move into the ion source region. This is a momentous task since a great variety of compounds are assayed by HPLC. In addition, many of the compounds are not very volatile (if at all) and are sensitive to thermal degradation, which is why we use HPLC in the first place.

Unlike GC-MS, there is no universal interface for removing the HPLC solvent. There are many different types of interfaces, though several have come to the forefront of use in recent years. Two of the most commonly used interfaces are the thermospray and the particle beam. Both provide different types of mass spectra and work for different types of compounds.

The **thermospray interface,** as its name suggests, removes heated solvent as it is sprayed into the ion source. A simple diagram of the thermospray interface is shown in Fig. 26-10. As the effluent leaves the HPLC instrument, it is pumped through a small piece of tubing into the thermospray probe. The effluent moves along the heated probe and exits at the tip as a spray. The solvent is removed from the spray by several vacuum pumps as small particles of sample plus solvent travel through the ion source to be ionized. The ionization process is considerably different from what occurs with GC-MS in that very little fragmentation occurs.

An important aspect of thermospray-MS (TS-MS)

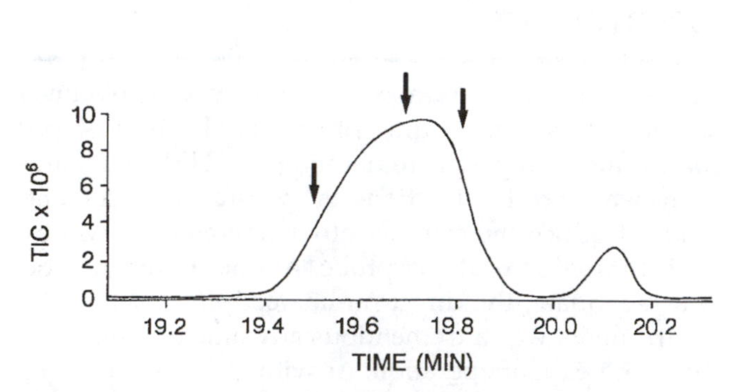

FIGURE 26-9.
Enlargement of the region 19.2–20.2 min from the TIC chromatogram shown in Fig. 26-7. Arrows indicate where mass spectra were obtained.

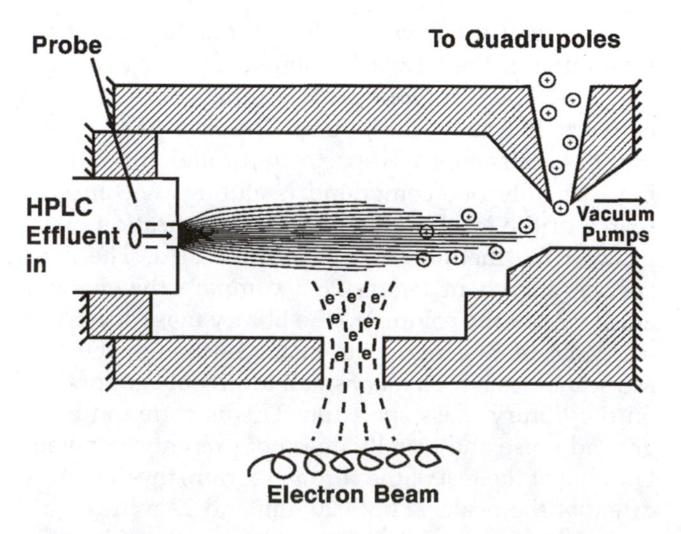

FIGURE 26-10.
Schematic of a thermospray LC-MS interface. Figure courtesy of Hewlett-Packard Co., Avondale, PA.

is that some type of volatile buffer must be present when the HPLC effluent exits the TS tip. The ionization is a reaction between the volatile buffer and the compounds as the solvent evaporates from the droplets in the source region. Ammonium acetate is commonly used; it produces ammonium ions that react in several ways. The ammonium ions can provide either H^+ or NH_4^+ to the molecule, which yields an $M + H^+$ parent ion or an $M + NH_4^+$ parent ion. In any case, the molecules do not fragment much, producing a very simple mass spectrum. The electron beam filament shown in Fig. 26-10 is used to enhance the ionization and increase the ions, although it is not necessary for many compounds.

The **particle beam** or **monodisperse aerosol generator** is another type of HPLC interface that is used for compounds that are slightly volatile. It functions very much like the jet separator used in GC-MS. As shown in Fig. 26-11, the design is somewhat different in that helium is added to the HPLC effluent as it enters the desolvation chambers. As the spray travels through the chambers and separators, solvent is removed, leaving behind a microscopic particle essentially free of solvent. The particles continue down a transfer tube and enter the MS source chamber, where ionization occurs by conventional electron impact. The resulting MS spectrum is essentially the same as that obtained with GC-MS. The major drawback of the particle beam interface is the requirement that compounds of interest have some volatility.

Improvements in LC-MS interfaces are continuously being made, although no universal interface for the various types of compounds is available yet. The method is now becoming routine, although it is still difficult to compare results when using different types of interfaces. As technological advances are made, we can expect the use of LC-MS to grow, since the MS is a universal detector for both qualitative and quantitative information.

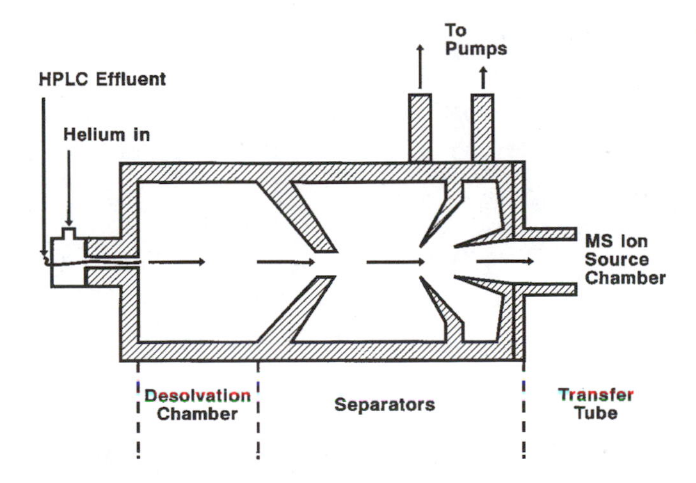

FIGURE 26-11.
Schematic of a particle beam LC-MS interface. Figure courtesy of Hewlett Packard Co., Avondale, PA.

26.6 APPLICATIONS

The use of mass spectrometry in the field of food science is well established, but many areas of improvement are still in their infancy. While GC-MS has been used for years, only recently have low-priced reliable units been available as standard lab instrumentation. Routine use of GC-MS can be expected to grow as more public and private labs have access to the instruments. LC-MS, on the other hand, is still in the developmental stages, as researchers look for better ways to interface the two instruments and improve the ionization process. Nonetheless, a mass spectrum can still be obtained for most compounds eluting off an HPLC column, making the technique very useful.

There are many different applications of mass spectrometry in food science. One of the most thorough treatments on this subject is the book by John Gilbert listed in the resource materials section of this chapter. Readers should consult this book concerning specifics on a particular food component or a certain type of food. The coverage is excellent, although somewhat outdated since many developments have occurred since 1986.

In considering the application of GC-MS or LC-MS in food systems, note that if a compound can be separated by a GC or LC method, then chances are good that a mass spectrometer can be used. For years, the mass spectrometer was used only in a qualitative manner, to check the purity of eluting peaks or for compound identification. With smaller units, the use of MS as a universal detector has gained wide acceptance. The advantage of utilizing the MS as a detector is that only certain ions need to be monitored, which makes it a selective detector. This technique is currently used extensively for pesticide analysis.

The use of **selected ion monitoring** (SIM) is especially helpful in LC-MS, where analysis is often limited by the lack of a suitable detector. Use of the SIM mode of detection is often the only way to detect some compounds eluting off the column. For example, fatty acids can be measured directly by HPLC, but unless the concentrations are high, an ultraviolet or refractive index detector will not pick them up. LC-MS will allow for the assay of trace amounts present in effluent. LC-MS has become especially helpful in the analysis of nonvolatile pesticides, amino acids, lipids, and sugars.

26.7 SUMMARY

Mass spectrometry is fairly simple when examined closely. The basic requirements are to (1) somehow get the sample into a ionizing chamber where ions are produced, (2) separate the ions formed by magnets or quadrupoles, (3) detect the m/z and amount, and (4) and output the data to some type of computer.

Since the qualitative and quantitative aspects of mass specs are so powerful, they are routinely coupled to GCs and HPLCs. The interface for GCs is very versatile and easy to use. However, interfacing an MS to an HPLC still presents problems because there is no universal interface. We will continue to see developments in the HPLC interface, although the technology now exists to analyze many different types of compounds. Future developments will expand the use of mass spectrometry to just about any type of chromatographic separation method.

ACKNOWLEDGMENT

Contribution No. 92-23-B from the Kansas Agircultural Experiment Station, Kansas State University, Manhattan, KS.

The author thanks his graduate student, Angelia Krizek, for her helpful suggestions and for drawing part of the figures for this chapter.

26.8 STUDY QUESTIONS

1. What are the basic components of a mass spectrometer?
2. What are the unique aspects of data that a mass spectrometer provides? How is this useful in the analysis of foods?
3. What is EI ionization? What is CI ionization?
4. What is a particle beam interface? What is a thermospray interface? What interface would you use for an HPLC run of a nonvolatile compound using a reversed-phase column, water, and methanol mobile phase?
5. What is the base peak on a mass spectrum? What is the parent ion peak?
6. What are the major ions (fragments) expected in the EI mass spectrum of ethanol (CH_3-CH_2-OH)?

26.9 RESOURCE MATERIALS

Davis, R., and Frearson, M. 1987. *Mass Spectrometry.* John Wiley & Sons, New York, NY. One of the best introductory texts on mass spectrometry. The authors start at a very basic level and slowly work through all aspects of MS, including ionization, fragmentation patterns, GC-MS, and LC-MS.

Gilbert, J. (Ed.) 1987. *Applications of Mass Spectrometry in Food Science.* Elsevier Science Publishing Co., New York, NY. This review book gives thorough coverage to all the uses of MS in food science. It is written for the researcher but is relatively easy to read. Several of the chapters contain sections on analysis of various food components.

Macrae, R. (Ed.) 1988. *HPLC in Food Analysis,* 2nd ed. Academic Press, Inc., New York, NY. In addition to being a comprehensive reference on the use of HPLC in food analysis, this book contains a section (Chapter 13) on the application of LC-MS. Both instrumentation and applications are discussed.

McLafferty, F.W. 1980. *Interpretation of Mass Spectra,* 3rd ed. University Science Books, Mill Valley, CA (available from Aldrich Chemical Company, Milwaukee, WI). A classic book on how molecules fragment in the ion source. Contains many examples of different types of molecules.

Pavia, D.L., Lampman, G.M., and Kriz, G.S., Jr. 1979. *Introduction to Spectroscopy.* Holt, Rinehart and Wilson, Philadelphia, PA. Chapter 6 is a good introduction for readers unfamiliar with mass spectrometry.

Silverstein, R.M., Bassler, G.C., and Morrill, T.C. 1981. *Spectrometric Identification of Organic Compounds,* 4th ed. John Wiley & Sons, New York, NY. An introductory text for students in organic chemistry. Chapter 2 presents mass spectrometry in an easily readable manner. Contains many examples of the mass spectra of organic compounds.

CHAPTER

27

Nuclear Magnetic Resonance and Electron Spin Resonance

Thomas M. Eads and Eugenia A. Davis

27.1 INTRODUCTION

Nuclear magnetic resonance (NMR) is a branch of spectroscopy that takes advantage of the exquisite sensitivity of magnetic atomic nuclei (nuclear spins) to their molecular or ionic environment. Electron spin resonance (ESR) is similar but deals with unpaired electrons (electron spins) that occur in organic free radicals (molecules) and in paramagnetic metal ions. The magnetic behavior of spins reflects molecular structure and molecular motion or ionic structure and ion movement. Thus NMR is used for analysis of chemical composition, physical state of food components, and food component mobility. ESR is used to detect free radicals that are naturally occurring in food, radicals produced in food by processing or irradiation, and paramagnetic ions. ESR is also used for measuring the physical state of molecules and molecular motion in foods and can be used to measure oxygen concentration. Another name for ESR is EPR, electron paramagnetic resonance.

Nuclei and electrons that are magnetic occur naturally in all matter and thus can be observed easily in foods. They are observed via their resonance with electromagnetic radiation, as in ultraviolet (UV), visible (Vis), or infrared (IR), except that the sample must first be immersed in a magnetic field. Because radio frequencies (used for NMR) and microwave frequencies (for ESR) penetrate food materials fairly well, detection is independent of optical transparency, unlike optical absorption (UV, Vis, IR), and provides information about material below the surface, unlike the corresponding reflectance spectroscopies. It is usually unnecessary to disrupt the sample, separate its components, or treat them chemically before analysis by NMR or ESR. Thus magnetic resonance is nondestructive and can often be used on samples as is. This is a very powerful advantage. The major tradeoffs for NMR are that it may not have the *sensitivity* of other spectroscopic methods and that it may not have the *selectivity* of chromatographic methods.

There are established magnetic resonance methods like moisture, oil, solid-to-liquid ratio, or solid content carried out on **benchtop NMR** instruments. Most are time-domain instruments that do not provide chemical information. A new class of benchtop instrument gives **NMR spectra** containing chemical information from liquid foods or from liquid phases in heterogeneous foods. Thus this chapter treats both time-domain NMR and frequency-domain NMR. Research in frequency-domain NMR is producing powerful new methods. While these have not yet been standardized, the food analyst can turn research methods into analytical protocols and even create his or her own NMR techniques. First, the analyst should understand the principles.

In this chapter, the basic vocabulary and principles of magnetic resonance are presented, benchtop NMR and ESR instruments are described, practical considerations are given, the main kinds of analyses are illustrated, and an introductory bibliography is given.

27.2 PRINCIPLES

Analysis by magnetic resonance requires awareness of basic principles. Results depend very much on instrumental settings and on sample condition. Magnetic resonance has a mathematical exposition, which we skip except for a few formulas that you will need. While not entirely accurate because it is so simplified, the following approach conveys indispensable concepts. More complete treatments are given in Resource Materials. The magnetic resonance measurement is summarized below.

1. The sample is first inserted into a probe immersed in a strong magnetic field created by a permanent magnet, electromagnet, or superconducting solenoid magnet.

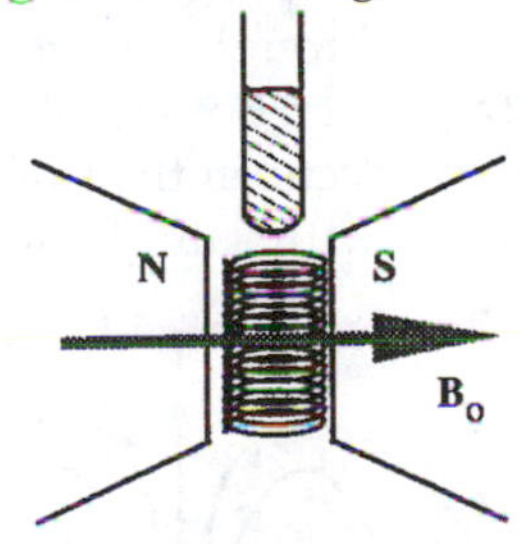

2. Nuclei (or electrons) **spin** like a top, and so they have **angular momentum P.** They also have charge. A spinning charge produces a **magnetic moment μ,** like a tiny bar magnet with a north and south pole. The magnetic moments of the spinning nuclei (or electrons) in the sample are at first randomly oriented but become aligned in the strong magnetic field. The sum of all the magnetic moments gives the sample a weak **magnetization M.**

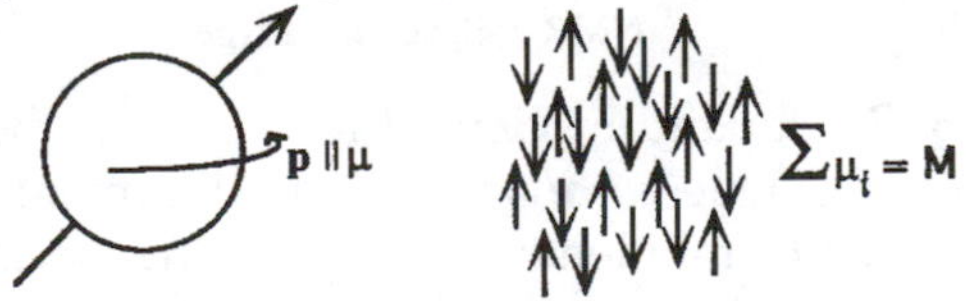

3. The magnetic moments also **precess** at high frequencies in the magnetic field: **radio frequencies** (megahertz, MHz) for nuclei and **microwave frequencies** in the radar range (gigahertz, GHz) for electrons. Precession is like the wobbling of a spinning top. The **pre-**

cession frequency ω_L is proportional to the **static field strength B_0.** The proportionality constant is the **gyromagnetic ratio** γ.

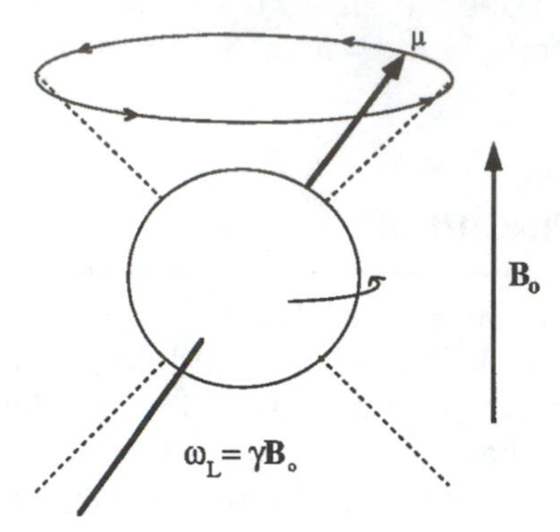

4. The **energy** of an **oscillating** magnetic field $B_1(t)$ can be absorbed by precessing nuclei (or electrons) if the oscillation is **on resonance,** that is, has the same frequency as precession. This is the classical description of magnetic resonance.
5. The oscillating field is produced by running an alternating current at radio frequencies through a wire **coil** (an inductor) that surrounds the sample (NMR) or by microwaves transmitted through a **wave guide** (ESR) into a **resonator cavity.** Both excitation and detection (see below) occur in the same circuit, which is bundled up with sample holder in the **probe.**

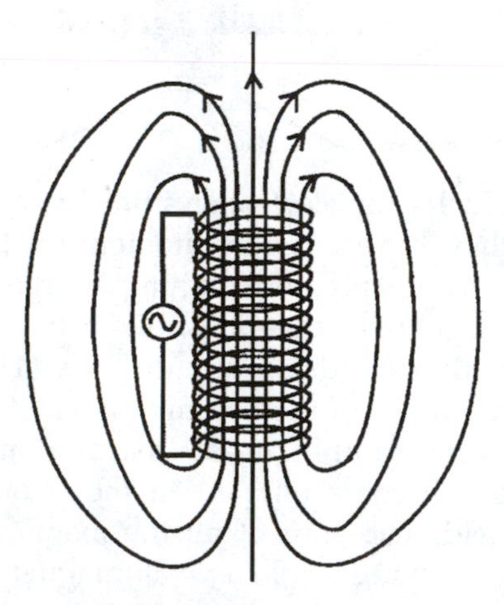

NMR coil, solenoid type

6. The quantum view of resonance is the following. Due to quantization, alignment of magnetic moments can be either parallel or antiparallel to B_0. Parallel is slightly favored, so there are now two subsets of magnetic moments with a very small energy difference, ΔE. Transitions can now occur, corresponding to reorientation of magnetic moments, if energy equal to ΔE is provided. This can come from $B_1(t)$, whose frequency ν must satisfy $\Delta E = h\nu$.

This ν turns out to be the same as the classical precession frequency $\nu_L = \omega_L/2\pi$.

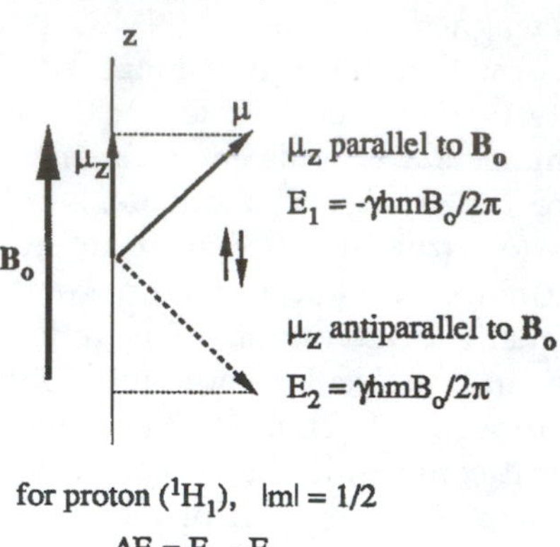

for proton ($^1H_1$), $|m| = 1/2$

$$\Delta E = E_2 - E_1$$
$$= \gamma h B/2\pi$$
$$= h\nu_{res}$$
$$\nu_{res} = \Delta E/h = \gamma B/2\pi$$

7. The oscillating magnetic field can be regarded as the sum of two counter-rotating fields, one of which can cause resonance (a). When the oscillating field is on, magnetic moments get **in phase** with each other. Now the sum over all of them is an appreciable magnetization, represented by a vector **M** (not shown). This magnetization follows the oscillating field in the **rotating frame.** The rotating frame (x′, y′, z) moves with $B_1(t)$(b).

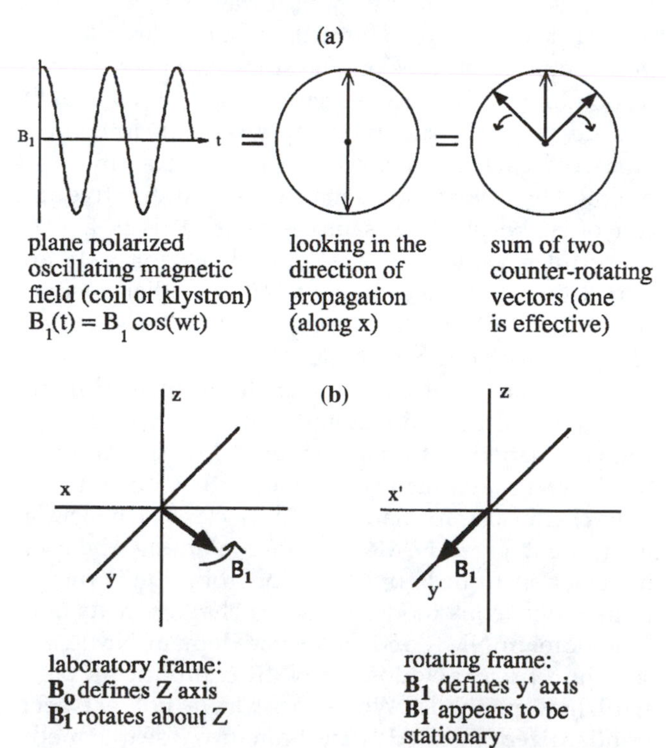

8. In **continuous wave NMR** and **ESR,** the excitation field is weak, but at resonance, nuclei (or

electrons) align with it (a) and the sample absorbs measurably. An **absorption spectrum** (b) is produced as the frequency of the oscillating field is swept. Alternately, the strength of the static field can be swept while holding the oscillating field frequency constant. The latter is common in ESR.

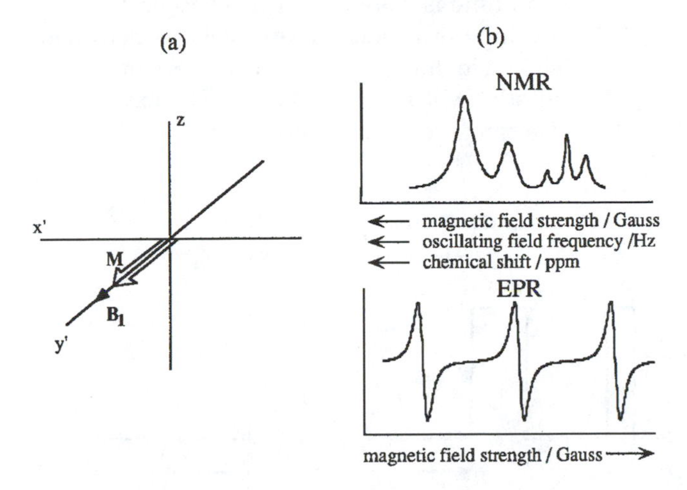

9. In **pulse NMR** and **pulse ESR,** the excitation field is strong and is applied in a short pulse (microseconds). The sample magnetization precesses *about it* instead of aligning. The angle of rotation $\Delta\theta$ is easily calculated or measured. The time to rotate is called the **pulse width,** Δt. A rotation of $\Delta\theta = 90°$ puts the magnetization onto the x′ axis. Detection is then in the **transverse plane** (x,y).

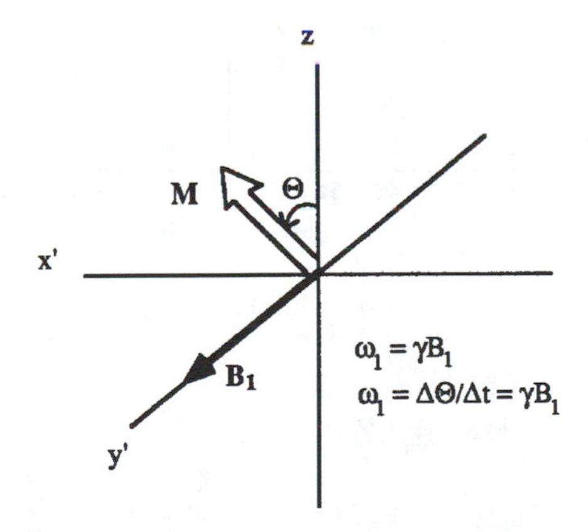

10. After the pulse, the magnetization continues to precess in the transverse plane because it still responds to the static field. This induces an electromotive force (emf) in the measuring coil. The principle of detection in pulse NMR is thus **electromagnetic induction.** For ESR, the perturbation of the electromagnetic energy in the **resonator cavity** surrounding the sample is measured. The NMR signal, a voltage V, oscillates at the precession frequency.

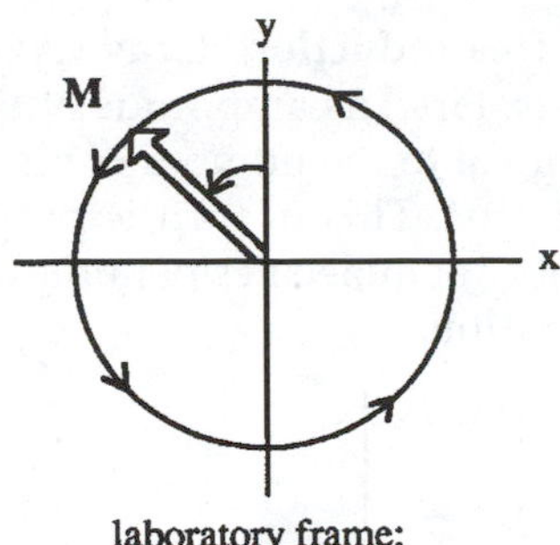

11. In NMR, the signal frequency provides chemical structure information because precession frequencies of nuclei are modified by **magnetic shielding** provided by electron clouds in the molecule (the **chemical shift effect**) and by **magnetic coupling** with nearby nuclei (the **spin-spin, J-coupling,** or **multiplet** effects). Splittings of unpaired electrons in ESR also occur due to coupling with nearby electrons and nuclei, and this sometimes enables signals from different free radicals and paramagnetic ions to be identified, but there is no real chemical shift effect in ESR.

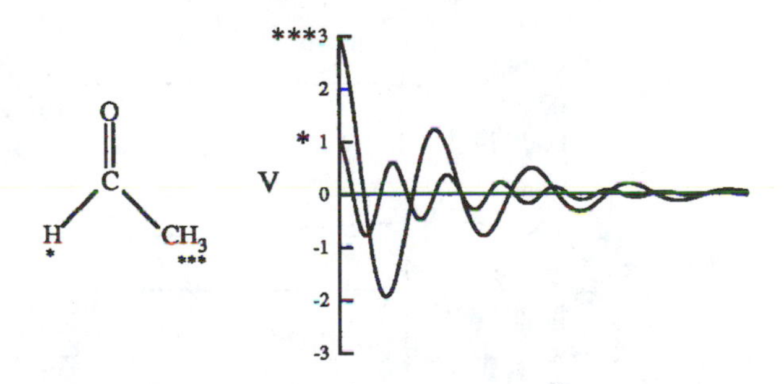

12. In pulse NMR, the magnetization dies away due to **nuclear relaxation processes.** At any instant, the magnetization has a **longitudinal component** (parallel to static field) and a **transverse component** (in the plane perpendicular to static field). Since the induction signal is detected in the transverse plane, the signal (see no. 11 above) reflects the decay of the transverse component of magnetization.

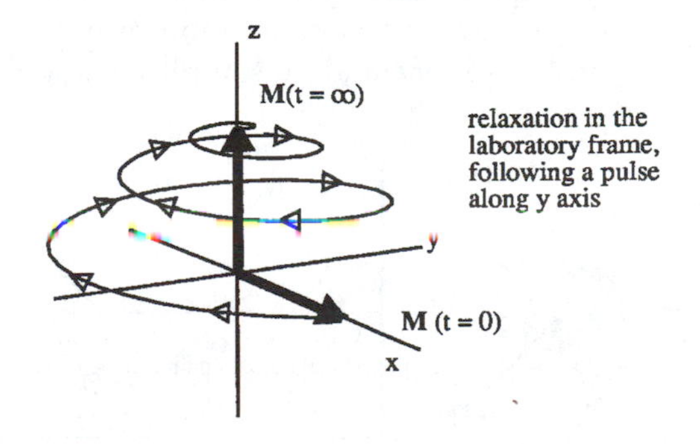

13. The **free induction decay** (FID) provides two things. First, its amplitude at time zero is proportional to the number of resonant nuclei (or electrons). This in turn is proportional to the number of molecules per unit volume, the **concentration.**

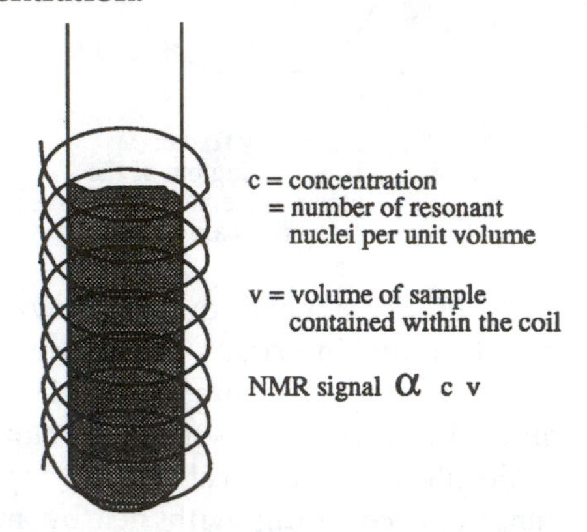

14. Second, the decay *rate* gives physical information because it is modified by **molecular motion.** In solidlike phases, motion is restricted and slow, producing a rapidly decaying signal. In liquidlike phases, motion is unrestricted and fast, yielding a slowly decaying signal. Note the differences in time scales in the two graphs.

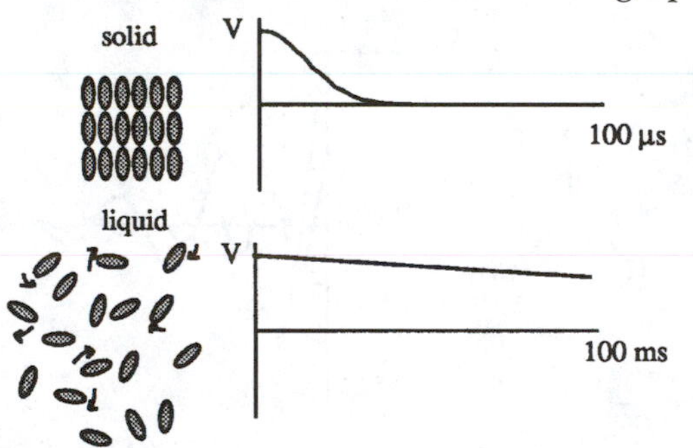

15. This is the basis for measurement of solid content, liquid content, and S/L ratio by NMR (e.g., solid fat content or SFC). The signal from a sample containing solid and liquid is the superposition of two decays. Such measurements do not distinguish the chemical identify of the component. The figure shows the direct method for SFC. In the indirect method, only the liquid intensity V_L is measured, and this is compared with a calibration graph obtained on a series of standard samples of known solid content.

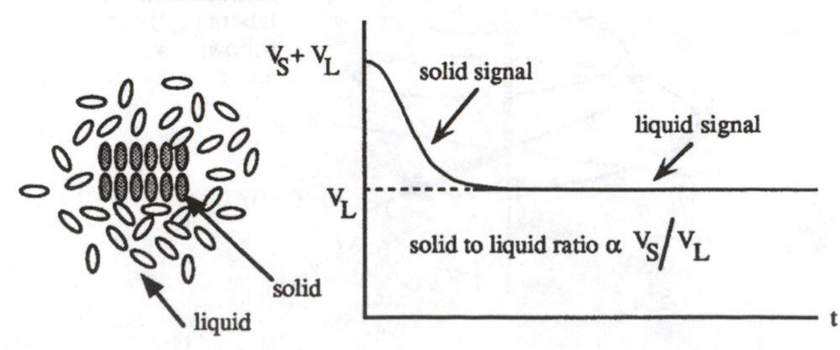

16. The decay rate of the transverse magnetization (the NMR signal) is called **transverse** or **spin-spin relaxation.** It contains two main contributions: that due to nuclear (or electron) relaxation and that due to **inhomogeneity** of the static magnetic field. The time constant describing the FID curve is called $\mathbf{T_2^*}$. The true nuclear relaxation time is called $\mathbf{T_2}$. It is obtained from the envelope of echoes in **spin echo** experiments (shown in the figure), which are very important in both NMR and ESR analysis. This experiment also can give solid-liquid ratios.

Carr-Purcell Meiboom Gill Spin Echo

(a) pulse sequence and NMR signal

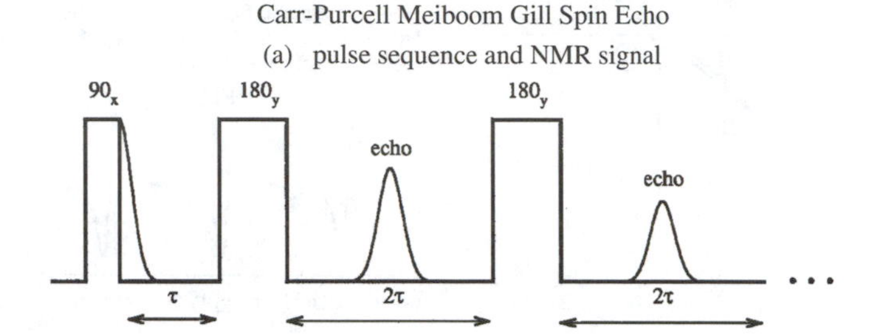

(b) formation of an echo

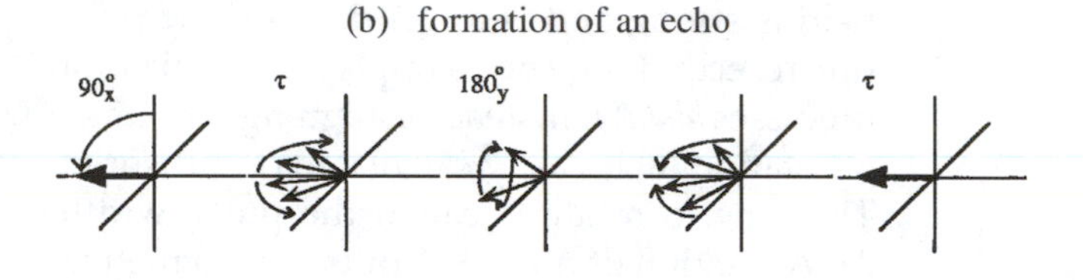

(c) plots of echo intensity decay

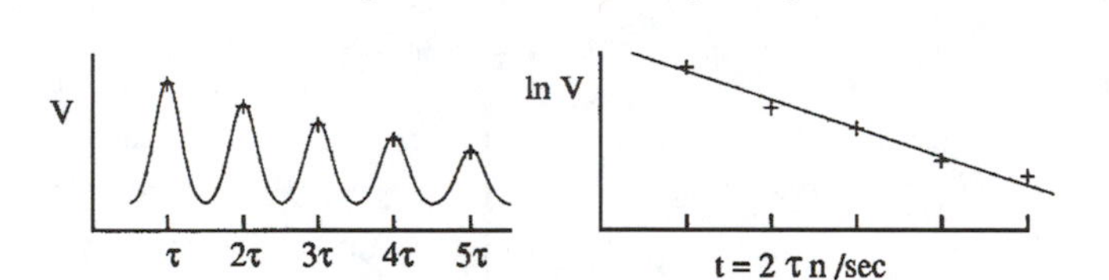

(d) calculation of T_2

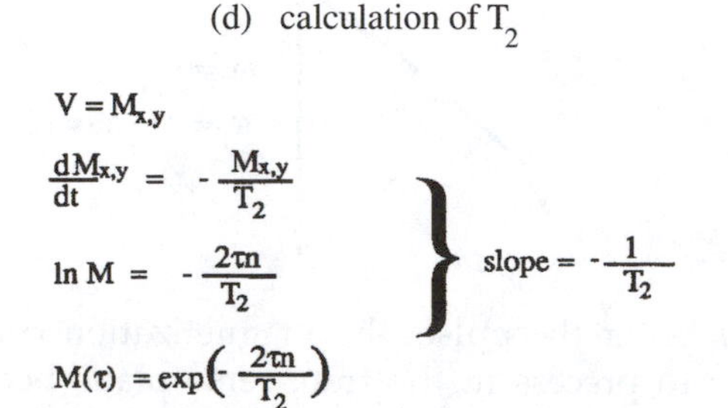

17. The time-domain signal in pulse NMR or ESR can be used *as is* for physical information but must be transformed for chemical information. The **magnetic resonance spectrum** is obtained by applying the **Fourier transform** (FT) to the time-domain signal. This is the difference between **time-domain** and **frequency-domain** (FT NMR) spectroscopy.

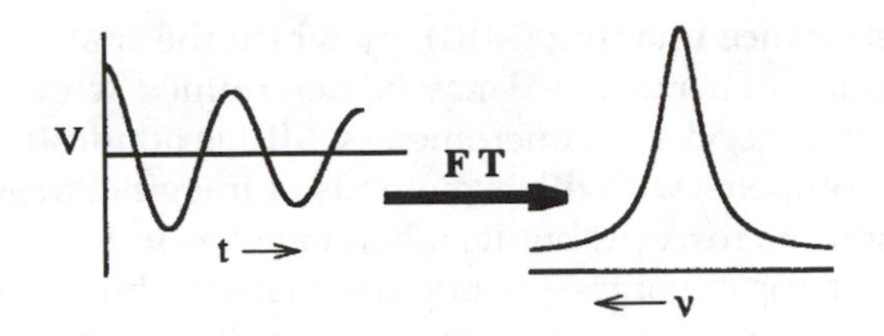

18. In NMR, the decay of a solid signal is fast, T_2 is short (tens of microseconds) (a), and the Fourier transform is a spectrum with a single **broad resonance** of width tens of kilohertz (b). It is not possible to get chemical information in this way since line widths are much larger than separations between resonances.

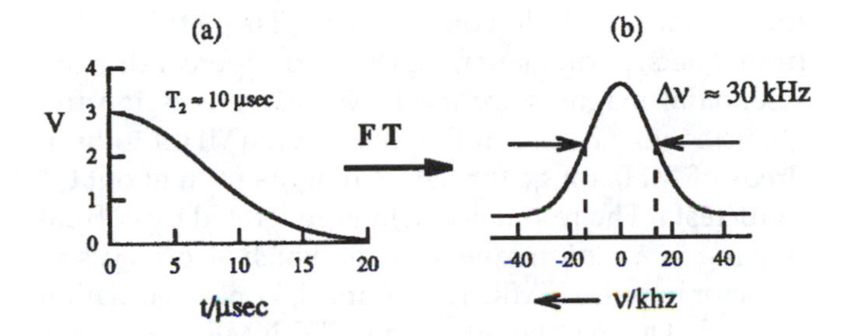

19. The decay of a liquid NMR signal is slow, T_2 is long (tenths of seconds to seconds) (a), and the Fourier transform is a spectrum with many **narrow resonances** of width tenths to tens of hertz (b). It is possible to get chemical information since line widths are smaller than separations between resonances.

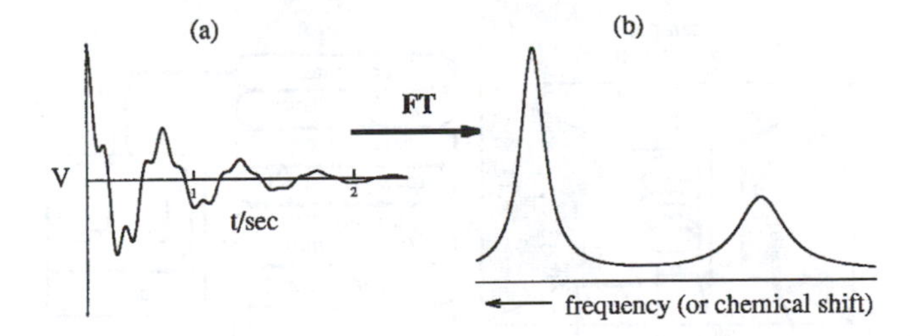

20. After a pulse, the **longitudinal** component of magnetization also returns to its equilibrium value along the static field direction. This occurs continuously during a continuous wave (CW) experiment. This is called **spin-lattice** or **longitudinal relaxation**, and the time constant for return is T_1. Since it is not in the transverse plane, longitudinal magnetization cannot be measured directly. Instead, the **inversion recovery** (shown) or **saturation recovery** experiment is used (not shown).

T_1 Measurement by Inversion Recovery

(a) pulse sequence and NMR signal

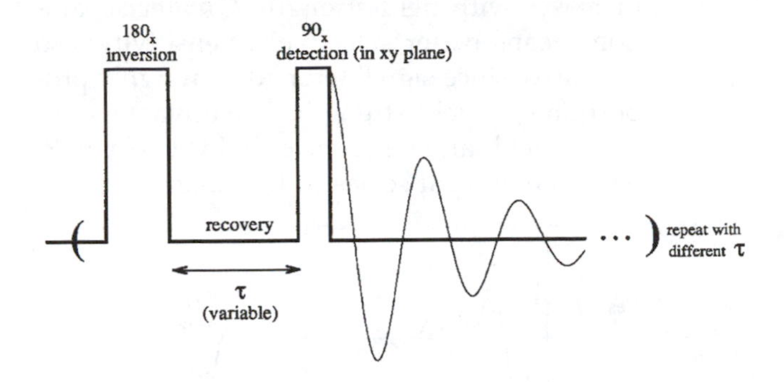

(b) path of longitudinal magnetization

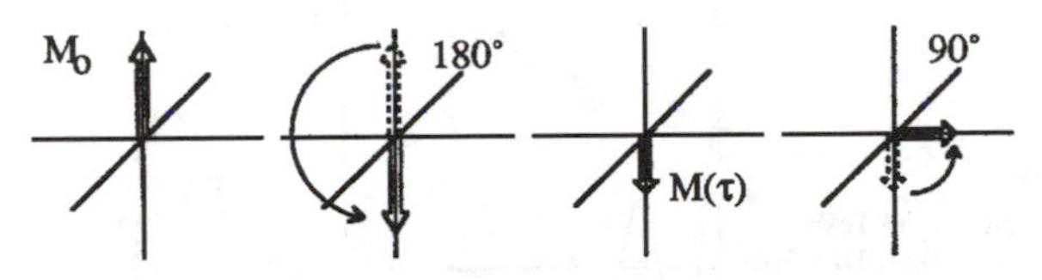

(c) plot of magnetization recovery

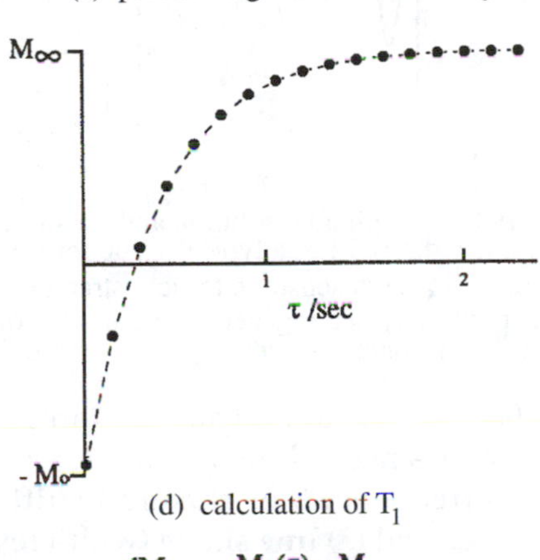

(d) calculation of T_1

$$\frac{dM_z}{dt} = -\frac{M_z(\tau) - M_o}{T_1}$$

$$M_z(\tau) = M_o\left\{1 - 2e^{-\left(\frac{\tau}{T_1}\right)}\right\}$$

21. If the sample has many components, each produces its own **pattern of resonances,** and with certain conditions met, they appear in proper proportions in the spectrum, enabling quantitative chemical analysis. When peaks are separated well enough, the method is called **high-resolution NMR.**

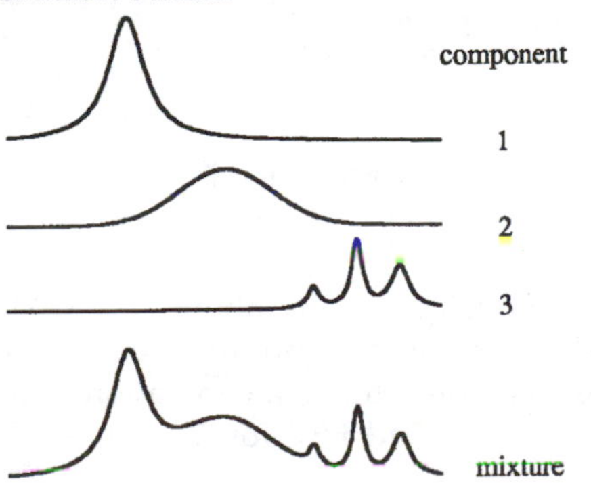

22. Since frequency is proportional to field strength, the separation between resonances increases with field strength. Greater separation means better selectively. Sensitivity also increases, since signal strength is roughly proportional to field strength. The advantages of higher fields and frequencies hold also for ESR. Note the vertical scales in the diagram.

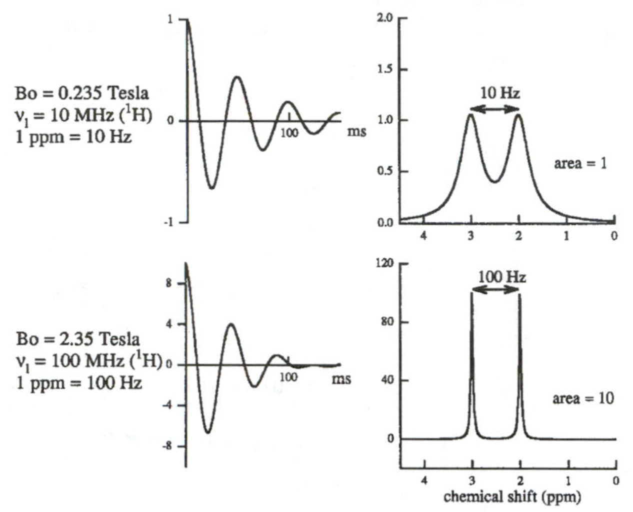

Effect of field strength on resolution and sensitivity. In each case, the linewidth is 5 Hz, a typical value for liquids in food. Linewidth of ^{1}H is not sensitive to field strength, unless the sample is physically heterogeneous. Linewidth of ^{13}C is more sensitive to field strength, especially at fields > 2.3 Tesla.

23. Each peak in a spectrum has three main characteristics needed in analysis: (1) **area** (amount) (2) **frequency** ν or **chemical shift** δ (chemical type), and (3) **line shape** (width reveals liquid, viscous, or solid; splittings reveal molecular structure). Line width is related to T_2* by the relation $\Delta\nu = 1/\pi T_2^*$.

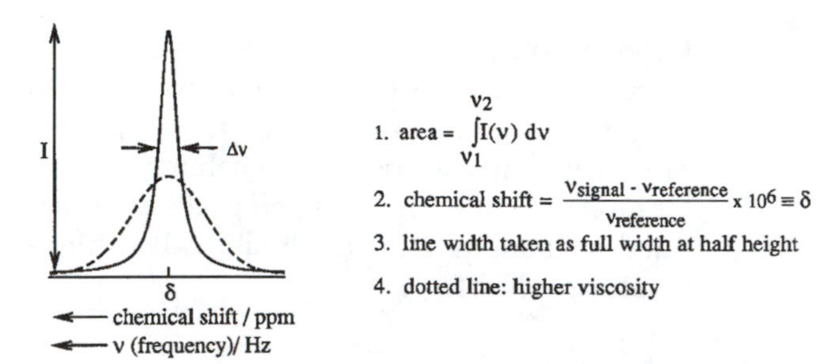

In its more sophisticated forms, NMR is an adjunct to chromatography and mass spectrometry for verification of chemical structure of isolated compounds of molecular weights up to about 20,000 kD. NMR can provide the total structure of a compound. NMR can be used as a detector in a separation scheme.

Both NMR and ESR form the basis for **magnetic resonance imaging** (MRI), by which the spatial distribution of a substance may be determined with resolution as good as 10 micrometers. MRI is nondestructive. It competes with other methods of imaging, including light microscopy, but its advantage lies in the fact that the basis of contrast is not just anatomy, but also flow, diffusion, or molecular motion. Analysis of food materials by MRI will become increasingly important.

27.3 INSTRUMENTATION

27.3.1 NMR Instruments

27.3.1.1 Overview

(See Fig. 27-1.) NMR can be a **continuous wave** (CW), **rapid scan** (a variation on CW), or **pulse** method. Electromagnets, permanent magnets, and superconducting solenoidal magnets are used. Nuclear precession frequencies $\gamma B_0/2\pi$ are in the megahertz (MHz) to hundreds of MHz range for field strengths from about 0.1 to 10 **Tesla.** The resonance field is generated by current flowing in a radiofrequency coil, which also serves as detector in pulse NMR. The sample is contained within the coil. During the sweep in CW NMR, resonance perturbs the energy in the probe/coil circuit. In pulse NMR, rotating magnetization induces electromotive force detected in the coil as voltage. This signal is displayed on a chart or video monitor. The **absorption** NMR spectrum is normally displayed. Demand for low-cost, modest capability, good-performance instrumentation has led to commercial development of

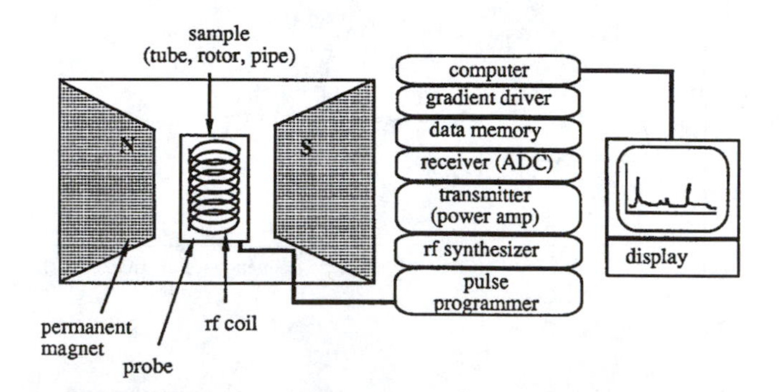

FIGURE 27-1.
Schematic diagram of a low-resolution pulse NMR analyzer (gradient coils not shown). Such an instrument would be capable of measuring solid-liquid ratio, moisture or oil content, moisture or oil diffusion, and solidity of solids via time-domain decays or frequency-domain low-resolution spectrum. With addition of homogeneous magnet, shims, and field-frequency lock, the spectrometer would become capable of identifying and quantifying major liquid-phase components such as water, oil, sugars, acids, and the like via the high-resolution NMR spectrum.

benchtop instruments that are far easier to use than the research versions and that are automated.

The major instrument categories are: **low resolution** (often improperly called "pulsed" or wide-line NMR) for solid-liquid ratio, moisture, or oil content; **high-resolution liquids** for chemical analysis of liquid-phase components in whole foods; **high-resolution solids** for chemical analysis of solid-phase components; and **magnetic resonance imaging** for drying, crystallization, flow, "anatomy," and so on.

The food analyst using NMR should ask the following questions: (1) What information do I want out of this sample? (2) What NMR techniques are required? (3) What instrumental capabilities are needed to perform the techniques (Table 27-1)? (4) How do I access the right instrument?

Research instrumentation can be expensive ($100,000 to $1,000,000), but benchtop NMR analyzers are available for $20,000 to $80,000, and the prices should decrease. NMR process sensors will cost even less. Magnetic resonance shows great promise for quality control and process monitoring in addition to analytical laboratory use.

27.3.1.2 Continuous Wave NMR (CW)

In the **frequency sweep** CW experiment, the strong field is held constant while the frequency of the oscillating field is varied. In the **field sweep** CW experiment, the field strength is varied while the frequency of the oscillating field is held constant. An electronic device measures energy absorbed in the circuit containing sample and radio frequency coil. The result is a spectrum.

In commercial CW instruments for benchtop solids analysis, the magnetic field is not homogeneous, and the sweep is wide. The solid resonance is broad (tens of kilohertz) and may be integrated. The result can be compared with a reference.

In CW instruments designed for liquids (becoming rare), the magnetic field is homogeneous to *at least* 1 part in 10 million (0.1 ppm), and the sweep is narrow. Liquid resonances occur at specific frequencies or chemical shifts (determined by comparison with a reference) and are narrow (tenths to tens of hertz). Integrals may be compared with a reference for quantitation.

27.3.1.3 Rapid Scan Correlation (RSC)

The rapid scan correlation experiment is similar to conventional CW, but its advantages are better resolution, higher signal-to-noise ratio, and greater speed of analysis. Commercial instruments can analyze liquid foods and liquid-phase components in heterogenous foods. However, if susceptibility broadening is severe, the spectra will suffer unacceptably from loss of resolution. Resolution may be recovered by **magic angle sample**

TABLE 27-1. Characteristics of NMR Spectrometers

Characteristic	*Capability*
Static magnetic field strength	Sensitivity, selectivity (resolution)
Field homogeneity	Resolution
Shim coils	Maximize resolution
Transmitter frequencies	Determines which nuclei are observable
Transmitter power	Pulse width and excitation bandwidth
Mode of excitation	CW, pulse
Decoupler capability	To undo J or dipolar coupling
Probe/coil configuration	Specific for wide-line, high resolution, decoupling, magic angle spinning, or other purpose
Signal mode	Phase referenced as in FT NMR or FT ESR; rectified as in some time-domain NMR
Pulse programming	Single pulse, spin-echo, decoupling, relaxation, signal averaging, other pulse sequences
Gradient coils	Required for diffusion or imaging
Data analysis program	Time-domain display, FT display, relaxation calculation, phasing, integration, subtraction, image construction, etc.
Sample spinning	Conventional (parallel, magic angle)
Sample control	Temperature, flow, automation, etc.
Connectivity with external data systems	LIMS, robots, controllers, etc.

spinning (MAS) (see section 27.4.2.3). Sometimes resolution is sufficient for analysis of total water, lipid, sugars, or another abundant species, even without MAS.

27.3.1.4 Pulse NMR: High Resolution and Low Resolution

Pulse NMR refers to excitation via a **hard pulse:** short time and large amplitude. The pulse has a selectable **carrier frequency** but results in **broad band excitation:** All resonances are excited simultaneously. The free induction signals from all components superimpose in the time domain but are sorted out according to frequency (or chemical shift) by the Fourier transform. The detector/receiver is characterized by its **bandwidth,** which must be greater than the spread in resonance frequencies. The signal is converted by an **analog-to-digital converter** and stored in **data memory** that holds thousands of time points. The time domain data may be analyzed immediately or Fourier-transformed into a spectrum.

In commercial benchtop pulse instruments designed for solids, the magnetic field is not homogeneous and the excitation bandwidth is larger (short, more powerful pulse). The solid resonance is broad (tens of kilohertz) and may be integrated and compared with a reference. The probe has a lower **Q-factor** (related to sensitivity) and short **acoustic ring down time,** the receiver has a large **bandwidth,** and data memory requirements are not stringent.

In commercial benchtop pulse instruments designed for liquids, the magnetic field is homogeneous and the excitation width is narrower (short, less powerful pulse). Liquid resonances are narrow (tenths to tens of hertz) and occur at specific frequencies (or chemical shifts). Resonances may be integrated and compared with references. The probe has a higher Q-factor, and data memory has to be large to provide good **digital resolution in the spectrum.**

Benchtop analytical instruments with both wide-line (solids) and high-resolution (liquids) capabilities have not yet appeared, but results from research show that the combination would be valuable.

27.3.2 Analytical ESR Instrumentation

ESR is usually a **field sweep** CW method (Fig. 27-2) using an electromagnet. Electron precession frequencies $\gamma B_0/2\pi$ are in the gigahertz (GHz) range for field strengths from about 0.3 to 0.6 Tesla. This corresponds to **X-band** ESR. The resonance field is the magnetic part of the electromagnetic radiation from a microwave source, usually a **klystron** or **gunn generator.** The sample is contained in the resonator cavity between the

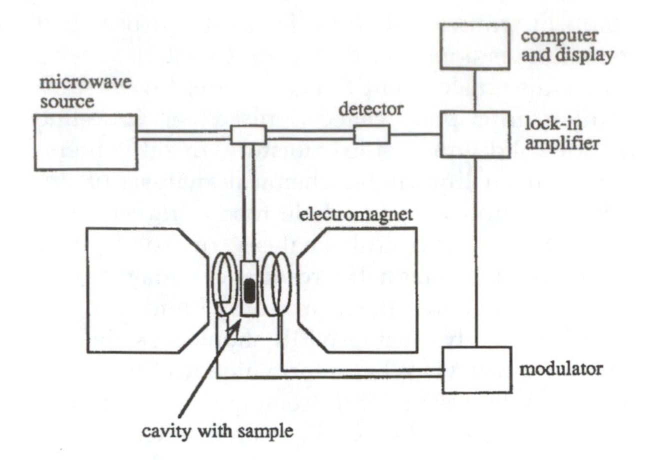

FIGURE 27-2.
Schematic diagram of an ESR spectrometer.

poles of the magnet. During the field sweep, resonance perturbs the microwave power level in the cavity. This is detected and the signal is displayed on a chart or video monitor. For sensitivity reasons, a method called **field modulation** is often applied. The result is that the **first derivative** of the ESR spectrum is displayed. Demand for low cost, modest capability, good performance instrumentation has led to commercial development of new benchtop instruments that are far easier to use than the research versions and that are automated. A primary food use of the benchtop ESR is detection of evidence of irradiation.

27.4 APPLICATIONS

27.4.1 Practical Aspects of NMR

27.4.1.1 Sample Requirements

The food sample can be virtually *anything*. This includes liquids (e.g., syrup, beverage, oil), semisolids (salad dressing, cheese, gel, emulsion, margarine, etc.), or solids (milk powder, sugar granules, chocolate, etc.). Sample size is limited by the magnet-coil combination. In most benchtop analyzers, the sample is between 0.1 cm^3 and 5 cm^3.

27.4.1.2 Observable Nuclei

The **observed nucleus** can be any magnetic isotope present in the sample, as long as the instrument can detect it. The most sensitive nuclei are hydrogen-1 (^1H, or **proton**) and phosphorous-31 (^{31}P) (the number refers to the **atomic mass**). These have large gyromagnetic ratios and high **natural isotopic abundances.** Most pulse time domain NMR analyzers detect only ^1H and have low field magnets that are not homogeneous.

New benchtop pulse FT NMR analyzers have strong magnets that are homogeneous and can detect other nuclei. ^{13}C is relatively insensitive due to a low isotopic abundance (1.3%) and small gyromagnetic ratio. However, the ^{13}C spectrum is more spread out (over 200 ppm) than the ^{1}H spectrum (10 ppm), so resolution is better even if sensitivity is not. Other NMR-active nuclei in foods include oxygen-17 (^{17}O), sodium-23 (^{23}Na), potassium-39 (^{39}K), and nitrogen-14 (^{14}N). These may someday be observable with benchtop instruments. Most others, such as calcium-43 (^{43}Ca) and magnesium-25 (^{25}Mg), cannot be observed without specific isotopic enrichment in the sample.

27.4.1.3 Magnets and Magnetic Field Strength

Benchtop analyzers use permanent magnets that are strong, small, and light, with enough space for a compact probe. Typical sample volume is about 1 cm^3. **Magnetic field strengths** of permanent magnets in benchtop analyzers vary from 0.235 to 2.35 Tesla. This corresponds to proton resonance frequencies from 10 to 100 MHz. The larger the field strength, the greater the signal (sensitivity) and the better the resolution (selectivity).

27.4.1.4 Magnetic Field Homogeneity

The homogeneity of the magnetic field in the region where the sample is placed is deliberately low (variation across the sample of 10 ppm or more) or high (less than 0.1 ppm). Low-homogeneity magnets are used for NMR analysis of solid and liquid content. No chemical information is obtainable. High-homogeneity magnets are used in high-resolution NMR to produce chemical information.

27.4.1.5 Probe Characteristics

Probes must be compact in order to fit into the magnet. The probe includes circuitry for **tuning** to resonance and **matching** to impedance, delivering the excitation pulse, and receiving the induction signal. The probe holds the sample and may permit **sample spinning** and sample environment control. In chemical analysis of liquid phases in foods, resolution (selectivity) may suffer from **susceptibility inhomogeneity,** a sample-dependent effect. Resolution is recovered when the sample is placed in a **rotor** and spun at rates between hundreds and thousands of hertz about an axis that makes the angle 54°44′ (the magic angle) with the static field. The technique is called **magic angle spinning** (MAS). It is not yet available on commercial benchtop instruments.

27.4.2 Examples of NMR Analysis of Foods

Table 27-2 is a partial listing of the *kinds* of analyses that may be performed using NMR. Note that each analysis listed may be applied to many kinds of food materials.

An important application involves the use of pulsed field gradients to determine particle size distribution in emulsions and translational diffusion of small molecules (moisture and fat) and of mobile polymers.

Time-domain solid content measurements work best from about 10 percent to 95 percent solids. Many foods, especially gels and thick liquids, have much lower solids. Alternative methods such as cross-relaxation must then be used (see section 27.4.2.4). These may be implemented by modification of some commercial benchtop instruments.

The following specific examples are limited to ^{1}H NMR, since commercial benchtop and in-line instruments with multinuclear capability are not yet available.

27.4.2.1 Solid or Liquid Content and Solid-Liquid Ratio

The most common use of NMR in food analysis is solid or liquid content. See section 27.2, items 14 and 15, for an illustration of the method of measuring solid content by NMR. The results of applying this method to a crystallizing fat are shown in Fig. 27-3.

27.4.2.2 Chemical Analysis of Liquid

It is possible to measure the composition of a liquid by NMR. In the example shown (see Fig. 27-4), the chemical functional groups of oil molecules produce a spectrum that has the signature of that oil. Further calculations can be made by measuring the area under each peak. The area, or integral, is proportional to the number of atoms with a particular chemical structure, such as the hydrogens near double bonds in fatty acyl groups in triacylglycerol molecules (fat molecules).

27.4.2.3 Chemical Analysis of Liquid Phases in a Heterogeneous Material

High-resolution NMR can be used to observe compounds that are in liquid phases. For heterogeneous semisolids like cheese, this requires use of magic angle spinning, as illustrated (Fig. 27-5).

TABLE 27-2. Examples of Food Analyses That Can Be Performed by NMR

Time-Domain NMR	*Frequency-Domain NMR (FT NMR)*
Solid fat content or solid-to-liquid ratio	Authentication of ingredients
Oil content	Spatial image of crystallization
Moisture or oil diffusion	Emulsion particle size distribution
Moisture content	Detection of adulteration in wines, juices
Crystallization	Fat crystal polymorphism
Oil saturation	Sugar crystal polymorphism
Crystalline, semisolid, mobile starch	Polysaccharide polymorphism
Plasticization of protein, polysaccharides	Rotational motion of liquid components
Solidity of solid phases	Postmortem phosphorous metabolism
Drying	pH via phosphorous shift
Hydration	Porosity (liquid spaces)
Solidification	Dissolved sugar composition
Melting	Liquid lipid analysis
Mobile species in frozen food	Oil average triacylglycerol molecular weight, saponification number, iodine value, etc.
	Organic acid composition

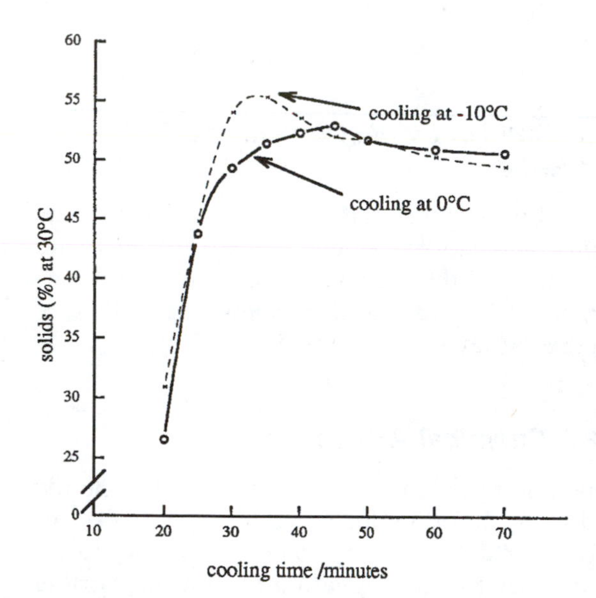

FIGURE 27-3.
Concept: Solid fat content varies during crystallization and solidification. This process can be monitored by NMR to compare different conditions of crystallization. *Sample*: Cocoa butter. *Spectrometer*: Bruker minispec p20. *Type*: Pulse time-domain. *Field strength*: 0.47 Tesla (20 MHz for ^{1}H). *Sample tube*: 10 mm o.d., wall 0.9 mm, sample height in tube 10 mm. *Discussion*: The indirect method of solid fat content (SFC) was used to obtain each point. Curve shows development of solid during crystallization of cocoa butter at two different temperatures. *Reference*: Adapted from Fig. 5 in Petersson, B., Anjou, K., and Sandstrom, L. 1985. Pulsed NMR method for solid fat content determination in tempering fats, part I: Cocoa butters and equivalents. *Fette Seifen Anstrichmittel* 87:225–230.

27.4.2.4 Solids and Solidity

Pulse NMR, or time-domain low-resolution NMR, is not the only way to observe solid phases. An alternative especially suitable for low solids foods is **cross-relaxation NMR.** In cross-relaxation NMR, the amount of rigidity of a solid phase is detected by its effect on the water signal. An example is the starch content of a ripening banana (Fig. 27-6).

27.4.3 Practical Aspects of ESR

27.4.3.1 Samples

Samples can be nearly anything. Volumes are generally small, typically 0.1 cm^3 but larger in benchtop ESR instruments. High-quality sample tubes (quartz) are usually needed. Sample shape and position affect results. Since water absorbs microwaves efficiently, detection efficiency is better the lower the H_2O content of the sample.

27.4.3.2 Sources of ESR Signals

The ESR can detect organic **free radicals** and **paramagnetic metal ions.** Free radicals are molecules containing an unpaired electron, such as carbon atom radicals in irradiated sugars. The radical must exist long enough to be detected. Most free radicals are unstable due to their chemical reactivity. They generally survive longer the drier and more solid the material. Sometimes they may be chemically trapped **(spin-trapping)** long enough to be detected.

It is possible to synthesize stable free radicals. They

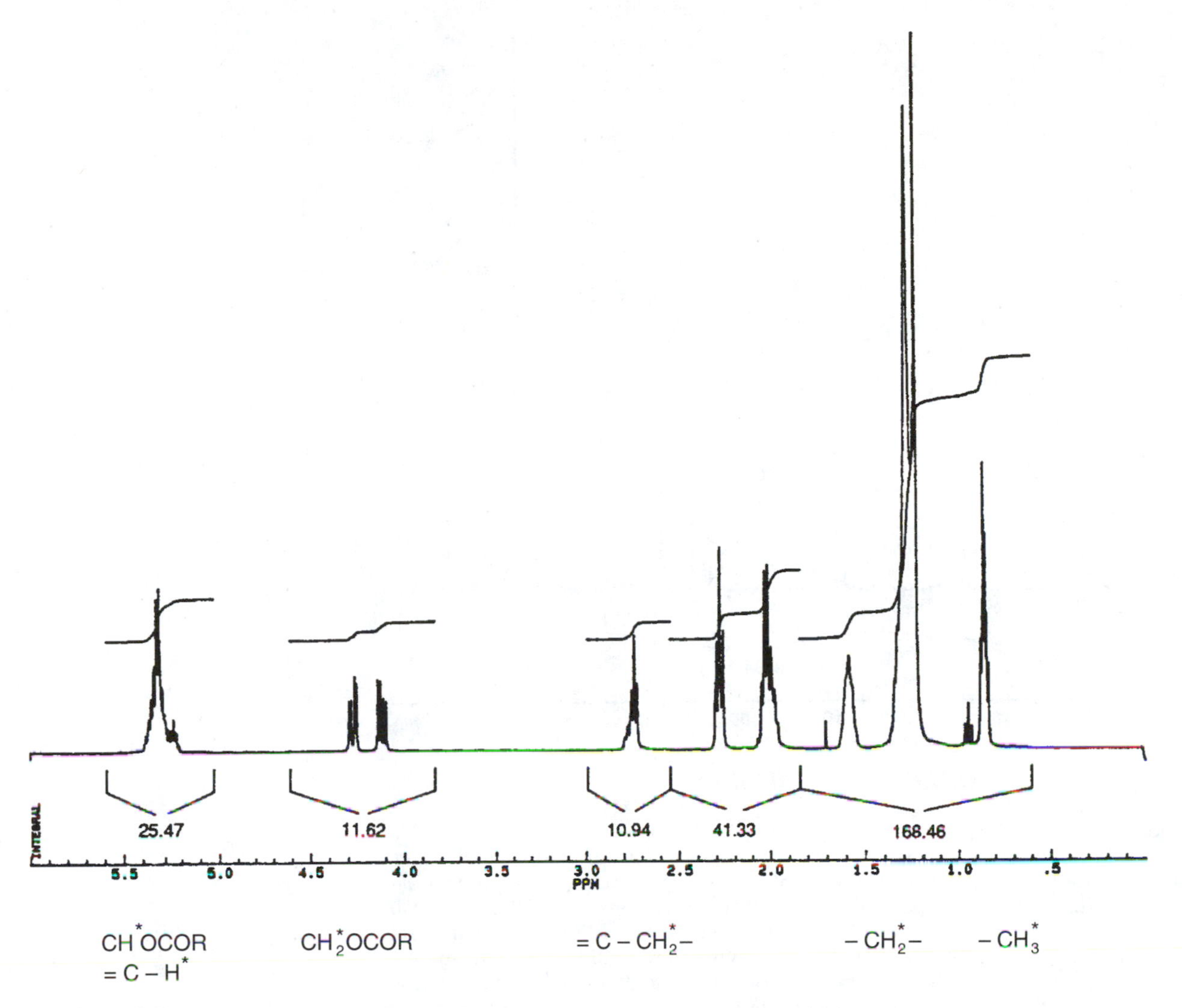

FIGURE 27-4.
Sample: Soybean oil. *Spectrometer*: Bruker MSL400. *Type*: Superconducting magnet, high resolution, Fourier transform, standard probe. *Field strength*: 9.4 Tesla (400 MHz for ^{1}H). *Sample tube*: 5 mm o.d., 0.3 mm wall, sample height in tube 20 mm. *Discussion*: FT NMR spectrum of pure soybean oil showing resonance clusters for all different chemical functional groups present in triacylglycerols (oil molecules). Integrals are shown (S-shaped curves) and the corresponding areas are given below each peak region. From these, the following may be calculated: number of olefinic (unsaturated) protons, average triacylglycerol molecular weight, iodine value, saponification value, average number of vinyl groups, and polyunsaturation factor. *Reference*: Reproduced with permission by American Oil Chemists' Society, from Eads, T.M., Multinuclear high resolution and wide line NMR methods for analysis of lipids, in *Analyses of Fats, Oils and Lipoproteins*, E.G. Perkins, ed. American Oil Chemists' Society, Champaign, IL, pp. 409–457, 1991. See also Schoolery, J.N., 1983. Applications of high resolution nuclear magnetic resonance to the study of lipids. In *Dietary Fats and Health*, E.G. Perkins & W.F. Visek, eds., American Oil Chemists' Society, Champaign, IL, Chapter 12, pp. 220–240.

can be solubilized in oil or water or attached covalently to specific food components like proteins, polysaccharides, or lipids. They can then be used to probe local molecular events during gelation, adsorption, membrane changes, denaturation, and so on. A probe in a liquidlike state produces narrower, better-resolved splitting patterns, while one in a solidlike state produces broader lines, up to a limit.

The origins of free radicals are those produced by irradiation with gamma rays (as used in food sterilization) or other ionizing radiation, and those produced by chemical reactions occurring naturally during aging of food (e.g., lipid oxidation); during processing (heating, grinding, drying, freezing, etc.); during addition of antioxidants or proxidants; or during browning (apples), greening (potatoes), reddening (meats), and other coloration reactions. The paramagnetic ions occurring in food are manganese (Mn), iron (Fe) and

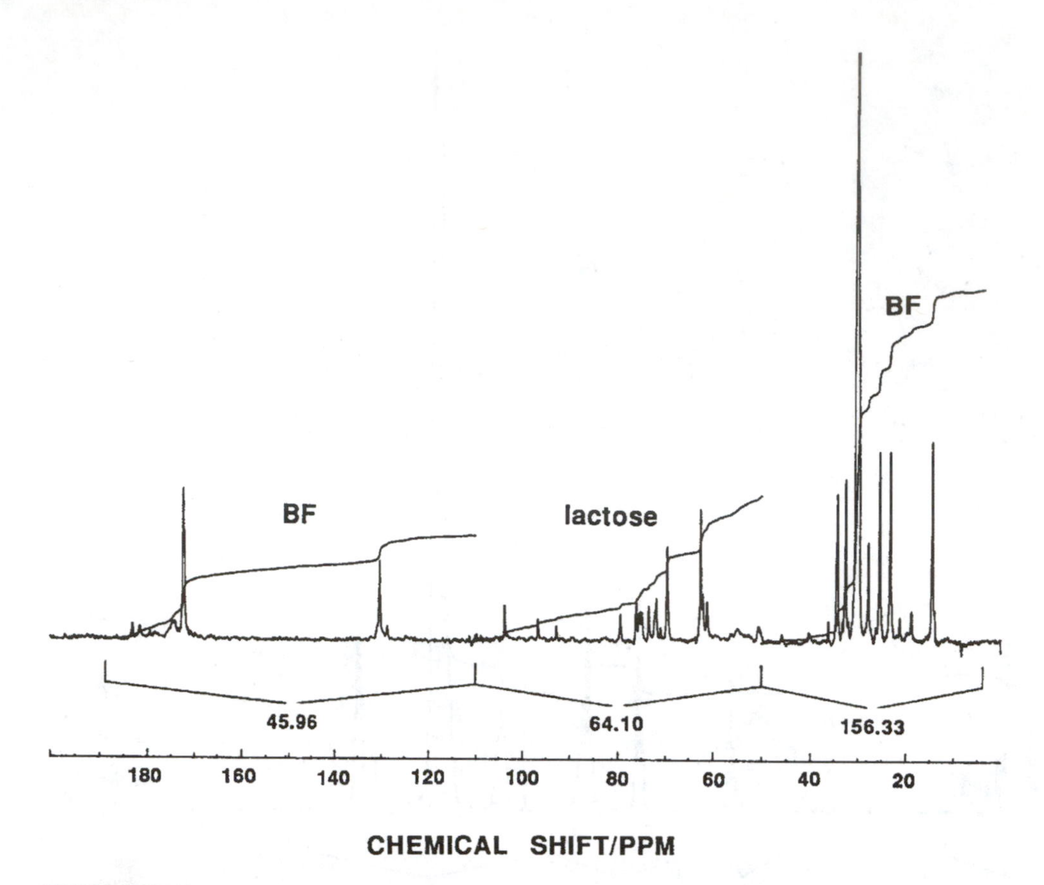

FIGURE 27-5.
Sample: Processed American cheese. *Spectrometer*: Bruker MSL400. *Type*: Superconducting magnet, high resolution, Fourier transform, with magic angle spinning (MAS) probe. *Field strength*: 9.4 Tesla (100 MHz for ^{13}C). *Sample rotor*: 7 mm o.d., 0.8 mm wall, 28 mm overall length, sample volume 0.44 ml. *Discussion*: Natural abundance ^{13}C NMR spectrum of intact piece of cheese shows sharp lines because of MAS; otherwise they would be less informative due to susceptibility broadening. Sharp resonances correspond to dissolved species (lactose in aqueous phase of protein gel) or to that fraction of butterfat (BF) that is liquid at room temperature. Values of integrals are shown. The corresponding proton (^1H) NMR spectrum (not shown) was also obtained, providing similar information and showing the water peak as well. *Reference*: Reproduced with permission of American Oil Chemists' Society, from Eads, T.M., Multinuclear high resolution and wide line NMR methods for analysis of lipids, in *Analyses of Fats, Oils and Lipoproteins*, E.G. Perkins, ed. American Oil Chemists' Society, Champaign, IL, pp. 409–457, 1991.

copper (Cu). These may occur naturally in an ESR-detectable state. Sometimes they must be released from the food matrix and brought to an appropriate oxidation state to be detected.

ESR line shapes may be integrated, and results may be expressed as ESR signal intensity per unit weight of sample. This makes quantitative analysis by ESR possible.

It is sometimes possible to tell what kind of substance is producing an ESR pattern from a food material. An unpaired electron may interact in characteristic ways with nearby electrons (as in metal ions whose electronic structures depend sensitively on ligands and oxidation state) or with nearby nuclei (as when an ESR triplet is formed when the radical is near a nitrogen-14 nucleus). If several ESR-active species are present, their signals will superimpose. Irradiation of food with gamma rays will often produce an ESR response where there were none before or add a new signal to the naturally occurring one.

27.4.3.3 Microwave Cavity

Sample environment control is possible in ESR probes. As for NMR, sample placement may then become tricky, and changing samples may become less convenient.

27.4.4 Examples of ESR Analysis of Foods

Table 27-3 is a partial listing of analyses that may be performed using ESR. Note that each analysis listed

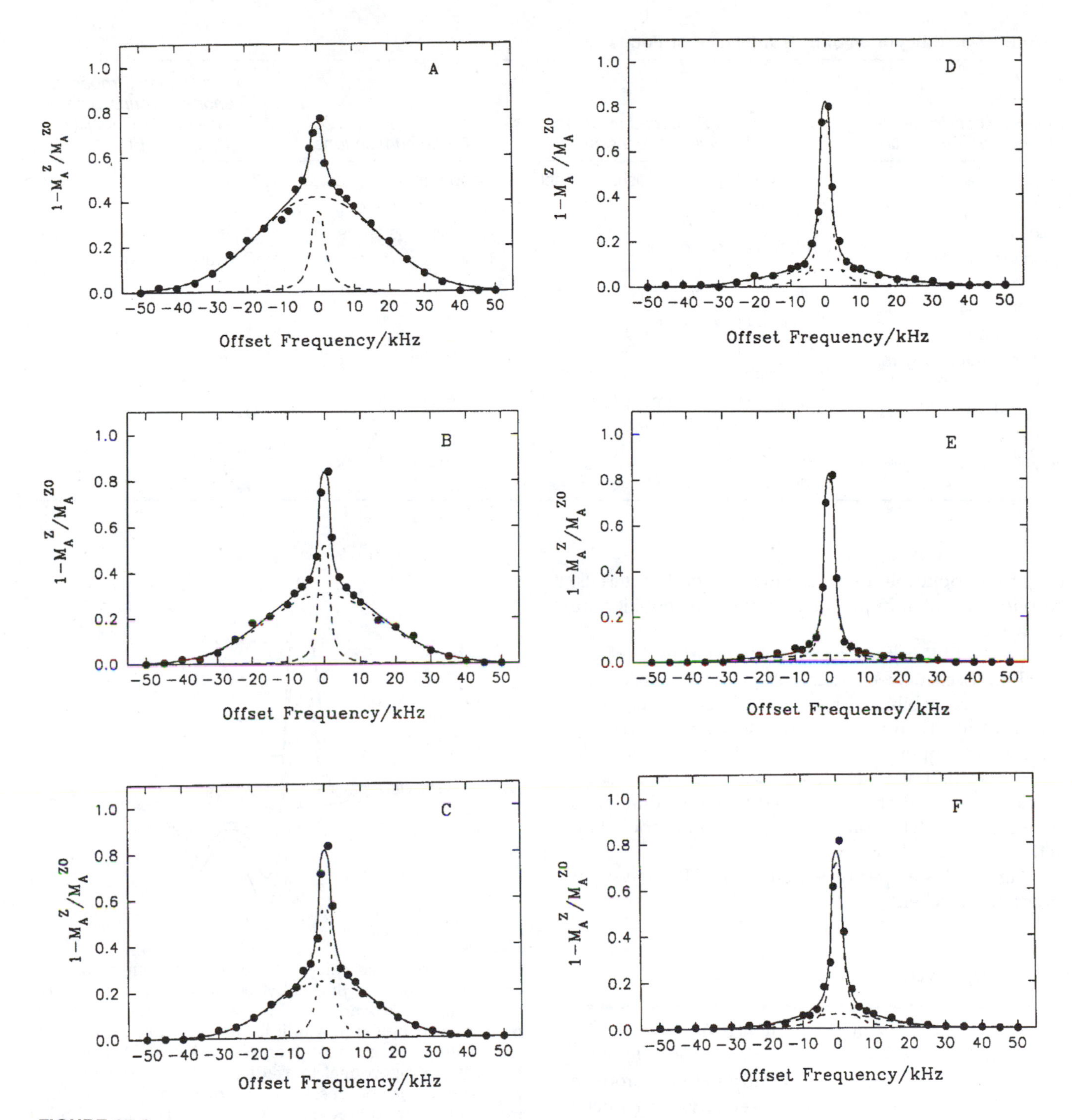

FIGURE 27-6.
Sample: Banana during ripening. *Spectrometer*: Nicolet NT200. *Type*: Superconducting magnet, high resolution, standard probe. *Field strength*: 4.7 Tesla (200 MHz for ^{1}H). *Sample tube*: 5 mm o.d., 0.3 mm wall, sample height 20 mm. *Discussion*: Cross relaxation NMR can detect solids at much lower levels than the pulsed NMR method (see Fig. 27-5 and theory section). Area under the curve gives amount of solids; width of curve gives solidity (wider is more rigid). Green banana has much starch (20% of fresh weight), which converts to sugars during ripening. (A), 1 day; (B) 2d; (C) 3d; (D) 4d; (E) 6d; (F) 8d. This valuable experiment can be carried out at low field and low resolution on a much cheaper, simpler instrument (not yet commercially available). *Reference*: Reprinted with permission from Ni, Q.X., and Eads, T.M. 1993. Analysis by proton NMR of changes in liquid phase and solid phase components during ripening of bananas, *J. Agric. Food Chem.* 41:1035–1040. Copyright 1993 American Chemical Society. Technique used is from Wu, J.Y., Bryant, R.G., and Eads, T.M. 1992. Detection of solidlike components in starch using cross-relaxation and Fourier transform wide-line ^{1}H NMR methods, *J. Agric. Food Chem.* 40, 449-455.

TABLE 27-3. Categories of ESR Analyses of Foods

Free Radicals Produced During Food Irradiation	*Free Radicals Produced in Chemical and Physical Processes*	*Paramagnetic Ions*	*Free Radical Probes as Reporter Groups Used to Investigate Mobility and/or Binding*
Meats containing bone	Lipid oxidation (meats, fish, vegetable oil, oilseeds, orange juice essence, etc.)	Copper (II) in soy sauce	Starch
Fish	Oxidation in polyphenols (teas, apples, etc.)	Mn in various foods	Gluten
Spices	Coffee aging, extraction	Cu in various foods	Whey protein concentrates
Food with shells attached	Vitamin E activity in oils	Fe in grains	Oil-water content
Cereal grains	Beer staling		
Frozen foods	Unsaturate fat oxidation of protein		
Carbohydrates (sugars, polysaccharides)	Browning reactions (many foods)		
Fats and oils	Greening reactions (potato)		
Fruits	Blueing reactions		
Vegetables	Reddening reactions (meat)		

might be applicable to other kinds of food materials. Applications of spin probes to molecular mobility are not listed.

Use of ESR in food analysis is important because: (1) High levels of free radicals and extensive oxidation are nutritive liabilities or signs of abuse or deterioration. (2) Not all countries permit sale of irradiated foodstuffs, and imports must be tested for exposure to ionizing radiation. (3) Evaluation of radiation dose and efficiency of irradiation is needed where irradiation is permitted. (4) Detection of antioxidant and prooxidant activity is needed in food research.

Fig. 27-7 is a specific example of ESR analysis of irradiated chicken.

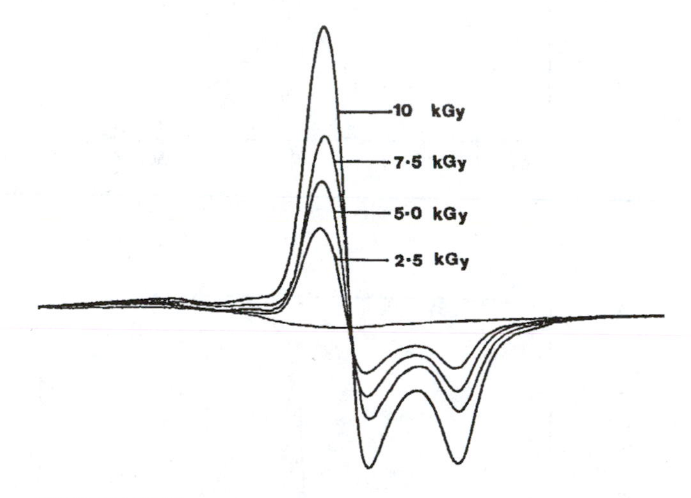

FIGURE 27-7.
Sample: Freeze-dried chicken bone powder from irradiated chicken. *Spectrometer*: Bruker ESP 300 with TE104 cavity. *Type*: electromagnet, X-band ESR. *Discussion*: Gamma irradiation induces formation of free radicals that are stable in fairly dry food materials (e.g., in bone of fresh chicken). ESR signal intensity is proportional to irradiation dose. *Reference*: Reproduced with permission from Bruker Instruments, Inc., from Gray, R., and Stevenson, M.H. 1992. The use of ESR spectroscopy for the identification of irradiated food. In *Bruker Report*, 1991–1992, Bruker Instruments, Billerica, MA.

27.5 SUMMARY

The principles of magnetic resonance are the same for NMR and ESR. A sample magnetization due to magnetic nuclei or electrons can be manipulated to produce a detectable signal. Due to the sensitivity of nuclei and electrons to molecular structure and motion, it is possible to extract chemical (compositional) and physical (solidity, liquidity) information from food samples. The instrumentation involves a magnet to polarize the spins; a transmitter of the resonant, oscillating magnetic field; a probe to hold the sample and to permit excitation and detection; a receiver; and a data system. There are many instruments; most are designed for a limited set of analyses. It is often possible to control sample environment and to automate analyses by robotics. Magnetic resonance methods are essentially nondestructive, except for preparing a sample that fits the probe. Potential issues in analysis by magnetic resonance are sensitivity and selectivity. Although magnetic resonance for analysis has a long history, its potential has barely been tapped, due mostly to expense and complexity. These problems are fast yielding to new technology and education of users. Magnetic resonance devices serve in food analytical labs and quality control labs and as sensors in food processing.

27.6 STUDY QUESTIONS

1. How are NMR and ESR similar or different in terms of the types of analyses for which they can be used?
2. a. Why do we apply a magnetic field to the nuclei or electrons?
   b. What would you expect to happen to sensitivity, selectivity, and resonance separation if you used a magnet whose field strength is ten times stronger?
   c. When can the energy of the oscillating magnetic field be absorbed by a sample?
3. You are interested in the chemistry and physics of baking. You wonder whether gluten binds amylose or amylopectin. You know that parts of proteins and polysaccharides are sometimes so mobile that they produce narrow NMR resonances. Binding would be expected to restrict these motions. Suggest a high-resolution NMR experiment to test whether binding takes place.
4. What must you do to convert a time-domain signal into a frequency-domain signal (spectrum)? What is the meaning of multiple hydrogen resonances in the spectrum?
5. Discuss the three main spectral characteristics that can help you determine amount, chemical type, and physical state of a particular component in your sample.
6. You are comparing a more hydrogenated, more saturated, and thus more solidlike competitor's margarine with yours, which is less hydrogenated, more unsaturated, and thus more liquidlike in texture. What two NMR techniques would you use to compare (a) solidity and (b) degree of saturation? What are the basic principles involved in these measurements? Draw the expected results. *Hint:* High-resolution liquids NMR for chemical analysis requires that the sample be in a liquid state.
7. ESR is especially attractive to follow molecular interactions. This is done by the proper selection of a free radical probe that dissolves in water or lipid or that adsorbs or binds to a molecule. If the free radical probe has a fatty acid tail that enables it to bind strongly with amylose, a very large, stiff starch polymer, what types of line shapes would you expect to see in the ESR spectra of (a) a solution of the probe; (b) the same solution with amylose added? Do you think you could quantitatively determine the amount that binds?

27.7 RESOURCE MATERIALS

Terminology

Axel, L., Margulis, A.R., and Meaney, T.F. 1984. Glossary of NMR terms. *Magn. Reson. Med.* 1:414–433.

Homans, S.W. 1989. *A Dictionary of Concepts in NMR.* Clarendon Press, Oxford, England.

Standard NMR Methods for Food

American Oil Chemists' Society (AOCS): NMR methods for SFC.

Association of Official Analytical Chemists International (AOAC) or other standard NMR methods for food.

Low-Resolution Time-Domain NMR of Foods

(There are dozens of articles and reviews on this topic. Very recent ones are cited).

Gambhir, P.N. 1992. Applications of low-resolution pulsed NMR to the determination of oil and moisture in oilseeds. *Trends in Food Science and Technology* 3:191–196.

Gribnau, M.C.M. 1992. Determination of solid/liquid ratios of fats and oils by low-resolution pulsed NMR. *Trends in Food Science and Technology* 3:186–190.

Bruker Spectrospin Ltr. There are many Bruker *minispec* instrument literature and application notes available (chocolate, cheese, oilseeds, margarine, beet pulp, etc.). Contact Bruker Spectrospin (Canada) Ltd., 555 Steeles Avenues, E. Milton, Ontario L9T 1Y6, Canada.

Introductory Pulse Fourier Transform NMR

Abraham, R.J., Fisher, J., and Loftus, P. 1988. *Introduction to NMR Spectroscopy.*Wiley & Sons, New York, NY. Graduate-level introduction to high-resolution NMR or liquids and solids. Text with problems.

Farrar, T.C., and Becker, E.D. 1971. *Pulse and Fourier Transform NMR, Introduction to Theory and Methods.* Academic Press, New York, NY. Classic introduction to basic theory and methods. Graduate level.

Friebolin, H. 1991. *Basic One- and Two-Dimensional NMR Spectroscopy.* VCH Publishers, New York, NY. Graduate level.

Hemminga, M.A. 1992. Introduction to NMR. *Trends in Food Science and Technology* 3:179–182. Undergraduate level.

James, T.L. 1975. *Nuclear Magnetic Resonance in Biochemistry: Principles and Applications.* Academic Press, New York, NY. Comprehensive introduction to one-dimensional NMR theory and practice, with many examples. Contains useful reference information. Undergraduate to graduate level.

Martin, M.L., Delpeuch, J.J., and Martin, G.J. 1980. *Practical NMR Spectroscopy.* Heyden & Son, Ltd., Philadelphia, PA. Experimental methods for liquids high-resolution NMR. Suitable for undergraduate level.

General and Selected Articles on NMR Foods

Belton, P.S., and Colquhoun, I.J. 1990. Nuclear magnetic resonance spectroscopy in food research. *Spectroscopy* 4:22–32. Covers new food applications outside of time-domain and pure high resolution. Undergraduate to graduate level.

Multiple articles and authors. 1992. Applications of nuclear magnetic resonance techniques in food research. Special issue. *Trends in Food Science and Technology* 3:177–250. Good overview. High resolution is missing. Undergraduate level.

Eads, T.M., and Croasmun, W.R. 1988. NMR applications to

fats and oils. *J. Am. Oil Chemists' Soc.* 65:79–83. Brief review. Undergraduate level.

Eads, T.M. 1991. Multinuclear high resolution and wide line NMR methods of analysis of lipids. In *Analyses of Fats, Oils and Lipoproteins*, E.G. Perkins (Ed.), 409–457. American Oil Chemists' Society, Champaign, IL. Graduate level.

Eads, T.M. 1991. New strategies for measuring molecular structure, dynamics, and composition of food materials by NMR. In *Frontiers in Carbohydrate Research 2*, R. Chandrasekaran (Ed.), 128–140. Elsevier, London.

Richardson, S.J., Baianu, I.C., and Steinberg, M.P. 1986. Mobility of water in wheat flour suspensions as studied by proton and oxygen-17 nuclear magnetic resonance. *J. Agric. Food Chem.* 34:17–23.

Richardson, S.J., Baianu, I.C., and Steinberg, M.P. 1987a. Mobility of water in corn starch suspensions determined by nuclear magnetic resonance. *Starch* 39:79–83.

Richardson, S.J., Baianu, I.C., and Steinberg, M.P. 1987b. Mobility of water in starch powders determined by nuclear magnetic resonance. *Starch* 39:198–203.

Richardson, S.J., Baianu, I.C., and Steinberg, M.P. 1987c. Mobility of water in sucrose solutions determined by deuterium and oxygen-17 nuclear magnetic resonance measurements. *J. Food Sci.* 52:806–809.

Umbach, S.L., Davis, E.A., Gordon, J. 1990. Effects of heat and water transport on the bagel-making process: Conventional and microwave baking. *Cereal Chem.* 67: 355–360.

Umbach, S.L., Davis, E.A., Gordon, J., and Callaghan, P.T. 1992. Water self-diffusion coefficients and dielectric properties determined for starch-gluten-water mixtures heated by microwave and by conventional methods. *Cereal Chem.* 69:637–642.

Williams, R.J.P. 1989. NMR studies of mobility within protein structure. *Eur. J. Biochem.* 183:479–497.

High Resolution NMR of Foods

(Few reviews are available. Some recent articles are cited. Original applications are easily obtained by computer search.)

Eads, T.M. 1991. Multinuclear high resolution and wide line NMR methods for analysis of lipids. In *Analyses of Fats, Oils and Lipoproteins*, E.G. Perkins (Ed.), 409–457. American Oil Chemists' Society, Champaign, IL.

Eads, T.M., and Bryant, R.G. 1986. High resolution proton NMR spectroscopy of milk, orange juice, and apple juice with efficient suppression of the water peak. *J. Agric. Food Chem.* 34:834–837.

Ni, Q.W., and Eads, T.M. 1992. Low speed magic angle spinning carbon-13 NMR of fruit tissue. *J. Agric. Food Chem.* 40:1507–1513.

Ni, Q.W., and Eads, T.M. 1993. Liquid phase composition of intact fruit tissue measured by high resolution proton NMR. submitted.

Rutar, V. 1989. NMR Studies of intact seeds, Ch. 4, in *Nuclear Magnetic Resonance in Agriculture*, P.E. Pfeffer and W.V. Gerasimowicz (Eds.). CRC Press, Inc., Boca Raton, FL.

NMR of Fats and Oils

Horman, I. 1984. In *Analysis of Foods and Beverages: Modern Techniques*, G. Charalambous (Ed.), 230–239. Academic Press, Orlando, FL.

Pollard, M. 1986. Nuclear magnetic resonance spectroscopy (high resolution), Ch. 9, in *Analysis of Oils and Fats*, R.J. Hamilton and J.B. Rossell (Eds.), 401–434. Elsevier, New York, NY.

Eads, T.M., and Croasmun, W.R. 1988. NMR applications to fats and oils. *J. Am. Oil Chemists' Soc.* 65:79–83. Brief review of the field. Undergraduate level.

Eads, T.M. 1991. Multinuclear high resolution and wide line NMR methods for analysis of lipids. In *Analyses of Fats, Oils and Lipoproteins*, E.G. Perkins (Ed.), 409–457. American Oil Chemists' Society, Champaign, IL.

Waddington, D. 1986. Applications of wide-line nuclear magnetic resonance in the oils and fats industry. In *Analysis of Oils and Fats*, R.J. Hamilton and J.B. Rossell (Eds.), 241–297. Elsevier, New York, NY.

Schoolery, J.N. Applications of high resolution nuclear magnetic resonance to the study of lipids, Ch. 12 in *Dietary Fats and Health*, E.G. Perkins and W.F. Visek (Eds.) American Oil Chemists' Society, Champaign, IL. Selected applications.

Lambelet, P. 1980. Determination of lipids in foods by pulsed nuclear magnetic resonance (NMR). *Mitt. Geb. Lebenmittelunters. Hyg.* 71:119–123.

Introductory ESR

Berliner, L.J. 1976. *Spin Labeling: Theory and Applications.* Academic Press, New York. Advanced, authoritative guide.

Berliner, L.J. 1979. *Spin Labeling II: Theory and Applications.* Academic Press, New York.

Knowles, P.F. 1976. *Magnetic Resonance of Biomolecules, an Introduction to the Theory and Practice of NMR and ESR in Biological Systems.* Wiley & Sons, New York, NY. Especially valuable are chapters 1, 2, 6, 7. Undergraduate to graduate level.

Swartz, H.M., Bolton, J.R., and Borg, D.C. *Biological Applications of Electron Spin Resonance.* 1972. Wiley & Sons, New York, NY. Especially chapters 1 and 2. Undergraduuate to graduate level.

Selected ESR Applications to Foods

(Many original articles can be easily retrieved by computer search.)

Biliaderis, C.C., and Vaughan, D.J. 1987. Electron spin resonance studies of starch-water-probe interactions. *Carbohydr. Polym.* 7:51.

Desrosiers, M.F., McLaughlin, W.L., Sheahen, L.A., Dodd, N.J.F., Lea, J.S., Evans, J.C., Rowlands, C.C., Raffi, J.J., and Agnel, J.P.L. 1990. Co-trial on ESR identification and estimates of gamma-ray and electron absorbed doses given to meat and bones. *Intl. J. Food Sci. Technol.* 25: 682–691.

Dodd, N.J.F., Lea, J.S., and Swallow, A.J. 1988. ESR detection of irradiated food. *Nature* 334:387.

Johnson, J.M., Davis, E.A., and Gordon, J. 1990. Lipid binding of modified corn starches studied by electron spin resonance. *Cereal Chem.* 67:236–240.

Johnson, J.M., Davis, E.A., and Gordon, J. 1990. Interactions of starch and sugar water measured by electron spin resonance and differential calorimetry. *Cereal Chem.* 67:286–291.

Pearce, L.E., Davis, E.A., Gordon, J., and Miller, W.G. 1985. Application of electron spin resonance techniques to model starch systems. *Food Microstruct.* 4:83–88.

Pearce, L.E., Davis, E.A., Gordon, J., and Miller, W.G. 1987. An electron spin resonance study of stearic acid interactions in model wheat starch gluten systems. *Food Microstruct.* 6:121–126.

Pearce, L.E., Davis, E.A., Gordon, J., and Miller, W.G. 1987. Stearic acid interactions as measured by electron spin resonance. *Cereal Chem.* 64:154–157.

Pearce, L.E., Davis, E.A., Gordon, J., and Miller, W.G. 1988. Electron spin resonance studies of isolated gluten systems. *Cereal Chem.* 65:55–58.

Schanen, P.A., Pearce, L.E., Davis, E.A., and Gordon, J. 1990. Hydration of whey protein-wheat starch systems as measured by electron spin resonance. *Cereal Chem.* 67: 124–128.

Schanen, P.A., Pearce, L.E., Davis, E.A., and Gordon, J. 1990. Lipid binding in whey protein-wheat starch systems as measured by electron spin resonance. *Cereal Chem.* 67:317–322.

Stevenson, M.H. 1991. Detection of irradiated food. *Food Sci. & Technol. Today* 5:69–73.

Emerging Methods

Bryant, R.G., Grad, J. and Eads, T.M. 1993. Nuclear cross relaxation spectroscopy. *Anal. Chem.* (Submitted.) Analysis of solid phases in heterogeneous materials including foods.

Wu, J.Y., Bryant, R.G., and Eads, T.M. 1992. Detection of solid-like domains in starch by cross-relaxation NMR spectroscopy. *J. Ag. Food Chem.* 40:449–455.

PART

IV

Chromatography

CHAPTER

28

Basic Principles of Chromatography

Mary Ann Rounds and S. Suzanne Nielsen

28.1 INTRODUCTION

The impact of chromatography has been very great on all areas of analysis and, therefore, on the progress of science in general. Chromatography differs from other methods of separation (e.g., fractional distillation, in which more or less the same operations and apparatus are always employed) in that a wide variety of materials, equipment, and techniques can be used. We have chosen to approach this extremely broad and complex field by first describing the general principles of extraction as a basis for understanding chromatography. This chapter will focus on **liquid chromatography,** subdivided as shown in Fig. 28-1. In view of their independent development and widespread use, gas chromatography and high-performance liquid chromatography are the topics of separate chapters. Supercritical-fluid chromatography, a recently utilized technique, is described briefly in section 28.3.2.

28.2 EXTRACTION

In its simplest form, extraction refers to the transfer of a solute from one liquid phase to another. Extraction in myriad forms is integral to food analysis—whether it be used for preliminary sample cleanup, concentration of the component of interest, or as the actual means of analysis. Extractions may be carried out by means of **batch, continuous,** or **countercurrent** processes.

28.2.1 Batch Extraction

In **batch extraction,** the solute is extracted from one solvent by shaking it with a second, immiscible solvent.

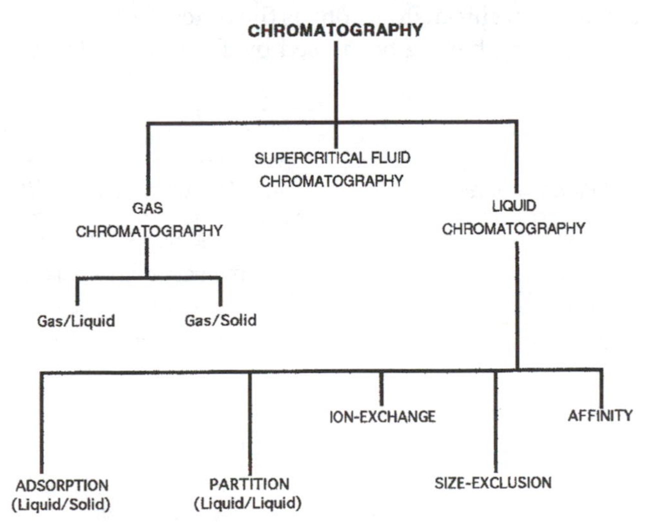

FIGURE 28-1.
A scheme for subdividing the field of chromatography.

The solute **partitions,** or distributes, itself between the two phases and, when equilibrium has been reached, the **partition coefficient,** K, is a constant.

$$K = \frac{\text{concentration of solute in Phase 1}}{\text{concentration of solute in Phase 2}} \quad (1)$$

The phases are allowed to separate, and the layer containing the desired constituent is removed, for example in a separatory funnel. In batch extraction, it is often difficult to obtain a clean separation of phases, owing to emulsion formation. Moreover, partition implies that a single extraction is usually incomplete.

28.2.2 Continuous Extraction

Continuous liquid/liquid extraction requires special apparatus, but is more efficient than batch separation. One example is the use of a Soxhlet extractor for extracting materials from solids. Solvent is recycled so that the solid is repeatedly extracted with fresh solvent. Other pieces of equipment have been designed for the continuous extraction of substances from liquids. Different extractors are used, depending on whether the nonaqueous solvent is heavier or lighter than water.

28.2.3 Countercurrent Extraction

Countercurrent distribution refers to a serial extraction process. It separates two or more solutes with different partition coefficients from each other by a series of partitions between two immiscible liquid phases. (Countercurrent distribution was once an important separation technique used by lipid chemists.) Liquid/liquid partition chromatography is a direct extension of countercurrent extraction; its modern version is known as countercurrent chromatography.

28.3 CHROMATOGRAPHY

28.3.1 Historical Perspective

Modern chromatography originated in the late nineteenth and early twentieth centuries from independent work by David T. Day, a distinguished American geologist and mining engineer, and Mikhail Tsvet, a Russian botanist. Day developed procedures for fractionating crude petroleum by passing it through Fuller's earth; Tsvet used a column packed with chalk to separate leaf pigments into colored bands. Since Tsvet recognized and correctly interpreted chromatographic processes and named the phenomenon **chromatography,** he is generally credited with its discovery.

After languishing in oblivion for years, chromatography began to evolve in the 1940s due to the development of column partition chromatography by Martin and Synge and the invention of paper chromatography. The first publication on gas chromatography (GC) appeared in 1952. By the late 1960s, GC, because of its importance to the petroleum industry, had developed into a sophisticated instrumental technique. Since early applications in the mid-1960s, high-performance liquid chromatography (HPLC), profiting from the theoretical and instrumental advances of GC, has extended the area of liquid chromatography into an equally sophisticated and useful method. Supercritical-fluid chromatography (SFC), demonstrated in 1962, is finally gaining popularity. Modern chromatographic techniques, including automated systems, are widely utilized in the characterization and quality control of food raw materials and food products.

28.3.2 General Terminology

Chromatography is a general term applied to a wide variety of separation techniques based on the partitioning or distribution of a sample **(solute)** between a **moving (mobile) phase** and a fixed or **stationary phase.** (Chromatography may be viewed as a series of equilibrations between the mobile and stationary phases). The relative interaction of a solute with these two phases is described by the **partition (K)** or **distribution (D) coefficient** (ratio of concentration of solute in stationary phase to concentration of solute in mobile phase). The mobile phase may be either a gas (GC) or liquid (LC) or a supercritical fluid. The stationary phase may be a liquid or, more usually, a solid. The field of chromatography can be subdivided or organized in several different ways, according to the physicochemical principles involved in the separation or according to the various techniques applied. Table 28-1 summarizes some of the chromatographic procedures or methods that have been developed on the basis of different mobile-stationary phase combinations. Inasmuch as the nature of interactions between solute molecules and the mobile or stationary phase differ, these methods have the ability to separate different kinds of molecules. (The reader is urged to review Table 28-1 again after having read this chapter.)

As previously stated, this chapter is devoted to the area of **liquid chromatography** (as performed at atmospheric pressure). **Gas chromatography** is covered in Chapter 30 and HPLC is the subject of Chapter 29. **Supercritical-fluid chromatography** (SFC) refers to chromatography performed at pressures and temperatures above the critical values of the mobile phase; a supercritical fluid (or compressed gas) is neither a liquid nor a typical gas. Consequently, SFC complements both GC and HPLC and can overcome some of the problems associated with each. Chapter 8 of *Chromatography,* 5th ed. (1992) provides detailed information on this technique.

28.3.3 Physicochemical Principles of Separation

In order to provide the reader with a broad understanding and better perspective of the field, we have chosen to describe first the physicochemical principles (illustrated in Fig. 28-2) that underlie liquid chromatographic separations, regardless of the specific techniques applied. Although it is more convenient to describe each of these phenomena separately, it must be emphasized that more than one mechanism may be involved in a given fractionation. For example, many cases of partition chromatography also involve adsorption.

28.3.3.1 Adsorption (Solid/Liquid) Chromatography

Adsorption chromatography is the oldest form of chromatography, having been used by Tsvet in 1903 in the

TABLE 28-1. Characteristics of Different Chromatographic Methods

Method	*Mobile/Stationary Phase*	*Retention Varies with*
Gas/liquid chromatography	gas/liquid	molecular size/polarity
Gas/solid chromatography	gas/solid	molecular size/polarity
Supercritical-fluid chromatography	supercritical fluid/solid	molecular size/polarity
Reversed-phase chromatography	polar liquid/nonpolar liquid or solid	molecular size/polarity
Normal-phase chromatography	less polar liquid/more polar liquid or solid	molecular size/polarity
Ion-exchange chromatography	polar liquid/ionic solid	molecular charge
Size-exclusion chromatography	liquid/solid	molecular size
Hydrophobic-interaction chromatography	polar liquid/nonpolar liquid or solid	molecular size/polarity
Affinity chromatography	water/binding sites	specific structure

From (4), used with permission.

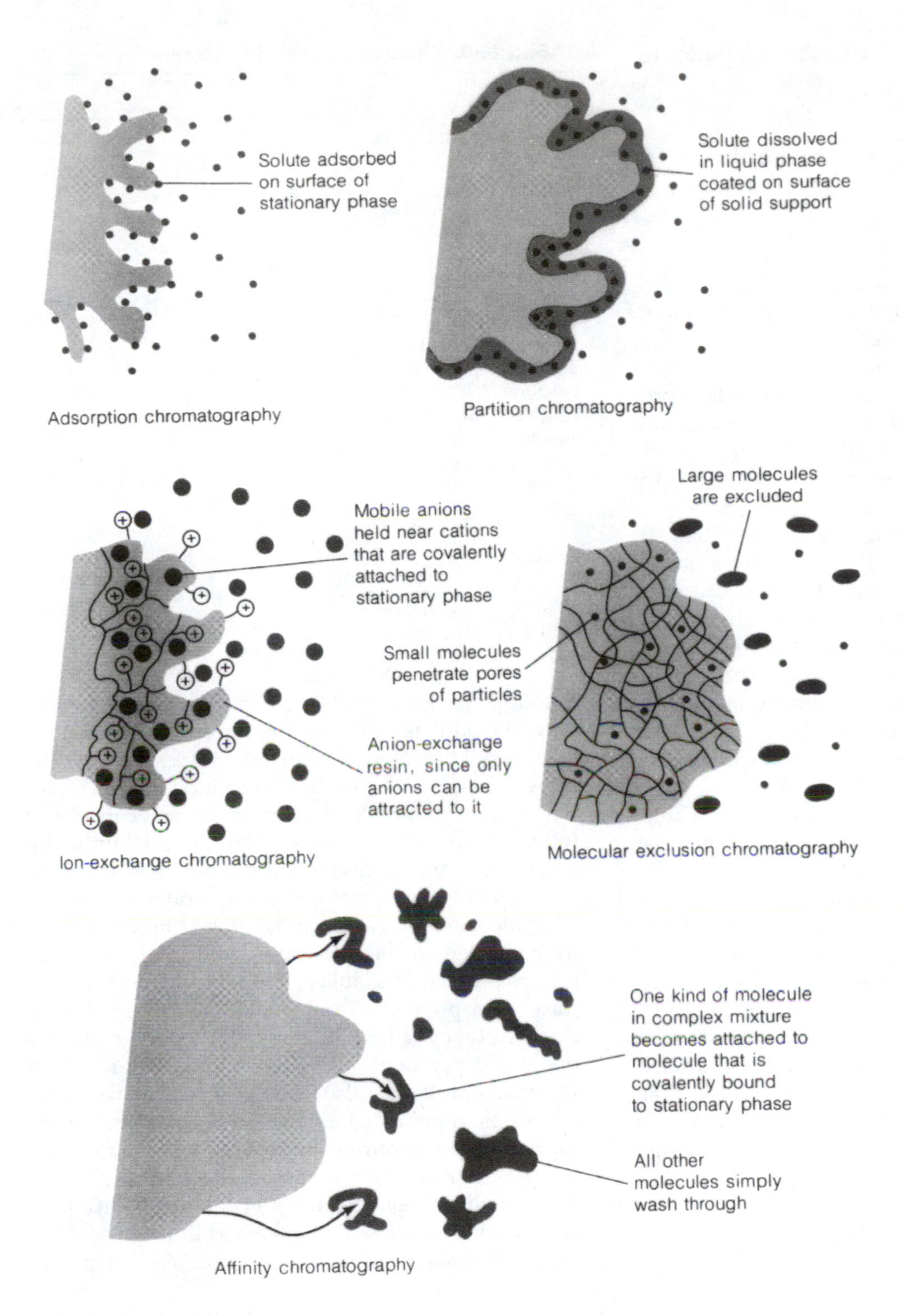

FIGURE 28-2.
Physicochemical principles of chromatography. From *Quantitative Chemical Analysis,* 3rd ed., by D.C. Harris. Copyright © 1991 by W. H. Freeman and Company. Reprinted with permission.

experiments that spawned modern chromatography. In this chromatographic mode, the stationary phase is a finely divided solid (to maximize the surface area), and the mobile phase may be either a gas or a liquid. (Gas/solid adsorption chromatography is discussed in Chapter 30.) The stationary phase (adsorbent) is chosen to permit differential interaction with the components of the sample to be resolved. The intermolecular forces thought to be primarily responsible for chromatographic adsorption include:

1. Van der Waal's forces
2. Electrostatic forces
3. Hydrogen bonds
4. Hydrophobic interactions

Sites available for interaction with any given substance are heterogeneous. Binding sites with greater affinities, the most active sites, tend to be populated first, so that additional solutes are less firmly bound. The net result is that adsorption is a concentration-dependent process, and the **adsorption** coefficient is *not* a constant (in contrast to the **partition** coefficient). Sample loads exceeding the adsorptive capacity of the stationary phase will result in relatively poor separation.

Classical adsorption chromatography utilizes silica (slightly acidic), alumina (slightly basic), charcoal (nonpolar), and a few other materials as the stationary phase. Both silica and alumina possess surface hydroxyl groups, and Lewis acid-type interactions determine their adsorption characteristics. The elution order of compounds from these adsorptive stationary phases can often be predicted on the basis of their relative polarities (Table 28-2). Compounds with the most polar functional groups are retained most strongly on polar adsorbents and, therefore, are eluted last. Nonpolar solutes are less well retained on polar adsorbents and are eluted first.

One model proposed to explain the mechanism of liquid/solid chromatography is that **solute** and **solvent** molecules are competing for active sites on the adsorbent. Thus, as relative adsorption of the mobile phase increases, adsorption of the solute must decrease. Solvents can be rated in order of their strength of adsorption on a particular adsorbent, such as silica. Such a **solvent strength** (or polarity) **scale** is called an **eluotropic series.** Table 28-3 is an example of such a series for alumina. (Silica has a similar rank ordering.)

TABLE 28-2. Compounds Class Polarity Scale[1]

Fluorocarbons
Saturated hydrocarbons
Olefins
Aromatics
Halogenated compounds
Ethers
Nitro compounds
Esters ≈ Ketones ≈ Aldehydes
Alcohols ≈ Amines
Amides
Carboxylic acids

From (6), used with permission.
[1]Listed in order of increasing polarity.

TABLE 28-3. Eluotropic Series for Alumina

Solvent
1-Pentane
Isooctane
Cyclohexane
Carbon tetrachloride
Xylene
Toluene
Benzene
Ethyl ether
Chloroform
Methylene chloride
Tetrahydrofuran
Acetone
Ethyl acetate
Aniline
Acetonitrile
2-Propanol
Ethanol
Methanol
Acetic acid

From (6), used with permission.

Eluotropic series provide the chromatographer with a way to modulate interaction between solutes and the stationary phase. Once an adsorbent has been chosen, solvents can be selected from the eluotropic series for that adsorbent. Mobile phase polarity can be increased (often by admixture of more polar solvents) until elution of the compound(s) of interest has been achieved.

Adsorption chromatography separates aromatic or aliphatic nonpolar compounds, such as lipids, primarily according to the type and number of functional groups present. The labile, fat-soluble chlorophyll and carotenoid pigments from plants have been studied extensively by adsorption chromatography (Tsvet's original experiment) utilizing columns. Adsorption chromatography has also been used for the analysis of fat-soluble vitamins. It is often used as a batch procedure for removal of impurities from samples prior to other analyses. For example, disposable solid-phase extraction cartridges (see Chapter 30) containing silica have been used for food analyses, such as lipids in soybean oil, carotenoids in citrus fruit, and vitamin E in grain.

28.3.3.2 Partition (Liquid/Liquid) Chromatography

28.3.3.2.1 Introduction In 1941, Martin and Synge undertook an investigation of the amino acid composition of wool, using a countercurrent extractor of 40 tubes with chloroform and water flowing in opposite directions. The efficiency of the extraction process was improved enormously by substituting a column of finely divided inert support material for holding one liquid phase (stationary phase) immobile, while allowing the second, immiscible solvent (mobile phase) to flow over it, thus providing intimate contact between the two phases. Solutes partitioned between the two liquid phases according to their partition coefficients; hence the name **partition chromatography.**

A partition system is manipulated by changing the nature of the two liquid phases, usually by combination of solvents or pH adjustment of buffers. Usually, the more polar of the two liquids is held stationary on the inert support and the less polar solvent is used to elute the sample components **(normal-phase chromatography).** Reversal of this arrangement, using a **nonpolar stationary phase** and a **polar mobile phase,** has come to be known as **reversed-phase chromatography** (see section 28.3.3.2.3)

Polar **hydrophilic substances,** such as amino acids, carbohydrates, and water-soluble plant pigments, are separable by **normal-phase** partition chromatography. **Lipophilic** compounds, such as lipids and fat-soluble pigments, may be resolved with **reversed-phase** systems. Liquid/liquid partition chromatography has been invaluable to carbohydrate chemistry. Column chromatography on finely divided cellulose has been used extensively in preparative chromatography of sugars and their derivatives. Paper chromatography (section 28.3.4.1) is still the simplest method to distinguish between various forms of sugars or phenolic compounds present in foods.

28.3.3.2.2 Coated Supports In its simplest form, the stationary phase for partition chromatography consists of a liquid coating on a solid matrix. The solid support should be as inert as possible and have a large surface area in order to maximize the amount of liquid held. Some examples of solid supports that have been used are silica, starch, cellulose powder, and glass beads. All are capable of holding a thin film of water, which serves as the stationary phase. It is important to note that materials prepared for **adsorption** chromatography must be **activated** by drying them to remove surface water. Conversely, some of these materials, such as silica gel, may be used for partition chromatography if they are deactivated by impregnation with water or the desired stationary phase. (The terms **silica gel** or **silicic acid** actually refer to hydrated silica precipitates, the properties of which can vary widely, depending on the method of preparation used. Depending on the amount of water held, silicic acid may act as an adsorbent or a partition support; usually, both properties contribute in varying degrees to the separations achieved.) One disadvantage of liquid/liquid chromatographic systems is that the liquid stationary phase is often stripped off. This problem can be overcome by chemically bonding the stationary phase to the support material, as described in the next section.

28.3.3.2.3 Bonded Supports The liquid stationary phase may be covalently attached to a support by a chemical reaction. These **bonded phases** have become very popular for HPLC use, and a wide variety of both polar and nonpolar stationary phases are now available. (Reactive silanols on the surface of silica gel may be derivatized with alkylamine, alkylnitrile, phenyl, or alkyl groups, as described in Chapter 29.) Again, chromatography performed with a mobile phase less polar than the bonded stationary phase is referred to as normal-phase chromatography. The use of a nonpolar bonded stationary phase (e.g., silica covered with C_8 or C_{18} groups) with a polar solvent (e.g., water/acetonitrile) is called reversed-phase chromatography. The latter has actually become the more widely used of the two methods. (Mechanisms other than partition are often involved in the separation.) Bonded-phase HPLC columns have greatly facilitated the analysis of vitamins in foods and feeds, as discussed in Chapter 3 of reference 7.

28.3.3.3 Ion-Exchange Chromatography

Ion exchange is a separation/purification process occurring naturally, for example, in soils, and utilized in water softeners and deionizers. Three types of separation may be achieved: (1) ionic from nonionic, (2) cationic from anionic, and (3) mixtures of similarly charged species. The first two cases are uncomplicated: One substance binds to the ion-exchange medium, whereas the other does not. Batch methods can be used for them, but chromatography is needed for the third category.

Ion-exchange chromatography may be viewed as a type of adsorption chromatography in which interactions between solute and stationary phase are primarily electrostatic in nature. The stationary phase (ion exchanger) contains fixed functional groups that are either negatively or positively charged (Fig. 28-3a). Exchangeable counterions preserve charge neutrality. A sample ion (or charged sites on large molecules) can exchange with the counterion to become the partner of the fixed charge. Ionic equilibrium is established as depicted in Fig. 28-3b. The functional group of the stationary phase determines whether cations or anions are exchanged. **Cation exchangers** contain covalently bound negatively charged functionalities, whereas **anion exchangers** contain bound positively charged groups. The chemical nature of these acidic or basic residues determines how stationary-phase ionization is affected by the mobile-phase pH.

The strongly acidic sulfonic acid moieties (RSO_3^-) of "strong"-cation exchangers are completely ionized at all pH values above 2. Strongly basic quaternary amine groups ($RNR'_3{}^+$) on "strong"-anion exchangers

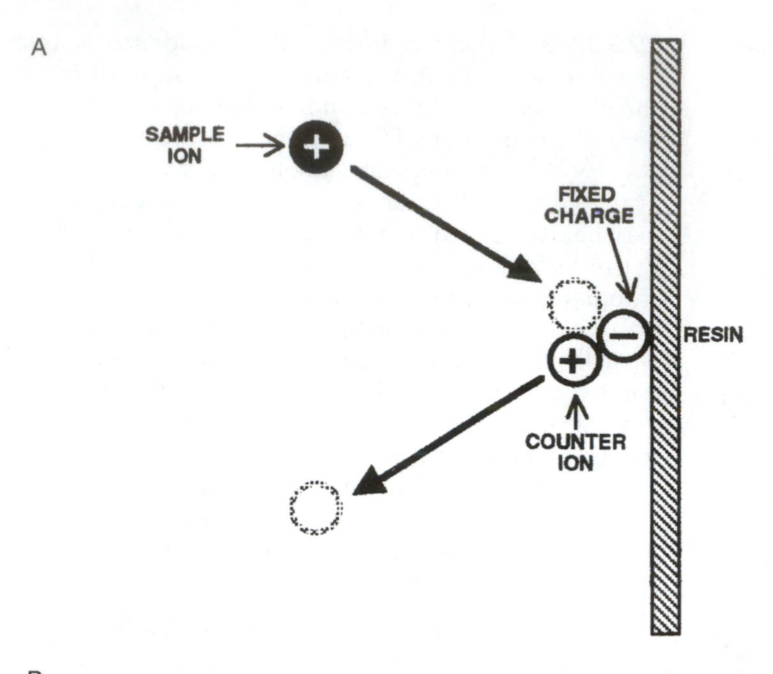

B

$$M^+ + (Na^+ \ {}^-O_3S\text{—}\text{Resin}) \rightleftharpoons (M^+ \ {}^-O_3S\text{—}\text{Resin}) + Na^+$$

M^+ = Cation

$$X^- + (Cl^- \ {}^+R_4N\text{—}\text{Resin}) \rightleftharpoons (X^- \ {}^+R_4N\text{—}\text{Resin}) + Cl^-$$

X^- = Anion

FIGURE 28-3.
The basis of ion-exchange chromatography. *A:* Schematic diagram of the ion-exchange process; *B:* ionic equilibria for cation- and anion-exchange processes. From (6), used with permission.

are ionized at all pH values below 10. Since maximum negative or positive charge is maintained over a broad pH range, the exchange or binding capacity of these stationary phases is essentially constant, regardless of mobile-phase pH. "Weak"-cation exchangers contain weakly acidic carboxylic acid functionalities (RCO_2^-); consequently, their exchange capacity varies considerably between ca. pH 4 to 10. Weakly basic anion exchangers possess primary, secondary, or tertiary amine residues ($R\text{-}NHR'_2{}^+$), which are deprotonated in moderately basic solution, thereby losing their positive charge and the ability to bind anions. Thus, one way of eluting solutes bound to an ion-exchange medium is to change the mobile-phase pH. A second way to elute bound solutes is to increase the ionic strength (e.g., use NaCl) of the mobile phase, to weaken the electrostatic interactions.

Chromatographic separations by ion exchange are based upon differences in affinity of the exchangers for the ions (or charged species) to be separated. The factors that govern **selectivity** of an exchanger for a particular ion include the ionic valence, radius, and concentration; the nature of the exchanger (including its displaceable counterion); and the composition and pH of the mobile phase. To be useful as an ion exchanger, a material must be both ionic in nature and highly permeable. Synthetic ion exchangers are thus crosslinked polyelectrolytes, and they may be inorganic (e.g., aluminosilicates) or, more commonly, organic compounds. Polystyrene, made by crosslinking styrene with divinyl benzene (DVB), may be modified to produce either anion- or cation-exchange resins (Fig. 28-4). Polymeric resins such as these are commercially available in a wide range of particle sizes and with different degrees of crosslinking (expressed as weight percent of DVB in the mixture). The extent of crosslink-

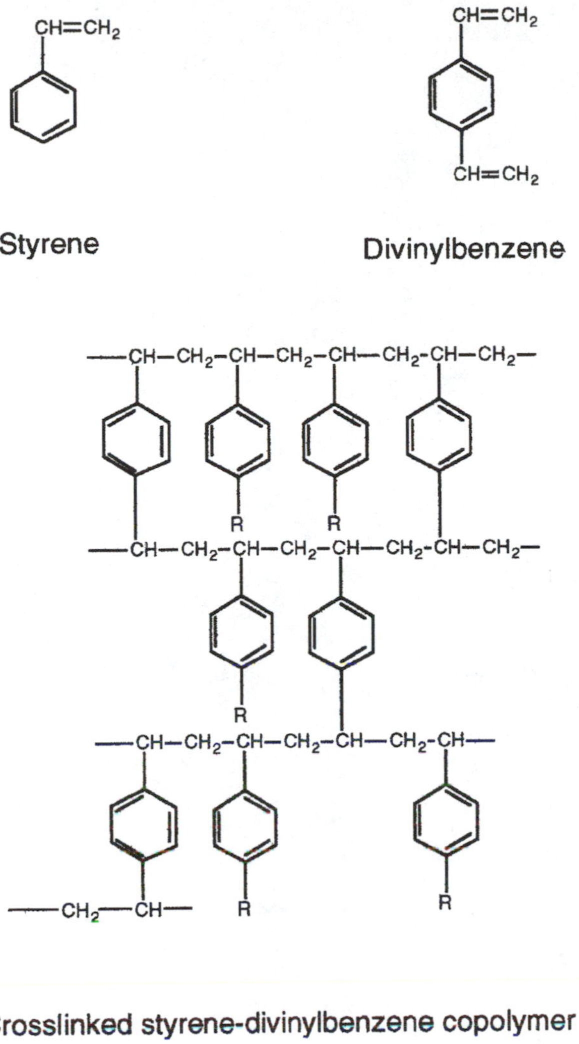

Styrene

Divinylbenzene

Crosslinked styrene-divinylbenzene copolymer

R=H, Plain polystyrene

R=$CH_2N^+(CH_3)_3Cl^-$, Anion-exchanger

R=$SO_3^-H^+$, Cation-exchanger

FIGURE 28-4.
Chemical structure of polystyrene-based ion-exchange resins.

ing controls the rigidity and porosity of the resin, which, in turn, determines its optimal use. Lightly crosslinked resins permit rapid equilibration of solute, but particles swell in water, thereby decreasing charge density and selectivity (relative affinity) of the resin for different ions. More highly crosslinked resins exhibit less swelling, higher exchange capacity, and selectivity, but longer equilibration times. The small pore size, high charge density, and inherent hydrophobicity of the older ion-exchange resins has limited their use to small molecules (M.W. < 500).

Ion exchangers based on polysaccharides, such as cellulose, dextran, or agarose, have proven very useful for the separation and purification of large molecules, such as proteins and nucleic acids. These materials, called **gels** because they are much softer than polystyrene resins, may be derivatized with strong or with weakly acidic or basic groups via OH moieties on the polysaccharide backbone (Fig. 28-5). They have much larger pore sizes and lower charge densities than the older synthetic resins.

Food-related applications of ion-exchange chromatography include the separation of amino acids, sugars, alkaloids, and proteins. Fractionation of amino acids in protein hydrolyzates was initially carried out by ion-exchange chromatography; automation of this process led to the development of commercially produced amino acid analyzers (see Chapter 15). Sugars, complexed with borate to form negatively charged species, can be separated on columns of strong-anion-exchange resin in the borate form. Many drugs, fatty acids, and the acids of fruit, being ionizable compounds, may be chromatographed in the ion-exchange mode.

28.3.3.4 Size-Exclusion Chromatography

Size-exclusion chromatography (SEC), also known as **molecular exclusion, gel permeation** (GPC), and **gel-filtration chromatography** (GFC), is probably the easiest mode of chromatography to perform and to understand. It is widely used in the biological sciences for the resolution of macromolecules, such as proteins and carbohydrates, but is also used for the fractionation and characterization of synthetic polymers. Unfortunately, nomenclature associated with this separation mode developed independently in the literature of the life sciences and in the field of polymer chemistry, resulting in inconsistencies.

In the **ideal SEC** system, molecules are separated solely on the basis of their size; no interaction between solutes and the stationary phase occurs. (In the event that solute/support interactions do occur, the separation mode is termed *nonideal SEC*.) The stationary phase in SEC consists of a column packing material that contains pores comparable in size to the molecules to be fractionated. Solutes too large to enter the pores travel with the mobile phase in the **interstitial space** (between particles) **outside the pores.** Thus, the largest molecules are eluted first from an SEC column. The volume of the mobile phase in the column, termed the **column void volume,** V_o, can be measured by chromatographing a very large (totally excluded) species, such as Blue Dextran, a dye of M.W. = 2×10^6.

As solute dimensions decrease, approaching those

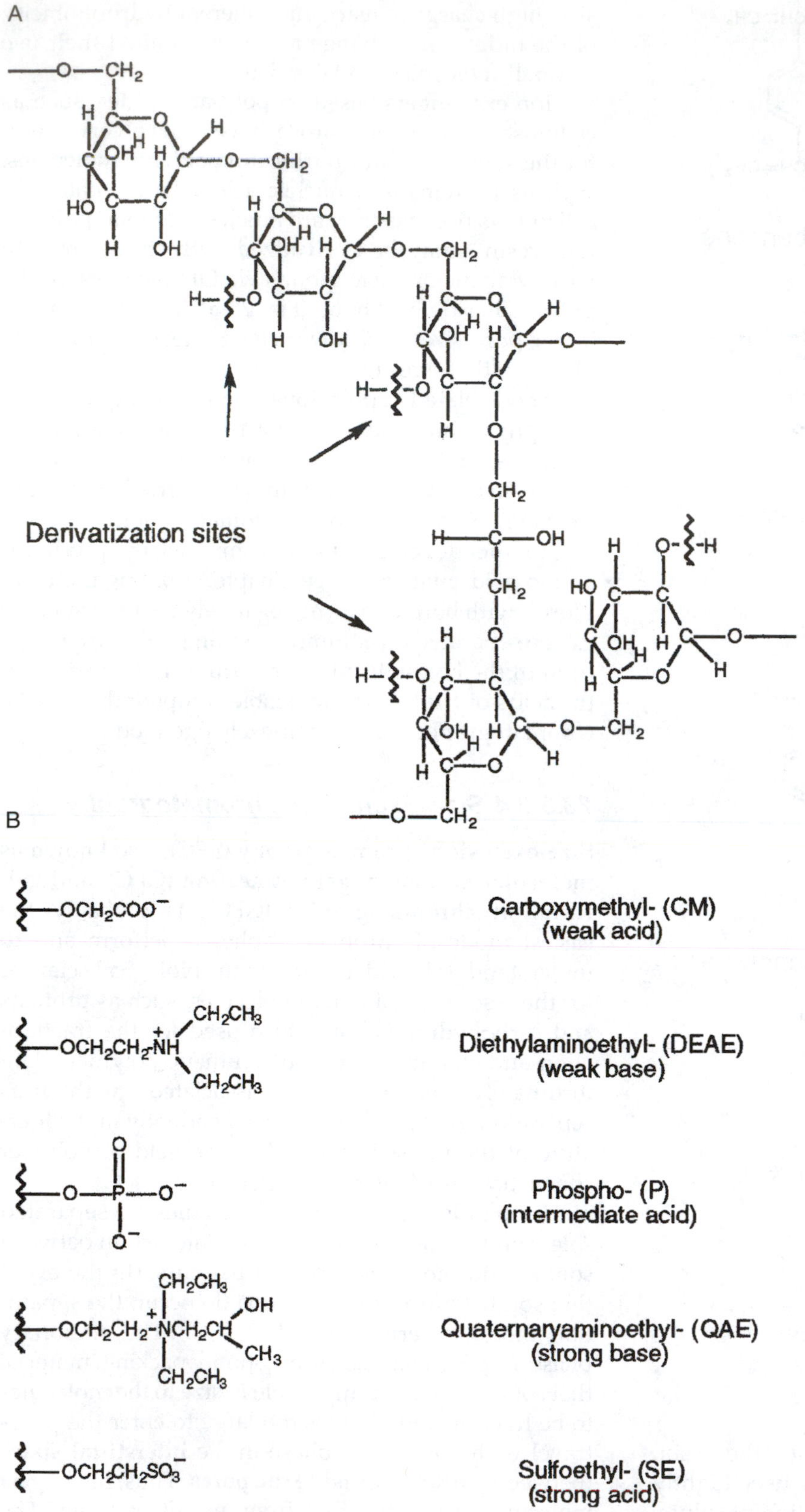

A

Derivatization sites

B

$-OCH_2COO^-$ Carboxymethyl- (CM) (weak acid)

$-OCH_2CH_2\overset{+}{N}H(CH_2CH_3)_2$ Diethylaminoethyl- (DEAE) (weak base)

$-O-P(=O)(O^-)-O^-$ Phospho- (P) (intermediate acid)

$-OCH_2CH_2\overset{+}{N}(CH_2CH_3)_2CH_2CH(OH)CH_3$ Quaternaryaminoethyl- (QAE) (strong base)

$-OCH_2CH_2SO_3^-$ Sulfoethyl- (SE) (strong acid)

FIGURE 28-5.
Chemical structure of one polysaccharide-based ion-exchange resin. *A:* Matrix of crosslinked dextran ("Sephadex," Pharmacia Biotech Inc., Piscataway, NJ); *B:* functional groups that may be used to impart ion-exchange properties to the matrix.

of the packing pores, molecules begin to diffuse into the packing particles and, consequently, are slowed down. Solutes of low molecular weight (e.g., glycyltyrosine) that have free access to all the available pore volume are eluted in the volume referred to as V_t. This value, which is equal to the column void volume, V_o, plus the volume of liquid inside the sorbent pores, V_i, is referred to as the **total permeation volume** of the packed column ($V_t = V_o + V_i$). These relationships are illustrated in Fig. 28-6. Solutes are ideally eluted between the void volume and the total liquid volume of the column. Because this volume is limited, only a relatively small number of solutes (ca. 10) can be completely resolved by SEC under ordinary conditions.

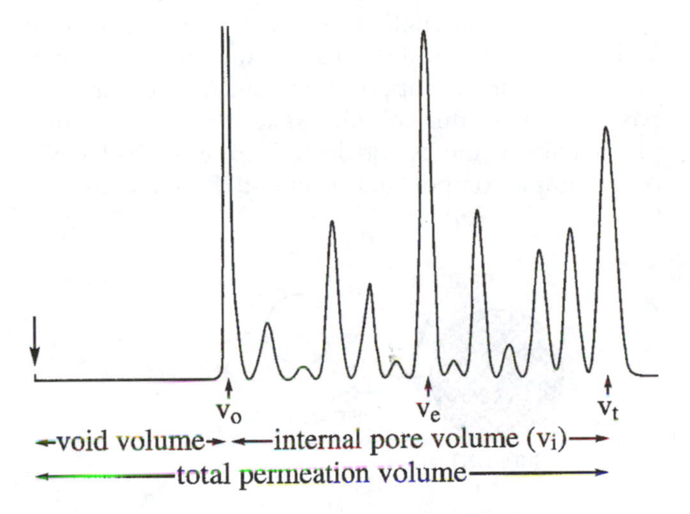

FIGURE 28-6.
Schematic elution profile illustrating some of the terms used in size-exclusion chromatography. Adapted from (4), with permission.

The behavior of a molecule in a size-exclusion column may be characterized in several different ways. Each solute exhibits an **elution volume,** V_e, as illustrated in Fig. 28-6. However, V_e depends on column dimensions and the way in which the column was packed. An expression frequently used to define solute behavior independent of these variables is:

$$K_{av} = (V_e - V_o)/(V_t - V_o) \quad (2)$$

where: K_{av} = available partition coefficient
V_e = elution volume of solute
V_o = column void volume
V_t = total permeation volume of column

The value of K_{av} calculated from experimental data for a solute chromatographed on a given SEC column defines the proportion of pores that can be occupied by that molecule. For a large, totally excluded species, such as Blue Dextran or DNA, $V_e = V_o$ and $K_{av} = 0$. For a small molecule with complete access to the internal pore volume, such as glycyltyrosine, $V_e = V_t$ and $K_{av} = 1$.

For each size-exclusion packing material, a plot of K_{av} versus the logarithm of the molecular weight for a series of solutes, similar in molecular shape and density, will give an S-shaped curve (Fig. 28-7). (In the case of proteins, K_{av} is actually better related to the Stokes radius, the average radius of the protein in solution.) The central, linear portion of this curve describes the **fractionation range** of the matrix, wherein maxi-

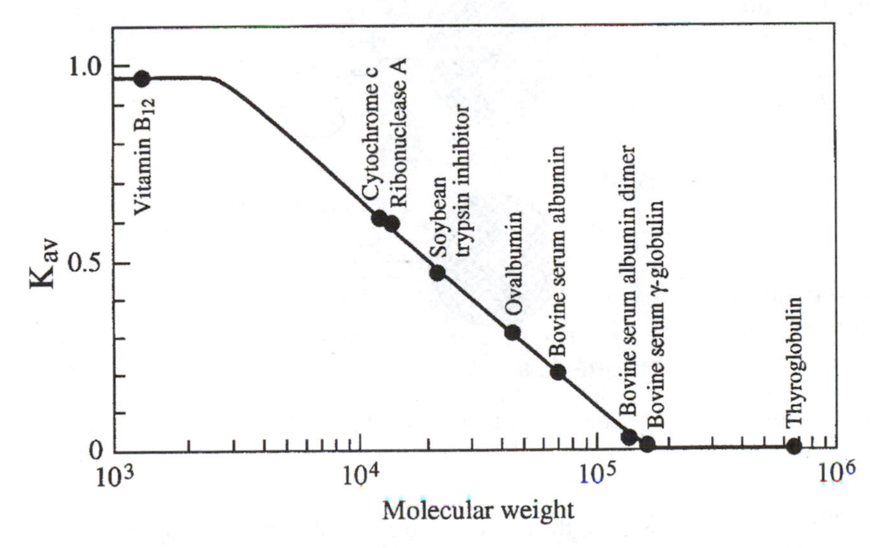

FIGURE 28-7.
Relationship between K_{av} and log (molecular weight) for globular proteins chromatographed on a column of Sephadex G-150 Superfine. Reproduced by permission of Pharmacia Biotech Inc., Piscataway, NJ.

mum separation among solutes of similar molecular weight is achieved. This correlation between solute elution behavior and molecular weight (or size) forms the basis for a widely used method for characterizing large molecules such as proteins and polysaccharides. A size-exclusion column is calibrated with a series of solutes of known molecular weight (or Stokes radius) to obtain a curve similar to that shown in Fig. 28-7. The **available partition coefficient,** K_{av}, for the unknown is then determined, and an estimate of molecular weight (or size) of the unknown is made by interpolation of the calibration curve.

Column packing materials for size-exclusion chromatography can be divided into two groups: semirigid **hydrophobic media** and **hydrophilic gels.** The former are usually derived from polystyrene and are used with organic mobile phases (GPC or nonaqueous SEC) for the separation of polymers, such as rubbers and plastics. Soft gels, polysaccharide-based packings, are typified by Sephadex, a crosslinked dextran (see Fig. 28-5a). These materials are available in a wide range of pore sizes and are useful for the separation of water-soluble substances in the molecular weight range 1–2.5 $\times$ 10^7. In selecting an SEC column packing, both the purpose of the experiment and size of the molecules to be separated must be considered. If the objective is to determine molecular weight, for example, a packing is chosen such that its fractionation range encompasses the anticipated molecular weight of the solute.

As discussed above, size-exclusion chromatography can be used, directly, to fractionate mixtures or, indirectly, as a means of obtaining information about a dissolved species. In addition to molecular weight estimations, SEC is used to determine the molecular weight distribution of natural and synthetic polymers, such as dextrans and gelatin preparations. Fractionation of biopolymer mixtures is probably the most widespread use of SEC, since the mild elution conditions employed rarely cause denaturation or loss of biological activity. It is also a fast, efficient alternative to dialysis for desalting solutions of macromolecules, such as proteins.

28.3.3.5 Affinity Chromatography

Affinity chromatography is unique in that separation is based on the specific, reversible interaction between a solute molecule and a ligand immobilized on the chromatographic stationary phase. While discussed here as a separate type of chromatography, affinity chromatography could be viewed as the ultimate extension of adsorption chromatography. Although the basic concepts of so-called biospecific adsorption were known as early as 1910, they were not perceived as potentially useful laboratory tools until ca. 1968.

Affinity chromatography usually involves immobilized biological materials as the stationary phase. These **ligands** can be antibodies, enzyme inhibitors, lectins, or other molecules that selectively and reversibly bind to complementary analyte molecules in the sample. Separation exploits the lock-and-key binding of biological systems. Although both ligands and the species to be isolated are usually biological macromolecules, the term **affinity chromatography** also encompasses other systems, such as separation of small molecules containing *cis*-diol groups via phenylboronic acid moieties on the stationary phase.

The principles of affinity chromatography are illustrated in Fig. 28-8. A ligand chosen because of the specificity and strength of interaction between itself and the molecule to be isolated (analyte) is immobilized on a suitable support material. As the sample is passed through this column, molecules that are complementary to the bound ligand are adsorbed while other sample components are eluted. Bound analyte is

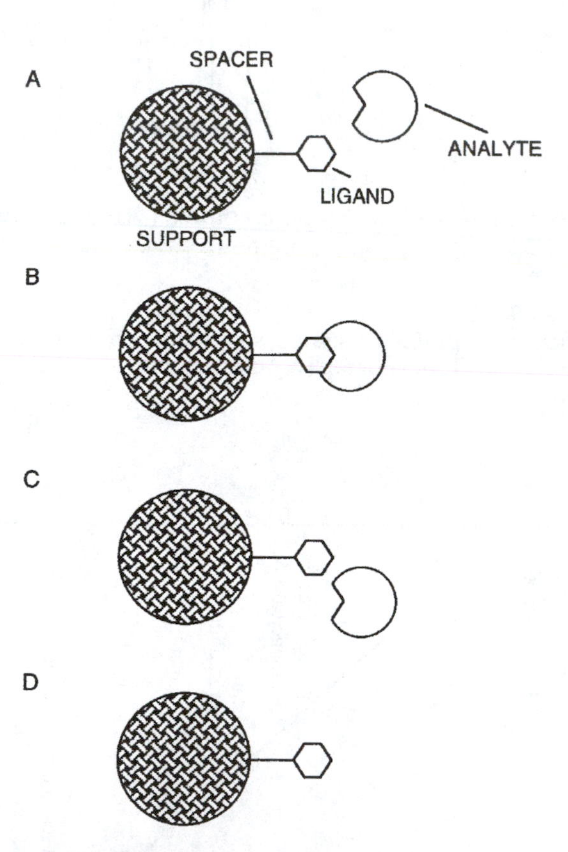

FIGURE 28-8.
Principles of bioselective affinity chromatography. (A) The support presents the immobilized ligand to the analyte to be isolated. (B) The analyte makes contact with the ligand and attaches itself. (C) The analyte is recovered by the introduction of an eluent, which dissociates the complex holding the analyte to the ligand. (D) The support is regenerated, ready for the next isolation. From (4), used with permission.

subsequently eluted via a change in the mobile-phase composition. (For example, changing the pH of the mobile phase often dissociates an enzyme-inhibitor complex.) After re-equilibration with the initial mobile phase, the stationary phase is ready to be used again. The ideal support for affinity chromatography should be a porous, stable, high-surface-area material that does not adsorb anything itself. Thus, polymers such as agarose, cellulose, dextran, and polyacrylamide are used, as well as controlled-pore glass.

Affinity ligands are usually attached to the support or matrix by covalent bond formation, and optimum reaction conditions often must be found empirically. Immobilization generally consists of two steps: **activation** and **coupling.** During the activation step, a reagent reacts with functional groups on the support, such as hydroxyl moieties, to produce an activated matrix. After removal of excess reagent, the ligand is coupled to the activated matrix. (Preactivated supports are now commercially available, and their availability has greatly increased the use of affinity chromatography). The coupling reaction most often involves free amino groups on the ligand, although other functional groups can also be used. When small molecules like phenylboronic acid are immobilized, a **spacer arm** (containing at least 4–6 methylene groups) is used to hold the ligand away from the support surface, enabling it to reach into the binding site of the analyte.

Ligands for affinity chromatography may be either **specific** or **general** (i.e., group specific). Specific ligands, such as antibodies, bind only one particular solute. General ligands, such as nucleotide analogs and lectins, bind to certain classes of solutes. For example, the lectin concanavalin A binds to all molecules that contain terminal glucosyl and mannosyl residues. Bound solutes can then be separated as a group or individually, depending upon the elution technique used. Some of the more common general ligands are listed in Table 28-4. Although less selective, general ligands are widely used because of their greater convenience.

Elution methods for affinity chromatography may be divided into **nonspecific** and **(bio)specific** methods. Nonspecific elution involves disrupting ligand analyte binding by changing the mobile-phase pH, ionic strength, dielectric constant, or temperature. If additional selectivity in elution is desired, for example in the case of immobilized general ligands, a biospecific elution technique is used. Free ligand, either identical to or different from the matrix-bound ligand, is added to the mobile phase. This free ligand competes for binding sites on the analyte. For example, glycoproteins bound to a concanavalin A (lectin) column can be eluted by using buffer containing an excess of lectin. In general, the eluent ligand should display greater affinity for the analyte of interest than the immobilized ligand.

In addition to protein purification, affinity chromatography may also be used to separate supramolecular structures such as cells, organelles, and viruses; concentrate dilute protein solutions; investigate binding mechanisms; and determine equilibrium constants. Affinity chromatography has been especially useful in the separation and purification of enzymes and glycoproteins. In the case of the latter, carbohydrate-derivatized adsorbents are used to isolate specific lectins, such as concanavalin A, and lentil or wheat-germ lectin. The lectin may then be coupled to agarose, such as concanavalin A- or lentil lectin-agarose, to provide

TABLE 28-4. General Affinity Ligands and Their Specificities

Ligand	*Specificity*
Cibacron Blue F3G-A dye, derivatives of AMP, NADH, and NADPH	Certain dehydrogenases via binding at the nucelotide binding site
Concanavalin A, lentil lectin, wheat germ lectin	Polysaccharides, glycoproteins, glycolipids, and membrane proteins containing sugar residues of certain configurations
Soybean trypsin inhibitor, methyl esters of various amino acids, D-amino acids	Various proteases
Phenylboronic acid	Glycosylated hemoglobins, sugars, nucleic acids, and other *cis*-diol-containing substances
Protein A	Many immunoglobin classes and subclasses via binding to the F_c region
DNA, RNA, nucleosides, nucleotides	Nucleases, polymerases, nucleic acids

a stationary phase for the purification of specific glycoproteins, glycolipids, or polysaccharides.

28.3.4 Chromatographic Techniques

The same general principles of chromatography apply, regardless of the specific method or technique used. Paper, thin-layer, and column liquid chromatography all utilize a liquid mobile phase, but the physical form of the stationary phase is quite different in each case. The remainder of this chapter will be devoted to these three techniques.

28.3.4.1 Paper Chromatography

Paper chromatography was introduced in 1944. Although adsorption by the paper itself has been utilized, paper generally serves only as a support for the liquid stationary phase **(partition chromatography).** To carry out this technique, the dissolved sample is applied as a small spot or streak one-half inch or more from the edge of a strip or square of filter paper (usually cellulose) and is allowed to dry. The strip is then suspended in a closed container, the atmosphere of which is saturated with the **developing solvent** (mobile phase), and the paper chromatogram is **developed.** The end closer to the sample is placed in contact with solvent, which then travels up or down the paper by capillary action (depending on whether **ascending** or **descending** development is used), separating sample components in the process. When the solvent front has traveled the length of the paper, the strip is removed from the developing chamber and the separated zones are detected by an appropriate method.

Physical, chemical, or biological methods may be used for the detection or **visualization** of separated compounds. The adsorption or emission of electromagnetic radiation can be measured directly by detectors. If physical methods such as these are not sufficient to identify the substances, chemical reactions **(derivatization)** may be employed, either before or after chromatography. Enzymatic tests are an example of biological detection.

The stationary phase in paper partition chromatography is usually water. However, the support may be impregnated with a nonpolar organic solvent and developed with polar solvents or water (reversed-phase paper chromatography). In the case of complex sample mixtures, a two-dimensional technique may be used. The sample is spotted in one corner of a square sheet of paper, and one solvent is used to develop the paper in one direction. The chromatogram is then dried, turned 90°, and developed again, using a second solvent of different polarity. Another means of improving resolution is the use of ion-exchange papers. Both paper that has been impregnated with ion-exchange resin and paper in which cellulose hydroxyl groups have been derivatized (with acidic or basic moieties) are commercially available.

The extent of migration of a compound on paper depends on the amount of the stationary phase compared to the amount of mobile phase within the same region and on the partition coefficient. It is measured in terms of the R_f value:

$$R_f = \frac{\text{Distance traveled by the center of the zone}}{\text{Distance traveled by the developer front}} \tag{3}$$

Unfortunately, R_f values are not always constant for a given solute/solvent/paper system and, thus, are not universal parameters. They also depend on temperature, the geometry of the paper, the distance of the solute spot from the solvent reservoir, and the duration and direction of development.

28.3.4.2 Thin-Layer Chromatography

Thin-layer chromatography (TLC), first described in 1938, has largely replaced paper chromatography because it is faster, more sensitive, and more reproducible. (Both of these techniques may be referred to as **planar chromatography.**) The resolution in TLC is greater than in paper chromatography because the particles on the plate are smaller and more regular than paper fibers. Experimental conditions can be easily varied to achieve separation and can be scaled up for use in column chromatography. (Thin-layer and column procedures are not necessarily interchangeable, however, due to differences such as the use of binders with TLC plates, vapor-phase equilibria in a TLC tank, etc.) Although TLC has been largely superseded by HPLC methods, both techniques have specific merits. The choice of which to use depends on the analytical problem at hand.

Thin-layer chromatography utilizes a thin (ca. 250 micrometers thick) layer of sorbent on a glass plate or commercially prepared sheet. If adsorption TLC is to be performed, the layer of sorbent (stationary phase) is **activated** by drying for a specified time and temperature. Following activation, the sample mixture (in carrier solvent) is applied as a spot near one end of the plate. After evaporation of carrier solvent, the plate is placed in a closed chamber with the end of the plate nearest the spot in the solvent at the bottom of the chamber. The solvent migrates up the plate by capillary action, and sample components are separated. After the TLC plate has been removed from the developing chamber and the solvent has been allowed to evaporate, the separated spots are made visible by an appropriate method. (For example, lipids may be detected by exposing a TLC plate to iodine vapor or by spraying it with a reagent that results in fluorescence of the lipid spots at 360 nm.) As previously described for paper chromatography, two-dimensional TLC may be used in the case of mixtures so complex that all components cannot be resolved by one solvent. **High-performance thin-layer chromatography** is a relatively new technique (analogous to HPLC) in which TLC plates are coated with smaller, more uniform particles of controlled porosity. This permits better separations in shorter times.

Although TLC applications have utilized partition chromatography (in which silica gel or cellulose serves simply as a matrix for a liquid film), most TLC relies on adsorptive properties of the thin layer. In reality, most applications probably do not involve pure adsorption or pure partition. Instead, one mechanism predominates, depending on experimental conditions. (The stationary phase in TLC is often called the **adsorbent,** even when it functions only as a support for a liquid phase in the partition mode.) An extensive table of applications of TLC to food analysis may be found in Chapter 5 of the book edited by Gruenwedel and Whitaker.

28.3.4.3 Column Liquid Chromatography

Column chromatography is the most useful method of separating compounds in a mixture. Fractionation of solutes occurs as a result of differential migration through a closed tube of stationary phase, and analytes can be monitored while the separation is in progress. The remainder of this chapter will discuss general procedures, theory, and the quantitation of data from column liquid chromatography.

28.3.4.3.1 General Procedures A system for *low-pressure* (i.e., performed at or near atmospheric pressure) column liquid chromatography is illustrated in Fig. 28-9. (While the procedure outlined below is applicable to column chromatography in general, the reader is referred to subsequent chapters for details specific to HPLC or GC).

Having selected a stationary and mobile phase suitable for the separation problem at hand, the analyst must first prepare the **stationary phase (resin, gel** or **packing material)** for use according to the supplier's instructions. (For example, the stationary phase must often be hydrated or preswelled in the mobile phase.) The prepared stationary phase is then packed into a column (usually glass), the length and diameter of which are determined by the amount of sample to be loaded, the separation mode to be used, and the degree of resolution required. Adsorption columns may be either dry- or wet-packed; other types of columns are wet-packed. The most common technique for wet-packing involves making a slurry of the adsorbent with the solvent and pouring this into the column. As the sorbent settles, excess solvent is drained off and additional slurry is added. This process is repeated until

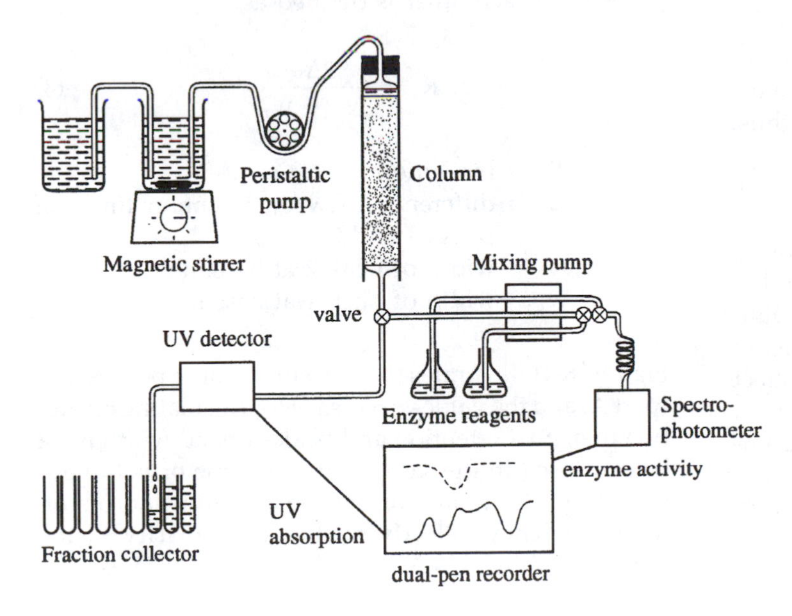

FIGURE 28-9.
A system for low-pressure column liquid chromatography. In this diagram, the column effluent is being split between two detectors in order to monitor both enzyme activity (at right) and UV absorption (at left). The two tracings can be recorded simultaneously by using a dual-pen recorder. Adapted from (8), with permission.

the desired bed height is obtained. (There is a certain art to pouring uniform columns, and no attempt is made to give details here.) If the packing solvent is different from the initial eluting solvent, the column must be thoroughly washed *(equilibrated)* with the starting mobile phase.

The sample to be fractionated, dissolved in a minimum volume of mobile phase, is applied in a layer at the top (or head) of the column. Classical or low-pressure chromatography utilizes only gravity flow or a peristaltic pump to maintain a flow of **mobile phase (eluent** or **eluting solvent)** through the column. In the case of a gravity-fed system, eluent is simply siphoned from a reservoir into the column. The flow rate is governed by the hydrostatic pressure, measured as the distance between the level of liquid in the reservoir and the level of the column outlet. If eluent is fed to the column by a peristaltic pump (see Fig. 28-9), the flow rate is determined by the pump speed and, thus, regulation of hydrostatic pressure is not necessary.

The process of passing the mobile phase through the column is called **elution,** and the portion that emerges from the outlet end of the column is sometimes called the **eluate** (or effluent). Elution may be **isocratic** (constant mobile-phase composition) or a **gradient** may be used. Gradient elution refers to changing the mobile phase (e.g., increasing solvent strength or pH) during elution in order to enhance resolution and decrease analysis time. The change may be continuous or stepwise. Gradients of increasing ionic strength are extremely valuable in ion-exchange chromatography. Gradient elution is commonly used for desorbing large molecules, such as proteins, which can undergo multiple-site interaction with a stationary phase. As elution proceeds, components of the sample are selectively retarded by the stationary phase according to one (or more) of the mechanisms discussed earlier and, thus, are eluted at different times.

The column effluent may be directed through a detector and then into receptacles, changed at intervals by a fraction collector. The detector response, in the form of an electrical signal, may be recorded (the **chromatogram**) and used for qualitative or quantitative analysis, as will be discussed in more detail later. The fraction collector may be set to collect eluate at specified time intervals or after a certain volume or number of drops has been collected. Components of the sample that have been chromatographically separated in this manner can then be further analyzed as needed.

28.3.4.3.2 Qualitative Analysis The volume of liquid required to elute a compound from a liquid chromatography column is called the **retention volume,** V_R. The associated time is the **retention time,** t_R. Comparing V_R or t_R to that of standards chromatographed under identical conditions often enables one to identify an unknown compound. (One should remember that different compounds may have identical retention times.) A related technique is to spike the unknown sample with a known compound and compare chromatograms of the original and spiked samples to see which peak has increased. In most cases, it will be necessary to collect the peak(s) of interest and establish their identity by another analytical method.

Often it is necessary to be able to compare chromatograms obtained from two different systems or columns. Differences in column dimensions, loading, temperature, flow rate, system dead-volume, and detector geometry may lead to discrepancies for uncorrected retention data. By subtracting the time required for the mobile phase or a nonretained solute (t_m or t_o) to travel through the column to the detector, one obtains an **adjusted retention time,** t'_R (or volume), as depicted in Fig. 28-10. The adjusted retention time (or volume) corrects for differences in system dead-volume; it may be thought of as the time the sample spends sorbed on the stationary phase.

A simple, reliable method for the identification of peaks is to use relative retention as expressed by the **separation factor,** α. Values for α (Fig. 28-10) depend only on temperature and the stationary phase and mobile phase used.

28.3.4.3.3 Separation and Resolution **1. Overview.** The goal of chromatography is to segregate components of a sample into separate bands or peaks as they migrate through the column. The **resolution** of two peaks from each other is defined as

$$R_S = \frac{2\Delta t}{w_2 + w_1} \tag{4}$$

where: R_S = resolution
Δt = difference between retention times of peaks 1 and 2
w_2 = width of peak 2 at baseline
w_1 = width of peak 1 at baseline

Figure 28-11 illustrates the measurement of peak width (part A) and the values necessary for calculating resolution (part B). (Retention and peak or band width must be expressed in the same units, i.e. time or volume.)

Chromatographic resolution is a function of column **efficiency, selectivity,** and the **capacity factor.** Mathematically, this relationship is expressed as

$$R_S = \underbrace{1/4\sqrt{N}}_{a}\ \underbrace{\left(\frac{\alpha - 1}{\alpha}\right)}_{b}\ \underbrace{\left(\frac{k'}{k' + 1}\right)}_{c} \tag{5}$$

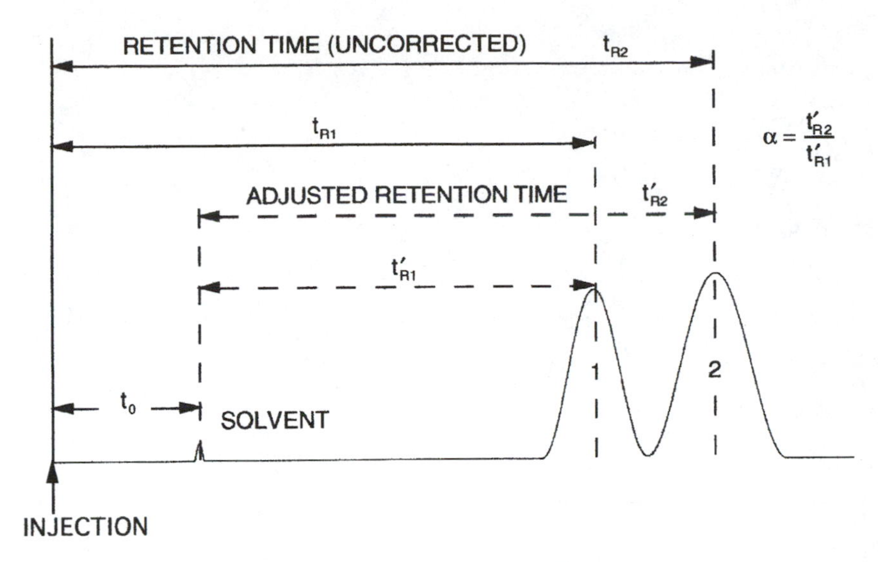

FIGURE 28-10.
Measurement of chromatographic retention. Adapted from (6), with permission.

where: a = the column **efficiency** term
b = the column **selectivity** term
c = the **capacity** term

These terms, and factors that contribute to them, will be discussed in the following paragraphs.

2. Efficiency. If faced with the problem of improving resolution, a chromatographer should first examine the **efficiency** of the column. An efficient column keeps the bands from spreading and/or gives narrow peaks. Column efficiency can be quantitated by

$$N = \left(\frac{t_R}{\sigma}\right)^2 = 16\left(\frac{t_R}{w}\right)^2 = 5.5\left(\frac{t_R}{w_{1/2}}\right)^2 \tag{6}$$

where: N = number of theoretical plates
t_R = retention time
σ = standard deviation for a Gaussian peak
w = peak width at baseline (w = 4σ)
$w_{1/2}$ = peak width at half-height

The measurement of t_R, w, and $w_{1/2}$ is illustrated in Fig. 28-11. (Retention volume may be used instead of t_R; in this case, band width is also measured in units of volume.) Although few peaks are actually Gaussian in shape, normal practice is to treat them as if they were. In the case of peaks that are incompletely resolved or slightly asymmetric, peak width at half-height is more accurate than peak width at baseline.

The value N calculated from the above equation is called the number of **theoretical plates.** The theoretical plate concept, borrowed from distillation theory, can best be understood by viewing chromatography as a series of equilibrations between mobile and stationary phases, analogous to countercurrent distribution. A column, then, would consist of N segments (theoretical plates) with one equilibration occurring in each. As a first approximation, N is independent of retention time and is, therefore, a useful measure of column performance. One method of monitoring column performance over time is to chromatograph a standard compound periodically, under constant conditions, and to compare the values of N obtained. Band broadening due to column deterioration will result in a decrease of N.

The number of theoretical plates is generally proportional to column length. Since columns are available in various lengths, it is useful to have a measure of column efficiency that is independent of column length. This may be expressed as

$$\text{HETP} = \frac{L}{N} \tag{7}$$

where: HETP = height equivalent of a theoretical plate
L = column length
N = number of theoretical plates

The so-called **height equivalent of a theoretical plate,** HETP, is sometimes more simply described as **plate height** (H). If a column consisted of discrete segments, HETP would be the height of each imaginary segment. Small plate height values (a large number of plates) indicate good efficiency of separation.

Obviously, plate theory is an oversimplification because chromatography is a continuous process. Columns are not divided into discrete segments and equilibration is not infinitely fast. Also, columns often

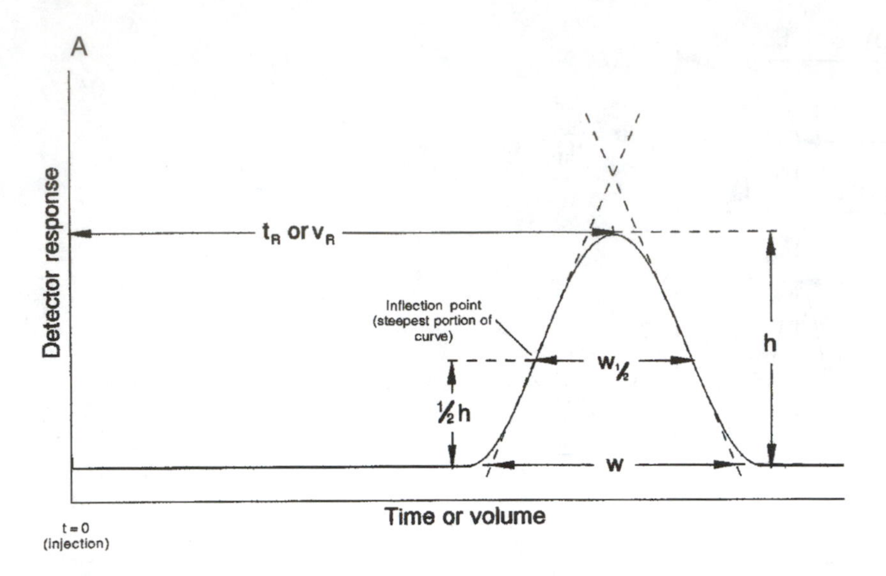

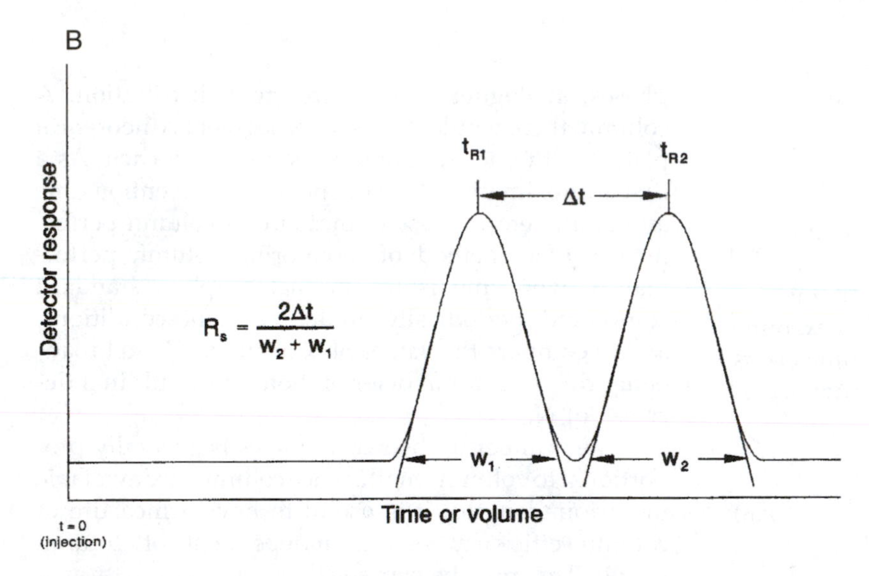

FIGURE 28-11.
Measurement of peak width and its contribution to resolution. A: Idealized Gaussian chromatogram, illustrating the measurement of w and $w_{1/2}$; B: the resolution of two bands is a function of both their relative retentions and peak widths. Adapted from (6), with permission.

behave as if they had a different number of plates for different solutes in a mixture. A more realistic description of the movement of solutes through a chromatography column takes into account the finite rate at which a solute can equilibrate itself between stationary and mobile phases. Thus, band shape depends on the rate of elution and is affected by solute diffusion. Any mechanism that causes a band of solute to broaden will increase HETP, consequently decreasing column efficiency. The various factors that contribute to plate height are expressed by the **Van Deemter equation:**

$$\text{HETP} = \text{A}u^{1/3} + \frac{\text{B}}{u} + \text{C}u \qquad (8)$$

where:

HETP = height equivalent to one theoretical plate
A, B, C = constants
u = mobile phase velocity

The constants A, B, and C are characteristic for a given column, mobile phase, and temperature. The $\text{A}u^{1/3}$ term refers to eddy diffusion or multiple flowpaths. Eddy diffusion arises from the different microscopic flowstreams that the mobile phase can take between particles in the column. Sample molecules can thus take different paths as well, depending on which flowstreams they follow. As a result, solute molecules spread from an initially narrow band to a broader area within the column. The B/u term of the Van Deemter equation refers to molecular diffusion along the length of the column, parallel to the direction of flow. Since the concentration of solute is lower at the edges of the solute zone than at the center, solute is always diffusing toward the edges of the zone. The more time a solute spends on the column, the greater will be its diffusive spreading. The third term in the Van Deemter equation, Cu, arises from the finite time required for solute to equilibrate itself between the mobile and stationary phases (rate of mass transfer). Ideally, each solute molecule is continually transferred into and out of the stationary phase. While in the stationary phase, however, a molecule is slowed down relative to the band center, which continues to migrate down the column. Conversely, when the molecule is in the mobile phase, it moves faster than the band center, since flow velocity exceeds band velocity. The net result is that the actual concentration of solute in the mobile phase is never in equilibrium with that in the immediately adjacent stationary phase, and the solute band is broadened. As shown by the Van Deemter equation, the mobile phase velocity, u, contributes to plate height in opposing ways—increasing the flow rate increases nonequilibrium and eddy diffusion (A and C terms) but decreases longitudinal diffusion (because a solute spends less time on the column). Optimizing the mobile-phase flow rate for a given column should reduce HETP and increase column efficiency.

3. Selectivity. Chromatographic resolution depends on **selectivity** as well as efficiency. Column selectivity refers to the distance, or relative separation, between two peaks and is given by

$$\alpha = \frac{t_{R2} - t_o}{t_{R1} - t_o} = \frac{t'_{R2}}{t'_{R1}} = \frac{K_2}{K_1} \quad (9)$$

where:

α = separation factor

t_{R1} and t_{R2} = retention times of components 1 and 2, respectively

t_o (or t_m) = retention time of unretained components (solvent front)

t'_{R1} and t'_{R2} = adjusted retention times of components 1 and 2, respectively

K_1 and K_2 = distribution coefficients of components 1 and 2, respectively

Retention times (or volumes) are measured as shown in Fig. 28-10. The time, t_o, can be measured by chromatographing a solute that is not retained under the separation conditions (i.e., travels with the solvent front). When this parameter is expressed in units of volume, V_o or V_m, it is known as the "dead-volume" of the system. Selectivity is a function of the stationary and/or mobile phase. For example, selectivity in ion-exchange chromatography is influenced by the nature and number of ionic groups on the matrix but can also be manipulated via pH and ionic strength of the mobile phase. Good selectivity is probably more important to a given separation than high efficiency (Fig. 28-12) since resolution is directly related to selectivity but is quadratically related to efficiency; that is, a fourfold increase in N is needed to double R_S (eq. 5).

4. Capacity Factor. The **capacity** or retention **factor,** k', is a measure of the amount of time a chromatographed species (solute) spends in/on the stationary phase relative to the mobile phase. The relationship between capacity factors and chromatographic retention (which may be expressed in units of either volume

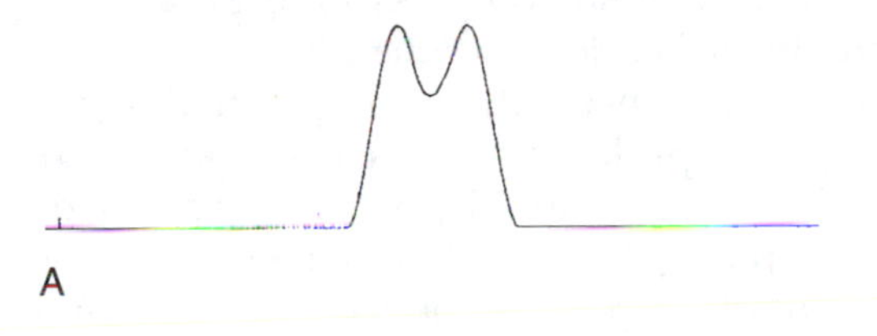

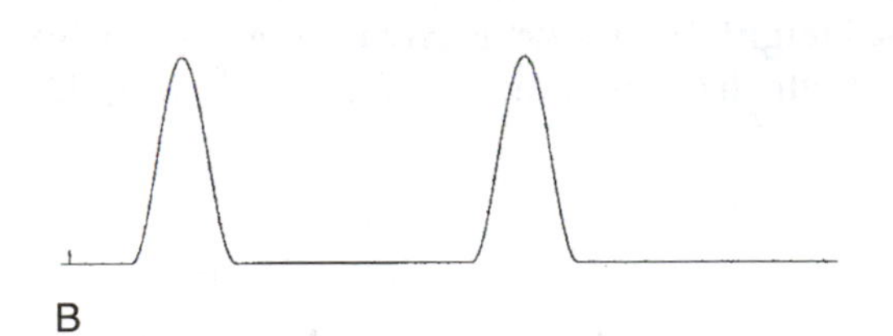

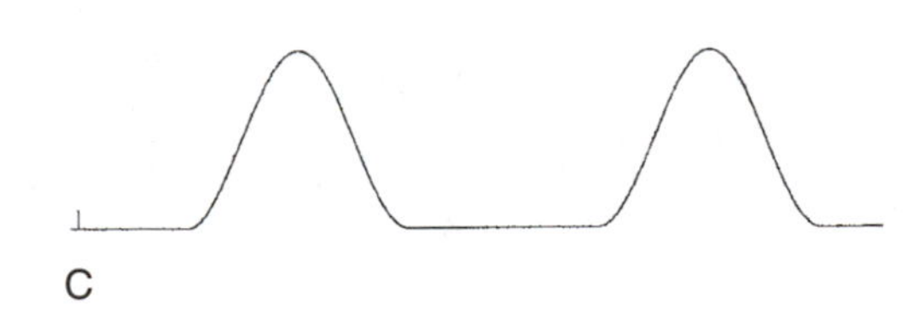

FIGURE 28-12.
Chromatographic resolution: efficiency *vs.* selectivity. *A:* Poor resolution; *B:* good resolution due to high column efficiency; *C:* good resolution due to column selectivity. From (6), used with permission.

or time, as previously discussed in section 28.3.4.3.2) is shown below:

$$k' = \frac{KV_S}{V_m} = \frac{V_R - V_m}{V_m} = \frac{t_R - t_o}{t_o} \quad (10)$$

where: k' = capacity factor
K = distribution coefficient of the solute
V_S = volume of stationary phase in column
V_m = volume of mobile phase
V_R = retention volume of solute
t_R = retention time of solute
t_o = retention time of unretained components (solvent front)

Small values of k′ indicate little retention, and components will be eluted close to the solvent front, resulting in poor separations. Large values of k′ result in improved separation but can also lead to broad peaks and long analysis times. On a practical basis, k′ values within the range of 1 to 15 are generally used. (In the equation for R_s, k′ is actually the average of k′ and k_2' for the two components separated.)

28.3.4.3.4 Quantitative Analysis Assuming that good chromatographic resolution and identification of sample components have been achieved, quantitation involves measuring peak height, area, or mass and comparing these data with those for standards of known concentration. (It should be remembered that a completely resolved peak is not necessarily equivalent to a pure substance. One peak can also represent several components that are not resolvable under the chromatographic conditions utilized.)

1. Peak Height versus Peak Area. There is much debate about which of these techniques is more useful. In general, peak height is less dependent on flow rate, but peak area is less affected by other instrumental and operator variations. Peak height is simply measured as the distance from baseline to peak maximum (see Fig. 28-11). Interpolation of the baseline from start to finish may be used to compensate for baseline drift (often a problem in gradient elution). Peak height measurement generally gives precisions of 1–2 percent and should not be used with visibly distorted peaks.

Several methods may be used to measure peak area, as depicted in Fig. 28-13. Since chromatographic peaks often approximate a triangle, area can be calculated by the formula for a triangle, $A = (w/2)\,h$ (Fig. 28-13a). The **width at half-height** ($w_{1/2}$) is used to reduce errors due to adsorption and tailing. This method should be used only for symmetrical peaks or those that have similar shapes. The area measured is less than the true area but is proportional to sample size, provided peaks are not badly distorted. Precision depends on the ratio of $h:w_{1/2}$, which preferably falls into a range of 2 to 10. A second area-measurement method is **triangulation,** which requires drawing lines tangent to the sides of the peak. Peak height and width are then measured as shown in Fig. 28-13b. Since this technique offers no greater accuracy than the previous method and is subject to operator error (drawing of the tangent lines), it is not recommended. A **planimeter** is a mechanical device that can be used to measure peak areas by tracing around their perimeters (Fig. 28-13c). The precision and accuracy of this method depend on the device itself and on operator skill. The planimeter technique can be more accurate than triangulation, especially if the peaks are skewed.

The **cut-and-weigh** method of quantitating peak mass requires carefully cutting out the chromatographic peak (or a copy of it) and weighing the paper

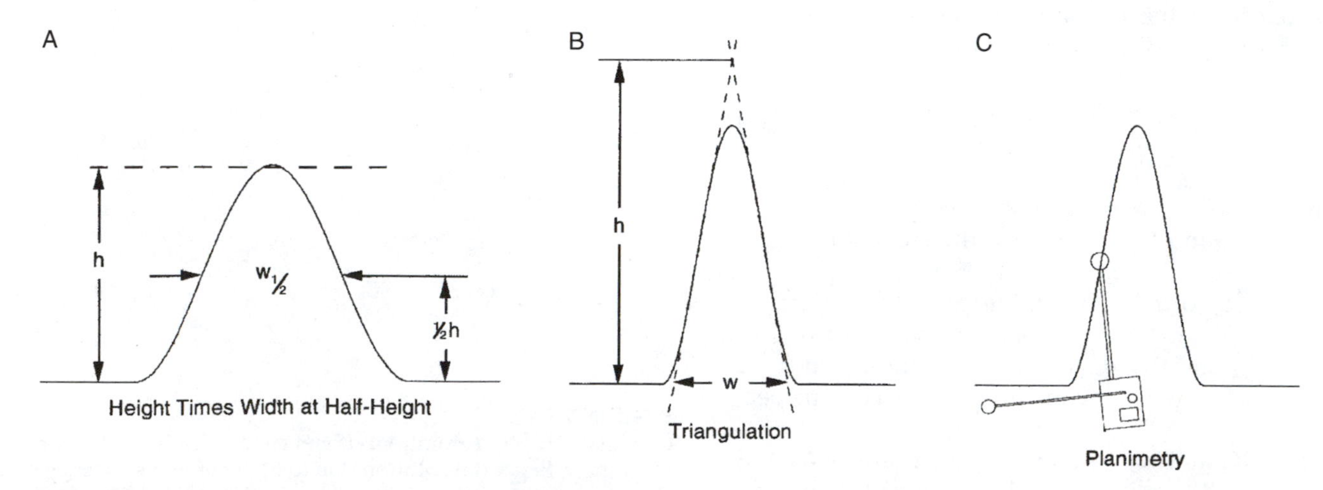

FIGURE 28-13.
Three different methods of estimating peak area. *A:* Peak height × width at half-height; *B:* triangulation; *C:* planimetry. From (6), used with permission.

on an analytical balance. The inaccuracy of cutting can be minimized by keeping the ratio of $h:w_{1/2}$ between 1 and 10. Homogeneity of the paper, moisture content, and the weight of the paper are all important factors, but this method is superior to triangulation techniques for irregularly shaped peaks.

Electronic integrators provide the chromatographer with highly precise and automatic conversion of detector output into numerical form. The only disadvantages of this method are that digital integrators are rather expensive and setting parameters often requires a fairly high level of operator sophistication.

2. External versus Internal Standards. Having quantitated sample peaks, one must compare these data with appropriate standards of known concentration in order to determine sample concentrations. Comparisons may be by means of **external** or **internal** standards. Comparison of peak height, area, or mass of unknown samples with standards injected separately, **external standards,** is common practice. Standard solutions covering the desired concentration range (preferably diluted from one stock solution) are chromatographed, and the appropriate data (peak height, area, or mass) plotted versus concentration to obtain a standard curve. An identical volume of sample is then chromatographed, and height, area, or mass of the sample peak is used to determine sample concentration via the standard curve (Fig. 28-14a). This absolute calibration method requires precise analytical technique and requires that detector sensitivity be constant from day to day if the calibration curve is to remain valid.

Use of the **internal standard** (relative or indirect) method can minimize errors due to sample preparation, apparatus, and operator technique. In this technique, a compound is utilized that is structurally related to, but is eluted independently of, compounds of interest in the sample to be analyzed. Basically, the amount of each component in the sample will be determined by comparing the height, area, or mass of that component peak to the height, area, or mass of the internal standard peak. However, variation in detector response between compounds of different chemical structure must be taken into account. One way to do this is by first preparing a set of standard solutions containing varying concentrations of the compound(s) of interest. Each of these solutions contains the same amount of the internal standard. These standards are chromatographed, and peak height, area, or mass is measured. Ratios of peak height, area, or mass (compound of interest/internal standard) are calculated and plotted against concentration to obtain calibration curves such as those shown in Fig. 28-14b. A separate response curve must be plotted for each sample component to be quantitated. Next, a known amount of internal standard is added to the unknown sample, and the sample is chromatographed. Peak height, area, or mass ratios (compound of interest/internal standard) are calculated and used to read the concentration of each relevant component from the appropriate calibration curve. The advantages of using internal standards are that injection volumes need not be accurately measured and the detector response need not remain constant since any change will not alter ratios. The main disadvantage of the internal standard technique is the difficulty of finding a standard that does not interfere chromatographically with components of interest in the sample.

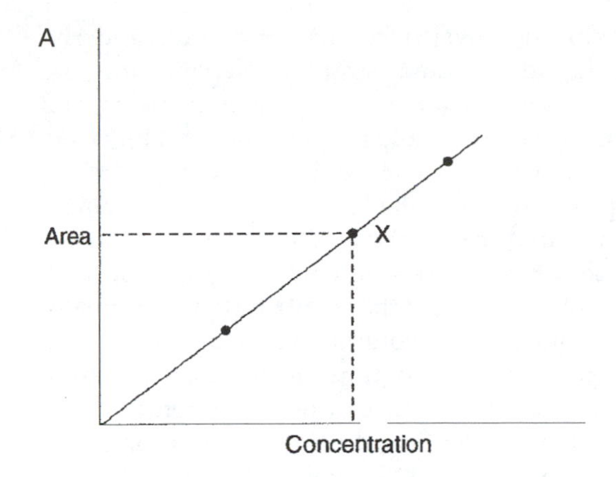

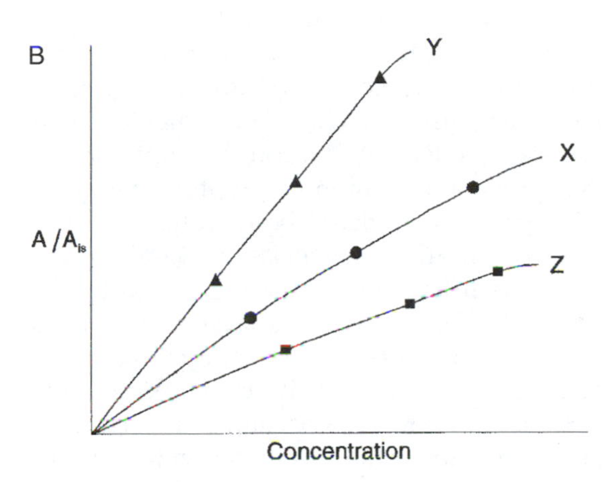

FIGURE 28-14.
Calibration curves for quantitation of a sample component, *x*. *A:* External standard technique; *B:* internal standard technique. Adapted from (6), with permission.

28.4 SUMMARY

Chromatography is a separation method based on the partitioning of a solute between a moving, or mobile, phase and a fixed, or stationary, phase. The mobile phase may be liquid, gas, or a supercritical fluid. The stationary phase may be a liquid or a solid. This chapter focuses on liquid chromatography, chromatography performed (at atmospheric pressure) with a liquid mobile phase. Basic physicochemical principles underlying all liquid chromatographic separations are adsorption, partition, ion exchange, size exclusion, and affinity. General terminology, and the specific techniques of paper, thin-layer, and column liquid chromatography are described. In the latter, sample components may be resolved into separate bands or peaks as they migrate through a column packed with stationary phase. Factors that contribute to chromatographic resolution are column efficiency, selectivity, and capacity. A chromatogram provides both qualitative and quantitative information via retention time (or volume) and peak height (area or mass) data.

For an introduction to the techniques of high-performance liquid chromatography and gas chromatography, the reader is referred again to Chapters 29 and 30 in this text. *Chromatography*, the standard work edited by E. Heftmann (1992 and earlier editions), is an excellent source of information on both fundamentals (Part A) and applications (Part B) of chromatography. Part B includes chapters on the chromatographic analysis of amino acids, proteins, lipids, carbohydrates, and phenolic compounds. In addition, *Fundamental* and *Applications Reviews* published (in alternating years) by the journal *Analytical Chemistry* relate new developments in all branches of chromatography, as well as their application to specific areas such as food. Recent books and general review papers are referenced, along with research articles published during the specified review period.

28.5 STUDY QUESTIONS

1. Differentiate batch, continuous, and countercurrent extraction, and explain how extraction relates to chromatography.
2. Describe the difference between adsorption and partition chromatography with respect to stationary phases. How does a solute interact with the stationary and mobile phase in each case?
3. Distinguish between normal-phase and reversed-phase chromatography by comparing the nature of the stationary and mobile phases and the order of solute elution.
4. What is the advantage of bonded supports over coated supports for partition chromatography?
5. You applied a mixture of proteins, in a buffer at pH 8.0, to an anion-exchange column. On the basis of some assays you performed, you know that the protein of interest adsorbed to the column.
   a. Does the anion-exchange stationary phase have a positive or negative charge?
   b. What is the overall charge of the protein of interest that adsorbed to the stationary phase?
   c. Is the isoelectric point of the protein of interest (adsorbed to the column) higher or lower than pH 8.0?
   d. What are the two most common methods you could use to elute the protein of interest from the anion-exchange column? Explain how each method works. (See also Chapter 15.)
6. Explain how you would use size-exclusion chromatography to estimate the molecular weight of a protein molecule. Include an explanation of what information must be collected and how it is used.
7. Would you use a polystyrene- or a polysaccharide-based stationary phase for work with protein? Explain your answer.
8. Explain the principle of affinity chromatography, why a spacer arm is used, and how the solute can be eluted.
9. Compare paper chromatography, thin-layer chromatography, and column liquid chromatography. Explain similarities and differences.
10. What is gradient elution from a column, and why is it often advantageous over isocratic elution?
11. Using the Van Deemter equation, HETP, and N, as appropriate, explain why the following changes may increase the efficiency of separation in column chromatography:
    a. changing the flow rate of the mobile phase
    b. increasing the length of the column
    c. reducing the inner diameter of the column
12. How can chromatographic data be used to quantitate sample components?
13. Why would you choose to use an internal standard rather than an external standard? Describe how you would select an internal standard for use.

28.6 REFERENCES

1. Dorsey, J.G., Foley, J.P., Cooper, W.T., Barford, R.A., and Barth, H.G. 1990. Liquid chromatography: Theory and methodology (Fundamental Review). *Anal. Chem.* 62:324R–356R.
2. Gruenwedel, D.W. and Whitaker, J.R. (Eds.) 1987. *Food Analysis Principles and Techniques*, Vol. 4, *Separation Techniques*. Marcel Dekker, Inc., New York, NY.
3. Harris, D.C. 1991. *Quantitative Chemical Analysis*, 3rd ed. W.H. Freeman and Co., New York, NY.
4. Heftmann, E. (Ed.) 1992. *Chromatography*, 5th ed. *Fundamentals and Applications of Chromatography and Related Differential Migration Methods.* Part A: *Fundamentals and Techniques.* Part B: *Applications.* Journal of Chromatography Library series volumes 51A and 51B. Elsevier, Amsterdam, The Netherlands.
5. Heftmann, E. (Ed.) 1975. *Chromatography: A Laboratory Handbook of Chromatographic and Electrophoretic Methods*, 3rd ed. Van Nostrand Reinhold, New York, NY.
6. Johnson, E.L., and Stevenson, R. 1978. *Basic Liquid Chromatography.* Varian Associates, Palo Alto, CA.
7. Lawrence, J.F. (Ed.) 1984. *Food Constituents and Food Residues: Their Chromatographic Determination.* Marcel Dekker, Inc., New York, NY.
8. Scopes, R.K. 1987. *Protein Purification: Principles and Practice*, 2nd ed. Springer-Verlag, New York, NY.
9. Snyder, L.R., and Kirkland, J.J. (Eds.) 1979. *Introduction to Modern Liquid Chromatography*, 2nd ed. John Wiley & and Sons, New York, NY.
10. Walters, R.R. 1985. Report on affinity chromatography. *Anal. Chem.* 57:1099A–1113A.

CHAPTER

29

High-Performance Liquid Chromatography

Mary Ann Rounds and Jesse F. Gregory, III

29.1 INTRODUCTION

High-performance liquid chromatography (HPLC) developed during the 1960s as a direct offshoot of classical column liquid chromatography through improvements in the technology of columns and instrumental components (pumps, injection valves, and detectors). Originally, HPLC was the acronym for *high-pressure liquid chromatography,* reflecting the high operating pressures generated by early columns. By the late 1970s, however, *high-performance liquid chromatography* had become the preferred term, emphasizing the effective separations achieved. In fact, newer columns and packing materials offer high performance at moderate pressure (although still high relative to gravity-flow liquid chromatography). *Advantages* of HPLC over traditional low-pressure column liquid chromatography include (1):

1. Speed (many analyses can be accomplished in 30 minutes or less)
2. Improved resolution (due, in part, to a wide variety of stationary phases)
3. Greater sensitivity (various detectors can be employed)
4. Reusable columns (although initially expensive, columns can be used for many analyses)
5. Ideal for ionic species and large molecules (substances of low volatility)
6. Easy sample recovery

High-performance liquid chromatography can be applied to the analysis of any compound with solubility in a liquid that can be used as the mobile phase. Unlike gas chromatography, sample components need not be volatile; derivatization, when used with HPLC, serves to enhance detectability of the analyte. Although most frequently employed as an **analytical** technique, HPLC may also be used in the **preparative** isolation of complex mixtures.

29.2 COMPONENTS OF AN HPLC SYSTEM

A schematic diagram of a basic HPLC system is shown in Fig. 29-1. The main components of this system—**pump, injector, column, detector,** and **recorder/data system**—will be discussed briefly in the sections below. (Additions of an autosampler or fraction collector are optional.) Connecting tubing, tube fittings, and the materials out of which components are constructed are also important since they influence system performance and longevity. References 1–4 include detailed discussions of HPLC equipment. Manufacturers can also be a source of practical information on HPLC instrumentation and fittings (5).

29.2.1 Pump

The role of the **HPLC pump** is to deliver the mobile phase through the system, typically at a flow rate of 1 ml/min, in a controlled, accurate, and precise manner. The two main types of pumps used are **constant pressure** and **constant volume.** Constant pressure pumps may be either of a reciprocating or a syringe-type design. Reciprocating pumps produce a pulsating flow, thereby requiring the addition of mechanical or electronic pulse dampers to suppress fluctuations. A mechanical **pulse damper** or **dampener** consists of a device (such as a deformable metal component or tubing filled with compressible liquid) that can change its volume in response to changes in pressure. Screw-driven syringe pumps produce pulseless flow but suffer the disadvantage of limited reservoir capacity. **Gradient elution systems** for HPLC may utilize low-pressure mixing, in which mobile phase components

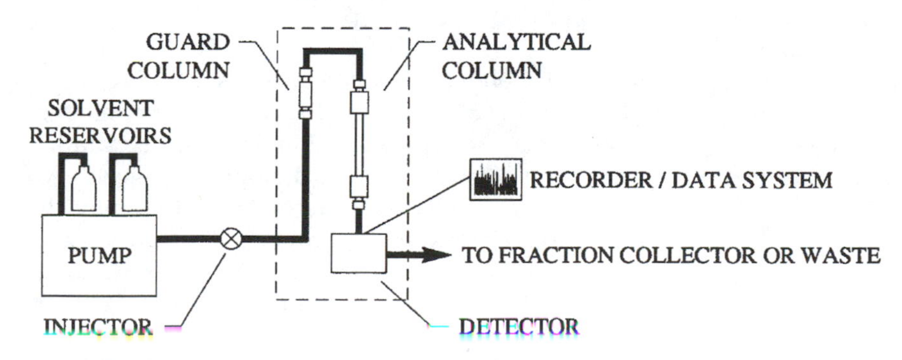

FIGURE 29-1.
Schematic representation of a system for high-performance liquid chromatography (not drawn to scale). Column(s) and detector may be thermostated, as indicated by the dashed line, for operation at elevated temperature.

are mixed before entering the high pressure pump, or high-pressure mixing, in which two or more independent, programmable pumps are used.

Most commercially available HPLC pumping systems and connecting lines are made out of grade ANSI 316 stainless steel, which can withstand the pressures generated and is resistant to corrosion by oxidizing agents, acids, bases, and organic solvents. Mineral acids and halide ions do attack stainless steel. Thus, the system should always be rinsed thoroughly with water if these substances have been used in the mobile phase. All HPLC pumps contain moving parts, such as check valves and pistons, and are quite sensitive to dust and particulate matter in the liquid being pumped. Therefore, it is advisable to filter the mobile phase using 0.45 or 0.22 μm porosity filters prior to use. Degassing HPLC eluents, by application of a vacuum and/or ultrasonication or by sparging with helium, is also recommended to prevent the problems that can be caused by air bubbles in a pump or detector.

29.2.2 Injector

The role of the injector is to place the sample into the flowing mobile phase for introduction onto the column. **Loop-type injection valves** are most frequently used in HPLC. A syringe is used to inject sample (at atmospheric pressure) into a loop connected between two ports of the multiport valve. When the valve is switched from "load" to "inject" positions, mobile phase (under higher pressure) is directed through the loop, thereby carrying sample onto the head of the column. Such systems are generally trouble free and afford good precision. Although injection volumes of 10 to 100 μl are most typical, both larger (e.g., 1–10 ml) and smaller (e.g., ≤2 μl) volumes of sample can be injected by utilizing special hardware.

Autosamplers may be used to automate the injection process. Samples are placed in uniform-size vials, sealed with a septum, and held in a revolving tray. A needle penetrates the septum to withdraw solution from the vial, and an electronically or pneumatically operated valve introduces it onto the column. Autosamplers can reduce the tedium and labor costs associated with routine HPLC analyses and improve assay precision. However, considerable time must be invested in initial setup, and these devices are seldom trouble free. Because samples may remain unattended for 12 to 24 hr prior to automatic injection, sample stability is a key factor to consider before purchasing this accessory.

29.2.3 Column

An HPLC column could be considered a tool for the separation of molecules. Since both external **hardware** and internal **packing material** are important, these two topics are discussed separately.

29.2.3.1 Column Hardware

An **HPLC column** is usually constructed of stainless steel tubing with terminators that allow it to be connected between the injector and detector of the system (Fig. 29-1). Columns have also been made from glass, fused silica, titanium, and PEEK (polyether ether ketone) resin. The most commonly used analytical HPLC column today has a diameter of 4.6 mm and a length of 10 to 25 cm and contains particles of 5 or 10 μm diameter (microparticulates). However, columns are commercially available in a wide variety of dimensions and formats to meet different application needs (2, 6).

Smaller (≤ 5 cm long) expendable columns called **guard columns** are often used to protect the analytical column from strongly adsorbed sample components. A guard column or cartridge is installed between the injector and analytical column via short lengths of capillary tubing or a special cartridge holder. Although the type of packing material in a guard column should be identical to that in the analytical column, larger particles (e.g., 30 to 40 μm) are frequently used because they allow the chromatographer to easily repack the guard column. A guard column or cartridge should always be repacked or replaced before its capacity is exceeded and contaminants break through to the analytical column. A silica-based guard column can also act as a **presaturator column** when harsh eluents are used. Columns containing large particles of bare silica may be placed between the HPLC pump and injector to presaturate the mobile phase with silica. This can help to slow down dissolution of silica-based packing materials during the use of aqueous mobile phases.

29.2.3.2 HPLC Column Packing Materials

The development of a wide variety of column packing materials has contributed substantially to the success and widespread use of HPLC.

29.2.3.2.1 General Requirements A **packing material** serves, first of all, to form the chromatographic bed. It may or may not be involved in the actual separation process, i.e. distribution of a solute between two phases. In classical liquid/liquid chromatography, the packing serves only to support the stationary phase, which is the liquid that resides in its pores (Chapter 28, section 28.3.3.2). In size-exclusion chromatography, it is important that there be no interaction between solutes and the column packing material since separation is accomplished, ideally, on the basis of differences in molecular size (Chapter 28, section 28.3.3.4). However, in adsorptive modes of chromatography, including ion-exchange and affinity, the column packing material simultaneously serves as both **support** and **stationary phase.** Additional requirements for HPLC column packing materials are (2):

1. Availability in a well-defined particle size, with a narrow particle size distribution
2. Sufficient mechanical strength to withstand pressure generated during packing and use
3. Good chemical stability

29.2.3.2.2 Silica-Based Column Packings **1. Porous Silica. Porous silica** meets the above criteria quite well and can be prepared in a wide range of particle and pore sizes (with a narrow particle size distribution). Both **particle size** and **pore diameter** are important with regard to HPLC separations. Small particles reduce the distance a solute must travel between stationary and mobile phases. This facilitates equilibration and results in good column efficiencies, meaning a large number of theoretical plates per unit of column length (Chapter 28, section 28.3.4.3.3). However, small particles also mean greater flow resistance; thus, higher pressure drops at equivalent flow velocity. Three, 5, or 10 μm particles are utilized in analytical columns. Larger particles may be used for industrial-scale preparative chromatography, since they are less expensive.

One half or more of the volume of porous silica consists of voids or **pores** (2). Choice of **pore diameter** is important, inasmuch as packing material surface area is inversely related to the mean pore diameter. In adsorptive modes, chromatographic retention (Chapter 28, section 28.3.4.3.2) increases as the amount of stationary phase surface area increases. Thus, use of the smallest possible pore diameter will maximize **surface area** and **sample capacity** (the amount of sample that can be separated on a given column). Packing materials with a pore diameter of 50 to 100 A and surface area of 200 to 400 m^2/g are used for low molecular weight (<500) solutes (2). For increasingly larger molecules, such as proteins and nucleic acids, it is necessary to use wider pore materials (pore diameter $\geq$ 300 A) so that internal surface is accessible to the solute (7).

Silica consists mainly of silicon dioxide, SiO_2, with each Si atom at the center of a tetrahedron. On the surface, one remaining valency is generally occupied by an -OH group, referred to as a **silanol.** These weakly acidic groups ($pK_a \sim 9$) (4) are rather reactive and can be utilized to modify the silica surface.

2. Bonded Phases. So-called **bonded phases** (Fig. 29-2a) are made by chemically bonding hydrocarbon groups to the surface of silica particles via surface silanols. Often, the silica is reacted with an organochlorosilane (4):

$$\geq Si{-}OH + Cl{-}\overset{R_1}{\underset{R_2}{Si}}{-}R_3 \longrightarrow \geq Si{-}O{-}\overset{R_1}{\underset{R_2}{Si}}{-}R_3 + HCl \quad (1)$$

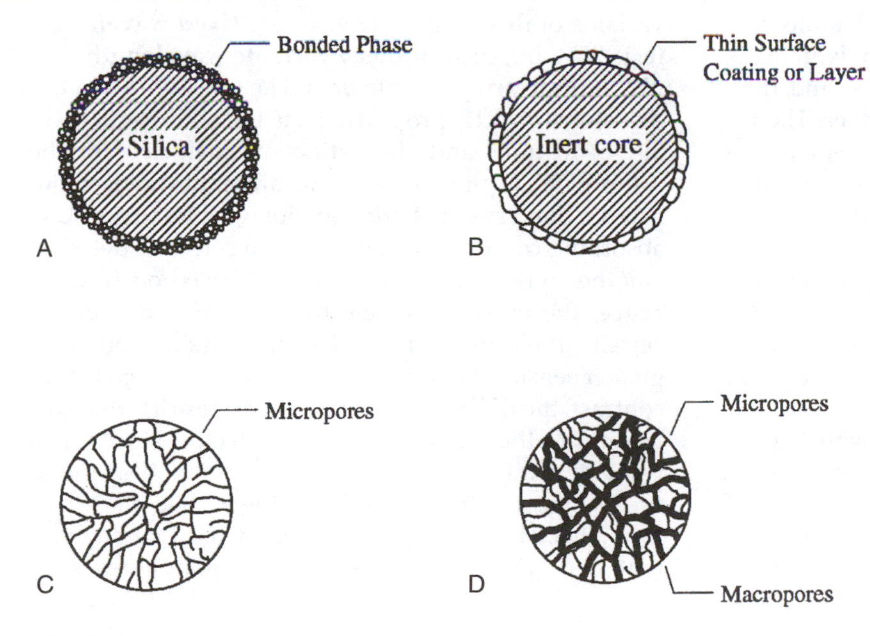

FIGURE 29-2.
Some types of packing materials utilized in HPLC. *A:* Bonded-phase silica; *B:* pellicular packing; *C:* microporous polymeric resin; *D:* macroporous polymeric resin. Adapted from (8), p. 621, by courtesy of Marcel Dekker, Inc.

Substituents R_1 and R_2 may be halides or methyl groups. The nature of R_3 determines whether the resulting bonded phase will exhibit normal-phase, reversed-phase, or ion-exchange chromatographic behavior. The siloxane (-Si-O-Si-) bond is stable in the pH range ca. 2 to 7.5.

The main disadvantage of silica and silica-based bonded-phase column packings is that the silica skeleton slowly dissolves in aqueous solutions, especially at pH > 8. Consequently, much effort has gone into the development of nonsilica HPLC packings. Other inorganic materials of greater pH stability that may be used are alumina, porous graphitic carbon, and hydroxyapatite.

3. Pellicular Packings. A **pellicular packing material** (Fig. 29-2b) is made by depositing a thin layer or coating onto the surface of an inert, usually nonporous, microparticulate **core.** Core material may be either inorganic or organic. Functional groups such as ion-exchange sites are then present at the surface only. The rigid core ensures good physical strength, whereas the thin stationary phase provides for rapid mass transfer and favorable column efficiency. A disadvantage is that the thin surface coating limits the number of interactive sites; consequently, binding capacity is low. Glass beads of limited porosity covered with chemically attached groups and silicas with extensive bonded-phase coverage also exhibit characteristics of pellicular packings. The distinction between bonded-phase and pellicular packings is not always clear cut (8).

Pellicular coatings on porous supports, such as macroporous silica or polystyrene, have proven to be quite useful for HPLC of large biological molecules such as proteins and oligonucleotides. A polyamine is physically adsorbed to the support surface and then crosslinked into a permanent polymeric layer. The resulting pellicular coating extends pH stability of underlying inorganic media and can mask the undesirable hydrophobicity of a polystyrene matrix (9).

29.2.3.2.3 Polymeric Column Packings Synthetic organic resins offer the advantages of good chemical stability and the possibility to vary interactive properties through direct chemical modification. Two categories of polymeric packing materials exist.

1. Microporous (Microreticular). Microporous or gel-type resins (Fig. 29-2c) are comprised of crosslinked copolymers in which the apparent porosity, evident only when the gel is in its swollen state, is determined by the degree of crosslinking. Styrene, crosslinked with 2 to 16 percent divinylbenzene, is an example of a microporous polymer. These gel-type packings undergo swelling and contraction with changes in the chromatographic mobile phase, which can result in bead fracture, poor mass transfer, and increased pressure drop and resistance to flow. Microporous polymers of less than ca. 8 percent crosslinking are not sufficiently rigid for HPLC use.

2. Macroporous (Macroreticular). Macroporous resins are highly crosslinked (e.g., ≥50%) and consist of a network of microspheric gel beads joined together to form a larger bead (Fig. 29-2d). Large permanent pores, ranging from 100 to 4000A or more in diameter, and high surface areas (≥100 m^2/g) are the result of interstitial spaces between the micro beads (8). Rigid microparticulate poly(styrene-divinylbenzene) packing materials of the macroporous type are now popular for HPLC use. They are stable from pH 1–14 and are available in a variety of particle and pore sizes. These resins can be used in unmodified form for reversed-phase chromatography or functionalized for use in other HPLC modes.

29.2.4 Detector

A **detector** translates concentration changes in the HPLC column effluent into electrical signals. Spectrochemical or electrochemical properties of solutes may be measured by a variety of instruments, each of which has advantages and disadvantages. The choice of which to use depends on solute type and concentration and detector sensitivity, linear range, and compatibility with the solvent and elution mode to be used. Most HPLC analyses are carried out using the **ultraviolet-visible** (UV-Vis) **absorption detector.** It measures the absorption of radiation in the wavelength range between 190 and 800 nm according to **Beer's law.** Some variants of this type of detector are **fixed wavelength** (e.g., 254 nm), **continuously variable wavelength,** and **photodiode-array** instruments. The sensitivity of a UV detector is directly proportional to the pathlength (usually 5–10 mm) and the extinction coefficient of the sample. It is quite sensitive to aromatics and conjugated π bond systems but does not respond to non-UV-absorbing compounds; hence, it is a selective detector. Another type of optical detector is based on **fluorescence,** the emission of electromagnetic radiation by certain molecules. This detection mode provides greater sensitivity and selectivity than UV detection. In contrast, the **differential refractive index (RI) detector** belongs to the universal class of detectors since it can respond to all solutes. Refractive index monitors measure change in refractive index of the mobile phase due to solutes. These detectors are widely used for analytes that do not contain UV-absorbing chromophores, such as carbohydrates and lipids. However, RI detectors suffer from sensitivity to changes in temperature and flow rates and cannot be used with gradient elution.

Electrochemical methods are also utilized for HPLC detection. **Amperometric detectors** measure the change in current as analyte is oxidized or reduced by the application of voltage across electrodes of the flow cell. Sensitivity and selectivity of amperometric detectors can sometimes be improved by the application of pulse techniques. **Pulsed amperometric detection** has recently been utilized in HPLC of carbohydrates. Another type of electrochemical detector is the **conductivity monitor,** which responds to the presence of ions eluting from the column. Ion-chromatography instruments generally use conductivity detection (section 29.3.3.2.1). Despite many attempts, there is not as yet a truly universal HPLC detector. However, since most detection methods are nondestructive, multiple monitors can be connected in series. There is also the possibility to couple spectrometers with HPLC. For example, mass spectrometers can be used as HPLC detectors (HPLC-MS).

Detection sensitivity or specificity can sometimes be enhanced by converting the analyte to a chemical derivative with different spectral or redox characteristics. An appropriate reagent can be added to the sample prior to injection, **pre-column derivatization,** or combined with column effluent before it enters the detector, **post-column derivatization.** Automated amino acid analyzers utilize post-column derivatization, usually with ninhydrin, for reliable and reproducible analyses of amino acids. Pre-column derivatization of amino acids with o-phthalaldehyde or similar reagents permits highly sensitive HPLC determination of amino acids using fluorescence detection.

29.2.5 Recorder/Data System

The results of a chromatographic separation are usually displayed on a strip-chart recorder. From this chart are obtained retention time (or volume) and peak height or area values. These data are used to identify and quantitate components of the sample, as discussed in Chapter 28. The addition of an electronic integrator can simplify data reduction, especially in the case of a large number of samples. Both accuracy and precision of analysis can be increased, assuming that operator-selected variables are optimized. Basically, retention time, peak area (or height), and percent of total peak area (or height) are determined for each component of the chromatogram and automatically printed out. Additional features may include the ability to calculate analyte levels based on multiple external or internal standards, compensation for nonlinear detector response, and the transfer of data to computers. As with other analytical instrumentation, computer-controlled HPLC systems are rapidly becoming more common. Software packages to aid data evaluation are commercially available (see Chapter 19, section 19.1.3.8.1 for some examples specific to HPLC analysis of pesticide residues). The field of computer-aided optimization of stationary/mobile phase systems is still under development (2).

29.3 SEPARATION MODES IN HPLC

The basic physicochemical principles underlying all liquid chromatographic separations—adsorption, partition, ion exchange, size exclusion, and affinity—are discussed in Chapter 28, and details will not be repeated here. The number of separation modes utilized in HPLC, however, is greater than that available to the classical chromatographer. This is due to the success of bonded phases, initially developed to facilitate classical liquid/liquid partition chromatography (Chapter 28, section 28.3.3.2.). In fact, reversed-phase HPLC is the most widely used separation mode in modern column liquid chromatography.

29.3.1 Normal Phase

29.3.1.1 Stationary and Mobile Phases

In **normal-phase** HPLC, the **stationary phase** is a **polar adsorbant** such as bare silica or silica to which polar nonionic functional groups—alcoholic hydroxyl, nitro, cyano (nitrile), or amino—have been chemically attached. Reactions similar to that described in section 29.2.3.2.2 are used, with terminal polar substituents being linked to the silica through a short hydrocarbon spacer (e.g., R_3 = $-CH_2CH_2CH_2NH_2$, aminopropyl) (8). These bonded phases are moderately polar and have advantages over bare silica: Solute-stationary phase interactions are less extreme and the surface is more uniform, resulting in better peak shapes.

The **mobile phase** for this mode consists of a **nonpolar solvent,** such as hexane, to which is added a more **polar modifier,** such as methylene chloride, to control solvent strength and selectivity. Solute retention, based on an adsorption/displacement process, can be modulated by varying the polarity or **solvent strength** of the mobile phase (Chapter 28, section 28.3.3.1). Solvent strength refers to the way a solvent affects the migration rate of the sample. **Weak solvents** increase retention (large k′ values), and **strong solvents** decrease retention (small k′ values). It is important to realize that the strength of a particular solvent depends solely on the chromatographic mode. For example, pentane is a weak solvent in normal-phase chromatography but a very strong solvent in the reversed-phase mode (10).

29.3.1.2 Applications of Normal-Phase HPLC

Normal-phase HPLC is best applied to the separation of compounds that are highly soluble in organic solvents, such as fat-soluble vitamins, or suffer from low stability in aqueous mobile phases, such as phospholipids. Compound classes (see Table 28-2), isomers, and highly hydrophilic species such as carbohydrates (see Chapter 10) may also be resolved by normal-phase chromatography. Although not as widely used as reversed-phase columns, amino and cyano bonded-phase HPLC columns are presently enjoying renewed popularity due to increased use for environmental applications and the separation of carbohydrates (6).

29.3.2 Reversed Phase

29.3.2.1 Stationary and Mobile Phases

More than 70 percent of all HPLC separations are carried out in the reversed-phase mode, which utilizes a **nonpolar stationary phase** and a **polar mobile phase.** The stationary phases most commonly employed are chemically **bonded phases** prepared by the reaction of silica surface silanols with an organochlorosilane as described in section 29.2.3.2.2. Usually, the R_3 group (eq. 1) is an octadecyl (C_{18}) chain [-$(CH_2)_{17}CH_3$]. **Octadecylsilyl** (ODS) **bonded phases** are one of the most popular reversed-phase packing materials. Shorter chain hydrocarbons such as octyl (C_8) or butyl (C_4) or phenyl groups can also be attached to the silica surface. The use of monochlorosilanes leads to monomeric bonded phases, where a monomolecular organic layer is formed on the silica surface. Reacting silica with a di- or trichlorosilane leads to the formation of a polymeric layer. In the reaction of silica surface silanol groups with bulky organosilanes, only ca. 50 percent of the -OH moieties are derivatized. Since they are weakly acidic and can possess a negative charge, these residual unreacted silanols can contribute significantly to undesirable band broadening and tailing of some solutes, especially amines. For this reason, the silica is sometimes subjected to a second reaction (endcapping) with a small silylating reagent such as trimethylchlorosilane.

Many silica-based reversed-phase columns are commercially available, and differences in their chromatographic behavior result from variation in the following (4):

1. Type of organic group bonded to the silica matrix, such as C_{18} versus phenyl
2. Chain length of organic moiety, such as C_8 versus C_{18}
3. Amount of organic moiety per unit volume of packing
4. Support particle size and shape
5. Matrix surface area and porosity
6. Bonded-phase surface topology, such as monomeric versus polymeric
7. Concentration of free silanols

Polymeric packing materials eliminate the problem of residual silanols, provide increased pH stability, and offer additional selectivity parameters. Highly cross-linked poly(styrene-divinylbenzene) may be used directly in the reversed-phase mode or can be modified with various hydrophilic or hydrophobic functional groups, including C_{18}.

Reversed-phase HPLC utilizes **polar mobile phases,** usually water mixed with methanol, acetonitrile, or tetrahydrofuran. Solutes are retained due to **hydrophobic interactions** with the **nonpolar stationary phase** and are eluted in order of increasing hydrophobicity (decreasing polarity). Increasing the polar (aqueous) component of the mobile phase increases solute retention (larger k′ values), whereas increasing the organic solvent content of the mobile phase decreases retention (smaller k′ values). Various additives can serve additional functions. An amine such as triethylamine may be added to the mobile phase to deactivate residual silanols. Aqueous buffers may be employed to suppress or otherwise control the ionization of sample components. Although ionogenic compounds can often be resolved without them, **ion-pair reagents** may be used to facilitate chromatography of ionic species on reversed-phase columns. These reagents are ionic surfactants, such as octanesulfonic acid, which can neutralize charged solutes and make them more lipophilic. Depending upon the concentration of ion-pairing agent, retention can be continuously varied from a reversed-phase process to an ion-exchange process. Hence, **ion-pair (paired-ion, ion-association) HPLC** is sometimes included in discussions of ion-exchange chromatography or may even be treated as an independent separation mode.

29.3.2.2 Applications of Reversed-Phase HPLC

As previously stated, reversed-phase HPLC is widely used. Only a few important food-related applications are mentioned here. Reversed-phase has been the HPLC mode most used for analysis of plant proteins. Cereal proteins, among the most difficult of these proteins to isolate and characterize, are now routinely analyzed by this method. In one recent study, a large number of wheat lines were analyzed in order to identify rye gene translocation. The chromatographic profiles obtained for complex mixtures of cereal proteins can also be correlated with functional properties (11). Both water- and fat-soluble vitamins can be analyzed by reversed-phase HPLC (see Chapter 17). The availability of fluorescence detectors has enabled researchers to quantitate very small amounts of the more than six possible forms of Vitamin B_6 (vitamers) in foods and biological samples. Figure 29-3 shows the separation of several of these vitamers in a rice bran extract (12). Ion-pair reversed-phase HPLC can be used to resolve carbohydrates on C_{18} bonded-phase columns. The constituents of soft drinks (caffeine, aspartame, etc.) can be rapidly separated using reversed-phase chromatography. Lipids, including triglycerides and cholesterol, and pigments, including chlorophylls, carotenoids, and anthocyanins (see Chapter 18), are other substances of interest to the food scientist that are amenable to reversed-phase HPLC.

29.3.3 Ion Exchange

29.3.3.1 Stationary and Mobile Phases

Packing materials for **ion-exchange HPLC** are often **functionalized organic resins,** namely sulfonated or aminated poly(styrene-divinylbenzene) (Chapter 28, section 28.3.3.3). Although microporous resins (of ≥8% crosslinking) can be used, **macroporous** (macroreticular) **resins** are much more satisfactory for HPLC columns due to their greater rigidity and permanent pore structure (section 29.2.3.2.3). **Pellicular packings** are also utilized, although a disadvantage is their limited ion-exchange capacity, i.e., the number of equivalents of ion-exchange moieties per unit weight of packing material.

Silica-based **bonded-phase ion exchangers** have also been developed. Since ionogenic groups, provided by the chemically bonded layer, tend to be at the surface, ion exchange is rapid and good separation efficiency is obtained. As with other silica-based packings, however, mobile phase pH must be somewhat restricted.

The **mobile phase** in ion-exchange HPLC is usually an **aqueous buffer,** and solute retention is controlled by changing mobile-phase ionic strength and/or pH. **Gradient elution** (gradually increasing ionic strength) is frequently employed. Ion exchange is similar to an adsorption process, wherein mobile-phase constituents compete with solutes for binding to sites on the stationary phase. Thus, increasing ionic strength of the mobile phase allows it to compete more effectively, and solute retention decreases. A change in mobile-phase pH can affect the sample or the stationary phase or both. The functional groups of "strong" ion exchangers remain ionized across a broad pH range. Consequently, the binding capacity of these stationary phases is essentially constant, regardless of mobile-phase pH, and changes in retention are due solely to alteration of solute charge. The ligand density or charge capacity of "weak" ion exchangers may be varied by manipulating mobile-phase pH (Chapter 28, section 28.3.3.3).

29.3.3.2 Applications of Ion-Exchange HPLC

Ion-exchange HPLC has many applications, ranging from the detection of simple inorganic ions to analysis of carbohydrates and amino acids to the preparative purification of proteins.

29.3.3.2.1 Ion Chromatography **Ion chromatography** is simply high-performance ion-exchange chromatography performed using a relatively **low-capacity stationary phase** (either anion or cation exchange) and, usually, a **conductivity detector.** (Other detection methods, e.g. electrochemical, can be employed.) All ions conduct an electric current; thus, measurement of electrical conductivity is an obvious way to detect ionic species. Because the mobile phase also contains ions, however, background conductivity can be relatively high. One step toward solving this problem is to use much lower capacity ion-exchange packing materials, so that more dilute eluents may be employed. In **non-suppressed** or **single-column ion chromatography,** the detector cell is placed directly after the column outlet and eluents are carefully chosen to maximize changes in conductivity as sample components elute from the column. **Suppressed ion chromatography** utilizes an eluent that can be selectively removed. Originally, this was accomplished by the addition of a second **suppressor column** between the analytical column and the conductivity detector. Today, suppressor columns have been replaced by ion-exchange membranes (2). Suppressed ion chromatography permits the use of more concentrated mobile phases and gradient elution. Ion chromatography can be used to determine inorganic anions and cations, transition metals, organic acids, amines, phenols, surfactants, and sugars. Some specific examples of ion chromatography applied to food matrices include the determination of organic and inorganic ions in milk; organic acids in coffee extract and wine; choline in infant formula; and trace metals, phosphates, and sulfites in foods. Figure 29-4 illustrates the simultaneous determination of organic acids and inorganic anions in coffee by ion chromatography.

29.3.3.2.2 Carbohydrates and Compounds of Biochemical Interest Both cation- and anion-exchange stationary phases are applied to HPLC of carbohydrates (as detailed in Chapter 10). Sugars and sugar derivatives can be separated on gel-type (microporous) polymeric

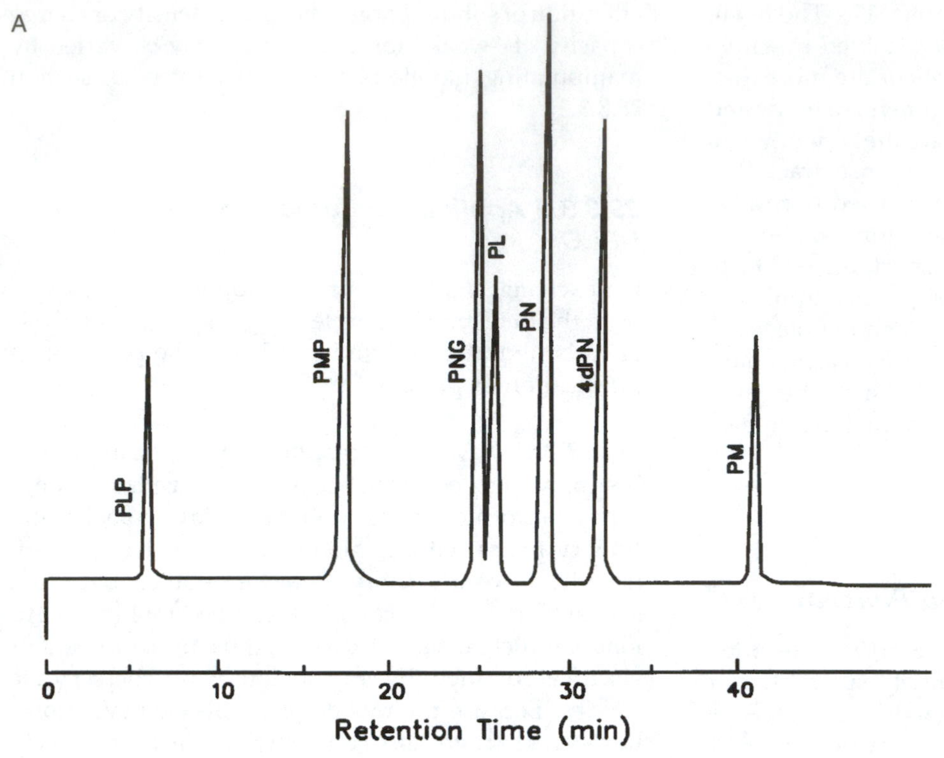

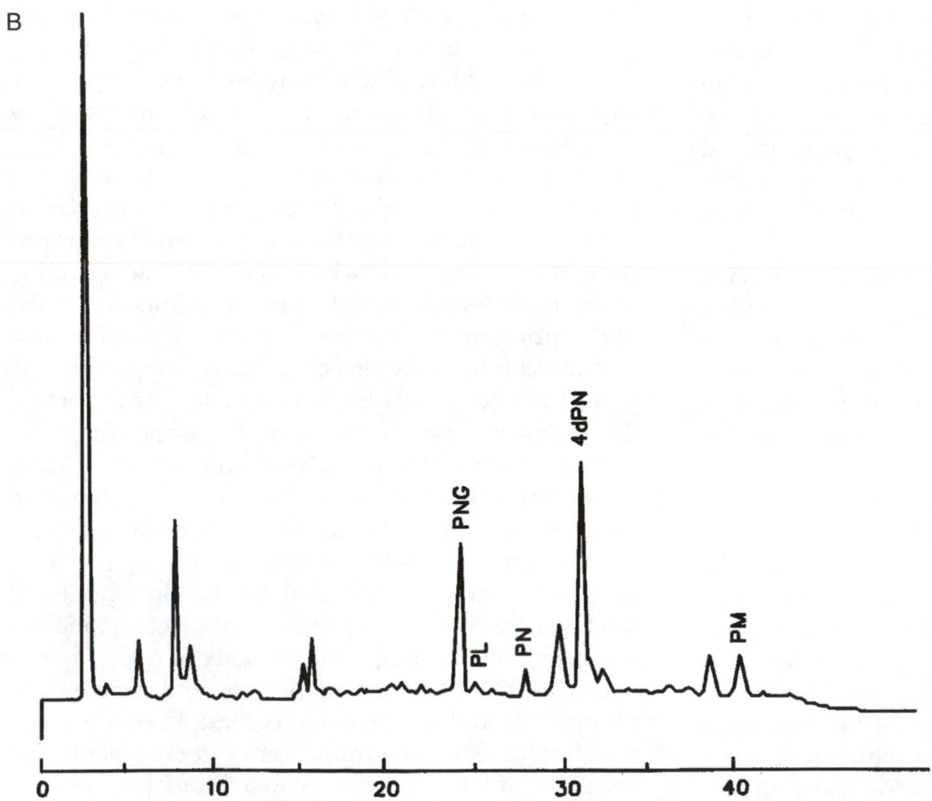

FIGURE 29-3.
Analysis of Vitamin B_6 compounds by reversed-phase HPLC with fluorescence detection. Some of the standard compounds (A) are present in a sample of rice bran extract (B). Sample preparation and analytical procedures are described in reference 12. Reprinted with permission from (12). Copyright 1991 American Chemical Society.

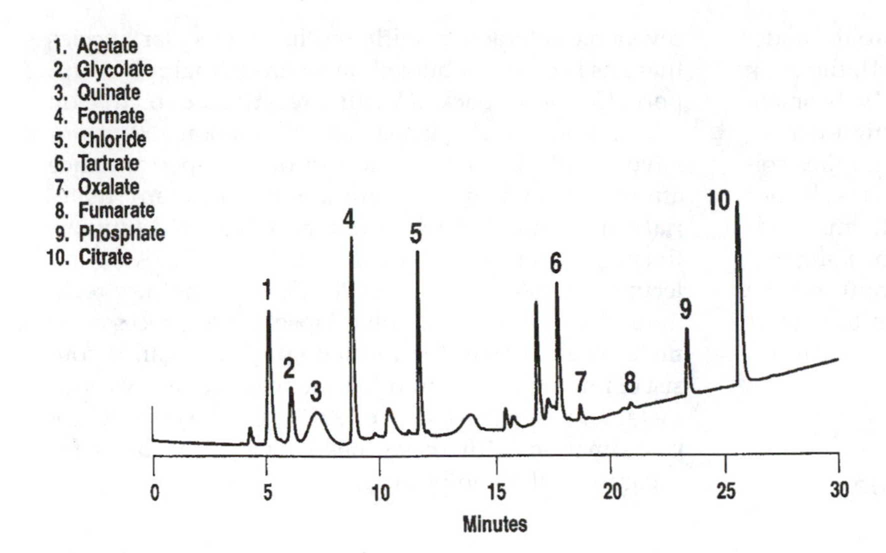

FIGURE 29-4.
Ion-chromatographic analysis of organic acids and inorganic anions in coffee. Ten anions (listed) were resolved on an IonPac AS5A column (Dionex) using a sodium hydroxide gradient and suppressed conductivity detection. Figure courtesy of Dionex Corp., Sunnyvale, CA.

cation-exchange resins in the H^+ form or loaded with Ca^{++} or other **metal counterions** (Fig. 29-5). Various separation mechanisms are involved, depending on the counterion associated with the resin, degree of crosslinking (4 vs. 8%), and type of carbohydrate. A review by Hicks (13) provides a good discussion of the application of cation-exchange resins to the separation of carbohydrates. These columns are generally operated at elevated temperatures (ca. 85°C) to obtain the highest efficiency and resolution. It is essential to perform adequate sample cleanup and avoid undue pressure on these packing materials.

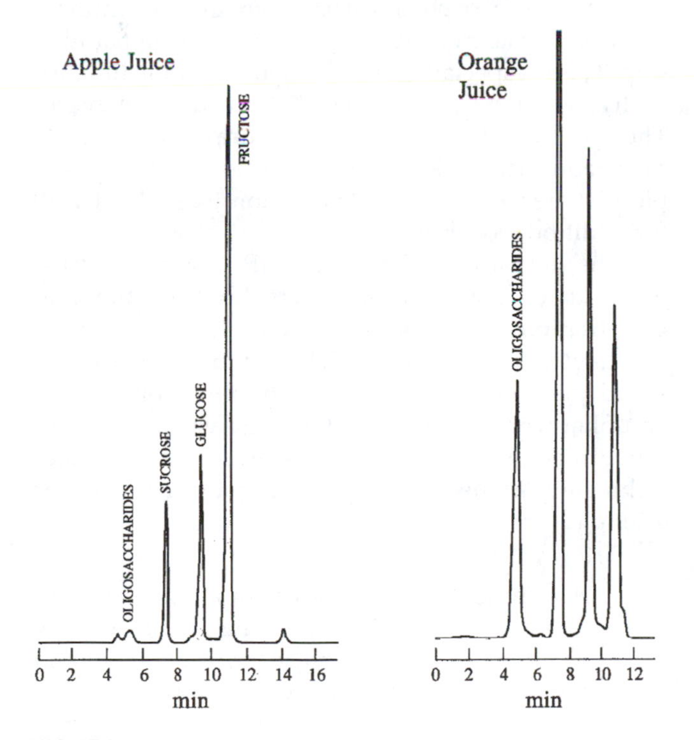

FIGURE 29-5.
Cation-exchange separation of sugars in fruit juice. Calcium-form column (Interaction CHO-620; 30 × 0.65 cm; 90°C) eluted with distilled water at 0.5 ml/min. Refractive index detection. Figure courtesy of Interaction Chromatography, San Jose, CA.

Carbohydrate analysis has benefited greatly by the relatively recent introduction of a technique that involves **anion-exchange** HPLC at high pH (≥12) and detection by a **pulsed amperometric detector.** Pellicular column packings, consisting of nonporous latex beads coated with a thin film of strong anion exchanger, provide the necessary fast exchange, high efficiency, and resistance to strong alkali. The advantage of separating carbohydrates by anion exchange is that retention and selectivity may be altered by changes in eluent composition. When metal-loaded cation-exchange resins are used, selectivity is a function of the stationary phase composition alone.

Amino acids have been successfully resolved on polymeric ion exchangers for more than 30 years (see Chapter 15). Ion exchange is one of the most effective modes for HPLC of proteins and has recently been recognized as valuable for the fractionation of peptides.

29.3.4 Size Exclusion

Size-exclusion chromatography fractionates solutes solely on the basis of their size. Due to the limited separation volume available to this chromatographic mode, as explained in Chapter 28 (section 28.3.3.4), the peak capacity of a size-exclusion column is relatively small, 10 to 13 peaks (see Fig. 28-6). Thus, the "high performance" aspect of HPLC, which generally implies columns with 2,000–20,000 theoretical plates, is really not applicable in the case of size exclusion. The main advantage gained from use of small particle packing materials is speed. Separation times of ≤60 min may be realized with 5–13 μm particles, compared to ≤24 hr separations using 30–150 μm particles in classical columns (2).

29.3.4.1 Column Packings and Mobile Phases

Size-exclusion packing materials or columns are selected so that matrix pore size matches the molecular weight range of the species to be resolved. Prepacked columns of microparticulate media are available in a wide variety of pore sizes. **Hydrophilic packings,** for use with water-soluble samples and aqueous mobile phases, may be surface-modified silica or methacrylate resins. **Poly(styrene-divinylbenzene) resins** are useful for nonaqueous size-exclusion chromatography of synthetic polymers.

The mobile phase in this mode is chosen for sample solubility, column compatability, and minimal solute-stationary phase interaction. Otherwise, it has little effect on the separation. Aqueous buffers are used for biopolymers, such as proteins and nucleotides, both to preserve biological activity and to prevent adsorptive interactions. Tetrahydrofuran or dimethylformamide is generally used for size-exclusion chromatography of polymer samples, in order to insure sample solubility.

29.3.4.2 Applications of High-Performance Size-Exclusion Chromatography

Hydrophilic polymeric size-exclusion packings are used for the rapid determination of **average molecular weights** and degree of polydispersity of polysaccharides, including amylose, amylopectin, pullulan, guar, and water-soluble cellulose derivatives. **Molecular weight distribution** can be determined directly from high-performance size-exclusion chromatography, if **low-angle laser light scattering** (LALLS) is used for detection (2).

29.3.5 Affinity

Affinity chromatography is based on the principle that the molecules to be purified can form a selective but reversible interaction with another molecular species that has been immobilized on a chromatographic support. Columns packed with **preactivated** media on which a ligand of interest can be immobilized *in situ* have greatly facilitated the use of high-performance affinity chromatography. Although almost any material can be immobilized on a suitably activated support, the major ligands are proteins, nucleic acids, dyes, and lectins (Chapter 28, section 28.3.3.5). An affinity technique that does not involve bioselective processes is **metal chelate affinity chromatography.** Ligands consist of immobilized iminodiacetic acid to which various metal ions, such as Cu^{++} or Zn^{++}, can be complexed. Coordination with these metal ions is the basis for separation of some proteins.

29.4 SAMPLE PREPARATION AND DATA EVALUATION

It is important to recognize that the success or failure of HPLC methods, as with applications of other analytical techniques, is ultimately dependent upon the adequacy of **sample preparation.** Other methodologies, such as extraction (solvent or solid phase), ultrafiltration, and precipitation, are generally needed to remove interfering materials prior to chromatography (14, 15). Variables such as extraction efficiency, analyte stability, and consistency of chemical or enzymatic pretreatment must be considered during each sample preparation step. This is especially critical when microconstituents, such as Vitamin B_6 compounds (12), are to be analyzed. The use of HPLC to analyze foods for pesticide, mycotoxin, and drug residues also requires fastidious sample preparation and data evaluation (see Chapter 19 for additional discussion).

The data acquired from an HPLC analysis must be evaluated from several aspects. **Identification** and **quantitation** of chromatographic peaks are discussed in Chapter 28 (sections 28.3.4.3.2 and 28.3.4.3.4). As stated previously, the fact that the retention time of an unknown and a standard are equivalent does not prove that the two compounds are identical. Other techniques are needed **to confirm peak identity.** For example:

1. A sample can be **spiked** with a small amount of added standard; only the height of the peak of interest should increase, with no change in retention time, peak width, or shape.
2. A diode array detector can provide **absorption spectra** of designated peaks. Although identical spectra do not prove identity, a spectral difference confirms that sample and standard peaks are different compounds.

3. In the absence of spectral scanning capability, other detectors, such as absorption or fluorescence, may be used in a **ratioing procedure.** Chromatograms of sample and standard are monitored at each of two different wavelengths. The ratio of peak areas at these wavelengths should be the same if sample and standard are identical.
4. Peaks of interest can be **collected** and subjected to additional chromatographic (different separation mode) or nonchromatographic (e.g., spectroscopic) analysis.

The **quantitative validity** of an HPLC analysis must also be confirmed. Standards should be included during each analytical session since detector response may vary from day to day. **Analyte recovery** should be checked periodically. This involves addition of a known quantity of standard to a sample (usually before extraction) and determination of how much is recovered during subsequent analysis. During routine analyses, it is highly desirable to include a control or **check sample,** a material of known composition. This material is analyzed parallel to unknown samples. When the concentration of analyte measured in the control falls outside an acceptible range, data from other samples analyzed during the same period must also be considered suspect. Carefully analyzed food samples and other substances are available from the National Institute of Standards and Technology (formerly the National Bureau of Standards) and other sources for use in this manner.

29.5 SUMMARY

High-performance liquid chromatography is a chromatographic technique of great versatility and analytical power. A basic HPLC system consists of a pump, injector, column, detector, and recorder/data system. The pump delivers mobile phase through the system. An injector allows sample to be placed into the flowing mobile phase for introduction onto the column. The HPLC column, connected between injector and detector, consists of stainless steel hardware filled with a packing material. Detectors used in HPLC include UV-Vis absorption, fluorescence, refractive index, amperometric, and conductivity. Detection sensitivity or specificity can sometimes be enhanced by chemical derivatization of the analyte. While a strip-chart recorder serves to record the basic results of a chromatographic separation, the use of electronic integrators and computer-controlled systems offers additional data-handling capability. A broad variety of column packing materials has contributed greatly to the widespread use of HPLC. These column packing materials may be categorized as silica-based (porous silica, bonded phases, pellicular packings) or polymeric (microporous, macroporous). The success of silica-based bonded phases has expanded the applications of normal-phase and reversed-phase modes of separation in HPLC. Separation can also be achieved by utilizing the principles of ion-exchange, size-exclusion, and affinity chromatography. HPLC is widely used for the analysis of small molecules and ions, such as sugars, vitamins, and amino acids, and has also been applied to the separation and purification of macromolecules, such as proteins, nucleic acids, and polysaccharides. As with other analytical techniques, careful sample preparation and confirming the identity of resolved species are essential to the success of HPLC methods.

29.6 STUDY QUESTIONS

1. Why might you choose to use HPLC rather than traditional low-pressure column chromatography?
2. What is a guard column and why is it used?
3. Give three general requirements for HPLC column packing materials. Describe and distinguish among porous silica, bonded phases, pellicular, and polymeric column packings, including the advantages and disadvantages of each type.
4. Explain the principle of operation for each of the various types of HPLC detectors. What factors would you consider in choosing an HPLC detector? Describe three different types of detectors and explain the principles of operation for each.
5. A sample containing compounds A, B, and C is analyzed with HPLC using the solid stationary phase of a C_{18} hydrocarbon chain attached to silica. A 1:5 solution of ethanol and H_2O was used as the mobile phase. A UV detector was used, and the following chromatogram was obtained.

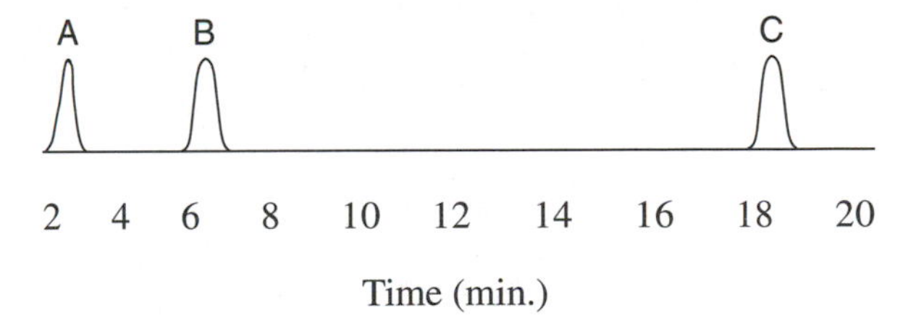

Assuming that the separation of compounds is based on their polarity,

a. Is this normal- or reversed-phase chromatography? Explain your answer.
b. Which compound is the most polar?
c. How would you change the mobile phase so Compound C would elute sooner, without changing the relative positions of Compounds A and B? Explain why this would work.

6. Ion chromatography has recently become a widely promoted chromatographic technique in food analysis. Describe ion chromatography and give at least two examples of its use.

7. Describe one application each for ion-exchange, size-exclusion, and affinity HPLC.

29.7 REFERENCES

1. Johnson, E.L., and Stevenson, R. 1978. *Basic Liquid Chromatography.* Varian Associates, Palo Alto, CA.
2. Heftmann, E. (Ed.) 1992. *Chromatography,* 5th ed. *Fundamentals and Applications of Chromatography and Related Differential Migration Methods.* Part A: *Fundamentals and Techniques.* Part B: *Applications.* Journal of Chromatography Library series volumes 51A and 51B. Elsevier, Amsterdam, The Netherlands.
3. Macrae, R. (Ed.) 1988. *HPLC in Food Analysis,* 2nd ed. Academic Press Ltd., New York, NY.
4. Krstulovic, A. M., and Brown, P.R. 1982. *Reversed-Phase High-Performance Liquid Chromatography.* John Wiley & Sons, New York, NY.
5. Upchurch, P. 1992. *HPLC Fittings,* 2nd ed. Paul Upchurch, Oak Harbor, WA.
6. Majors, R.E. 1991. Trends in HPLC column usage. LC · GC. 9:686.
7. Gooding, K.M., and Regnier, F. E. (Eds.) 1990. *HPLC of Biological Macromolecules: Methods and Applications.* Chromatographic Science series volume 51. Marcel Dekker, Inc., New York, NY.
8. Unger, K.K. 1990. *Packings and Stationary Phases in Chromatographic Techniques.* Marcel Dekker, Inc., New York, NY.
9. Rounds, M.A., Rounds, W. D., and Regnier, F. E. 1987. Poly(styrene-divinylbenzene)-based strong anion-exchange packing material for high-performance liquid chromatography of proteins. *J. Chromatogr.* 397:25–38.
10. Heftmann, E. (Ed.) 1975. *Chromatography: A Laboratory Handbook of Chromatographic and Electrophoretic Methods,* 3rd ed. Van Nostrand Reinhold, New York, NY.
11. Dorsey, J.G., Foley, J.P., Cooper, W.T., Barford, R.A., and Barth, H.G. 1992. Liquid chromatography: Theory and methodology (Fundamental Review). *Anal. Chem.* 64: 353R–389R.
12. Gregory, J. F., and Sartain, D. B. 1991. Improved chromatographic determination of free and glycosylated forms of vitamin B_6 in foods. *J. Agric. Food Chem.* 39:899–905.
13. Hicks, K.B. 1988. High-performance liquid chromatography of carbohydrates, in *Advances in Carbohydrate Chemistry and Biochemistry,* Vol. 46. R. S. Tipson and D. Horton (Eds.), 17–329. Academic Press, Inc., San Diego, CA.
14. Lim, C. K. (Ed.) 1986. *HPLC of Small Molecules: A Practical Approach.* IRL Press Ltd., Oxford, England.
15. Lawrence, J. F. (Ed.) 1984. *Food Constituents and Food Residues: Their Chromatographic Determination.* Marcel Dekker, Inc., New York, NY.

CHAPTER

30

Gas Chromatography

Gary A. Reineccius

30.1 INTRODUCTION

The first publication on gas chromatography (GC) was in 1952 (1), while the first commercial instruments were manufactured in 1956. James and Martin (1) separated fatty acids by gas chromatography, collected the column effluent, and titrated the individual fatty acids for quantitation. Gas chromatography has advanced greatly since that early work and is now considered to be a mature field that is approaching theoretical limitations.

The types of analysis that can be done by GC are very broad. GC has been used for the determination of fatty acids, triglycerides, cholesterol and other sterols, gases, solvent analysis, water, alcohols, and simple sugars, as well as oligosaccharides, amino acids and peptides, vitamins, pesticides, herbicides, food additives, antioxidants, nitrosamines, polychlorinated biphenyls (PCBs), drugs, flavor compounds, and many more. The fact that GC has been used for these various applications does not necessarily mean that it is the best method—often better choices exist. Gas chromatography is ideally suited to the analysis of thermally stable volatile substances. Substances that do not meet these requirements (e.g., sugars, oligosaccharides, amino acids, peptides, and vitamins) are more suited to analysis by a technique such as high-performance liquid chromatography (HPLC) or supercritical-fluid chromatography (SFC). Yet gas chromatographic methods appear in the literature for these substances.

This chapter will discuss sample preparation for gas chromatography, gas chromatographic hardware, columns, and chromatographic theory as it uniquely applies to gas chromatography. Texts devoted to gas chromatography in general (2–4) and food applications in particular (5) should be consulted for more detail.

30.2 SAMPLE PREPARATION FOR GAS CHROMATOGRAPHY

30.2.1 Introduction

One cannot generally directly inject a food product into a GC without some sample preparation. The high temperatures of the injection port will result in the degradation of nonvolatile constituents and create a number of false GC peaks corresponding to the volatile degradation products formed. Additionally, very often the constituent of interest must be isolated from the food matrix simply to permit concentration such that it is at detectable limits for the GC. Thus, one must generally do some type of sample preparation, component isolation, and concentration prior to GC analysis.

Sample preparation often involves grinding, homogenization, or otherwise reducing particle size. There is substantial documentation in the literature showing that foods may undergo changes during storage and sample preparation. Many foods contain active enzyme systems that will alter the composition of the food product. This is very evident in the area of flavor work (6–8). Inactivation of enzyme systems via high temperature-short time thermal processing, sample storage under frozen conditions, drying the sample, or homogenization with alcohol may be necessary (see Chapter 3).

Microbial growth or chemical reactions may also occur in the food during sample preparation. Chemical reactions will often result in the formation of volatiles that will again give false peaks on the GC. Thus, the sample must be maintained under conditions such that degradation does not occur. Microorganisms are often inhibited by chemical means (e.g., sodium fluoride), thermal processing, drying, or frozen storage.

30.2.2 Isolation of Solutes from Foods

The isolation procedure may involve headspace analysis (static or dynamic), distillation, preparative chromatography (e.g., solid-phase extraction, column chromatography on silica gel), simple solvent extraction, or some combination of these basic methods. The procedure used will depend on the food matrix as well as the compounds to be analyzed. The primary considerations are to isolate the compounds of interest (e.g., flavor compounds, PCBs, pesticides) from nonvolatile food constituents (e.g., carbohydrates, proteins, vitamins) or those that would interfere with gas chromatography (e.g., lipids). Some of the chromatographic methods that might be applied to this task have been discussed in the basic chromatography chapter (Chapter 28) of this text. Methods for the isolation of volatile substances (e.g., headspace analysis, distillation, and extraction methods) have not been discussed in this text, so they will be covered briefly as they pertain to the isolation of components for gas chromatographic analysis.

It should be emphasized that the isolation procedure used is critical in determining the results obtained. An improper choice of method or poor technique at this step negates the best gas chromatographic analysis of the isolated solutes. The influence of isolation technique on gas chromatographic results is evident from Fig. 30-1. All of the chromatograms (presented as bar charts) shown in this figure were obtained from the analysis of the *same aqueous model system*. The gross differences in GC profile are simply due to the biases introduced by the selectivity of the isolation method. These biases are discussed in the sections that follow.

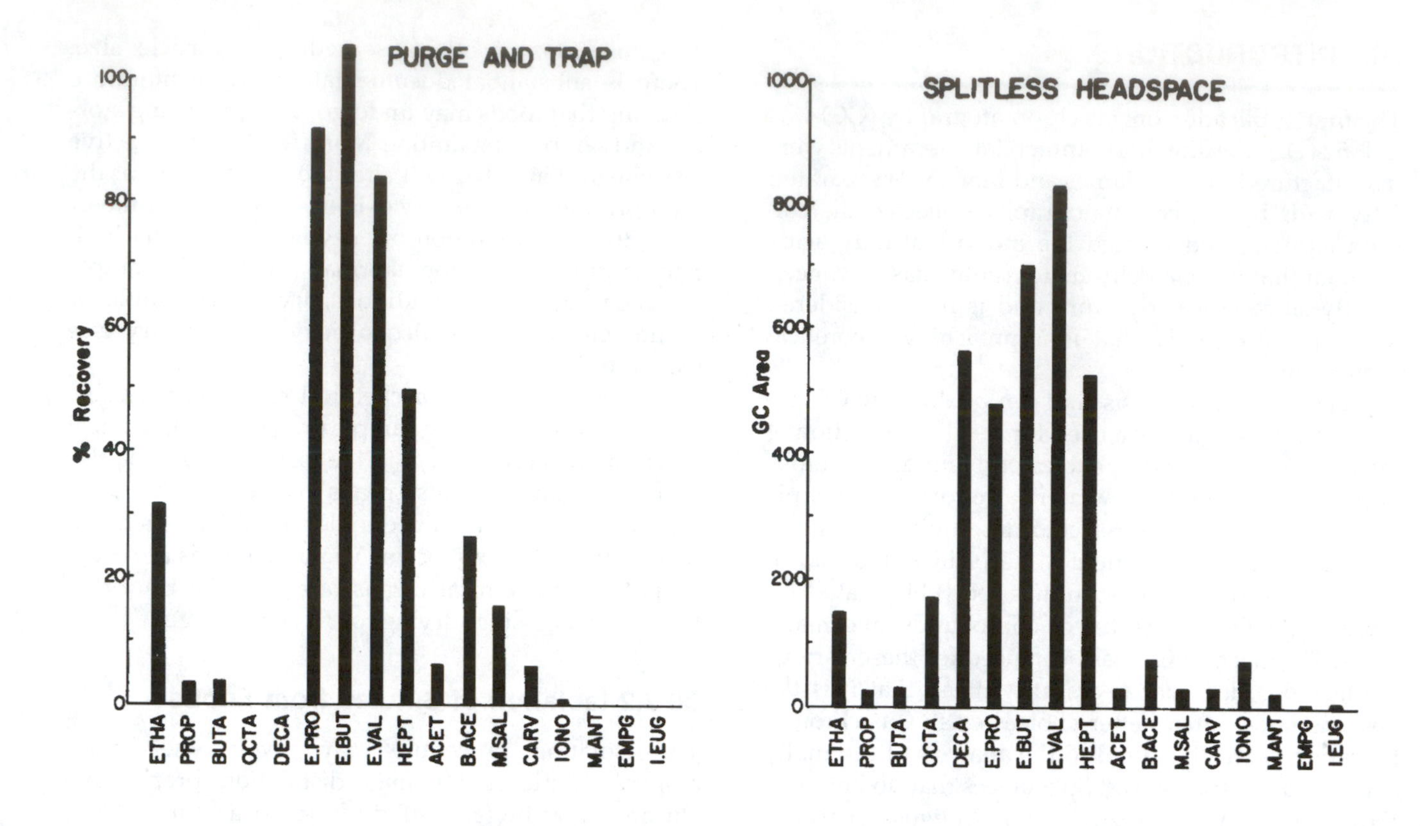

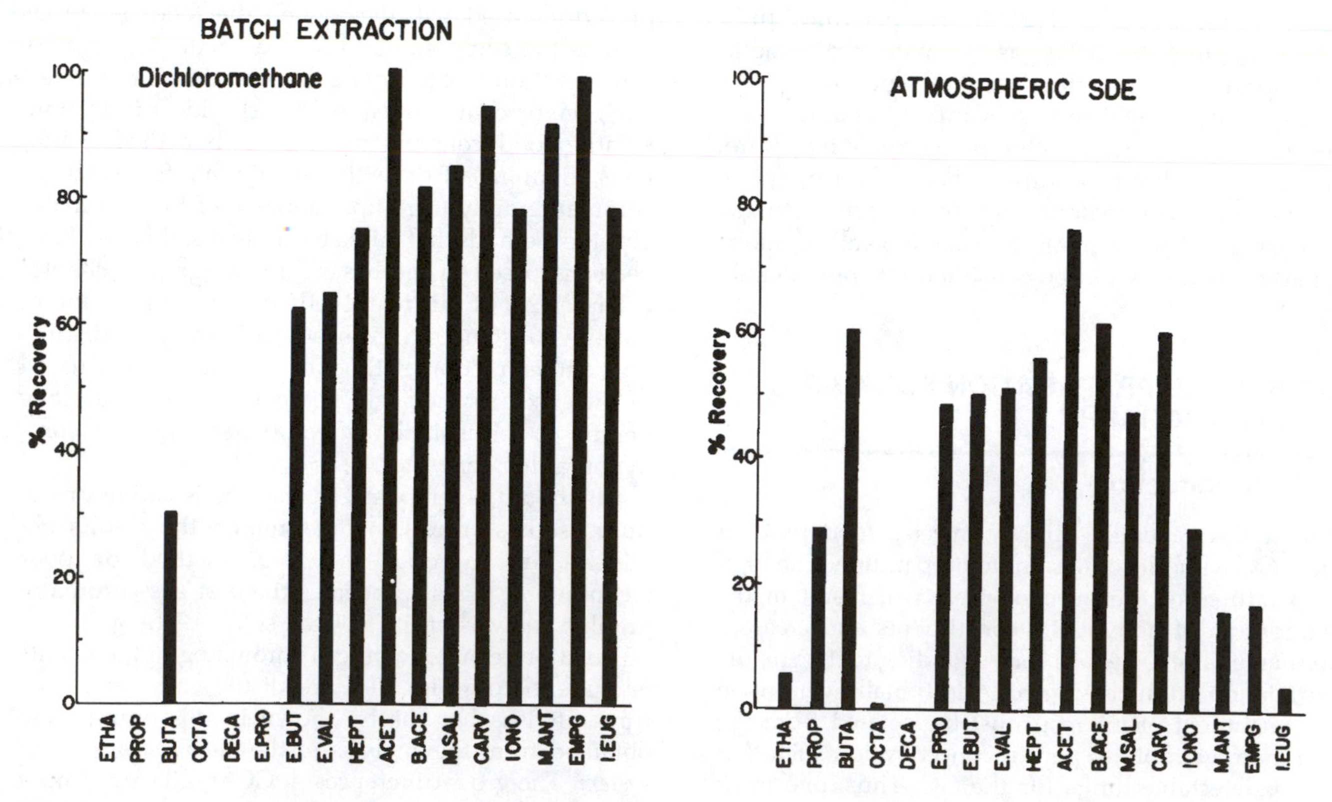

FIGURE 30-1.
Comparison of methods for the isolation of volatile flavor compounds from aqueous model systems. (All bars should be of equal height if recovered at 100%.) Adapted from (9), with permisson.

30.2.2.1 Headspace Methods

One of the simplest methods of isolating volatile compounds from foods is by direct injection of the headspace vapors above a food product. Unfortunately, this method does not provide the sensitivity needed for trace analysis (Fig. 30-1). Instrumental constraints typically limit headspace injection volumes to 5 ml or less. Therefore, only volatiles present in the headspace at concentrations greater than 10^{-7}g/L headspace would be at detectable levels (using a flame ionization detector).

Direct headspace sampling has been used extensively where rapid analysis is necessary and major component analysis is satisfactory. Examples of method applications include measurement of hexanal as an indicator of oxidation (10, 11), and 2-methyl propanal, 2 methyl butanal, and 3-methyl butanal as indicators of nonenzymatic browning (12). The determination of volatiles in packaging materials may also be approached by this method.

Headspace concentration techniques (often called **dynamic headspace** or **purge and trap methods**) have found wide usage in recent years. This concentration method may involve simply passing large volumes of headspace vapors through a cryogenic trap or, alternatively, a more complicated extraction and/or adsorption trap. A simple **cryogenic trap** offers some advantages and disadvantages. A cryo trap (if properly designed and operated) will collect headspace vapors irrespective of compound polarity and boiling point. However, water is typically the most abundant volatile in a food product, and, therefore, one collects an aqueous distillate of the product aroma. This distillate must be extracted with an organic solvent, dried, and then concentrated for analysis. These additional steps add analysis time and provide opportunity for sample contamination. Recently, the use of adsorbent traps has become the most common means to concentrate headspace vapors.

Adsorbent traps offer the advantages of providing a water-free flavor isolate (trap material typically has little affinity for water) and are readily automated. The adsorbent initially used for headspace trapping was charcoal. The charcoal was either solvent extracted (CS_2) or thermally desorbed with backflushing (inert gas) to recover the adsorbed volatiles. The use of synthetic porous polymers as headspace trap material now dominates. Initially, Tenax^R (a porous polymer very similar to the skeleton of ion exchange resins) was most commonly used; however, combinations of Tenax^R and other polymers are now seeing greater application. These polymers exhibit good thermal stability and reasonable capacity. Adsorbent traps are generally placed in a closed system and loaded, desorbed, and so on via the use of automated multiport valving systems. The automated closed system approach provides reproducible GC retention times and quantitative precision necessary for some studies. The primary disadvantage of adsorbent traps is their differential adsorption affinity and limited capacity. Buckholz et al. (13) have shown that the most volatile peanut aroma constituents will break through two Tenax^R traps in series after purging at 40 mL/min for only 15 min. Therefore, the GC profile may only poorly represent the actual food composition due to biases introduced by the purging and trapping steps (Fig. 30-1).

30.2.2.2 Distillation Methods

Distillation processes are quite effective at isolating volatile compounds from foods for GC analysis (Fig. 30-1). Product moisture or outside steam is used to heat and codistill the volatiles from a food product. This means that a very dilute aqueous solution of volatiles results, and a solvent extraction must be performed on the distillate in order to permit concentration for analysis. The distillation method most commonly used today is some modification of the original Nickerson-Likens distillation head. In this apparatus, a sample is boiled in one side flask and an extracting solvent in another. The product steam and solvent vapors are intermixed and condensed; the solvent extracts the organic volatiles from the condensed steam. The solvent and extracted distillate return to their respective flasks and are distilled to again extract the volatiles from the food. While this method is convenient and efficient, artifacts from solvents used in extraction, antifoam agents, steam supply (contaminated water), thermally induced chemical changes, and leakage of contaminated laboratory air into the system may contaminate the volatile isolate.

30.2.2.3 Solvent Extraction

Solvent extraction is often the preferred method for the recovery of volatiles from foods (Fig. 30-1). Recovery of volatiles will depend upon solvent choice and the solubility of the solutes being extracted. Solvent extraction typically involves the use of an organic solvent (unless sugars, amino acids, or some other water-soluble components are of interest). Extraction with organic solvents limits the method to the isolation of volatiles from fat-free foods (e.g., wines, some breads, fruit and berries, some vegetables, and alcoholic beverages), or an additional procedure must be employed to separate the extracted fat from the isolated volatiles (e.g., a chromatographic method). Fat will otherwise interfere with subsequent concentration and GC analysis.

Solvent extractions may be carried out in quite elaborate equipment, such as supercritical CO_2 extractors, or can be as simple as a batch process in a separatory funnel. Batch extractions can be quite efficient if multiple extractions and extensive shaking are used (14). The continuous extractors (liquid/liquid) are more efficient but require more costly and elaborate equipment.

30.2.2.4 Solid-Phase Extraction

The extractions discussed above involved the use of two immiscible phases (water and an organic solvent). However, a newer and very rapidly growing alternative to such extractions is solid-phase extraction (15, 16). In this technique, a liquid sample (most often aqueous based) is passed through a column (2–10 mL volume) filled with chromatographic packing or a Teflon^R filter disk that has the chromatographic packing imbedded in it. The chromatographic packing may be any of a number of different materials (e.g., ion-exchange resins or a host of different reversed- or normal-phase HPLC column packings).

When a sample is passed through the cartridge or filter, solutes that have an affinity for the chromatographic phase will be retained on the phase while those with little or no affinity will pass through. The phase is next rinsed with water, perhaps a weak solvent (e.g., pentane) if a reversed-phase technique is used, and then a stronger solvent (e.g., dichloromethane). The strong eluant is chosen such that it will remove the solutes of interest.

Solid-phase extraction has numerous advantages over traditional liquid/liquid extractions including: (1) Less solvent is required; (2) it is faster; (3) less glassware is needed (less cost and potential for contamination); (4) better precision and accuracy; (5) minimal solvent evaporation for further analysis (e.g. gas chromatography); and (6) it is readily automated.

30.2.2.5 Direct Injection

It is theoretically possible to analyze some foods by direct injection of the food into a gas chromatograph. Assuming one can inject 2–3 μL samples into a GC and the GC has a detection limit of 0.1 ng (0.1 ng/2μL), one could detect volatiles in the sample at concentrations greater than 50 ppb. Problems with direct injection arise due to thermal degradation of the nonvolatile food constituents, damage to the GC column, decreased separation efficiency due to water in the food sample, contamination of the column and injection port by nonvolatile materials, and reduced column efficiency due to slow vaporization of volatiles from the food (injection port temperatures are reduced to minimize thermal degradation of the nonvolatile food constituents). Despite these concerns, direct injection is commonly used to determine the flavor quality of vegetable oils (17, 18). A relatively large volume of oil (50–100 μL) can be directly injected into an injection port of a GC that has been packed with glass wool. Since vegetable oils are reasonably thermally stable and free of water, this method is particularly well suited to oil analysis.

There are numerous other approaches for the isolation of volatiles from foods. Some are simple variations of these methods, while others are unique. Several review articles are available that provide a more complete view of methodology (19–21).

30.2.3 Sample Derivatization

The compounds one wishes to determine must be thermally stable under the GC conditions employed. Thus, for some compounds (e.g., pesticides, aroma compounds, PCBs, and volatile contaminants) the analyst can simply isolate the components of interest from a food as discussed above and directly inject them into the GC. For compounds that are thermally unstable, too low in volatility (e.g., sugars and amino acids), or yield poor chromatographic separation due to polarity (e.g., phenols or acids), a derivatization step must be included prior to GC analysis (see also Chapters 10 and 13). A listing of some of the reagents used in preparing volatile derivatives for GC is given in Table 30-1. The conditions of use for these reagents are often specified by the supplier or can be found in the literature.

30.3 GAS CHROMATOGRAPHIC HARDWARE AND COLUMNS

The major parts of a GC are the **gas supply** and **regulators, injection port, oven, column, detector, electronics,** and **recorder/data handling system** (Fig. 30-2).

The hardware as well as operating parameters used in any GC analysis must be accurately and completely recorded. The information that must be included is presented in Table 30-2.

30.3.1 Gas Supply System

The gas chromatograph will require at least a supply of carrier gas, and most likely, gases for the detector (e.g., hydrogen and air for a flame ionization detector). The gases used must be of high purity and all regulators, gas lines, and fittings of good quality. The regulators should have stainless steel rather than polymer diaphragms since polymers will give off volatiles that may contribute peaks to the analytical run. All gas lines must be clean and contain no residual drawing

TABLE 30-1. Reagents Used for Making Volatile Derivatives of Food Components for GC Analysis

Reagent	*Chemical Group*	*Food Constituent*
Silyl reagents	Hydroxy, amino carboxylic acids	Sugars, sterols, amino acids
Trimethylchlorosilane/ hexamethyldisilazane		
BSA [N,O-bis(trimethylsilyl) acetamide]		
BSTFA [N,O-bis (trimethylsilyl) trifluoroacetimide]		
t-BuDMCS (t-Butyldimethylchlorosilyl/ imidazole)		
Esterifying reagents	Carboxylic acids	Fatty acids, amines, amino acids, triglycerides, wax esters, phospholipids, cholesteryl esters
Methanolic HCl		
Methanolic sodium methoxide		
N,N-Dimethylformamide dimethyl acetal		
Boron trifluoride (or trichloride)/ methanol		
Miscellaneous		
Acetic anhydride/pyridine	Alcoholic and phenolic	Phenols, aromatic hydroxyl groups, alcohols
N-trifluoroacetylimidizole/N-heptafluorobutyrlimidizole	Hydroxy and amines	Same as above
Alkylboronic acids	Polar groups on neighboring atoms	
O-alkylhydroxylamine	Compounds containing both hydroxyl and carbonyl groups	Ketosteroids, prostaglandins

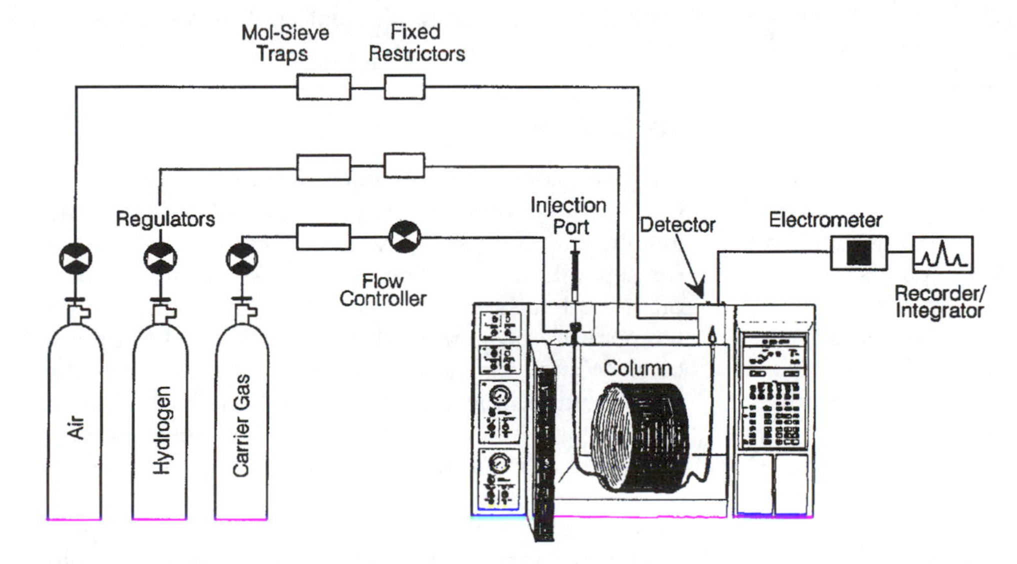

FIGURE 30-2.
Diagram of a gas chromatographic system. Figure courtesy of Hewlett-Packard Co., Analytical Customer Training, Atlanta, GA.

TABLE 30-2. Gas Chromatographic Hardware and Operating Conditions to be Recorded for All GC Separations

Parameter	*Description*
Sample	Name and injection volume
Injection	Type of injection (e.g., split versus splitless) and conditions (injection port flow rates)
Column	Length, diameter (material-packed columns), and manufacturer
Packing/phase	Packed columns—solid support; size mesh; coating; loading (%) Capillary columns—phase material and thickness
Temperatures	Injector; detector; oven and any programming information
Carrier gas	Flow rate (velocity) and type
Detector	Type
Data output	Attenuation and chart speed

oil. The gas lines should have traps in line to remove any moisture and contaminants from the incoming gases. These traps must be periodically replaced to maintain effectiveness.

30.3.2 Injection Port

30.3.2.1 Hardware

The injection port serves the purpose of providing a place for sample introduction, its vaporization, and possibly some dilution and splitting. Liquid samples make up the bulk of materials analyzed by GC, and they are always done by syringe injection (manual or automated). The injection port contains a soft septum that provides a gas-tight seal but can be penetrated by a syringe needle for sample introduction.

The sample must be vaporized in the injection port in order to pass through the column for separation. This vaporization can occur quickly by flash evaporation (standard injection ports) or slowly in a more gentle manner (temperature programmed injection port or on-column injection). The choice depends upon the thermal stability of the analytes.

The injection port may serve the additional function of splitting the injection so that only a portion of the analyte goes on the column. Capillary columns have limited capacity, and the injection volume may have to be reduced to permit efficient chromatography.

Due to the various sample as well as instrumental requirements, there are several different designs of injection ports available. These include the **standard heated port** (split or splitless, Fig. 30-3), **temperature programmed,** and **on-column injectors.** The standard injection port is operated about 20° warmer than the maximum column oven temperature. The sample may be diluted with carrier gas to accomplish a split (1:50–1:100 preferred), whereby only a small portion of the analyte goes on the column or a splitless injection mode whereby all of the analyte goes on the column. Splitless injection requires the use of a sample solvent that has a boiling point about 20°C above the initial column temperature. The solvent must recondense in the column for acceptable chromatography.

For temperature-programmed injection ports, the sample is introduced into an ambient temperature port and then it is temperature programmed to some desired temperature. On-column is a technique whereby the sample is actually introduced into the column whose temperature is at that of the GC oven or that of the room. Either technique is desired for temperature-sensitive analytes.

30.3.2.2 Sample Injection

Samples may be introduced into the injection port using a **manual syringe technique** or an **automated sampling system.** Manual sample injection is generally the largest single source of poor precision in GC analysis. Ten μL syringes are usually chosen since they are more durable than the microsyringes, and sample injection volumes typically range from 1 to 3 μL. These syringes will hold about 0.6 μL in the needle and barrel (this is in addition to that measured on the barrel). Thus the amount of sample that is injected into the GC depends upon the proportion of this 0.6 μL that is included in the injection and the ability of the analyst to accurately read the desired sample volume on the syringe barrel. This can be quite variable for the same analyst and be grossly different between analysts.

30.3.3 Oven

The oven controls the temperature of the column. In GC, one takes advantage of both an interaction of the analyte with the stationary phase and the boiling point for separation of compounds. Thus, the injection is often made at a lower oven temperature and is then temperature programmed to some elevated temperature. While analyses may be done isothermally, compound elution time and resolution are extremely dependent upon temperature, so temperature-programmed runs are most common. It should be obvious that higher temperatures will cause the sample to elute faster and, therefore be at a cost of resolution.

Oven temperature program rates can range from as little as 0.1°C/min to the maximum temperature heating rate that the GC can provide. A rate of 2 to 10°C/min is most common.

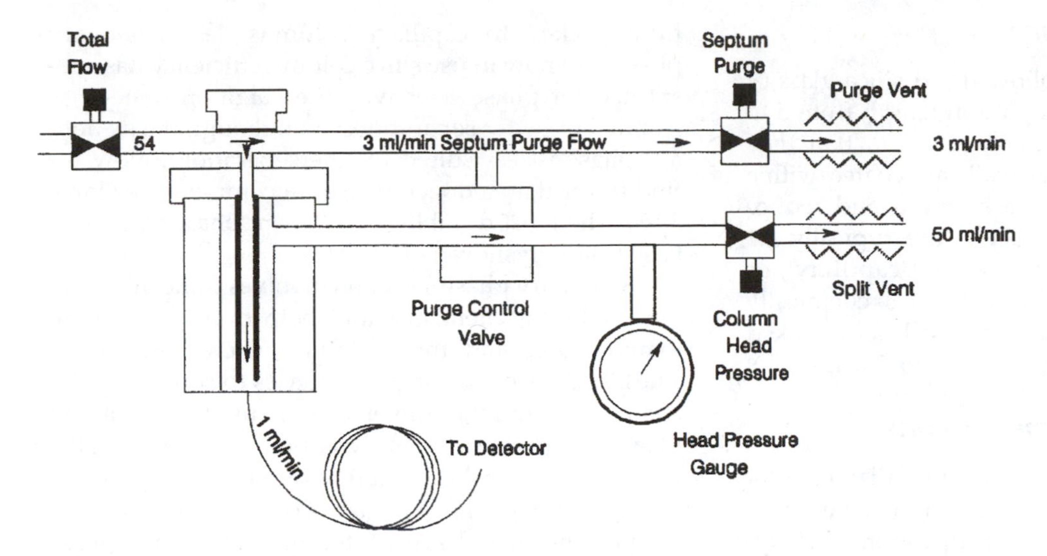

FIGURE 30-3.
Schematic of a GC injection port. Figure courtesy of Hewlett-Packard Co., Analytical Customer Training, Atlanta, GA.

30.3.4 Column and Stationary Phases

The GC column may be either **packed** or **capillary.** Early chromatography was done on packed columns, but the advantages of capillary chromatography so greatly outweigh those of packed column chromatography that few packed column instruments are sold any longer (Fig. 30-4). While some use HRGC (high-resolution gas chromatography) to designate capillary GC, GC today means capillary chromatography to most individuals.

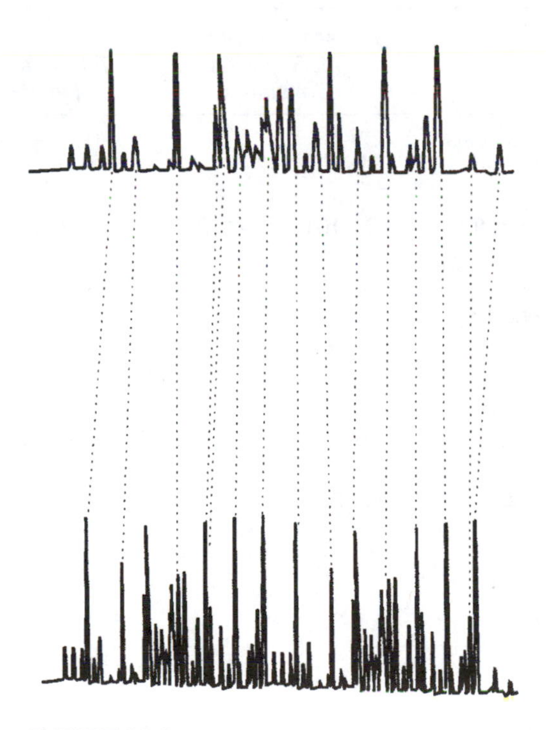

FIGURE 30-4.
Comparison of gas chromatographic separation of perfume base using packed *(top)* and capillary columns *(bottom)*. Bottom figure courtesy of Hewlett-Packard Co., Analytical Customer Training, Atlanta, GA.

30.3.4.1 Packed Columns

The packed column is most commonly made of stainless steel or glass and may range from 1.6 to 12.7 mm in outer diameter and be 0.5 to 5.0 m long (generally 2–3 m). It is packed with a granular material consisting of a "liquid" coated on an allegedly inert solid support. The **solid support** is most often diatomaceous earth (skeletons of algae) that has been purified, possibly chemically modified (e.g., silane treated), and then sieved to provide a definite mesh size (60/80, 80/100, or 100/120).

The liquid loading is usually applied to the solid support at 1–10 percent by weight of the solid support. While the liquid coating can be any one of the approximately 200 available, the most common are silicone-based phases (methyl, phenyl, or cyano substituted) and Carbowax (ester-based).

The liquid phase as well as the loading is determined by the analysis desired. The choice of liquid is typically such that it is of similar polarity as the analytes to be separated (see section 30.4.4.5 for more discussion). Loading influences time of analysis (retention time is proportional to loading), resolution (generally improved by increasing phase loading, within limits), and bleed. The liquid coatings are somewhat volatile and will be lost from the column at high temperatures (this is dependent upon the phase itself). This results in an increasing baseline **(column bleeding)** during temperature programming.

30.3.4.2 Capillary Columns

The capillary column is a hollow fused silica glass (< 100 ppm impurities) tube ranging in length from 5 to 100 m. The walls are so thin, ca. 25 μm, that they are flexible. The column outer walls are coated with a polyamide material to enhance strength and reduce breakage. Column inner diameters are typically 0.1 mm **(microbore),** 0.2–0.32 mm **(normal capillary),** or 0.53 mm **(megabore).** Liquid coating is chemically bonded to the glass walls and internally crosslinked at phase thicknesses ranging from 0.1 to 5 μm.

30.3.4.3 Gas-Solid Chromatography

Gas-solid chromatography is a very specialized area of chromatography accomplished without using a liquid phase—the analyte interaction is with a synthetic polymer such as Poropak or Chromosorb (trade names of polymers based on vinyl benzene). This material has been applied both to packed and capillary columns. Separations usually involve water or other very volatile materials.

30.3.4.4 Stationary Phases

As many as 200 different liquid phases have been developed for gas chromatography. As GC has changed from packed to capillary columns, less stationary phases are now in use since column efficiency has substituted for phase selectivity (i.e., high efficiency has resulted in better separations even though the stationary phase is less suited for the separation). Now we find fewer than a dozen phases in common use (Table 30-3). The most durable and efficient phases are those based on polysiloxane (-Si-O-Si-).

Stationary phase selection involves some intuition, knowledge of chemistry, and help from the column manufacturer and the literature. There are general rules, such as choosing polar phases to separate polar compounds and the converse or phenyl-based column phase to separate aromatic compounds. However, the high efficiency of these columns often results in separation even though the phase is not optimal. For example, a 5 percent phenyl substituted methyl silicone phase applied to a capillary column will separate polar as well as nonpolar compounds and is a commonly used phase coating.

30.3.5 Detectors

There are numerous detectors available for GC, each offering certain advantages in either sensitivity (e.g., electron capture) or selectivity (e.g., atomic emission

TABLE 30-3. Common Stationary Phases

Composition	*Polarity*	*Applications*[1]	*Phases with Similar McReynolds Constants*[2]	*Temperature Limits*[3]
100% dimethyl polysiloxane (gum)	Nonpolar	Phenols, hydrocarbons, amines, sulfur compounds, pesticides, PCBs	OV-1 SE-30	−60°C–325°C
100% dimethyl polysiloxane (fluid)	Nonpolar	Amino acid derivatives, essential oils	OV-101, SP-2100	0°C–280°C
5% diphenyl 95% dimethyl polysiloxane	Nonpolar	Fatty acids, methyl esters, alkaloids, drugs, halogenated compounds	SE-52 OV-23 SE-54	−60°–325°C
7% cyanopropyl 7% phenyl polysiloxane	Intermediate	Drugs, steroids, pesticides	OV-1701	−20°C–280°C
50% phenyl, 50% methyl polysiloxane	Intermediate	Drugs, steroids, pesticides, glycols	OV-17	60°–240°C
50% cyanopropylmethyl, 50% phenylmethyl polysiloxane	Intermediate	Fatty acids, methyl esters, alditol acetates	OV-225	60°C–240°C
50% trifluoropropyl polysiloxane	Intermediate	Halogenated compounds, + aromatics	OV-210	45°C–240°C
Polyethylene glycol-TPA modified	Polar	Acids, alcohols, aldehydes, acrylates, nitriles, ketones	OV-351 SP-1000	60°C–240°C
Polyethylene glycol	Polar	Free acids, alcohols, esters, essential oils, glycols, solvents	Carbowax 20M	60°C–220°C

[1]Specific application notes from column suppliers provide information for choosing a specific column.
[2]McReynolds constants are used to group stationary phases together on the basis of separation properties.
[3]Stationary phases have both upper and lower temperature limits. Lower temperature limit is often due to a phase change (liquid to solid) and upper temperature limit to a volatilization of phase.

detector). The most common detectors are the **flame ionization** (FID), **thermal conductivity** (TCD), **electron capture** (ECD), **flame photometric** (FPD), and **photoionization** (PID) detectors. The operating principles and food applications of these detectors are discussed below (more detail can be found in reference 22). The characteristics of these detectors are summarized in Table 30-4.

30.3.5.1. Thermal Conductivity Detector (TCD)

30.3.5.1.1 Operating Principles As the carrier gas passes over a hot filament (tungsten), it cools the filament at a certain rate depending on carrier gas velocity and composition. The temperature of the filament determines its resistance to electrical current. As a compound elutes with the carrier gas, the cooling effect on the filament is typically less, resulting in a temperature increase in the filament and an increase in resistance that is monitored by the GC electronics. Older style TCDs used two detectors and two matching columns; one system served as a reference and the other as the analytical system. Newer designs use only one detector (and column), which employs a carrier gas switching value to pass alternately carrier gas or column effluent though the detector (Fig. 30-5). The signal is then a change in cooling of the detector as a function of which gas is passing through the detector—from the analytical column or carrier gas supply (reference gas flow).

The choice of carrier gas is important since differences between its thermal properties and the analytes determines response. While hydrogen is the best choice, helium is most commonly used since hydrogen is flammable.

30.3.5.1.2 Applications The most valuable properties of this detector are that it is *universal* in response and nondestructive to the sample. Thus, it is used in food applications where there is no other detector that will adequately respond to the analytes (e.g., water, permanent gases, CO, or CO_2) or when the analyst wishes to recover the separated compounds for further analysis (e.g., trap the column effluent for infrared, NMR, or sensory analysis). It does not find broad use because it is relatively insensitive, and often the analyst desires specificity in detector response to remove interfering compounds from the chromatogram.

30.3.5.2 Flame Ionization Detector (FID)

30.3.5.2.1 Operating Principles As compounds elute from the analytical column, they are burned in a hydrogen flame (Fig. 30-6). A potential (often 300 volts) is applied across the flame. The flame will carry a current across the potential which is proportional to the organic ions present in the flame from the burning of an organic compound. The current flowing across the flame is amplified and recorded. The FID responds to organics on a weight basis. It gives virtually no response to H_2O, NO_2, CO_2, H_2S and limited response to many other compounds. Response is best with compounds containing C-C or C-H bonds.

30.3.5.2.2 Applications The food analyst is most often working with organic compounds, which this detector responds well to. Its very good sensitivity, wide linear range in response (necessary in quantitation), and dependability make this detector the choice for most food work. Thus, this detector is used for virtually all food analyses where a specific detector is not desired or

TABLE 30-4. Characteristics of Detectors for Gas Chromatography

Characteristic	*TCD*[1]	*FID*[2]	*ECD*[3]	*FPD*[4]	*PID*[5]
Specificity	Very little; detects almost anything, including H_2O; called the "universal detector"	Most organics	Halogenated compounds and those with nitro or conjugaged double bonds	Organic compounds with S or P (determined by which filter is used)	Depends on ionization energy of lamp relative to bond energy of solutes
Sensitivity limits	ca. 400 pg; relatively poor; varies with thermal properties of compound	10–100 pg for most organics; very good	0.05–1 pg; excellent	2 pg for S and 0.9 pg for P compounds; excellent	1–10 pg depending on compound and lamp energy; excellent
Linear Range	10^4—poor; response easily becomes nonlinear	10^6–10^7; excellent	10^4—poor; 10^4 for P; poor	10^3 for S	10^7—excellent

[1]Thermal conductivity detector
[2]Flame ionization detector
[3]Electron capture detector
[4]Flame photometric detector
[5]Photoionization detector

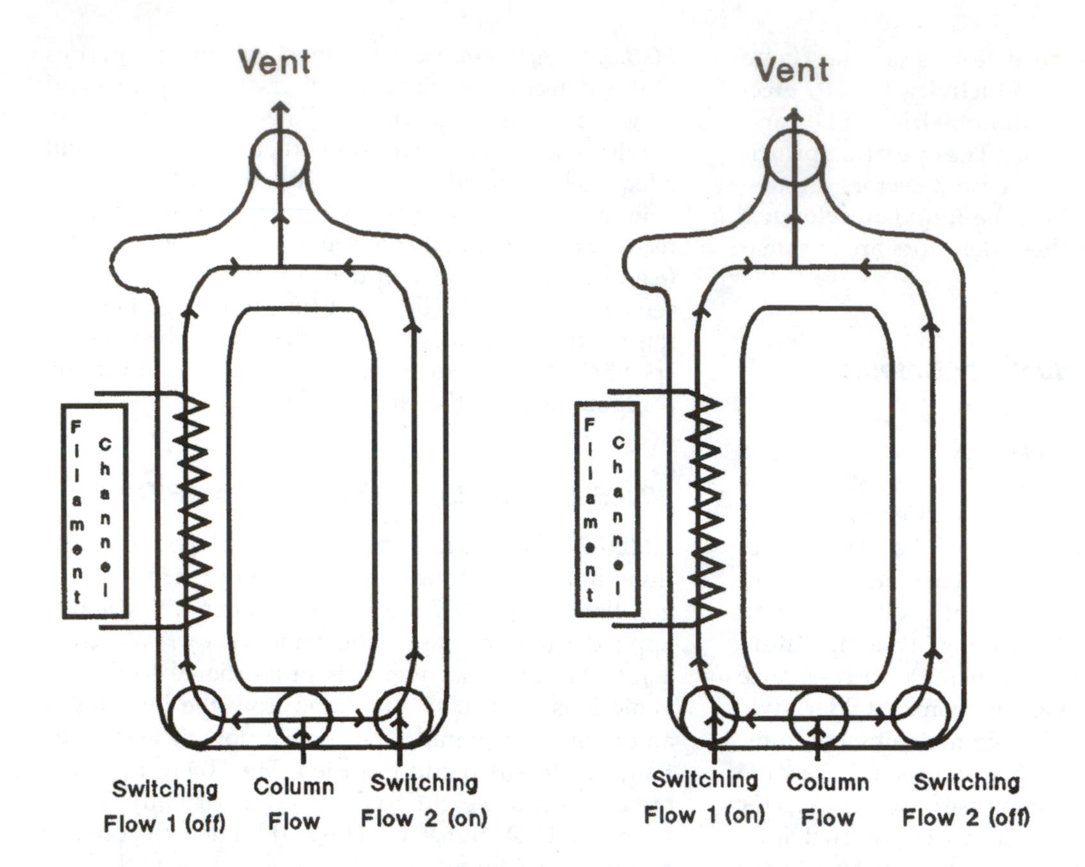

FIGURE 30-5.
Schematic of the thermal conductivity detector. Figure courtesy of Hewlett-Packard Co., Analytical Customer Training, Atlanta, GA.

sample destruction is acceptable (column eluant is burned in flame). This includes, for example, flavor studies, fatty acid analysis, carbohydrate analysis, sterols, contaminants in foods, and antioxidants.

30.3.5.3 Electron Capture Detector (ECD)

30.3.5.3.1 Operating Principles The electron capture detector contains a radioactive foil coating that emits electrons as it undergoes decay (Fig. 30-7). The electrons are collected on an anode, and the standing current is monitored by instrument electronics. As an analyte elutes from the GC column, it passes between the radioactive foil and the anode. Compounds that capture electrons reduce the standing current and thereby give a measurable response.

30.3.5.3.2 Applications In food applications, the ECD has found its greatest use in determining pesticide residues (see Chapter 19). The specificity and sensitivity of this detector make it ideal for this application.

30.3.5.4 Flame Photometric Detector (FPD)

30.3.5.4.1 Operating Principles The FPD detector works by burning all analytes eluting from the analytical column and then measuring specific wavelengths of light that are emitted from the flame using a filter and photometer (Fig. 30-8). The wavelengths of light that are suitable in terms of intensity and uniqueness are characteristic of sulfur and phosphorous. Thus this detector is specific for these two elements.

30.3.5.4.2 Applications The FPD has found its major food applications in the determination of organophosphorus pesticides and volatile sulfur compounds in general. The determination of sulfur compounds has typically been in relation to flavor studies.

30.3.5.5 Photoionization Detector (PID)

30.3.5.5.1 Operating Principles The photoionization detector uses UV irradiation (usually 10.2 eV) to ionize analytes eluting from the analytical column (Fig. 30-

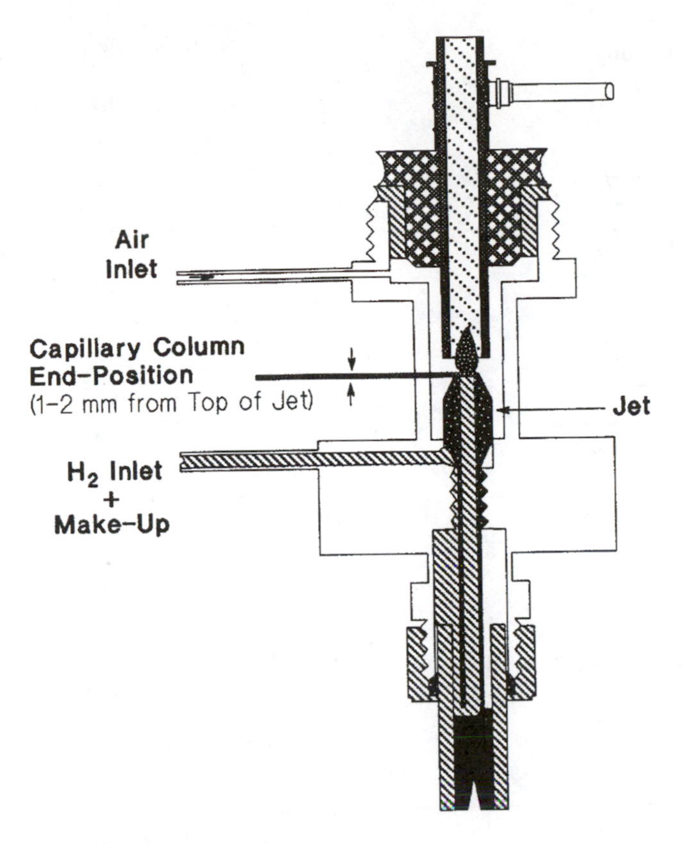

FIGURE 30-6.
Schematic of the flame ionization detector designed for use with capillary columns. Figure courtesy of Hewlett-Packard Co., Analytical Customer Training, Atlanta, GA.

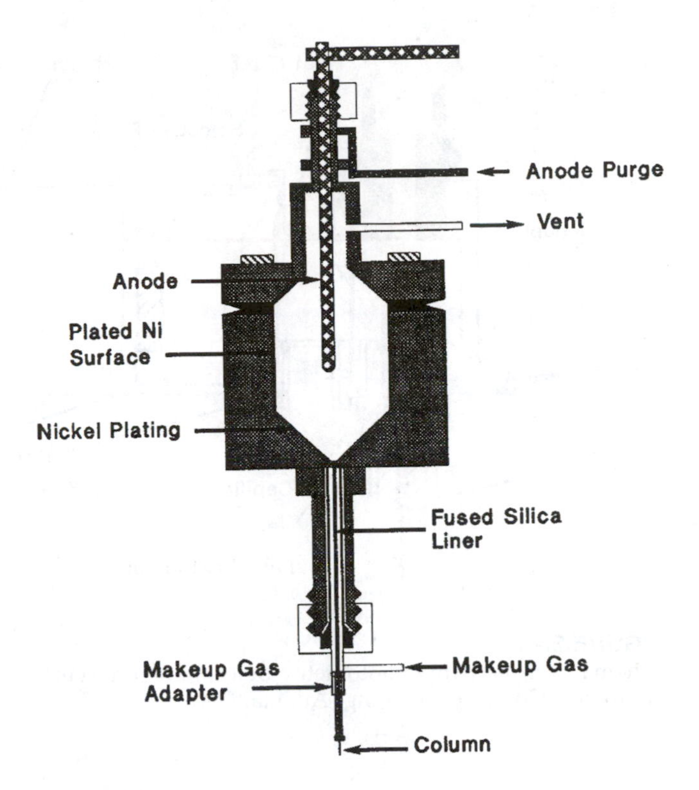

FIGURE 30-7.
Schematic of the electron capture detector. Figure courtesy of Hewlett-Packard Co., Analytical Customer Training, Atlanta, GA.

9). The ions are accelerated by a polarizing electrode to a collecting electrode. The small current formed is magnified by the electrometer of the GC to provide a measurable signal.

This detector offers the advantages of being quite sensitive and nondestructive and may be operated in a selective response mode. The selectivity comes from being able to control the energy of ionization, which will determine the classes of compounds that are ionized and thus detected.

30.3.5.5.2 Applications The PID finds primary use in analyses for which excellent sensitivity is required from a nondestructive detector. This is most often a flavor application in which the analyst wishes to smell the GC effluent to determine the sensory character of the individual GC peaks. While this detector might find broader use, the widespread availability of the FID (which is suitable for most of the same applications) meets most of these needs.

30.3.5.6 Miscellaneous Detectors

As was noted earlier in this section, there are numerous other detectors available for GC. These detectors tend to be specific detectors that find use in limited applications (e.g., electrolytic detectors or nitrogen phosphorous detectors).

30.3.5.7 Hyphenated Gas Chromatographic Techniques

Hyphenated gas chromatographic techniques are those that combine GC with another major technique. Examples are **GC-AED** (atomic emission detector), **GC-FTIR** (Fourier transform infrared), and **GC-MS** (mass spectrometry). While all of the techniques are established methods of analysis in themselves, they become powerful tools when combined with a technique such as GC. GC provides the separation and the hyphenated technique the detector. GC-MS has long been known to be a most valuable tool for the identification of volatile compounds (see Chapter 26). The MS, however, may perform the task of serving as a specific detector for the GC by selectively focusing on ion fragments unique to the analytes of interest. The analyst can detect and quantify components without their gas chromatographic resolution in this manner. The same statements can be made about

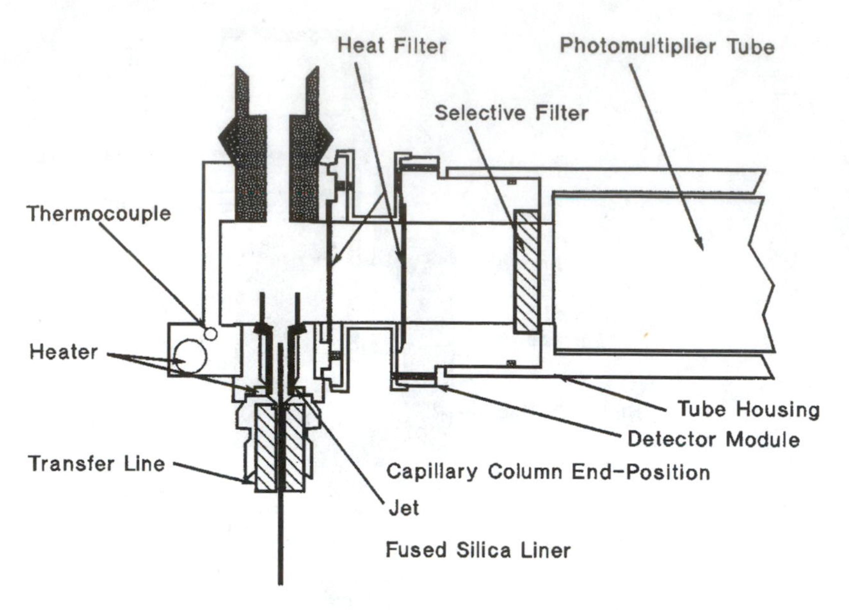

FIGURE 30-8.
Schematic of the flame photometric detector. Figure courtesy of Hewlett-Packard Co., Analytical Customer Training, Atlanta, GA.

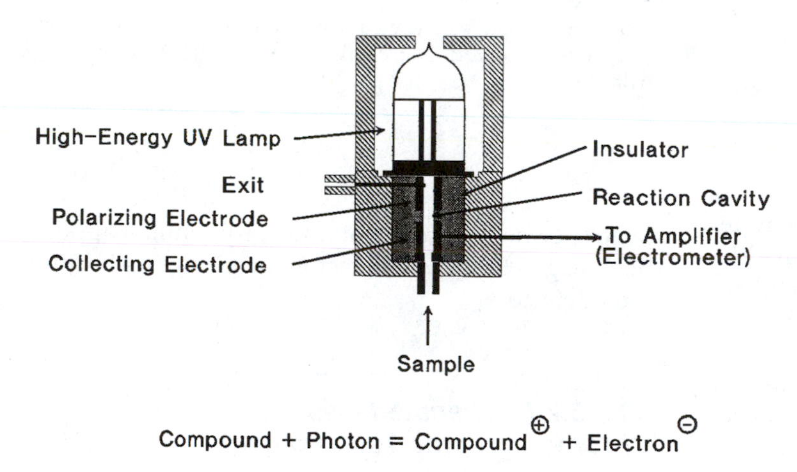

FIGURE 30-9.
Schematic of the photoionization detector. Figure courtesy of Hewlett-Packard Co., Analytical Customer Training, Atlanta, GA.

GC-FTIR (see Chapter 24). The FTIR can readily serve as a GC detector.

A relatively new combination is GC-AED. In this technique, the GC column effluent enters a microwave-generated Helium plasma that excites the atoms present in the analytes. The atoms emit light at their characteristic wavelengths, and this emission is monitored using a diode ray detector similar to that used in HPLC. This results in a very sensitive and specific elemental detector. This detector will find very broad application as it enters the field.

30.4 CHROMATOGRAPHIC THEORY

30.4.1 Introduction

The principles of chromatographic separations and chromatographic theory are discussed in Chapter 28 (sections 28.3.3 and 28.3.4.3.3, respectively). Gas chromatography may depend upon several types (or principles) of chromatography for separation. For example, size-exclusion chromatography is used in the separation of permanent gases such as N_2, O_2, and H_2. A variation of size exclusion is used to separate chiral compounds on cyclodextrin-based columns; one enantiomorphic form will fit better into the cavity of the cyclodextrin than will the other form, resulting in separation. Adsorption chromatography is used to separate very volatile polar compounds (e.g., alcohols, water, and aldehydes) on porous polymer columns (e.g., Tenax^R phase). Partition chromatography is the workhorse for gas chromatographic separations. There are over 200 different liquid phases that have been developed for gas chromatographic use over time. Fortunately, the vast majority of separations can be accomplished with only a few of these phases, and the other phases have fallen into disuse. Gas chromatography depends not only upon adsorption, partition, and/or size exclusion for separation, but also upon solute boiling point for additional resolving powers. Thus, the separations accomplished are based on several properties of the solutes. This gives gas chromatography virtually unequaled resolution powers as compared to other types of chromatography (e.g., HPLC, paper, or thin-layer chromatography).

A brief discussion of chromatographic theory will follow. The purpose of this additional discussion is to apply this theory to gas chromatography to optimize separation efficiency so that analyses can be done faster, less expensively, or with greater precision and accuracy. If one understands the factors influencing resolution in GC, one can optimize the process and gain in efficiency of operation.

30.4.2 Separation Efficiency

A good separation has narrow-based peaks and ideally, but not essential to quality of data, baseline separation of compounds. This is not always achieved. Peaks broaden as they pass through the column—the more they broaden, the poorer is the separation and efficiency. As discussed in Chapter 28 (section 28.3.4.3.3) a measure of this broadening is **height equivalent to a theoretical plate** (HEPT). This term is derived from N, the number of plates in the column, and L, the length of the column. A good packed column might have N = 5,000, while a good capillary column should have about 3,000 to 4,000 plates per meter for a total of 100,000 to 500,000 plates depending on column length. HETP will range from about 0.1 to 1 mm for good columns.

30.4.2.1 Carrier Gas Flow Rates and Column Parameters

Several factors influence column efficiency (peak broadening). These are related by the **Van Deemter equation** (1): (HEPT values should be small.)

$$\text{HEPT} = A + B/U + CU \tag{1}$$

where: HETP = height equivalent to a theoretical plate
A = eddy diffusion
B = band broadening due to diffusion
U = velocity of the mobile phase
C = resistance to mass transfer

A is eddy diffusion; this is a spreading of the analytes in the column due to the carrier gas having various pathways or nonuniform flow (Fig. 30-10). In packed column chromatography, poor uniformity in solid support size or poor packing results in channeling and multiple pathways for carrier flow, which results in spreading of the analyte in the column. Thus, improved efficiency is obtained by using the high-performance solid supports and commercially packed columns.

In capillary chromatography, the A term is relatively very small. However, as the diameter of the capillary column increases, the flow properties deteriorate, and band spreading occurs. The most efficient capillary columns have small diameters (0.1 mm), and efficiency decreases rapidly as one goes to megabore columns (Fig. 30-11). Megabore columns are only slightly more

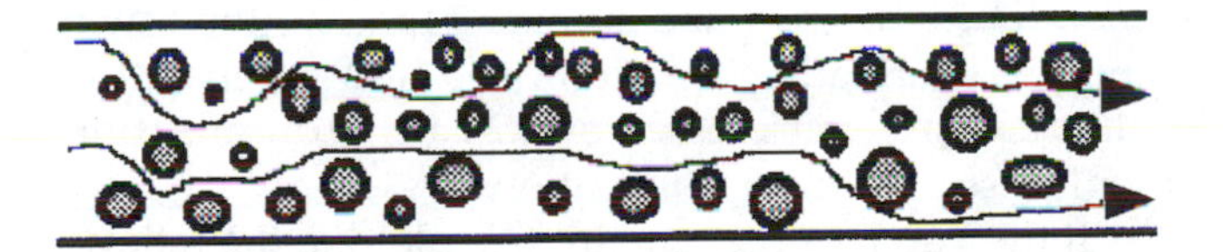

FIGURE 30-10.
Illustration of flow properties that lead to large eddy diffusion (term A).

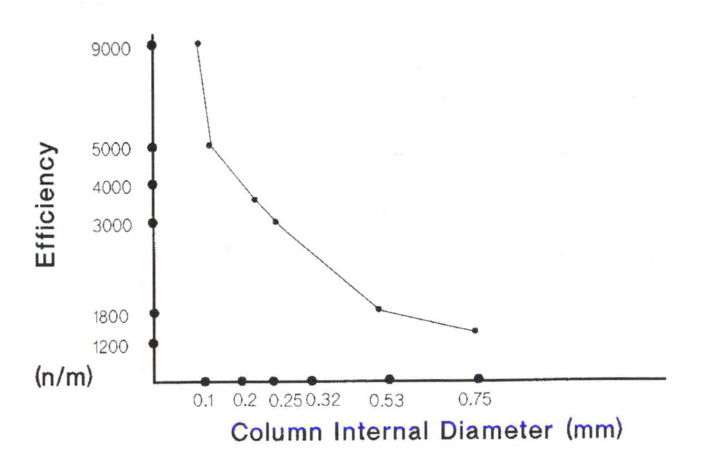

FIGURE 30-11.
The influence of column diameter on column efficiency (plates/meter). Figure courtesy of Hewlett-Packard, Analytical Customer Training, Atlanta, GA.

efficient than packed columns. While column efficiency increases as we go to smaller columns, column capacity decreases rapidly. Microbore columns are easily overloaded (capacity may be 1–5 ng per analyte), resulting again in poor chromatography. Thus, column diameter is generally chosen as 0.2 to 0.32 mm to compromise efficiency with capacity.

B is band broadening due to diffusion; solutes will go from a high to a low concentration (Fig. 30-12). U is velocity of the mobile phase. Thus, very slow flow rates result in large amounts of diffusion band broadening, and faster flow rates minimize this term.

C is resistance to mass transfer. If the flow (U) is too fast, the equilibrium between the phases is not established, and poor efficiency results (Fig. 30-13). This can be visualized in the following way: If one molecule of solute is dissolved in the stationary phase and another is not, the undissolved molecule continues to move through the column while the other is retained. This results in band spreading within the column. Other factors that influence this term are thickness of the stationary phase and uniformity of coating on the phase support. Thick films give greater capacity (ability to handle larger amounts of a solute) but at a cost in terms of band spreading (efficiency of separation) since thick films provide more variation in diffusion properties in and out of the stationary phase. Thus phase thickness is a compromise between maximizing separation efficiency and sample capacity (too much sample—overloading a column—destroys separation ability). Phase thicknesses of 0.25 to 1 μm are commonly used for most applications.

If the Van Deemter equation is plotted, the following figure results (Fig. 30-14). We see an optimum in flow rate due to the opposing effects of the B and the C terms. It should be noted that the GC may not be operated at a carrier flow velocity yielding maximum efficiency (lowest HETP). Analysis time is directly proportional to carrier gas flow velocity. If the analysis time can be significantly shortened by operating above the optimum flow velocity and adequate resolution is still obtained, velocities well in excess of optimum should be used.

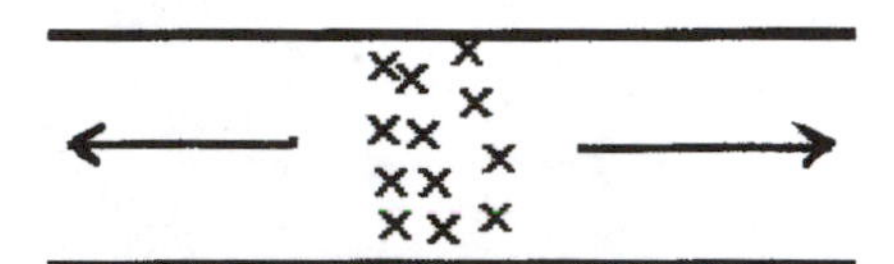

FIGURE 30-12.
Illustration of the diffusion term (B) influencing column efficiency. (U is velocity of the mobile phase.)

 **is slower than** 

FIGURE 30-13.
Illustration of the mass transfer term (C) influencing column efficiency.

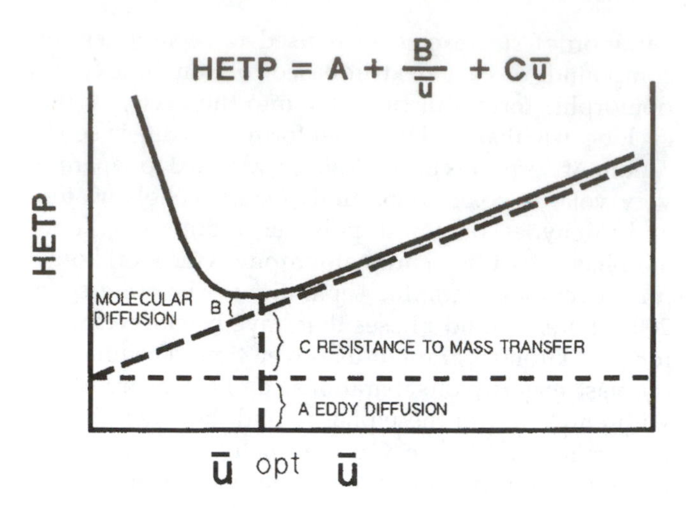

FIGURE 30-14.
The relationship between carrier gas flow rate (U) and column efficiency (HETP)—Van Deemter equation (optimum U is noted). Figure courtesy of Hewlett-Packard Co., Analytical Customer Training, Atlanta, GA.

30.4.2.2 Carrier Gas Type

The relationship between HETP and carrier gas flow velocity is strongly influenced by carrier gas choice (Fig. 30-15). Nitrogen is the most efficient but has an optimum at such a low flow velocity (long analysis

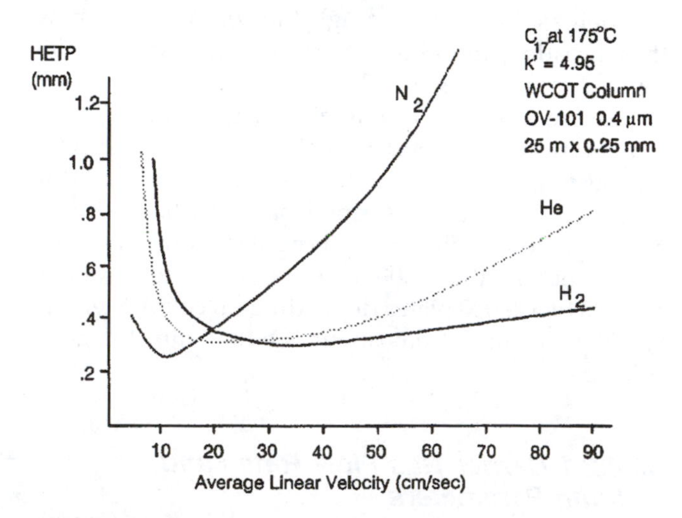

FIGURE 30-15.
Influence of carrier gas type and flow rate on column efficiency. Figure courtesy of Hewlett-Packard Co., Analytical Customer Training, Atlanta, GA.

time) that it is a poor choice since analysis times are much longer than required. Helium is the next best choice and is the most commonly used carrier. Hydrogen, however, is generally the best choice since it offers high efficiency but small dependency on flow velocity. One can operate the GC with very high flow velocities (short analysis time) and yet lose little in terms of separation efficiency. Hydrogen is not commonly used as a carrier gas because it is flammable.

30.4.2.3 Summary of Separation Efficiency

In summary, an important goal of analysis is to achieve the necessary separation in the minimum amount of time. The following factors should be considered:

1. In general, small diameter columns (packed or capillary) should be used since separation efficiency is strongly dependent on column diameter. While small diameter columns will limit column capacity, limited capacity can often be compensated for by increasing phase thickness. Increased phase thickness will also decrease column efficiency but to a lesser extent than increasing column diameter.
2. Lower operating temperatures should be used—if elevated column temperatures are required in order for the compounds of interest to elute, use a shorter column if resolution is adequate.
3. One should keep columns as short as possible (analysis time is directly proportional to column length—resolution is proportional to the square root of length).
4. Use hydrogen as the carrier gas if the detector permits. Some detectors have specific carrier gas requirements.
5. Operate the GC at the maximum carrier gas velocity that provides resolution.

The pyramid shown in Fig. 30-16 summarizes the compromises that must be made in choosing the analytical column and gas chromatographic operating conditions. One cannot get all four of the properties under any given operating conditions and column choices. For example, resolution will be at the cost of capacity (e.g., small bore columns and thin film coating) and speed (e.g., low operating temperatures and long columns). Capacity will be at a cost of resolution (e.g., wide bore columns with thick films) and speed (e.g., slow carrier velocities and low temperatures). The choice of column and operating parameters must consider the needs of the analyst and the compromises involved in these choices.

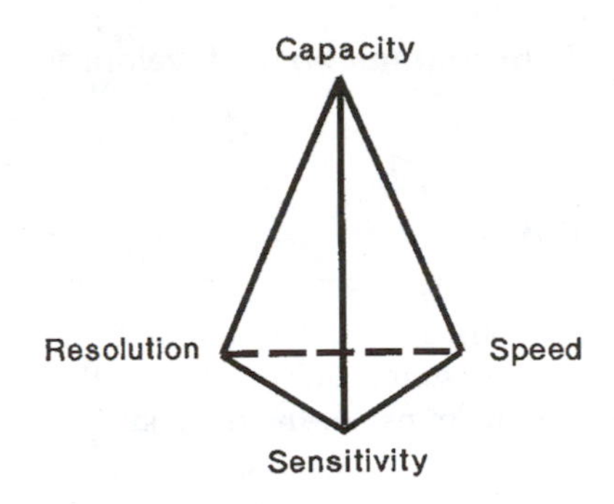

FIGURE 30-16.
Relationships among column capacity, efficiency, resolution, and analysis speed. Figure courtesy of Hewlett-Packard Co., Analytical Customer Training, Atlanta, GA.

30.5 SUMMARY

Gas chromatography has found broad application in both the food industry and academia. It is exceptionally well suited to the analysis of volatile thermally stable compounds. This is due to the outstanding resolving properties of the method and the wide variety of detectors that can provide either sensitivity or selectivity in analysis.

Sample preparation generally involves the isolation of solutes from foods, which may be accomplished by headspace analysis, distillation, preparative chromatography (including solid-phase extraction), or extraction (liquid/liquid). Some solutes can then be directly analyzed, while others must be derivatized prior to analysis.

The GC consists of a gas supply and regulators (pressure and flow control), injection port, column and column oven, detector, electronics, and a data recording and processing system. The analyst must be knowledgeable about each of these GC components: carrier and detector gases; injection port temperatures and operation in split, splitless, temperature-programmed, or on-column modes; column choices and optimization (gas flows and temperature profile during separation); and detectors (TCD, FID, NPD, ECD, FPD, and PID). The characteristics of these GC components and an understanding of basic chromatographic theory are essential to balancing the properties of resolution, capacity, speed, and sensitivity.

Unlike most of the other chromatographic techniques, GC has reached the theoretical limits in terms of both resolution and sensitivity. Thus, this method will not change significantly in the future other than minor innovations in hardware or associated computer software.

GC as a separation technique has been combined with AED, FTIR, and MS as detection techniques to make GC an even more powerful tool. Such hyphenated techniques are likely to continue to be developed and refined.

30.6 STUDY QUESTIONS

1. For each of the following methods to isolate solutes from food prior to GC analysis, describe the procedure, the applications, and the cautions in use of the method:
   a. headspace methods
   b. distillation methods
   c. solvent extraction
2. What is solid-phase extraction and why is it advantageous over traditional liquid/liquid extractions?
3. Why must sugars and fatty acids be derivatized before GC analysis, while pesticides and aroma compounds need not be derivatized?
4. Why is the injection port of a GC at a higher temperature than the oven temperature?
5. Differentiate packed columns from capillary columns (microbore and megabore) with regard to physical characteristics, column capacity, and column efficiency.
6. You are doing GC with a packed column and notice that the baseline rises from the beginning to the end of each run. Explain a likely cause for this increase.
7. The most common detectors for GC are FID, TCD, ECD, FPD, and PID. Differentiate each of these with regard to the operating principles. Also, indicate below which detector(s) fits the description given.
   a. ____ least sensitive
   b. ____ most sensitive
   c. ____ least specific
   d. ____ greatest linear range
   e. ____ nondestructive to sample
   f. ____ commonly used for pesticides
   g. ____ commonly used for volatile sulfur compounds
8. What types of chromatography does GC rely upon for separation of compounds?
9. A fellow lab worker is familiar with HPLC for food analysis but not with GC. As you consider each component of a typical chromatographic system (and specifically the components and conditions for GC and HPLC systems), explain GC to the fellow worker by comparing and contrasting it to HPLC. Following that, state in general terms the differences among the types of samples appropriate for analysis by GC versus HPLC, and give several examples of food constituents appropriate for analysis by each. (See also Chapter 29.).

30.7 REFERENCES

1. James, A.T., and Martin, A.J.P. 1952. Gas-liquid chromatography: The separation and microestimation of volatile fatty acids from formic acid to dodecanoic acid. *Biochem. J.* 50:679.
2. Jennings, W.G. 1981. *Applications of Glass Capillary Chromatography.* Marcel Dekker, Inc., New York, NY.
3. Jennings, W.G. 1987. *Analytical Gas Chromatography.* Academic Press, Inc., Orlando, FL.
4. Schomburg, G. 1990. *Gas Chromatography: A Practical Course.* Weinheim, New York, NY.
5. Gordon, M.H. 1990. *Principles and Applications of Gas Chromatography in Food Analysis.* E. Horwood, New York, NY.
6. Drawert, F., Heimann, W., Enberger, R., and Tressl, R. 1965. Enzymatische Veranderung des naturlichen Apfelaromass bei der Aurfarbeitung. *Naturwissenschaften.* 52:304.
7. Fleming, H.P., Fore, S.P., and Goldblatt, L.A. 1968. The formation of carbonyl compounds in cucumbers. *J. Food Sci.* 33:572.
8. Kazeniak, S.J., and Hall, R.M. 1970. Flavor chemistry of tomato volatiles. *J. Food Sci.* 35:519.
9. Leahy, M.M., and Reineccius, G.A. 1984. Comparison of methods for the analysis of volatile compounds from aqueous model systems, in *Analysis of Volatiles: New Methods and Their Application,* P. Schreier (Ed.). DeGruyter Publ., Berlin, Germany.
10. Sapers, G.M., Panasiuk, O., and Talley, F.B. 1973. Flavor quality and stability of potato flakes: Effects of raw material and processing. *J. Food Sci.* 38:586.
11. Seo, E.W., and Joel, D.L. 1980. Pentane production as an index of rancidity in freeze-dried pork. *J. Food Sci.* 45:26.
12. Buttery, R.G., and Teranishi, R. 1963. Measurement of fat oxidation and browning aldehydes in food vapors by direct injection gas-liquid chromatography. *J. Agric. Food Chem.* 11:504.
13. Buckholz, L.L., Withycombe, D.A., and Daun, H. 1980. Application and characteristics of polymer adsorption method used to analyze flavor volatiles from peanuts. *J. Agric Food Chem.* 28:760.
14. Reineccius, G.A., Keeney, P.A., and Weiseberger, W. 1972. Factors affecting the concentration of pyrazines in cocoa beans. *J. Agric. Food Chem.* 20:202.
15. Markel, C., Hagen, D.F., and Bunnelle, V.A. 1991. New technologies in solid-phase extraction. *LC-GC.* 9:332.
16. Majors, R.E. 1986. Sample preparation for HPLC and GC using solid-phase extraction. *LC-GC.* 4:972.
17. Dupuy, H.P., Fore, S.P., and Goldbatt, L.A. 1971. Elution and analysis of volatiles in vegetable oils by gas chromatography. *J. Am. Oil Chem. Soc.* 48:876.
18. Legendre, M.G., Fisher, G.S., Fuller, W.H., Dupuy, H.P., and Rayner, E.T. 1979. Novel technique for the analysis of volatiles in aqueous and nonaqueous systems. *J. Am. Oil Chem. Soc.* 56:552.
19. Reineccius, G.A., and Anandaraman, S. 1984. Analysis of volatile flavors in foods, in *Food Constituents and Food Residues: Their Chromatographic Determination,* J. Whitaker and K. Steward (Eds.). AVI Publishing, Westport, CT.
20. Widmer, H.M. 1990. Recent developments in instrumental analysis, in *Flavor Science and Technology,* Y. Bessière and A.F. Thomas (Eds.), p. 181. John Wiley & Sons, Chichester, England.
21. Parliment, T.H. 1986. Sample preparation techniques for gas-liquid chromatographic analysis of biologically derived aromas, in *Biogeneration of Aromas,* T.H. Parliment and R. Croteau (Eds.), p. 34. American Chemical Society, Washington, DC.
22. Buffington, R., and Wilson, M.K. 1987. *Detectors for Gas Chromatography.* Hewlett-Packard Corp., Avondale, PA.

P A R

V

Other Methods and Instrumentation

CHAPTER

31

The pH Meter and the Use of Ion-Selective Electrodes

Dick H. Kleyn

31.1 INTRODUCTION

While the measurement of hydrogen ion concentration by use of the pH meter is a common technique in analytical laboratories, the food analyst often has little more than a vague understanding of the theory of this electroanalytical technique. Furthermore, the opportunity to measure the concentrations of other cations and anions with **ion-selective electrodes** (ISE) is likely to be a completely new discovery. Thus, the purpose of this chapter is to introduce students to the theory of potentiometry, using the concept of pH as the means of illustration. Finally, attention is given to the use of other ion-selective electrodes in electroanalytical chemistry.

31.2 PRINCIPLES AND PROCEDURES

31.2.1 Theory of Potentiometry

Potentiometry is one of the two main categories of electroanalytical techniques, being classified as voltammetry at zero current. The other technique is known as **voltammetry** and is classified as voltammetry at finite current. pH measurement is an example of potentiometry, while polarography is an example of voltammetry.

The basic principle of potentiometry involves the use of an electrolytic cell composed of two electrodes dipped into a test solution. A voltage (**electromotive force**-EMF) develops, which is related to the ionic concentration of the solution. This EMF is measured under conditions such that an infinitessimal current (10^{-12} amperes or less) is drawn during measurement. Hence, the classification *voltammetry at zero current* is used for potentiometry. When an appreciable current is drawn, concentration changes in the solution surrounding the electrodes will be produced, the measured potential corresponding to a different system from that present initially. In addition, irreversible changes may occur in either of the two electrodes.

The early potentiometers compared EMFs in terms of resistance. These instruments used a moving coil galvanometer as a balance detector, establishing the point of balance between an unknown EMF and an opposing known EMF. The galvanometer was replaced by a vacuum tube amplifier capable of registering minute amounts of current, thus permitting measurement of the EMF differences. The vacuum tube instruments have been replaced by transistors for measuring the EMF difference because they operate with very little current.

31.2.2 Concept of pH

The Bronsted-Lowry theory of neutralization is based upon the following definitions for acid and base:

Acid: A substance capable of donating protons. It must contain hydrogen.

Base: A substance capable of accepting protons.

Neutralization is the reaction of an acid with a base:

$$HCl + NaOH \rightarrow NaCl + H_2O \qquad (1)$$

Acids form hydrated protons called **hydronium ions** (H_3O^+) and bases form **hydroxide ions** (OH^-) in aqueous solutions:

$$H_3O^+ + OH^- \rightleftarrows 2H_2O \qquad (2)$$

At any temperature, the product of the **molar concentrations** (moles/liter) of H_3O^+ and OH^- is a constant referred to as the **ion product constant for water** (K_w):

$$[H_3O^+]\,[OH^-] = K_w \qquad (3)$$

K_w varies with the temperature. For example, at 25°C, $K_w = 1.04 \times 10^{-14}$, but at 100°C, $K_w = 58.2 \times 10^{-14}$.

The above concept of K_w leads to the question of what the concentrations of $[H_3O^+]$ and $[OH^-]$ are in pure water. Experimentation has revealed that the concentration of $[H_3O^+]$ is approximately 1.0×10^{-7}M, as is that of the $[OH^-]$ at 25°C. Because the concentrations of these ions is equal, pure water is referred to as being neutral.

Suppose that a drop of acid (proton donator) is added to pure water. The $[H_3O^+]$ concentration would increase. However, the K_w would remain constant (1.0×10^{-14} at 25°C), revealing a decrease in the $[OH^-]$ concentration. Conversely, if a drop of base is added to pure water, the $[H_3O^+]$ would decrease while the $[OH^-]$ would increase, maintaining the K_w of 1.0×10^{-14} at 25°C.

How did the term pH derive from the above considerations? In approaching the answer to this question, one must observe the concentrations of $[H_3O^+]$ and $[OH^-]$ in various foods, as shown in Table 31-1. The numerical values found in Table 31-1 for $[H_3O^+]$ and $[OH^-]$ are bulky and led a Swedish chemist, S.L.P. Sorensen, to develop the pH system in 1909.

pH is defined as the logarithm of the reciprocal of the hydrogen ion concentration. It may also be defined as the negative logarithm of the molar concentration of hydrogen ions. Thus, a $[H_3O^+]$ concentration of 1×10^{-6} is expressed simply as pH 6. The $[OH^-]$ concentration is expressed as pOH and would be pOH 8 in this case, as shown in Table 31-2.

TABLE 31-1. Concentration of $[H_3O^+]$ and $[OH^-]$ in Various Foods at 25°

Food	*$[H_3O^+]$*[1]	*$[OH^-]$*[1]	*K_w*
Cola	2.24×10^{-3}	4.66×10^{-12}	1×10^{-14}
Grape juice	5.62×10^{-4}	1.78×10^{-11}	1×10^{-14}
SevenUp	3.55×10^{-4}	2.82×10^{-11}	1×10^{-14}
Schlitz beer	7.95×10^{-5}	1.26×10^{-10}	1×10^{-14}
Pure water	1.00×10^{-7}	1.00×10^{-7}	1×10^{-14}
Tap water	4.78×10^{-9}	2.09×10^{-6}	1×10^{-14}
Milk of magnesia	7.94×10^{-11}	1.26×10^{-4}	1×10^{-14}

From (9), used with permission. Copyright 1971 American Chemical Society.
[1]Moles per liter.
Calculating the pH of the cola:

Step 1. Substitute the [H+] into the pH equation:
$pH = -\log(H+)$
$pH = -\log(2.24 \times 10^{-3})$

Step 2. Separate 2.24×10^{-3} into two parts; determine the logarithm of each part:
$\log 2.24 = 0.350$
$\log 10^{-3} = -3$

Step 3. Add the two logs together since adding logs is equivalent to multiplying the two numbers:
$0.350 + (-3) = 2.65$

Step 4. Place the value into the pH equation:
$pH = -(-2.65)$
$pH = 2.65$

While the use of pH notation is simpler from the numerical standpoint, it is a confusing concept in the minds of many scientists. One must remember that it is a logarithmic value and that a change in one pH unit is actually a tenfold change in the concentration of $[H_3O^+]$.

It is important to understand that pH and titratable acidity are not the same. This point may be illustrated by comparing the dissociations of two acidic solutions.

$$HCl \rightleftarrows H^+ + Cl^- \quad (4)$$

$$CH_3COOH \rightleftarrows H^+ + CH_3COO^- \quad (5)$$

Thus N/10 solutions of the above acids have the same strength as measured by titration. However, the HCl solution is much stronger (active) in terms of $[H_3O^+]$. The reaction of the HCl solution is pH 1.02, while that of the CH_3COOH is pH 2.89 at 25°C.

31.2.3 The pH Meter: Its Major Components

31.2.3.1 General Information

The pH meter is a good example of a potentiometer. Four major parts of the pH system are always needed: (1) reference electrode, (2) indicator electrode (pH sensitive), (3) voltmeter or amplifier that is capable of measuring small EMF differences in a circuit of very high resistance, and (4) the sample being analyzed (Fig. 31-1).

One notes that there are two electrodes involved in the measurement. Each of these electrodes tends to send its ions into solution (electrolytic solution pressure). Also, the ions in solution tend to react with the electrode (activity). The combination of these two factors produces the so-called **electrode potential.**

Hydrogen ion concentration (activity) is determined by the voltage that develops between the two electrodes. The **Nernst equation** relates the electrode response to the activity.

$$E = E^\circ + 2.3\frac{RT}{NF}\log A \quad (6)$$

where:
E = actual electrode potential
E° = a constant, the sum of several potentials in the system
R = universal gas constant, 8.313 joules/degree/g mole wt.
F = Faraday constant, 96,490 coulombs per g equiv. wt.
T = absolute temperature (°Kelvin)
N = charge of the ion
A = activity of the ion being measured

At 25°C, the relationship of 2.3RT/F is calculated to be 0.0591, as follows:

$$\frac{2.3 \times 8.316 \times 298}{96{,}490} = 0.0591 \text{ volts} \quad (7)$$

TABLE 31-2. Relationship of [H+] versus pH and [OH−] versus pOH at 25°C.

$[H^+]$[1]	pH	$[OH^-]$[1]	pOH
1×10^0	0	1×10^{-14}	14
10^{-1}	1	10^{-13}	13
10^{-2}	2	10^{-12}	12
10^{-3}	3	10^{-11}	11
10^{-4}	4	10^{-10}	10
10^{-5}	5	10^{-9}	9
10^{-6}	6	10^{-8}	8
10^{-7}	7	10^{-7}	7
10^{-8}	8	10^{-6}	6
10^{-9}	9	10^{-5}	5
10^{-10}	10	10^{-4}	4
10^{-11}	11	10^{-3}	3
10^{-12}	12	10^{-2}	2
10^{-13}	13	10^{-1}	1
10^{-14}	14	10^0	0

From (9), used with permission. Copyright 1971 American Chemical Society.

[1]Moles per liter. Note that the product of $[H^+][OH^-]$ is always 1×10^{-14}.

Calculation of [H+] of a beer with pH 4.30:

Step 1. Substitute numbers into the pH equation:

$$pH = -\log [H+]$$
$$4.30 = -\log [H+]$$
$$-4.30 = \log [H+]$$

Step 2. Divide the −4.30 into two parts so that the first part contains the decimal places and second part the whole number:

$$-4.30 = 0.70 - 5 = \log [H+]$$

Step 3. Find the antilogs:

antilog of 0.70 = 5.0

antilog of $-5 = 10^{-5}$

Step 4. Multiply the two antilogs to get [H+]:

$$5 \times 10^{-5} = [H+]$$
$$[H+] = 5 \times 10^{-5} \text{ Molar}$$

Thus, voltage produced by the electrode system is a linear function of the pH, the electrode potential being essentially +60 millivolts (0.059 volts) for each change of one pH unit. At neutrality (pH 7), the electrode potential is zero millivolts. At pH 6, the electrode potential is +60 millivolts, while at pH 4, the electrode potential is +180 millivolts. Conversely, at pH 8, the electrode potential is −60 millivolts.

It must be recognized that the above relationship of millivolts versus pH exists only at 25°C: it is temperature dependent. For example, at 0°C, the electrode potential is 54 millivolts, while at 100°C it is 70 millivolts. Modern pH meters have a sensitive attentuator (temperature compensator) built into them in order to account for this effect of temperature.

31.2.3.2 Reference Electrode

The **reference electrode** is needed to complete the circuit in the pH system. This half cell is one of the most troublesome parts of the pH meter—problems in obtaining pH measurements are often traced to a faulty reference electrode.

The **saturated calomel electrode** (Fig. 31-1) is the most common reference electrode. It is based upon the following reversible reaction:

$$Hg_2Cl_2 + 2e^- \rightleftarrows 2Hg + 2\,Cl^- \quad (8)$$

The $E°_{25°C}$ for the saturated KCl salt bridge is +0.2444 volts; the Nernst equation for the reaction is as follows:

$$E = E° - \frac{0.059}{2} \log [Cl^-]^2 \quad (9)$$

Thus, one observes that the potential is dependent upon the chloride ion concentration, which is easily regulated by the use of saturated KCl solution in the electrode.

A calomel reference electrode has three principal parts: (1) a platinum wire covered with a mixture of calomel (Hg_2Cl_2), (2) a filling solution (saturated KCl), and (3) a permeable junction through which the filling solution slowly migrates into the sample being measured. Junctions are made of ceramic material or fibrous material. A sleeve junction may also be used. Because these junctions tend to clog up, causing a slow, unstable response and inaccurate results, one electrode manufacturer has introduced a free-flowing junction wherein electrolyte flowing from a cartridge is introduced at each measurement.

A less widely used reference electrode is the **silver-silver chloride electrode.** Because the calomel electrode is unstable at high temperatures (>80°C) or in strongly basic samples (pH >9), a silver-silver chloride electrode must be used for such application. It is a very reproducible electrode based upon the following reaction:

$$AgCl(s) + e^- \rightleftarrows Ag(s) + Cl^- \quad (10)$$

The internal element is a silver-coated platinum wire, the surface silver being converted to silver chloride by hydrolysis in hydrochloric acid. The filling solution is a mixture of 4M KCl, saturated with AgCl that is used to prevent the AgCl surface of the internal element from dissolving. The permeable junction is usually of the porous ceramic type. Due to the relative insolubility of AgCl, this electrode tends to clog more readily than the calomel reference electrode. However, it is possible to obtain a double-junction electrode in which a separate inner body holds the Ag/AgCl internal element, electrolyte, and ceramic junction. An outer body containing a second electrolyte and junction isolates the inner body from the sample.

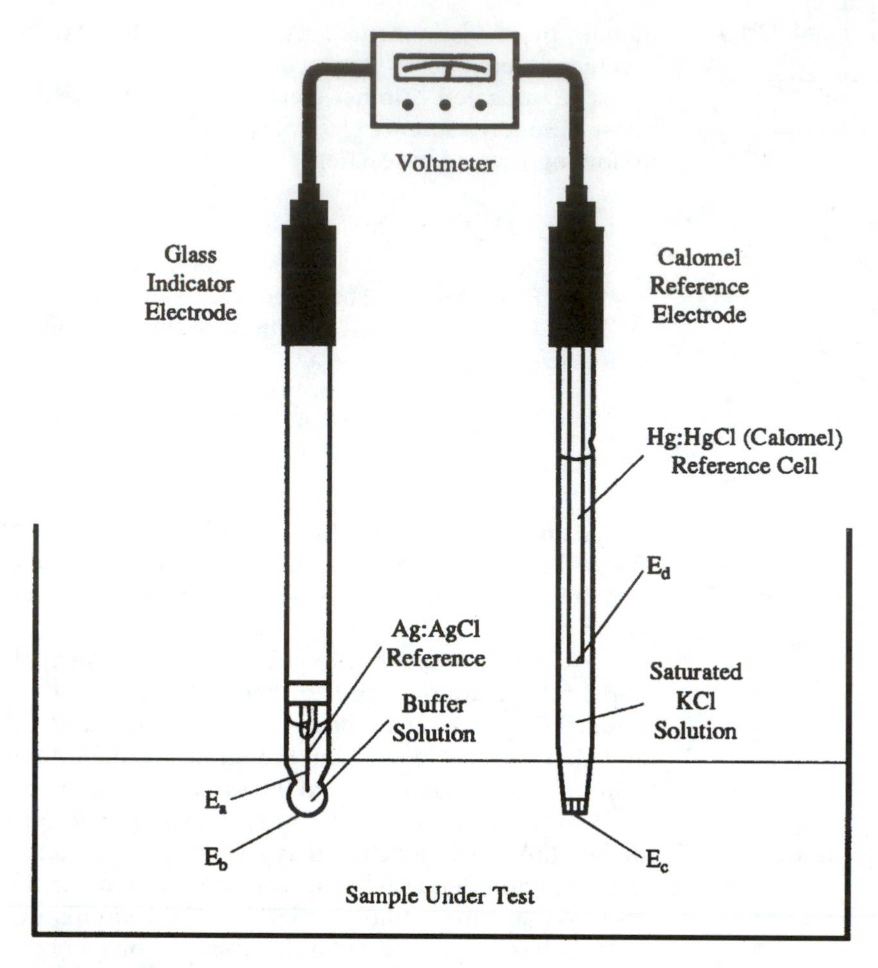

FIGURE 31-1.
The measuring circuit of the potentiometric system.
E_a: contact potential between Ag:AgCl electrode and inner liquid. E_a is independent of pH of the test solution but is temperature dependent.
E_b: potential developed at the pH-sensitive glass membrane. E_b varies with the pH of the test solution and also with temperature. In addition to this potential, the glass electrode also develops an asymmetry potential, which depends upon the composition and shape of the glass membrane. It also changes as the electrode ages.
E_c: diffusion potential between saturated KCl solution and test sample. E_c is essentially independent of the solution under test.
E_d: contact potential between calomel portion of electrode and KCl salt bridge. E_d is independent of the solution under test but is temperature dependent.
From (4), used with permission.

31.2.3.3 Indicator Electrode

The **indicator electrode** most commonly used in measuring pH today is referred to as the **glass electrode.** Prior to its development, the hydrogen electrode and the quinhydrone electrode were used.

The history of the glass electrode goes back to 1875, when it was suggested by Lord Kelvin that glass was an electrical conductor. Cremer discovered the glass electrode potential 30 years later when he observed that a thin glass membrane placed between two aqueous solutions exhibited an electrical potential sensitive to changes in acidity. Subsequently, the reaction was shown to be dependent upon the hydrogen ion concentration. These observations were of great importance in the development of the pH meter. One of the early pioneers in development of this instrumentation was Dr. Arnold O. Beckman.

What is the design of the glass electrode? This electrode (Fig. 31-1) also has three principal parts: (1) a silver-silver chloride electrode with a mercury connection that is needed as a lead to the potentiometer, (2) a buffer solution consisting of 0.01N HCl, 0.09N KCl, and acetate buffer used to maintain a constant pH (E_a), and (3) a small pH-sensitive glass membrane whose potential (E_b) varies with the pH of the test solution. The composition of this pH-sensitive glass membrane is 72.2 percent SiO_2, 6.4 percent CaO, and 21.4 percent Na_2O (mole percent).

In using the glass electrode as an indicator electrode in pH measurements, the measured potential (measured against the calomel electrode) is directly proportional to the pH as discussed earlier, $E = E° + 0.059$ pH.

Conventional glass electrodes are suitable for measuring pH in the range of pH 1–9. However, this electrode is sensitive to higher pH, especially in the presence of sodium ions. Thus, equipment manufacturers have developed modern glass electrodes that are usable over the entire pH range of 0–14 and feature a very low sodium ion error, such as <0.01 pH at 25°C.

31.2.3.4 Combination Electrodes

Combination (reference + indicator) electrodes have been introduced in recent years and have great application when the volume of sample is limited or the shape of the sample vessel is restricted in either size or shape. However, there is still merit in the utilization of individual electrodes, allowing one to meet varying requirements for a particular analysis. In addition, the use of other ion-selective electrodes is growing in the analysis of foods. Glass-body combination electrodes yield excellent results over the pH range of 0–11 and a temperature range of 5°–50°C. They are available with either a silver-silver chloride or a calomel reference element.

31.2.3.5 Guidelines for Use of pH Meter

It is very important that the pH meter be operated and maintained properly. One should always follow the specific instructions provided by the instrument manufacturer. For maximum accuracy, the meter should be standardized using two buffers **(two-point calibration).** Select two buffers of pH values about 3 pH units apart, bracketing that of the anticipated sample pH. Buffers of pH 7 and either pH 4 or pH 10, depending on whether samples to be measured are acidic or basic, are generally used. First, follow manufacturer's instructions for **one-point calibration;** rinse thoroughly with distilled water and blot dry. Immerse electrode in the second buffer (pH 4, for example) and perform a second standardization **(sloping).** This time, the pH meter **slope control** is used to adjust the reading to the correct value of the second buffer. Repeat these two steps, if necessary, until a value within 0.1 pH unit of the correct value of the second buffer is displayed. If this cannot be achieved, the instrument is not in good working condition. Electrodes should be checked, remembering that the reference electrode is more likely in need of attention. One should always follow the electrode manufacturer's specific directions for storage of a pH electrode. In this way, the pH meter is always ready to be used and the life of the electrodes is prolonged. One precaution that should be followed pertains to a calomel reference electrode. The storage solution level should always be at least 2 cm below the saturated KCl solution level in the electrode in order to prevent diffusion of storage solution into the electrode (Fig. 31-2).

31.2.4 Other Ion-Selective Electrodes

31.2.4.1 General Information

The concept of measuring $[H^+]$ has been considered in the previous sections. One must question whether this application of potentiometry can be applied to the measurement of other ions. It is only in recent years that much attention has been given to this question. Indeed, many electrodes have been developed for the direct measurement of various cations and anions, such as bromide, calcium, chloride, fluoride, potassium, sodium, and sulfide. There are even electrodes available for measuring dissolved gases, such as ammonia and carbon dioxide. While some of these methods are limited in their application due to interference from other ions, it is often possible to overcome this problem by pH adjustment, reduction of the interferent, or removal of it by complexing or precipitation.

Varying the composition of the glass in a glass electrode is one means of changing the sensitivity of the glass membrane to other ions. An electrode membrane containing 71 percent SiO_2, 11 percent Na_2O, and 18 percent Al_2O_3 is sensitive to potassium.

A typical **glass membrane sodium-indicating electrode** operates in the range of 1 to 10^{-6} molar or 23,000 to 0.023 ppm. Interferences from silver, lithium, potassium, and ammonium ions are a possibility. Response time is less than 30 seconds. Combination

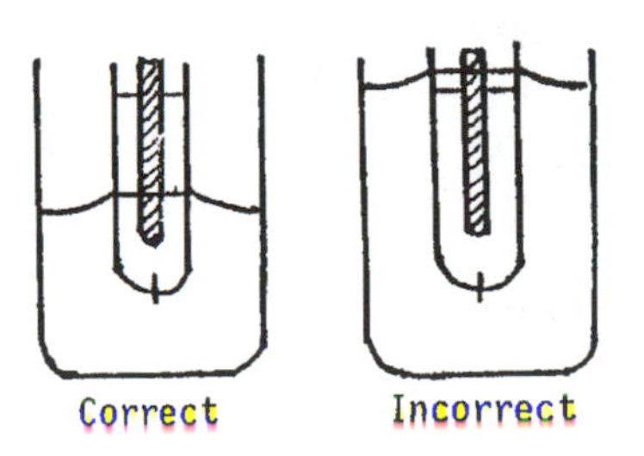

FIGURE 31-2.
Correct and incorrect depth of calomel electrodes in solutions. Reprinted with permission from (9). Copyright 1971 American Chemical Society.

polymer-body combination sodium ion-selective electrodes are also available, a calomel reference half-cell being used in this system.

Solid-state ion-selective electrodes are also available. These electrodes do not use a glass-sensitive membrane. Instead, the active membrane consists of a single inorganic crystal treated with a rare earth. The fluoride electrode serves as a good example, consisting of a crystal of lanthanum chloride treated with europium, which permits ionic charge transport and lowered electrical resistance. Fluoride concentrations of 0.02 ppm may be detected with this electrode. Other commonly used solid-state ion-selective electrodes are available. For example, bromide can be detected at concentrations of 0.04 ppm and chloride at 0.178 ppm. Response time for all the solid-state electrodes is less than 30 seconds. These electrodes are also subject to interferences from various anions.

In addition to the various glass membrane and solid-state ion-selective electrodes, it should be noted that there are other types of these electrodes, such as precipitate-impregnated, liquid-liquid membrane, and even enzyme electrodes. **Gas-sensing electrodes** also are gaining in use. These electrodes possess a gas-permeable membrane and a combination pH electrode with internal buffer solution. Upon passing through the membrane, the gas dissolves in a thin layer of buffer solution that surrounds the combination pH electrode. The dissolved gas causes the pH of the solution to change, and the combination electrode detects this change. Ammonia, carbon dioxide, and sulfur dioxide can be measured by these type electrodes.

31.2.4.2 Activity versus Concentration

In using ion-selective electrodes, the concept of activity versus concentration must be considered. **Activity** is a measure of chemical reactivity, while **concentration** is a measure of all forms (free and bound) of ions in solution. Due to interactions of ions with themselves and with the solvent, the effective concentration or activity is, in general, lower than the actual concentration. Activity and concentration are related by the following equation:

$$A = VC \qquad (11)$$

where: A = activity
V = activity coefficient
C = concentration

The **activity coefficient** is a function of **ionic strength.** Ionic strength is a function of the concentration of, and the charge on, all ions in solution.

By adjusting the ionic strength for all test samples and standards to a nearly constant level, the Nernst equation can be used to relate electrode response to the concentration of the measured species. In practice, both samples and calibrating standards are adjusted to a high but constant ionic strength. An **ionic-strength adjustment buffer** is used for this purpose. It is a solution of neutral or noninterfering ions that raises the total ionic strength of the solution to the level at which the effects of other ions are canceled. These buffers can also be used to control pH, remove ionic interferences, and limit chemical interferences arising from association and complexation. Thus, to measure accurately the concentration of ionic species using an ion-selective electrode, the following requirements must be met:

1. Maintain a constant reference potential.
2. Operate a constant temperature.
3. Adjust ionic strength.
4. Adjust pH.
5. Remove electrode interferences.
6. Eliminate method interferences.

31.2.4.3 Calibration Curves

It is common practice to develop a **calibration curve** in working with ion-selective electrodes. The two electrodes (indicator and reference) are immersed in a series of solutions of known concentration. The **electrode potential** (millivolts) developed in these standard solutions is recorded and plotted (on semi-log paper) against the **logarithms of the standard concentrations** (Fig. 31-3). Upon analysis of a test sample, the observed millivolt reading for the electrode potential is used to determine the concentration by referring to the calibration curve.

The calibration curve has a linear region at which the electrode has a constant response to changes in concentration, fitting the Nernst equation ($E = E° - 0.059$ [Ion]). Note also the nonlinear region of the curve at low concentrations. The total ionic strength and the concentrations of interfering ions are among the factors that determine the lowest level of activity that can be detected in practical applications. Examples of calibration curves for various gases are found in Figure 31-4.

31.2.4.4 Other ISE Methodologies

Although a calibration curve is the most common means of using ion-selective electrodes, there are other applications. For example, in a titration the ISE may be employed to detect the **equivalence point of the titration.** The ISE may be responsive to either the sample species (S titration) or the titrant (T titration), the latter titration probably being of more common use.

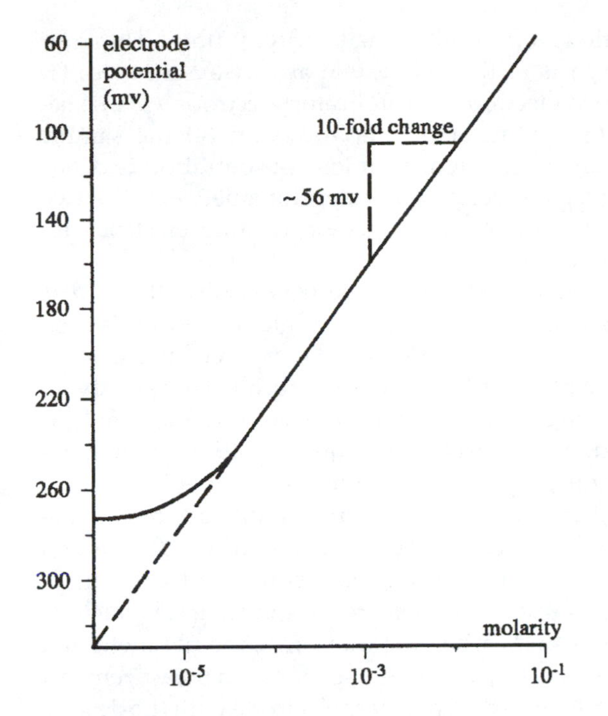

FIGURE 31-3.
A typical calibration curve. From (3), used with permission.

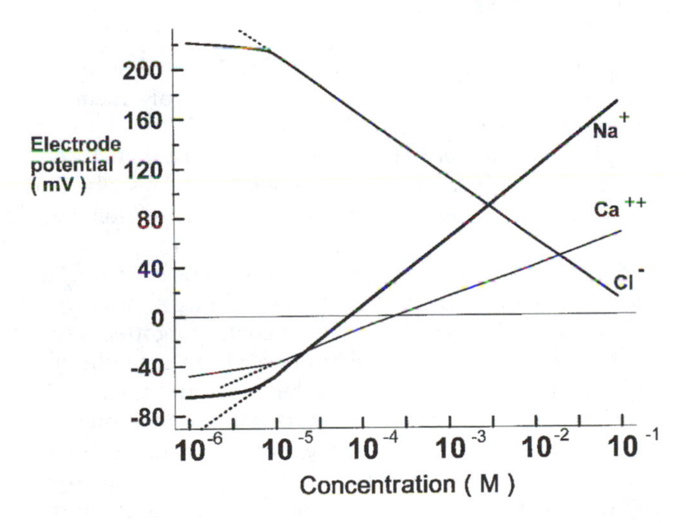

FIGURE 31-4.
Examples of ion-selective electrode calibration curves for ions important in foods. Figure courtesy of Phoenix Electrode Co., Houston, TX.

In the T titration, little change in electrode potential occurs as titrant is added because it is reacting with the sample species. However, when all of the sample species has reacted with the titrant, a large increase in the electrode potential occurs, revealing the equivalence point of the titration (Fig. 31-5).

Finally, the method of **standard addition** shall be

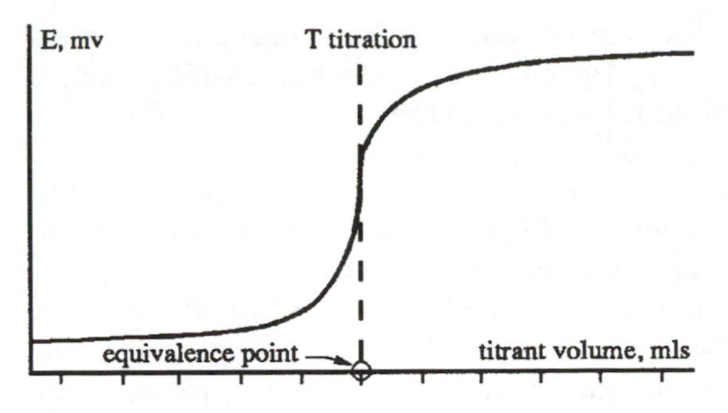

FIGURE 31-5.
A typical T-type titration. From (3), used with permission.

considered, as it does have application in ISE methodology. The electrodes (indicator and reference) are immersed in the sample, and the initial electrode potential is determined. Then an aliquot containing a known concentration of the measured species is added (standard addition) to the sample, and a second measurement of electrode potential is determined. These measured values in electrode potential may then be used to determine the concentration of the active species in the original sample. This method may not require the use of an ionic strength adjustment buffer. It is of great value when only a few samples are to be measured and time does not permit the development of a calibration curve. It also eliminates complex unknown background effects.

31.3 APPLICATIONS

31.3.1 pH Measurement

The measurement of pH has many uses in food science; the safety of many foods is pH dependent, and the desired flavor of a food often is determined by the $[H^+]$ concentration. Thus, pH measurements are routinely conducted in both the laboratory and the processing plant of the food scientist.

31.3.2 Acidimetry

One application of pH measurement is in acidimetry. For example, it is possible to determine the equivalence point of a titration of a weak acid with a strong base by measuring pH changes at several points in the titration, followed by plotting pH versus titration values. This titration is an example of a T titration (Fig. 31-5) with the glass electrode (H^+ sensitive).

The pH meter is sometimes used to determine the endpoint of a titration. For example, it is possible to use an indicator (phenolphthalein) to determine the acidity of milk since one can readily see the color change (colorless to pink) at the endpoint. However, with highly colored foods, such as chocolate ice cream or strawberry frozen yogurt, it is impossible to detect this endpoint. Thus, the pH meter may be used instead of an indicator.

The tips of the pH electrodes are immersed in the unknown liquid, which is being stirred with a magnetic stirrer. Care must be given to keep the electrode tips above the magnet in order to avoid breaking them. Titrant (sodium hydroxide) is then added slowly, and the endpoint is detected when the pH reaches the range where phenolphthalein experiences a color change (pH 8.0–9.6).

31.3.3 Ion-Selective Electrodes

The pH meter with both a pH scale and millivolt scale may also be used for other ion-selective electrode analyses. One simply replaces the glass electrode with the ISE of choice and follows instructions for the determination.

Some examples of applications of these electrodes are salt and nitrate in processed meats, salt content of butter and cheese, calcium in milk, sodium in low-sodium ice cream, carbon dioxide in soft drinks, potassium and sodium levels in wine, and nitrate in canned vegetables. An ISE method applicable to foods containing <100 mg sodium/100 grams is an official method of the Association of Official Analytical Chemists International (AOAC Method 976.25). This method employs a sodium combination ISE, pH meter, magnetic stirrer, and a special type of graph paper for plotting a standard curve. Obviously, there are many other applications, but the above serve to demonstrate the versatility of this valuable measuring tool.

A major *advantage* in the use of ion-selective electrodes lies in the ability to measure many anions and cations directly. Such measurements are relatively simple compared to most other analytical techniques, particularly because the pH meter may be used as the voltmeter. Analyses are independent of sample volume when making direct measurements, while turdidity, color, and viscosity are all of no concern.

A major *disadvantage* in the use of ion-selective electrodes is their inability to measure below 2–3 ppm, although there are some electrodes that are sensitive down to 1 part per billion. At low levels of measurement (below 10^{-4} molar), the electrode response time is slow. Finally, some electrodes have had a high rate of premature failure or a short operating life and possible excessive noise characteristics.

31.4 SUMMARY

Potentiometry is classified as voltammetry at zero current because electrode potentials are measured under conditions of infinitessimal current flow. The four major parts of the pH system are always needed: (1) reference electrode, (2) indicator electrode (pH sensitive), (3) voltmeter or amplifier, and (4) the sample being analyzed. Hydrogen ion concentration is determined by the voltage that develops between the two electrodes, the Nernst equation relating electrode response to activity.

The saturated calomel electrode is usually used as the reference electrode, while the glass electrode is used as an indicator electrode for pH measurements. Combination electrodes (reference + indicator) are also available, being of value when the volume of sample is limited. These electrodes require continuous maintenance for proper operation of the pH meter.

Other ion-selective electrodes (ISE) are now available for the direct measurement of various cations and anions, such as sodium, potassium, and calcium. It is also possible to measure dissolved gases such as ammonia and carbon dioxide. Since the pH meter has a millivolt scale, it may be used for such measurements by simply replacing the glass electrode with the desired ISE.

31.5 STUDY QUESTIONS

1. Explain the theory of potentiometry and the Nernst equation as they relate to being able to use a pH meter to measure H^+ concentration.
2. Explain the difference between a saturated calomel electrode and a silver-silver chloride electrode; describe the construction of a glass electrode and a combination electrode.
3. You return from a two-week vacation and ask your lab technician about the pH of the apple juice sample you gave him or her before you left. Having forgotten to do it before, the technician calibrates a pH meter with one standard buffer stored next to the meter and then reads the pH of the sample of unpasteurized apple juice immediately after removing it from the refrigerator (4°C), where it has been stored for two weeks. Explain the reasons why this stated procedure could lead to inaccurate or misleading pH values.
4. Explain the principles of using an ion-selective electrode to measure the concentration of a particular inorganic element in food. Explain how an ion-selective electrode works and why electrode potential can be correlated to concentration when one is really measuring activity and not concentration.
5. Your lab technician forgot to add the appropriate ionic strength adjustor (ISA) solution to samples and standards when preparing solutions for analysis with an ion-selective electrode. Should you tell the technician to proceed with the samples as already prepared, or go back and prepare samples and standards with ISA solution? Explain your answer, with reference to the principles of using an ISE to quantitate ions.

6. To measure accurately the concentration of a particular element with an ion-selective electrode, ionic strength of the sample being analyzed is only one of the factors that must be controlled. List the other things one must do (i.e., factors to control, consider, or eliminate) for an accurate measure of concentration by the ISE method.
7. You have decided to purchase an ion-selective electrode to monitor the sodium content of foods produced by your plant. List the advantages this would have over the atomic absorption/emission method or the Mohr/Volhard titration method. List the problems and disadvantages of ISE that you should anticipate. (See also Chapters 9 and 25.)
8. Calibration curves (i.e., standard curves) are used in (a) ultraviolet-visible spectroscopy, (b) atomic emission spectroscopy, and with (c) ion-selective electrodes. For each method, state what factors are plotted against each other, and state what type of curve is expected (i.e., linear or nonlinear, positive or negative slope). (See also Chapters 23 and 25.)

31.6 PRACTICE PROBLEMS

1. Vinegar has an [H+] of 1.77×10^{-3}M. What is the pH? What is the major acid found in vinegar? Its structure?
2. Orange juice has an [H+] of 2.09×10^{-4}M. What is the pH? What is the major acid found in orange juice? What is its structure?
3. A sample of yogurt has a pH of 3.95. What is the [H+] concentration? What is the major acid found in yogurt? What is its structure?
4. An apple pectin gel has a pH of 3.30. What is the [H+] concentration? What is the major acid found in apples? What is its structure?

Answers

1. 2.75; acetic acid;

```
   H
   |
H—C—COOH
   |
   H
```

2. 3.68; citric acid;

```
  COOH    COOH    COOH
   |       |       |
H—C———————C———————C—H
   |       |       |
   H       OH      H
```

3. 1.1×10^{-4}M; lactic acid;

```
        OH
        |
 CH3—C—COOH
        |
        H
```

4. 5.0×10^{-4}M; malic acid;

```
    COOH    COOH
     |       |
HO—C———————C—H
     |       |
     H       H
```

31.7 RESOURCE MATERIALS

1. Beckman Instruments. 1980. *The Beckman Handbook of Applied Electrochemistry.* Fullerton, CA.
2. Brumblay, R.U. 1972. *Quantitative Analysis,* revised ed. Harper & Row Publishers, New York, NY.
3. Comer, J. 1986. Ion selective and oxygen electrodes. Presented at IFT Short Course, "Instrumental Methods for Quality Assurance and Research," Dallas, TX, June 18–20.
4. Dicker, D.H. 1969. The Laboratory pH Meter. *American Laboratory,* February.
5. Fisher, J.E. 1984. Measurement of pH. *American Laboratory,* p. 54–60, June.
6. Fisher Scientific Co. 1989. *Electrode Handbook,* 5th ed. Pittsburgh, PA.
7. Hach Company. 1988. *Hach Systems for Electrochemical Analysis.* Loveland, CO.
8. Kenkel, J. 1988. *Analytical Chemistry for Technicians.* Lewis Publishers, Inc., Chelsea, MI.
9. Pecsok, R.L., Chapman, K., and Ponder, W.H. 1971. *Modern Chemical Technology,* Vol. 3, revised ed. American Chemical Society, Washington, DC.
10. Pomeranz, Y., and Meloan, C.E. 1987. *Food Analysis: Theory and Practice,* 2nd ed. Van Nostrand Reinhold, New York, NY.

CHAPTER

32

Application of Enzymes in Food Analysis

Joseph R. Powers

32.1 INTRODUCTION

Enzymes are protein catalysts that are capable of very great specificity and reactivity under physiological conditions. Enzymatic analysis is the measurement of compounds with the aid of added enzymes or the measurement of endogenous enzyme activity to give an indication of the state of a biological system including foods. The fact that enzyme catalysis can take place under relatively mild conditions allows for measurement of relatively unstable compounds not amenable to some other techniques. In addition, the specificity of enzyme reactions can allow for measurement of components of complex mixtures without the time and expense of complicated chromatographic separation techniques.

There are several uses of enzyme analyses in food science and technology. In several instances, enzyme activity is a useful measure for adequate processing of a food product. The thermal stability of enzymes has been used extensively as a measure of heat treatment; for example, peroxidase activity is used as a measure of adequacy of blanching of vegetable products. Enzyme activity assays are also used by the food technologist to assess potency of enzyme preparations used as processing aids.

The food scientist can also use commercially available enzyme preparations to measure constituents of foods that are enzyme substrates. For example, glucose content can be determined in a complex food matrix containing other monosaccharides by using readily available enzymes. A corollary use of commercially available enzymes is to measure enzyme activity as a function of enzyme inhibitor content in a food. Organophosphate insecticides are potent inhibitors of the enzyme acetylcholinesterase, and hence the activity of this enzyme in the presence of a food extract is a measure of organophosphate insecticide concentration in the food. Also of interest is the measurement of enzyme activity associated with food quality. For example, catalase activity is markedly increased in milk from mastitic udders. Catalase activity also parallels the bacterial count in milk. Another use of enzyme assays to determine food quality is estimation of protein nutritive value by monitoring the activity of added proteases on food protein samples (see Chapter 16). Enzymes can be used to measure the appearance of degradation products such as trimethylamine in fish during storage. Enzymes are also used as preparative tools in food analysis. Examples include the use of amylases and proteases in fiber analysis (Chapter 11) and the enzymatic hydrolysis of thiamine phosphate esters in vitamin analysis.

In order to successfully carry out enzyme analyses in foods, an understanding of certain basic principles of enzymology is necessary. After a brief overview of these principles, some examples of the use of enzymatic analyses in food systems are examined here.

32.2 PRINCIPLES

32.2.1 Enzyme Kinetics

32.2.1.1 Overview

Enzymes are biological catalysts that are proteins. A catalyst increases the rate (velocity) of a thermodynamically possible reaction. The enzyme does not modify the equilibrium constant of the reaction, and the enzyme catalyst is not consumed in the reaction. Because enzymes affect rates (velocities) of reactions, some knowledge of **enzyme kinetics** (study of rates) is needed for the food scientist to effectively use enzymes in analysis. To measure the rate of an enzyme-catalyzed reaction, typically one mixes the enzyme with the substrate under specified conditions (pH, temperature, ionic strength, etc.) and follows the reaction by measuring the amount of product that appears or by measuring the disappearance of substrate. Consider the following as a simple representation of an enzyme-catalyzed reaction:

$$S + E \rightleftarrows ES \rightarrow P + E \tag{1}$$

where: S = substrate
E = enzyme
ES = enzyme-substrate complex
P = product

The time course of an enzyme-catalyzed reaction is illustrated in Fig. 32-1. The formation of the enzyme substrate complex is very rapid and is not normally seen in the laboratory. The brief time in which the enzyme-substrate complex is initially formed is on the millisecond scale and is called the **pre-steady state period.** The slope of the linear portion of the curve following the pre-steady state period gives us the **initial velocity** (v_0). After the pre-steady state period, a **steady state period** exists in which the concentration of the enzyme-substrate complex is constant. A time course needs to be established experimentally by using a series of points or a continuous assay to establish the appropriate time frame for the measurement of the initial velocity.

The rate of the enzyme-catalyzed reaction depends on the concentration of the enzyme and also depends on the substrate concentration. With a fixed enzyme concentration, increasing substrate concentration will result in an increased velocity (see Fig. 32-2). As substrate concentration increases further, the increase in velocity slows until, with a very large concentration of substrate, no further increase in velocity is noted.

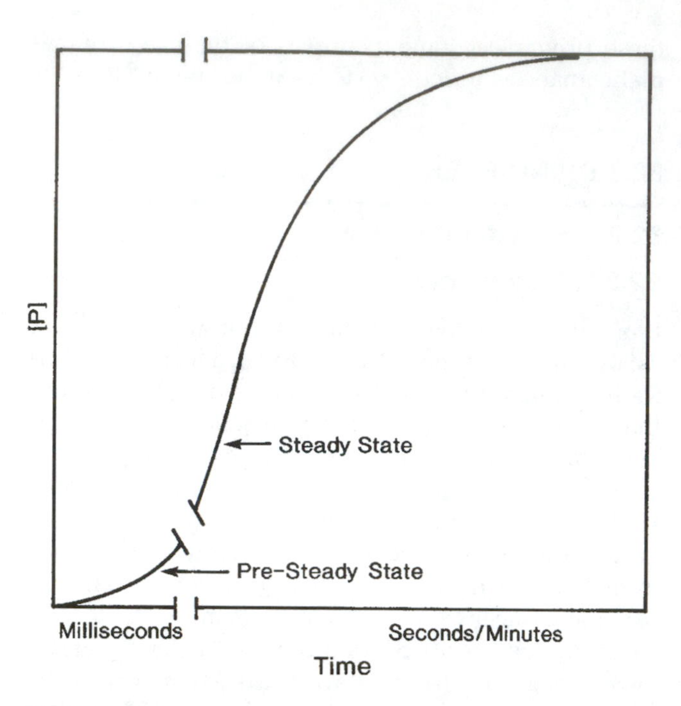

FIGURE 32-1.
Time course for a typical enzyme-catalyzed reaction showing the pre-steady state and steady state periods. [P] = product concentration.

The velocity of the reaction at this very large substrate concentration is the **maximum velocity** (V_m) of the reaction under the conditions of that particular assay. The substrate concentration at which one-half V_m is observed is defined as the **Michaelis constant** or K_m. K_m is an important characteristic of an enzyme. It is an indication of the relative binding affinity of the enzyme for a particular substrate. The lower the K_m, the greater the affinity of the enzyme for the substrate.

If we examine relationships that hold in the steady state period, the Michaelis-Menten equation can be derived for the simplified enzyme catalyzed reaction:

$$E + S \underset{k_{-1}}{\overset{k_1}{\rightleftharpoons}} ES \overset{k^2}{\rightarrow} E + P \qquad (2)$$

where: k_1, k_{-1}, k_2 = reaction rate constants for reactions indicated

In the steady state, the rate of change in enzyme substrate complex concentration is zero: $dES/dt = 0$ and:

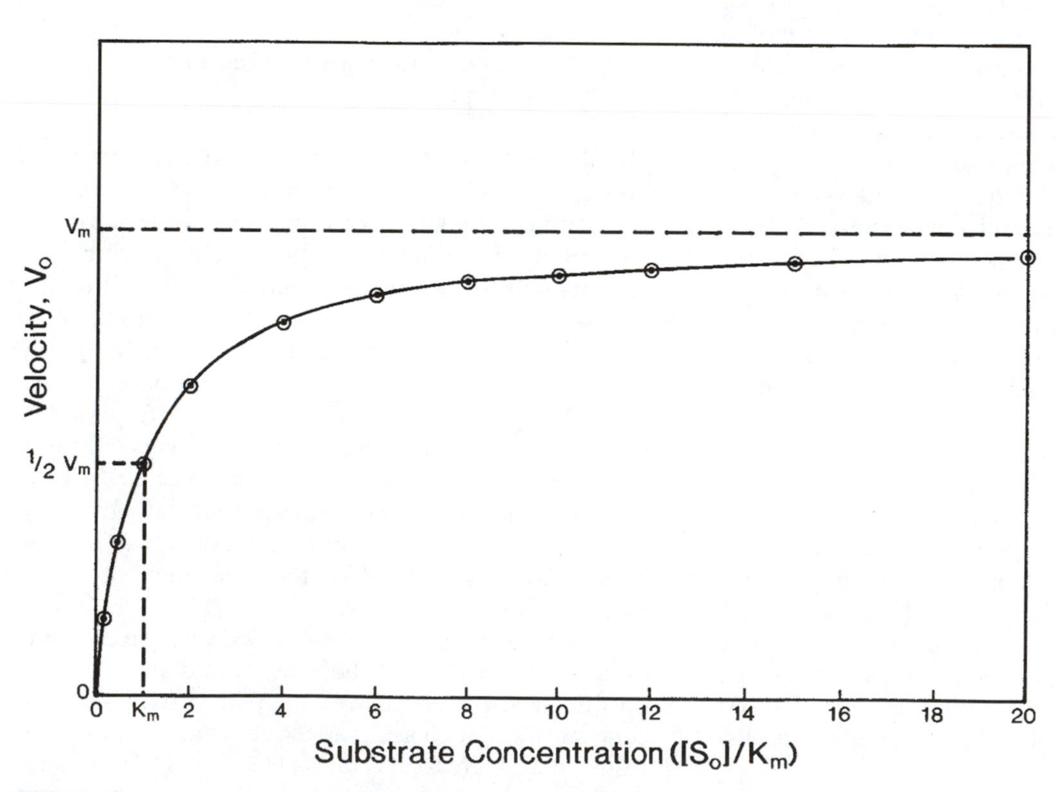

FIGURE 32-2.
Effect of substrate concentration on the rate of an enzyme-catalyzed reaction. Plotted according to the Michaelis-Menten equation.

$$\text{Rate of disappearance of ES} = k_{-1}[ES] + k_2[ES] \quad (3)$$

$$\text{Rate of appearance of ES} = k_1[E][S] \quad (4)$$

$$\text{then } k_1[E][S] = k_{-1}[ES] + k_2[ES] \quad (5)$$

$$[E_o] = [E] + [ES] \quad (6)$$

where: E_o = total enzyme
E = free enzyme
ES = enzyme substrate complex

$$\text{Substituting } [E] = [E_o] - [ES] \quad (7)$$

$$k_1([E_o] - [ES])[S] = k_{-1}[ES] + k_2[ES] = (k_{-1} + k_2)[ES] \quad (8)$$

Rearranging and solving for [ES]:

$$ES = \frac{k_1[E_o][S]}{k_1[S] + (k_{-1} + k_2)} = \frac{[E_o][S]}{\frac{k_{-1} + k_2}{k_1} + [S]} \quad (9)$$

If the collection of rate constants in the denominator is defined as the Michaelis constant (K_m):

$$K_m = \frac{k_{-1} + k_2}{k_1} \quad (10)$$

Note that K_m is *not* affected by enzyme *or* substrate concentration.

Then:

$$[ES] = \frac{[E_o][S]}{K_m + [S]} \quad (11)$$

If we define the velocity (v_o) of the enzyme-catalyzed reaction as:

$$v_o = k_2[ES] \quad (12)$$

Then:

$$v_o = \frac{k_2[E_o][S]}{K_m + [S]} \quad (13)$$

When all the enzyme is saturated—all substrate binding sites on the enzyme are occupied—at the large substrate concentrations in Fig. 32-2 we have maximum velocity, V_m: All of E_o is in the ES form and

$$k_2[ES] = k_2[E_o] \text{ at } S >> K_m \text{ and:}$$

$$v_o = \frac{V_m[S]}{K_m + [S]} \quad (14)$$

This is the **Michaelis-Menten equation,** the equation for a right hyperbola; the data plotted in Fig. 32-2 fit such an equation. A convenient way to verify this equation is to simply remember that $v_o = ½V_m$ when $[S] = K_m$. Therefore, by simple substitution

$$1/2V_m = \frac{V_mK_m}{K_m + K_m} = \frac{V_m}{2} \quad (15)$$

32.2.1.2 Order of Reactions

The velocity of an enzyme-catalyzed reaction increases as substrate concentration increases (see Fig. 32-2). A **first order** reaction with respect to substrate concentration is obeyed in the region of the curve where substrate concentration is small ($S << K_m$). This means that the velocity of the reaction is directly proportional to the substrate concentration in this region. When the substrate concentration is further increased, the velocity of the reaction no longer increases linearly, and the reaction is **mixed order.** This is seen in the figure as the curvilinear portion of the plot. If substrate concentration is increased further, the velocity asymptotically approaches the maximum velocity (V_m). In this linear, nearly zero slope portion of the plot, the velocity is independent of substrate concentration. However, note that at large substrate concentrations ($S >> K_m$), the velocity is directly proportional to enzyme concentration ($V_m = k_2 [E_o]$). Thus, in this portion of the curve where $S >> K_m$, the rate of the reaction is **zero order** with respect to substrate concentration (is independent of substrate concentration) but first order with respect to enzyme concentration.

If we are interested in measuring the amount of **enzyme** in a reaction mixture, we should, if possible, work at substrate concentrations so that the observed velocity approximates V_m. At these substrate concentrations, enzyme is directly rate limiting to the observed velocity. Conversely, if we are interested in measuring substrate concentration by measuring initial velocity, we must be at substrate concentrations less than K_m in order to have a rate directly proportional to substrate concentration.

32.2.1.3 Determination of Michaelis Constant (K_m) and V_m

To properly design an experiment in which velocity is zero order with respect to substrate and first order with respect to enzyme concentration, or conversely an experiment in which we would like to measure rates that are directly proportional to substrate concentration, we must know the K_m. The most popular method for determining K_m is the use of a **Lineweaver-Burk plot.** The reciprocal of the Michaelis-Menten equation is:

$$\frac{1}{v_o} = \frac{K_m}{V_m[S]} + \frac{1}{V_m} \qquad (16)$$

This equation is that of a straight line y = mx + b where m = slope and b = y intercept. A plot of substrate concentration versus initial velocity as shown in Fig. 32-2 can be transformed to a linear form via use of the reciprocal eq. 16 and Fig. 32-3 (Lineweaver-Burk plot) results. The intercept of the plotted data on the y (vertical) axis is $1/V_m$ while the intercept on the x (horizontal) axis is $-1/K_m$. The slope of the line is K_m/V_m. Consequently, both K_m and V_m can be obtained using this method.

A disadvantage of the Lineweaver-Burk plot is that the data with the inherently largest error, collected at very low substrate concentrations and consequently low rates, tend to direct the drawing of a best fit line. An alternative method of plotting the data is the **Eadie-Hofstee method.** The Michaelis-Menten equation can be rearranged to give:

$$v_o = V_m - \frac{v_o K_m}{[S]} \qquad (17)$$

Equation 17 is also the equation of a straight line, and when v_o versus $v_o/[S]$ are plotted, the slope of the line is $-K_m$, the y intercept is V_m, and the x intercept is V_m/K_m. Data plotted by this method give a more even spacing of the data than the Lineweaver-Burk method.

32.2.2 Factors that Affect Enzyme Reaction Rate

The velocity of an enzyme-catalyzed reaction is affected by a number of factors, including enzyme and substrate concentrations, temperature, pH, ionic strength, and the presence of inhibitors and activators.

32.2.2.1 Effect of Enzyme Concentration

The velocity of an enzyme-catalyzed reaction will depend on the enzyme concentration in the reaction mixture. The expected relationship between enzyme activity and enzyme concentration is shown in Fig. 32-4. Doubling the enzyme concentration will double the rate of the reaction. If possible, determination of enzyme activity should be done at concentrations of substrate much greater than K_m. Under these conditions, a zero-order dependence of the rate with respect to substrate concentration and a first-order relationship between rate and enzyme concentration exist. It is critical that the substrate concentration is saturating during the entire period the reaction mixture is sampled and the amount measured of product formed or substrate

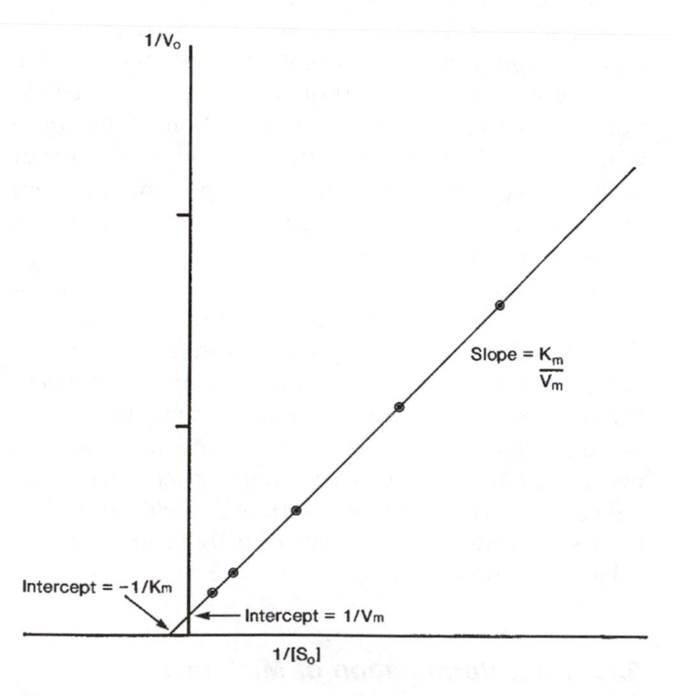

FIGURE 32-3.
Plot of substrate-velocity data by the Lineweaver-Burk method.

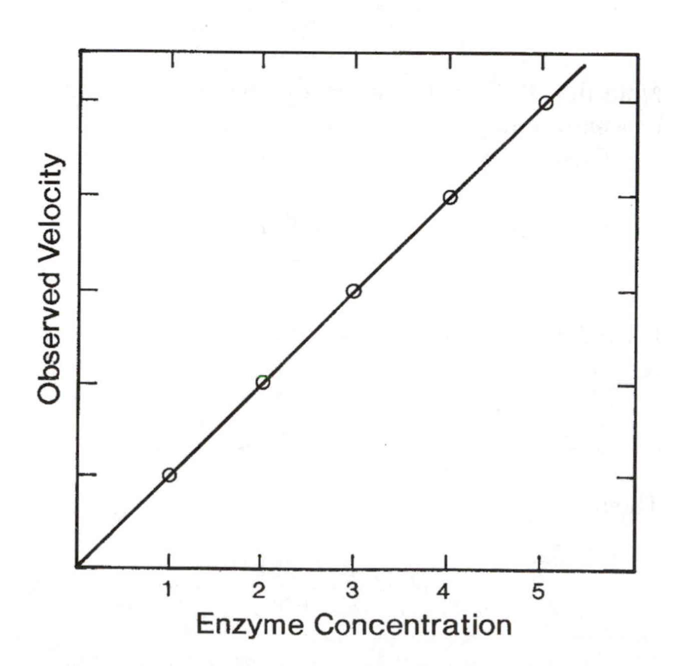

FIGURE 32-4.
Expected effect of enzyme concentration on observed velocity of an enzyme-catalyzed reaction.

disappearing is linear over the period during which the reaction is sampled. The activity of the enzyme is obtained as the slope of the linear part of the line of a plot of product or substrate concentration versus time.

If a large number of samples are to be assayed, a single aliquot is often taken at a single time. This can be risky and will give good results only if the time at which the sample is taken falls on the linear portion of a plot of substrate concentration or product concentration versus time of reaction (see Fig. 32-5). The plot becomes nonlinear if the substrate concentration falls below the concentration needed to saturate the enzyme, if the increase in concentration of product produces a significant amount of back reaction, or if the enzyme loses activity during the time of the assay. Normally, one designs an experiment in which enzyme concentration is estimated such that no more than 5 to 10 percent of the substrate has been converted to product within the time used for measuring the initial rate. In the example shown in Fig. 32-5, by sampling at the single point, a, an underestimation of the rate is made for curves 3 and 4. A better method of estimating rates is to measure initial rates of the reactions, in which the change in substrate or product concentration is determined at times as close as possible to time zero. This is shown in Fig. 32-5 by the solid lines drawn tangent to the slopes of the initial parts of the curves. The slope of the tangent line gives the initial rate.

Sometimes it is not possible to carry out enzyme assays at $[S] >> K_m$. The substrate may be very expensive or relatively insoluble or K_m may be large (i.e., $K_m > 100$ mM). Enzyme concentration can also be estimated at substrate concentrations much less than K_m. When substrate concentration is much less than K_m, the substrate term in the denominator of the Michaelis-Menten equation can be ignored and $v = (V_m[S])/K_m$, which is the equation for a first-order reaction with respect to substrate concentration. Under these conditions, a plot of product concentration versus time gives a nonlinear plot (Fig. 32-6). A plot of log $([S_o]/[S])$ versus time gives a straight-line relationship (Fig. 32-6 inset). The slope of the line of the log plot is directly related to the enzyme concentration. When the slope of a series of these log plots is further plotted as a function of enzyme concentration, a straight-line relationship should result. If possible, the reaction should be followed continuously or aliquots removed at frequent time intervals and the reaction allowed to proceed to greater than 10 percent of the total reaction.

32.2.2.2 Effect of Substrate Concentration

The substrate concentration velocity relationship for an enzyme-catalyzed reaction in which enzyme con-

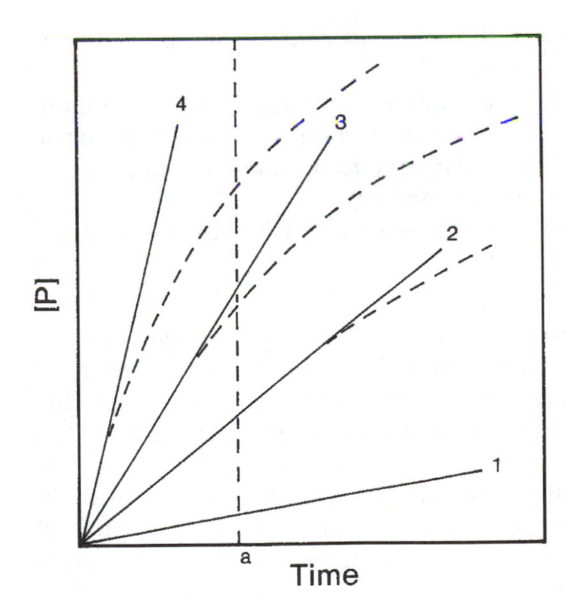

FIGURE 32-5.
Effect of enzyme concentration on time course of an enzyme-catalyzed reaction. The dashed lines are experimentally determined data with enzyme concentration increasing from 1–4. The solid lines are tangents drawn from the initial slopes of the experimental data. If a single time point, a, is used for data collection, a large difference between actual data collected and that predicted from initial rates is seen.

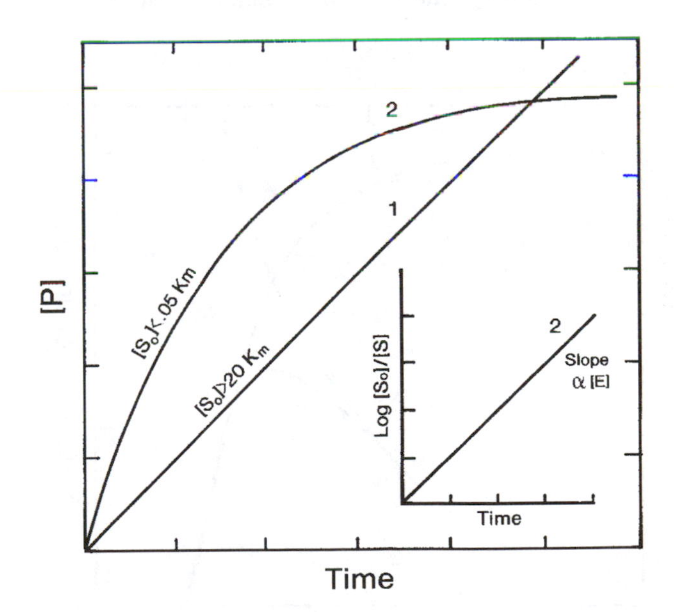

FIGURE 32-6.
Product concentration [P] formed as a function of time for an enzyme-catalyzed reaction. Line 1 is linear, indicating a zero-order reaction with respect to substrate concentration [S]. The slope of line 1 is directly related to enzyme concentration. Line 2 is nonlinear. A replot of line 2 data, plotting log S_o/S versus time, is linear (insert), indicating the reaction is first order with respect to substrate concentration. The slope of the replot is directly related to enzyme concentration.

centration is constant is shown in Fig. 32-2. As noted before, the rate of the reaction is first order with respect to substrate concentration when [S] << K_m. At [S] >> K_m, the reaction is zero order with respect to substrate concentration and first order with respect to [E]. At substrate concentrations between the first-order and zero-order regions, the enzyme-catalyzed reaction is mixed order with respect to substrate concentration. However, when initial rates are obtained, a linear relationship between v_o and E_o should be seen.

32.2.2.3 Environmental Effects

32.2.2.3.1 Effect of Temperature on Enzyme Activity Temperature can affect observed enzyme activity in several ways. Most obvious is that temperature can affect the stability of enzyme and also the rate of the enzyme-catalyzed reaction. Other factors in enzyme-catalyzed reactions that may be considered include the effect of temperature on the solubility of gases that are either products or substrates of the observed reaction and the effect of temperature on pH of the system. A good example of the latter is the common buffering species Tris (tris [hydroxymethyl] aminomethane), for which the pK_a changes 0.031 per 1°C change.

Temperature affects both the stability and the activity of the enzyme, as shown in Fig. 32-7. At relatively low temperatures, the enzyme is stable. However, at

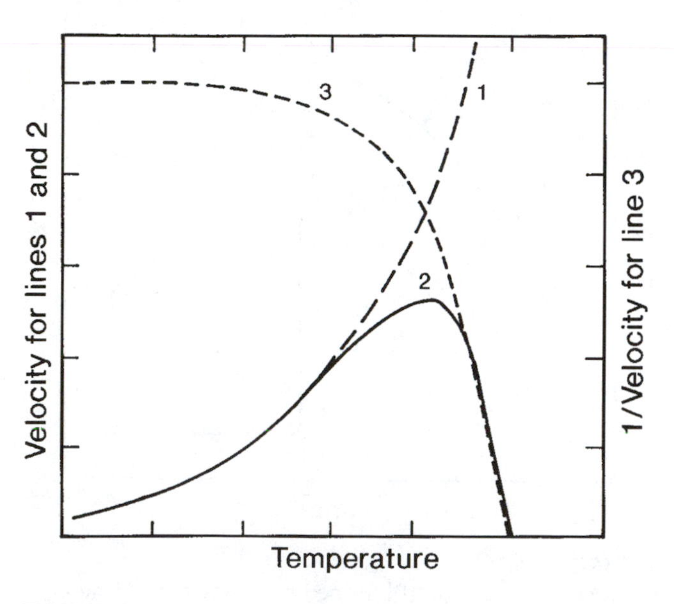

FIGURE 32-7.
Effect of temperature on velocity of an enzyme-catalyzed reaction. Temperature effect on substrate to product conversion is shown by line 1. Line 3 shows effect of temperature on rate of enzyme denaturation (right hand y-axis is for line 3). The net effect of temperature on the observed velocity is given by line 2, and the temperature optimum is at the maximum of line 2.

higher temperatures, denaturation dominates, and a markedly reduced enzyme activity represented by the negative slope portion of line 2 is observed. Line 1 of Fig. 32-7 shows the effect of temperature on the velocity of the enzyme-catalyzed reaction. The velocity is expected to increase as the temperature is increased. As shown by line 1, the velocity approximately doubles for every 10°C rise in temperature. The net effect of increasing temperature on the rate of conversion of substrate to product (line 1) and on the rate of the denaturation of enzyme (line 3) is line 2 of Fig. 32-7. The temperature optimum of the enzyme is at the maximum point of line 2. The temperature optimum is not a unique characteristic of the enzyme. The optimum applies instead to the entire system because type of substrate, pH, salt concentration, substrate concentration, and time of reaction can affect the observed optimum. For this reason, investigators should fully describe a system in which the effects of temperature on observed enzyme activity are reported.

The data of line 2 of Fig. 32-7 can be plotted according to the **Arrhenius equation:**

$$k = Ae^{-Ea/RT} \tag{18}$$

which can be written:

$$\log k = \log A - \frac{E_a}{2.3RT} \tag{19}$$

where: k = a specific rate constant at some temperature, T (°K)
E_a = activation energy, the minimum amount of energy a reactant molecule must have to be converted to product
R = gas constant
A = a frequency factor (pre-exponential factor)

The positive slope on the left side (high temperature) of the **Arrhenius plot** (Fig. 32-8) gives a measure of the **activation energy** (E_a) for the denaturation of the enzyme. Note that a small change in temperature has a very large effect on the rate of denaturation. The slope on the right side of Fig. 32-8 gives a measure of the E_a for the transformation of substrate to product catalyzed by the enzyme. If the experiment is carried out under conditions in which V_m is measured ([S] >> K_m), then the activation energy observed will be for the catalytic step of the reaction.

32.2.2.3.2 Effect of pH on Enzyme Activity The observed rate of an enzyme-catalyzed reaction is greatly affected by the pH of the medium. Enzymes have pH optima and commonly have bell-shaped curves for

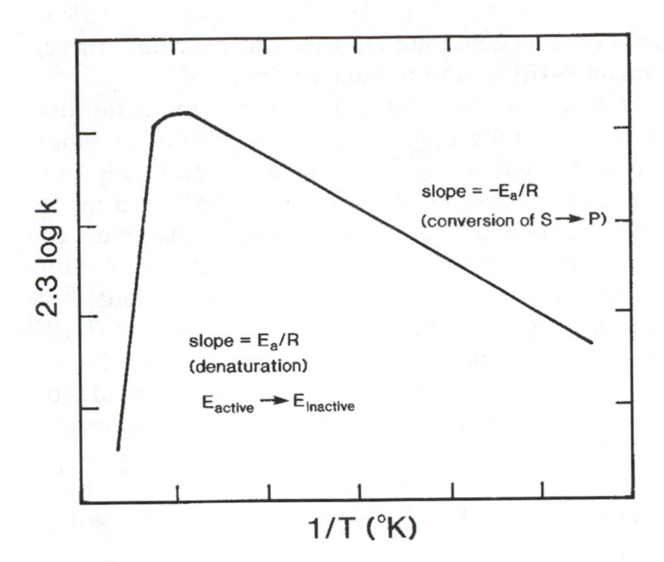

FIGURE 32-8.
Effect of temperature on rate constant of an enzyme-catalyzed reaction. The data are plotted 2.3 log k versus 1/T (°K) according to the Arrhenius equation, $k = Ae^{-Ea/RT}$.

activity versus pH (Fig. 32-9). This pH effect is a manifestation of the effects of pH on enzyme stability and on rate of substrate to product conversion and may also be due to changes in ionization of substrate.

The rate of substrate to product conversion is affected by pH because pH may affect binding of substrate to enzyme and the ionization of catalytic groups such as carboxyl or amino groups that are part of the enzyme's active site. The stability of the tertiary or quaternary structure of enzymes is also pH dependent and affects the velocity of the enzyme reaction, especially at extreme acidic or alkaline pHs. The pH for maximum stability of an enzyme does not necessarily coincide with the pH for maximum activity of that same enzyme. For example, the proteolytic enzymes trypsin and chymotrypsin are stable at pH 3, while they have maximum activity at pH 7 to 8.

To establish the pH optimum for an enzyme reaction, the reaction mixture is buffered at different pHs and the activity of the enzyme is determined. To determine pH enzyme stability relationships, aliquots of the enzyme are buffered at different pH values and held for a specified period of time (e.g., 1 hr). The pH of the aliquots is then adjusted to the pH optimum and each aliquot is assayed. The effect of pH on enzyme stability is thus obtained. These studies are helpful in establishing conditions for handling the enzyme and also may be useful in establishing methods for controlling enzyme activity in a food system. Note that pH stability and the pH optimum for the enzyme activity are not true constants. That is to say, these may vary with particular source of enzyme, the specific substrate used, the temperature of the experiment, or even the buffering species used in the experiment. In the use of enzymes for analysis, it is not necessary that the reaction be carried out at the pH optimum for activity, or even at a pH at which the enzyme is most stable, but it is critical to maintain a fixed pH during the reaction (i.e., use buffer) and to use the same pH in all studies to be compared.

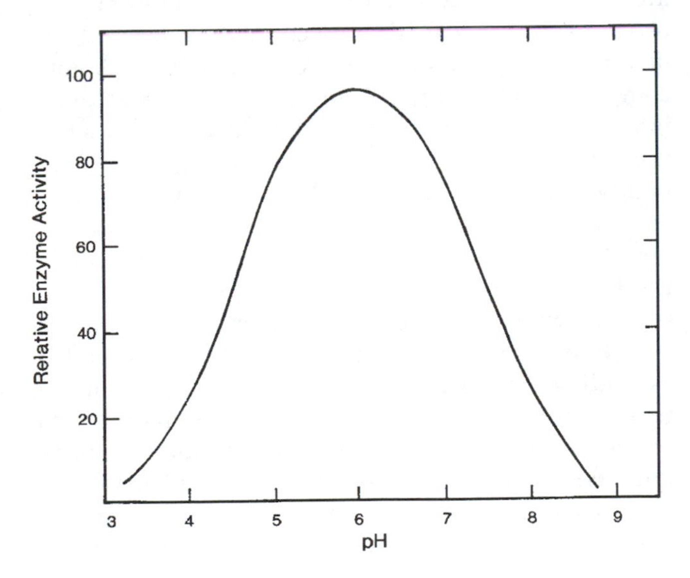

FIGURE 32-9.
Typical velocity-pH curve for an enzyme-catalyzed reaction. The maximum on the curve is the optimum for the system and can vary with temperature, specific substrate, and enzyme source.

32.2.2.4 Activators and Inhibitors

32.2.2.4.1 Activators Some enzymes contain, in addition to a protein portion, small molecules that are **activators of the enzyme.** Some enzymes show an absolute requirement for a particular inorganic ion for activity, while others show increased activity when small molecules are included in the reaction medium. These small molecules can play a role in maintaining the conformation of the protein, or they may form an essential component of the active site, or they may form part of the substrate of the enzyme.

In some cases, the activator forms a nearly irreversible association with the enzyme. These nonprotein portions of the enzyme are called **prosthetic groups.** The amount of enzyme activator complex formed is equal to the amount of activator present in the mixture. In these cases, activator concentration can be estimated up to concentrations equal to total enzyme concentration by simply measuring enzyme activity.

In most cases, dissociation constants for an enzyme activator complex are within the range of enzyme concentration. Dissociable nonprotein parts of enzymes are categorized as coenzymes. When this type of activator is added to enzyme, a curvilinear relationship similar to a Michaelis-Menten plot results, making difficult the determination of an unknown amount of activator. A reciprocal plot analogous to a Lineweaver-Burk plot can be constructed using standards and unknown activator concentrations estimated from such a plot.

An example of an essential activator is the pyridine coenzyme NAD^+. NAD^+ is essential for the oxidation of ethanol to acetaldehyde by alcohol dehydrogenase:

$$\text{ethanol} + NAD^+ \underset{}{\overset{\text{alcohol dehydrogenase}}{\rightleftharpoons}} \text{acetaldehyde} + NADH + H^+ \quad (20)$$

In the reaction, NAD^+ is reduced to NADH and can be considered a second substrate. Another example of an activator of an enzyme is the chloride ion with alpha-amylase. In this case, alpha-amylase has some activity in the absence of chloride. With saturating levels of chloride, the alpha-amylase activity increases about fourfold. Other anions, including F^-, Br^-, and I^-, also activate alpha-amylase. These anions must not be in the reaction mixture if alpha-amylase stimulation is to be used as a method of determining chloride concentration.

32.2.2.4.2 Inhibitors An **enzyme inhibitor** is a compound that when present in an enzyme-catalyzed reaction medium decreases the enzyme activity. Enzyme inhibitors can be categorized as **irreversible** or **reversible** inhibitors. Enzyme inhibitors include inorganic ions, such as Pb^{2+} or Hg^{2+}, which can react with suphydryl groups on enzymes to inactivate the enzyme, compounds that resemble substrate, and naturally occurring proteins that specifically bind to enzymes (such as protease inhibitors found in legumes).

1. Irreversible Inhibitors. When the dissociation constant of the inhibitor enzyme complex is very small, the decrease in enzyme activity observed will be directly proportional to the inhibitor added. The speed at which the irreversible combination of enzyme and inhibitor reacts may be slow, and the effect of time on the reduction of enzyme activity by the addition of inhibitor must be determined to ensure complete enzyme-inhibitor reaction. For example, the amylase inhibitor found in many legumes must be preincubated under specified conditions with amylase prior to measurement of residual activity in order to accurately estimate inhibitor content (1).

2. Reversible Inhibitors. Most inhibitors exhibit a dissociation constant such that both enzyme and inhibitor are found free in the reaction mixture. Several types of reversible inhibitors are known: **competitive, noncompetitive,** and **uncompetitive.**

Competitive inhibitors usually resemble the substrate structurally and compete with substrate for binding to the active site of the enzyme, and only one molecule of substrate or inhibitor can be bound to the enzyme at one time. An inhibitor can be characterized as competitive by adding a fixed amount of inhibitor to reactions at various substrate concentrations and by plotting the resulting data by the Lineweaver-Burk method and noting the effect of inhibitor relative to that of control reactions in which no inhibitor is added. If the inhibitor is competitive, the slope and x intercept of the plot with inhibitor are altered while the y intercept ($1/V_m$) is unaltered. It can be shown that the ratio of the **uninhibited initial velocity** (v_o) to the **inhibited initial velocity** (v_i) gives:

$$\frac{v_o}{v_i} = \frac{[I]K_m}{K_i(K_m + [S])} \quad (21)$$

where: K_i = Dissociation constant of the enzyme-inhibitor complex
$[I]$ = Concentration of competitive inhibitor

Thus, a plot of v_o/v_i versus inhibitor concentration will give a straight-line relationship. From this plot the concentration of a competitive inhibitor can be found (2).

A **noncompetitive inhibitor** binds to enzyme independent of substrate and is bound outside the active site of the enzyme. A noncompetitive inhibitor can be identified by its effect on the rate of enzyme-catalyzed reactions at various substrate concentrations and the data plotted by the Lineweaver Burk method. A noncompetitive inhibitor will affect the slope and the y intercept as compared to the uninhibited system while the x intercept, $1/K_m$, is unaltered. Analogous to competitive inhibitors, a standard curve of v_o/v_i versus inhibitor concentration may be prepared and used to determine the concentration of a noncompetitive inhibitor (2).

Uncompetitive inhibitors bind only to the enzyme substrate complex. Uncompetitive inhibition is noted by adding a fixed amount of inhibitor to reactions at several substrate concentrations and plotting the data by the Lineweaver-Burk method. An uncompetitive inhibitor will affect both the x and y intercepts of the Lineweaver-Burk plot as compared to the uninhibited system, while maintaining an equal slope to the uninhibited system (i.e., a parallel line will result). A plot of v_o/v_i versus inhibitor concentration can be prepared to be used as a standard curve for the determination of the concentration of an uncompetitive inhibitor (2).

32.2.3 Methods of Measurement

32.2.3.1 Overview

For practical enzyme analysis, it is necessary to be familiar with the methods of measurement of the reaction. Any physical or chemical property of the system that relates to substrate or product concentration can be used to follow an enzyme reaction. A wide variety of methods are available to follow enzyme reactions, including **absorbance spectrometry, fluorimetry, manometric methods, titration, isotope measurement,** and **viscosity.** A good example of the use of spectrophotometry as a method for following enzyme reactions is use of the spectra of the pyridine coenzyme NAD(H) and NADP(H), in which there is a marked change in absorbance at 340 nm upon oxidation/reduction (Fig. 32-10). Many methods depend on the increase or decrease in absorbance at 340 nm when these coenzymes are products or substrates in a coupled reaction.

An example of using several methods to measure the activity of an enzyme is in the assay of alpha-amylase activity (3). Alpha-amylase cleaves starch at alpha 1,4 linkages in starch and is an endoenzyme. An endoenzyme cleaves a polymer substrate at internal linkages. This reaction can be followed by a number of methods, including reduction in viscosity, increase in reducing groups upon hydrolysis, reduction in color of the starch iodine complex, and polarimetry. However, it is difficult to differentiate the activity of alpha-amylase from beta-amylase using a single assay.

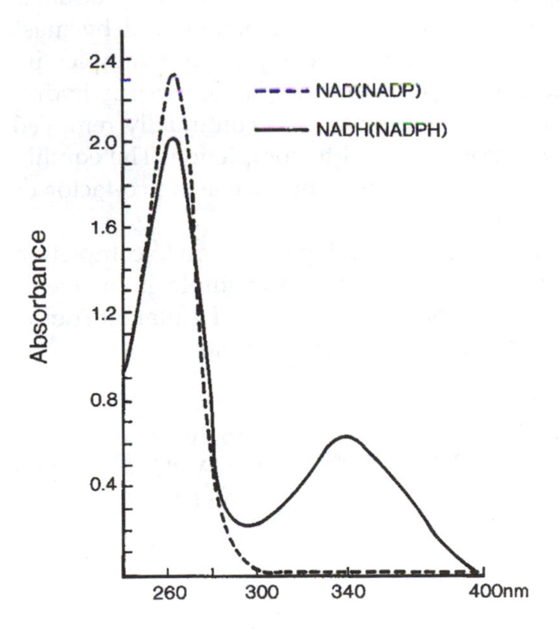

FIGURE 32-10.
Absorption curves of NAD(P) and NAD(P)H; λ = wavelength. Many enzymatic analysis methods are based on the measurement of an increase or decrease in absorbance at 340 nm due to NAD(H) or NADP(H).

Beta-amylase cleaves maltose from the nonreducing end of starch. While a marked decrease in viscosity of starch or reduction in iodine color would be expected to occur due to alpha-amylase activity, beta-amylase can also cause changes in viscosity and iodine color if in high concentration. In order to establish whether alpha-amylase or beta-amylase is being measured, the analyst must determine the change in number of reducing groups as a basis of comparison. Because alpha-amylase is an endoenzyme, hydrolysis of a few bonds near the center of the polymeric substrate will cause a marked decrease in viscosity, while hydrolysis of an equal number of bonds by the exoenzyme, beta-amylase, will have little effect on viscosity.

In developing an enzyme assay, it is wise to first write out a complete, balanced equation for the particular enzyme-catalyzed reaction. Inspection of the products and substrates for chemical and physical properties that are readily measurable with available equipment will often result in an obvious choice of method for following the reaction in the laboratory.

If one has options in methodology, one should select the method that is able to monitor the reaction continuously, is most sensitive, and is specific for the enzyme-catalyzed reaction.

32.2.3.2 Coupled Reactions

Enzymes can be used in assays via coupled reactions. **Coupled reactions** involve using two or more enzyme reactions so that a substrate or product concentration can be readily followed. In using a coupled reaction, there is an **indicator reaction** and a **measuring reaction.** For example:

$$\underset{\text{measuring reaction}}{S1 \xrightarrow{E1} P1} \tag{22}$$

$$\underset{\text{indicating reaction}}{P1 \xrightarrow{E2} P2} \tag{23}$$

The role of the indicating enzyme (E2) is to produce P2, which is readily measurable and, hence, is an indication of the amount of P1 produced by E1. Alternatively the same sequence can be used in measuring S1, the substrate for E1. When a coupled reaction is used to measure the activity of an enzyme (e.g., E1 above), it is critical that the indicating enzyme E2 not be rate limiting in the reaction sequence: The measuring reaction must always be rate determining. Consequently, E2 activity should be much greater than E1 activity for an effective assay. Coupled enzyme reactions can have problems with respect to pH of the system if the pH optima of the coupled enzymes are quite different. It may be necessary to allow the first reaction (e.g., the measuring reaction catalyzed by E1 above, eq. 22) to proceed for a time and then arrest the reaction by heating to denature E1. The pH is adjusted, the indicating enzyme (E2, eq. 23) added, and the reaction completed. If an endpoint method is used with a coupled system, the requirements for pH compatibility are not as stringent as for a rate assay because an extended time period can be used to allow the reaction sequence to go to completion.

32.3 APPLICATIONS

As described above, certain information is needed prior to using enzyme assays analytically. In general, knowledge of K_m, time course of the reaction, the enzyme's specificity for substrate, the pH optimum and pH stability of the enzyme, and effects of temperature on the reaction and stability of the enzyme are desirable. Many times this information is available from the literature. However, a few preliminary experiments may be necessary, especially in the case of experiments in which velocities are measured. A time course to establish linearity of product formation or substrate consumption in the reaction is a necessity. An experiment to show linearity of velocity of the enzyme reaction to enzyme concentration is recommended (see Fig. 32-5).

32.3.1 Substrate Assays

The following is not an extensive compendium of methods for the measurement of food components by enzymatic analysis. Instead, it is meant to be representative of the types of analyses possible. The reader can consult handbooks published by the manufacturers of enzyme kits, for example, Boehringer-Mannheim (4), the review article by Whitaker (2), and the series by Bergmeyer (5) for a more comprehensive guide to enzyme methods applicable to foods.

32.3.1.1 Sample Preparation

Because of the specificity of enzymes, sample preparation prior to enzyme analysis is often minimal and may involve only extraction and removal of solids by filtration or centrifugation. Regardless, due to the wide variety of foods that might be encountered by the analyst using enzyme assays, a check should be made of the extraction and enzyme reaction steps by standard addition of known amounts of analyte to the food and extract, and measuring recovery of that standard. If the standard additions are fully recovered, this is a positive indication that the extraction is complete, that sample does not contain interfering substances that require removal prior to the enzymatic analysis, and that the reagents are good. In some cases, interfering substances are present but can be readily removed by precipitation or adsorption. For example, polyvinylpolypyrrolidone (PVPP) powder can be used to decolorize juices or red wines. With the advent of small syringe mini-columns (e.g., C18, silica, and ion exchange cartridges), it is also relatively easy and fast to attain group separations to remove interfering substances from a sample extract.

32.3.1.2 Total Change/Endpoint Methods

While substrate concentrations can be determined in rate assays when the reaction is first order with respect to substrate concentration ($S << K_m$), substrate concentration can also be determined by the total change or endpoint method. In this method, the enzyme-catalyzed reaction is allowed to go to completion so that concentration of product, which is measured, is directly related to substrate. An example of such a system is the measurement of glucose using glucose oxidase and peroxidase, described below.

In some cases, an equilibrium is established in an endpoint method in which there is a significant amount of substrate remaining in equilibrium with product. In these cases, the equilibrium can be altered. For example, in cases in which a proton-yielding reaction is used, alkaline conditions (increase in pH) can be used. Trapping agents can also be used, in which product is effectively removed from the reaction, and by mass action the reaction goes to completion. Examples include the trapping of ketones and aldehydes by hydrazine. In this way, the product is continually removed and the reaction is pulled to completion. The equilibrium can also be displaced by increasing co-factor or co-enzyme concentration.

Another means of driving a reaction to completion is a regenerating system (5). For example, in the measurement of glutamate, with the aid of glutamate dehydrogenase, the following can be done:

$$\text{glutamate} + NAD^+ + H_2O \xrightleftharpoons{\text{glutamate dehydrogenase}} \alpha\text{-ketoglutarate} + NADH + NH_4^+ \quad (24)$$

$$\text{pyruvate} + NADH + H^+ \xrightarrow{\text{lactate dehydrogenase}} NAD^+ + \text{lactate} \quad (25)$$

In this system, NADH is recycled to NAD^+ via lactate dehydrogenase until all the glutamate to be measured is consumed. The reaction is stopped by heating to denature the enzymes present, a second aliquot of glutamate dehydrogenase and NADH is added, and the α-ketoglutarate (equivalent to the original glutamate) measured via decrease in absorbance at 340 nm. An example in which the same equilibrium is displaced in the measurement of glutamate is as follows:

$$\text{glutamate} + NAD^+ + H_2O \xrightleftharpoons{\text{glutamate dehydrogenase}} \alpha\text{-ketoglutarate} + NADH + NH_4^+ \quad (26)$$

$$NADH + INT \xrightarrow{\text{diaphorase}} NAD^+ + \text{formazan} \quad (27)$$

INT, iodonitrotetrazolium chloride, is a trapping reagent for the NADH product of the glutamate dehydrogenase catalyzed reaction. The formazan formed is measurable colorimetrically at 492 nm.

32.3.1.3 Specific Applications

32.3.1.3.1 Measurement of Sulfite Sulfite is a food additive that can be measured by several techniques, including titration, distillation followed by titration, gas chromatography, and colorimetric analysis. Sulfite can also be specifically oxidized to sulfate by the commercially available enzyme sulfite oxidase (SO):

$$SO_3^= + O_2 + H_2O \xrightarrow{SO} SO_4^= + H_2O_2 \quad (28)$$

The H_2O_2 product can be measured by several methods, including the use of the enzyme NADH-peroxidase.

$$H_2O_2 + NADH + H^+ \xrightarrow{\text{NADH-peroxidase}} 2H_2O + NAD^+ \quad (29)$$

The amount of sulfite in the system is equal to the NADH oxidized, which is determined by decrease in absorbance at 340 nm. Ascorbic acid can interfere with the assay but can be removed by using ascorbic acid oxidase (6).

32.3.1.3.2 Colorimetric Determination of Glucose The combination of the enzymes glucose oxidase and peroxidase can be used to specifically measure glucose in a food system (7). Glucose is preferentially oxidized by glucose oxidase to produce gluconolactone and hydrogen peroxide. The hydrogen peroxide plus o-dianisidine in the presence of peroxidase produces a yellow color that absorbs at 420 nm (eqs. 30 and 31). This assay is normally carried out as an endpoint assay, and there is stoichiometry between the color formed and the amount of glucose in the extract, which is established with a standard curve. Because glucose oxidase is quite specific for glucose, it is a useful tool in determining the amount of glucose in the presence of other reducing sugars.

$$\beta\text{-D-glucose} + O_2 \xrightarrow{\text{glucose oxidase}} \delta\text{-gluconolactone} + H_2O_2 \quad (30)$$

$$H_2O_2 + \text{o-diansidine} \xrightarrow{\text{peroxidase}} H_2O + \text{oxidized dye (colored)} \quad (31)$$

32.3.1.3.3 Starch/Dextrin Content Starch and dextrins can be determined by enzymatic hydrolysis using amyloglucosidase, an enzyme that cleaves alpha 1,4 and alpha 1,6 bonds of starch, glycogen, and dextrins, liberating glucose. The glucose formed can be subsequently determined enzymatically. Glucose can be determined by the previously described colorimetric method, in which glucose is oxidized by glucose oxidase and coupled to a colored dye via reaction of the glucose oxidase product, hydrogen peroxide, with peroxidase. An alternative method of measuring glucose enzymatically is by coupling hexokinase (HK) and glucose-6-phosphate dehydrogenase (G6PDH) reactions:

$$\text{glucose} + ATP \xrightarrow{HK} \text{glucose-6-phosphate} + ADP \quad (32)$$

$$\text{glucose-6-phosphate} + NADP^+ \xrightarrow{G6PDH} \text{6 phospho-gluconate} + NADPH + H^+ \quad (33)$$

The amount of NADPH formed is measured by absorbance at 340 nm and is a stoichiometric measure of the glucose originating in the dextrin or starch hydrolyzed by amyloglucosidase. Note that hexokinase catalyzes the phosphorylation of fructose as well as glucose. The determination of glucose is specific because of the specificity of the second reaction catalyzed by glucose-6-phosphate dehydrogenase in which glucose-6-phosphate is the substrate.

This assay sequence can be used to detect the dextrins of corn syrup used to sweeten a fruit juice product. A second assay would be needed, however, without treatment with amyloglucosidase to account for the glucose in the product. The glucose determined in that assay would be subtracted from the result of the assay in which amyloglucosidase is used.

The same hexokinase glucose-6-phosphate dehydrogenase sequence used to measure glucose can also be used to measure other carbohydrates in foods. For example, lactose and sucrose can be determined via specific hydrolysis of these disaccharides by beta-galactosidase and invertase respectively, followed by the use of the earlier described hexokinase, glucose-6-phosphate dehydrogenase sequence.

32.3.1.3.4 Determination of D-Malic Acid in Apple Juice Two stereo isomeric forms of malic acid exist. L-malic acid occurs naturally, while the D form is normally not found in nature. Synthetically produced malic acid is a mixture of these two isomers. Consequently, synthetic malic acid can be detected by a determination of D-malic acid. One means of detecting the malic acid is through the use of the enzyme decarboxylating D-malate dehydrogenase (8). Decarboxylating D-malate dehydrogenase (DMD) catalyzes the conversion of D-malic acid as follows:

$$\begin{matrix}\text{D-malic acid} \\ + \text{NAD}^+\end{matrix} \xrightarrow{\text{DMD}} \begin{matrix}\text{pyruvate} + CO_2 \\ + \text{NADH} + H^+\end{matrix} \qquad (34)$$

The reaction can be followed by the measurement of NADH photometrically. Because CO_2 is a product of this reaction and escapes, the equilibrium of the reaction lies to the right and the process is irreversible. This assay is of value because the addition of synthetic D/L malic acid can be used to illegally increase the acid content of apple juice and apple juice products.

32.3.2 Enzyme Activity Assays

32.3.2.1 Peroxidase Activity

Peroxidase is found in most plant materials and is reasonably stable to heat. A heat treatment that will destroy all peroxidase activity in a plant material is usually considered to be more than adequate to destroy other enzymes and most microbes present. In vegetable processing, therefore, the adequacy of the blanching process can be monitored by following the disappearance of peroxidase activity (9). Peroxidase catalyzes the oxidation of guaiacol (colorless) in the presence of hydrogen peroxide to form tetraguaiacol (yellow brown) and water (eq. 35). Tetraguaiacol has an absorbance maximum around 450 nm. Increase in absorbance at 450 nm can be used to determine the activity of peroxidase in the reaction mixture.

$$H_2O_2 + \text{guaiacol} \xrightarrow{\text{peroxidase}} \begin{matrix}\text{tetraguaiacol} \\ + H_2O \\ \text{(colored)}\end{matrix} \qquad (35)$$

32.3.2.2 Lipoxygenase

Recently it has been pointed out that lipoxygenase may be a more appropriate enzyme to measure the adequacy of blanching of vegetables than peroxidase (10). Lipoxygenase refers to a group of enzymes that catalyzes the oxidation by molecular oxygen of fatty acids containing a cis, cis, 1,4-pentadiene system producing conjugated hydroperoxide derivatives:

$$(\text{—CH}=\text{CH—CH}_2\text{—CH}=\text{CH—}) + O_2$$

$$\xrightarrow{\text{lipoxygenase}} \underset{\underset{\underset{\text{H}}{|}}{\underset{\text{O}}{|}}}{\underset{\text{O}}{|}}{(\text{—C}}\text{—CH}=\text{CH—CH}=\text{CH—}) \; \text{(conjugated)} \qquad (36)$$

A variety of methods can be used to measure lipoxygenase activity in plant extracts. The reaction can be followed by measuring loss of fatty acid substrate, oxygen uptake, occurrence of the conjugated diene at 234 nm, or the oxidation of a cosubstrate such as carotene (11). All these methods have been used, and each has its advantages. The oxygen electrode method is widely used and replaces the more cumbersome manometric method. The electrode method is rapid and sensitive and gives continuous recording. It is normally the method of choice for crude extracts, but secondary reactions involving oxidation must be corrected for or eliminated. Zhang et al. (12) have reported the adaptation of the O_2 electrode method to the assay of lipoxygenase in green bean homogenates without extraction. Due to the rapidity of the method (<3 min including the homogenization), on-line process control using lipoxygenase activity as a control parameter for optimization of blanching of green beans is a real possibility. The formation of conjugated diene fatty acids with a chromophore at 234 nm can be followed continuously. However, optically clear mixtures are necessary. Bleaching of carotenoids has also been used as a measure of lipoxygenase activity. However, the stoichiometry of this method is uncertain, and all lipoxygenases do not have equal carotenoid bleaching activity. Williams et al. (10) have developed a semiquantitative spot test assay for lipoxygenase in which I^- is oxidized to I_2 in the presence of the linoleic acid hydroperoxide product and the I_2 detected as an iodine starch complex.

32.3.2.3 Alkaline Phosphatase Assay

Alkaline phosphatase is a relatively heat-stable enzyme found in raw milk. The thermal stability of alkaline phosphatase in milk is greater than the non-spore forming microbial pathogens present in milk. The phosphatase assay is applied to dairy products to determine whether pasteurization has been done properly and to detect the addition of raw milk to pasteurized milk. The common phosphatase test is based on the phosphatase-catalyzed hydrolysis of disodiumphenyl phosphate liberating phenol (13). The phenol product is measured colorimetrically after reaction with CQC (2,6-dichloroquinonechloroimide) to form a blue indophenol. The indophenol is extracted into n-butanol and measured at 650 nm. This is an example of a physical separation of product to allow the ready measurement of an enzyme reaction. Recently, a fluorometric assay has been suggested and has been commercialized for measurement of alkaline phosphatase in which the rate of fluorophore production can be monitored directly without butanol extraction used to measure indophenol when phenylphosphate is used as substrate (14). The fluorometric assay was shown to give greater repeatability compared to the standard assay in which phenylphosphate is used as substrate and was capable of detecting 0.05 percent raw milk in a pasteurized milk sample.

32.3.2.4 Alpha-Amylase Activity in Malt

Amylase activity in malt is a critical quality parameter. The amylase activity in malt is often referred to as diastatic power and refers to the production of reducing substances by the action of alpha- and beta-amylases on starch. The measurement of diastatic power involves digestion of soluble starch with a malt infusion (extract) and following increase in reducing substances by measuring reduction of Fehling's solution or ferricyanide. Specifically measuring alpha-amylase activity (often referred to as dextrinizing activity) in malt is more complicated and is based on using a limit dextrin as substrate. Limit dextrin is prepared by action of beta-amylase (free of alpha-amylase activity) on soluble starch. The beta-amylase clips maltose units off the nonreducing end of the starch molecule until an alpha 1,6-branch point is encountered. The resulting product is a beta-limit dextrin that serves as the substrate for the endo cleaving alpha-amylase. A malt infusion is added to the previously prepared limit dextrin substrate and aliquots removed periodically to a solution of dilute iodine. The alpha-amylase activity is measured by changed color of the starch iodine complex in the presence of excess beta-amylase used to prepare the limit dextrin. The color is compared to a colored disc on a comparator. This is continued until the color is matched to a color on a comparator. The time to reach that color is dextrinizing time and is a measure of alpha-amylase activity, a shorter time representing a more active preparation.

32.3.2.5 Rennet Activity

Rennet, an extract of bovine stomach, is used as a coagulating agent in cheese manufacture. Most rennet activity tests are based on noting the ability of a preparation to coagulate milk. For example, 12 percent nonfat dry milk is dispersed in a 10 mM calcium chloride solution and warmed to 35°C. An aliquot of the rennet preparation is added and the time of milk clotting observed visually. The activity of the preparation is calculated in relationship to a standard rennet. As opposed to coagulation ability, rennet preparations can also be evaluated for proteolytic activity by measuring the release of a dye from azocasein (casein to which a dye has been covalently attached). In this assay, the rennet preparation is incubated with 1 percent azocasein. After the reaction period, the reaction is stopped by adding trichloroacetic acid. The trichlorolacetic acid precipitates the protein that is not hydrolyzed. The small fragments of colored azocasein produced by the hydrolysis of the rennet are left in solution and absorbance read at 345 nm (15, 16). This assay is based on the increase in solubility of a substrate upon cleavage by an enzyme.

32.3.3 Bio-Sensors/Immobilized Enzymes

The use of immobilized enzymes as analytical tools is currently receiving increased attention. An immobilized enzyme in concert with a sensing device is an example of a bio-sensor. A bio-sensor is a device comprised of a biological sensing element (e.g., enzyme, antibody, etc.) coupled to a suitable transducer (e.g., optical, electrochemical, etc.). Immobilized enzymes, because of their stability and ease of removal from the reaction, can be used repeatedly, thus eliminating a major cost in enzyme assays. The most widely used enzyme electrode is the glucose electrode in which glucose oxidase is combined with an oxygen electrode to determine glucose concentration (17). When the electrode is put into a glucose solution, the glucose diffuses into the membrane where it is converted to gluconolactone by glucose oxidase with the uptake of oxygen. The oxygen uptake is a measure of the glucose concentration. Glucose can also be measured by the action of glucose oxidase with the detection of hydrogen peroxide, in which the hydrogen peroxide is detected amperometrically at a polarized electrode (18). A large number of enzyme electrodes (bio-sensors) have been reported in the literature recently. For example, a glycerol sensor, in which glycerol dehydrogenase was immobilized, has been developed for the determination of glycerol in wine (19). NADH produced by the enzyme was monitored with a platinum electrode.

32.4 SUMMARY

Enzymes, due to their specificity and sensitivity, are valuable analytical devices for quantitating compounds that are enzyme substrates, activators, or inhibitors. In enzyme-catalyzed reactions, the enzyme and substrate are mixed under specific conditions (pH, temperature, ionic strength, substrate, and enzyme concentrations). Changes in these conditions can affect the reaction rate of the enzyme and thereby the outcome of the assay. The enzymatic reaction is followed by measuring either the amount of product generated or the disappearance of the substrate. Applications for enzyme analyses will increase as a greater number of enzymes are purified and become commercially available. In some cases, gene amplification techniques will make enzymes available that are not naturally found in great enough amounts to be used analytically. The measurement of enzyme activity is useful in assessing food quality and as an indication of the adequacy of heat processes such as pasteurization and blanching. In the future, as in-line process control (to maximize efficiencies and drive quality developments) in the food industry becomes more important, immobilized enzyme sensors, along with microprocessors, will likely play a prominent role.

32.5 STUDY QUESTIONS

1. The Michaelis-Menten equation mathematically defines the hyperbolic nature of a plot relating reaction velocity to substrate concentration for an enzyme-mediated reaction. The reciprocal of this equation gives the Lineweaver-Burk formula and a straight-line relationship as shown below.
   a. Define what v_o, K_m, V_m, and [S] refer to in the Lineweaver-Burk formula.
   b. Based on the components of the Lineweaver-Burk formula, label the y axis, x axis, slope, and y intercept on the plot.
   c. What factors that control or influence the rate of enzyme reactions affect K_m and V_m?

$$\frac{1}{v_o} = \frac{K_m}{V_m}\frac{1}{[S]} + \frac{1}{V_m}$$

   a. v_o –
      K_m –
      V_m –
      [S] –

   b.

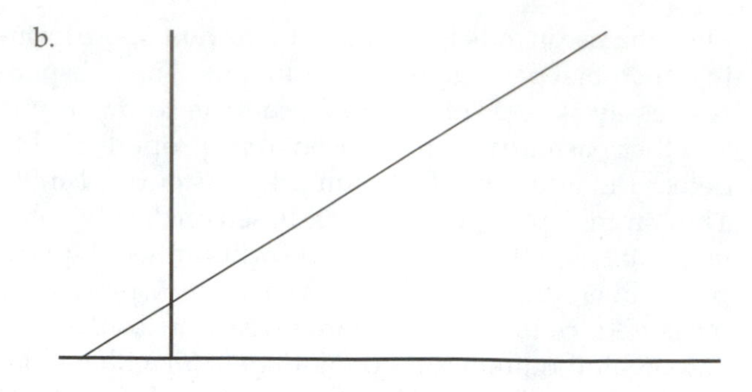

   c. K_m –
      V_m –
2. Explain, on a chemical basis, why extremes of pH and temperature can reduce the rate of enzyme-catalyzed reactions.
3. Differentiate among competitive, noncompetitive, and uncompetitive enzyme inhibitors.
4. You believe that the food product you are working with contains a specific enzyme inhibitor. Explain how you would quantitate the amount of enzyme inhibitor (I) present in an extract of the food. The inhibitor (I) in question can be purchased commercially in a purified form from a chemical company. The inhibitor is known to inhibit the specific enzyme E, which reacts with the substrate S to generate product P, which can be quantitated spectrophotometrically.
5. What methods can be used to quantitate enzyme activity in enzyme-catalyzed reactions?
6. What is a coupled reaction, and what are the concerns in using coupled reactions to measure enzyme activity? Give a specific example of a coupled reaction used to measure enzyme activity.
7. Explain how D-malic acid can be quantitated by an enzymatic method to test for adulteration of apple juice.
8. Why is the enzyme peroxidase often quantitated in processing vegetables?
9. Explain the purpose of testing for phosphatase activity in the dairy industry, and explain why it can be used in that way.
10. Explain how glucose can be quantitated using a specific immobilized enzyme.

32.6 REFERENCES

1. Powers, J.R., and Whitaker, J.R. 1977. Effect of several experimental parameters on combination of red kidney bean *(Phaseolus vulgaris)* α-amylase inhibitor with porcine pancreatic α-amylase. *J. Food Biochem* 1:239.
2. Whitaker, J.R. 1985. Analytical uses of enzymes, in *Food Analysis. Principles and Techniques*. Vol. 3. *Biological Techniques*. D. Gruenwedel and J.R. Whitaker (Eds.), 297–377. Marcel Dekker, Inc., New York, NY.
3. Bernfeld, P. 1955. Amylases, α and β. *Methods in Enzymology* 1:149.
4. Boehringer-Mannheim. 1987. *Methods of Biochemical Analysis and Food Analysis*. Boehringer Mannheim Gmb H Mannheim, W. Germany.

5. Bergmeyer, H.U. 1983. *Methods of Enzymatic Analysis*. Academic Press, Inc., New York, NY.
6. Beutler, H. 1984. A new enzymatic method for determination of sulphite in food. *Food Chem.* 15:157.
7. Raabo, E., and Terkildsen, T.C. 1960. On the enzyme determination of blood glucose. *Scand. J. Clin. Lab. Invest.* 12:402.
8. Beutler, H., and Wurst, B. 1990. A new method for the enzymatic determination of D-malic acid in foodstuffs. Part I: Principles of the Enzymatic Reaction. *Deutsche Lebensmittel-Rundschau* 86:341.
9. USDA. 1975. Enzyme inactivation tests (frozen vegetables). Technical inspection procedures for the use of USDA inspectors. Agricultural Marketing Service, U.S. Dept. of Agriculture, Washington, DC.
10. Williams, D.C., Lim, M.H., Chen, A.O., Pangborn, R.M., and Whitaker, J.R. 1986. Blanching of vegetables for freezing—Which indictor enzyme to use. *Food Technol.* 40(6):130.
11. Surrey, K. 1964. Spectrophotometric method for determination of lipoxidase activity. *Plant Physiol.* 39:65.
12. Zhang, Q., Cavalieri, R.P., Powers, J.R., and Wu, J. 1991. Measurement of lipoxygenase activity in homogenized green bean tissue. *J. Food Sci.* 56:719.
13. Druchrey, I., Kleyn, D.H., and Murthy, G.K. 1985. Phosphatase methods, in *Standard Methods for the Examination of Dairy Products*, 15th ed. G.H. Richardson (Ed.), p. 311. American Public Health Association, Washington, DC.
14. Rocco, R. 1990. Fluorometric determination of alkaline phosphatase in fluid dairy products: Collaborative study. *J. Assoc. Off. Anal. Chem.* 73:842.
15. Christen, G.L., and Marshall, R.T. 1984. Selected properties of lipase and protease of *Pseudomonas fluorescens* 27 produced in 4 media. *J. Dairy Sci.* 67:1680.
16. Kim, S.M., and Zayas, J.F. 1991. Comparative quality characteristics of chymosin extracts obtained by ultrasound treatment. *J. Food Sci.* 56:406.
17. Guilbault, G.G., and Lubrano, G.J. 1972. Enzyme electrode for glucose. *Anal. Chim. Acta* 60:254.
18. Shimizu, Y., and Morita, K. 1990. Microhole assay electrode as a glucose sensor. *Anal. Chem.* 62:1498.
19. Matsumoto, K. 1990. Simultaneous determination of multicomponent in food by amperometric FIA with immobilized enzyme reactions in a parallel configuration, in *Flow Injection Analysis (FIA) Based on Enzymes or Antibodies*, R.D. Schmid (Ed.). GBF Monographs, Vol. 14, pp. 193.

CHAPTER

33

Immunoassays

Deborah E. Dixon

33.1 INTRODUCTION

Immunological methods are finding widespread application in food analysis. The classical methods used in food analysis consisted of agglutination and gel precipitation reactions and gave way to the isotopic assays, such as **radioimmunoassay** (RIA). This assay provided excellent sensitivity, but the need for expensive equipment and radioisotopes presented a drawback to its widespread use. Evolution of enzyme-labeled reagents and development of **enzyme immunoassays** (EIAs) overcame the undesirable features (e.g., potential health hazards from exposure to radioactivity) posed by RIAs. **Enzyme-linked immunosorbent assays** (ELISAs) have found widespread use in the food industry. They can be used to detect desirable as well as undesirable substances. Questions have recently been raised about the safety of the food supply. A great deal of attention has been placed on the detection of undesirable substances such as pesticides, drug residues, hormones, growth promoters, microbial toxins, either mycotoxins or enterotoxins, natural intoxicants such as alkaloids and other undesirable additives. However, there are a large number of chemicals present in foods that are natural components (e.g., carbohydrates, proteins, fats, flavor, color, minerals), as well as additives added intentionally to enhance processing. Concentrations of one or more of these may need to be monitored on a regular basis. Microbes may produce beneficial compounds for food processing (fermentation for preservation). They may also be the source of harmful mycotoxins and enterotoxins.

Immunological methods and most definitely ELISAs provide sensitivity, specificity, speed, and cost effectiveness that many of the classical microbiological and chemical methods of analysis lack. These methods require extensive sample preparation (e.g., extraction and clean-up procedures, concentration and separation steps), trained personnel, and expensive equipment, and both are costly and time consuming. Immunoassays lend themselves to routine analysis of large numbers of samples. They can be used in the field, where qualitative screening is often desired. They can also be used quantitatively to obtain a value of how much of the analyte is present in the sample. ELISAs can be used in the private sector as well as government laboratories as research tools. They can also find application in routine surveillance and quality control testing.

Until the mid 1970s, immunoassays were developed with polyclonal antibodies. Milstein and Kohler (1) developed hybridomas that secrete monoclonal antibodies. Monoclonal antibodies have now been developed for use in food analysis.

Principles of immunoassays as well as their applications are discussed in the following sections.

33.2 PRINCIPLES AND PROCEDURES

33.2.1 Immunological Definitions

33.2.1.1 Antibodies

Antibodies are members of the family of immunoglobulins. These proteins are slightly glycosylated and exhibit a number of important and diagnostic features (2, 3). There are two types of antibodies: **polyclonal antibodies** and **monoclonal antibodies.** Briefly, to produce **polyclonal antibodies,** a properly selected antigen is injected into the host animal. The animal's immune system will recognize the antigen as a foreign substance and respond to it. The resulting antibodies are a mixture. There are a number of different determinants or one repeating determinant. There are various determinants to which the immune cells respond and result in a mixture of antibodies to those determinants. If a successful preparation is made, the population will contain some antibodies whose affinity and avidity are great for the foreign protein.

Monoclonal antibodies are secreted by hybridomas that are created by fusing hyperimmunized spleen cells, usually from mice, to myeloma cells, usually of mouse origin. The secreted antibodies are homogeneous, since all the cells in the culture that secrete the antibodies originated from one cell (1).

33.2.1.2 Antigen

The term **antigen** is more complex in meaning than is the term antibody. The antigen may be described as the substance to which the antibody binds. The work of Landsteiner (4) demonstrated that the antibodies bind to and have specificity for fairly small chemical moieties. Kabat estimated the chemical group may be the size of a penta saccharide (5), while another study estimated it to be as large as a tetrapeptide (6). Therefore, an antigen can be a large soluble protein, a mammalian cell, or an organism (e.g., virus or bacteria). The actual site of antibody binding should be called a **determinant** (7) or **epitope** (8).

33.2.1.3 Hapten

The term **multivalent** can be used to describe antigens made of proteins or bacteria. Small hormones and haptens are examples of **univalent antigens.** A **hapten** is a small molecule of less than 1,000 Daltons. It is nonimmunogenic in its own right and must be chemically linked to proteins (*in vivo* or *in vitro*) to produce antibodies. When a small molecule is attached to a large carrier molecule [e.g., bovine serum albumin (BSA) or keyhole limpet hemocyanin (KLH)], the immunogen is called a **conjugate antigen.**

33.2.2 Methodology for Immunoassays

Immunoassays (IAs) comprise a test format that is antibody based. They can be used to detect antigens or antibodies. They can be developed for detection of large or small molecules.

33.2.2.1 Isotopic Immunoassays

In an **isotopic immunoassay,** hapten or antigen can be measured. The assay is based on competition for antibody between a radioactive indicator antigen and its unlabeled counterpart in the test sample. As the amount of unlabeled antigen in the test sample increases, less labeled antigen is bound. The concentration of antigen in the test sample can be determined from comparison with a standard calibration curve prepared with known concentrations of purified antigen (9).

33.2.2.1.1 Radioimmunoassay (RIA) The sensitivity of the RIA is largely limited by the amount of radioactivity that can be introduced into the radiolabeled antigen (9). Levels as low as 1 ng can be detected when carrier-free radioactive iodine ^{125}I is used as an extrinsic label. Precipitates usually do not result, due to the extremely low concentrations used. There are several procedures that can be used to separate free and bound indicator antigen. In the classical method, complexes of antibody bound to radiolabeled antigen are precipitated with antiserum prepared against the antibody moiety (anti-species antibody).

There are also some solid-phase assays that alleviate need for the second anti-species antibody (Fig. 33-1). In one method, antibodies to the antigen are adsorbed onto the plastic (e.g., polystyrene) tube. Next, radiolabeled antigen binds specifically to the adsorbed antigens and can be counted. When unlabeled antigen that competes is also present, less radiolabeled antigen can be proportionally bound. The unbound fraction that is left can be removed. This method is cost effective, quick, and highly sensitive. For example, it is possible to detect less than 0.001 mcg of antigen when a tube is coated with 1 mcg of antibody (10).

33.2.2.2 Nonisotopic Immunoassays

Nonisotopic immunoassays are different from isotopic immunoassays mainly due to the type of label used, the means of endpoint detection, and the possibility of eliminating a separation step. Two types of nonisotopic immunoassays are **fluoroimmunoassays** (FIAs) and **enzyme immunoassays** (EIAs).

33.2.2.2.1 Fluoroimmunoassays **Fluorescein** and **rhodamine** are commonly used for labeling molecules. There are three types of fluoroimmunoassays: (1) **non-**

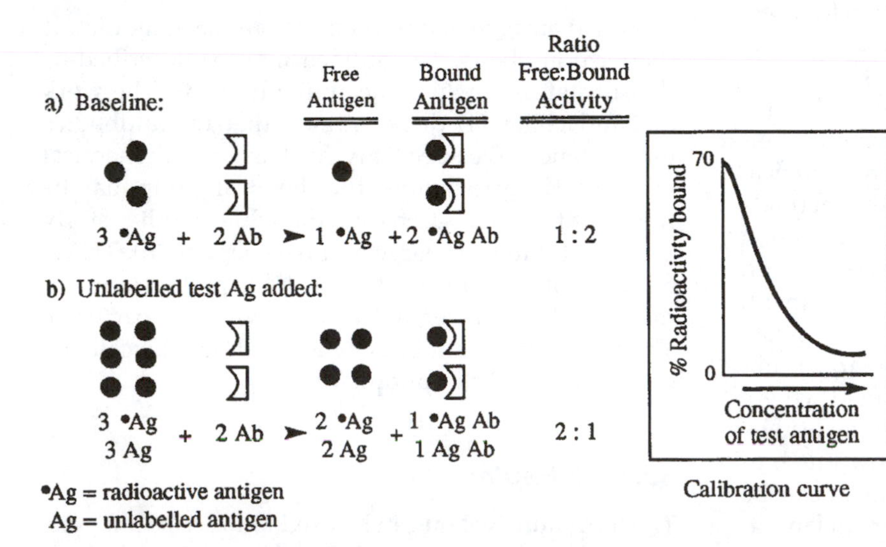

FIGURE 33-1.
Radioimmunoassay methodology. Principle of radioimmunoassay (simplified by assuming a very highly avid antibody and one combining site per antibody molecule). (a) If we add 3 mol of radiolabeled Ag (•) to 2 mol of Ab, 1 mol of Ag will be free and 2 bound to Ab. The ratio of the counts of free to bound will be 1:2. (b) If we now add 3 mol of unlabelled Ag (•) plus 3 mol of Ag to the Ab, again only 2 mol of total Ag will be bound, but since the Ab cannot distinguish labeled from unlabeled Ag, half will be radioactive. The remaining antigen will be free and the ratio of free:bound radioactivity changes to 2:1. This ratio will vary with the amount of unlabeled Ag added, and this enables a calibration curve to be constructed. From (11), used with permission.

separation fluoroimmunoassays that require no separation step of bound from unbound product, (2) **polarization fluoroimmunoassays** that operate based on antibody binding to enhance the signal, and (3) **quenching fluoroimmunoassays** that depend on a decrease in signal from the bound fraction (Fig. 33-2). The quenching of activity is attributed to the antibody's ability to affect the excitation or the emission from the labeled small molecule (12).

33.2.2.2.2 Enzyme Immunoassays (EIAs) EIAs employ enzyme labels and are divided into two categories: homogeneous and heterogeneous. **Homogeneous assays** require no separation of unreacted reagents because the immune reaction affects the enzyme activity. **Heterogeneous assays** have separation steps. ELISA, a type of heterogeneous assay, requires washing between each step to remove unbound reagents. Enzyme labels widely used include alkaline phosphatase, glucose oxidase, and horseradish peroxidase. These enzymes catalyze reactions that cause substrates to degrade and form a colored product that can be read either spectrophotometrically or visually by eye. Depending on format, either antibody or antigen is adsorbed onto a solid phase, which can be polystyrene tubes, polystyrene microtiter wells, or membranes (i.e., nitrocellulose and nylon) (13).

1. Sandwich ELISA. This format enables detection of large antigens (bacterial, viral, and other large proteins). Two preparations of antibodies are used, one to coat the solid phase and one onto which the enzyme is attached (Fig. 33-3). These antibodies can have the same specificity or can be directed against separate

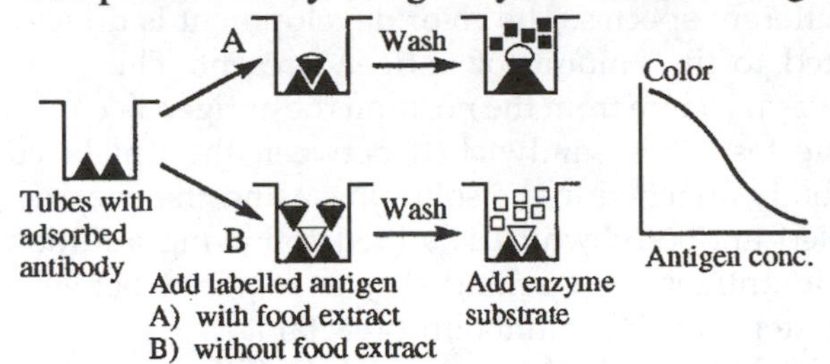

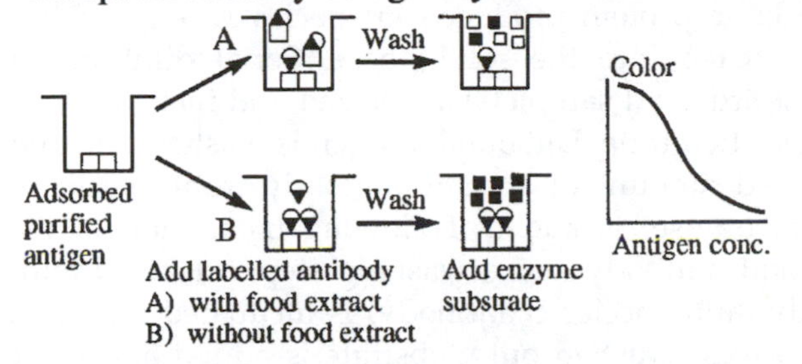

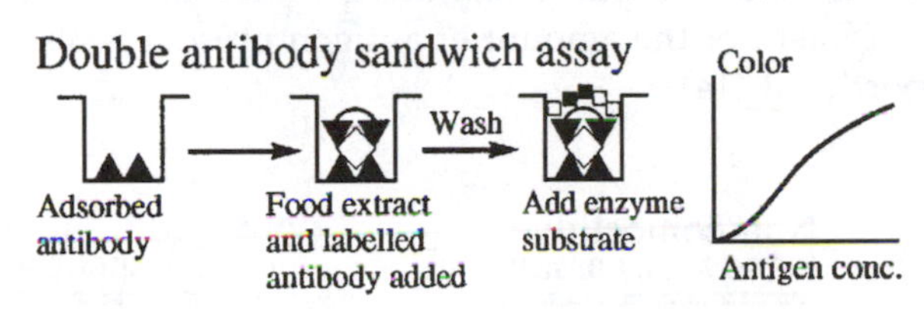

FIGURE 33-3.
Sandwich ELISA methodology. Enzyme-immunoassay formats used for quantification of food protein antigens. From (15), used with permission. Illustration from *Development and Application of Immunoassay for Food Analysis*, J.H. Rittenburg, Ed., copyright © 1990 by Elsevier Science Publishers Ltd., reprinted by permission of publisher.

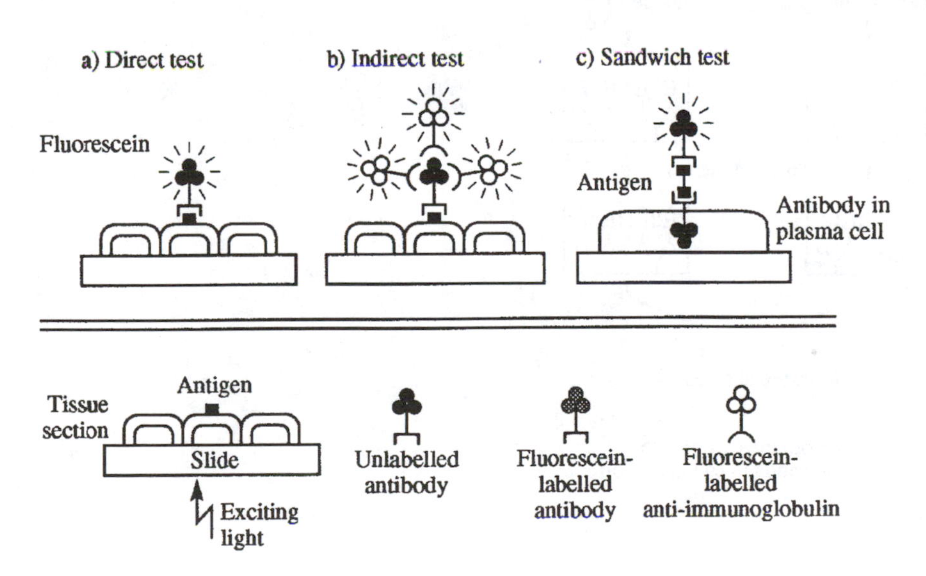

FIGURE 33-2.
Immunofluorescence methodology. The basis of fluorescence antibody tests for identification of tissue antigens or their antibodies. ● = fluorescein labeled. From (11), used with permission.

antigenic determinants. They can be raised in the same or different species. The color development is directly related to the amount of antigen present. The assay derives its name from the position the antigen occupies in the test. It is sandwiched between the unlabeled antibody attached to the solid phase and the enzyme-labeled antibody, which is added following addition of the antigen. A washing step is required between each step to remove unbound reactants.

An **indirect double-sandwich ELISA** may be developed. Here the solid-phase antibodies specific to the large protein produced in species A (e.g., rabbit) are coated into the solid phase. Serial dilutions of standards and sample(s) are added and incubated for a specified time. Unbound antigen is washed out, and a fixed amount of specific antibody from species B (e.g., mouse) is added. Following incubation of the second antibody and a washing step, a labeled antibody (anti-species B antibody) is incubated and then the excess washed out. Substrate is added and color development of the amount of antigen present is also proportional (14).

2. Competitive Direct ELISA. In this format, free hapten competes with an enzyme-labeled hapten for a number of limited antibody sites attached onto a solid phase. Unbound reactants are washed before substrate is added. Following substrate addition, the color produced is indirectly related to the amount of hapten in the test sample (Fig. 33-4).

3. Competitive Indirect ELISA. In this format, antibody is free in solution, and competition occurs between free hapten and solid-phase antigen bound to the plastic or membrane. Following an incubation step, unreacted reagents are washed away, and an antispecies antibody labeled with enzyme is added and incubated. After another washing step, substrate is added. The color produced is indirectly related to the amount of hapten present in the sample (Fig. 33-5).

33.2.2.3 Other Immunochemical Methods

33.2.2.3.1 Agglutination The principle of the **passive hemagglutination test** is based on the least amount of

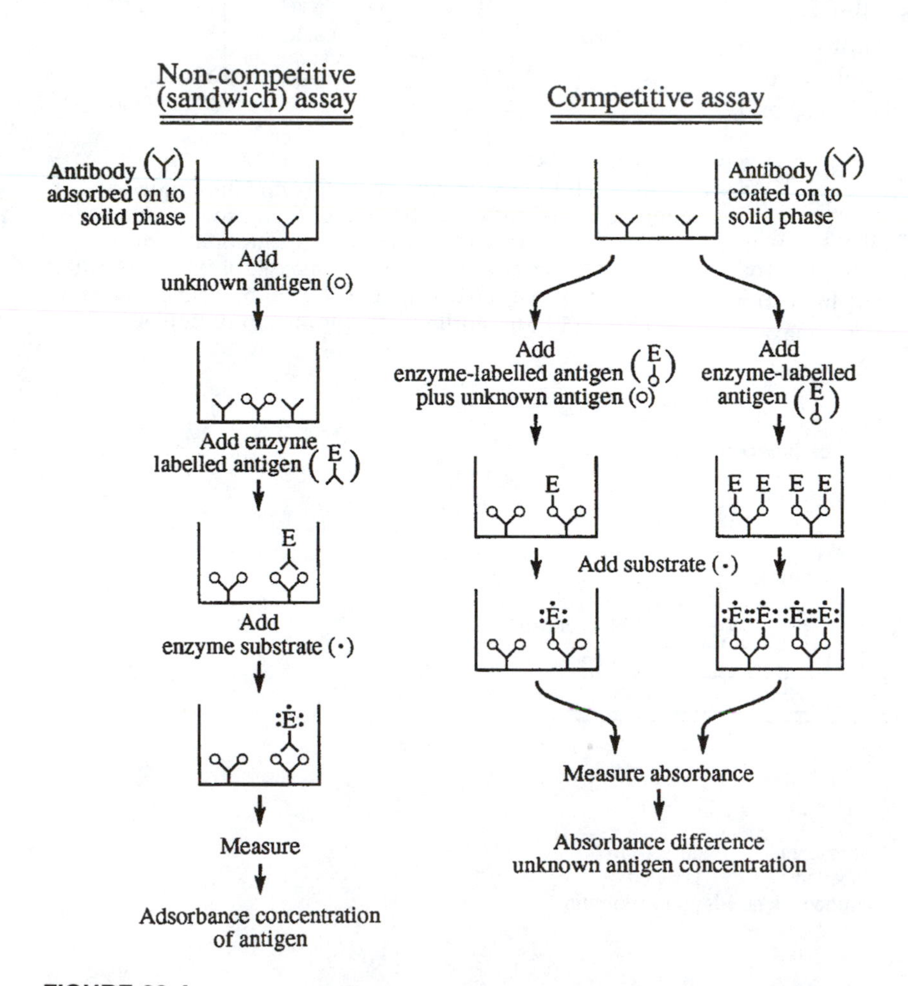

FIGURE 33-4.
Examples of direct competitive and noncompetitive enzyme-linked immunosorbent assays. From (15), used with permission. Illustration from *Development and Application of Immunoassay for Food Analysis*, J.H. Rittenburg, Ed., copyright © 1990 by Elsevier Science Publishers Ltd., reprinted by permission of publisher.

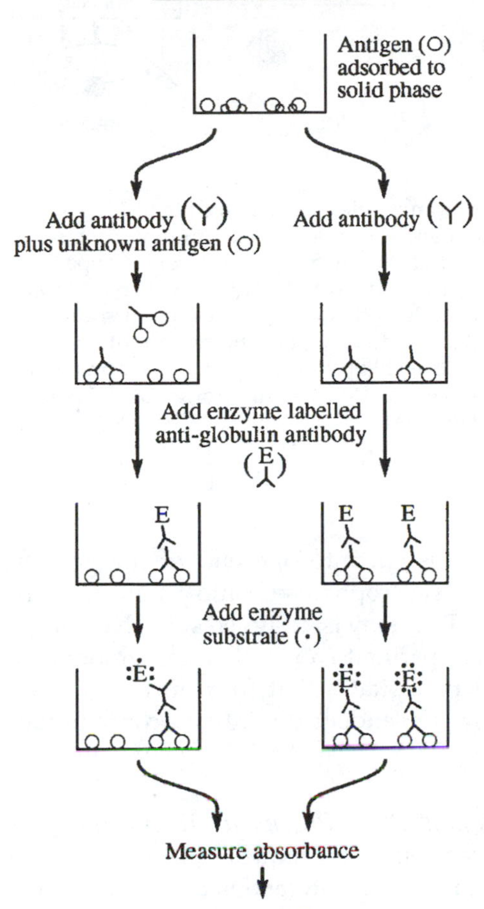

FIGURE 33-5.
Example of an indirect competitive enzyme-linked immunosorbent assay. From (15), used with permission. Illustration from *Development and Application of Immunoassay for Food Analysis*, J.H. Rittenburg, Ed., copyright © 1990 by Elsevier Science Publishers Ltd., reprinted by permission of publisher.

soluble antigen required to inhibit agglutination of red blood cells (Fig. 33-6). This is the concentration of antigen in the last tube that will give a wide ring agglutination pattern (known as a **mat**). This concentration is the amount in the last tube that will still cause a mat formation (16).

Methods based on the agglutination are semiquantitative procedures. It is typical to use a twofold dilution scheme. Therefore, the procedures can only yield results that reflect the dilution sequence. If the interval between antigen concentration were greater, then the inherent error would also be greater. However, the narrower the range, the more accurate the assay (17).

Hemagglutination assays are easy to perform and possess the desired sensitivity. Unfortunately, they require large quantities of antiserum, and some antibody

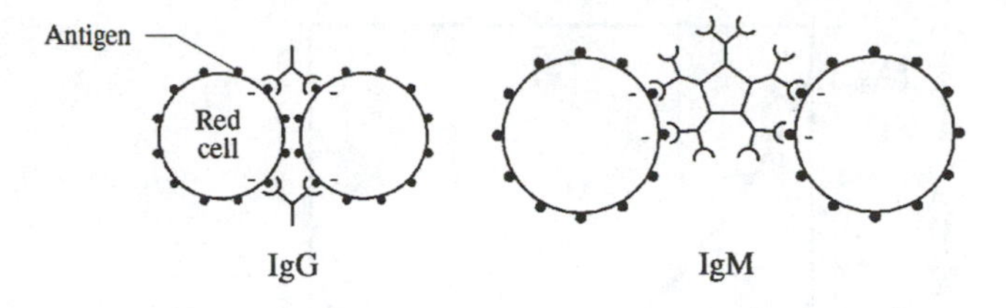

FIGURE 33-6.
Hemagglutination methodology. Mechanism of agglutination of antigen-coated particles by antibody crosslinking to form large macroscopic aggregates. If red cells are used, several crosslinks are needed to overcome the electrical charge at the cell surface. IgM is superior to IgG as an agglutinator because of its multivalent binding and because the charged cells are further apart. From (11), used with permission.

preparations do not react with red blood cells. The biggest potential disadvantage is that there may be nonspecific interaction from other proteins also capable of causing agglutination (18). Latex particles can also be used for agglutination assays.

33.2.2.3.2 Immunodiffusion **1. Single Radial Immunodiffusion.** The polyclonal antiserum prepared against the desired antigen is uniformly dispensed in an agar gel (19). Standards of known concentration of the antigen and sample extracts are placed into the appropriate wells cut into the agar. Diffusion of the standards and samples into the agar cause formation of precipitin rings. At any given time, the diameter of the rings is proportional to the initial antigen concentrations in the wells (Fig. 33-7). Microgram quantities of antigen can be detected with this method. The method requires large quantities of antiserum and is not ideal for analysis of sparingly soluble antigens. However, gels that contain urea do enable analysis of such proteins (e.g., gluten) (20).

2. Double Diffusion. Double diffusion assays are not used quantitatively; they enable the user to obtain information about the immunochemical relationship (similarities and dissimilarities) of several antigens (21). An extract may be screened against several different antisera by placing the extract in the center well and the antisera in peripheral wells (Fig. 33-8).

3. Immunoelectrophoresis. This procedure is carried out by electrophoretically separating the antigen to provide resolution of different antigenic components in the mixture prior to immunodiffusion against the antiserum (22). The antigen is electrophoresed in a gel. A trough is cut parallel to the direction of the separation. An antiserum is added to the trough and the separated antigens diffuse toward one another, which results in formation of precipitates (Fig. 33-9).

4. Rocket Immunoelectrophoresis. In this method, the antiserum is dispensed into the agarose gel (23). The antigen standards and sample extracts are added

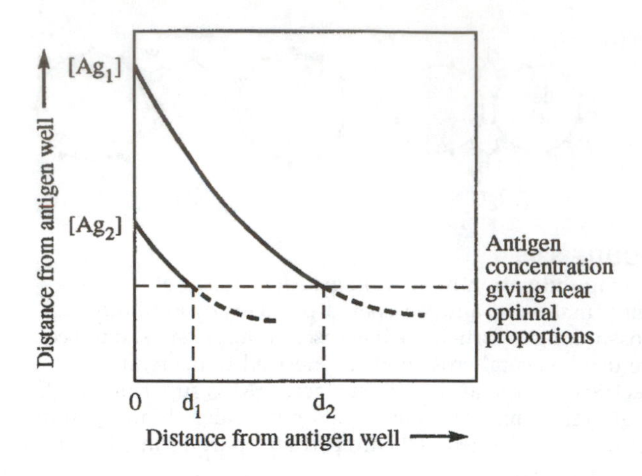

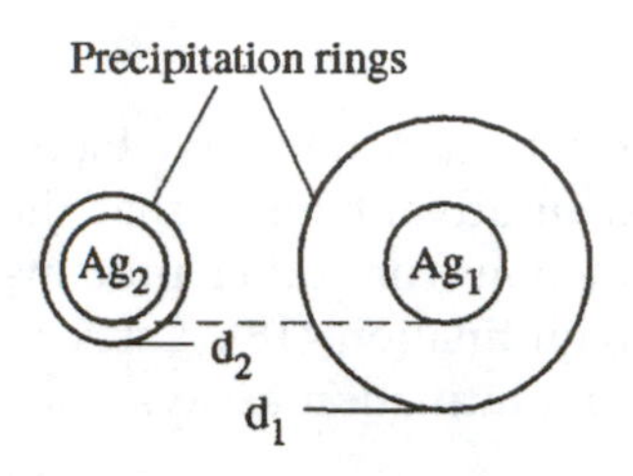

FIGURE 33-7.
Single radial immunodiffusion: relation of antigen concentration to size of precipitation ring formed. Antigen at the higher concentration (Ag_1) diffuses further from the well before it falls to the level giving precipitation with antibody near optimal proportions. From (11), used with permission.

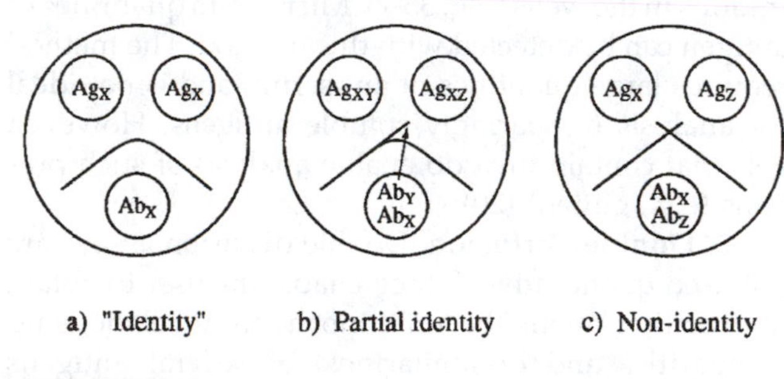

FIGURE 33-8.
Double immunodiffusion methodology. (a) Line of confluence obtained with two antigens that cannot be distinguished by the antiserum used. (b) Spur formation by partially related antigens having a common determinant x but individual determinants y and z reacting with a mixture of antibodies directed against x and y. The antigen with determinants x and z can only precipitate antibodies directed to x. The remaining antibodies (Ab_y) cross the precipitin line to react with the antigen from the adjacent well, which has determinant y giving rise to a spur over the precipitin line. (c) Crossing over of lines formed with unrelated antigens. From (11), used with permission. Illustration from *Development and Application of Immunoassay for Food Analysis*, J.H. Rittenburg, ed., copyright © 1990. Elsevier Science Publishers, Ltd (original publisher). Chapman & Hall, Ltd (current publisher).

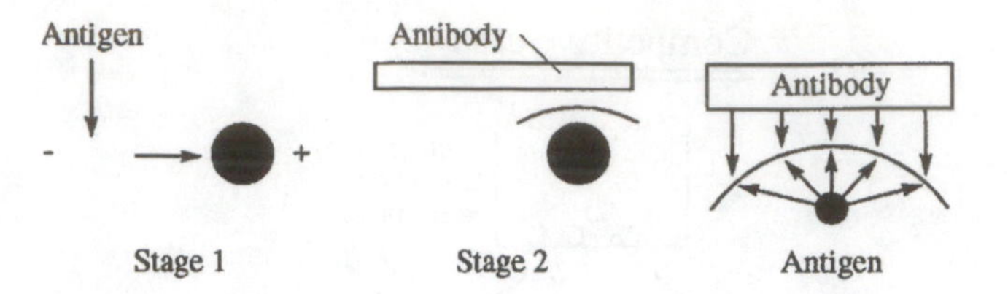

FIGURE 33-9.
The principle of immunoelectrophoresis. Stage 1: Electrophoresis of antigen in agar gel. Antigen migrates to hypothetical position shown. Stage 2: Current stopped. Trough cut in agar and filled with antibody. Precipitin arc formed. Because antigen theoretically at a point source diffuses radially, and antibody from the trough diffuses with a plane front, they meet in optimal proportions for precipition along an arc. The arc is closest to the trough at the point where antigen is in highest concentration. From (11), used with permission.

to the small wells cut into one end of the gel. The antigens then electrophorese onto the antibody-containing gel. The assay is designed such that antigen migrates with a sparing to no antibody migration. The rocket-shaped precipitates that form in the gel have heights that are dependent on antigen concentration (Fig. 33-10).

33.2.2.3.3 Quantitative Precipitin Techniques For quantitative precipitin techniques (24), an insoluble complex forms following interaction between a soluble

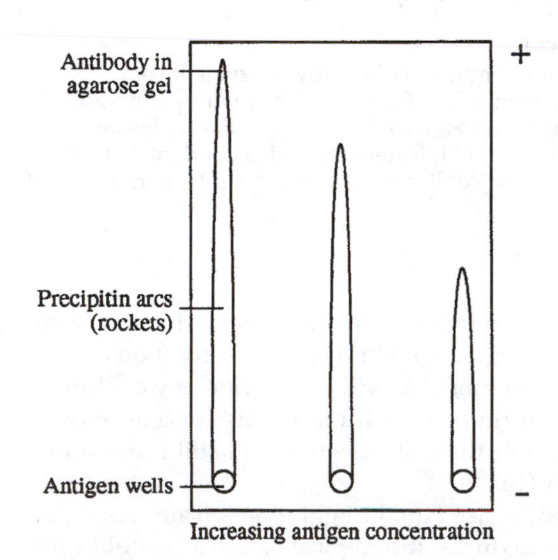

FIGURE 33-10.
Rocket electrophoresis. Antigen, in this case human serum albumin, is electrophoresed into gel containing antibody. The distance from the starting well to the front of the rocket-shaped arc is related to antigen concentration. In the example shown, human serum albumin is present at relative concentrations from left to right: 3, 2, and 1. From (47), used with permission.

antigen and specific antibodies (Fig. 33-11). This method works in solution, not in a gel. The amount of precipitate is subsequently analyzed by protein assay. A fair amount of sensitivity is achievable, but the assay requires long times to obtain maximal precipitate formation. Generation of biphasic antigen concentration versus precipitate curves is a potential problem.

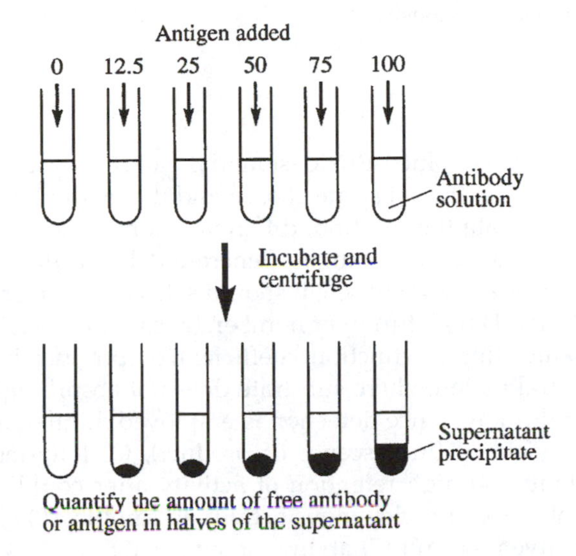

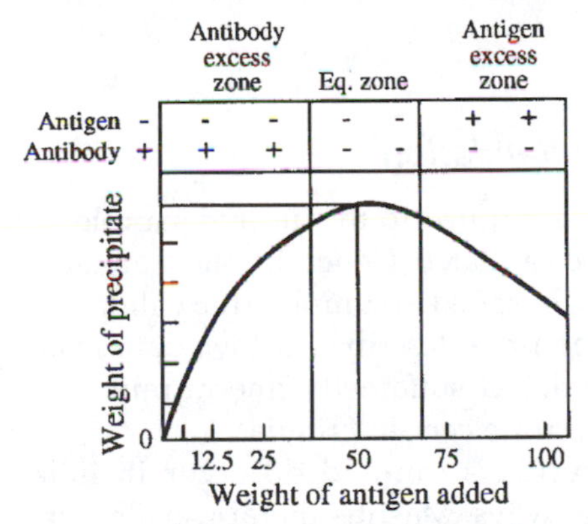

FIGURE 33-11.
Qualitative precipitin reaction. A series of tubes containing a standard quantity of antibody is prepared and to these, various amounts of antigen are added. Following incubation and centrifugation, the precipitate and supernatant are separated and the amount of antibody and antigen in equal halves of the supernatant quantified. The antibody content of the serum can be calculated from the equivalence point where no antigen or antibody is present in the supernatant. All the antigen added is therefore complexed in the precipitate with all the antibody available and the antibody content is given by subtracting the weight of antigen added from the weight of the antibody–antigen precipitated. From (15), used with permission. Illustration from *Development and Application of Immunoassay for Food Analysis*, J.H. Rittenburg, Ed., copyright © 1990. Elsevier Science Publishers, Ltd (original publisher). Chapman & Hall Ltd (current publisher).

33.2.2.3.4 Immunoaffinity Columns An immunoaffinity column is constructed by attaching antibodies with specificity for a certain analyte to a solid support (e.g., gel matrix). A sample is extracted and passed through the column. The analyte of interest binds to the antibodies and can be eluted and quantitated using fluorescent derivatization or instrumental quantitation [e.g., high-performance liquid chromatography (HPLC)].

33.2.3 Considerations for Immunoassay Development

It is not possible, due to the scope and length of this chapter, to describe in any great detail all the considerations for development of all the immunoassays. It is appropriate, however, to provide in brief detail the steps required to develop one such method. Since development and application of ELISAs for detection of small molecules important in the food industry are constantly on the rise, an overview of the general procedure for developing a direct competitive ELISA is provided, along with the procedure to validate such an assay.

33.2.3.1 Overview

The key tasks required for ELISA development for haptens are as follows: (1) Prepare a suitable immunogen, (2) immunize host animal (e.g., mouse or rabbit), (3) obtain test bleeds to titer antisera for specific antibodies, (4) develop an assay for optimizing (balancing) of antibodies and enzyme conjugate, (5) apply the test to the desired sample matrix, and (6) validate the method.

In order to elicit an immune response, the hapten must be chemically linked (*in vitro* or *in vivo*) to a carrier molecule, namely a protein. If a reactive group is not present, one must be added onto a portion of the molecule. Amino or carboxyl groups are added, which then enables the molecule to be linked to the carrier protein via an amino or carboxyl group. Commonly used carriers are BSA, KLH, and bovine gamma globulin (BGG). Reports of hemisuccinate (26) and oxime (27) derivatizations appear in the literature. Periodate is a coupling method also cited (28). The mixed anhydride reaction (29), N-hydroxysuccinimide reaction (30), two-step glutaraldehyde reaction (31–33), carbodiimide (34), and diazotization (35) have been used to couple the derivitized hapten to the carrier molecule. The advantages and disadvantages of use of these methods are described in (36).

Once a titer is observed in the test serum and the reaction is shown to be specific, the reagents are optimized to generate the best possible curve in the desired range of analyte concentration. The reagents are often first standardized in pure solutions of the hapten, and then the desired sample matrix is applied to the assay. Samples that would be analyzed for any specific analyte could be one or more of the following: tissue (e.g.,

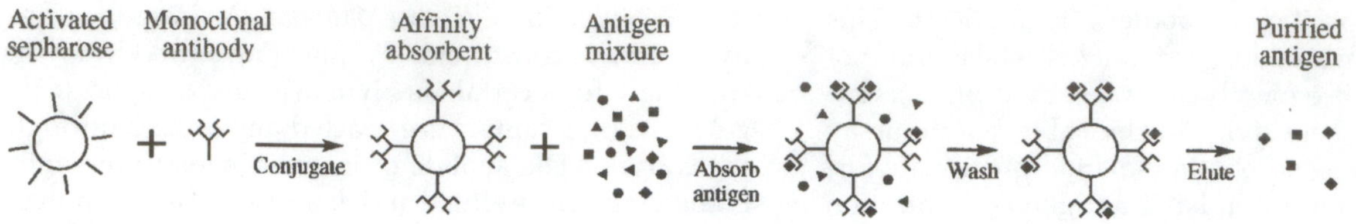

FIGURE 33-12.
Affinity chromatography. A column is filled with Sepharose-linked antibody. The antigen mixture is poured down the column. Only the antigen binds and is released, by change in pH, for example. An antigen-linked affinity column will purify antibody, obviously. From (11), used with permission.

liver, kidney, muscle), dairy products (e.g., milk, yogurt and cheese), feeds, cereal grains, or processed foods.

When an assay is optimized, the antibody concentration and enzyme-labeled conjugate concentrations are set up so that the assay operates within the linear part of the curve (absorbance versus concentration of hapten). Many assays are designed to work around a concentration that is regulated. For example, aflatoxins are regulated by the U.S. Food and Drug Administration (FDA) at 20 ppb (37). An assay for detection of this mycotoxin, whether it is used to screen for the presence or absence of the toxin, or quantitatively to determine an actual concentration, would need to accurately detect aflatoxins above and below the 20 ppb concentration.

In order to construct an assay that provides the desired sensitivity, the antibody and enzyme conjugate must work together. When an assay is constructed time is a key factor. Usually, an assay will be developed to work in the shortest time possible necessary to generate the desired color and sensitivity. Color development is extremely important when an assay is designed to be read visually. Distinct color must be observable at the control concentration as well as above and below that level.

An assay needs to be developed in a manner so that little or no color development is due to nonspecific color in the test. There are various ways antibody can be loaded onto the solid phase, and various ways plastic sites not coated with antibody can be blocked with agents to decrease or eliminate nonspecific binding of enzyme conjugate.

Optimization of an assay requires a discussion in much greater detail. Therefore, the reader is encouraged to review additional discussions on the subject (13, 14).

ELISAs can be developed using microtiter wells (e.g., polystyrene) or membranes (e.g., nitrocellulose, nylon, polypropylene) as solid phases. The membranes are immobilized onto dipsticks, cups, or disks. Choices for enzymes include horseradish peroxidase, alkaline phosphatase, glucose oxidase, and β-galactosidase. Depending on the substrate chosen and the method used to terminate the reaction, different colors (e.g., green, blue, and yellow) can be generated. Regardless of which enzyme is chosen, it should satisfy a number of criteria: (1) high turnover number, (2) easily detectable product (high extinction coefficient of product in a spectral region where substrate does not absorb light, and if fluorescence detection is employed, high quantum yield of fluorescence of product), (3) long-term stability, (4) high retention of activity after coupling, (5) absence of endogenous activity in sample, (6) cost effectiveness, and (7) abundant supply (38).

33.2.3.2 Method Validation

Once optimized and applied to the desired sample(s), the method must be validated. Generally, there are nine criteria that must be satisfied to complete the validation (39). As these criteria are described below, data from an ELISA used to detect sulfamethazine in milk (40) will be used to illustrate certain criteria.

Sulfonamide residues can and do occur in milk from any of several ways. Mastitis therapy, deliberate feeding, inadvertent feeding, and the use of sulfamethazine-containing boluses to prevent infection in cows that have calved are some of the common ways for sulfonamide residues to occur in milk. Surveys and studies have been performed that showed that 50 percent or more of the milk samples analyzed contained measurable amounts of sulfamethazine residues (32).

33.2.3.2.1 Limit of Detection (Chemical Sensitivity) This is the smallest quantity or concentration of an analyte that can be reliably distinguished from background in the test.

When determining the limit of detection for a direct competitive ELISA, the following equation can be used:

$$\text{Limit of detection} = \frac{X_o - 2SD}{X_o} \times 100\% \quad (1)$$

where: X_o = mean of the absorbance value for 0 parts per billion (ppb)
2SD = 2 times the standard deviation

The mean of the 0 ppb value is a composite value obtained from the average of a number of standard curves. For example, the limit of detection for an ELISA used to detect sulfamethazine in milk is 0.7 ppb (Table 33-1). The test enables the detection of sulfamethazine well below the safe level set by the FDA (41).

33.2.3.2.2 Crossreactivity (Chemical Specificity) This is the extent to which the assay responds to only the specified analyte and not to other compounds or substances in the sample.

The sulfamethazine antibody was found to be highly specific for sulfamethazine and did not bind to closely related sulfonamides (Table 33-2). There was 14 percent crossreactivity with sulfamerazine, which is not surprising since sulfamerazine only lacks one methyl group that sulfamethazine possesses. The antibody did not recognize penicillin or chlortetracycline, which are commonly used in combination with sulfonamides.

33.2.3.2.3 Reproducibility This is the ability of the assay to duplicate results in repeat determinations. It is the opposite of variability in the assay.

1. **Intra-Assay Variability.** This is the variability between replicate determinations in the same assay. Intra-assay variability for the sulfamethazine ELISA ranged from 6.02 to 7.70 percent (Table 33-3), with overall variability being 6.89 percent. This assay is highly repeatable.

2. **Inter-Assay Variability.** This is the variability between replicate determinations from different groups. Inter-assay variability for the sulfamethazine ELISA ranged from 2.84 percent to 17.44 percent, with an overall value of 8.54 percent. A coefficient of variation (CV) below 10 percent is very good. Chemical methods often have CVs of ±50 percent when measuring in the ppb range (42).

33.2.3.2.4 Reference Correlation This is the degree of closeness of the linear relationship between the results obtained using the ELISA and a reference assay [e.g., HPLC or thin-layer chromatography (TLC) versus ELISA] over the range of the test.

When 64 milk samples were analyzed for sulfamethazine by the ELISA and HPLC, a correlation of 94.9 percent was obtained (Table 33-4). Excellent agreement between immunoassays and chemical methods is possible.

33.2.3.2.5 Sensitivity This is the ability of the test to detect positive samples as positive. It is the percent positivity in a population of true positives.

When 74 raw milk samples were analyzed for sul-

TABLE 33-1. Limit of Detection for the Direct Competitive ELISA for Sulfamethazine in Milk

Concentration of Sulfamethazine (ppb)						
ABSORBANCE OF 650 nm						
0	*2.5*	*5*	*10*	*25*	*50*	
0.735	0.620	0.544	0.367	0.209	0.122	
0.774	0.614	0.540	0.377	0.205	0.124	
0.720	0.568	0.521	0.347	0.196	0.117	
0.724	0.560	0.493	0.352	0.192	0.137	
0.704	0.560	0.467	0.364	0.215	0.137	
0.734	0.594	0.510	0.421	0.239	0.174	
0.732	0.586	0.513	0.371	0.209	0.135	Mean

$$\text{Limit of Detection} = \frac{X_o - 2SD}{X_o} \times 100\%$$

$$= \frac{0.732 - 2(0.0215)}{0.732} \times 100\%$$

$$= 94.1\%$$

94.1 percent absorbance corresponds to 0.7 ppb concentration of sulfamethazine when graphed on logit-log paper.

From (40), used with permission.

TABLE 33-2. Percent Relative Crossreactivity of Anti-Sulfamethazine Polyclonal Antibody Determined Using Sulfamethazine and Other Drugs

Compound	*Concentration Required to Inhibit 50% of Antibody Binding (ppb)*	*Percent Crossreactivity*[1]
sulfamethazine	90	100
sulfamerazine	630	14
sulfathiazole	>50,000	<0.18
sulfapyridine	>50,000	<0.18
sulfaquinoxaline	>50,000	<0.18
sulfachloropyridazine	>50,000	<0.18
sulfadimethoxine	>50,000	<0.18
sulfanilamide	>50,000	<0.18
chlorotetracycline HCl	>50,000	<0.18
procaine penicillin	>50,000	<0.18
p-amino benzoic acid	>50,000	<0.18

From (40), used with permission.

$$^{1}\text{Relative percent crossreactivity} = \frac{\text{Concentration of sulfamethazine required to inhibit 50\% of antibody binding (ppb)}}{\text{Concentration of other antimicrobial agent required to inhibit 50\% of antibody binding (ppb)}} \times 100$$

TABLE 33-3. Reproducibility of the Direct Competitive ELISA for Detection of Sulfamethazine Residues in Milk

	ng SULFAMETHAZINE/ml (ABSORBANCE at 650 nm)						
Replicate	*0*	*2.5*	*5*	*10*	*25*	*50*	$\bar{x}$
1A	0.735	0.620	0.544	0.367	0.209	0.122	
1B	0.720	0.568	0.521	0.347	0.196	0.117	
1C	0.774	0.614	0.540	0.377	0.205	0.124	
1D	0.724	0.560	0.493	0.352	0.192	0.137	
1E	0.720	0.554	0.535	0.416	0.215	0.127	
1F	0.751	0.649	0.556	0.419	0.244	0.190	
$\bar{x}$	0.737	0.594	0.532	0.380	0.210	0.136	
S.D.	0.020	0.035	0.020	0.028	0.017	0.025	
C.V.	2.66	5.89	3.76	7.37	8.10	18.4	7.70
2A	0.704	0.632	0.540	0.429	0.282	0.176	
2B	0.720	0.598	0.520	0.381	0.211	0.130	
2C	0.722	0.587	0.522	0.380	0.226	0.147	
2D	0.751	0.632	0.500	0.393	0.208	0.142	
2E	0.734	0.594	0.510	0.421	0.239	0.174	
$\bar{x}$	0.726	0.609	0.518	0.401	0.233	0.154	
S.D.	0.016	0.019	0.013	0.020	0.027	0.018	
C.V.	2.20	3.12	2.50	4.99	11.59	11.69	6.02
3A	0.767	0.647	0.483	0.340	0.221	0.162	
3B	0.725	0.665	0.546	0.434	0.297	0.206	
3C	0.769	0.697	0.561	0.405	0.231	0.155	
3D	0.754	0.596	0.458	0.370	0.224	0.134	
$\bar{x}$	0.754	0.639	0.500	0.387	0.243	0.200	
S.D.	0.018	0.026	0.053	0.036	0.031	0.164	
C.V.	2.39	4.07	10.60	9.30	12.75	2.60	6.95
$\bar{x}$ Singles	0.738	0.611	0.522	0.388	0.227	0.149	
S.D. Singles	0.021	0.034	0.028	0.030	0.028	0.026	
C.V. Singles	2.84	5.56	5.36	7.73	12.33	17.44	8.54

From (40), used with permission.
$\bar{x}$ = mean
S.D. = Standard Deviation
C.V. = Coefficient of Variation

TABLE 33-4. Reference Correlation of High-Performance Liquid Chromatography (HPLC) and Direct Competitive Enzyme-Linked Immunosorbent Assay (ELISA) Analyzing Raw Milk Samples for Sulfamethazine Residues

Sample	HPLC (ppb)	ELISA (ppb)
1	5.3 (5.25)	3.9
2	9.0	7.3
3	13.43	12.4
4	20.55	19.4
5	<1.0	2.0
6	4.20	3.5
7	7.82	8.7
8	16.37	18.6
9	<1.0	0
10	3.42	3.5
11	7.70	8.0
12	15.65	19.6
13	<1.0	0
14	4.94	4.6
15	8.41	8.7
16	15.17	19.2
17	<1.0	0
18	4.14	4.1
19	8.77	7.2
20	16.48	13.4
21	<1.0	0
22	4.52	5.9
23	9.43	9.1
24	18.57	19.6
25	1.7	0
26	5.94	0
27	10.48	8.1
28	19.67	18.0
29	<1.0	0.7
30	5.03	4.0
31	9.38	8.2
32	15.68	18.1
33	<1.0	1.2
34	4.45	5.2
35	8.26	7.9
36	15.96	14.7
37	1.26	0
38	5.43	4.8
39	9.38	12.6
40	17.12	18.1
41	<1.0	0
42	4.75	5.0
43	9.09	9.0
44	16.70	14.1
45	<1.0	0
46	4.73	3.4
47	8.70	7.7
48	16.68	14.2
49	<1.0	1.7
50	5.03	4.6
51	8.84	8.9
52	17.14	16.7
53	1.04	0
54	5.17	3.7
55	9.19	7.3
56	17.70	12.9
57	<1.0	0
58	4.91	3.7
59	9.60	6.8
60	17.91	12.7
61	<1.0	2.6
62	5.36	5.5
63	9.49	13.0
64	17.7	22.9

Correlation Coefficient = .949
Intercept = .529
Slope = 1.003

From (40), used with permission.

famethazine by ELISA, the sensitivity was determined to be 20/21 = 95.2 percent (>10 ppb sulfamethazine is positive).

33.2.3.2.6 Specificity This is the ability of the test to detect negative samples as negative. It is the percent test negativity in a population of true negatives.

When the 74 raw milk samples were analyzed for sulfamethazine by ELISA, the specificity was determined to be 52/53 = 98.1 percent (<10 ppb sulfamethazine is negative).

33.2.3.2.7 Overall Accuracy This is the combined or total ability of the test to correctly detect positive and negative samples (sensitivity and specificity = overall accuracy).

The overall accuracy for the sulfamethazine ELISA was determined to be 20/21 + 52/53 = 72/74 = 97.3 percent, which demonstrates that ELISAs can produce highly accurate results.

33.2.3.2.8 Stability This is the usable shelf life of the kit for specified storage conditions. It can be determined by accelerated aging studies and confirmed by real-time testing. This validation point, is, of course, important if a method is to be commercialized into a test kit format (43). Many ELISAs that are commercially available in kits are stable for 6 months to 1 year when stored at 4°C. Some tests can be stored at room temperature.

33.2.3.2.9 Ruggedness It is common for ruggedness testing to be included in the validation. An example of ruggedness testing for an ELISA would be to determine test performance over a range of temperatures (e.g., extreme temperatures). This might be a concern when a test is used in the field in a setting exposed to seasonal temperature fluctuations (e.g., slaughterhouse).

33.3 APPLICATIONS

Immunoassays and particularly ELISAs are finding increasing application every year for food analysis. A review article of this nature cannot even begin to summarize, let alone discuss in any great deal, each type of immunoassay and how it is being used. There are numerous books and lengthy review articles that describe the drawbacks to using chemical methods (e.g., TLC, HPLC, gas chromatography–mass spectrometry) as well as older immunoassays (e.g., gel diffusion, agglutination) for routine sample analysis (15, 44–46). These articles and books also describe in great detail the development and application of ELISAs for analysis of foods, whether the samples comprise food components, are naturally occurring contaminants, or are by-products of crop or livestock production. The literature describes the potential interferences in samples and discusses which samples are more difficult to analyze.

The references cited above discuss not only assay development and application but also issues concerning use of ELISAs by regulatory agencies for sample analysis. There are also discussions concerning implementation of ELISA testing for uses other than research purposes (e.g., quality control testing, routine sample analysis, regulatory testing, field versus laboratory setting).

Also provided in the references are lists of tests that can be purchased commercially for testing various commodities. Integration of ELISA kits and other immunoassays into routine testing programs is an ever-increasing occurrence. In many instances, immunoassays are used for initial screening, and then classical microbiological tests are used for confirmation. There are more and more immunoassays to choose from that enable the user to obtain quick and accurate results for a modest price.

33.4 SUMMARY

It is obvious upon review of this chapter that a great number of immunoassays exist that can be used for analysis of large and small molecules. All have advantages and potential disadvantages that must be considered when selecting a method for analysis. Numerous factors must also be considered in developing and validating any immunoassay method.

In an era when questions are raised on a regular basis that focus on the safety of the food supply, when consumers are aware and concerned about the food they are buying, when large numbers of samples need to be tested on a routine basis, the need is great and the application of immunoassays, mainly ELISAs, is timely. While immunoassays were once a tool used only for analysis of clinical samples, development and use of immunoassays for detection of food constituents, additives, natural contaminants, growth promoters, and the like are becoming more and more commonplace in the testing scenario. With the advent of commercial kits, testing can be handled efficiently and effectively for a minimal cost. It is an area of explosive growth, with newer, more sensitive, and more specific methods available for sample analysis. Consumers can rest assured that implementation of immunological testing tools will monitor the safety of their food supply in a more effective manner than has ever before been possible.

33.5 STUDY QUESTIONS

1. a. What are antibodies?
   b. Describe the difference between monoclonal antibodies and polyclonal antibodies.
2. a. What is an antigen?
   b. What is a hapten?
   c. Describe the procedure that must be done to prepare a conjugate antigen.
3. a. What does the acronym RIA stand for?
   b. Describe the method as well as the potential advantages and disadvantages of using this method.
4. a. What is a homogeneous enzyme immunoassay?
   b. What is a heterogenous assay?
   c. Is an ELISA a homogeneous or heterogeneous assay?
5. a. What does the acronym ELISA stand for?
   b. Describe the three types found in the text.
   c. Why can this type of assay be used in field-type situations?
6. Describe how an immunoaffinity column could be used to purify a substance from a food extract.
7. List the key tasks required for development of an ELISA for detection of a low molecular weight compound.
8. Describe the eight criteria required to validate a method.

3.6 REFERENCES

1. Kohler, F., and Milstein, C. 1975. Continuous cultures of predefined specificity. *Nature* 256:495–497.
2. Nisonoff, A., Hooper, J.E., and Spring, S.B. (Eds.) 1975. *The Antibody Molecule.* Academic Press, New York.
3. Gally, J.A. 1973. Structure of immunoglobulins, Ch. 2, in *The Antigens,* M. Sela (Ed.) pp. 161–175. Academic Press, New York.
4. Landsteiner, K. (Ed.) 1962. *The Specificity of Serological Reactions,* 3rd ed., Dover, New York, NY.
5. Kabat, E.A. 1956. Heterogeneity in extent of the combining regions of human anti-dextran. *J. Immunol.* 77: 377–385.
6. Schechter, B., Schechter, I., and Sela, M. 1970. Specific fractionation of antibodies to peptide determinants. *Immunochem.* 7:587–599.
7. Sela, M. 1969. Antigenicity: Some molecular aspects. *Science* 166:1365–1370.
8. Jerne, N.K. 1960. Immunological speculations. *Annu. Rev. Microbiol.* 14:341–345.
9. Luft, R., and Yalow, R.S. 1974. Radioimmunoassay meth-

odology and application, in *Physiology and Clinical Studies*. George Thieme Verlag, Stuttgart.
10. Eisen, H. (Ed.) 1980. *Immunology: An Introduction to Molecular and Cellular Principles of the Immune Responses*, 2nd ed. Harper & Row, Hagerstown, MD.
11. Roitt, I., Costoff, J.R., and Male, D.K. 1989. Immunological techniques, Ch. 5, in *Essential Immunology*, Blackwell Scientific Publications Ltd., Oxford, England.
12. Munro, A.J., Landon, J., and Share, E.J. 1982. The basis of immunoassays for antibiotics. *Anti. Microb. Chemother.* 9:423–432.
13. Burdon, R.H., and Van Knippenberg, P.H. (Eds.) 1985. *Practice and Theory of Enzyme Immunoassays*, Vol. 15. Elsevier, New York, NY.
14. Voller, A., Bidwell, D., and Bartlett, A. 1979. *The Enzyme-Linked Immunosorbent Assay (ELISA)*. Dynatech Labs., Inc., Chantilly, VA.
15. Rittenburg, J.H. (Ed.) 1990. *Development and Application of Immunoassay for Food Analysis*. Elsevier Applied Science, New York, NY.
16. Steiner, S.J. 1981. The development of a model immunological assay system for the detection of antibiotic residues. Ph.D. dissertation. Rutgers University, New Brunswick, NJ.
17. Dixon, D.E., Steiner, S.J., and Katz, S.E. 1986. Immunological approaches, Ch. 11, in *Modern Analysis of Antibiotics*, Vol. 27, A. Azsolalos (Ed.), pp. 415–431. Marcel Dekker, Inc., New York.
18. Kang'ethe, E.K. 1990. Use of immunoassays in monitoring meat protein additives, Ch. 5, in *Development and Application of Immunoassays for Food Analysis*, J.H. Rittenburg (Ed.), pp. 127–139. Elsevier Applied Science, New York, NY.
19. Mancini, G., Carbonara, A.O., and Heremans, J.F. 1965. Immunochemical quantitation of antigens by single radial immunodiffusion. *Immunochem.* 2:235–254.
20. Ouchterlony, O. 1949. Antigen-antibody reactions in gels. *Acta Pathol. Microbiol. Scand.* 26:507–515.
21. Skerritt, J.H. 1990. Immunoassays of non-meat protein additives in food, Ch. 4, in *Development and Applications of Immunoassays for Food Analysis*, J.H. Rittenburg (Ed.), pp. 81–125. Elsevier Applied Science, New York, NY.
22. Grabar, P., and Williams, C.P. 1953. Méthode permettant létude conjugée des propriétés électrophorétiques et immunochimiques d'un mélange de proteines. *Biochim. Biophys. Acta* 10:193–194.
23. Laurell, C.B. 1966. Quantitative estimation of proteins by electrophoresis in agarose gel containing antibodies. *Anal Biochem.* 15:45–52.
24. Heidelberger, M., and Kendall, F.E. 1932. Quantitative studies on the precipitin reaction. Determination of small quantities of a specific poly saccharides. *J. Exp. Med.* 55:555–561.
25. Garvey, J.S., Cremer, N.E., and Sussdorf, D.H. 1977. *Methods in Immunology*, 3rd ed. W.A. Benjamin, Inc., Advanced Book Program, London, England.
26. Casale, W.L., Pestka, J.J., and Hart, L.P. 1988. Enzyme-linked immunosorbent assay employing monoclonal antibody specific for deoxynivalenol (vomitoxin) and several analogues. *J. Agri. Food Chem.* 36:663–668.
27. Dixon, D.E., Warner, R.L., Ram, B.P., Hart, L.P., and Pestka, J.J. 1987. Hybridoma cell line production of a specific monoclonal antibody to the mycotoxins zearalenone and α-zearalenol. *J. Agric. Food Chem.* 353:122–126.
28. Berkowitz, D.B., and Webert, P.W. 1986. Enzyme immunoassay based survey of prevalence of gentamicin in serum of marketed animals. *J. Assoc. Anal. Chem.* 69: 437–440.
29. Campbell, G.S., Mageau, R.P., Schwab, B., and Johnston, R.W. 1984. Detection and quantitation of chloramphenicol by competitive enzyme-linked immunosorbent assay. *Antimicrob. Agents and Chemother.* 25:205–211.
30. Schmidt, D.J., Clarkson, C.E., Swanson, T.A., Egger, M.L., Carlson, R.E., Van Emon, J.E., and Karu, A.E. 1990. Monoclonal antibodies for immunoassay of avermectins. *J. Agric. Food Chem.* 38:1763–1770.
31. Dixon-Holland, D.E., and Katz, S.E. 1989. Use of a competitive direct enzyme-linked immunosorbent assay on detection of sulfamethazine residues in swine tissue and urine. *J. Assoc. Off. Anal. Chem.* 71:1137–1140.
32. Dixon-Holland, D.E., and Katz, S.E. 1989. Direct competitive enzyme-linked immunosorbent assay on detection of sulfamethazine residues in milk. *J. Assoc. Offic. Anal. Chem.* 71:447–450.
33. Dixon-Holland, D.E., and Katz, S.E. 1991. A competitive direct enzyme-linked immunosorbent screening assay for the detection of sulfonamide contamination of animal feeds. *J. Assoc. Offic. Anal. Chem.* 74:784–789.
34. Fleeker, J.R., and Lovelt, L.J. 1985. Enzyme immunoassay for screening sulfamethazine residues in swine blood. *J. Assoc. Offic. Anal. Chem.* 68:172–174.
35. Singh, P., Ram, B.P., and Sharkov, N. 1989. Enzyme immunoassay for screening of sulfamethazine in swine. *J. Agric. Food Chem.* 37:109–113.
36. Erlanger, B.F. 1973. Principles and methods for the preparation of drug protein conjugates for immunological studies. *Pharmacol. Reviews* 25:271–277.
37. Porterfield, R.I., and Cupone, J.J. 1984. Accelerated aging studies. *MD and DI*, April, p. 45.
38. Albini, B., and Andres, A. 1978. Immunoelectronmicroscopy, Ch. 30, in *Principles of Immunology*, N.R. Rose, F. Milgram, and C. Van Oss (Eds.), 2nd ed. MacMillan, Los Angeles, CA.
39. O'Rangers, J.J. 1990. Development of drug residue immunoassays, Ch. 4, in *Immunochemical Methods for Environmental Analysis*, J.M. Van Emon and R.O. Mumma (Eds.), pp. 25–37. American Chemical Society, Washington, D.C.
40. Dixon-Holland, D.E. 1990. Unpublished data. Research and Development, Neogen Corporation, Lansing, MI.
41. Fed. Regist. 1977. 426-2211, U.S. Government Printing Office, Washington, D.C.
42. Horwitz, W. 1981. Review of analytical foods and feeds. I. Review in Methodology. *J. Assoc. Off. Anal. Chem.* 64:104–130.
43. Fed. Regist. 1990. Unavoidable Contaminants in Food for Human Consumption and Food Packaging Material, 21 CFR109. U.S. Government Printing Office, Washington, D.C.
44. Dixon-Holland, D.E. 1992. Immunological methods, in *Encyclopedia of Food Science and Technology*, H.Y. Hui (Ed.), pp. 1452–1475. Wiley and Sons, New York, NY.
45. Samarajeewa, U., Wei, C.I., Huang, T.S., and Marshall,

M.R. 1991. Application of immunoassay in the food industry. *Critical Reviews in Food Science and Nutrition* 29:403–434.

46. Van Emon, J.M., and Mumma, R.O. (Eds.) 1990. *Immunochemical Methods for Environmental Analysis*. American Chemical Society, Washington, D.C.

47. Weir, D.M., Herzenberg, L.A., Blackwell, C., and Herzenberg, Leonore A. (Eds.). 1986, *Handbook of Experimental Immunology in Four Volumes, Vol. 1: Immunochemistry*. 4th ed., p. 32.30. Blackwell Scientific Publications, Oxford, England.

CHAPTER

34

Thermal Analysis

Eugenia A. Davis

34.1 INTRODUCTION

Thermal analysis is a broad term that encompasses numerous techniques by which chemical or physical changes of a substance, either alone or in the presence of other substances, are measured as a sample is subjected to a controlled temperature program versus time. Food products or food components are chemically complex mixtures. They may include small molecules such as sugars with dextro- or levorotatory forms, and the chemical properties, solubilities, and reactivities can be quite different in their different forms. Sometimes the molecular constituents can form different phases in water at different water contents and temperatures. An example would be the micellar states of surfactants. Other constituents are natural polymers, such as amylose and amylopectin from starch, or actin and myosin from muscle tissue.

The ultimate chemical change that can be monitored in a food system by thermal analysis is the total combustion of a product in order to determine its mineral and caloric contents. This is usually determined by bomb calorimetry and is achieved by uncontrolled heating of the material in an O_2 atmosphere once this material is ignited.

There are many methods of thermal analysis that involve highly programmed temperature regimes, during which measurements can include:

1. Temperatures of transition (temperature when a phase change or chemical reaction begins, reaches a peak, or ends)
2. Heat capacity changes
3. Weight losses or gains
4. Energies of transition or enthalpic changes (ΔH)
5. Dimensional changes
6. Viscoelastic property changes during phase changes or chemical reactions
7. Changes in electrical polarization
8. Evolved gases

The most popular modern thermal analysis techniques are those that follow dynamically (a sequence) of physicochemical changes that a substance undergoes during heating or cooling. **Differential scanning calorimetry** (DSC), **differential thermal analysis** (DTA), and **thermogravimetric analysis** (TGA) are three common dynamic thermoanalytical techniques. More recently, **dynamic mechanical thermal analysis** (DMTA) has been developed as well. There are specialized forms of DSC, such as modulated DSC and polarization DSC, that are considered to be beyond the scope of this chapter and are mentioned to alert the reader to their existence.

TGA measures changes in weight of a sample as a function of temperature; both losses and gains can occur. TMA measures changes in penetration, extension, expansion, or contraction as a function of temperature. DMTA responds to polymer chain movement at a fixed or changing temperature. The mechanical damping measures the energy lost as the material flexes with a certain amplitude. The resonant frequency gives the elastic modulus. With proper instrumentation, the technique can be very sensitive. Such measurements are important in studies where maintaining or losing rigidity is important.

DTA and DSC have been the most commonly used methods of thermal analysis in food science. They measure the differential temperature or heat flow to or from a sample versus a reference material, and this is displayed as a function of temperature or time. These techniques can differentiate between endothermic and exothermic reactions. The types of **thermal transitions** and what they can represent are summarized below:

1. Endothermic curves usually relate to physical rather than chemical changes.
   a. A sharp endotherm is indicative of crystalline rearrangement, fusion, or solid state transitions for pure materials.
   b. A broad transition is indicative of transitions that relate to dehydration, temperature dependent phase behavior, or polymer melt.
2. Exothermic transitions that relate to reactions without decomposition can be caused by a decrease in enthalpy of a phase or chemical system.
   a. Narrow exotherms can result from crystallization (ordering) of a metastable system, whether undercooled organic, inorganic, amorphous polymer or liquid, or annealing of stored energy resulting from mechanical stress.
   b. Broad exotherms can result from solid-solid phase transitions, chemical reactions, polymerization, or curing of polymers.
3. Exothermic transitions that relate to decomposition can be narrow or broad, depending on kinetic behavior.
4. A change in heat capacity of a material at a glass transition can be seen simply as a small change in heat capacity (change in the base line) with no well-defined peak being produced.

Because DSC and DTA are very commonly used in thermal analysis of foods, the remainder of this chapter will focus mainly on these techniques (especially DSC).

34.2 PRINCIPLES AND PROCEDURES

34.2.1 General Principle of Calorimetry

Calorimetry involves the measurement of temperature or heat, more specifically, the determination of the temperature or the quantity of heat absorbed or given off when a definite amount of material undergoes a specific physicochemical change. The chemical changes in foods are coupled with energy transformations:

1. **Oxidation** is characterized by a net amount of heat being released during a temperature rise in the sample. These are called **exothermic** reactions.
2. **Hydrolysis** is characterized by little or no heat evolution. Reactions are almost **isothermic.**
3. **Reduction** is characterized by a net amount of heat absorbed during a temperature rise. Reactions are **endothermic** (energy is taken up by the sample). The net amount of heat absorbed or released by a sample can be measured quantitatively in a **calorimeter.** Units of thermal energy include:

$$\begin{aligned} 1 \text{ cal/g} &= 1.8 \text{ Btu/lb} \\ &= 0.001 \text{ Kcal (Cal)/g} \\ &= 4.186 \text{ J/g} \end{aligned} \quad (1)$$

Temperature is measured in °F, °C, or K:

$$\begin{aligned} T(°C) &= [T(°F) - 32] \cdot 5/9; \; T(K) \\ &= T(°C) + 273.16 \end{aligned} \quad (2)$$

34.2.2 Dynamic Thermal Analyzers

Dynamic thermal analyzers are special calorimeters (e.g., DSCs, DTAs), where a test sample and an inert material are used. Both test and reference samples are heated or cooled at the same time under identical conditions. The temperature of the test sample will either be higher or lower than that of the reference, depending on whether the reaction or change is net exothermic or endothermic, respectively. If no difference in temperature exists between test sample and reference, then an isothermal state exists. When a nearly isothermal state exists, a small temperature differential between test sample and reference may indicate slight differences in heat capacity and thermal conductivity or differences in the weight and density of the two materials. Thus, **dynamic thermal analysis** can monitor a property change in a sequence. Such changes can include phase transitions, molecular conformational changes, interactions with other constituents, or pyrolytic degradation.

As mentioned earlier, until recently the most common dynamic thermal analyzers used in food science are differential analyzers such as DSCs and DTAs. Therefore, it is important to explain the differences between them, as well as the type of data each one can deliver.

Standardization of DSC and DTA nomenclature began in 1965 with the **International Confederation for Thermal Analysis** (ICTA). ICTA also developed a set of standards with well-characterized melting temperatures, such as indium, melts, and heat capacities, such as sapphire crystals, to calibrate thermal analysis equipment, and the U.S. National Bureau of Standards (now called the National Institute of Standards and Technology) began marketing these standards in 1971. ICTA has defined DTA (1) as "a technique for recording the difference in temperature between a substance and reference material against either time or temperature as the two specimens are subjected to identical temperature regimes in an environment heated or cooled at a controlled rate." The data are obtained as a DTA curve with an ordinate axis in **temperature difference** (ΔT) between sample and reference materials. The abscissa is **time or temperature.** A downward peak is an endotherm, while an upward peak is an exotherm. Even though this is the ICTA's recommendation to record data, one often sees the thermographs inverted. Therefore, care must be taken to determine whether a downward peak is endothermic or exothermic.

The amount of heat or energy that is absorbed or evolved to attain a physical or chemical change can be calculated from the area between the curve and an appropriate baseline. In DTA, there is a slight difference in heat capacity and thermal conductivity between sample and reference that causes temperature detections to be slightly different, with increasing temperature even when there is not a physical or chemical change occuring. Therefore, the DTA cannot be used to obtain heat capacity data directly. Sometimes, a calibration constant can be used as a function of temperature with the use of computerized transformations to achieve such information.

The ICTA definition of DSC (1) is "a technique for recording the energy necessary to establish a zero temperature difference between a substance and a reference material against either time or temperature, as the two specimens are subjected to identical temperature regimes in an environment heated or cooled at a controlled rate." Based on this definition, one can see that the sample and the reference cells must have separate heating elements and temperature sensors. In order to maintain the same temperature between sample and reference, the resistance of the temperature sensor must be changed to influence the rate by which heat is supplied to the test and reference, respectively. The DSC curve represents, on the ordinate axis, the **rate of energy absorption** by the sample relative to the reference. Like for DTA, the abscissa is **time** or **temperature.** The heat capacity of the sample will add algebraically to the rate of the heat enthalpy for the endothermic or exothermic process being monitored.

In Fig. 34-1 are found the schematics of the temperature sensors and resistance heat sources in the sample

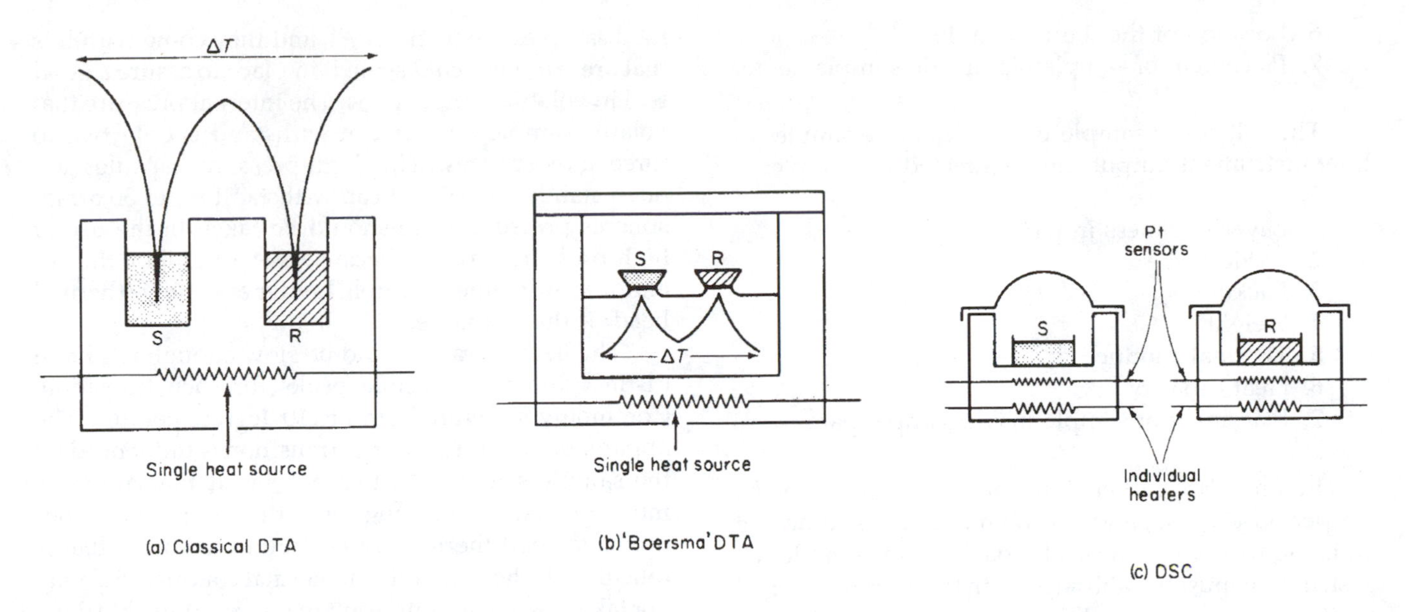

FIGURE 34-1.
Schematics of typical configurations in differential thermal analyzers: (a) classical differential thermal analyzer (DTA), (b) Boersma DTA, (c) differential scanning calorimeter analyzer (DSC). From (2), used with permission. Figure courtesy of Perkin-Elmer Corporation, Norwalk, CT.

and reference areas of the calorimeter (2). The first DTA was made by Robert-Austen in 1899. It took about 50 years to design a practical DTA and determine that a DTA peak gave a direct measure of the heat required to affect a physicochemical change in a material (3).

The general design of the classical DTA can be found in Fig. 34-1a, where the thermocouples are embedded in the sample and a single heat source is used. Boersma (4) modified the classical DTA. The differences can be seen in Fig. 34-1b, where the thermocouples are embedded in the blocks containing the sample and reference cells, but still having one heater.

In 1964, Perkin-Elmer introduced the first commercial DSC (5). The sample and reference areas can be seen schematically in Fig. 34-1c. This technique involves recording the difference in energy flux necessary to establish a **zero temperature difference** between a substance and reference material against either time or temperature when both are heated and cooled. The heating or cooling takes place at a controlled rate. Separate heating and temperature sensing devices are used to attain this (usually temperature sensitive resistors control the rate of heat flow). The exact rates that the heat is supplied to the test and reference areas, in order to achieve a given heating rate, depend on the heat capacities of the sample and reference.

Proponents of DSC or DTA analyses argue as to the relative advantages of one over the other, although with updated instrumentation containing microprocessors, computerized transformations can be made to give similar data. DTAs ideally cover wider temperature ranges, while DSCs are better at very low heating rates. In favor of DSC, one instrument constant can be used across all temperatures since sample and reference are maintained at the same temperature. To calculate enthalpy changes from DTA data, a calibration constant is needed, which is a function of temperature and is influenced by the heat capacities of the sample and reference materials. The temperature of transition can be accurately determined by DTA. For DSC, the temperature and **enthalpy** (ΔH) of transition can be directly measured since two heaters, for sample and reference, maintain equal temperature. The rate of differential heat input (dH/dt) plotted versus temperature or time can be monitored for the whole transition. The remainder of this chapter will primarily describe the DSC instrumentation and applications.

34.2.3 Differential Scanning Calorimeters

In order to properly use a DSC, it is important that the experimental conditions be standardized for each series of experiments. The following are some of the conditions that influence instrument output:

1. Furnace atmosphere
2. Size and shape of the furnace
3. Sample pan size, material, and its resistance to corrosion
4. Wire and bead size of the thermal junction
5. Heating rate of scan

6. Location of the thermocouple
7. Placement of sample pan inside sample holder

The following sample properties are examples of how instrument output can be affected by sample:

1. Layer thickness in pan
2. Particle size
3. Packing
4. Weight
5. Thermal conductivity
6. Heat capacity
7. Placement of sample inside sample pan

In order to gather, interpret, and calculate the proper onset, peak, and end temperatures and heat of transition, as well as the heat capacity of the sample, the instrument must be calibrated with the aforementioned well-characterized standards, such as indium. Indium has a ΔH of fusion of 28.4 J/g, m.p. 156.64°C (6, 7). This information can be used to calculate the calibration constant of the instrument from the following equation:

$$K_R = \Delta H_{indium} \cdot W_{indium}/A_{indium} \qquad (3)$$

where:
K_R = calibration constant at a given scan rate, R
ΔH_{indium} = enthalphy per gram of the heat of fusion for indium
W_{indium} = weight of indium in the sample pan
A = area of indium heat of fusion peak for the weight of indium used on the DSC used

Then the enthalpy of sample is:

$$\Delta H_{sample} = (K_R)(A_{sample})/W_{sample} \qquad (4)$$

where: A_{sample} = area of the transition peak for the sample at scan rate R
W_{sample} = weight of the sample

In this way, the area under the peak for an unknown transition can be used to calculate the ΔH of the transition, since in DSC work the area does not include the heat capacity.

The usual sample size (6–12 mg) can be placed in either small (up to 20 mg) volatile or nonvolatile sample pans (usually sealed volatile sample pans are used in food science work) or stainless steel capsules that can withstand high pressure build-up inside them, such as that caused by the volatilization of water. The high-pressure capsule can accommodate larger sample sizes (up to 45 mg). The pans most commonly used with food samples are made of Al, and they come with lids that are crimped (cold sealed) in place to assure a good seal in volatile sample pans. The internal pressure that volatile sample holders can withstand is only two to three atmospheres. The high-pressure capsules are often stainless steel and can withstand up to 30 atmospheres pressure. Care should be taken in the use of high-pressure capsules because they can injure the operator or instrument sample holder assembly (thermal head) if they explode.

The heating rate should be slow enough to obtain distinct and reproducible peaks for each transition. Commonly, scans are from 1 to 10 degrees per min. The apparent temperature of the transition is influenced by the sample size due to thermal lag. If the sample is much greater than 20 mg, or if the sample pan does not have good thermal contact with the holder that is touching the heater, uniform thermal conductivity may not take place. This will result in nonreproducible data. Also, if the entire assembly is dirty, the data may be indicative not of the sample but of the contaminants.

Once the scan is completed, the remaining problem is proper interpretation of the data obtained.

34.2.4 Data Interpretation

The following examples of DSC heating curves, adopted from the Perkin-Elmer instruction manual for the DSC-2 (8), show differences due to changes in **heat capacity** (Cp) and the peak thermal transitions: Fig. 34-2A shows no change in heat capacity across the transition; Fig. 34-2B shows a broad transition (where the baseline may not be flat); Fig. 34-2C shows a transition with a change in heat capacity; Fig. 34-2D shows a change in heat capacity during a glass transition, and defines its transition temperature (some researchers report the onset temperature as for other transitions); Fig. 34-2E shows how an increase in sample size affects the transition peak temperature (hence it is better to report); Fig. 34-2F compares a primary thermogram with its first derivative (dashed curve). Some researchers use the first derivative curve to better interpret onset, peak, and end temperatures.

Fig. 34-3 is a composite stylized set of thermal output. The directions of the endothermic and exothermic heat flows are labeled on the y axis, and the temperature increases from left to right. This output can be explained by a rise due to Cp as the scan starts from the isothermal state, followed by a ΔCp due to a glass transition; then a broad endothermic peak due to temperature-dependent phase changes, and ending with an exothermic peak.

One major controversy in the interpretation of DSC curves relates to the onset temperature, the meaning of the peak temperature, and the determination of the

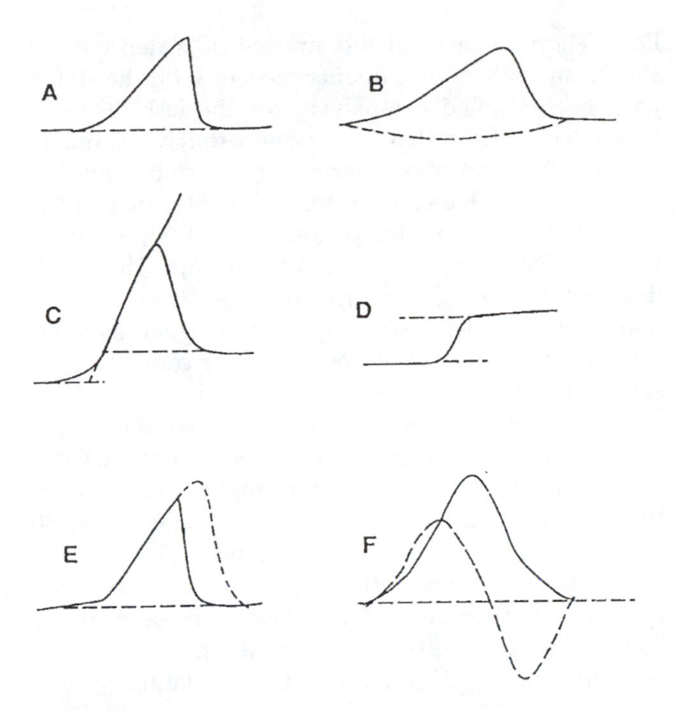

FIGURE 34-2.
Stylized curves one might find in differential scanning calorimetry (DSC) scans. (A) Curve with no change in heat capacity, (B) a broad transition (baseline may not be flat), (C) a transition with a concomitant change in heat capacity, (D) heat capacitiy change during a glass transition, (E) the effect of an increase in sample size (dashed line) on the transition, (F) comparision of a primary thermogram and its first derivative (dashed line). Adapted from (8). Figure courtesy of Perkin-Elmer Corporation, Norwalk, CT.

baseline. Figure 34-4 shows a stylized thermal peak that might result from a DSC heating scan. ICTA defines the area under the curve as ABCA. An open question is whether the onset temperature is at A or D. This author prefers to report the first deviation from the baseline (A) instead of the extrapolated onset temperature (D). The precise value would depend on whether there is a sharp versus a broad transition, or whether there is a single transition. The peak temperature for melts has been interpreted as the point when all the material has melted; however, it has been shown from other types of measurements that the peak temperature does not indicate either the maximum rate or the end of the transition being monitored.

Most researchers agree onset temperature is more significant than peak temperature, since peak temperature is greatly influenced by scan rate and sample size and does not always relate to a specific physical change. It is for these reasons that, although reporting peak temperature to compare to other reports, this author feels uncomfortable to interpret the meaning.

34.3 APPLICATIONS

DSC is widely applied in the food industry to interpret water, starch, protein, lipid, and carbohydrate interactions. There will be numerous studies in this area, and a few examples are given below.

The amount of freezable versus nonfreezable water in the system can be evaluated. Figure 34-5 shows an

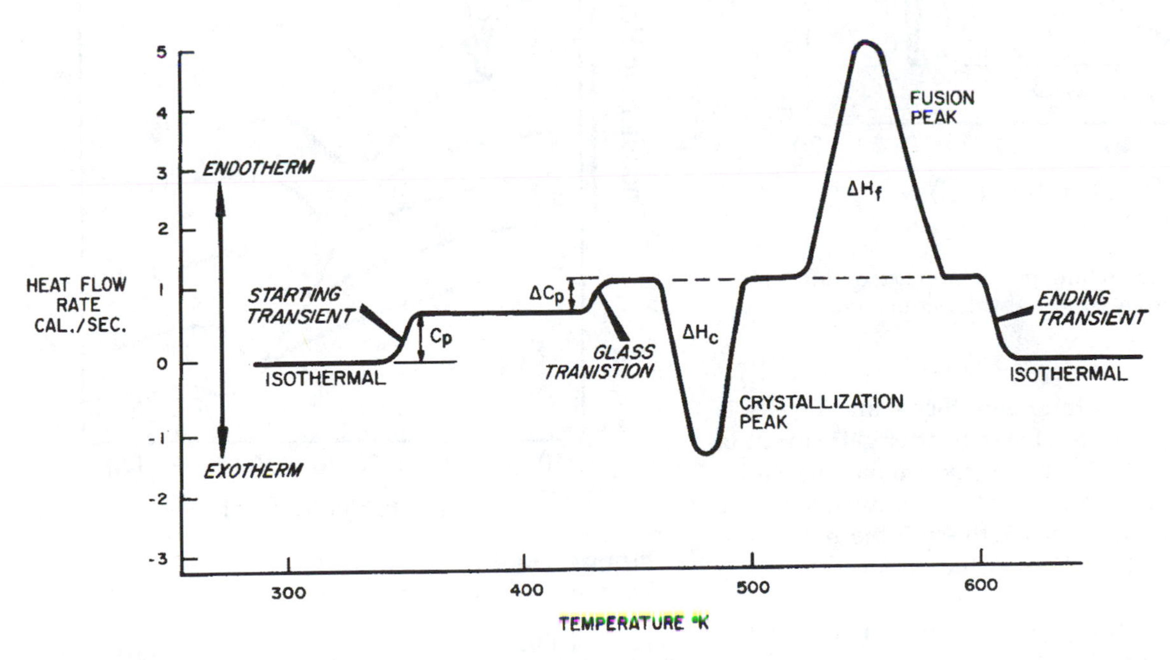

FIGURE 34-3.
A set of thermal transitions as they might appear when a sample is heated. From (8), used with permission. Figure courtesy of Perkin-Elmer Corporation, Norwalk, CT.

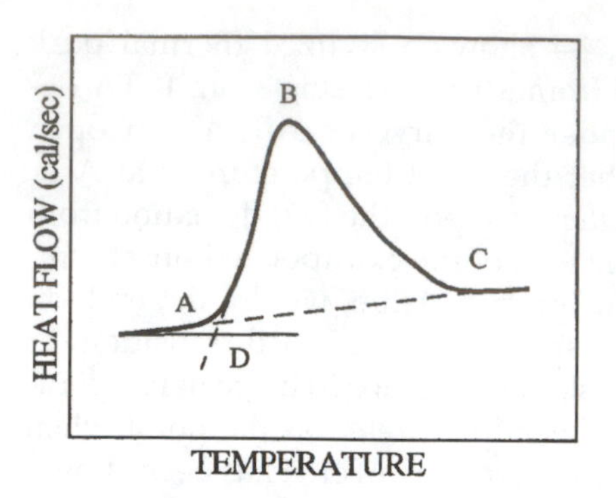

FIGURE 34-4.
A stylized thermal peak indicating possible ways to report thermal transition temperature (A → D).

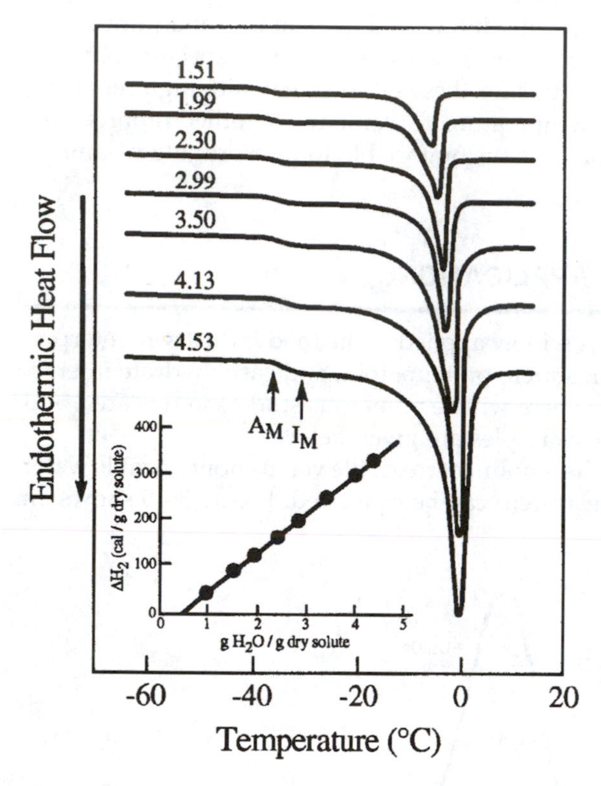

FIGURE 34-5.
Thermograms of ice melting in the presence of different amounts starch. From (9), used with permission.

example of the ice-melting endotherm and how it is affected by the presence of starch, after gelatinization has taken place (9). The larger the ice-melting endotherm, the greater the amount of free water in the system. As would be expected, the endotherm for pure ice is larger than that of ice in the presence of starch after gelatinization. Also, there might be a shift in the onset temperature due to some contribution of bound water.

Starches can be studied in many different ways by DSC. The influences of the amount of water, type of starch, and other components present with the starch have been studied extensively for the last 20 years. Examples of the enthalpy of potato starch gelatinization as a function of percent moisture can be found in Fig. 34-6 (10). One can see from Fig. 34-6 that as the volume fraction of moisture goes below 0.81, the original peak decreases and a shoulder develops. The shoulder is shifted to higher temperatures as it becomes the main peak. Finally, the enthalpy of the higher temperature endotherm decreases as the water content of the sample is further reduced.

An example of using DSC to study glass transitions of wheat starch can be seen in Fig. 34-7, from the work of Zeleznak and Hoseney (11). In thermograms of native wheat starch at various water contents below 30 percent, one can see onset temperature (T_o) shifting to higher temperatures with decreasing water content, and glass transitions (T_g) can readily be seen above 20°C, below 20.1 percent water content.

Chemical modification affects the gelatinization of

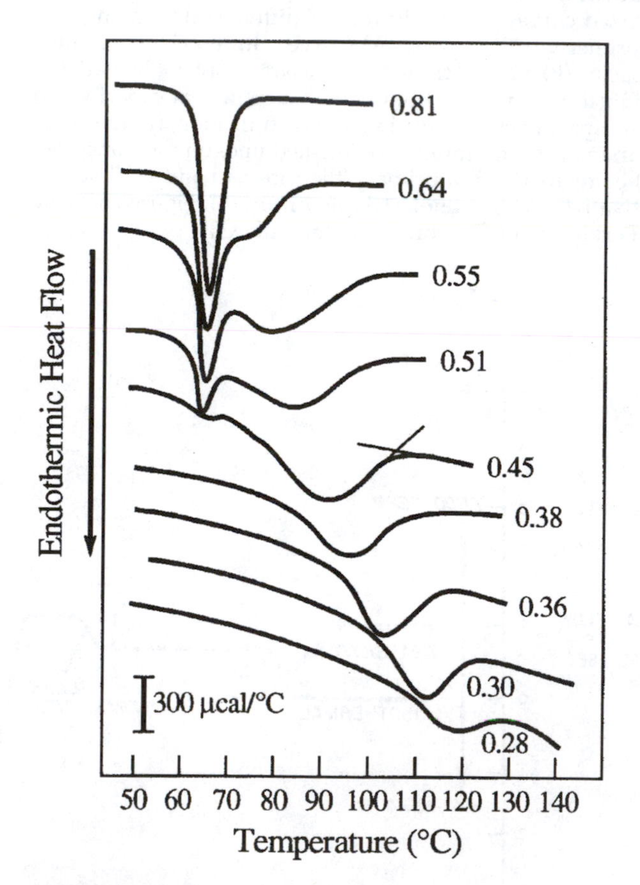

FIGURE 34-6.
The process of potato starch gelatinization with different volume fractions of water. From (10), used with permission. Reprinted from *Biopolymers*, J.W. Donovan. Copyright © 1979. John Wiley & Sons, Inc. Reprinted by permission of John Wiley & Sons, Inc.

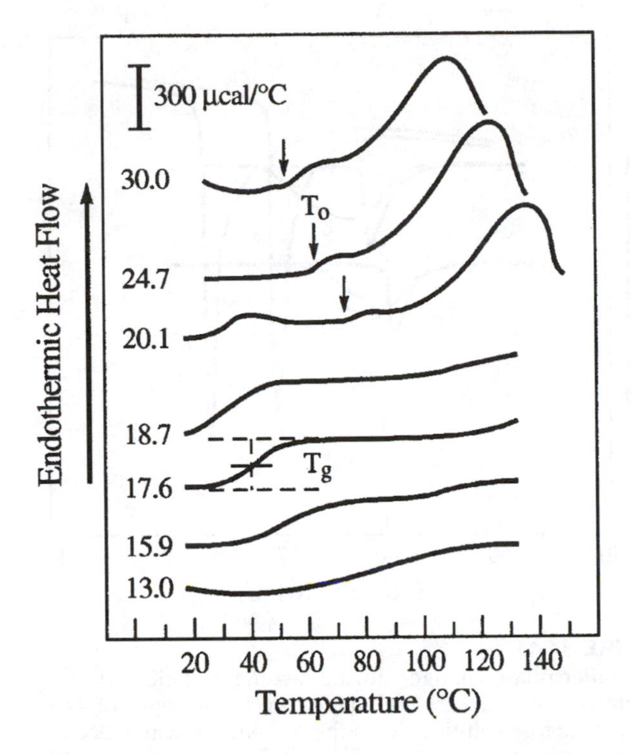

FIGURE 34-7.
Thermograms of wheat starch and the glass transition at different percentages of water. From (11), used with permission.

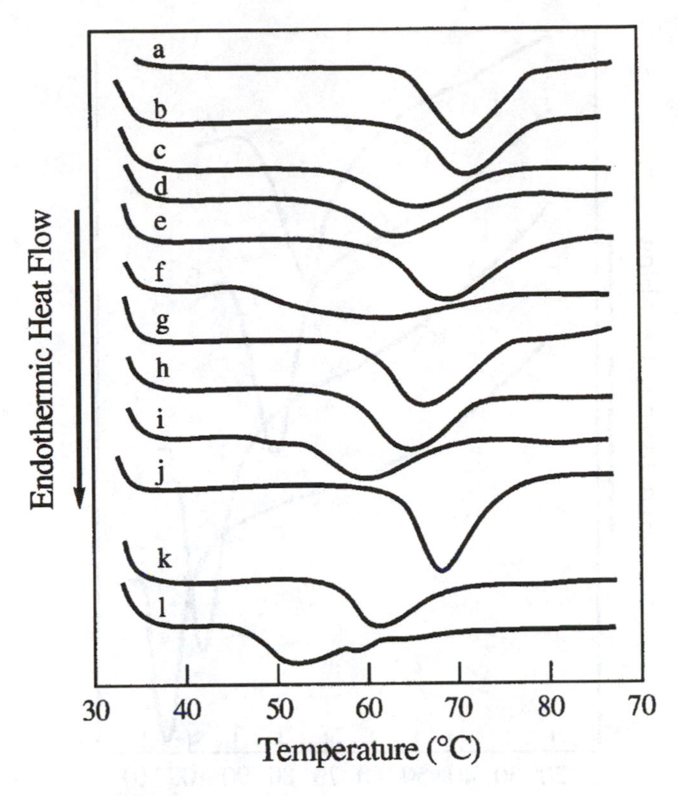

FIGURE 34-8.
Heating of waxy and normal corn starch with 1:2 starch to water: (a) unmodified waxy corn starch, (b) octenylsuccinate, 0.015 degree of substitution (DS), (c) quarternary ammonium, 0.036 DS, (d) quarternary ammonium, 0.052 DS, (e) phosphate, 0.0087 DS, (f) phosphate, 0.026 DS, (g) acetate 0.286 DS, (h) hydroxypropyl, 0.055 DS, (i) hydroxypropyl, 0.133 DS, (j) unmodified normal corn starch, (k) and (l) normal corn starch modified with hydroxypropyl at 0.055 and 0.148 DS. From (12), used with permission.

native starches, as can be seen from Fig. 34-8, where waxy and normal corn starch have been phosphorylated, octenylsuccinylated, hydroxypropylated, acetylated, or quarternary ammoniated (12). Also, corn starch gelatinization can be delayed by addition of different concentrations of sucrose, as seen from the work of Johnson et al. (13) in Fig. 34-9.

Polymorphism of mono- and triglycerides has been recognized for many years. Such molecules play an important role in the functionality of emulsifiers in foods. Generally, triglycerides have been shown to exist in an α, β, and β′ forms. Within those categories one can find polymorphic forms. The physical and thermodynamic behavior identifies the various polymorphic forms. For example, Simpson and Hagemann (14) have shown that tristearin can exist in an α, β, and β'_1, β'_2. From Fig. 34-10, one can see that tristearin, when heated, gives different DSC curves. An exothermic reaction takes place during its conversion from the α to the β phase at 55°C. However, β'_1, β'_2 convert to the β phase at 64°C differently. The β'_2 shows a small endothermic peak at 61°C prior to the phase change at 64°C, while the β'_1 does not. If the triglyceride is all in the β phase prior to heating, no conversion is seen by DSC. Therefore, Simpson and Hagemann could not show a DSC scan for the β form during heating.

Emulsifier phase changes can be studied by DSC either alone or in the presence of other components (15). In Fig. 34-11, one can see saturated monoglyceride transitions without water or sucrose (Fig. 34-11, a); in the presence of 42 percent sucrose solution (Fig. 34-11, b); and in the presence of pure water (Fig. 3-11, c). Figure 34-11, a′, b′, and c′ are rescans of Fig. 34-11, a, b, and c after cooling.

Meat scientists use DSC to study the thermal denaturation of muscle. An example of such an application can be seen in Fig. 34-12 where post-rigor beef and cod muscle can be seen to differ in their thermograms (5). The three peaks refer to denaturation of the head, tail, and a flexible region of myosin. These denaturations are strongly influenced by ionic strength and other factors.

Further applications of DSC to food research can be found in numerous articles. Biliaderis has published a review that gives numerous examples of interest to food scientists (16).

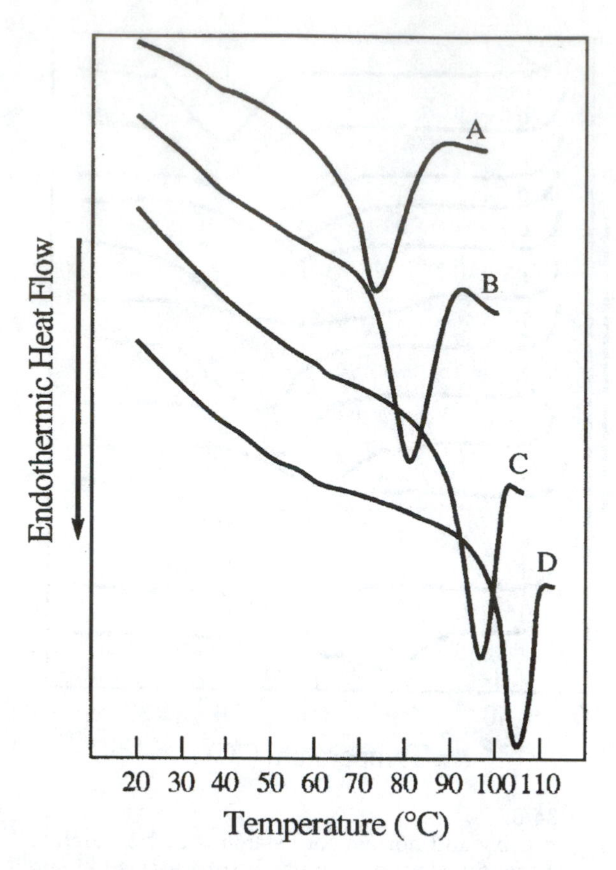

FIGURE 34-9.
Thermograms of corn starch in the presence of differing concentrations of sucrose in water (ratios of starch:sucrose:water): A, 1:0.5:2; B, 1:1:2; C, 1:2:2; D, 1:3:2. From (13), used with permission.

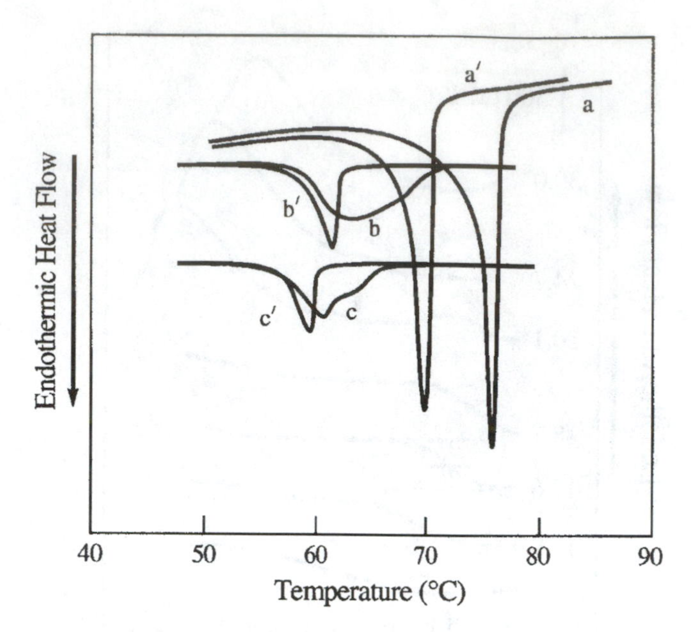

FIGURE 34-11.
Emulsifier phase changes during heating by DSC. (a) Saturated monoglyceride alone; (b) in the presence of 42 percent sucrose solution; (c) in the presence of water. Second heating scans also are shown (a′–c′). From (15), used with permission.

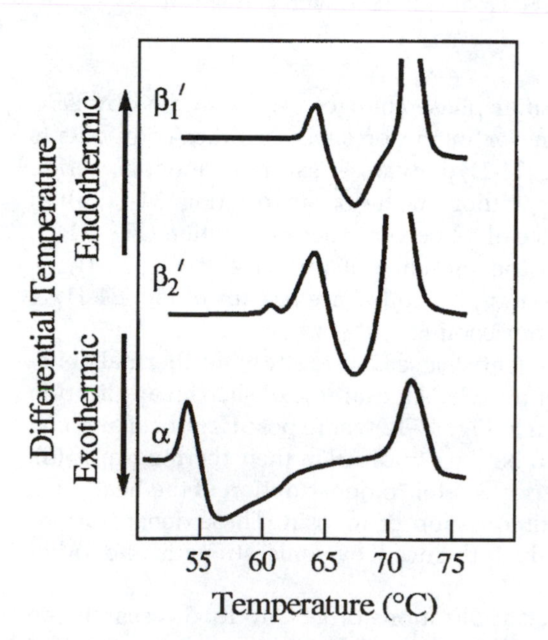

FIGURE 34-10.
DSC heating curves for tristearin phases showing phase transitions. The initial phase state of each curve is (a) β'_1, (b) β'_2, and (c) α. From (14), used with permission.

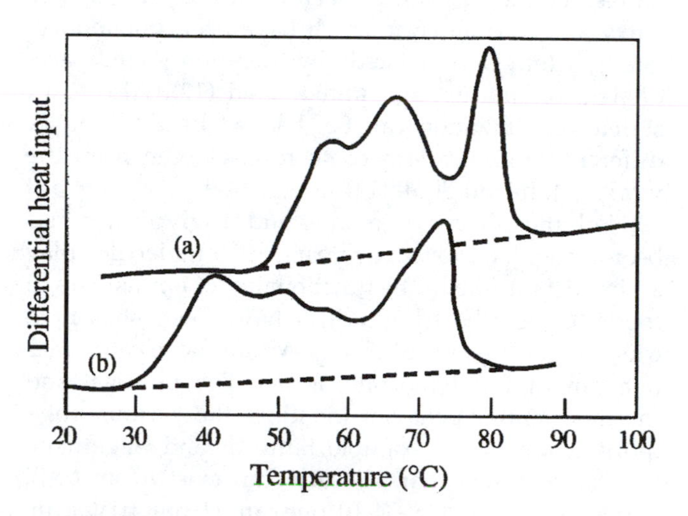

FIGURE 34-12.
Myosin denaturation in (a) beef and (b) cod muscle (post-rigor state) during heating by DSC. From (5), used with permission. Reprinted from *Food Gels*, Rodger, G.W. and Wilding, P. 1990. Ch. 9, Muscle proteins. P. Harris (Ed.), pp. 361–400. Elsevier Applied Science, New York, NY (original publishers). Chapman & Hall, London, England (current publishers).

34.4 SUMMARY

Thermal analysis techniques of various types are evolving to capture events that take place dynamically during heating under specialized experimental conditions. Some examples of the type of measurements that can be made include changes in heat capacity, weight, energies of transitions, dimensional changes, and viscoelastic and other energy and rheologically related properties. Differential scanning calorimetry is an example of a commonly used thermal analytical technique used in food science. By this method, energies involving endothermic and exothermic reactions can be measured, as well as changes in heat capacity that may or may not be be associated with a chemical reaction. Definitions of differential thermal analyzers developed by the International Confederation for Thermal Analysis are given. The general principles by which differential thermal analyzers operate, schematics of differential analyzers, examples of how data can be interpreted, and examples of data obtained from food science–related research are given in this chapter. The application examples include water, starch, protein, lipid, and carbohydrate transformations and interactions.

34.5 STUDY QUESTIONS

1. What types of information can you obtain from thermal analysis work? What types of thermal analyzers might you use for each of the thermal analyses listed in the text?
2. Define differential scanning calorimetry and how it differs from differential thermal analysis.
3. The glass transition temperature (transformation from the glassy to rubbery state) is needed for an extruded product that involves a change in heat capacity for an extruded product in order to obtain an indication of the molecular state of the system as it might relate to a crispy versus rubbery product. What type of data can you get to determine this by thermal analysis?
4. What sample constraints are needed to optimize DSC or DTA signals? Can you quantify the amount of energy required for a molecule to undergo a thermal transition?
5. You have a delay in starch gelatinization by adding a large amount of sucrose to your starch solution. Sketch and discuss the DSC-type curves you expect to obtain for the two starch systems and why.
6. How might the data in question 5 above differ from those where water is limiting?
7. You have a mixture of an emulsifier, protein, and water. Sketch what type of DSC curve you might obtain if you scanned from −30°C to 100°C and why.

34.6 REFERENCES

1. Mackenzie, R.C. 1969. Nomenclature in thermal analysis. *Talanta*. 16:1227–1230.
2. Perkin-Elmer Corporation. 1970. *Thermal Analysis Newsletter,* No. 9. Perkin-Elmer Corporation, Norwalk, CT.
3. Kerr, P.F., and Kulp, J.L. 1948. Multiple differential thermal analysis. *Am. Mineral*. 33:387–419.
4. Boersma, S.L. 1955. A theory of differential thermal analysis and new methods of measurement and interpretation. *J. Am. Ceram. Soc*. 38:281–284.
5. Rodger, G.W., and Wilding, P. 1990. Muscle proteins, Ch. 9, in *Food Gels,* P. Harris (Ed.), pp. 361–400. Elsevier Applied Food Science Series, Elsevier Applied Science, New York, NY.
6. Mortimer, C.E. 1975. *Chemistry: A Conceptual Approach*. Van Nostrand Reinhold, New York, NY.
7. Pope, M.I., and Judd, M.D. 1977. *Differential Thermal Analysis: A Guide to the Technique and Its Application*. Heydon and Sons Ltd., London, England.
8. Perkin-Elmer Corporation. 1981. *Instructions for Model DSC-2c Differential Scanning Calorimeter User's Manual*. Manual 0990–9806. Perkin-Elmer Corporation, Norwalk, CT.
9. Gekko, K., and Satake, I. 1981. Differential scanning calorimetry of unfreezable water in water-protein-polyol systems. *Agric. Biol. Chem*. 45:2209–2217.
10. Donovan, J.W. 1979. Phase transitions of the starch-water systems. *Biopolymers* 18:263–275.
11. Zeleznak, K.J., and Hoseney, R.C. 1984. The glass transition in starch. *Cereal Chem*. 64:121–124.
12. Miller, L.A., Gordon, J., and Davis, E.A. 1991. Dielectric and thermal transition properties of chemically modified starches during heating. *Cereal Chem*. 68:441–448.
13. Johnson, J.M., Davis, E.A., and Gordon, J. 1990. Interactions of starch and sugar water measured by electron spin resonance and differential scanning calorimetry. *Cereal Chem*. 67:286–291.
14. Simpson, T.D., and Hagemann, J.W. 1982. Evidence of two β' phases in tristearin. *J. Amer. Oil Chem. Soc*. 59: 169–171.
15. Cloke, J.D., Gordon, J., and Davis, E.A. 1983. Enthalpy changes in model cake systems containing emulsifiers. *Cereal Chem*. 60:143–146.
16. Biliaderis, C.G. 1983. Differential scanning calorimetry in food research—a review. *Food Chem*. 10:239–265.

Index